Perimeter and Area of Plane Figures

1. Triangle

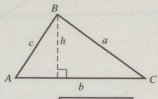

Perimeter: $P = a + b + c$

Area: $A = \dfrac{1}{2}bh$

2. Rectangle

Perimeter: $P = 2l + 2w$

Area: $A = lw$

3. Parallelogram

Perimeter: $P = 2a + 2b$

Area: $A = bh$

4. Trapezoid

Perimeter: $P = a + b + c + d$

Area: $A = \dfrac{1}{2}h(a + b)$

5. Circle

Circumference: $C = 2\pi r$

Area: $A = \pi r^2$

Volume and Surface Area of Solid Figures

1. Rectangular prism

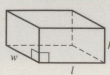

Volume: $V = lwh$

Surface area: $S = 2lw + 2lh + 2wh$

2. Right circular cylinder

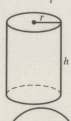

Volume: $V = \pi r^2 h$

Surface area: $S = 2\pi r^2 + 2\pi rh$

3. Sphere

Volume: $V = \dfrac{4}{3}\pi r^3$

Surface area: $S = 4\pi r^2$

4. Right circular cone

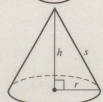

Volume: $V = \dfrac{1}{3}\pi r^2 h$

Surface area: $S = \pi r^2 + \pi rs$

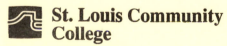

NCTIONS,

MODELING, FUNCTIONS, AND GRAPHS

MODELING, FUNCTIONS, AND GRAPHS

Algebra for College Students

Second Edition

Katherine Yoshiwara
Los Angeles Pierce College

Bruce Yoshiwara
Los Angeles Pierce College

Irving Drooyan
Los Angeles Pierce College, Emeritus

PWS Publishing Company

I(T)P **An International Thomson Publishing Company**

Boston • Albany • Bonn • Cincinnati • Detroit • London • Madrid • Melbourne • Mexico City • New York • Paris
San Francisco • Singapore • Tokyo • Toronto • Washington

PWS PUBLISHING COMPANY

20 Park Plaza, Boston, MA 02116-4324

 This book is printed on recycled, acid-free paper.

I(T)P™

International Thomson Publishing
The trademark ITP is used under license.

For more information, contact:

PWS Publishing Co.
20 Park Plaza
Boston, MA 02116

International Thomson Publishing Europe
Berkshire House I68-I73
High Holborn
London WC1V 7AA
England

Thomas Nelson Australia
102 Dodds Street
South Melbourne, 3205
Victoria, Australia

Nelson Canada
1120 Birchmount Road
Scarborough, Ontario
Canada M1K 5G4

International Thomson Editores
Campos Eliseos 385, Piso 7
Col. Polanco
11560 México D.F., Mexico

International Thomson Publishing GmbH
Königswinterer Strasse 418
53227 Bonn, Germany

International Thomson Publishing Asia
221 Henderson Road
#05-10 Henderson Building
Singapore 0315

International Thomson Publishing Japan
Hirakawacho Kyowa Building, 31
2-2-1 Hirakawacho
Chiyoda-ku, Tokyo 102
Japan

Library of Congress Cataloging-in-Publication Data

Yoshiwara, Katherine
    Modeling, functions, and graphs : algebra for college students /
Katherine Yoshiwara, Bruce Yoshiwara, Irving Drooyan.—2nd ed.
        p.   cm.
    Includes index.
    ISBN 0-534-94560-0
    1. Algebra.   I. Yoshiwara, Bruce.   II. Drooyan, Irving.
III. Title.
QA154.2.Y67   1995
512.9—dc20
                                          95-38377
                                            CIP

Sponsoring Editor: David Dietz
Editorial Assistant: Ange Mlinko
Production Coordinator: Robine Andrau
Marketing Development Manager: Marianne C. P. Rutter
Manufacturing Coordinator: Marcia A. Locke
Interior/Cover Designer: Julia Gecha

Interior Illustrator: Rolin Graphics/Scientific Illustrators
Typesetter: Modern Graphics Inc.
Cover Art: Vytas Sakalas, "Improving Relations with the Rectangle #37." Used with permission of the artist.
Cover Printer: Coral Graphic Services, Inc.
Text Printer/Binder: Quebecor Printing/Hawkins

Printed and bound in the United States of America
95 96 97 98 99 — 10 9 8 7 6 5 4 3 2 1

Contents

v

Preface

*T*he secondary mathematics curriculum is undergoing fundamental changes. One of the primary forces for change is the national renewal effort outlined in the National Council of Teachers of Mathematics Standards and already well under way at the college level as the calculus reform movement. A second major influence is the ready availability of powerful calculators and computers. *Modeling, Functions, and Graphs,* Second Edition, incorporates the benefits of technology and the philosophy of the reform movement and applies them to the material covered in a typical intermediate algebra course.

MODELING

"Investigations are designed to show students how the mathematical techniques they are learning can be applied to study and understand new situations."

The ability to model problems or phenomena by algebraic expressions and equations is the ultimate goal of any algebra course. Through simple applications, we motivate students to develop the necessary skills and techniques of algebra. In several chapters we introduce new material in the form of interactive "Investigations" that give students an opportunity to explore open-ended modeling problems. These Investigations can be used in class as guided explorations or as projects for small groups. They are designed to show students how the mathematical techniques they are learning can be applied to study and understand new situations.

FUNCTIONS

"While the formal study of functions is usually the content of precalculus, it is not too early to begin building an intuitive understanding of functional relationships in the preceding algebra courses."

The fundamental concept underlying calculus and related disciplines is the notion of function; students should acquire a good understanding of functions before they embark on the study of calculus or other college-level mathematics. While the formal study of functions is usually the content of precalculus, it is not too early to begin building an intuitive understanding of functional relationships in the preceding algebra courses. These ideas are useful not only in calculus but in practically any field students may pursue. After some preliminary review of variables in Chapter 1, we begin working with functions in Chapter 2, Linear Models. In Chapter 3 we consider applications of linear models, and in Chapter 4 we study quadratic models. We have deliberately delayed the introduction of formal function notation and terminology until Chapter 5 in order to concentrate on the meaning behind the notation.

In all our work with functions and modeling, we employ the now-celebrated "Rule of Four"—that is, that all problems should be considered using algebraic, numerical, graphical, and verbal methods. It is the connections among these approaches that we have endeavored to establish in this course. At this level it is crucial that students learn to write an algebraic expression from a verbal description, recognize trends in a table of data, and extract and interpret information from the graph of a function.

In order to get a sense of the text's approach to modeling and functions, see Section 2.1 (linear models), Exercises 1–10; Section 3.4 (scatter plots), Exercises 13–24; Section 4.1 (quadratic models), Exercises 1–6; Section 5.5 (functions as mathematical models), Exercises 1–12 and 33–36; and Section 7.2 (exponential functions), Exercises 47–52 and 63–66. These exercises are designed to help students understand the link between observed phenomena and mathematical models.

"At this level it is crucial that students learn to write an algebraic expression from a verbal description, recognize trends in a table of data, and extract and interpret information from the graph of a function."

GRAPHS

No tool for conveying information about a system is more powerful than a graph. Yet many students have trouble progressing from a point-wise understanding of graphs to a more global view. By taking advantage of graphing calculators, we examine a large number of examples and study them in more detail than is possible when every graph must be plotted by hand. We consider more realistic models in which calculations by more traditional methods are difficult or impossible.

We have incorporated graphing calculators into the text wherever they can be used to enhance understanding. Calculator use is not simply an add-on but in many ways shapes the organization of the material. The text includes some instructions for the TI-82 graphing calculator, but these can easily be adapted to any other graphing utility. We have not attempted to use all the features of the calculator or to teach calculator use for its own sake but in all cases have let the mathematics suggest how technology should be used.

In order to get a sense of the text's approach to graphing calculators, see Section 2.6 (solving equations graphically), Exercises 17–30; Section 6.6 (distance formula), Exercises 17–22; Section 8.3 (rational functions), Exercises 19–30; and Section 8.6 (motion problems), Exercises 21–24.

"No tool for conveying information about a system is more powerful than a graph. Yet many students have trouble progressing from a point-wise understanding of graphs to a more global view."

CONTENT

Modeling, Functions, and Graphs, Second Edition, includes all of the material found in a typical intermediate algebra course, although the order of the topics is nontraditional in some instances. Chapter 1, Fundamentals, reviews algebraic expressions and the solution of simple equations, recalls some

"Although the order of the topics is nontraditional in some instances, *Modeling, Functions, and Graphs,* Second Edition, includes all of the material found in a typical intermediate algebra course."

useful facts from geometry, and introduces the graphing calculator. Chapters 2 and 3 consider linear equations and their applications, including systems of equations and inequalities and the rudiments of curve fitting. Chapter 4 deals with quadratic models. The definition of functions is introduced in Chapter 5, along with the graphs of some basic functions, direct and inverse variation, and some examples of modeling with functions.

The second half of the text continues the study of various families of functions and their algebraic properties. Chapter 6, Powers and Roots, includes the laws of exponents, radical notation and the distance formula, as well as radical equations and their applications. Chapter 7 studies exponential and logarithmic models, and Chapter 8 considers polynomial and rational functions. Chapter 9 covers sequences and series, and Chapter 10, Additional Topics, includes the conic sections, inverse functions, some additional graphing techniques, and matrix solution of linear systems. A review of polynomial products and factoring, a note on the structure of the real number system, and an introduction to complex numbers are included as an appendix.

EXERCISE SETS

The exercise sets reflect our focus on modeling, functions, and graphs. We have provided a wide range of open-ended problems that emphasize mathematical modeling using tables of values, algebraic expressions, and graphs. We have also included practice exercises for each new skill, so that instructors may decide how much drill on manipulation is appropriate for their classes. Each chapter concludes with a set of review exercises that students can use as a practice test for the chapter. Answers to the odd-numbered exercises in each section and to all exercises in the chapter reviews are provided.

ANCILLARIES

For instructors: An **Instructor's Manual** provides answers to the even-numbered exercises as well as comments on pedagogy for each section of the text and additional ideas and tips for using graphing calculators in the classroom. Printed and computerized **test banks** contain about fifteen hundred test items for IBM and Macintosh computers.

For students: A **Student's Solutions Manual** provides solutions to all odd-numbered exercises, and a **Graphing Calculator Guidebook** teaches keystrokes and commands for the TI-81 and TI-85 graphing calculators, to correspond with coverage in the text. **MathQuest,** a software tutorial for Windows and Macintosh, reinforces manipulative skills through drill. Prepared specifically for *Modeling, Functions, and Graphs,* MathQuest enables students to work exercises and examples corresponding to selected sections, sharpening their algebra skills for applications in the text. Also excellent for algebra review, **Intermediate Algebra Videos** from Educational Video Resources are available to adopters.

ACKNOWLEDGMENTS

We benefited greatly from the comments and suggestions of the many instructors who reviewed drafts of the manuscript. They include

Rob Biagini-Komas
College of San Mateo

Alice Burstein
Middlesex Community-Technical College

D. Sarah Garland
University of Alaska at Fairbanks

Gary S. Kersting
Sacramento City College

Bobby M. Righi
Seattle Central Community College

Fundamentals

$\mathcal{Y}$ou may have heard the saying, "Mathematics is the language of science." Traditionally, anyone who wanted to understand physics or engineering needed to understand mathematics first. Today, however, the availability of powerful computers and calculators has allowed professionals in not only the physical sciences but in practically every discipline to take advantage of mathematical methods. Mathematical techniques are now used in a wide array of fields from business to sociology to analyze data, identify trends, and predict the effects of change. At the heart of these quantitative methods are the concepts and skills of algebra. In this course you will use some of the skills you learned in elementary algebra to solve problems and to study the behavior of a variety of phenomena.

1.1 INTRODUCTION

We will begin with a few examples of how we can use algebra to investigate numerical situations.

Investigation #1

Delbert is offered a job as a salesman for a company that manufactures restaurant equipment. He will be paid $1000 per month plus a 6% commission on his sales. The sales manager informs Delbert that he can expect to sell about $8000 worth of equipment per month. To help him decide whether to accept the job, Delbert does a few calculations.

1. Based on the sales manager's estimate, what monthly income can Delbert expect from this job? What annual salary would that provide?

2. What would Delbert's monthly salary be if he sold, on the average, only $5000 worth of equipment per month? What if he sells $10,000 worth per month? Compute monthly salaries based on several other possible sales totals, and record your findings in Table 1.1.

Sales	Incomes
5,000	
8,000	
10,000	
12,000	
15,000	
18,000	
20,000	
25,000	
30,000	
35,000	

Table 1.1

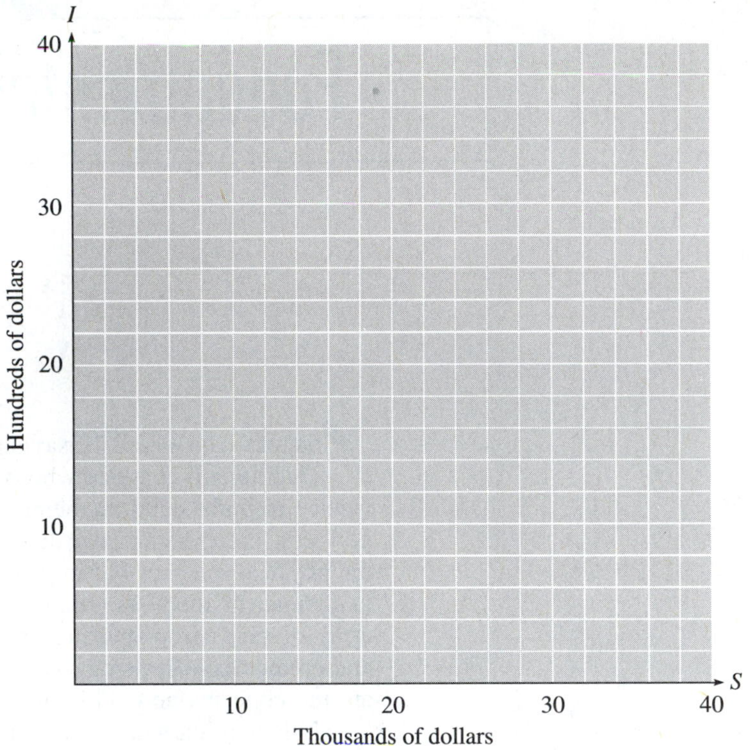

Figure 1.1

3. Plot your data points on a graph, using the sales figures (S) on the horizontal axis and income (I) on the vertical axis, as shown in Figure 1.1. Connect the data points to show Delbert's monthly income for all possible monthly sales totals.

4. Add two new data points to Table 1.1 by reading values from your graph.

5. Write an algebraic expression for Delbert's monthly income (I) in terms of his monthly sales (S). Use the verbal description in the problem to help you:

"He will be paid $1000 per month plus a 6% commission on his sales."
Income = _____

6. Test your formula from part (5) to see if it gives the same results you recorded in Table 1.1.

7. Now use your formula to find out what monthly sales total Delbert would need in order to have a monthly income of $2500.

8. Each increase of $1000 in monthly sales increases Delbert's monthly income by _____ .

9. Summarize the results of your work: In your own words, describe the relationship between Delbert's monthly sales and his monthly income. Include in your discussion a description of your graph.

 Investigation #2

Francine is the program director for the City Park Summer Program. This week she is planning the soccer league for 7- to 9-year-olds. In this league each team plays every other team once before the play-offs. The number of games Francine must schedule (not including the play-off games) depends on how many teams she allows in the league. To get a feel for the numbers involved, Francine does some calculations.

1. How many games will be needed if there is only one team in the league? (This event is not very likely, but Francine is looking for a pattern so she considers all possibilities.) How many games will be needed if there are two teams? Start recording your findings in Table 1.2.

2. The number of games gets harder to calculate as the number of teams increases. Francine discovers that she can "model" the problem by drawing one dot for each team in the league and then connecting each dot to every other dot with a straight line. The number of lines represents the number of games needed. The diagrams for three teams and for four teams are shown in Figure 1.2. Use similar diagrams (see Figure 1.3) to fill in Table 1.2 up to eight teams. (Be careful that no three dots "line up.")

Teams	Games
1	
2	
3	
4	
5	
6	
7	
8	

Table 1.2

Figure 1.2

Five Teams   **Six Teams**   **Seven Teams**   **Eight Teams**

Figure 1.3

3. Plot the data points on the graph in Figure 1.4, using the number of teams (n) on the horizontal axis and the number of games (G) on the vertical axis. For this problem we cannot really connect the points because n cannot have fractional values. (Remember that n is the number of teams.) However, we will draw a faint curve through the points anyway to help us see a pattern. Do the points lie on a straight line?

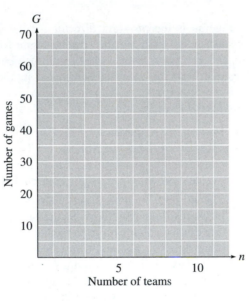

Figure 1.4

n	G	ΔG
1	0	—
2	1	1
3	3	2
4		
5		
6		
7		
8		
9		
10		
11		
12		

Table 1.3

4. Can you see a quick way to obtain more entries in the table? Here is a hint: Add another column to your table, as shown in Table 1.3. We will call this column ΔG, which means "change in G." As you move down the table, how much does G increase at each step? The first few values of ΔG are filled in for you. Do you see a pattern? Use this pattern to fill in the table up to $n = 12$.

5. Francine would like to have a formula that tells her directly (in terms of n) how many games will be needed if there are n teams in the league. To find such a formula, she goes back to her dot and line diagrams. Consider the diagram for four teams in Figure 1.5(a).

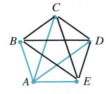

Figure 1.5   (a) Four Teams   (b) Five Teams

Step 1 Pick any team, say Team *A* as shown. How many games does Team *A* play? _____

Now, each team plays the same number of games. How many teams are there? _____

That makes a total of how many games? _____ × _____ = _____ .

But be careful! We have counted each game twice! (Notice that the line from Team *A* to Team *B* is the same as the line from Team *B* to Team *A*.) We must divide our previous answer by _____ , which gives us _____ games.

(Does this agree with the value in Table 1.3?)

Step 2 Try the same reasoning on the diagram for five teams shown in Figure 1.5(b).

Pick any team, say Team *A* as shown. How many games does Team *A* play? _____

Each team plays the same number of games. How many teams are there? _____

That makes a total of how many games? _____ × _____ = _____ .

Again, we have counted each game twice. We must divide our answer by _____ , which gives us _____ games.

(Does this agree with the value in Table 1.3?)

Step 3 Now try to generalize your work: Suppose there are *n* teams in the league. How many games will any particular team play? (How many teams are left to be opponents?) _____

Since each team plays the same number of games, multiply the number of games by the number of teams: _____ × _____ . Each game is counted twice, so we must divide by _____ .

The formula for the number of games needed in a league of *n* teams is _____ .

6. Use the formula to calculate the number of teams that can be in the league if the maximum number of games Francine can schedule is 160. (Do you remember how to solve quadratic equations? If not, refer to Chapter 4.)

7. Summarize your work: In your own words, describe the relationship between the number of teams in the league and the number of games Francine must schedule. Include in your discussion a description of your graph.

Investigation #3

A contagious disease whose spread is unchecked can devastate a confined population. For example, in the early sixteenth century, Spanish troops introduced smallpox into the Aztec population in Central America. The resulting epidemic contributed significantly to the fall of Montezuma's empire.

Suppose an outbreak of cholera follows severe flooding in an isolated town of 5000 people. Initially (Day 0), 40 people are infected. Every day after that, 25% of those still healthy fall ill.

1. On the first day, how many people are still healthy? _____
   How many will fall ill on the first day? _____
   What is the total number of people infected after the first day? _____

2. Check your results against Table 1.4. Subtract the total number of infected residents from 5000 to find the number of healthy residents at the beginning of the second day. Then fill in the rest of the table for 10 days. (Round off decimal results to the nearest whole number.)

Day	Number Healthy	New Patients	Total Infected
0	5000	40	40
1	4960	1240	1280
2			
3			
4			
5			
6			
7			
8			
9			
10			

Table 1.4

3. Use the last column of Table 1.4 to plot the total number of infected residents (I) against time (t). (Consider the values of I you must plot and choose a suitable scale for the vertical axis in Figure 1.6.) Connect your data points with a smooth curve.

4. Do the values of I approach some largest value? Draw a dotted horizontal line at that value of I. Will the values of I ever exceed that value?

5. What is the first day on which at least 95% of the population is infected?

6. Look back at Table 1.4. What is happening to the number of new patients each day as time goes on? How is this phenomenon reflected in the graph in Figure 1.6? How would the graph look if the number of new patients every day were a constant?

7. Summarize your work: In your own words, describe how the number of residents infected with cholera changes with time. Include a description of your graph.

In the three investigations just presented, we used a variety of mathematical techniques to build an understanding of the situation. These

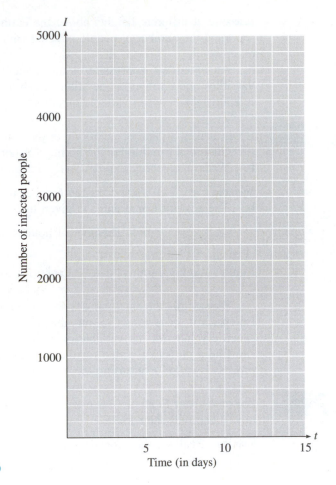

Figure 1.6

techniques included performing numerical calculations, describing relationships between variables and algebraic expressions, solving equations or formulas, using a model to represent an abstract idea, plotting data on a graph, and using the graph to identify trends or make predictions. The study of mathematics and its applications depends on understanding and applying these basic skills.

1.2   ALGEBRAIC EXPRESSIONS

In the next three sections, we will review some algebraic skills: writing expressions, solving equations in one variable, and using formulas.

From your previous algebra courses, you are familiar with the use of letters, or **variables,** to stand for unknown numbers in equations or formulas. Variables are also used to represent numerical quantities that change over time or in different situations. For example, p might stand for the atmospheric pressure at different heights above the earth's surface. Or, we might let N represent the number of people infected with cholera t days after the start of an epidemic.

An **algebraic expression** is any meaningful combination of numbers, variables, and symbols of operation. Algebraic expressions are used to express relationships between variable quantities.

Example 1

Loren makes $6 an hour working at the campus bookstore.
  a. Choose a variable for the number of hours Loren works per week.
  b. Write an algebraic expression for the amount of Loren's weekly earnings.

Solutions   **a.**   Let h stand for the number of hours Loren works per week.
  b.   The amount Loren earns is given by

$$6 \times (\text{number of hours Loren worked})$$

or $6 \cdot h$.
Loren's weekly earnings can be expressed as $6h$.

We say that the algebraic expression $6h$ represents the amount of money Loren earns *in terms of* the number of hours she works. If we substitute specific values for the variable or variables in an algebraic expression, we can find a numerical value for the quantity the expression represents. This process is called **evaluating** the expression.

Example 2

If Loren from Example 1 works for 16 hours in the bookstore one week, how much will she earn?

Solution   Evaluate the expression $6h$ for $h = 16$:

$$6h = 6(\mathbf{16}) = 96$$

Loren will make $96.

If we also assign a variable to Loren's weekly earnings, say w, we can write an equation relating the two variables:

$$w = 6h$$

An **equation** is a mathematical statement that two expressions are equal. Equations relating two variables are particularly useful. If we know the value of one of the variables, we can find the corresponding value of the other variable by solving the equation.

Example 3

How many hours does Loren need to work next week if she wants to earn $225?

Solution   In mathematical language we know that $w = 225$,   and we would like to know the value of h. We will substitute the value for the known variable into our equation and then solve for the unknown variable.

$$w = 6h \qquad \text{Substitute 225 for } w.$$

$$225 = 6h \qquad \text{Divide both sides by 6.}$$

$$\frac{225}{6} = \frac{6h}{6} \qquad \text{Simplify.}$$

$$37.5 = h$$

According to our calculations, Loren must work 37.5 hours in order to earn $225. In reality, Loren will probably have to work for 38 hours, because most employers do not pay for portions of an hour's work. ▭

Order of Operations

Algebraic expressions often involve more than one arithmetic operation. To resolve any ambiguities over how such expressions should be evaluated, we follow the **order of operations** that you learned in elementary algebra.

> **ORDER OF OPERATIONS**
> **1.** Simplify any expressions within grouping symbols (parentheses, brackets, square root bars, or fraction bars). Start with the innermost grouping symbols and work outward.
> **2.** Evaluate all powers.
> **3.** Perform multiplications and divisions in order from left to right.
> **4.** Perform additions and subtractions in order from left to right.

We will consider several examples illustrating how to apply the order of operations when writing and evaluating algebraic expressions.

Example 4

April sells environmentally friendly cleaning products. Her income consists of $200 per week plus a commission of 9% of her sales.
a. Choose variables to represent the unknown quantities, and write an algebraic expression for April's weekly income in terms of her sales.
b. Find April's income for a week in which she sells $350 worth of cleaning products.

Solutions a. Let I represent April's total income for the week, and let S represent the total amount of her sales. We can translate the information from the problem into mathematical language as follows:

"Her income consists of $200 . . . plus . . . 9% of her sales."
$$I \quad = \quad \$200 \quad + \quad 0.09 \quad S$$

Thus $I = 200 + 0.09S$.

b. We want to evaluate our expression from part (a) with $S = 350$. We substitute 350 for S to find

$$I = 200 + 0.09\,(\mathbf{350})$$

According to the order of operations, we should perform the multiplication before the addition. Thus to simplify our expression for I, we begin by computing 0.09 (350).

$$I = 200 + 0.09\,(350) \qquad \text{Multiply 0.09 (350) first.}$$
$$= 200 + 31.5 \qquad\qquad \text{Add.}$$
$$= 231.5$$

April's income for the week is $231.50.

On a scientific or a graphing calculator, we can enter the expression from Example 4(b) just as it is written:

$$200 \boxed{+} \; 0.09 \; \boxed{\times} \; 350 \; \boxed{\text{ENTER}}$$

and the calculator will perform the operations in the correct order—multiplication first.

In some situations we may want to write an algebraic expression in which an addition or subtraction should be performed before a multiplication. We use parentheses to "override" the multiplication-first rule. Compare these two expressions:

The sum of four times x and ten	$4x + 10$
Four times the sum of x and ten	$4(x + 10)$

In the first expression, we perform the multiplication $4 \times x$ first, but in the second expression we perform the addition $x + 10$ first because it is enclosed in parentheses.

Example 5

Economy Parcel Service charges $2.80 per pound to deliver a package from Pasadena to Cedar Rapids. Andrew wants to mail a painting that weighs 8.3 pounds, plus whatever packing material he uses.
a. Choose variables to represent the unknown quantities and write an expression for the cost of shipping Andrew's painting.
b. Find the shipping cost if Andrew uses 2.9 pounds of packing material.

Solutions **a.** Let C stand for the shipping cost and let w stand for the weight of the packing material. Andrew must find the total weight of his package first and then multiply by the shipping charge. The total weight of the package is $8.3 + w$ pounds. We use parentheses around this expression to show that it should be computed first and the sum should be multiplied by the shipping charge of $2.80 per pound. Thus

$$C = 2.80\,(8.3 + w)$$

b. Evaluate the formula from part (a) with $w = 2.9$.

$$C = 2.80\,(8.3 + w) \qquad \text{Substitute 2.9 for } w.$$
$$= 2.80\,(8.3 + \mathbf{2.9}) \qquad \text{Add inside parentheses.}$$
$$= 2.80\,(11.2) \qquad \text{Multiply.}$$
$$= 31.36$$

The cost of shipping the painting is \$31.36.

On a calculator we enter the expression for C in the order it appears, including the parentheses. (Experiment to see whether your calculator requires you to enter the ⊠ symbol after 2.80.) The keying sequence

$$2.80 \; \boxed{\times} \; \boxed{(} \; 8.3 \; \boxed{+} \; 2.9 \; \boxed{)} \; \boxed{\text{ENTER}}$$

gives the correct result, 31.36.

 COMMON ERROR

If we omit the parentheses, the calculator will perform the multiplication before the addition. Thus the keying sequence

$$2.80 \; \boxed{\times} \; 8.3 \; \boxed{+} \; 2.9$$

gives an incorrect result for Example 5(b). (The sequence

$$8.3 \; \boxed{+} \; 2.9 \; \boxed{\times} \; 2.80$$

does not work either!)

We have seen that the location (or absence) of parentheses can drastically alter the meaning of an algebraic expression. In Example 6, notice how the location of the parentheses changes the value of the expression.

Example 6

a. $\begin{aligned} 5 - 3 \cdot 4^2 &= 5 - 3 \cdot 16 \\ &= 5 - 48 = -43 \end{aligned}$

b. $\begin{aligned} 5 - (3 \cdot 4)^2 &= 5 - 12^2 \\ &= 5 - 144 = -139 \end{aligned}$

c. $\begin{aligned} (5 - 3 \cdot 4)^2 &= (5 - 12)^2 \\ &= (-7)^2 = 49 \end{aligned}$

d. $\begin{aligned} (5 - 3) \cdot 4^2 &= 2 \cdot 16 \\ &= 32 \end{aligned}$

 COMMON ERROR

In the expression $5 - 12^2$, which appears in Example 6(b), the exponent 2 applies only to 12, not to -12. Thus $5 - 12^2 \neq 5 + 144$.

The order of operations mentions other grouping devices besides parentheses: fraction bars and square root bars. Notice how the placement of the fraction bar affects the expressions in Example 7.

Example 7

a. $\dfrac{1+2}{3\cdot 4} = \dfrac{3}{12}$

$\qquad\quad = \dfrac{1}{4}$

b. $1 + \dfrac{2}{3\cdot 4} = 1 + \dfrac{2}{12}$

$\qquad\qquad = 1 + \dfrac{1}{6} = \dfrac{7}{6}$

c. $\dfrac{1+2}{3}\cdot 4 = \dfrac{3}{3}\cdot 4$

$\qquad\qquad = 1\cdot 4 = 4$

d. $1 + \dfrac{2}{3}\cdot 4 = 1 + \dfrac{8}{3}$

$\qquad\qquad = \dfrac{3}{3} + \dfrac{8}{3} = \dfrac{11}{3}$

Using a Calculator   Because your calculator cannot use a fraction bar as a grouping symbol, *you must insert parentheses around any expression that appears above or below a fraction bar.* Here are the keying sequences for each expression in Example 7.

a. $\dfrac{1+2}{3\cdot 4}$ $\boxed{(}\,1\,\boxed{+}\,2\,\boxed{)}\,\boxed{\div}\,\boxed{(}\,3\,\boxed{\times}\,4\,\boxed{)}$

b. $1 + \dfrac{2}{3\cdot 4}$ $1\,\boxed{+}\,2\,\boxed{\div}\,\boxed{(}\,3\,\boxed{\times}\,4\,\boxed{)}$

c. $\dfrac{1+2}{3}\cdot 4$ $\boxed{(}\,1\,\boxed{+}\,2\,\boxed{)}\,\boxed{\div}\,3\,\boxed{\times}\,4$

d. $1 + \dfrac{2}{3}\cdot 4$ $1\,\boxed{+}\,2\,\boxed{\div}\,3\,\boxed{\times}\,4$

Similarly, your calculator does not use a square root bar as a grouping symbol. When using a calculator, *you must insert parentheses around any expression that appears under a square root symbol.*

Example 8   If an object falls from a height of h meters, then the time, t, it will take to

reach the ground is given in seconds by the formula $t = \sqrt{\dfrac{h}{4.9}}$.   How long will it take a marble to fall to the ground from the top of the Sears Tower in Chicago, which is 443.2 meters tall?

Solution   In this problem the formula is given, so we need only evaluate the expression for $h = 443.2$.   Thus

$$t = \sqrt{\frac{443.2}{4.9}}$$

To simplify the expression, we perform the division first, then take the square root of the quotient. On a graphing calculator, we use the keying sequence

$$\boxed{\sqrt{}}\;\boxed{(}\;443.2\;\boxed{\div}\;4.9\;\boxed{)}\;\boxed{\text{ENTER}}$$

and the calculator returns the result 9.510466844. The marble takes approximately 9.51 seconds to reach the ground.

 COMMON ERRORS

In Example 8 note that if we omit the parentheses and key in the sequence

$$\boxed{\sqrt{}}\;443.2\;\boxed{\div}\;4.9\;\boxed{\text{ENTER}}$$

the calculator will compute $\dfrac{\sqrt{443.2}}{4.9},$ which is not what we want.

Also note that on a scientific calculator, we enter the $\boxed{\sqrt{}}$ key *after* the radicand, as follows:

$$\boxed{(}\;443.2\;\boxed{\div}\;4.9\;\boxed{)}\;\boxed{\sqrt{}}$$

Example 9

Use a calculator to simplify each expression. Then explain how the calculator performed the operations.

a. $-8\,(6.2)^2$ **b.** $\dfrac{\sqrt{72}}{4 + 2.5}$

Solutions   a.   Use the keying sequence $\boxed{(-)}\;8\;\boxed{\times}\;6.2\;\boxed{x^2}\;\boxed{\text{ENTER}}$. (Notice that we use the gray negative key, $\boxed{(-)}$, not the blue subtraction key, $\boxed{-}$, to enter -8.) Your calculator performs the power first, to get $(6.2)^2 = 38.44,$ then multiplies the result by -8 to get $-8\,(38.44) = -307.52.$

b.   Use the keying sequence $\boxed{\sqrt{}}\;72\;\boxed{\div}\;\boxed{(}\;4\;\boxed{+}\;2.5\;\boxed{)}\;\boxed{\text{ENTER}}$. Notice that we must insert parentheses around the sum $4 + 2.5$. Your calculator performs the operations in the following order.

$$\frac{\sqrt{72}}{4 + 2.5} = \frac{\sqrt{72}}{6.5} \qquad \text{\textcolor{teal}{Perform operations below the fraction bar.}}$$

$$= \frac{8.485281374}{6.5} \qquad \text{\textcolor{teal}{Simplify the square root.}}$$

$$= 1.305427904 \approx 1.31 \qquad \text{\textcolor{teal}{Divide.}}$$

Exercise 1.2

Write algebraic expressions to describe each situation, and evaluate for the given values.

1. Jim was 27 years old when Ana was born.

   a. Write an expression for Jim's age, j, in terms of Ana's age, a.

   b. Use your expression to find Jim's age when Ana is 22 years old.

2. Rani wants to replace the wheels on her in-line skates. New wheels cost $6.59 each.

   a. Write an expression for the total cost, C, of new wheels in terms of the number of wheels, n, Rani must replace.

   b. Use your expression to find the total cost if Rani must replace eight wheels.

3. Helen decides to drive to visit her father. The trip is a distance of 1260 miles.

   a. Write an expression for the total number of hours, h, Helen must drive in terms of her average driving speed, r.

   b. Use your expression to find how long Helen must drive if she averages 45 miles per hour.

4. Radio station KPUB needs to raise $75,000 to cover its operating costs for the next 6 months.

   a. Write an expression for the amount of money, n, the station still needs in terms of the amount, r, they have already raised.

   b. Use your expression to find how much money they still need after they have raised $28,500.

5. Dress fabric is sold (from bolts with a standard width) for $5.79 per yard.

   a. Write an expression for the number of yards, y, of fabric in terms of the number of feet, f, of the fabric. (There are 3 feet in a yard.)

   b. Write an expression for the cost of the fabric in terms of the number of feet.

   c. Find the cost of 10 feet of fabric.

6. Yoon can maintain a running speed of 12 miles per hour.

   a. Write an expression for Yoon's running time (in hours), t, in terms of the distance, d, he runs.

   b. Write an expression for Yoon's time in minutes in terms of the distance he runs.

   c. How many minutes will it take Yoon to run 2.7 miles?

7. The area of a circle is equal to π times the square of its radius.

   a. Write an expression for the area, A, of a circle in terms of its radius, r.

   b. Find the area of a circle whose radius is 5 centimeters.

8. To estimate the speed (in miles per hour) at which an automobile was moving on a dry concrete road when the brakes were applied, multiply the length (in feet) of its skid mark by 24 and take the square root of the resulting product.

   a. Write an expression for the speed, s, of the car in terms of the length, l, of the skid mark.

   b. How fast was a car traveling if its skid mark was 160 feet long?

9. Farshid wants to compute the average of his homework scores. He has a total of 198 points from the 20 assignments he turned in. However, he missed some of the assignments entirely (and therefore has no points on any of them).

   a. Write an expression for the total number of assignments, n, in terms of how many Farshid missed, m.

   b. Write an expression for Farshid's average homework score, a, in terms of the number of assignments he missed.

   c. Find Farshid's average score if he missed five assignments.

10. Roma and four friends together bought a single lottery ticket and agreed to split any winnings evenly. The ticket was a $10,000 winner and in addition qualified for the Big Spin, a cash prize depending on the outcome of a spin of a prize wheel.

a. Write an expression for their total winnings, *w*, in terms of the prize, *p*, from the Big Spin.

b. Write an expression for Roma's share, *s*, of the winnings in terms of the prize from the Big Spin.

c. Find Roma's share if the prize from the Big Spin is $50,000.

11. The sales tax in the city of Preston is 7.9%.

a. Write an expression for the amount of sales tax, *t*, in terms of the price, *p*, of the item.

b. Find the total bill, *b*, for an item (price plus tax) in terms of the price of the item.

c. Find the total bill for an item whose price is $490.

12. A savings account pays (at the end of 1 year) 6.4% interest on the amount deposited.

a. Write an expression for the interest, *I*, earned in 1 year in terms of the amount deposited, *P*.

b. Find the total amount (initial deposit plus interest), *A*, in the account after 1 year in terms of the amount deposited.

c. Find the total amount in the account after 1 year if $350 was deposited.

13. Your best friend moves to another state and the long distance phone call costs $1.97 plus $0.39 for each minute.

a. Write an expression for the cost, *C*, of a long distance phone call in terms of the number of minutes, *m*, of the call.

b. Find the cost of a 27-minute phone call.

14. Arenac Airlines charges 47 cents per pound on its flight from Omer to Pinconning, both for passengers and for luggage. Mr. Owsley wants to take the flight with 15 pounds of luggage.

a. Write an expression for the cost, *C*, of the flight in terms of Mr. Owsley's weight, *w*.

b. Find the cost if Mr. Owsley weighs 162 pounds.

15. Juan buys a 50-pound bag of rice and consumes about 0.4 pound per week.

a. Write an expression for the amount of rice, *c*, Juan has consumed in terms of the number of weeks, *w*, since he bought the bag.

b. Write an expression for the amount of rice, *r*, Juan has left in terms of the number of weeks since he bought the bag.

c. Find the amount of rice Juan has left after 6 weeks.

16. Trinh is bicycling down a mountain road that loses 500 feet in elevation for each 1 mile of road. She started at an elevation of 6300 feet.

a. Write an expression for the elevation, *l*, Trinh has lost in terms of the distance, *d*, she has cycled.

b. Write an expression for Trinh's elevation, *e*, in terms of the number of miles she has cycled.

c. Find Trinh's elevation after she has cycled 9 miles.

17. Leon's truck gets 20 miles per gallon of gasoline. The gas tank holds 14.6 gallons of gasoline.

a. Write an expression for the amount, *u*, of gasoline used in terms of the number of miles driven, *m*.

b. Write an expression for the amount, *g*, of gasoline remaining in the tank in terms of the number of miles driven since the tank was last filled.

c. Find the amount of gasoline left in the tank when Leon has driven 110 miles.

18. As part of a charity fund-raising event, Sanaz found sponsors to pay her $16 for each hour that she skates. In addition, Sanaz will win a scholarship if she can raise $500.

a. Write an expression for the number of hours, *s*, Sanaz skates in terms of the amount of money, *m*, she raises.

b. Write an expression for the number of hours, *h*, Sanaz has left to skate to earn the scholarship in terms of the amount of money she has already raised.

c. Find the number of hours she has left to skate if she has already raised $380.

Simplify each expression according to the order of operations.

19. $\dfrac{3(6-8)}{-2} - \dfrac{6}{-2}$ **20.** $\dfrac{5(3-5)}{2} - \dfrac{18}{-3}$

21. $6[3 - 2(4 + 1)]$ **22.** $5[3 + 4(6 - 4)]$

23. $(4 - 3)[2 + 3(2 - 1)]$

24. $(8 - 6)[5 + 7(2 - 3)]$

25. $64 \div (8[4 - 2(3 + 1)])$

26. $27 \div (3[9 - 3(4 - 2)])$

27. $5[3 + (8 - 1)] \div (-25)$

28. $-3[-2 + (6 - 1)] \div 9$

29. $[-3(8 - 2) + 3] \cdot [24 \div 6]$

30. $[-2 + 3(5 - 8)] \cdot [-15 \div 3]$

31. -5^2 **32.** $(-15)^2$ **33.** $(-3)^4$

34. -3^4 **35.** -4^3 **36.** $(-4)^3$

37. $(-2)^5$ **38.** -2^5

39. $\dfrac{4 \cdot 2^3}{16} + 3 \cdot 4^2$ **40.** $\dfrac{4 \cdot 3^2}{6} + (3 \cdot 4)^2$

41. $\dfrac{3^2 - 5}{6 - 2^2} - \dfrac{6^2}{3^2}$ **42.** $\dfrac{3 \cdot 2^2}{4 - 1} + \dfrac{(-3)(2)^3}{6}$

43. $\dfrac{(-5)^2 - 3^2}{4 - 6} + \dfrac{(-3)^2}{2 + 1}$

44. $\dfrac{7^2 - 6^2}{10 + 3} - \dfrac{8^2 \cdot (-2)}{(-4)^2}$

Use a calculator to simplify each expression.

45. $\dfrac{-8398}{26 \cdot 17}$ **46.** $\dfrac{-415.112}{8.58 + 18.73}$

47. $\dfrac{112.78 + 2599.124}{27.56}$ **48.** $\dfrac{202,462 - 9510}{356}$

49. $\sqrt{24 \cdot 54}$ **50.** $\sqrt{\dfrac{1216}{19}}$

51. $\dfrac{116 - 35}{215 - 242}$ **52.** $\dfrac{842 - 987}{443 - 385}$

53. $\sqrt{27^2 + 36^2}$ **54.** $\sqrt{13^2 - 4 \cdot 21 \cdot 2}$

55. $\dfrac{-27 - \sqrt{27^2 - 4(4)(35)}}{2 \cdot 4}$

56. $\dfrac{13 + \sqrt{13^2 - 4(5)(-6)}}{2 \cdot 5}$

Write an algebraic expression that corresponds to each graphing calculator keying sequence.

57. a. 2 $\boxed{+}$ 3 $\boxed{\div}$ 4 $\boxed{\text{ENTER}}$

    b. $\boxed{(}$ 2 $\boxed{+}$ 3 $\boxed{)}$ $\boxed{\div}$ 4 $\boxed{\text{ENTER}}$

58. a. 72 $\boxed{\div}$ $\boxed{(}$ 6 $\boxed{\times}$ 2 $\boxed{)}$ $\boxed{\text{ENTER}}$

    b. 72 $\boxed{\div}$ 6 $\boxed{\times}$ 2 $\boxed{\text{ENTER}}$

59. a. $\boxed{(-)}$ 23 $\boxed{x^2}$ $\boxed{\text{ENTER}}$

    b. $\boxed{(}$ $\boxed{(-)}$ 23 $\boxed{)}$ $\boxed{x^2}$ $\boxed{\text{ENTER}}$

60. a. 96 $\boxed{\div}$ 2 $\boxed{\wedge}$ 3 $\boxed{\text{ENTER}}$

    b. $\boxed{(}$ 96 $\boxed{\div}$ 2 $\boxed{)}$ $\boxed{\wedge}$ 3 $\boxed{\text{ENTER}}$

61. a. $\boxed{\sqrt{\ }}$ $\boxed{(}$ 9 $\boxed{+}$ 16 $\boxed{)}$ $\boxed{\text{ENTER}}$

    b. $\boxed{\sqrt{\ }}$ 9 $\boxed{+}$ 16 $\boxed{\text{ENTER}}$

62. a. $\boxed{(-)}$ 5 $\boxed{\times}$ 2 $\boxed{x^2}$ $\boxed{\text{ENTER}}$

    b. $\boxed{(}$ $\boxed{(-)}$ 5 $\boxed{\times}$ 2 $\boxed{)}$ $\boxed{x^2}$ $\boxed{\text{ENTER}}$

Evaluate each expression for the given values of the variable. Use your calculator where appropriate.

63. $\dfrac{5(F - 32)}{9}$; $F = 212$

64. $\dfrac{a - 4s}{1 - r}$; $r = 2$, $s = 12$,   and $a = 4$

65. $P + Prt$; $P = 1000$, $r = 0.04$,   and $t = 2$

66. $R(1 + at)$; $R = 2.5$, $a = 0.05$,   and $t = 20$

67. $\dfrac{1}{2} gt^2 - 12t$; $g = 32$ and $t = \dfrac{3}{4}$

68. $\dfrac{Mv^2}{g}$; $M = \dfrac{16}{3}$, $v = \dfrac{3}{2}$,   and $g = 32$

69. $\dfrac{32(V - v)^2}{g}$; $V = 12.78$, $v = 4.26$, and $g = 32$

70. $\dfrac{32(V - v)^2}{g}$; $V = 38.3$, $v = -6.7$, and $g = 9.8$

1.3 LINEAR EQUATIONS

In the last section, we wrote simple equations in two variables to model a problem. In this section and the next, we review some techniques for solving equations that you learned in elementary algebra.

Finding the solutions of many equations involves generating simpler equations that have the same solutions. (If the new equation is simple enough, we can can see its solution right away. Then we know the solution of the original equation as well.) Equations that have identical solutions are called **equivalent equations**. For example,

$$3x - 5 = x + 3 \tag{1}$$

and

$$2x = 8$$

are equivalent equations because the solution to each equation is 4. Often we can find simpler equivalent equations by analyzing what operations were performed on the variable and then "undoing" those operations in reverse order.

Solving Linear Equations

A great variety of practical problems can be solved with **linear**, or first-degree, equations. These are equations that can be written so that every term is either a constant or a constant times the variable. Equation (1) is an example of a linear equation. Recall the following rules for generating equivalent equations.

TO GENERATE EQUIVALENT EQUATIONS

1. Add or subtract the *same* number to *both* sides of an equation.

2. Multiply or divide *both* sides of an equation by the *same* number (except zero).

An application of either of these rules will produce a new equation that is equivalent to the old one, and thus preserve the solution. We apply the rules with the goal of "isolating" the variable on one side of the equation.

Example 1

Solve the equation $3x - 5 = x + 3$.

Solution We first collect all the variable terms on one side of the equation and the constant terms on the other side.

$$3x - 5 - x = x + 3 - x \qquad \text{Subtract } x \text{ from both sides.}$$

$$2x - 5 = 3 \qquad \text{Simplify.}$$

$$2x - 5 + 5 = 3 + 5 \qquad \text{Add 5 to both sides.}$$

$$2x = 8 \qquad \text{Simplify.}$$

$$\frac{2x}{2} = \frac{8}{2} \qquad \text{Divide both sides by 2.}$$

$$x = 4 \qquad \text{Simplify.}$$

The solution is 4. (You can check the solution by substituting 4 into the original equation and showing that a true statement results.)

The following steps should enable you to solve any linear equation. Of course, you may not need all the steps for a particular equation.

TO SOLVE A LINEAR EQUATION

1. First, simplify each side of the equation separately.

   a. Apply the distributive law to remove parentheses.

   b. Collect like terms.

2. By adding or subtracting appropriate terms to both sides of the equation, get all the variable terms on one side and all the constant terms on the other.

3. Divide both sides of the equation by the coefficient of the variable.

Example 2

Solve $3(2x - 5) - 4x = 2x - (6 - 3x)$.

Solution Begin by simplifying each side of the equation.

$$3(2x - 5) - 4x = 2x - (6 - 3x) \qquad \text{Apply the distributive law.}$$

$$6x - 15 - 4x = 2x - 6 + 3x \qquad \text{Combine like terms on each side.}$$

$$2x - 15 = 5x - 6$$

Next, get all the variable terms on the left side of the equation and all the constant terms on the right side.

$$2x - 15 - 5x + 15 = 5x - 6 - 5x + 15 \qquad \text{Add} - 5x + 15 \text{ to both sides.}$$

$$-3x = 9$$

Finally, divide both sides of the equation by the coefficient of the variable.

$$-3x = 9 \qquad \text{Divide both sides by } -3.$$

$$x = -3$$

The solution is -3.

Problem Solving Most forms of problem solving involve translating a real-life problem into some structured system, say a computer programming language, or in our case, algebraic expressions. We can then use the techniques of algebra to solve the mathematical problem and translate that solution back into the context of the original problem. Learning to extract the relevant information from a situation and to create a model for the problem are among the most important mathematical skills you can acquire.

Here are some guidelines for problem solving with algebraic equations.

> **GUIDELINES FOR PROBLEM SOLVING**
> **Step 1** Identify the unknown quantity and assign a variable to represent it.
> **Step 2** Find some quantity that can be expressed in two different ways, and write an equation by setting the expressions equal.
> **Step 3** Solve the equation.
> **Step 4** Interpret your solution to answer the question in the problem.

In Step 1 it is a good idea to begin by writing an English phrase to describe the quantity you are looking for. Be as specific as possible in your description—if you are going to write an equation about this quantity, you must understand its properties! Remember that your variable must represent a *numerical* quantity. For example, x can represent the *speed* of a train, but not just "the train."

Writing an equation is the hardest part of the problem. Note that the "quantity" mentioned in Step 2 will probably *not* be the same unknown quantity you are looking for, but the algebraic expressions you write *will* involve your variable. For example, if your variable represents the *speed* of a train, your equation might be about the *distance* the train traveled.

In the rest of this section, we consider some common applications of linear equations in one variable.

Supply and Demand

Sometimes we can obtain an equation by knowing something about the subject under discussion. For example, in economics we often hear about the law of supply and demand. The higher the price of a product, the more its manufacturers will be willing to supply, but consumers will buy more of the product the lower its price is set. The price at which the demand for a product equals the supply is called the *equilibrium price*.

Example 3

The Coffee Connection finds that when it charges p dollars for a pound of coffee, it can sell $800 - 60p$ pounds per month. On the other hand, at a price of p dollars a pound, International Food and Beverage will supply the Coffee Connection with $175 + 40p$ pounds of coffee per month. What price should the Coffee Connection charge for a pound of coffee so that its monthly inventory will sell out?

Solution

Step 1   We are looking for the equilibrium price, p.

Step 2   The Coffee Connection would like the demand for its coffee to equal its supply. We equate the expressions for supply and for demand to obtain the equation

$$800 - 60p = 175 + 40p$$

Step 3   Solve the equation. We want all terms containing the variable, p, on one side of the equation and all constant terms on the other side. To accomplish this, we add $60p$ to both sides of the equation and subtract 175 from both sides to obtain

$$800 - 60p + 60p - 175 = 175 + 40p + 60p - 175$$

$$625 = 100p \qquad \text{Divide both sides by 100.}$$

$$6.25 = p$$

Step 4   The Coffee Connection should charge $6.25 per pound for its coffee.

Percent Problems

Problems involving percent provide many examples of linear equations. Recall the basic formula for computing percentages:

$$P = rW$$
the **P**art (or percentage) = the percent **r**ate $\times$ the **W**hole amount

Also recall that a *percent increase* or *percent decrease* is calculated as a fraction of the *original* amount. For example, suppose you make $6.00 an hour now, but next month you are expecting a 5% raise. You new salary should be

$$\underset{\substack{\text{Original} \\ \text{salary}}}{\$6.00} + \underset{\text{Increase}}{0.05(\$6.00)} = \underset{\substack{\text{New} \\ \text{salary}}}{\$6.30}$$

Example 4

The price of housing in urban areas increased 4% over the past year. If a certain house costs $100,000 today, what was its price last year?

Solution

Step 1   Let c represent the cost of the house last year.

Step 2   Express the current price of the house in two different ways. During the past year, the price of the house increased by 4%, or $0.04c$. Its

current price is thus

$$(1)c + 0.04c = c(1 + 0.04) = 1.04c$$

Original cost Price increase

This expression is equal to the value given for the current price of the house:

$$1.04c = 100,000$$

Step 3 To solve this equation, we divide both sides by 1.04 to find

$$c = \frac{100,000}{1.04} = 96,153.846$$

Step 4 To the nearest cent, the cost of the house last year was $96,153.85.

COMMON ERROR

In Example 4 it would be incorrect to calculate last year's price by subtracting 4% of $100,000 from $100,000 to get $96,000. (Do you see why?)

Weighted Averages Recall that we find the **average**, or **mean**, of a set of values by adding up the values and dividing the sum by the number of values. Thus the average, $\bar{x}$, of the numbers $x_1, x_2, \ldots , x_n$ is given by

$$\bar{x} = \frac{x_1 + x_2 + \cdots + x_n}{n}$$

In a **weighted average**, the numbers being averaged occur with different frequencies, or are "weighted" differently in their contribution to the average value. For instance, suppose a biology class of 12 students takes a 10-point quiz. Two students receive 10s, three receive 9s, five receive 8s, and two receive scores of 6. The average score earned on the quiz is then

$$x = \frac{2(10) + 3(9) + 5(8) + 2(6)}{12} = 8.25$$

The numbers in color are called the **weights**—in this example they represent the number of times each score was counted. Note that n, the total number of scores, is equal to the sum of the weights:

$$12 = 2 + 3 + 5 + 2$$

Example 5

Kwan's grade in his accounting class will be computed as follows: Tests count for 50% of the grade, homework counts for 20%, and the final exam counts for 30%. If Kwan has an average of 84 on tests and 92 on homework, what score does he need on the final exam to earn a grade of 90?

Solution

Step 1 Let x represent the final exam score Kwan needs.

Step 2 Kwan's grade is the weighted average of his test, homework, and final exam scores:

$$\frac{\mathbf{0.50}(84) + \mathbf{0.20}(92) + \mathbf{0.30}x}{1.00} = 90 \tag{2}$$

(Notice that the sum of the weights is 1.00, or 100% of Kwan's grade.) Multiply both sides of the equation by 1.00 (the sum of the weights) to get

$$0.50(84) + 0.20(92) + 0.30x = 1.00(90) \tag{3}$$

Step 3 Solve the equation. Simplify the left side first.

$$60.4 + 0.30x = 90 \qquad \text{Subtract 60.4 from both sides.}$$
$$0.30x = 29.6 \qquad \text{Divide both sides by 0.30.}$$
$$x = 98.7$$

Step 4 Kwan needs a score of 98.7 on the final exam to earn a grade of 90.

In Example 5 notice that we rewrote the formula for a weighted average, Equation (2), in a simpler form, Equation (3). In this form the formula says

$$w_1x_1 + w_2x_2 + \cdots + w_nx_n = W\bar{x}$$

where W is the sum of the weights, or

The sum of the weighted values equals the sum of the weights times the average value.

This form is particularly useful in solving problems involving mixtures.

Example 6

The vet advised Delbert to feed his dog Rollo with kibble that is no more than 8% fat. Rollo likes Juicy Bits, which are 15% fat. LeanMeal is much more expensive but is only 5% fat. How much LeanMeal should Delbert mix with 50 pounds of Juicy Bits to make a mixture that is 8% fat?

Solution

Step 1 Let p represent the number of pounds of LeanMeal needed.

Step 2 In this problem we want the weighted average of the fat contents in the two kibbles to be 8%. The weights are the number of pounds of each kibble we use. It is often useful to summarize the given information in a table such as Table 1.5.

	% Fat	Total Pounds	Pounds of Fat
Juicy Bits	15%	50	0.15(50)
LeanMeal	5%	p	0.05p
Mixture	8%	50 + p	0.08(50 + p)

Table 1.5

The amount of fat in the mixture must come from adding the amounts of fat in the two ingredients. This gives us the equation

$$0.15\,(50) + 0.05p = 0.08\,(50 + p)$$

Note that this equation is an example of the formula for weighted averages.

Step 3 Simplify each side of the equation, then solve.

$$7.5 + 0.05p = 4 + 0.08p$$
$$3.5 = 0.03p$$
$$p = 116.\overline{6}$$

Step 4 Delbert must mix 116⅔ pounds of LeanMeal with 50 pounds of Juicy Bits to make a mixture that is 8% fat.

Formulas A **formula** is any equation that relates several variable quantities. For example, the equation

$$P = 2l + 2w \tag{4}$$

is a formula that gives the perimeter of a rectangle in terms of its length and width. Formulas are used in almost every discipline that deals with numerical information.

Sometimes a formula is not given to us in the form that will be most useful for our purposes. For instance, suppose we have a fixed amount of wire fencing to enclose an exercise area for rabbits, and we would like to see what dimensions are possible for different rectangles with that perimeter. In this case it would be more useful to have a formula for, say, the length of the rectangle in terms of its perimeter and its width. We can find such a formula by solving Equation (4) for l in terms of P and w.

$$2l + 2w = P \qquad \text{Subtract 2w from both sides.}$$
$$2l = P - 2w \qquad \text{Divide both sides by 2.}$$
$$l = \frac{P - 2w}{2}$$

The result is a new formula that gives the length of a rectangle in terms of its perimeter and its width.

Example 7

The formula $5F = 9C + 160$ relates the temperature in degrees Fahrenheit, F, to the temperature in degrees Celsius, C. Solve the formula for C in terms of F.

Solution Begin by isolating the term that contains C.

$$5F = 9C + 160 \qquad \text{Subtract 160 from both sides.}$$

$$5F - 160 = 9C \qquad \text{Divide both sides by 9.}$$

$$\frac{5F - 160}{9} = C$$

The formula for C in terms of F is

$$C = \frac{5F - 160}{9}, \quad \text{or} \quad C = \frac{5}{9}F - \frac{160}{9}$$

Exercise 1.3

Solve each linear equation.

1. $3x + 5 = 26$
2. $2 + 5x = 37$
3. $3(z + 2) = 37$
4. $2(z - 3) = 15$
5. $3y - 2(y - 4) = 12 - 5y$
6. $5y - 3(y + 1) = 14 + 2y$
7. $0.8w - 2.6 = 1.4w + 0.3$
8. $4.8 - 1.3w = 0.7w + 2.1$
9. $0.25t + 0.10(t - 4) = 11.60$
10. $0.12t + 0.08(t + 10,000) = 12,000$

Solve each problem by writing and solving an equation.

11. Celine's boutique carries a line of jewelry made by a local artists' co-op. If she charges p dollars for a pair of earrings, she finds that she can sell $200 - 5p$ pairs per month. On the other hand, the co-op will provide her with $56 + 3p$ pairs of earrings when she charges p dollars per pair. What price should Celine charge so that the demand for earrings will equal her supply?

12. Curio Electronics sells garage door openers. If it charges p dollars per unit, it sells $120 - p$ openers per month. The manufacturer will supply $20 + 2p$ openers at a price of p dollars each. What price should Curio Electronics charge so that its monthly supply will meet its demand?

13. Roger sets out on a bicycle trip at an average speed of 16 miles per hour. Six hours later his wife finds his patch kit on the dining room table. If she heads after him in the car at 45 miles per hour, how long will it be before she catches him?

   a. What are we asked to find in this problem? Assign a variable to represent it.

   b. Write an expression in terms of your variable for the distance Roger's wife drives.

   c. Write an expression in terms of your variable for the distance Roger has cycled.

   d. Write an equation and solve it.

14. Kate and Julie set out in their sailboat on a straight course at 9 miles per hour. Two hours later their mother becomes worried and sends their father after them in the speedboat. If their father travels at 24 miles per hour, how long will it be before he catches them?

a. What are we asked to find in this problem? Assign a variable to represent it.

b. Write an expression in terms of your variable for the distance Kate and Julie sailed.

c. Write an expression in terms of your variable for the distance their father traveled.

d. Write an equation and solve it.

15. The reprographics department has a choice of two new copying machines. One sells for $20,000 and costs $0.02 per copy to operate. The other sells for $17,500, but its operating costs are $0.025 per copy. The repro department decides to buy the more expensive machine. How many copies must they make before the higher price is justified?

a. What are we asked to find in this problem? Assign a variable to represent it.

b. Write expressions in terms of your variable for the total cost incurred by each machine.

c. Write an equation and solve it.

16. Annie needs a new refrigerator and can choose between two models of the same size. One model sells for $525 and costs $0.08 per hour to run. A more energy efficient model sells for $700 but runs for $0.05 per hour. If Annie buys the more expensive model, how long will it be before she starts saving money?

a. What are we asked to find in this problem? Assign a variable to represent it.

b. Write expressions in terms of your variable for the total cost incurred by each refrigerator.

c. Write an equation and solve it.

17. The population of Midland has been growing at an annual rate of 8% over the past 5 years. Its present population is 135,000.

a. Assuming the same rate of growth, what do you predict the population of Midland will be next year?

b. What was the population of Midland last year?

18. For the past 3 years, the annual inflation rate has been 6%. This year a steak dinner at Benny's costs $12.

a. Assuming the same rate of inflation, what do you predict the price of a steak dinner will be next year?

b. What did a steak dinner cost last year?

19. Virginia took a 7% pay cut when she changed jobs last year. What percent pay increase must she receive this year in order to match her old salary of $24,000? (*Hint:* What was Virginia's salary after the pay cut?)

20. Clarence W. Networth took a 16% loss in the stock market last year. What percent gain must he realize this year in order to restore his original holdings of $85,000? (*Hint:* What was the value of Clarence's stock holdings after the loss?)

21. Delbert's test average in algebra is 77. If the final exam counts for 30% of the grade and the test average counts for 70%, what must Delbert score on the final exam to have a term average of 80?

22. Harold's batting average for his first 40 at-bats is .385. What batting average did he maintain over his last 90 at-bats if his season average was .350?

23. A horticulturist needs a fertilizer that is 8% potash, but she can find only fertilizers that contain 6% and 15% potash. How much of each should she mix to obtain 10 pounds of 8% potash fertilizer?

a. What are we asked to find in this problem? Assign a variable to represent it.

b. Write algebraic expressions in terms of your variable for the amounts of each fertilizer the horticulturist uses. Use Table 1.6.

Pounds of Fertilizer	% Potash	Pounds of Potash

Table 1.6

c. Write expressions for the amount of potash in each batch of fertilizer.

d. Write two different expressions for the amount of potash in the mixture. Now write an equation and solve it.

24. A sculptor wants to cast a bronze statue from an alloy that is 60% copper. He has 30 pounds of a 45% alloy. How much 80% copper alloy should he mix with it to obtain the 60% copper alloy?

a. What are we asked to find in this problem? Assign a variable to represent it.

b. Write algebraic expressions in terms of your variable for the amounts of each alloy the sculptor uses. Use Table 1.7.

Pounds of Alloy	% Copper	Pounds of Copper

Table 1.7

c. Write expressions in terms of your variable for the amount of copper in each batch of alloy.

d. Write two different expressions for the amount of copper in the mixture. Now write an equation and solve it.

25. Lacy's Department Stores wants to keep the average salary of its employees under $19,000 per year. If the downtown store pays its 4 managers $28,000 per year and its 12 department heads $22,000 per year, how much can it pay its 30 clerks?

a. What are we asked to find in this problem? Assign a variable to represent it.

b. Write algebraic expressions for the total amounts Lacy's pays its managers, its department heads, and its clerks.

c. Write two different expressions for the total amount Lacy's pays in salaries each year.

d. Write an equation and solve it.

26. Federal regulations require that 60% of all vehicles manufactured next year comply with new emission standards. Major Motors can bring 85% of their small trucks in line with the standards but only 40% of their automobiles. If Major Motors plans to manufacture 20,000 automobiles next year, how many trucks will they have to produce in order to comply with the federal regulations?

a. What are we asked to find in this problem? Assign a variable to represent it.

b. Write algebraic expressions for the number of trucks and the number of cars that will meet emission standards.

c. Write two different expressions for the total number of vehicles that will meet the standards.

d. Write an equation and solve it.

Solve each formula for the specified variable.

27. $v = k + gt$,   for t

28. $S = 3\pi d + \pi a$,   for d

29. $S = 2w(w + 2h)$,   for h

30. $A = P(1 + rt)$,   for r

31. $P = a + (n - 1)d$,   for n

32. $R = 2d + h(a + b)$,   for b

33. $A = \pi rh + \pi r^2$,   for h

34. $A = 2w^2 + 4lw$,   for l

1.4   OTHER TYPES OF EQUATIONS

Many other kinds of equations are useful in problem solving. In this section we consider a few simple examples.

Simple Quadratic Equations   A **quadratic equation** is one of the form

$$ax^2 + bx + c = 0$$

where a, b, and c are constants, and a is not zero. In other words, a term containing the square of the variable appears in the equation. We will study quadratic equations in detail in Chapter 4; for now consider an example like

$$x^2 = 16$$

The variable, x, has been squared, so it seems reasonable that to solve the equation we can do the opposite, or take square roots of both sides of the equation. Doing so gives us

$$x = \pm \sqrt{16}$$
$$= \pm 4$$

and you can verify that both 4 and -4 are solutions of $x^2 = 16$. Note that 16, like every positive number, has two square roots, so the equation has two solutions.

Example 1

If a cat falls off a tree branch 20 feet above the ground, its height t seconds later is given by $h = 20 - 16t^2$.
a. What is the height of the cat 0.5 second later?
b. How long does the cat have to get in position to land on its feet before it reaches the ground?

Solutions **a.** In this question we are given the value of t and asked to find the corresponding value of h. To do this we should evaluate the formula for $t = 0.5$. We substitute 0.5 for t into the formula and simplify.

$$h = 20 - 16\,(\mathbf{0.5})^2 \qquad \text{Perform the power.}$$
$$= 20 - 16\,(0.25) \qquad \text{Multiply, then subtract.}$$
$$= 20 - 4 = 16$$

The cat is 16 feet above the ground after 0.5 second.
   A calculator keying sequence for the evaluation above is

$$20 \;\boxed{-}\; 16 \;\boxed{\times}\; 0.5 \;\boxed{x^2}\; \boxed{\text{ENTER}}$$

b. We would like to find the value of t when the height, h, is known. We substitute $h = 0$ into the equation to obtain

$$0 = 20 - 16\,t^2$$

To solve this equation for t, we first isolate t^2 on one side of the equation.

$$16\,t^2 = 20 \qquad\qquad \text{Divide by 16.}$$
$$t^2 = \frac{20}{16} = 1.25$$

Now take the square roots of both sides of the equation to find

$$t = \pm \sqrt{1.25} \approx \pm 1.118$$

We discard the negative solution, so the cat has approximately 1.12 seconds to be in position for landing.

On a graphing calculator, we can solve the equation above in two steps:

20 ÷ 16 ENTER   (Dividing by 16)

√ ANS ENTER   (Taking square roots)   ▭

Notice that in Example 1(a) we are given a value of t and asked to find h, and in Example 1(b) we are given a value of h and asked to find t.

Simple Radical Equations

A **radical equation** is one in which the variable appears under a square root (or other radical). For example,

$$\sqrt{x + 3} = 4$$

is a radical equation. As you might guess, we can solve this equation by first squaring both sides in order to "undo" the radical:

$$(\sqrt{x + 3})^2 = 4^2$$

$$x + 3 = 16$$

We can now complete the solution as usual to find $x = 13$.   The solution checks, because substituting 13 for x into the original equation results in a true statement: $\sqrt{13 + 3} = 4$.

We should always check the solution to a radical equation, because it is possible to introduce false or extraneous solutions when we square both sides of the equation. For example, the equation

$$\sqrt{x} = -5$$

has no solution, since the square root of x is never a negative number. However, if we try to solve the equation by squaring both sides, we find

$$(\sqrt{x})^2 = (-5)^2$$

$$x = 25$$

Since $\sqrt{25}$ does not equal -5, 25 is not a solution to the original equation, $\sqrt{x} = -5$.

Example 2

When a car brakes suddenly, its speed can be estimated from the length of the skid marks it leaves on the pavement. A formula for the car's speed in miles per hour is $v = \sqrt{24d}$, where the length of the skid marks, d, is given in feet.

a. If a car leaves skid marks 80 feet long, how fast was the car traveling when the driver applied the brakes?

b. How far will a car skid if its driver applies the brakes while traveling 80 miles per hour?

Solutions **a.** To find the velocity of the car, we evaluate the formula with $d = 80$.

$$v = \sqrt{24\,(\mathbf{80})} \qquad \text{Substitute 80 for } d.$$

$$= \sqrt{1920} \qquad \text{Multiply inside the radical.}$$

$$\approx 43.8178046 \qquad \text{Take the square root.}$$

The car was traveling approximately 44 miles per hour.

A graphing calculator keying sequence for the evaluation above is

$$\boxed{\sqrt{}}\ \boxed{(}\ \boxed{24}\ \boxed{\times}\ \boxed{80}\ \boxed{)}\ \boxed{\text{ENTER}}$$

b. We would like to find the value of d when the value of v is known. We substitute $v = 80$ into the equation to obtain

$$\mathbf{80} = \sqrt{24\,d}$$

Since d appears under a square root in the equation, we first square both sides to get

$$80^2 = (\sqrt{24\,d}\,)^2$$

or

$$6400 = 24d \qquad \text{Divide by 24.}$$

$$266.\overline{6} = d$$

You can check that this value for d works in the original equation. Thus the car will skid approximately 267 feet.

On a graphing calculator, we can solve the equation above by keying in

$$80\ \boxed{x^2}\ \boxed{\div}\ 24\ \boxed{\text{ENTER}} \qquad\qquad \boxed{}$$

Fractional Equations   If the variable in an equation appears in the denominator of a fraction, we should first multiply both sides of the equation by the denominator. This will bring the variable to the numerators of each term. For instance, to solve the equation

$$\frac{5}{x + 2} = 3$$

we multiply both sides of the equation by $x + 2$ to obtain

$$(x + 2)\,\frac{5}{x + 2} = 3\,(x + 2)$$

or

$$5 = 3\,(x + 2)$$

We can now proceed as usual to complete the solution. Simplify the right side by applying the distributive law to get

$$5 = 3x + 6 \qquad \text{Subtract 6 from both sides.}$$

$$-1 = 3x \qquad \text{Divide both sides by 3.}$$

$$x = \frac{-1}{3}$$

We are not quite finished with this equation. Whenever an equation involves variables in the denominator of a fraction, we must exclude from consideration any values that cause that denominator to equal zero. For our example the value -2 must be excluded, because the fraction $\dfrac{5}{x+2}$ is undefined when $x = -2$. After solving the equation algebraically, we must check that the value we obtain for the solution is not one of these excluded values. Since our value, $x = \dfrac{-1}{3}$, does not cause the denominator in the original equation to equal zero, it is in fact the solution of the equation.

Example 3

The weight of an object decreases as its distance from earth increases. If an object weighs W pounds on earth, then its weight w at a distance d miles from the center of the earth is given by

$$w = \frac{(3963)^2 W}{d^2}$$

(3963 is the radius of the earth in miles). An astronaut who weighs 125 pounds on earth weighs 75 pounds on a space station in orbit around the earth. How far is the space station above the center of the earth?

Solution Substitute 125 for W and 75 for w in the equation and solve for d. Note that the variable d appears in the denominator of a fraction. We must exclude the value $d = 0$ from this problem, since if $d = 0$ the fraction is undefined.

$$75 = \frac{(3963)^2\,(\mathbf{125})}{d^2} \qquad \text{Multiply both sides by } d^2.$$

$$75\,d^2 = (3963)^2\,(125) \qquad \text{Divide by 75.}$$

$$d^2 = \frac{(3963)^2\,(175)}{75} \qquad \text{Take square roots of both sides.}$$

$$d = \pm\sqrt{\frac{(3963)^2\,(175)}{75}} \approx \pm\,6053.58$$

The station is approximately 6054 miles above the center of the earth, or $6054 - 3963 = 2091$ miles above the surface of the earth. ⬜

In the solution to Example 3, notice that we did not perform the numerical calculations at each step. Instead we wrote down the operations needed, but we actually performed all the calculations in the last step. If you are working with a scientific or graphing calculator it is easy to key in all the calculations at once, so this approach can save time and effort. A keying sequence for this expression is

☑ [( ] 3963 [x^2] [×] 175 [÷] 75 [ )] [ENTER]

Proportions   Two variables are said to be **proportional** if the ratio of their corresponding values is always the same. For example, if Loren makes $6 an hour, then her earnings are proportional to the number of hours she works. Consider Table 1.8.

Hours Worked	Earnings	Ratio: $\dfrac{\text{Earnings}}{\text{Hours Worked}}$
5	$30	$\dfrac{30 \text{ dollars}}{5 \text{ hours}} = 6$ dollars per hour
12	$72	$\dfrac{72 \text{ dollars}}{12 \text{ hours}} = 6$ dollars per hour
30	$180	$\dfrac{180 \text{ dollars}}{30 \text{ hours}} = 6$ dollars per hour

Table 1.8

The ratio of Loren's earnings to the number of hours she works is always equal to 6. (This ratio is called the **constant of proportionality**.)

A **proportion** is a special kind of fractional equation in which each side is a ratio: $\dfrac{a}{b} = \dfrac{c}{d}$. We can use proportions to solve problems about proportional variables. If we multiply both sides of a proportion by the common denominator of the two fractions, bd, we get an equivalent equation without fractions. Thus

$$(bd)\frac{a}{b} = (bd)\frac{c}{d}$$

or

$$ad = bc$$

This fundamental property can be used to rewrite any proportion.

> **FUNDAMENTAL PROPERTY OF PROPORTIONS**
>
> If $\dfrac{a}{b} = \dfrac{c}{d}$,   then $ad = bc$.

Example 4

Solve the proportion $\dfrac{3.6}{2.1} = \dfrac{4.8}{v}$.

Solution   Rewrite the proportion using the fundamental property.

$$3.6\,v = (2.1)\,(4.8)$$

Now solve the equation as usual.

$$v = \frac{(2.1)\,(4.8)}{3.6} = 2.8$$

When using your calculator, you can key in this expression for v just as it appears.

We may know that two variables are proportional without knowing the constant of proportionality. For example, a car's gas mileage, the ratio of miles driven to gallons of gasoline used, is fairly constant under normal conditions.

Example 5

Jan's car uses 8 gallons of gas to travel 140 miles. How many gallons will be required for a trip of 450 miles?

Solution
Step 1   Let g represent the number of gallons of gas needed for the trip.
Step 2   Identify the ratio that remains constant: in this problem it is gas mileage, or the ratio $\dfrac{\text{miles}}{\text{gallons}}$. Set up two ratios for gas mileage, one using the given values and another using the unknown quantity. Make sure you use the same quantity (miles in this case) in the numerator of each ratio. The two ratios are equal, yielding the proportion

$$\frac{140 \text{ miles}}{8 \text{ gallons}} = \frac{450 \text{ miles}}{g \text{ gallons}}$$

Step 3   Apply the fundamental property of proportions above and solve to obtain

$$140g = 8(450)$$

$$g = \frac{8(450)}{140} \approx 25.71$$

Step 4   Jan will need approximately 25.7 gallons of gas for the trip.

Formulas We can use the equation-solving techniques discussed above to solve formulas for one variable in terms of the others. Here are some examples that involve fractions, radicals, and squared variables.

Example 6

Solve the formula $p = \dfrac{v}{q + v}$ for v.

Solution Because the variable we want appears in the denominator, we must first multiply both sides of the equation by that denominator, $q + v$:

$$(q + v)\, p = \frac{v}{q + v}\,(q + v)$$

or

$$(q + v)\, p = v$$

Apply the distributive law on the left side, and collect all terms that involve v on one side of the equation.

$$qp + vp = v \qquad \text{Subtract } vp \text{ from both sides.}$$

$$qp = v - vp$$

We cannot combine the two terms containing v because they are not like terms. However, we can *factor out* v, so that the right side is written as a single term containing the variable v. We can then complete the solution.

$$qp = v\,(1 - p) \qquad \text{Divide both sides by } 1 - p.$$

$$\frac{qp}{1 - p} = v$$

$\qquad\qquad\qquad\qquad\qquad\qquad\qquad\qquad\qquad\qquad$ ▭

Example 7

Solve the formula $V = \dfrac{1}{3}\pi r^2 h$ for r.

Solution Since the variable we want is squared, use extraction of roots, First, multiply both sides by 3 to clear the fractions.

$$3V = \pi r^2 h \qquad \text{Divide both sides by } \pi h.$$

$$\frac{3V}{\pi h} = r^2 \qquad \text{Take square roots.}$$

$$\pm\sqrt{\frac{3V}{\pi h}} = r$$

$\qquad\qquad\qquad\qquad\qquad\qquad\qquad\qquad\qquad\qquad$ ▭

Example 8

Solve the formula $t = \sqrt{1 + s^2}$ for s.

Solution Since the variable we want is under a radical, square both sides of the equation.

$$t^2 = 1 + s^2 \qquad \text{Subtract 1 from both sides.}$$

$$t^2 - 1 = s^2 \qquad \text{Take square roots.}$$

$$s = \pm\sqrt{t^2 - 1}$$

$\qquad\qquad\qquad\qquad\qquad\qquad\qquad\qquad\qquad\qquad$ ▭

Exercise 1.4

Solve each quadratic equation.

1. $9x^2 = 25$ 2. $4x^2 = 9$ 3. $2x^2 = 14$

4. $3x^2 = 15$ 5. $4x^2 - 24 = 0$

6. $3x^2 - 9 = 0$ 7. $\dfrac{2x^2}{3} = 4$ 8. $\dfrac{3x^2}{5} = 6$

9. If you invest P dollars in an account that pays compound interest at annual rate r, the amount of money, A, in the account after 2 years will be given by the formula

$$A = P(1 + r)^2$$

To the nearest tenth of a percent, what interest rate will you require if you want an investment of $1600 to grow to $2000 in 2 years?

10. A machinist wants to make a metal section of pipe that is 80 millimeters long and has an interior volume of 9000 cubic millimeters. If the pipe is 2 millimeters thick, its interior volume is given by the formula

$$V = \pi(r - 2)^2 h$$

where h is the length of the pipe and r is its outside radius. What should the outside radius of the pipe be?

Solve each radical equation.

11. $\sqrt{x} - 5 = 3$ 12. $\sqrt{x} - 4 = 1$

13. $\sqrt{y + 6} = 2$ 14. $\sqrt{y - 3} = 5$

15. $4\sqrt{z} - 8 = -2$

16. $-3\sqrt{z} + 14 = 8$

17. $5 + 2\sqrt{6 - 2w} = 13$

18. $8 - 3\sqrt{9 + 2w} = -7$

19. The period of a pendulum is the time it takes for the pendulum to complete one entire swing, say from far left to far right and back to far left. The greater the length, L, of the pendulum, the longer its period, T. In fact, if L is measured in feet, then the period is given in seconds by

$$T = 2\pi \sqrt{\dfrac{L}{32}}$$

Suppose you are standing in the Convention Center in Portland, Oregon, and you time the period of its Foucault pendulum (the longest in the world). Its period is approximately 10.54 seconds. Approximately how long is the pendulum?

20. If you are flying in an airplane at an altitude of h miles, then on a clear day you can see a distance of d miles to the horizon, where

$$d = 89.4 \sqrt{h}$$

At what altitude will you be able to see for a distance of 100 miles? How high is that in feet?

Solve each fractional equation.

21. $\dfrac{6}{w + 2} = 4$ 22. $\dfrac{12}{r - 7} = 3$

23. $9 = \dfrac{h - 5}{h - 2}$ 24. $-3 = \dfrac{v + 1}{v - 6}$

25. $\dfrac{15}{s^2} = 8$ 26. $\dfrac{3}{m^2} = 5$

27. $4.3 = \sqrt{\dfrac{18}{y}}$ 28. $6.5 = \dfrac{52}{\sqrt{z}}$

29. The total weight, S, that a beam can support is given in pounds by

$$S = \dfrac{182.6wh^2}{l}$$

where w is the width of the beam in inches, h is its height in inches, and l is the length of the beam in feet. A beam over the doorway in an interior wall of a house must support 1600 pounds. If the beam is 4 inches wide and 9 inches tall, how long can it be?

30. If two appliances are connected in parallel in an electrical circuit, the total resistance, R, in the circuit is given by

$$R = \frac{a\,b}{a + b}$$

where a and b are the resistances of the two appliances. If one appliance has a resistance of 18 ohms and the total resistance in the circuit is measured at 12 ohms, what is the resistance of the second appliance?

Solve each proportion.

31. $\dfrac{3}{4} = \dfrac{y + 2}{12 - y}$

32. $\dfrac{-3}{4} = \dfrac{y - 7}{y + 14}$

33. $\dfrac{50}{r} = \dfrac{75}{r + 20}$

34. $\dfrac{30}{r} = \dfrac{20}{r - 10}$

For each problem state the ratio that remains constant. Then set up a proportion and solve.

35. If the taxes on a house worth \$120,000 are \$2700, what would the taxes be on a house assessed at \$275,000 at the same tax rate?

36. If a typical household in the Midwest uses 83 million BTUs of electricity annually and pays \$1236, how much will a household that uses 70 million BTUs annually spend for energy?

37. Your rich uncle leaves \$100,000 to be divided between you and his daughter, Myrtle, in the ratio of 3 to 5. How much will you get?

38. The school district receives \$320,000 to be divided between maintenance and new equipment in the ratio of 4 to 5. How much should be allocated to maintenance?

39. The scale on a map of Michigan uses $\frac{3}{8}$ inch to represent 10 miles. If Isle Royale is $1\frac{11}{16}$ inches long on the map, what is the actual length of the island?

40. A photographer plans to enlarge a photograph that measures 8.3 centimeters by 11.2 centimeters to produce a poster that is 36 centimeters wide. How long will the poster be?

41. The Forest Service tags 200 perch and releases them into Spirit Lake. One month later it captures 80 perch and finds that 18 of them are tagged. What is the Forest Service's estimate of the original perch population of the lake?

42. The Wildlife Commission tags 30 Canada geese at one of its migratory feeding grounds. When the geese return, the commission captures 45 geese, of which 4 are tagged. What is the commission's estimate of the number of geese that use the feeding ground?

Solve each formula for the specified variable.

43. $S = \dfrac{a}{1 - r}$, for r

44. $I = \dfrac{E}{r + R}$, for R

45. $H = \dfrac{2xy}{x + y}$, for x

46. $M = \dfrac{ab}{a + b}$, for b

47. $F = \dfrac{mv^2}{r}$, for v

48. $A = \dfrac{\sqrt{3}}{4}s^2$, for s

49. $F = \dfrac{Gm_1m_2}{d^2}$, for d

50. $F = \dfrac{kq_1q_2}{r^2}$, for r

51. $T = 2\pi\sqrt{\dfrac{L}{g}}$, for L

52. $T = 2\pi\sqrt{\dfrac{m}{k}}$, for m

53. $r = \sqrt{t^2 - s^2}$, for s

54. $c = \sqrt{a^2 - b^2}$, for b

In this section we review some of the information you will need from geometry. We assume you are familiar with the formulas for area and perimeter of common geometric figures; you can find these formulas listed at the front of the book.

Right Triangles and the Pythagorean Theorem

A **right triangle** is a triangle in which one of the angles is a right angle, or 90°. Because the sum of the three angles in *any* triangle is 180°, the other two angles in a right triangle must have a sum of 180° − 90°, or 90°. For instance, if we know that one of the angles in a right triangle is 37°, then the remaining angle must be 90° − 37°, or 53°, as shown in Figure 1.7.

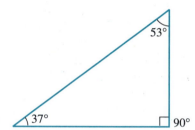

Figure 1.7

Example 1

In a right triangle, the medium-sized angle is 15° less than twice the smallest angle. Find the sizes of the three angles in Figure 1.8.

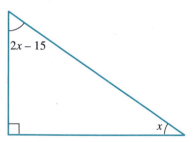

Figure 1.8

Solution

Step 1   Let x stand for the size of the smallest angle. Then the medium-sized angle must be $2x - 15$.

Step 2   Since the right angle is the largest angle, the sum of the smallest and medium-sized angles must be the remaining 90°. Thus

$$x + (2x - 15) = 90$$

Step 3 Solve the equation. Begin by simplifying the left side.

$$3x - 15 = 90 \qquad \text{Add 15 to both sides.}$$

$$3x = 105 \qquad \text{Divide both sides by 3.}$$

$$x = 35$$

Step 4 The smallest angle is 35°; the medium-sized angle is $2(35°) - 15°$, or 55°.

In a right triangle (see Figure 1.9), the longest side is opposite the right angle and is called the **hypotenuse**. Ordinarily, even if we know the lengths of two sides of a triangle, it is not easy to find the length of the third side (to solve this problem we need trigonometry). But for the special case of a right triangle, there is an equation that relates the lengths of the three sides. This property of right triangles was known to many ancient cultures, and we know it today by the name of a Greek mathematician, Pythagoras, who provided a proof of the result.

PYTHAGOREAN THEOREM

In a right triangle, if c stands for the length of the hypotenuse and a and b stand for the lengths of the other two sides, then

$$a^2 + b^2 = c^2$$

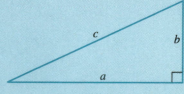

Figure 1.9

Example 2

The hypotenuse of a right triangle is 15 feet long. The third side is twice the length of the shortest side. Find the lengths of the other two sides.

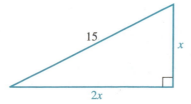

Figure 1.10

Solution

Step 1 Let x represent the length of the shortest side, so that the third side has length $2x$ (see Figure 1.10).

Step 2 Substituting these expressions into the Pythagorean theorem, we find

$$x^2 + (2x)^2 = 15^2$$

Step 3   This is a quadratic equation with no linear term, so we simplify and then isolate x^2.

$$x^2 + 4x^2 = 225 \qquad \text{Combine like terms.}$$

$$5x^2 = 225 \qquad \text{Divide both sides by 5.}$$

$$x^2 = 45$$

Taking square roots of both sides yields

$$x = \pm\sqrt{45} \approx \pm 6.708203932$$

Step 4   Since a length must be a positive number, we have that the shortest side has length approximately 6.71 feet, and the third side has length 2 (6.71), or approximately 13.42 feet.

Isosceles and Equilateral Triangles

Recall also that an **isosceles** triangle is one that has at least two sides of equal length. In an isosceles triangle, the angles opposite the equal sides, called the **base angles**, are equal in measure, as shown in Figure 1.11(a). In an **equilateral** triangle [Figure 1.11(b)], all three sides have equal length and all three angles have equal measure.

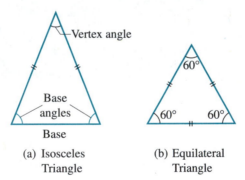

Figure 1.11

(a) Isosceles Triangle

(b) Equilateral Triangle

Inequalities and the Triangle Inequality

The longest side in a triangle is always opposite the largest angle, and the shortest side is opposite the smallest angle. It is also true that the sum of the lengths of any two sides of a triangle must be greater than the third side, or else the two sides will not meet to form a triangle! This fact is called the **triangle inequality.**

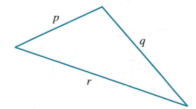

Figure 1.12

In Figure 1.12 we must have that

$$p + q > r$$

where p, q, and r are the lengths of the sides of the triangle. Recall that the symbol > is called an **inequality symbol,** and the statement $p + q > r$ is called an **inequality**. There are four basic inequality symbols:

> > "is greater than"
> < "is less than"
> $\geq$ "is greater than or equal to"
> $\leq$ "is less than or equal to"

Inequalities that include the symbol > or < are called **strict inequalities;** those that include $\geq$ or $\leq$ are called **nonstrict**.

You may also recall that inequalities have one curious feature: If we multiply or divide both sides of an inequality by a negative number, the direction of the inequality must be reversed. For example, if we multiply both sides of the inequality

$$2 < 5$$

by -3, we get

$$-3\,(2) > -3\,(5) \qquad \text{Inequality symbol changes from} < \text{to} >.$$

or

$$-6 > -15$$

Because of this property the rules for solving linear equations must be revised slightly for solving linear inequalities.

> ### RULES FOR SOLVING LINEAR INEQUALITIES
> **1.** We may add or subtract the same number to both sides of an inequality without changing its solutions.
> **2.** We may multiply or divide boths sides of an inequality by a *positive* number without changing its solutions.
> **3.** If we multiply or divide both sides of an inequality by a *negative* number, we must *reverse the direction of the inequality symbol.*

Example 3

Solve the inequality $4 - 3x \geq -17$.

Solution Use the rules above to isolate x on one side of the inequality.

$$4 - 3x \geq -17 \qquad \text{Subtract 4 from both sides.}$$

$$-3x \geq -21 \qquad \text{Divide both sides by } -3.$$

$$x \leq 7$$

Notice that we reversed the direction of the inequality when we divided by -3. Any number less than or equal to 7 is a solution of the inequality.

Now we can use the triangle inequality to discover information about the sides of a triangle.

Example 4

Two sides of a triangle have lengths 7 inches and 10 inches. What can you say about the length of the third side (see Figure 1.13)?

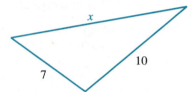

Figure 1.13

Solution   Let x represent the length of the third side of the triangle. By the triangle inequality, we must have that

$$x < 7 + 10 \qquad \text{or} \qquad x < 17$$

Looking at another pair of sides, we must also have that

$$10 < x + 7 \qquad \text{or} \qquad x > 3$$

Thus the third side must be greater than 3 inches but less than 17 inches long.

Similar Triangles

Two triangles are said to be **similar** if their corresponding angles are equal. Two similar triangles will have the same shape, but not necessarily the same size. One of the triangles will be an enlargement or a reduction of the other, so that their corresponding sides are proportional. In other words, for similar triangles the ratios of the corresponding sides are equal (see Figure 1.14).

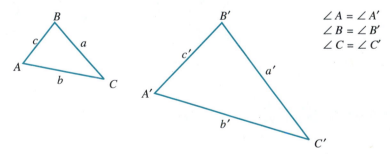

Figure 1.14

If any two pairs of corresponding angles of two triangles are equal, then the third pair must also be equal, since in both triangles the sum of the angles is 180°. Thus to show that two triangles are similar, we need only show that two pairs of angles are equal.

Example 5

The roof of an A-frame ski chalet forms an isosceles triangle with the floor (see Figure 1.15). The floor of the chalet is 24 feet wide, and the ceiling is 20 feet tall at the center. If a loft is built at a height of 8 feet from the floor, how wide will the loft be?

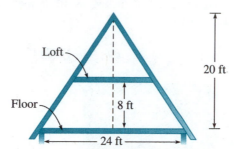

Figure 1.15

Solution From Figure 1.16 we can show that $\triangle ABC$ is similar to $\triangle ADE$. Both triangles include $\angle A$, and since $\overline{DE}$ is a parallel to $\overline{BC}$, $\angle ADE$ is equal to $\angle ABC$. Thus the triangles have two pairs of equal angles and are therefore similar triangles.

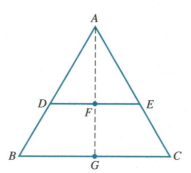

Figure 1.16

Step 1 Let w stand for the width of the loft.

Step 2 First note that if $FG = 8$, then $AF = 12$. Because $\triangle ABC$ is similar to $\triangle ADE$, the ratios of their corresponding sides (or corresponding altitudes) are equal. In particular,

$$\frac{w}{24} = \frac{12}{20}$$

Step 3 Solve the proportion for w. Begin by "cross-multiplying."

$$20w = (12)(24) \qquad \text{Apply the fundamental principle.}$$

$$w = \frac{288}{20} = 14.4 \qquad \text{Divide by 20.}$$

Step 4 The floor of the loft will be 14.4 feet wide.

Volume and Surface Area

The **volume** of a three-dimensional object measures its capacity, or how much space it encloses. Volume is measured in cubic units, such as cubic inches or cubic meters. The volume of a rectangular prism, or box, is given by the product of its length, width, and height. For example, the volume of the box of length 4 inches, width 3 inches, and height 2 inches shown in Figure 1.17 is

$$V = lwh = (4)\,(3)\,(2) = 24 \text{ cubic inches}$$

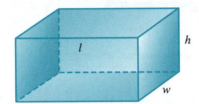

Figure 1.17

Formulas for the volumes of other common objects can be found at the front of the book.

Example 6

A cylindrical can must have a height of 6 inches but can have any reasonable radius.

a. Write an algebraic expression for the volume of the can in terms of its radius.

b. If the volume of the can should be approximately 170 cubic inches, what should its radius be?

Solutions a. The formula for the volume of any right circular cylinder is $V = \pi r^2 h$. If the height of the cylinder is 6 inches, then $V = \pi r^2 (6)$, or $V = 6\pi r^2$ (see Figure 1.18).

b. Substitute 170 for V and solve for r.

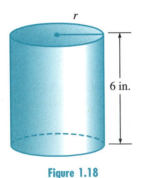

Figure 1.18

$$170 = 6\pi r^2 \qquad\qquad \text{Divide both sides by } 6\pi.$$

$$r^2 = \frac{170}{6\pi} \qquad\qquad \text{Take square roots.}$$

$$r = \sqrt{\frac{170}{6\pi}} \approx 3.00312$$

Thus the radius of the can should be approximately 3 inches. A calculator keying sequence for the expression above is

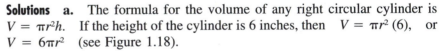

The **surface area** of a solid object is the sum of the areas of all the exterior faces of the object. It measures the amount of paper that would be needed to cover the object entirely. Since it is an area, it is measured in square units.

Example 7

Write a formula for the surface area of a closed box in terms of its length, width, and height (see Figure 1.19).

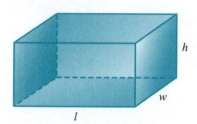

Figure 1.19

Solution The box has six sides; we must find the area of each side and add them together. The top and bottom of the box both have area *lw*, so together they contribute 2*lw* to the surface area. The back and front of the box both have area *lh*, so they contribute 2*lh* to the surface area. Finally, the left and right sides of the box both have area *wh*, so they add 2*wh* to the surface area. Thus the total surface area is

$$S = 2lw + 2lh + 2wh$$

Formulas for the surface areas of other common solids can be found at the front of the book.

Exercise 1.5

Solve each problem involving properties of triangles.

1. One angle of a triangle is 10° larger than another, and the third angle is 29° larger than the smallest. How large is each angle?

2. One angle of a triangle is twice as large as the second angle, and the third angle is 10° less than the larger of the other two. How large is each angle?

3. One acute angle of a right triangle is twice the other acute angle. How large is each acute angle?

4. One acute angle of a right triangle is 10° less than three times the other acute angle. How large is each acute angle?

5. The vertex angle of an isosceles triangle is 20° less than the sum of the equal angles. How large is each angle?

6. The vertex angle of an isosceles triangle is 30° less than one of the equal angles. How large is each angle?

7. The perimeter of an isosceles triangle is 42 centimeters and its base is 12 centimeters long. How long are the equal sides?

8. The altitude of an equilateral triangle is $\dfrac{\sqrt{3}}{2}$ times its base. If the perimeter of an isosceles triangle is 18 inches, what is its area?

Solve each problem by using the Pythagorean theorem.

9. The size of a TV screen is the length of its diagonal. If the width of a 35-inch TV screen is 28 inches, what is its height?

Figure 1.20

10. How high on a building will a 25-foot ladder reach if its base is 15 feet away from the bottom of the wall?

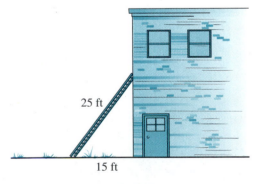

Figure 1.21

11. If a 30-meter pine tree casts a shadow of 30 meters, how far is the tip of the shadow from the top of the tree?

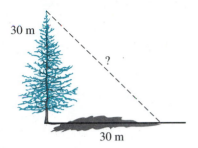

Figure 1.22

12. A baseball diamond is a square whose sides are 90 feet in length. Find the straight-line distance from home plate to second base.

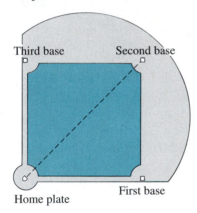

Figure 1.23

13. What size square can be inscribed in a circle of radius 8 inches?

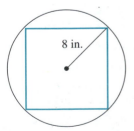

Figure 1.24

14. What size rectangle can be inscribed in a circle of radius 30 feet if the length of the rectangle must be three times its width?

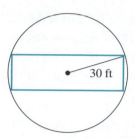

Figure 1.25

Solve each inequality.

15. $3x - 2 > 1 + 2x$ **16.** $2x + 3 \leq x - 1$

17. $\dfrac{-2x - 6}{-3} > 2$ **18.** $\dfrac{-2x - 3}{2} \leq -5$

19. $\dfrac{2x - 3}{3} \leq \dfrac{3x}{-2}$ **20.** $\dfrac{3x - 4}{-2} > \dfrac{-2x}{5}$

21. If two sides of a triangle are 6 feet and 10 feet long, what can you say about the length of the third side?

22. If one of the equal sides of an isosceles triangle is 8 millimeters long, what can you say about the length of the base?

Solve each problem involving similar triangles.

23. A 6-foot man stands 12 feet from a lamppost. His shadow is 9 feet long. How tall is the lamppost?

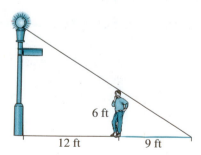

Figure 1.26

24. A rock climber estimates the height of a cliff she plans to scale as follows: She places a mirror on the ground so that she can just see the top of the cliff in the mirror while she stands straight. (The

angles 1 and 2 formed by the light rays are equal.) She then measures the distance from the spot where she is standing to the mirror (2 feet) and the distance from the mirror to the base of the cliff. If she is 5 feet 6 inches tall, how high is the cliff?

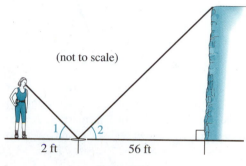

Figure 1.27

25. A conical tank is 12 feet deep and the diameter of the top is 8 feet. If the tank is filled with water to a depth of 7 feet, what is the area of the exposed surface of the water?

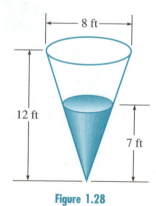

Figure 1.28

26. A florist fits a cylindrical piece of foam into a conical vase that is 10 inches high and measures 8 inches across the top. If the radius of the foam cylinder is $2\frac{1}{2}$ inches, how tall should the cylinder be to reach just to the top of the vase? (See Figure 1.29.)

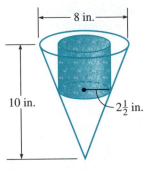

Figure 1.29

27. To measure the distance across the river shown in Figure 1.30, stand at *A* and sight across the river to a convenient landmark at *B*. Then measure the distances *AC, CD,* and *DE*. How wide is the river if *AC* = 20 feet, *CD* = 13 feet, and *DE* = 58 feet?

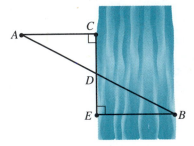

Figure 1.30

28. To measure the distance *EC* across the lake shown in Figure 1.31, stand at *A* and sight point *C* across the lake, then mark point *B*. Then sight to point *E* and mark point *D* so that *DB* is parallel to *CE*. If *AD* = 25 yards, *AE* = 60 yards, and *BD* = 30 yards, how wide is the lake?

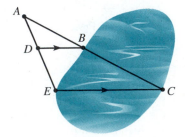

Figure 1.31

Use the formulas at the front of the book to find the volumes and surface areas of the given objects.

29. a. How much helium (in cubic meters) is needed to inflate a spherical balloon to a radius of 1.2 meters?

   b. How much gelatin (in square centimeters) is needed to coat a spherical pill whose radius is 0.7 centimeter?

30. a. How much storage space is in a rectangular box whose length is 12.3 inches, width is 4 inches, and height is 7.3 inches?

   b. How much marine sealer will be needed to paint a rectangular wooden storage locker with length 6.2 feet, width 5.8 feet, and height 2.6 feet?

31. a. How much grain can be stored in a cylindrical silo whose radius is 6 meters and height is 23.2 meters?

   b. How much paint is needed to cover a cylindrical storage drum whose radius is 15.3 inches and height is 4.5 inches?

32. a. A conical pile of sand is 8.1 feet high and has a radius of 4.6 feet. How much sand is in the pile?

   b. How much plastic is needed to line a conical funnel with a radius of 16 centimeters and a slant height of 42 centimeters?

Use the formulas for surface area and volume to answer the questions.

33. A cardboard packing box must have length 20 inches and width 16 inches.

   a. Write a formula for the surface area of the box in terms of its height.

   b. What is the maximum possible height for the box if its surface area should be no more than 1216 square inches?

34. We can find the area of a trapezoid by computing the product of its height and the average of its two bases.

   a. Write a formula for the area of a trapezoid whose height is 30 centimeters.

b. If the area of the trapezoid in part (a) is 1020 square centimeters and one base is 36 centimeters long, find the length of the other base.

35. A conical coffee filter is 8.4 centimeters tall.

  a. Write a formula for the filter's volume in terms of its radius.

  b. If the volume of the filter is 302.4 cubic centimeters, what is its radius?

36. A large bottle of shampoo is 20 centimeters tall and cylindrical in shape.

  a. Write a formula for the volume of the bottle in terms of its radius.

  b. What radius should the bottle have if it must hold 240 milliliters of shampoo? (A milliliter is equal to 1 cubic centimeter.)

1.6   GRAPHS

Graphs are one of the most useful tools for studying mathematical relationships. Constructing a graph can provide an overview of a quantity of data and perhaps help us identify trends or unexpected occurrences. Being able to interpret a graph can help us answer any number of questions about the data.

The data in Table 1.9 show the atmospheric pressure, on a certain day, at different altitudes above the surface of the earth. Meteorologists regularly collect such data by attaching to a weather balloon a device called a radio-sonde, which is equipped with a barometer and a radio transmitter. Altitudes are given in feet, and atmospheric pressures are given in inches of mercury.

Altitude	0	5,000	10,000	20,000	30,000	40,000	50,000
Pressure	29.7	24.8	20.5	14.6	10.6	8.5	7.3

Table 1.9

We observe a generally decreasing trend in pressure as the altitude increases, but it is difficult to say anything more precise about the relationship between pressure and altitude. A clearer picture emerges if we plot the data on a graph. To do this we use two perpendicular number lines called **axes.** We use the horizontal axis for the values of the first variable, altitude, and the vertical axis for the values of the second variable, pressure.

The entries in Table 1.9 are called **ordered pairs** in which the **first component** is the altitude and the **second component** is the atmospheric pressure measured at that altitude. For example, the first two entries can be represented by (0, 29.7) and (5000, 24.8). We plot the points whose **coordinates** are given by the ordered pairs as shown in Figure 1.32(a).

If we plot all the ordered pairs from Table 1.9, we can connect them with a smooth curve as shown in Figure 1.32(b). In doing this we are actually "estimating" the pressures that correspond to altitudes between those given, such as 15,000 feet or 37,000 feet. However, for many physical situations, variables are related so that one changes "smoothly" with respect to the other. We will assume this occurs in most of the modeling we do.

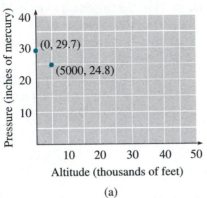

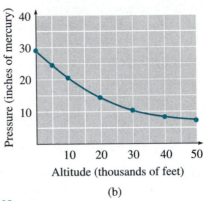

(a)

(b)

Figure 1.32

Example 1

From the graph in Figure 1.32(b), estimate
a.   the atmospheric pressure measured at an altitude of 15,000 feet.
b.   the altitude at which the pressure is 12 inches of mercury.

Solutions   **a.**   Notice that the point on the graph in Figure 1.33 with first coordinate 15,000 has a second coordinate of approximately 17.4. Hence we estimate the pressure at 15,000 feet to be 17.4 inches of mercury.

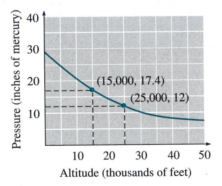

Figure 1.33

b.   Notice that the point on the graph with second coordinate 12 has a first coordinate of approximately 25,000, so an atmospheric pressure of 12 inches of mercury occurs at about 25,000 feet.

By using the graph of the data in Figure 1.32(b), we can obtain information about the relationship between altitude and pressure that would be difficult or impossible to obtain from the data alone.

Example 2

a.   For what altitudes is the pressure less than 18 inches of mercury?
b.   How much does the pressure decrease as the altitude increases from 15,000 feet to 25,000 feet?

c. For which 10,000-foot increase in altitude does the pressure change most rapidly?

Solutions **a.** From the graph in Figure 1.32(b), we see that the pressure has dropped to 18 inches of mercury at about 14,000 feet and that it continues to decrease as the altitude increases. Therefore, the pressure is less than 18 inches of mercury for altitudes greater than 14,000 feet.

b. At 15,000 feet the pressure is approximately 17.4 inches of mercury, and at 25,000 feet it is 12.3 inches. This represents a decrease in pressure of 17.4 − 12.3, or 5.1, inches of mercury.

c. By studying the graph, we see that the pressure decreases most rapidly at low altitudes, so we conclude that the greatest drop in pressure occurs between 0 and 10,000 feet. ▭

Graphs of Equations

In Example 1 we used a graph to illustrate a collection of data given by a table. Graphs can also help us analyze models given in the form of equations. Let's first review some facts about solutions of equations in two variables.

An equation in two variables, such as $y = 2x + 3$, is said to be *satisfied* if the variables are replaced by a pair of numbers that make the statement true. The pair of numbers is called a **solution** of the equation and is usually written as an ordered pair (x, y). The first number in the pair is the value of x, and the second number is the value of y.

To find a solution of a given equation, we assign any number to one of the variables and then solve the resulting equation for the second variable. Of course, by choosing different values for x, we can find many different solutions to the equation. For example, to obtain solutions to the equation

$$y = 2x + 3 \tag{5}$$

we might choose the values $-2, 0$, and 1 for x. When we substitute these x-values into the equation, we find a corresponding y-value for each.

$$\text{When } x = -2, \quad y = 2\,(-2) + 3 = -1$$

$$\text{When } x = 0, \quad y = 2\,(0) + 3 = 3$$

$$\text{When } x = 1, \quad y = 2\,(1) + 3 = 5$$

Thus the ordered pairs $(-2, -1)$, $(0, 3)$, and $(1, 5)$ are three solutions of $y = 2x + 3$.

We can also substitute values for y and see if any x-values make the equation true. For example, if we let $y = 10$, we have

$$10 = 2x + 3 \qquad \text{Subtract 3 from both sides.}$$

Solving this equation for x, we find

$$7 = 2x$$

or $x = 3.5$.   This means that the ordered pair (3.5, 10) is another solution of Equation (5). Because we could have used any value for x (or for y), Equation (5) has infinitely many solutions.

Since an equation in two variables may have infinitely many solutions, we obviously cannot list them all. However, we can display the solutions on a graph. To do this we use a **Cartesian** (or **rectangular**) **coordinate system**, as shown in Figure 1.34. Every point in the plane can be located by its coordinates, and every ordered pair of coordinates corresponds to a unique point in the plane, called its graph.

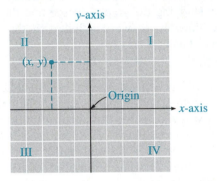

Figure 1.34

The **graph of an equation** is the graph of all the solutions of the equation. Thus a particular point is included in the graph of an equation if the coordinates of the point satisfy the equation. If the coordinates of a point do not satisfy the equation, then the point is not part of the graph. We can think of the graph of an equation as a picture of the solutions of the equation. For example, the graph of Equation (5),

$$y = 2x + 3$$

is shown in Figure 1.35.

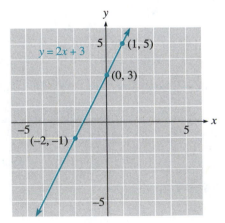

Figure 1.35

The graph does not display *all* the solutions of Equation (5), but it shows important features like the intercepts on the x- and y-axes and enough of a picture for us to infer how the rest of the curve looks (in this case, a straight line). Since there is a solution corresponding to every real number x, the graph extends infinitely in either direction, as indicated by the arrows.

Example 3

In each case use the graph of $y = 0.5x^2 - 2$ in Figure 1.36 to decide whether the given ordered pair is a solution of the equation. Verify your answers algebraically.

a. $(-4, 6)$ **b.** $(3, 0)$

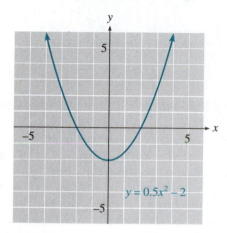

Figure 1.36

Solutions **a.** Since the point $(-4, 6)$ does lie on the graph, the ordered pair $x = -4$, $y = 6$ is a solution of $y = 0.5x^2 - 2$. We can verify this by substituting -4 for x:

$$0.5\,(-4)^2 - 2 = 0.5\,(16) - 2$$
$$= 8 - 2 = 6$$

b. Since the point $(3, 0)$ does not lie on the graph, the pair $x = 3$, $y = 0$ is not a solution of $y = 0.5x^2 - 2$. We substitute 3 for x to verify this:

$$0.5\,(3)^2 - 2 = 0.5\,(9) - 2$$
$$= 4.5 - 2 = 2.5 \neq 0$$

Using a Graphing Calculator

A graphing calculator is a versatile tool for exploring the graphs of equations. We will describe procedures for the TI-82 graphing calculator.

We begin by "setting the graphing window," which corresponds to drawing the x- and y-axes and choosing a scale for each when we graph by hand. Press the [ZOOM] key and then the number [6]. We have selected the "standard graphing window," which displays values from -10 to 10 on both axes.

Next we must enter the equation we wish to graph. Let's begin with the graph of

$$y = 2x + 3 \tag{6}$$

We press the $\boxed{Y =}$ key, then type in 2X + 3 after "$Y_1 =$." (Use the $\boxed{X, T, \theta}$ key to enter the X.) Finally, we press $\boxed{\text{GRAPH}}$ to display the graph. The graph looks like Figure 1.37.

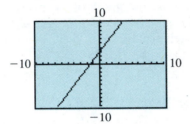

Figure 1.37

Example 4

Use a graphing calculator to graph the given equations.
a. $y = 2x + 3$ **b.** $y = 2x - 3$ **c.** $y = 2x$

Solutions   Press $\boxed{Y =}$ and type in 2X + 3 after "$Y_1 =$." To enter the second equation, press the down arrow $\boxed{\nabla}$ and enter 2X − 3 after "$Y_2 =$." (Be careful to use the subtraction key, $\boxed{-}$, not the negation key, $\boxed{(-)}$, when typing in this equation.) Press $\boxed{\nabla}$ again and enter 2X after "$Y_3 =$." Finally, choose the standard graphing window by pressing $\boxed{\text{ZOOM}}$ $\boxed{6}$. The calculator will draw the three graphs in the order they were entered. (Make sure you know which is which!) What do you notice about these three graphs in Figure 1.38?

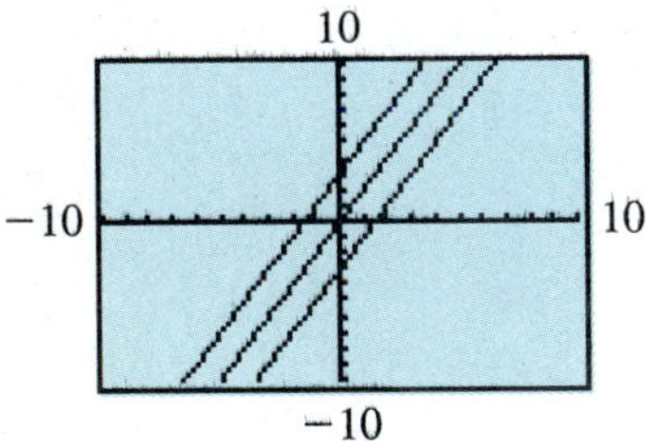

Figure 1.38

Here is a summary of the procedure for entering a graph in the standard window.

> **TO GRAPH AN EQUATION**
> **1.** Press $\boxed{Y =}$ and enter the equation(s) you wish to graph.
> **2.** Select the standard graphing window by pressing $\boxed{\text{ZOOM}}$ and then $\boxed{6}$.

When you want to erase the graphs, press $\boxed{Y =}$ and then press $\boxed{\text{CLEAR}}$. This deletes the expression following Y_1. Use the $\boxed{\nabla}$ key and the $\boxed{\text{CLEAR}}$ key to delete the rest of the equations on the display.

Trace Key   Graphing calculators have a "trace" feature to make it easy to read coordinates of points on the graph. Once the $\boxed{\text{TRACE}}$ button is pressed on the TI-82 graphing calculator, a "bug" begins flashing on the display. For example, let's graph the equation $y = x^2 - 2$. Press $\boxed{Y=}$ and enter X² − 2 after "$Y_1 =$." The keying sequence looks like this:

$$\boxed{X, T, \theta} \quad \boxed{x^2} \quad \boxed{-} \quad \boxed{2}$$

Next press $\boxed{ZOOM}$ $\boxed{6}$ to see the graph. (Notice that this graph is not a straight line.) Now press $\boxed{TRACE}$ to activate the bug. The coordinates of the bug, $x = 0$ and $y = 2$, appear at the bottom of the display.

We use the $\boxed{\triangleleft}$ and $\boxed{\triangleright}$ "arrow" keys to make the bug move along the graph. On the TI-82 calculator, you will notice that the $\boxed{TRACE}$ key does not locate exactly the point with x-coordinate 0.2: The bug jumps from $x = 0$ to $x = 0.21276596$, as shown in Figure 1.39. This jump occurs because the calculator screen cannot display *every* point. The size of the jumps in x-values depends on the window you choose.

The "ZOOM Decimal" window shows us a smaller piece of the graph, but the x-values change by 0.1 unit each time you press the arrow keys. Press $\boxed{ZOOM}$ $\boxed{4}$ to choose the **ZDecimal** option. The **ZDecimal** window displays x-values from -4.7 to 4.7 and y-values from -3.1 to 3.1. The graph of $y = x^2 - 2$ in this window is shown in Figure 1.40.

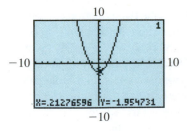

Figure 1.39

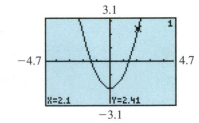

Figure 1.40

Example 5

Use the $\boxed{TRACE}$ key to approximate the missing coordinates for points on the graph of $y = x^2 - 2$.
a. $(2.1, ?)$ **b.** $(?, -1.2)$

Solutions **a.** First follow the discussion above to display Figure 1.40 on your calculator. Press $\boxed{TRACE}$ and use the left and right arrow keys to move the bug to the point with x-coordinate 2.1. The y-coordinate of this point is 2.41. Our answer is $(2.1, 2.41)$.
b. Use the left and right arrow keys to move the bug as close as possible to the point with y-coordinate -1.2. On the TI-82, this point has $y = -1.19$. The x-coordinate of this point is 0.9. But there is another point with y-coordinate -1.19. Use the left arrow key to move the bug to the point $(-0.9, -1.19)$. Thus our answers are $(0.9, -1.2)$ and $(-0.9, -1.2)$.

We can use the equation of the graph, $y = x^2 - 2$, to see how accurate our approximations are. To leave the graphing window and get back to the calculation or "Home" screen, press $\boxed{\text{2nd}}$ $\boxed{\text{QUIT}}$. First we will find the y-coordinate corresponding to $x = 2.1$ by evaluating $x^2 - 2$ at $x = 2.1$.   Enter

$$2.1 \ \boxed{x^2} \ \boxed{-} \ 2 \ \boxed{\text{ENTER}}$$

which gives us 2.41. Thus the exact coordinates of the point in Example 5(a) are (2.1, 2.41).

To find the x-coordinate(s) corresponding to $y = -1.2$, we solve the equation $-1.2 = x^2 - 2$ for x.

$$-1.2 = x^2 - 2$$
$$0.8 = x^2$$
$$x = \pm\sqrt{0.8} \approx \pm 0.89$$

Add 2 to both sides: -1.2 $\boxed{+}$ 2 $\boxed{\text{ENTER}}$
Take square roots: $\boxed{\text{2nd}}$ $\boxed{\sqrt{x}}$ $\boxed{\text{2nd}}$ $\boxed{\text{ANS}}$

Rounding this result to one decimal place, we approximate the points in Example 5(b) as $(0.9, -1.2)$ and $(-0.9, -1.2)$.

Example 6

At what point do the graphs of $y = 7.5 - 2.5x$ and $y = x + 0.5$ intersect?

Solution Graph the two equations in the standard graphing window as follows. Press $\boxed{\text{Y} =}$ and clear the display if necessary. Then enter the given equations at "$Y_1 =$" and "$Y_2 =$." Finally, press $\boxed{\text{ZOOM}}$ $\boxed{6}$ to see the graphs. Now use $\boxed{\text{TRACE}}$ to read the coordinates of the intersection point. Use the left and right arrow keys to move along the graph; use the up and down arrow keys to move from one graph to the other. (Note that a 1 or 2 appears in the upper right corner to tell you which graph the bug is on.)

We cannot locate the intersection point exactly in the standard window, so we try the **ZDecimal** window by pressing $\boxed{\text{ZOOM}}$ $\boxed{4}$. Now the $\boxed{\text{TRACE}}$ feature gives the coordinates of the intersection point as (2, 2.5), as shown in Figure 1.41. You can verify that these are the exact coordinates of the intersection point by substituting $x = 2$, $y = 2.5$ into each equation.

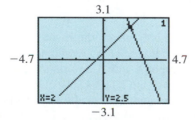

Figure 1.41

Example 7

Use the $\boxed{\text{TRACE}}$ key to approximate the coordinates of the turning points of

$$y = 0.04 \ (x - 1) \ (x + 4) \ (x - 5) + 1.3$$

Solution First graph the equation in the ZDecimal graphing window. After pressing $\boxed{Y =}$, enter the equation just as it is written. The graph should appear as in Figure 1.42. The turning points are the "peaks" and "valleys" where the graph changes from increasing to decreasing or vice versa.

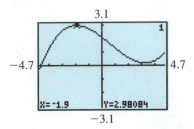

Figure 1.42

Activate the $\boxed{\text{TRACE}}$ and press the left arrow until the bug coincides with the peak of the graph. This point has a larger y-coordinate than the points just left or right of it. Its coordinates are $(-1.9, 2.98084)$, or about $(-1.9, 3)$. Now press the right arrow until the bug shows the smallest y-coordinate in the valley, 0.15828. At this point the x-coordinate is 3.3. Rounded to one decimal place, our approximation for the second turning point is (3.3, 0.2).

Here is a review of the keystrokes we have used.

> **SOME USEFUL KEYS**
> 1. Use the $\boxed{\text{CLEAR}}$ key and the up and down arrows to erase equations from the $\boxed{Y =}$ display.
> 2. Use $\boxed{\text{ZOOM}}$ $\boxed{6}$ for the standard graphing window; use $\boxed{\text{ZOOM}}$ $\boxed{4}$ for the **ZDecimal** window.
> 3. Use the $\boxed{\text{TRACE}}$ key and the left and right arrows to read the coordinates of points on a graph; use the up and down arrows to move to a different graph.
> 4. Press $\boxed{\text{2nd}}$ $\boxed{\text{QUIT}}$ to leave the graphing window and return to the Home screen.

Choosing a Graphing Window The standard graphing window is helpful in many instances. However, it is not always the best choice for our purposes—sometimes important features of the graph are hard to see in the standard window.

Imagine that your calculator display is an actual window and that you are looking through it. The graph lies on a plane on the other side of the window. The plane is infinite, but we can see only a portion of it through our window.

What happens when we try to graph the equation $y = 20$? Try it! Your answer should be "Nothing," because the graph of $y = 20$ does not appear in our standard graphing window, which includes only points whose *x*- and *y*-coordinates are between −10 and 10. We need to define a new graphing window in order to see the graph of $y = 20$.

To change the graphing window, press the ⌐WINDOW⌐ key, and you will see a display like the one shown in Figure 1.43. These are the settings for the standard graphing window. "Xmin" and "Xmax" give the coordinates of the left and right edges of the window; "Ymin" and "Ymax" give the coordinates of the top and bottom edges. The scales used on the *x*- and *y*-axes are indicated by "Xscl" and "Yscl," which give the distance between tick marks on each axis.

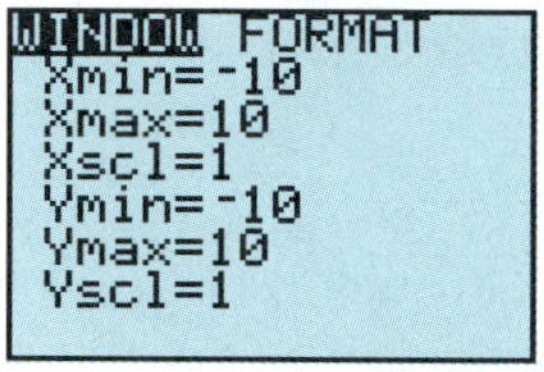

Figure 1.43

```
WINDOW FORMAT
Xmin=-10
Xmax=10
Xscl=1
Ymin=-10
Ymax=10
Yscl=1
```

To see the settings for the ZOOM Decimal window, press ⌐ZOOM⌐ ⌐4⌐, and then press ⌐WINDOW⌐ again. You can see that the ZOOM Decimal window shows a much smaller portion of the graph: from −4.7 to 4.7 on the *x*-axis, and from −3.1 to 3.1 on the *y*-axis. The ZOOM Decimal window is an example of a "friendly" graphing window, that is, one in which the ⌐TRACE⌐ picks out points with "nice" coordinates.

Now return to the standard window in preparation for graphing $y = 20$. (Press ⌐ZOOM⌐ ⌐6⌐ and then ⌐WINDOW⌐.) We will change only the *y*-value settings for our graph. Use the down arrow to move the cursor and change Ymin to 0, Ymax to 25, and Yscl to 5. Press ⌐GRAPH⌐ again and study the result. Your graph should look like Figure 1.44. Use the ⌐TRACE⌐ key to find the coordinates of some points on the graph.

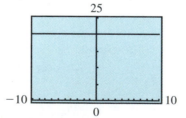

Figure 1.44

Choosing a good graphing window for an equation is not always easy. It will become easier when you know more about some standard equations and their graphs. For now we will have to depend on intuition and trial and error.

Example 8

Graph the equation $y = \sqrt{x}$.

Solution   We press the ⌐Y =⌐ key, then ⌐CLEAR⌐ to erase our earlier work. Next we enter ⌐√⌐ ⌐X, T, θ⌐ and finally press ⌐GRAPH⌐. With our current graphing window, we see only a small portion of the graph (see Figure 1.45). Since the graph of $y = \sqrt{x}$ involves only non-negative numbers, we can improve our graph by defining a new graphing window. Press the $\boxed{\text{WINDOW}}$ key and modify the range variables as follows: Xmin = 0, Xmax = 20, Ymin = 0, and Y max = 5. When we $\boxed{\text{GRAPH}}$ the same equation with the new graphing window, we get something like Figure 1.46.

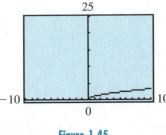

Figure 1.45

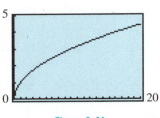

Figure 1.46

Exercise 1.6

Answer the questions about each graph.

1. Figure 1.47 shows the graph of the temperatures recorded during a winter day in Billings, Montana.

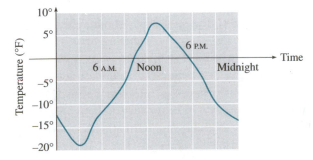

Figure 1.47

a. What were the high and low temperatures recorded during the day?

b. During what time intervals was the temperature above 5°F? Below −5°F?

c. Estimate the temperatures at 7:00 A.M. and 2:00 P.M. At what time(s) was the temperature approximately 0°F? Approximately −12°F?

d. How much did the temperature increase between 3:00 A.M. and 6:00 A.M.? Between 9:00 A.M. and noon? How much did the temperature decrease between 6:00 P.M. and 9:00 P.M.?

e. During which 3-hour interval did the temperature increase most rapidly? Decrease most rapidly?

2. Figure 1.48 shows a graph of the altitude of a commercial jetliner during its flight from Denver to Los Angeles.

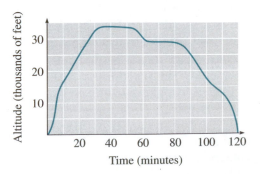

Figure 1.48

a. What was the highest altitude the jet achieved? At what time(s) was this altitude recorded?

b. During what time intervals was the altitude greater than 10,000 feet? Below 20,000 feet?

c. Estimate the altitudes 15 minutes into the flight and 35 minutes into the flight. At what time(s) was the altitude approximately 16,000 feet? 32,000 feet?

d. How many feet did the jet climb during the first 10 minutes of flight? Between 20 minutes and 30 minutes? How many feet did the jet descend between 100 minutes and 120 minutes?

e. During which 10-minute interval did the jet ascend most rapidly? Descend most rapidly?

3. Figure 1.49 shows the gas mileage achieved by an experimental model automobile at different speeds.

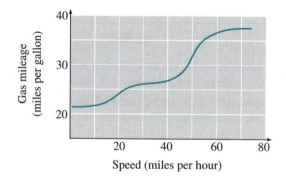

Figure 1.49

a. Estimate the gas mileage achieved at 43 miles per hour.

b. Estimate the speed at which a gas mileage of 34 miles per gallon is achieved.

c. At what speed is the best gas mileage achieved? Do you think the gas mileage will continue to improve as the speed increases? Why or why not?

d. The data illustrated by the graph were collected under ideal test conditions. What factors might affect the gas mileage if the car were driven under more realistic conditions?

4. Figure 1.50 shows the average height of young women aged 0 to 18 years.

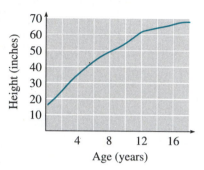

Figure 1.50

a. Estimate the average height of 5-year-old girls.

b. Estimate the age at which the average young woman is 50 inches tall.

c. At what age does the average woman achieve her maximum height? Do you think that height will continue to increase as age increases? Why or why not?

d. The data recorded in the graph reflect the average heights for young women at given ages. What factors might affect the heights of specific individuals?

5. Figure 1.51 shows the speed of a car during a 1-hour journey.

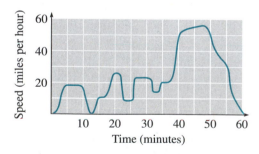

Figure 1.51

a. When did the car stop at a traffic signal?

b. During what time interval did the car drive in stop-and-go city traffic?

c. During what time interval did the car travel on the freeway?

6. Figure 1.52 shows the fish population of a popular fishing pond.

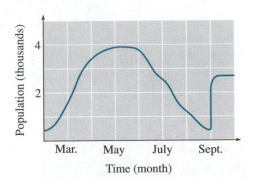

Figure 1.52

a. During what months do the young fish hatch?
b. During what months is fishing allowed?
c. When does the park service restock the pond?

a. Use the graph (Figures 1.53–1.60) to find the missing component in each solution of the equation.
b. Verify your answers algebraically.

7. $s = 2t + 4$

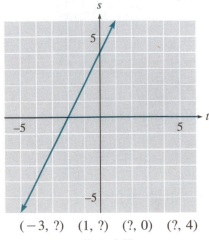

$(-3, ?)$ $(1, ?)$ $(?, 0)$ $(?, 4)$

Figure 1.53

8. $s = -2t + 4$

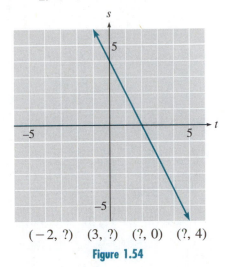

$(-2, ?)$ $(3, ?)$ $(?, 0)$ $(?, 4)$

Figure 1.54

9. $w = v^2 + 2$

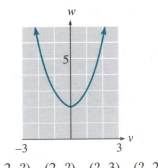

$(-2, ?)$ $(2, ?)$ $(?, 3)$ $(?, 2)$

Figure 1.55

10. $w = v^2 - 4$

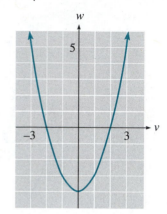

$(-1, ?)$ $(3, ?)$ $(?, 0)$ $(?, -4)$

Figure 1.56

12. $p = \dfrac{1}{m + 1}$

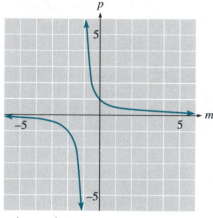

$\left(\dfrac{-3}{2}, ?\right)$ $(3, ?)$ $(?, -1)$ $(?, 2)$

Figure 1.58

11. $p = \dfrac{1}{m - 1}$

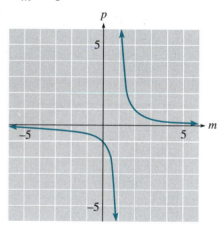

$(-1, ?)$ $\left(\dfrac{1}{2}, ?\right)$ $\left(?, \dfrac{1}{3}\right)$ $(?, -1)$

Figure 1.57

13. $y = x^3$

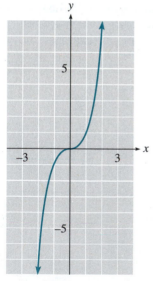

$(-2, ?)$ $\left(\dfrac{1}{2}, ?\right)$ $(?, 0)$ $(?, -1)$

Figure 1.59

14. $y = \sqrt{x} + 4$

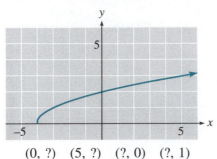

$(0, ?)$ $(5, ?)$ $(?, 0)$ $(?, 1)$

Figure 1.60

Use a graphing calculator to graph each equation. Use the standard graphing window.

15. $y = x - 2$

16. $y = 5 - x$

17. $y = x^2 - 9$

18. $y = 5 - x^2$

19. $y = x^3 - 8x$

20. $y = 2 + 4x - x^3$

21. $y = x^4 - 5x^2 + 3$

22. $y = 1 + 6x^2 - x^4$

Graph each set of equations in the same window. Compare the graphs.

23. $y = x - 3,\ \ y = -x - 3$

24. $y = 2x + 1,\ \ y = -2x + 1$

25. $y = x^2,\ \ y = x^2 + 4,\ \ y = x^2 - 4$

26. $y = 3 - x^2,\ \ y = 1 - x^2,\ \ y = -2 - x^2$

27. $y = x,\ \ y = x^2,\ \ y = x^3$

28. $y = x^2,\ \ y = 0.5x^2,\ \ y = 0.1x^2$

29. $y = \sqrt{x},\ \ y = \sqrt{x - 4},\ \ y = \sqrt{x + 4}$

30. $y = \sqrt{x},\ \ y = -\sqrt{x},\ \ y = \sqrt{-x}$

Graph each equation in the "friendly" window.

Xmin $= -9.4$ Xmax $= 9.4$ Xscl $= 1$
Ymin $= -9.3$ Ymax $= 9.3$ Yscl $= 1$

Use the TRACE **key to answer the questions. Round your answers to one decimal place.**

31. Find all the points on the graph of $y = x^2 + 8x + 9$ that have y-coordinate -3.

32. Find all the points on the graph of $y = \sqrt{36 - x^2}$ that have x-coordinate -2.

33. Find all points on the graph of $y = \sqrt{x^2 - 6}$ that have x-coordinate 5.

34. Find all points on the graph of $y = 3 + 4x - x^2$ that have y-coordinate -2.

35. Estimate the coordinates of the point where the graphs of $y = 2x - 6$ and $y = 0.5x + 2$ intersect.

36. Estimate the coordinates of the point where the graphs of $y = -1.75x - 6$ and $y = 0.6x - 1$ intersect.

37. Estimate the coordinates of the "turning points" on the graph of $y = 2 + 4x - x^3$.

38. Estimate the coordinates of the "turning points" on the graph of $y = 0.5x^3 - 4x$.

Graph each equation

a. using the standard window

b. using the suggested window.

Explain how the window alters the appearance of the graph in each case.

39. a. $y = \sqrt{1 - x^2}$

    b. Xmin $= -2$,   Xmax $= 2$
       Ymin $= -2$,   Ymax $= 2$

40. a. $y = \dfrac{1}{x^2 + 10}$

    b. Xmin $= -5$,   Xmax $= 5$
       Ymin $= 0$,   Ymax $= 0.5$

41. a. $y = 5x^2 + 20$

    b. Xmin $= -3$,   Xmax $= 3$
       Ymin $= 20$,   Ymax $= 80$

42. a. $y = -20 - (x + 20)^2$

    b. Xmin $= -25$,   Xmax $= -15$

       Ymin $= -25$,   Ymax $= -15$

43. a. $y = (x - 8)(x + 6)(x - 15)$

    b. Xmin $= -10$,   Xmax $= 20$

       Ymin $= -250$,   Ymax $= 750$

44. a. $y = 200x^3$

    b. Xmin $= -5$,   Xmax $= 5$

       Ymin $= -10{,}000$,   Ymax $= 10{,}000$

Chapter 1 Review

Write algebraic expressions to describe each situation.

1. Romina is taking the train to visit her grandfather in Canada, a journey of 2400 miles.

a. The train travels at an average speed of 48 miles per hour. Write an expression for the distance the train has traveled after t hours.

b. Write an expression for the distance Romina has left to go after traveling for t hours.

2. The employees of Griswell Accounting will all receive a 7% raise at the end of this year.

a. Write an expression for the amount of each employee's raise in terms of his or her current salary, s.

b. Write an expression for each employee's salary next year in terms of s, the current salary.

3. The pressure, P, exerted by a gas is given by a constant k times the temperature, T, of the gas, divided by the volume, V, it occupies.

a. Write an algebraic expression for the pressure, P.

b. Determine the pressure in pounds per square inch exerted by 200 cubic inches of a gas at 400° Kelvin if the value of the gas constant is 20.

4. The expansion in length, E, that a section of highway will experience on a very hot day is given by a constant k times the length, L, of the section times the difference between the present temperature, T, and 65° Fahrenheit.

a. Write an algebraic expression for the expansion, E.

b. Determine the amount that a 1000-foot section of highway will expand at 105° if the value of the constant is 0.000012.

Simplify.

5. -6^2 **6.** $(-6)^2$

7. $(4 - 2)[3 - 2(3 - 4)]$

8. $2[1 + (6 - 2)] \div (-4)$

9. $\dfrac{2 \cdot 3^2}{6} - 3 \cdot 2^2$

10. $-2^3 + 3\left[\dfrac{5^2 + 3}{4 - (-3)}\right] + (-3)^2$

Use a calculator to simplify each expression.

11. $\sqrt{18^2 - 4(3)(-6)}$ **12.** $\dfrac{26.8 - 32.4}{-16(25.5)}$

Evaluate.

13. $\dfrac{1}{2}gt^2 - 6t$, for $g = 32$, $t = 2$

14. $\dfrac{a - ar^n}{1 - r}$, for $a = 2.1$, $r = 0.5$, $n = 3$

Solve.

15. $2x - 6 = 4x - 8$ **16.** $0.40x = 240$

17. $2(x - 3) + 4(x + 2) = 6$

18. $2x - (x + 3) = 2(x + 1)$

19. $0.30(y + 2) = 2.10$

20. $0.06y + 0.04(y + 1000) = 60$

21. $5x^2 = 30$ **22.** $3x^2 - 6 = 0$

23. $\dfrac{2x^2}{5} - 7 = 9$ **24.** $\dfrac{3x^2 - 8}{4} = 10$

25. $\dfrac{y + 3}{y + 5} = \dfrac{1}{3}$ **26.** $\dfrac{y}{6 - y} = \dfrac{1}{2}$

27. $\dfrac{6}{z^2 - 1} = 2$ **28.** $\dfrac{z^2 + 2}{z^2 - 2} = 3$

29. $2\sqrt{w} - 5 = 21$ **30.** $16 - 3\sqrt{w} = -5$

31. $12 - \sqrt{5v + 1} = 3$

32. $3\sqrt{17 - 4v} - 8 = 19$

Solve each formula for the indicated variable.

33. $3N = 5t - 3c$, for t

34. $C = 10 + 2p - 2t$, for t

35. $S = 2\pi (R - r)$,   for R

36. $9C = 5 (F - 32)$,   for F

37. $V = C\left[1 - \dfrac{t}{n}\right]$,   for t

38. $\dfrac{p}{q} = \dfrac{r}{q + r}$,   for q

39. $t = \sqrt{\dfrac{2v}{g}}$,   for g

40. $r = \sqrt{\dfrac{A}{\pi}}$,   for A

Solve each inequality.

41. $3x + 1 < 2 - 3(x - 1)$

42. $2(x - 2) - 3(x + 3) \le -24$

43. $\dfrac{-x - 2.3}{2} > 1.2$ **44.** $\dfrac{-2x + 4}{3} \ge \dfrac{-3x}{2}$

For each problem
a. identify the unknown quantity and assign a variable to represent it.
b. write an equation that models the problem.
c. solve the equation and answer the question in the problem.

45. Two retired tennis pros go into business to market their own brand of tennis racket. Their initial outlay is $36,000, and the rackets cost $22 each to manufacture.

a. If they sell the rackets for $40 each, how many must they sell before they break even?

b. How many must they sell in order to clear $10,000 apiece in the first year?

46. A cosmetics manufacturer expands her product line to include bubble bath. She spends $16,000 on start-up costs and markets the bubble bath for $8 per bottle.

a. If the bubble bath costs $5 per bottle to produce, how many bottles must the manufacturer sell to break even?

b. If the manufacturer can afford a loss no greater than $10,000 on the new product, how many bottles must she sell?

47. The length of a table tennis table is 4 feet longer than its width, and its perimeter is 28 feet. Find the dimensions of the table.

48. The length of a tennis court for singles is 24 feet longer than twice its width, and its perimeter is 210 feet. Find the dimensions of the tennis court.

49. One angle of a triangle is twice as large as another angle, and the third angle is 12° more than the sum of the other two. Find the size of each angle.

50. The smallest angle of a triangle is 25° less than the second angle and 50° less than the third angle. Find the size of each angle.

51. The length of each of the two equal sides of an isosceles triangle is 15.6 centimeters greater than the length of the third side. The perimeter of the triangle is 65.7 centimeters. Find the length of each side.

52. The longest side of a triangle is three times the length of the shortest side, and the third side is 72.4 centimeters longer than the shortest side. Find the length of each side if the perimeter is 272.9 centimeters.

53. Last year the Dean of Admissions at Beckworth College selected 75% of all applicants for the freshman class. How many students applied if the college accepted 600 students?

54. A part-time secretary takes home 80% of his total salary. What is his total salary if he takes home $120 per week?

55. A sporting goods supplier contracts to manufacture 1710 baseball bats. If 5% of all the bats produced are discarded because of defects, how many bats should the supplier make in order to produce 1710 acceptable ones?

56. An electronics firm always ships 2% additional transistors over the number ordered to cover any defective transistors that might be included. How many transistors were ordered if 22,440 were shipped?

57. A solar energy firm experienced a 12% increase in sales this year on top of a 9% increase last year. What is the total percent increase in its sales over 2 years ago?

58. A suburban school district experienced a 15% increase in enrollment last year and anticipates an 8% increase this year. What is the total percent increase in its enrollment over 2 years ago?

59. How many liters of a 20% sugar solution should be added to 40 liters of a 32% solution in order to obtain a 28% solution?

60. An automobile radiator contains 8 quarts of a 40% antifreeze solution. How many quarts of this solution should be drained from the radiator and replaced with water if the resulting solution is to be 20% antifreeze?

61. A recipe for applesauce bread calls for $1\frac{1}{2}$ cups of flour and $\frac{1}{3}$ cup of honey. How much honey should be used if the flour is increased to $2\frac{1}{2}$ cups?

62. The instructions for planting an azalea call for $\frac{3}{4}$ pound of peat moss mixed with $1\frac{1}{3}$ pounds of nitrohumus. How many pounds of nitrohumus will you need to make azalea mix with a 7-pound bag of peat moss?

63. On an illustrated map of Paris, 4.5 centimeters represents 1 kilometer. If the distance on the map from your hotel to the Louvre is about 10 centimeters, how long a walk will it be?

64. If 2.5 pounds of tin are required to make 12 pounds of a certain alloy, how many pounds of tin are needed to make 90 pounds of the alloy?

65. A thin strip of sheet aluminum is bent into the shape shown in Figure 1.61. What was the length of the strip before it was bent?

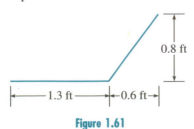

0.8 ft

1.3 ft 0.6 ft

Figure 1.61

66. A roof truss has the dimensions shown in Figure 1.62. What is the length of the rafter?

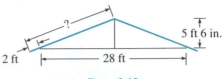

?

5 ft 6 in.

2 ft 28 ft

Figure 1.62

67. You decide to use geometry to find the height of an elm tree in your front yard. You ask a friend who is 6 feet 6 inches tall to stand next to the tree. The friend casts a shadow 14.2 feet long, and the tree casts a shadow 39.4 feet long. How tall is the tree?

68. A surveyor set up two markers on either side of a small lake and made the measurements shown in Figure 1.63. Use her measurements to find the distance across the lake.

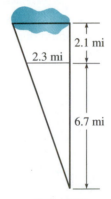

2.1 mi

2.3 mi

6.7 mi

Figure 1.63

69. A wooden packing crate has a square base, 6 feet by 6 feet. If the wood for the crate costs 10 cents per square foot, how tall a crate can be built for $16.80?

70. A redwood planter has a square cross section, 8 inches by 8 inches. If the redwood costs $\frac{1}{2}$ cent per square inch, how long a planter can be built for $5.44?

71. The volume of a pyramid with a square base is given by $V = \frac{1}{3} s^2 h$, where s is the side of the base and h is the height. The Great Pyramid of Cheops in Egypt is 160 yards tall and its base measures 250 yards on each side (see Figure 1.64). Find the volume of the Great Pyramid.

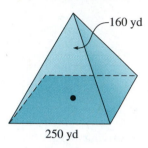

Figure 1.64

72. a. Use the Pythagorean theorem to find the altitude of one of the triangular faces of the Great Pyramid described in Exercise 71.

b. Find the surface area of the Great Pyramid, not including the base.

Use the graphs to answer the questions.

73. Figure 1.65 shows a graph of the snow level at a ski resort in Colorado during a 2-week period in January 1995.

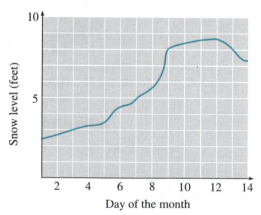

Figure 1.65

a. What was the highest snow level recorded?
b. On how many days was the snow level above 4 feet?
c. By how much did the snow level increase between January 7 and January 9?
d. During which daily interval did the snow level increase most rapidly? Decrease most rapidly?

74. Figure 1.66 shows a graph of the maximum daily temperatures at the ski resort in Exercise 73 over the same 2-week period.

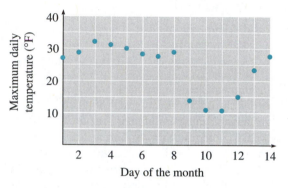

Figure 1.66

a. Estimate the highest maximum temperature and the lowest maximum temperature.
b. On how many days was the maximum temperature below 20°F?
c. On what day did a cold front pass over the resort?
d. By approximately how much did the maximum temperature drop on the day the cold front passed through?

a. Use the graph to find the missing component in each solution of the equation.
b. Verify your answers algebraically.

75. $y = 2 - x^3$

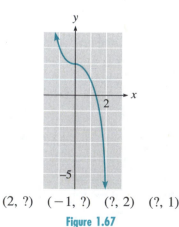

$(2, ?)$ $(-1, ?)$ $(?, 2)$ $(?, 1)$

Figure 1.67

76. $y = \sqrt{1 - x}$

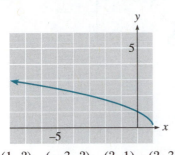

$(1, ?)$ $(-3, ?)$ $(?, 1)$ $(?, 3)$

Figure 1.68

Use the TRACE feature on your graphing calculator to estimate the coordinates of the points on each graph. Round your answers to one decimal place.

77. Graph $y = 0.5x^4 - 5x^2 + 4.5$.

a. Find all the points on the graph that have y-coordinate 0.

b. Find all the points on the graph that have y-coordinate 8.

c. How many points on the graph have y-coordinate -2?

d. Find the "turning points" of the graph.

78. Graph on the same screen

$y = 0.5 (x - 1)^2 + 2$ and
$y = -(x + 1)^2 + 8$

a. What is the minimum point (with the smallest y-coordinate) of the first graph?

b. What is the maximum point (with the largest y-coordinate) of the second graph?

c. Find the intersection points of the two graphs.

Linear Models

In this chapter we investigate how to describe a relationship between two variables by using equations, graphs, and tables of values. This process is called **mathematical modeling**. The simplest and perhaps most useful type of model is a linear model.

2.1 SOME EXAMPLES OF LINEAR MODELS

In each of the following examples, notice how the relationship between the variables, which is initially described in words, can also be represented by

1. A table of values,

2. An algebraic equation,

3. A graph.

Example 1

Yumiko's long-distance telephone company charges a $3 access fee for a call to Tokyo and $2 for each minute of the call. (A fraction of a minute is charged as the corresponding fraction of $2.)

a. Write an equation that expresses the cost of a call to Tokyo in terms of the length of the call.

b. Graph the equation in (a).

Solutions **a.** Let t represent the length of the call in minutes and let C represent the cost of the call. Then

$$C = 3 + 2t \qquad (t \geq 0)$$

b. Choose several values for t and calculate the corresponding values for C to obtain the ordered pairs shown in the following table.

t	C
0	3
1	5
2	7
3	9

(t, C)
(0, 3)
(1, 5)
(2, 7)
(3, 9)

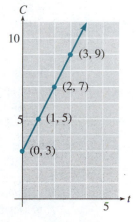

Figure 2.1

The graphs of these points lie on a straight line, as shown in Figure 2.1. Notice that the line extends infinitely in only one direction, since negative values of t do not make sense here.

Example 2

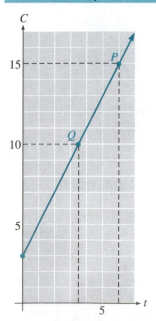

Figure 2.2

Use the equation $C = 3 + 2t$ from Example 1 to answer the following questions. Then show how the answers can be found by using the graph.
a. How much will a 6-minute call to Tokyo cost Yumiko?
b. How long can Yumiko talk for $10?

Solutions **a.** Substitute $t = 6$ into the equation to find

$$C = 3 + 2(6) = 15$$

A 6-minute call will cost $15. The point P on the graph in Figure 2.2 represents the cost of a 6-minute call to Tokyo. The value on the C-axis at the same height as point P is 15, so a 6-minute call costs $15.
b. Substitute $C = 10$ into the equation and solve for t.

$$10 = 3 + 2t$$
$$7 = 2t$$
$$t = 3.5$$

For $10 Yumiko can talk for $3\frac{1}{2}$ minutes. The point Q on the graph represents a $10 call. The value on the t-axis below point Q is 3.5, so $10 will buy a 3.5-minute phone call.

To draw a graph that is useful in analyzing a problem, we must choose scales for the axes that reflect the magnitudes of the variables involved.

Example 3

In 1970 a three-bedroom house in Midville cost $30,000. The price of a home has increased by an average of $4000 per year since then.
a. Write an equation that expresses the price of a three-bedroom house in Midville in terms of the number of years since 1970.
b. Graph the equation in (a).
c. Use the graph to calculate the increase in the price of the home from 1982 to 1989. Illustrate the increase on your graph.

Solutions   a. Let P represent the cost of the house t years after 1970. Then

$$P = 30,000 + 4000t \qquad (t \geq 0)$$

b. Choose several values for t and calculate the corresponding values for P to obtain the following ordered pairs.

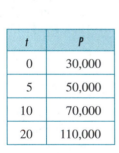

t	P	(t, P)
0	30,000	(0, 30,000)
5	50,000	(5, 50,000)
10	70,000	(10, 70,000)
20	110,000	(20, 110,000)

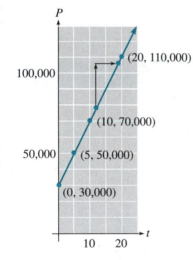

Figure 2.3

To graph the equation we scale the horizontal axis, or t-axis, in 5-year intervals and the vertical axis, or P-axis, in intervals of $10,000, then plot the points we found.

    c. Find the points on the graph corresponding to 1982 to 1989. These points lie above $t = 12$ and $t = 19$ on the t-axis. Now find the values on the P-axis corresponding to the two points. The values are $P = 78,000$ in 1982 and $P = 106,000$ in 1989. The increase in price is the difference of the two P-values.

$$\text{increase in price} = 106,000 - 78,000$$
$$= 28,000$$

The price of the home increased $28,000 between 1982 and 1989. The increase is indicated by the arrows in Figure 2.3.   ▭

Perhaps you noticed that the graphs in Examples 1–3 were portions of straight lines. In fact, the graph of any equation that can be written in the form

$$ax + by = c$$

where a and b are not both equal to zero, is a straight line. For this reason such equations are called **linear equations.** (Note that the equation in Example 1, $C = 3 + 2t$, can be written equivalently as $C - 2t = 3$. The equation in Example 3, $P = 30,000 + 4000t$, can be written as $P - 4000t = 30,000$.) The graphs in the preceding examples are *increasing graphs.* As we move along the graph from left to right (that is, in the direction of increasing t), the second coordinate of points on the graph increases as well. Example 4 illustrates a *decreasing graph.*

Example 4

Silver Lake has been polluted by industrial waste products. The concentration of toxic chemicals in the water is currently 285 parts per million (ppm). Local environmental officials would like to reduce the concentration by 15 ppm each year.
a. Write an equation that expresses the concentration of toxic chemicals t years from now.
b. Graph the equation in (a).

Solutions a. Let C represent the concentration of toxic chemicals t years from now. Then

$$C = 285 - 15t \qquad (t \geq 0) \tag{1}$$

b. Choose several values for t and calculate the corresponding values for C to obtain the following ordered pairs. The graph is shown in Figure 2.4.

t	C
0	285
5	210
10	155
15	60

(t, C)
(0, 285)
(5, 210)
(10, 155)
(15, 60)

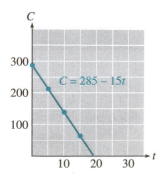

Figure 2.4   ▭

In Figure 2.4 notice that we extend the graph until it reaches the horizontal axis, but no farther; points with negative C-coordinates have no meaning for the problem.

Intercepts   Consider the graph of the equation

$$3y - 4x = 12$$

shown in Figure 2.5. The points at which the graph crosses the axes are called the **intercepts** of the graph. The coordinates of these points are relatively easy to find, since the y-coordinate of the x-intercept is zero and the x-coordinate of the y-intercept is zero. The intercepts are usually significant for the model as well.

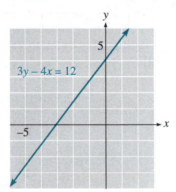

Figure 2.5

Example 5

a. Find the intercepts of the graph in Example 4.
b. What is the significance of the intercepts to the problem?

Solutions   a.   To find the C-intercept of the graph, set t equal to zero in Equation (1) on page 71.

$$C = 285 - 15\,(0) = 285$$

The C-intercept is the point $(0, 285)$, or simply 285. To find the t-intercept, set C equal to zero in Equation (1).

$$0 = 285 - 15t \qquad \text{Add } 15t \text{ to both sides.}$$

$$15t = 285 \qquad \text{Divide both sides by 15.}$$

$$t = 19$$

The t-intercept is the point $(19, 0)$, or simply 19.
b.   The C-intercept represents the concentration of toxic chemicals in Silver Lake now: When $t = 0$, $C = 285$, so the concentration is currently 285 ppm. The t-intercept represents the number of years it will take for the

concentration of toxic chemicals to drop to zero: When $C = 0$, $t = 19$, so it will take 19 years for the pollution to be eliminated entirely.

⬜

Since we really need only two points to determine the graph of a linear equation, we might as well find the intercepts first and use them to draw the graph. It is always a good idea to find a third point as a check.

Example 6

a.   Find the x- and y-intercepts of the graph of $1500x - 1800y = 9000$.
b.   Use the intercepts to graph the equation. Find a third point as a check.

Solutions   **a.**   To find the x-intercept, set $y = 0$.

$$1500x - 1800(0) = 9000 \qquad \text{Simplify.}$$

$$1500x = 9000 \qquad \text{Divide both sides by 1500.}$$

$$x = 6$$

The x-intercept is the point $(6, 0)$. To find the y-intercept, set $x = 0$.

$$1500(0) - 1800y = 9000 \qquad \text{Simplify.}$$

$$-1800y = 9000 \qquad \text{Divide both sides by } -1800.$$

$$y = -5$$

The y-intercept is the point $(0, -5)$.
b.   Plot the two intercepts, $(6, 0)$ and $(0, -5)$. Draw the line through them, as shown in Figure 2.6. Now find another point and check that it lies on this line. We choose $x = 2$ and solve for y.

$$1500(2) - 1800y = 9000$$

$$3000 - 1800y = 9000$$

$$-1800y = 6000$$

$$y = -3.\overline{3}$$

Plot the point $(2, -3\frac{1}{3})$. Since this point lies on the line, we can be reasonably confident that our graph is correct.

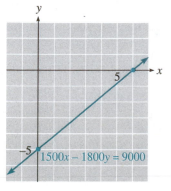

Figure 2.6

Lines Through the Origin

If the graph of a linear equation passes through the origin, then the x-intercept and the y-intercept are the same point, namely $(0, 0)$. In this case, of course, we must locate at least one other point in order to sketch the graph.

Example 7

Mariel is planning a driving tour of the western states for her summer vacation. Mariel's car gets an average of 25 miles per gallon of gas.
a. Write an equation that relates the number of gallons of gas she will need to the number of miles she plans to drive.
b. Graph the equation in (a).

Solutions   **a.**   Let m represent the number of miles Mariel plans to drive and let g represent the number of gallons of gas she will need. Then

$$g = \frac{m}{25} \qquad (m \geq 0)$$

b. When $m = 0$, $g = 0$, so the point $(0, 0)$ is both the m-intercept and the g-intercept of the graph. To find some other points on the graph, choose several convenient values for m. (Values for m are "convenient" if they are divisible by 25.)

m	g
100	4
250	10
400	16

(m, g)
(100, 4)
(250, 10)
(400, 16)

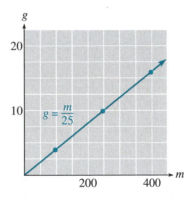

Figure 2.7

To graph the equation, we scale the horizontal axis, or m-axis, in intervals of 50 miles, and the vertical axis, or g-axis, in units of 5 gallons, as shown in Figure 2.7. (These choices are arbitrary. You can use any scale that results in a reasonable graph.) ▭

The examples in this section model some simple linear relationships between two variables. Such relationships, in which the value of one variable is determined by the value of the other, are called **functions.** We will study various kinds of functions throughout the course.

Exercise 2.1

In Exercises 1–10 write and graph a linear equation for each situation. Then answer the questions.

1. Frank plants a dozen corn seedlings, each 6 inches tall. With plenty of water and sunlight they will grow approximately 2 inches per day.

   a. Write an equation that expresses the height, h, of the seedlings in terms of the number of days, t, since they were planted.

   b. Graph the equation.

   c. How tall is the corn after 3 weeks?

   d. How long will it be before the corn is 6 feet tall?

2. In the desert the temperature at 6 A.M., just before sunrise, was 65°F. The temperature rose about 5 degrees every hour until it reached its maximum value at about 5 P.M.

   a. Write an equation that expresses the temperature, T, in the desert in terms of the number of hours, h, since 6 A.M.

   b. Graph the equation.

   c. How hot is it at noon?

   d. When will the temperature be 110°F?

3. On October 31, Betty and Paul fill their 250-gallon heating fuel oil tank. Beginning in November they use an average of 15 gallons per week of heating fuel oil.

   a. Write an equation that expresses the amount of oil, A, in the tank in terms of the number of weeks, w, since October 31.

   b. Graph the equation.

   c. How much did the amount of fuel oil in the tank decrease between the third week and the eighth week? Illustrate this amount on the graph.

   d. When will the tank contain more than 175 gallons of fuel oil? Illustrate this on the graph.

4. Leon's camper has a 20-gallon tank, and he gets 12 miles to the gallon. (Note that getting 12 miles to the gallon is the same as using 1/12 gallon per mile.)

   a. Write an equation that expresses the amount of gasoline, g, in Leon's fuel tank in terms of the number of miles, m, he has driven.

   b. Graph the equation.

   c. How much gasoline will Leon use between 8 A.M., when his odometer reads 96 miles, and 9 A.M., when his odometer reads 144 miles? Illustrate this amount on the graph.

   d. If Leon has less than 5 gallons of gas left, how many miles has he driven? Illustrate this on the graph.

5. Phil and Ernie buy a used photocopier for $800 and set up a copy service on their campus. For each hour that the copier runs continuously, Phil and Ernie make $40.

   a. Write an equation that expresses Phil and Ernie's profit (or loss), P, in terms of the number of hours, t, they run the copier.

   b. Find the intercepts and sketch the graph.

   c. What is the significance of the intercepts to Phil and Ernie's profit?

6. A deep-sea diver is taking some readings at a depth of 400 feet. She begins ascending at 20 feet per minute. (Consider a depth of 400 feet as an altitude of -400 feet.)

   a. Write an equation that expresses the diver's altitude, h, in terms of the number of minutes, m, elapsed.

   b. Find the intercepts and sketch the graph.

   c. What is the significance of the intercepts to the diver's depth?

7. The owner of a gas station has $4800 to spend on unleaded gas this month. Regular unleaded costs him $0.60 per gallon, and premium unleaded costs $0.80 per gallon.

   a. Write an equation that relates the amount of regular unleaded gasoline, x, he can buy and the amount of premium unleaded, y. (Your equation will have the form $ax + by = c$.)

b. Find the intercepts and sketch the graph.

c. What is the significance of the intercepts to the amount of gasoline?

8. Five pounds of body fat is equivalent to 16,000 calories. Carol can burn 600 calories per hour bicycling and 400 calories per hour swimming.

a. Write an equation that relates the number of hours, x, of cycling and, y, of swimming Carol needs to perform in order to lose 5 pounds. (Your equation will have the form $ax + by = c$.)

b. Find the intercepts and sketch the graph.

c. What is the significance of the intercepts to Carol's exercise program?

9. A real estate agent receives a salary of $10,000 plus 3% of her total sales for the year.

a. Write an equation that expresses the agent's salary, I, in terms of her total annual sales, s.

b. Graph the equation. (*Hint:* Use multiples of $100,000 for total annual sales.)

c. Use your equation to determine what annual sales result in salaries between $16,000 and $22,000 for the year. Confirm your answer by using the graph.

d. Use your graph to calculate the increase in the agent's salary if her sales increase from $500,000 to $700,000. Use arrows to illustrate the increase on your graph.

10. Under a proposed graduated income tax system, a single taxpayer whose taxable income is between $13,000 and $23,000 would pay $1500 plus 20% of the amount of income over $13,000.

a. Write an equation that expresses the taxes, T, owed to the amount of income, I, over $13,000.

b. Graph the equation.

c. Use your equation to determine what incomes result in taxes between $2500 and $3100. Confirm your answer by using the graph.

d. Use your graph to calculate the increase in taxes corresponding to an increase in income from $20,000 to $22,000. Use arrows to illustrate the increase on your graph.

In Exercises 11–22 graph each equation by the intercept method.

11. $x + 2y = 8$

12. $2x - y = 6$

13. $3x - 4y = 12$

14. $2x + 6y = 6$

15. $\dfrac{x}{9} - \dfrac{y}{4} = 1$

16. $\dfrac{x}{5} + \dfrac{y}{8} = 1$

17. $\dfrac{2x}{3} + \dfrac{3y}{11} = 1$

18. $\dfrac{8x}{7} - \dfrac{2y}{7} = 1$

19. $20x = 30y - 45,000$

20. $30x = 45y + 60,000$

21. $0.4x + 1.2y = 4.8$

22. $3.2x - 0.8y = 12.8$

Graph each equation in Exercises 23–28.

23. $x + y = 0$

24. $x - y = 0$

25. $2x - y = 0$

26. $2x + y = 0$

27. $x = 3y$

28. $x = -3y$

29. A freight train travels at a constant speed of 50 miles per hour.

a. Write an equation that expresses the distance the train has traveled in terms of the number of hours elapsed.

b. Graph the equation.

30. A computer programmer is paid $25 an hour.

a. Write an equation that expresses the programmer's wages in terms of the number of hours she works.

b. Graph the equation.

31. Sandra works for the post office as a mail carrier. On a typical day she can deliver mail to 24 houses per hour.

a. Write an equation that expresses the number of hours in Sandra's shift in terms of the number of houses on her route.

b. Graph the equation.

32. Jim volunteers at the blood bank. He can process an average of 20 blood samples per hour.

a. Write an equation that expresses the number of hours Jim works in terms of the number of blood samples the bank receives.

b. Graph the equation.

2.2 SLOPE

Using Ratios for Comparison

Which is more expensive: a 64-ounce bottle of Velvolux dish soap that costs $3.52, or a 60-ounce bottle of Rainfresh dish soap that costs $3.36?

You are probably familiar with the notion of comparison shopping. To determine which dish soap is the better buy, we compute the unit price, or price per ounce, for each bottle. The unit price for Velvolux is

$$\frac{352 \text{ cents}}{64 \text{ ounces}} = 5.5 \text{ cents per ounce}$$

and the unit price for Rainfresh is

$$\frac{336 \text{ cents}}{60 \text{ ounces}} = 5.6 \text{ cents per ounce}$$

The Velvolux costs less per ounce, so it is the better buy. By computing the price of each brand for *the same amount of soap* we can easily compare their unit prices.

In many situations a ratio, similar to a unit price, can provide a basis for comparison. Example 1 uses a ratio to measure a rate of growth.

Example 1

Which grow faster: Hybrid A wheat seedlings, which grow 11.2 centimeters in 14 days, or Hybrid B seedlings, which grow 13.5 centimeters in 18 days?

Solution   We compute the growth rate for each strain of wheat. Growth rate is expressed as a ratio, $\frac{\text{centimeters}}{\text{days}}$, or centimeters per day. The growth rate for Hybrid A is

$$\frac{11.2 \text{ centimeters}}{14 \text{ days}} = 0.8 \text{ centimeter per day}$$

and the growth rate for Hybrid B is

$$\frac{13.5 \text{ centimeters}}{18 \text{ days}} = 0.75 \text{ centimeter per day}$$

Because their rate of growth is larger, we see that the Hybrid A seedlings grow faster.

By computing the growth of each strain of wheat seedling over *the same unit of time,* a single day, we have a basis for comparison. We can use this same idea, finding a common basis for comparison, to measure and compare the steepness of an incline.

Measuring Steepness

Imagine you are an ant carrying a heavy burden along one of the two paths shown in Figure 2.8. Which path is more strenuous? Most ants would agree that the steeper path is more difficult. But what exactly is "steepness"? It

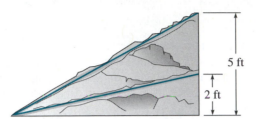

Figure 2.8

is not merely the gain in altitude, because even a gentle incline will reach a great height eventually. Steepness measures how sharply the altitude increases. An ant finds the second path more difficult, or steeper, because it rises 5 feet while the first path rises only 2 feet *over the same horizontal distance.*

To compare the steepness of two inclined paths, we compute the ratio of change in altitude to change in horizontal distance for each path.

Example 2

Which is steeper: Stony Point trail, which climbs 400 feet over a horizontal distance of 2500 feet, or Lone Pine trail, which climbs 360 feet over a horizontal distance of 1800 feet?

Solution   For each trail we compute the ratio of vertical gain to horizontal distance. For Stony Point trail the ratio is

$$\frac{400 \text{ feet}}{2500 \text{ feet}} = 0.16$$

and for Lone Pine trail the ratio is

$$\frac{360 \text{ feet}}{1800 \text{ feet}} = 0.20$$

Lone Pine trail is steeper, since it has a vertical gain of 0.20 foot for every foot traveled horizontally. Or in more practical units, Lone Pine trail rises 20 feet for every 100 feet of horizontal distance, whereas Stony Point trail rises only 16 feet over a horizontal distance of 100 feet.

Note that to compare the steepness of the two trails in Example 2 it is not enough to know which trail has the greater gain in elevation overall. Instead we compare their elevation gains over *the same horizontal distance.* Using the same horizontal distance provides a basis for comparison. The two trails are illustrated in Figure 2.9. We show the trails on a coordinate system where horizontal distance is measured in the *x*-direction and vertical distance is measured in the *y*-direction. Each trail appears as a line on the graph.

The ratio that we computed in Example 2,

$$\frac{\text{change in elevation}}{\text{change in horizontal position}}$$

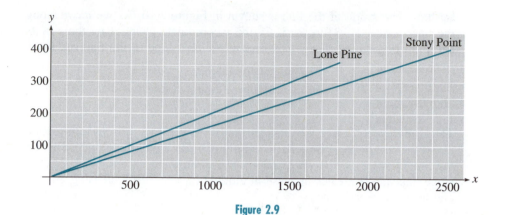

Figure 2.9

appears on the graphs in Figure 2.9 as $\dfrac{\text{change in } y\text{-coordinate}}{\text{change in } x\text{-coordinate}}$. This ratio is used to measure the steepness of a line; the larger the ratio, the steeper the line. For example, as we travel along the line representing Stony Point trail, we move from the point $(0, 0)$ to the point $(2500, 400)$. The y-coordinate changes by 400 and the x-coordinate changes by 2500, giving the ratio 0.16 that we found in Example 2. We call this ratio the **slope** of the line.

> The **slope** of a line is the ratio $\dfrac{\textbf{change in } y\textbf{-coordinate}}{\textbf{change in } x\textbf{-coordinate}}$ as we move from one point to another on the line.

Example 3

Compute the slope of the line that passes through points A and B in Figure 2.10.

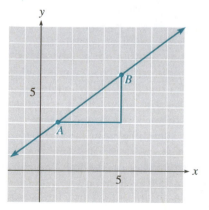

Figure 2.10

Solution   The graph of the line is shown in Figure 2.10. As we move along the line from A $(1, 3)$ to B $(5, 6)$, the y-coordinate changes by three units and the x-coordinate changes by four units. The slope of the line is thus

$$\frac{\text{change in } y\text{-coordinate}}{\text{change in } x\text{-coordinate}} = \frac{3}{4}$$

Thus the slope of a line is a *number* that we can compute from information about the line. The slope tells us how much the y-coordinate of points on the line increases when we increase the x-coordinate by one unit. A larger slope indicates a greater increase in altitude (at least for increasing graphs) and hence a steeper line.

Definition of Slope   Because the ratio that defines slope, $\dfrac{\text{change in } y\text{-coordinate}}{\text{change in } x\text{-coordinate}}$, is cumbersome to write out, we use a shorthand notation. The symbol Δ (the Greek letter "delta") is used in mathematics to denote "change in." In particular, Δy means "change in y-coordinate," and Δx means "change in x-coordinate." (See Figure 2.11.) We also use the letter m to stand for slope. With these symbols we can write the definition of slope as follows.

The **slope** of a line is given by $m = \dfrac{\Delta y}{\Delta x}$ $(\Delta x \neq 0)$.

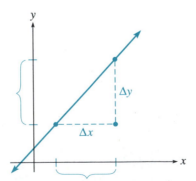

Figure 2.11

Although we have only considered examples in which Δx and Δy are positive numbers, they can also be negative. We say that Δx is positive if we move to the right and negative if we move to the left as we travel from one point to another on the line. We say that Δy is positive if we move up and negative if we move down.

Example 4

Compute the slope of the line that passes through the points $P(-4, 2)$ and $Q(5, -1)$ shown in Figure 2.12. Illustrate Δy and Δx on the graph.

Solution The graph of the line is shown in Figure 2.12. As we move from point $P(-4, 2)$ to point $Q(5, -1)$, we move three units down in the y-direction, so $\Delta y = -3$. We then move nine units to the right in the x-direction, so $\Delta x = 9$. Thus the slope is

$$m = \frac{\Delta y}{\Delta x} = \frac{-3}{9} = \frac{-1}{3}$$

Δy and Δx are labeled on the graph.

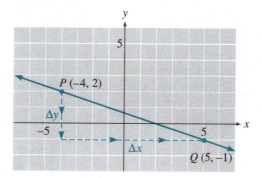

Figure 2.12

Notice that the line graphed in Example 4, which *decreases* as we move from left to right, has a negative slope. This is still true if we compute the slope by moving from point Q to point P instead of from P to Q. (See Figure 2.13.) In that case our computation looks like this:

$$m = \frac{\Delta y}{\Delta x} = \frac{3}{-9} = \frac{-1}{3}$$

The *ratio* of change in y to change in x is still the same. In general, we can move from point to point in either direction to compute the slope, as long as we are consistent between Δy and Δx.

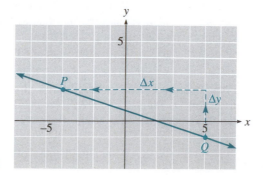

Figure 2.13

Lines Have Constant Slope

How do we know which two points to choose when we want to compute the slope of a line? It turns out that it doesn't matter.

Example 5

a. Graph the line $4x - 2y = 8$.
b. Compute the slope of the line using the x-intercept and y-intercept.
c. Compute the slope of the line using the points (4, 4) and (1, −2).

Solutions a. To graph the equation, find the x- and y-intercepts. Set x equal to zero to find

$$4\,(0) - 2y = 8$$
$$y = -4$$

The y-intercept is (0, −4). Set y equal to zero to find

$$4x - 2\,(0) = 8$$
$$x = 2$$

The x-intercept is (2, 0). The graph is shown in Figure 2.14.

b. If we move from (0, −4) to (2, 0) along the line, we move four units up and two units to the right; that is, $\Delta y = 4$ and $\Delta x = 2$. The slope is

$$m = \frac{\Delta y}{\Delta x} = \frac{4}{2} = 2$$

c. If we move from (4, 4) to (1, −2) along the line, we move six units down and three units to the left; that is, $\Delta y = -6$ and $\Delta x = -3$. The slope is

$$m = \frac{\Delta y}{\Delta x} = \frac{-6}{-3} = 2$$

the same as the value we calculated in part (b). Both calculations are illustrated on the graph in Figure 2.14.

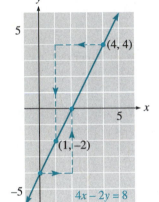

Figure 2.14

Example 5 illustrates an important property of lines: they have constant slope. This means that no matter what two points we use to calculate the slope, we will always get the same result. We will see later that lines are the only graphs that have this property.

Meaning of Slope

In Example 1 on page 68 we graphed the equation $C = 3 + 2t$, showing the cost of a long-distance call to Tokyo in terms of its length. The graph is reproduced in Figure 2.15. We can choose any two points on the line to compute its slope. Using points P and Q as shown,

$$m = \frac{\Delta C}{\Delta t} = \frac{6}{3} = 2$$

The slope of the line is 2.

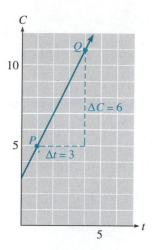

Figure 2.15

What is the significance of this value for the problem setting, namely, the cost of a call to Tokyo? Recall that the symbol Δ means "change in." Thus the expression

$$\frac{\Delta C}{\Delta t} = \frac{6}{3} \quad \text{stands for} \quad \frac{\text{change in cost}}{\text{change in time}} = \frac{6 \text{ dollars}}{3 \text{ minutes}}$$

In other words, if we increase the length of the call by 3 minutes, the cost of the call increases by $6. We might also say that the cost of the call increases at a rate of $2 per minute. For this problem the slope is a ratio that gives the *rate of increase* in the cost of a telephone call, in dollars per minute.

In general, the slope of a line measures the rate of change of y with respect to x. In different situations this rate might be interpreted as a rate of growth or a speed. A negative slope might represent a rate of decrease or a rate of consumption. Interpreting the slope of a graph can give us valuable information about the variables involved. In fact, one branch of calculus is concerned with just this skill: finding slopes and using them to analyze mathematical models of problems.

Example 6

The graph in Figure 2.16 shows the distance traveled by a driver for a cross-country trucking firm in terms of the number of hours she has been on the road.

a. Compute the slope of the graph.
b. What is the significance of the slope to the problem?

Solutions **a.** To compute the slope of the graph, choose any two points on the line, say (2, 100) and (4, 200). These points are labeled G and H, respectively, in Figure 2.16. As we move from G to H, we find

$$m = \frac{\Delta D}{\Delta t} = \frac{100}{2} = 50$$

The slope of the line is 50.
b. Consider the meanings of ΔD and Δt in the computation of the slope. The best way to understand these symbols is to include the units in the calculation.

$$\frac{\Delta D}{\Delta t} \quad \text{means} \quad \frac{\text{change in distance}}{\text{change in time}}$$

or

$$\frac{\Delta D}{\Delta t} = \frac{100 \text{ miles}}{2 \text{ hours}} = 50 \text{ miles per hour}$$

The slope represents the trucker's average speed or velocity.

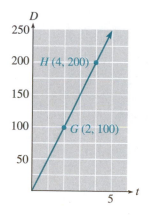

Figure 2.16

Exercise 2.2

Compute ratios to answer the questions in Exercises 1–10.

1. Carl runs 100 meters in 10 seconds. Anthony runs 200 meters in 19.6 seconds. Who has the faster average speed?

2. A spiked volleyball travels 6 feet in 0.04 second. A pitched baseball travels 66 feet in 0.48 second. Which ball travels faster?

3. Øksendahl's premium ice cream costs $6.98 for 30 ounces. Fran and Jenny's brand sells 32 ounces for $7.19. Which brand costs more per ounce?

4. Farm Fresh eggs sell for $1.56 a dozen. The company also sells packages of 18 eggs for $2.36. Which package has the lower price per egg?

5. On his 512-mile round trip to Las Vegas and back, Corey needed 16 gallons of gasoline. He used 13 gallons of gasoline on a 429-mile trip to Los Angeles. On which trip did he get better fuel economy?

6. Kendra needs $4\frac{1}{2}$ gallons of Luke's Brand primer to cover 1710 square feet of wall. She uses $5\frac{1}{3}$ gallons of Slattery's Brand primer for 2040 square feet of wall. Which brand covered more wall per gallon?

7. Which is steeper: Stone Canyon Drive, which rises 840 feet over a horizontal distance of 1500 feet, or Highway 33, which rises 1150 feet over a horizontal distance of 2000 feet?

8. The top of Romeo's ladder is on Juliet's window sill, which is 11 feet above the ground, and the bottom of the ladder is 5 feet from the base of the wall. Is the incline of this ladder as steep as a firefighter's ladder that rises a height of 35 feet over a horizontal distance of 16 feet?

9. Which is steeper: the truck ramp for Acme Movers, which rises 4 feet over a horizontal distance of 9 feet, or a toy truck ramp, which rises 3 centimeters over a horizontal distance of 7 centimeters?

10. Grimy Gulch Pass rises 0.6 mile over a horizontal distance of 26 miles. Kazuo's driveway rises 3 inches over a horizontal distance of 140 inches. Which is steeper?

In Exercises 11–16 compute the slope of the line through the indicated points.

11.

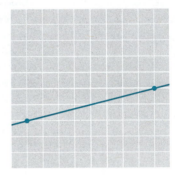

Figure 2.17

12.

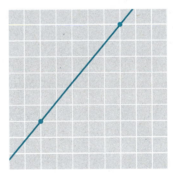

Figure 2.18

13.

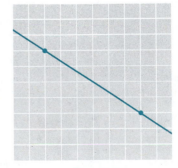

Figure 2.19

14.

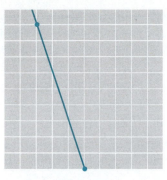

Figure 2.20

15.

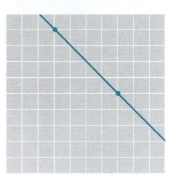

Figure 2.21

16.

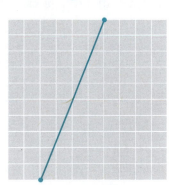

Figure 2.22

17. A line has slope 7/3. Find the vertical change associated with each horizontal change along the line.
a. 6   **b.** 10   **c.** -24

18. A line has slope $-4/5$. Find the horizontal change associated with each vertical change along the line.
a. 2   **b.** -12 **c.** 5

19. Residential staircases are usually built with a slope of 70%, or 7/10. If the vertical distance between stories is 10 feet, how much horizontal space does the staircase require?

20. A straight section of highway in the Midwest maintains a grade (or slope) of 4%, or 1/25, for 12 miles. How much does your elevation change as you travel the road?

In Exercises 21–26 graph each line by the intercept method and then use the intercepts to compute the slope.

21. $3x - 4y = 12$

22. $2y - 5x = 10$

23. $2y + 6x = -18$

24. $9x + 12y = 36$

25. $\dfrac{x}{5} - \dfrac{y}{8} = 1$

26. $\dfrac{x}{7} - \dfrac{y}{4} = 1$

Which of the tables in Exercises 27 and 28 represent variables that are related by a linear equation? (*Hint:* Which relationships have constant slope?)

27. a.

x	y
2	12
3	17
4	22
5	27
6	32

b.

t	P
2	4
3	9
4	16
5	25
6	36

c.

h	w
-6	20
-3	18
0	16
3	14
6	12

d.

t	d
5	0
10	3
15	6
20	12
25	24

28. a.

r	E
1	5
2	5/2
3	5/3
4	5/4
5	1

b.

s	t
10	6.2
20	9.7
30	12.6
40	15.8
50	19.0

c.

w	A
2	−13
4	−23
6	−33
8	−43
10	−53

d.

x	C
0	0
2	5
4	10
8	20
16	40

29. A temporary typist's paycheck (before deductions) in dollars is given by $S = 8t$, where t is the number of hours he worked.

a. Graph the equation.

b. Using two points on the graph, compute the slope $\dfrac{\Delta S}{\Delta t}$, including units.

c. What is the significance of the slope in terms of the typist's paycheck?

30. The distance in miles covered by a cross-country competitor is given by $d = 6t$, where t is the number of hours she runs.

a. Graph the equation.

b. Using two points on the graph, compute the slope $\dfrac{\Delta d}{\Delta t}$, including units.

c. What is the significance of the slope in terms of the cross-country runner?

For each graph in Exercises 31–40,
a. choose two points and compute the slope (including units).
b. explain what the slope measures in the context of the problem.

31. Figure 2.23 shows the distance, d, traveled (in meters) by a train t minutes after the train passes an observer.

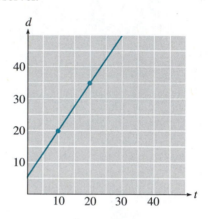

Figure 2.23

32. Figure 2.24 shows the altitude (in feet), a, of a skier t minutes after getting on a ski lift.

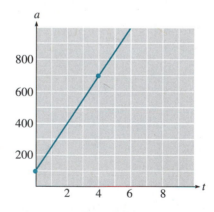

Figure 2.24

after a new drill is installed.

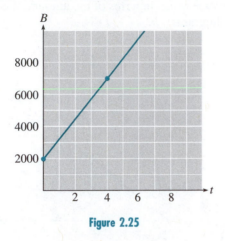

Figure 2.25

34. Figure 2.26 shows the amount of garbage (in tons), *G,* that has been deposited at a dump site *t* years after new regulations go into effect.

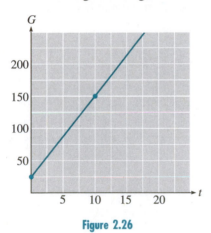

Figure 2.26

35. Figure 2.27 shows the amount of emergency

water (in liters), *W,* remaining in a southern Californian household *t* days after an earthquake.

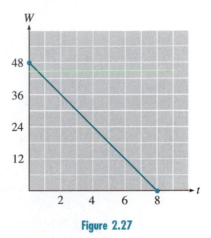

Figure 2.27

36. Figure 2.28 shows the amount of money (in dollars), *M,* in Tammy's bank account *w* weeks after she loses all sources of income.

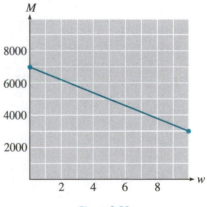

Figure 2.28

37. Figure 2.29 shows the length in inches, *i*, corresponding to various lengths in feet, *f*.

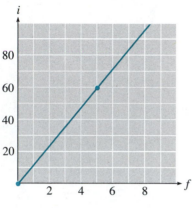

Figure 2.29

38. Figure 2.30 shows the number of ounces, *z*, that correspond to various weights measured in pounds, *p*.

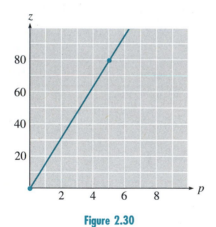

Figure 2.30

39. Figure 2.31 shows the cost (in dollars), *C*, of coffee beans in terms of the amount of coffee (in kilograms), *b*.

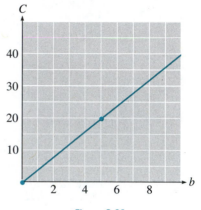

Figure 2.31

40. Figure 2.32 shows Tracey's earnings (in dollars), *E*, in terms of the number of hours, *h*, she babysits.

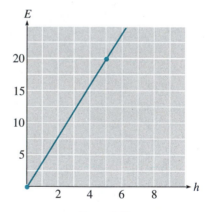

Figure 2.32

2.3   USING A GRAPHING CALCULATOR

Recall that a linear equation is one that can be written in the form

$$ax + by = c$$

where a, b, and c are constants, and a and b are not both zero. Some examples of linear equations are

$$2x + 5y = 10 \qquad 4x = 3 - 8y$$

$$y = \frac{2}{3}x - 1 \qquad 0 = x - 2 - y$$

The order of the terms is not important as long as each variable has an exponent of one and no variables appear in the denominators of fractions.

We can use a graphing calculator to graph any of the linear equations we have encountered. However, the calculator will only accept an equation written in the form   "$y = $ (expression in x)."   For example, suppose you would like to use your graphing calculator to graph the equation

$$2x - y = 5$$

Before you can enter the equation, you must solve for y in terms of x.

$$2x - y = 5 \qquad\qquad \text{Subtract } 2x \text{ from both sides.}$$

$$-y = -2x + 5 \qquad \text{Divide both sides by } -1.$$

$$y = 2x - 5$$

Now press $\boxed{Y=}$ and enter $2X - 5$ after $Y_1 =$. Select the standard graphing window by pressing $\boxed{\text{ZOOM}}$ $\boxed{6}$. The graph is shown in Figure 2.33.

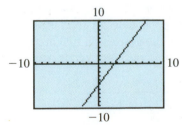

Figure 2.33

Example 1

Use a graphing calculator to graph the equation $3x + 2y = 16$.

Solution   First solve the equation for y in terms of x.

$$3x + 2y = 16 \qquad\qquad \text{Subtract } 3x \text{ from both sides.}$$

$$2y = -3x + 16 \qquad \text{Divide both sides by 2.}$$

$$y = -\frac{3}{2}x + 8$$

$$\text{or} \quad y = -1.5x + 8$$

Press $\boxed{\text{Y =}}$ and clear any equations on the display. Enter $-1.5\,\text{X} + 8$ after $\text{Y}_1 =$. Choose the standard graphing window by pressing $\boxed{\text{ZOOM}}$ $\boxed{6}$. The graph is shown in Figure 2.34.

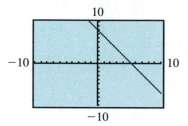

Figure 2.34

Choosing a Graphing Window

Now consider the equation

$$6x - 5y = 90 \tag{2}$$

Solving for y, we first subtract $6x$ from both sides of the equation.

$$-5y = 90 - 6x \qquad \text{Divide both sides by } -5.$$

$$y = \frac{90}{-5} - \frac{6x}{-5} \qquad \text{Simplify.}$$

$$y = -18 + \frac{6}{5}x$$

If we enter this equation in the graphing calculator, we obtain the graph shown in Figure 2.35. In the standard window we see only a small piece of the graph in the lower corner of the screen. We need to choose a window that shows more of the graph. How do we decide on an appropriate window for a given graph? The window should show the essential features of the graph and give the viewer a good idea of its overall shape. The most informative points on the graph of a line are the x- and y-intercepts. Both intercepts should be visible in the graphing window.

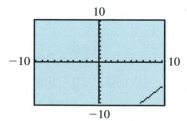

Figure 2.35

To graph a linear equation, choose scales for the axes so that both intercepts can be shown on the graph.

You can check that the intercepts of the graph of Equation (2) are $(15, 0)$ and $(0, -18)$. Thus a good choice for the window might be

$$Xmin = -20 \qquad Xmax = 20$$
$$Ymin = -20 \qquad Ymax = 20$$

This window is shown in Figure 2.36. If we wanted to concentrate on the portion of the graph in the fourth quadrant, we could choose the window settings

$$Xmin = 0 \qquad Xmax = 20$$
$$Ymin = -20 \qquad Ymax = 0$$

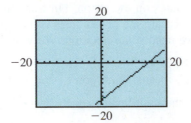

Figure 2.36

Example 2   Choose an appropriate window and graph the equation $10y - 15x = 6$.

Solution   First solve for y in terms of x to find

$$10y = 15x + 6 \qquad \text{Add 15x to both sides.}$$
$$y = 1.5x + 0.6 \qquad \text{Divide both sides by 10.}$$

Press $\boxed{Y =}$ and enter this equation.

To decide on a good window, we find the x- and y-intercepts of the graph. Setting x equal to zero, we find $y = 0.6$,   and when $y = 0$ we find $x = -0.4$.   The intercepts are thus $(-0.4, 0)$ and $(0, 0.6)$. For this graph we want to choose a very small window so we can distinguish the intercepts from the origin, so we will set x- and y-limits slightly larger than the intercept values. Figure 2.37 shows the graph in the window

$$Xmin = -1 \qquad Xmax = 1$$
$$Ymin = -1 \qquad Ymax = 1$$

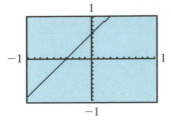

Figure 2.37

We have also set Xscl and Yscl equal to 0.1, so that the tick marks occur at intervals of 0.1 unit.

A Formula for Slope   In Section 2.2 we defined the slope of a line to be the ratio

$$m = \frac{\Delta y}{\Delta x}$$

Recall that this means the ratio $\dfrac{\text{change in } y\text{-coordinate}}{\text{change in } x\text{-coordinate}}$ as we move from one point to another on the line. This ratio tells us how fast the y-coordinate increases or decreases as we increase the x-coordinate, which gives us a way to measure how steep the line is. So far we have computed Δy and Δx by counting squares on the graph, but this method is not always practical. All we really need are the coordinates of two points on the graph.

First we need some notation. We will use *subscripts* to refer to the two points:

P_1 means "first point" and P_2 means "second point."

We denote the coordinates of P_1 by (x_1, y_1) and the coordinates of P_2 by (x_2, y_2).

Now consider a specific example. The line that passes through the two points P_1 (2, 9) and P_2 (7, -6) is shown in Figure 2.38. Note that we can find Δx by subtracting the x-coordinate of the first point from the x-coordinate of the second point:

$$\Delta x = 7 - 2 = 5$$

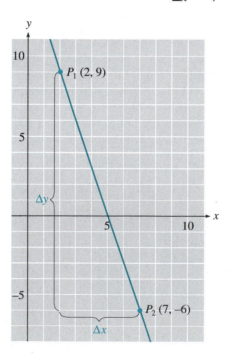

Figure 2.38

In general, we have

$$\Delta x = x_2 - x_1$$

and similarly

$$\Delta y = y_2 - y_1$$

These formulas work even if some of the coordinates are negative; in our example

$$\Delta y = y_2 - y_1 = -6 - 9 = -15$$

By counting squares down from P_1 to P_2, we see that Δy is indeed -15.

We now have a formula for the slope of a line that works even if we do not have a graph.

SLOPE FORMULA

The slope of the line passing through the points P_1 (x_1, y_1) and P_2 (x_2, y_2) is given by

$$m = \frac{\Delta y}{\Delta x} = \frac{y_2 - y_1}{x_2 - x_1} \qquad (x_2 \neq x_1)$$

Example 3

Graph the equation $y = -2.6x - 5.4$. Use $\boxed{\text{TRACE}}$ to find the coordinates of two points on the line, and compute its slope.

Solution For this problem use the following window settings:

$$\text{Xmin} = -5 \qquad \text{Xmax} = 4.4$$

$$\text{Ymin} = -20 \qquad \text{Ymax} = 15$$

The graph is shown in Figure 2.39. Press $\boxed{\text{TRACE}}$ and use the left and right arrows to choose two points on the graph, say $(2, -10.6)$ and $(-1, -2.8)$. (Remember that you can use any two points on the line to calculate the slope,

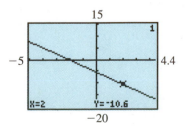

Figure 2.39

so if you choose different points you should get the same values for m.) Now substitute the coordinates of the two points into the slope formula.

$$m = \frac{y_2 - y_1}{x_2 - x_1} = \frac{-2.8 - (-10.6)}{-1 - 2}$$

You can let the calculator do the calculations for you: Press $\boxed{\text{CLEAR}}$ and enter

$$\boxed{(}\;\boxed{(-)}\;2.8\;\boxed{+}\;10.6\;\boxed{)}\;\boxed{\div}\;\boxed{(-)}\;3\;\boxed{\text{ENTER}}$$

The slope is $m = -2.6$. You can verify that this answer is correct by positioning the "bug" at the y-intercept, $(0, -5.4)$. Now move the bug one unit to the right, to the point $(1, -8)$. The y-coordinate has changed by -2.6 units, from -5.4 to -8. In other words, for a one-unit change in x there is a corresponding -2.6-unit change in y, or

$$m = \frac{\Delta y}{\Delta x} = \frac{-2.6}{1} = -2.6$$

Equations and Graphs

Perhaps you noticed that the slope of the line in Example 3, -2.6, is the same as the coefficient of x in its equation, $y = -2.6x - 5.4$. This is not a coincidence. In fact, the equation of a line can tell us quite a bit about how the graph should look. We can explore this idea by comparing the graphs of several equations in the same window.

As an example we will graph the equations

$$y = 0.5x + 8$$

$$y = 0.5x - 8$$

$$y = 0.5x$$

in the same window. Press $\boxed{\text{Y} =}$ and enter the three equations as follows: Enter .5X + 8 after "$Y_1 =$" and then press the down arrow and enter .5X − 8 after "$Y_2 =$". Press the down arrow again and enter .5X after "$Y_3 =$". We then choose the standard graphing window by pressing $\boxed{\text{ZOOM}}$ and then $\boxed{6}$. The graphs are shown in Figure 2.40.

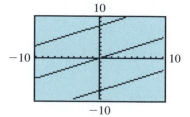

Figure 2.40

The lines appear to be parallel, or to have the same slope. We can verify that this is true by computing the slope of each line. To make the calculations easier, we will switch to a new graphing window, the Integer window. Press ZOOM 8, then press ENTER to choose the **ZInteger** window.

Your graphs should now look like Figure 2.41. Notice that when you activate the TRACE in the Integer window, you get points whose x-coordinates are integers.

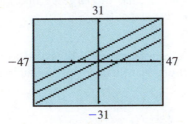

Figure 2.41

Choose two points on one line and calculate its slope. You should get a slope of 0.5. Now press the up arrow key to move to another line and calculate its slope in the same way. You should find that all three lines have slope 0.5, which is the coefficient of x in all three equations.

Now use TRACE to locate the y-intercept of each graph. You should find that the y-intercept of Y_1 is (0, 8), the y-intercept of Y_2 is (0, −8), and the y-intercept of Y_3 is (0, 0). In each case the y-intercept is the same as the constant term in the equation of the line.

Example 4

a. Graph $y = 0.2x + 5$, $y = 1.2x + 5$, and $y = 2x + 5$ in the **ZInteger** window.

b. Find the slope and y-intercept of each line. What do you conclude?

Solutions   a. Press Y = and enter the three equations after Y_1, Y_2, and Y_3, using the down arrow key to start a new equation. Choose the Integer graphing window by pressing ZOOM 8, and ENTER. The graphs are shown in Figure 2.42.

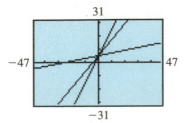

Figure 2.42

b. Choose two points on each line and compute its slope. You should find that the slope of Y_1 is 0.2, the slope of Y_2 is 1.2, and the slope of Y_3 is 2.

In each case the slope is the same as the coefficient of x in the equation. You should also find that all three lines have the same y-intercept, (0, 5). Note that 5 is the constant term in each equation. It appears that the slope of a line is given by the coefficient of x in its equation, and the y-intercept is given by the constant term. You should also observe that the steepest of the three lines, Y_3, has the largest slope, namely 2, and the least steep, Y_1, has the smallest slope, 0.2.

In the exercises you will be asked to conduct similar investigations of linear equations and their graphs, including lines that have negative slopes.

Exercise 2.3

In Exercises 1–12 solve each equation for y in terms of x.

1. $2x + y = 6$

2. $y - 3x + 5 = 0$

3. $8 - y + 3x = 0$

4. $4x - y = 2$

5. $x + 2y = 5$

6. $7 - 2y = x$

7. $3x - 4y = 6$

8. $4x + 3y = 6$

9. $0.2x + 0.5y = 1$

10. $1.2x - 4.2y = 3.6$

11. $7x + 3y = y - 32$

12. $4y - 8x = 78 - 2y$

For each equation in Exercises 13–24,
a. find the x- and y-intercepts.
b. use the intercepts to select a suitable graphing window. (Both intercepts should be visible and distinguishable.)
c. solve the equation for y.
d. graph the resulting equation using the window you specified.

13. $x + y = 100$

14. $x - y = -72$

15. $25x - 36y = 1$

16. $43x + 71y = 1$

17. $\dfrac{y}{12} - \dfrac{x}{47} = 1$

18. $\dfrac{x}{8} + \dfrac{y}{21} = 1$

19. $-2x = 3y + 84$

20. $7x = 91 - 13y$

21. $y - 42 = \dfrac{1}{3}(x - 12)$

22. $y + 27 = \dfrac{3}{2}(x - 6)$

23. $y - 3 = \dfrac{-13}{22}(x + 5)$

24. $y + 7 = \dfrac{-15}{17}(x - 8)$

In Exercises 25–34 find the slope of the line segment joining each pair of points.

25. $(-3, 2), (2, 14)$ **26.** $(-4, -3), (1, 9)$

27. $(2, -3), (-2, -1)$

28. $(5, -4), (-1, 1)$

29. $(-6, -3), (-6, 3)$ **30.** $(-2, 9), (4, 9)$

31. $(7.6, -4.2), (-3.1, 6.8)$

32. $(-9.7, -2.1), (-3.5, -6.2)$

33. $\left(\dfrac{3}{4}, -\dfrac{1}{8}\right), \left(\dfrac{5}{6}, -\dfrac{1}{2}\right)$

34. $\left(-\dfrac{7}{5}, -\dfrac{1}{3}\right), \left(\dfrac{1}{10}, -\dfrac{4}{3}\right)$

Find the slope of each line in Exercises 35–40.

35.

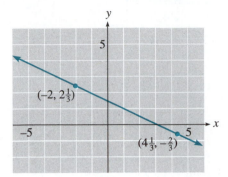

Figure 2.43

36.

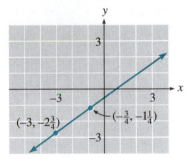

Figure 2.44

37.

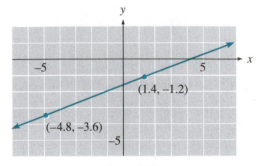

Figure 2.45

38.

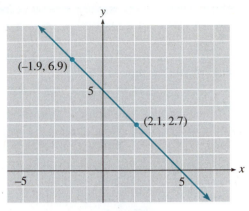

Figure 2.46

39.

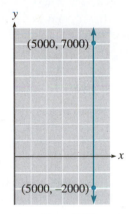

Figure 2.47

40.

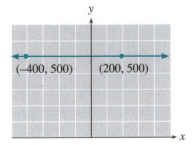

Figure 2.48

For Exercises 41–44 set the graphing window as indicated and graph the equation. For convenience, set both Xscl and Yscl to 0.
a. Use TRACE to read off coordinates of two convenient points and compute the slope of the line.
b. Use TRACE again to find the coordinates of the y-intercept.

41. Xmin $= -4.7$,   Xmax $= 4.7$
Ymin $= -10$,   Ymax $= 20$
$y = 2.5x + 6.25$

42. Xmin $= -4$,   Xmax $= 5.4$
Ymin $= -20$,   Ymax $= 15$
$y = -4.2x - 3.7$

43. Xmin $= -0.4$,   Xmax $= 9$
Ymin $= -10$,   Ymax $= 100$
$y = -8.4x + 63$

44. Xmin $= -7.2$,   Xmax $= 2.2$
Ymin $= -200$,   Ymax $= 10$
$y = -28x - 182$

45. The table gives the radius and circumference of various circles rounded to three decimal places.
a. If we plot the data, will the points lie on a straight line?
b. What familiar number does the slope turn out to be? (*Hint:* Recall a formula from geometry.)

r	C
4	25.133
6	37.699
10	62.832
15	94.248

46. The table gives the side and the diagonal of various squares rounded to three decimal places.
a. If we plot the data, will the points lie on a straight line?
b. What familiar number does the slope turn out to be? (*Hint:* Draw a picture and use the Pythagorean theorem.)

s	d
3	4.243
6	8.485
8	11.314
10	14.142

47. The formula $F = \dfrac{9}{5}C + 32$ converts the temperature in degrees Celsius (°C) to degrees Fahrenheit (°F).
a. Choose appropriate WINDOW settings and graph the equation $y = \dfrac{9}{5}x + 32$.
b. What is the Fahrenheit temperature when it is 10°C?
c. What is the Celsius temperature when it is -4°F?
d. Calculate the slope of the graph.
e. What is the meaning of the slope in this problem?
f. What is the significance of the C-intercept? of the F-intercept?

48. If the temperature on the ground is 70°F, the formula $T = 70 - \dfrac{3}{820}h$ gives the temperature at an altitude of h feet.
a. Choose appropriate WINDOW settings and graph the equation $y = 70 - \dfrac{3}{820}h$.
b. What is the temperature at an altitude of 4100 feet?
c. At what altitude is the temperature 34°F?
d. Calculate the slope of the graph.
e. What is the meaning of the slope in this problem?
f. What is the significance of the h-intercept? of the T-intercept?

In Exercises 49–58 set your window as follows: First press ZOOM 6 and then press ZOOM 8 ENTER to obtain the ZInteger window centered at the origin.
a. Graph the given equations in the same graphing window.
b. Note the slope and y-intercept for each graph.
c. Write a paragraph describing your observations.

49. $y = \dfrac{1}{2} x$, $\quad y = 2x$, $\quad y = -2x$

50. $y = x$, $\quad y = -x$, $\quad y = 0.1 x$

51. $y = 2x$, $\quad y = 3x$, $\quad y = 4x$

52. $y = -2x$, $\quad y = -4x$, $\quad y = -8x$

53. $y = -x$, $\quad y = \dfrac{-1}{2} x$, $\quad y = \dfrac{-1}{4} x$

54. $y = \dfrac{1}{3} x$, $\quad y = \dfrac{2}{3} x$, $\quad y = \dfrac{4}{3} x$

55. $y = -2x$, $\quad y = -2x + 10$, $y = -2x - 25$

56. $y = 1.5x - 18$, $\quad y = 1.5x$, $\quad y = 1.5x + 18$

57. $y = \dfrac{2}{3} x - 12$, $\quad y = \dfrac{2}{3} x - 24$, $y = \dfrac{2}{3} x - 36$

58. $y = -0.8x + 4$, $\quad y = -0.8x + 10$, $y = -0.8x + 20$

2.4 EQUATIONS OF LINES

Slope-Intercept Form

In Exercise 49 above you graphed three lines on the same set of axes:

$$y = \dfrac{1}{2} x \qquad y = 2x \qquad y = -2x$$

In Figure 2.49 you can see that all three of these lines pass through a common point (the origin) but that each line has a different slope. If we calculate the slope of each line, we find the following.

For l_1, $y = \dfrac{1}{2} x$, the slope is

$$m = \dfrac{1}{2}$$

For l_2, $y = 2x$, the slope is

$$m = 2$$

For l_3, $y = -2x$, the slope is

$$m = -2$$

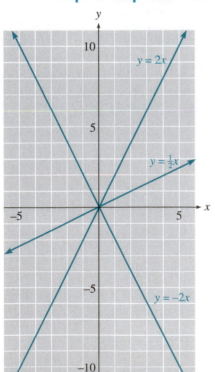

Figure 2.49

In each of these examples, the slope is the same as the coefficient of x in the equation of the line. This should come as no surprise when we recall that the slope measures how steep the line is. Think of the slope as a scale factor that tells us how many units y should increase (or decrease) for each unit that we increase x.

Now consider the graphs of the three lines shown in Figure 2.50.

$$y = 2x - 2 \qquad y = 2x + 1 \qquad y = 2x + 3$$

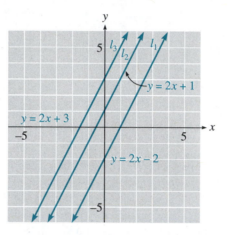

Figure 2.50

These three lines have the same slope, but they have different x- and y-intercepts. In fact, if we calculate the y-intercept of each line, we find that it is given by the constant term in the equation.

For l_1 the y-intercept is

$$y = 2(0) - 2 = -2$$

For l_2 the y-intercept is

$$y = 2(0) + 1 = 1$$

For l_3 the y-intercept is

$$y = 2(0) + 3 = 3$$

We have discovered that the equation of a line gives us direct information about the graph of the line. If we write the equation in the form

$$y = mx + b$$

then the coefficient of x is the slope of the line and the constant term is its y-intercept. This form for the equation of a line is called the **slope-intercept** form.

SLOPE-INTERCEPT FORM

If we write the equation of a line in the form

$$y = mx + b$$

then m is the **slope** of the line and b is the **y-intercept**.

Any linear equation (except, as we shall see, equations of vertical lines) can be put into slope-intercept form simply by solving for y. Once that is done we can find the slope and y-intercept without any further calculation by reading off their values from the equation.

Example 1

a.   Write the equation $3x + 4y = 6$ in slope-intercept form.
b.   Specify the slope of the line and its y-intercept.

Solutions   **a.**   Solve the equation for y.

$$3x + 4y = 6 \qquad \text{Subtract } 3x \text{ from both sides.}$$

$$4y = -3x + 6 \qquad \text{Divide both sides by 4.}$$

$$y = \frac{-3x}{4} + \frac{6}{4} \qquad \text{Simplify.}$$

$$y = \frac{-3}{4}x + \frac{3}{2}$$

b.   The slope is the coefficient of x and the y-intercept is the constant term. Thus

$$m = \frac{-3}{4} \qquad \text{and} \qquad b = \frac{3}{2}$$

Slope-Intercept Method of Graphing

We can graph a line whose equation is given in slope-intercept form without having to make a table of values. We first plot the y-intercept $(0, b)$ and then use the definition of slope to find a second point on the line. Starting at the y-intercept, we move Δy units in the y-direction and then Δx units in the x-direction. (See Figure 2.51.) The point at this location lies on the graph.

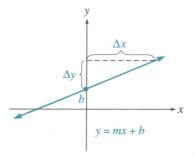

Figure 2.51

Example 2   Graph the line $y = \dfrac{4}{3}x - 2$.

Solution   Since the equation is given in slope-intercept form, we see that the slope of the line is $m = \dfrac{4}{3}$ and its y-intercept is $b = -2$.   We begin by plotting the y-intercept, $(0, -2)$. We then use the slope to find another point on the line. We have

$$m = \frac{\Delta y}{\Delta x} = \frac{4}{3}$$

so starting at $(0, -2)$ we move four units in the y-direction and three units in the x-direction, to arrive at the point $(3, 2)$. Finally, we draw the line through these two points. (See Figure 2.52.)

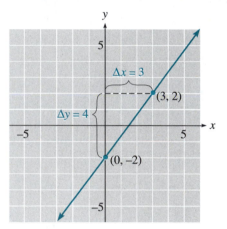

Figure 2.52

The slope of a line is a fraction and can be written in many equivalent ways. In Example 2 the slope is equal to $\dfrac{8}{6}$, $\dfrac{12}{9}$, and $\dfrac{-4}{-3}$. We can use any of these fractions to locate a third point on the line as a check. If we use $m = \dfrac{\Delta y}{\Delta x} = \dfrac{-4}{-3}$,   we move down four units and left three units from the y-intercept to find the point $(-3, -6)$ on the line.

We can also use the slope-intercept form to determine the equation of a line from its graph. First note the value of the y-intercept from the graph and calculate the slope using two convenient points. Then use these values to write the slope-intercept form of the equation.

Example 3

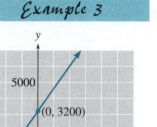

Figure 2.53

Find an equation for the line shown in Figure 2.53.

Solution The line crosses the y-axis at the point $(0, 3200)$, so the y-intercept is 3200. To calculate the slope of the line, locate another convenient point, say $(20, 6000)$, and compute

$$m = \frac{\Delta y}{\Delta x} = \frac{6000 - 3200}{20 - 0}$$

$$= \frac{2800}{20} = 140$$

The slope-intercept form of the equation, with $m = 140$ and $b = 3200$, is $y = 140x + 3200$.

Point-Slope Form

There is only one line that passes through a given point and has a given slope. This means that if we know just one point on a line and the slope, we can sketch the graph and we can find an equation for the line. For example, consider a line of slope $\frac{-3}{4}$ that passes through the point $(1, -4)$. To graph the line, we first plot the given point, $(1, -4)$. Then we use the slope to find another point on the line. Since $m = \frac{-3}{4} = \frac{\Delta y}{\Delta x}$, starting from $(1, -4)$ we move down three units and then four units to the right, to find the point $(5, -7)$. Draw the line through these two points, as shown in Figure 2.54.

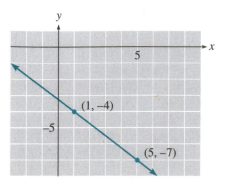

Figure 2.54

To find an equation for this line, we will use the formula for slope,

$$m = \frac{y_2 - y_1}{x_2 - x_1}$$

We can substitute the values for the slope, m, and for the coordinates of the given point, (x_1, y_1). For the second point, (x_2, y_2), we will use the variable point (x, y). We can do this because we can compute the slope using *any* two points on the line.

Example 4

Find an equation for the line that passes through $(1, -4)$ and has slope $\dfrac{-3}{4}$.

Solution   We know the slope, $m = \dfrac{-3}{4}$,   and a point, $(x_1, y_1) = (1, -4)$. We use the variable point (x, y) for the second point. Substituting these values into the slope formula gives us

$$\frac{-3}{4} = \frac{y - (-4)}{x - 1}$$

To solve for y, we first multiply both sides by $x - 1$.

$$(x - 1)\,\frac{-3}{4} = \frac{y - (-4)}{x - 1}\,(x - 1)$$

$$\frac{-3}{4}\,(x - 1) = y + 4 \qquad\qquad \text{Apply the distributive law.}$$

$$\frac{-3}{4}\,x + \frac{3}{4} = y + 4 \qquad\qquad \text{Subtract 4 from both sides.}$$

$$\frac{-3}{4}\,x - \frac{13}{4} = y \qquad\qquad \frac{3}{4} - 4 = \frac{3}{4} - \frac{16}{4} = \frac{13}{4} \quad \square$$

When we use the slope formula in this way to find the equation of a line, we will always substitute a variable point (x, y) for the second point. Thus if we like we can modify the slope formula for this purpose and write

$$m = \frac{y - y_1}{x - x_1}$$

We can even go one step further and clear the fraction to obtain

$$(x - x_1)\,m = \frac{y - y_1}{x - x_1}\,(x - x_1)$$

or

$$(x - x_1)\,m = y - y_1$$

This equation is just another version of the slope formula, but in this form it is usually called the **point-slope form** for a linear equation.

> **POINT-SLOPE FORM**
>
> The equation of the line that passes through the point (x_1, y_1) and has slope m is
>
> $$y - y_1 = m\,(x - x_1)$$

Equation of a Line Through Two Points

Of course, we can easily graph a line that passes through two given points. We can find the equation of the line in two steps: First find the slope of the line using the two points, and then use the point-slope formula.

Example 5

Find an equation for the line whose graph includes the points $(2, 2)$ and $(-4, 1)$.

Solution   First find the slope. You can use either point for (x_1, y_1) and the other point for (x_2, y_2).

$$m = \frac{y_2 - y_1}{x_2 - x_1} = \frac{1 - 2}{-4 - 2}$$

$$= \frac{-1}{-6} = \frac{1}{6}$$

Next use the point-slope formula to find the equation of the line. You can use either point for (x_1, y_1). Using $(2, 2)$ for (x_1, y_1) yields

$$y - 2 = \frac{1}{6}(x - 2)$$ Apply the distributive law.

$$y - 2 = \frac{1}{6}x - \frac{1}{3}$$ Add 2 to both sides.

$$y = \frac{1}{6}x + \frac{5}{3}$$ $\dfrac{-1}{3} + 2 = \dfrac{-1}{3} + \dfrac{6}{3} = \dfrac{5}{3}$ ▭

The technique illustrated in Example 5 is very useful for finding linear models when we have at least two data points.

Example 6

Ms. Randolph bought a new car in 1990. In 1992 the car was worth $9000, and in 1995 it was valued at $4500.

a.   Assuming that the depreciation is linear, that is, that the value of the car decreases by the same amount each year, write an equation that expresses the value of Ms. Randolph's car in terms of the number of years she has owned it.

b.   Interpret the slope as a rate of change.

c.   Find the value of the car when it was new.

Solutions   a.   Let t represent the number of years that Ms. Randolph has owned her car, and let V represent its value after t years. Then the two ordered pairs (2, 9000) and (5, 4500) are the coordinates of points of the graph of V versus t. (See Figure 2.55.) To find its equation, first compute the slope:

$$m = \frac{V_2 - V_1}{t_2 - t_1}$$

$$= \frac{9000 - 4500}{2 - 5}$$

$$= \frac{4500}{-3} = -1500$$

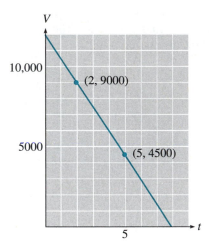

Figure 2.55

Then use the point-slope formula with either point, say (2, 9000), and $m = -1500$:

$$V - V_1 = m\,(t - t_1) \qquad \text{Substitute the given values.}$$
$$V - 9000 = -1500\,(t - 2) \qquad \text{Apply the distributive law.}$$
$$V - 9000 = -1500t + 3000 \qquad \text{Add 9000 to both sides.}$$
$$V = -1500t + 12{,}000$$

b.   The slope represents the change in the value of the car per year. Thus

$$m = \frac{\Delta V}{\Delta t} = \frac{-1500 \text{ dollars}}{1 \text{ year}}$$

so the car depreciated at a rate of $1500 per year.

c. The car was new when $t = 0$, so we substitute 0 for t to find

$$V = -1500\,(0) + 12,000$$

$$= 12,000$$

The car was worth $12,000 when new.

Exercise 2.4

In Exercises 1–12
a. write each equation in slope-intercept form.
b. state the slope and *y*-intercept of the line.

1. $3x + 2y = 1$ **2.** $3x - y = 7$

3. $x - 3y = 2$ **4.** $5x - 4y = 0$

5. $\dfrac{1}{4}x + \dfrac{3}{2}y = \dfrac{1}{6}$ **6.** $\dfrac{7}{6}x - \dfrac{2}{9}y = 3$

7. $4.2x - 0.3y = 6.6$

8. $0.8x + 0.004y = 0.24$

9. $y + 29 = 0$ **10.** $y - 37 = 0$

11. $250x + 150y = 2450$

12. $80x - 360y = 6120$

In Exercises 13–20
a. graph the line with the given slope and *y*-intercept.
b. write an equation for the line.

13. $m = 3$ and $b = -2$

14. $m = -4$ and $b = 1$

15. $m = -2$ and $b = 4$

16. $m = 5$ and $b = -3$

17. $m = \dfrac{5}{3}$ and $b = -6$

18. $m = -\dfrac{3}{4}$ and $b = -2$

19. $m = -\dfrac{1}{2}$ and $b = 3$

20. $m = \dfrac{2}{3}$ and $b = -4$

Exercises 21–30: Find an equation for each graph in Exercises 31–40 in Section 2.2.

In Exercises 31–36
a. estimate the slope and vertical intercept of each line.
b. Using your estimates from (a), write an equation for the line.

31.

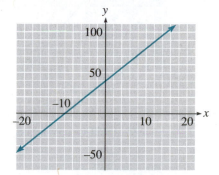

Figure 2.56

32.

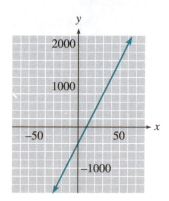

Figure 2.57

33.

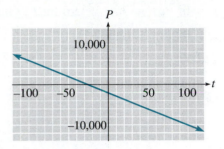

Figure 2.58

34.

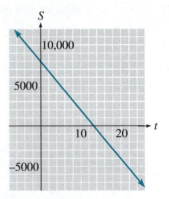

Figure 2.59

35.

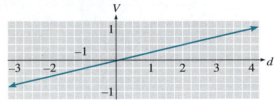

Figure 2.60

36.

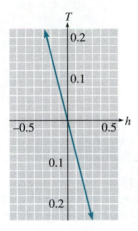

Figure 2.61

In Exercises 37–42 find an equation for the line that passes through the given point and has the given slope.

37. $(2, -5);\quad m = -3$

38. $(-6, -1);\quad m = 4$

39. $(2, -1);\quad m = \dfrac{5}{3}$

40. $(-1, 2);\quad m = -\dfrac{3}{2}$

41. $(-6.4, -3.5),\quad m = -0.27$

42. $(7.2, -1.3);\quad m = 1.55$

In Exercises 43–48 find an equation for the line whose graph includes the two given points.

43. $(-4, 2), (3, 3)$

44. $(5, -1), (2, -3)$

45. $(-2, -6), (2, 5)$

46. $(-1, 3), (4, -3)$

47. $(15.3, 9.6), (-2.4, -10.8)$

48. $(11.2, -18.3), (5.1, -6.8)$

In Exercises 49–54
a. find a linear equation relating the variables.
b. graph the equation.
c. state the slope of the line, including units, and explain its meaning in the context of the problem.

49. It cost a bicycle company $9000 to make 50 touring bikes in its first month of operation and $15,000 to make 125 bikes during its second month. Express its production costs, C, in terms of the number, b, of bikes made.

50. Under ideal conditions Andrea's Porsche can travel 312 miles on a full tank (12 gallons of gasoline) and 130 miles on 5 gallons. Express the distance, d, Andrea can drive in terms of the amount of gasoline, g, she buys.

51. On an international flight a passenger may check two bags, each weighing 70 kilograms, or 154 pounds, and one carry-on bag weighing 50 kilograms, or 110 pounds. Express the weight,

p, of a bag in pounds in terms of its weight, k, in kilograms.

52. A radio station in Detroit, Michigan, reports the high and low temperatures in the Detroit/Windsor area as 59°F and 23°F, respectively. A station in Windsor, Ontario, reports the same temperatures as 15°C and −5°C. Express the Fahrenheit temperature, F, in terms of the Celsius temperature, C.

53. When Harold and Nancy leave their motel at 8 A.M. on the second day of their summer vacation, they are 265 miles from Los Angeles. When they stop for lunch at 1 P.M., they are 590 miles from Los Angeles. Express their distance, d, from Los Angeles on the second day in terms of the time, t, they have driven.

54. Flying lessons cost $645 for an 8-hour course and $1425 for a 20-hour course. Both prices include a fixed insurance fee. Express the cost, C, of flying lessons in terms of the length, h, of the course.

2.5 **ADDITIONAL PROPERTIES OF LINES**

Horizontal and Vertical Lines

Two special cases of linear equations are worth noting. First, an equation such as

$$y = 4$$

can be thought of as an equation in two variables,

$$0x + y = 4$$

For each value of x, this equation assigns the value 4 to y. Thus any ordered pair of the form $(x, 4)$ is a solution of the equation. For example,

$$(-1, 4), \qquad (2, 4), \qquad \text{and} \qquad (4, 4)$$

are all solutions of the equation. If we draw a straight line through these points, we obtain the graph shown in Figure 2.62.

The other special case of a linear equation is of the type

$$x = 3$$

which may be thought of as an equation in two variables,

$$x + 0y = 3$$

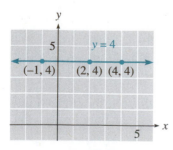

Figure 2.62

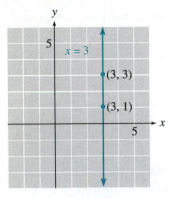

Figure 2.63

Here only one value is permissible for x, namely 3, whereas any value may be assigned to y. Thus any ordered pair of the form $(3, y)$ is a solution of this equation. If we choose two solutions, say $(3, 1)$ and $(3, 3)$, and draw a straight line through these two points, we have the graph shown in Figure 2.63. In general, we have the following results.

> The graph of $x = k$ (k a constant) is a **vertical** line.
> The graph of $y = k$ (k a constant) is a **horizontal** line.

Example 1

a.   Graph $y = 2$.   **b.**   Graph $x = -3$.

Solutions   **a.**

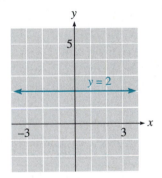

Figure 2.64

b.

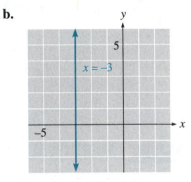

Figure 2.65

Now let's compute the slopes of the two lines in Example 1. Choose two points on the graph of $y = 2$,   say $(-5, 2)$ and $(4, 2)$. Use these points to compute the slope.

$$m = \frac{y_2 - y_1}{x_2 - x_1} = \frac{2 - 2}{4 - (-5)}$$

$$= \frac{0}{9} = 0$$

The slope of the horizontal line $y = 2$ is zero. In fact, the slope of any horizontal line is zero, because the y-coordinates of all the points on the line are equal. Thus

$$m = \frac{y_2 - y_1}{x_2 - x_1} = \frac{0}{x_2 - x_1} = 0$$

On a vertical line the x-coordinates of all the points are equal. For example, two points on the line $x = -3$ are $(-3, 1)$ and $(-3, 6)$. Using these points to compute the slope, we find

$$m = \frac{y_2 - y_1}{x_2 - x_1} = \frac{6 - 1}{-3 - (-3)}$$

$$= \frac{5}{0}$$

which is undefined. The slope of any vertical line is undefined because the expression $x_2 - x_1$ equals zero.

> The slope of a **horizontal** line is **zero**.
> The slope of a **vertical** line is **undefined**.

Parallel and Perpendicular Lines

Consider the graphs of the equations

$$y = \frac{2}{3}x - 4$$

$$y = \frac{2}{3}x + 2$$

shown in Figure 2.66. The lines have the same slope, $\frac{2}{3}$, but different y-intercepts. Since slope measures the steepness, or inclination, of a line, lines with the same slope are parallel.

> Two lines with slopes m_1 and m_2 are **parallel** if and only if
>
> $$m_1 = m_2$$

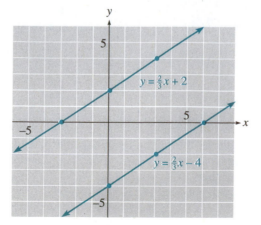

Figure 2.66

Example 2

Are the graphs of the equations $3x + 6y = 6$ and $y = -\dfrac{1}{2}x + 5$ parallel?

Solution   The lines are parallel if their slopes are equal. We can find the slope of the first line by putting its equation into slope-intercept form. Solve for y:

$$3x + 6y = 6 \qquad \text{Subtract } 3x \text{ from both sides.}$$

$$6y = -3x + 6 \qquad \text{Divide both sides by 6.}$$

$$y = \frac{-3x}{6} + \frac{6}{6} \qquad \text{Simplify.}$$

$$y = -\frac{1}{2}x + 1$$

The slope of the first line is $m_1 = -\frac{1}{2}$.   The equation of the second line is already in slope-intercept form; its slope is $m_2 = -\frac{1}{2}$.   Thus $m_1 = m_2$, so the lines are parallel.

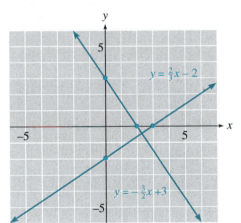

Figure 2.67

Now consider the graphs of the equations

$$y = \frac{2}{3}x - 2$$

$$y = -\frac{3}{2}x + 3$$

shown in Figure 2.67. The lines appear to be perpendicular. The relationship between the slopes of perpendicular lines is not as easy to see as the relationship for parallel lines. For this example, $m_1 = \frac{2}{3}$ and $m_2 = -\frac{3}{2}$. Note that

$$m_2 = -\frac{3}{2} = \frac{-1}{\frac{2}{3}} = \frac{-1}{m_1}$$

This relationship holds for any two perpendicular lines with slopes m_1 and m_2, as long as $m_1 \neq 0$ and $m_2 \neq 0$.

> Two lines with slopes m_1 and m_2 are **perpendicular** if
>
> $$m_2 = \frac{-1}{m_1}$$

We say that m_2 is the *negative reciprocal* of m_1.

Example 3

Are the graphs of $3x - 5y = 5$ and $2y = \frac{10}{3}x + 3$ perpendicular?

Solution Find the slope of each line by putting the equation into slope-intercept form. For the first line,

$$5y = 3x - 5 \qquad \text{Divide both sides by 5.}$$

$$y = \frac{3}{5}x - 1$$

so $m_1 = \frac{3}{5}$. For the second line,

$$y = \frac{5}{3}x + \frac{3}{2}$$

so $m_2 = \frac{5}{3}$. Now the negative reciprocal of m_1 is

$$\frac{-1}{m_1} = \frac{-1}{\frac{3}{5}} = \frac{-5}{3}$$

but $m_2 = \frac{5}{3}$. Thus $m_2 \neq \frac{-1}{m_1}$, so the lines are not perpendicular.

Example 4

Show that the triangle with vertices A $(0, 8)$, B $(6, 2)$, and C $(-4, 4)$ is a right triangle.

Solution We will show that two of the sides of the right triangle are perpendicular. The line segment $\overline{AB}$ has slope

$$m_1 = \frac{2 - 8}{6 - 0} = \frac{-6}{6} = -1$$

and the line segment $\overline{AC}$ has slope

$$m_2 = \frac{4 - 8}{-4 - 0} = \frac{-4}{-4} = 1$$

Since

$$\frac{-1}{m_1} = \frac{-1}{-1} = 1 = m_2$$

the sides $\overline{AB}$ and $\overline{AC}$ are perpendicular, and the triangle is a right triangle. (See Figure 2.68.)

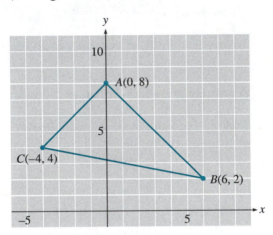

Figure 2.68

Consider the graph of $4x - 2y = 6$ shown in Figure 2.69. Can we find the equation of the line that is parallel to this line but passes through the point (1, 4)? If we can find the slope of the desired line, we can use the slope-intercept formula to find its equation.

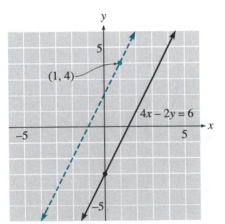

Figure 2.69

Since the line we want is parallel to the given line, they must have the same slope. To find the slope of the given line, we write its equation in slope-intercept form:

$$4x - 2y = 6 \qquad \text{Subtract } 4x \text{ from both sides.}$$

$$-2y = -4x + 6 \qquad \text{Divide both sides by } -2.$$

$$y = \frac{-4x}{-2} + \frac{6}{-2} \qquad \text{Simplify.}$$

$$y = 2x - 3$$

The slope of the given line is $m_1 = 2$. Since the unknown line is parallel to this line, its slope is also 2. Now we know the slope of the desired line, $m = 2$, and one point on the line, (1, 4). Substituting these values into the point-slope formula will give us the equation.

$$y - y_1 = m (x - x_1)$$

$$y - 4 = 2 (x - 1) \qquad \text{Apply the distributive law.}$$

$$y - 4 = 2x - 2 \qquad \text{Add 4 to both sides.}$$

$$y = 2x + 2$$

Example 5

Find an equation for the line that passes through the point (1, 4) and is perpendicular to the line $4x - 2y = 6$.

Solution We follow the same strategy as in the discussion above: First find the slope of the desired line, then use the point-slope formula to write its equation. The line we want is perpendicular to the given line, so its slope is the negative reciprocal of $m_1 = 2$, the slope of the given line. Thus

$$m_2 = \frac{-1}{m_1} = \frac{-1}{2}$$

Now use the point-slope formula with $m = \frac{-1}{2}$ and $(x_1, y_1) = (1, 4)$.

$$y - y_1 = m (x - x_1)$$

$$y - 4 = \frac{-1}{2} (x - 1) \qquad \text{Apply the distributive law.}$$

$$y - 4 = \frac{-1}{2} x + \frac{1}{2} \qquad \text{Add 4 to both sides.}$$

$$y = \frac{-1}{2} x + \frac{9}{2} \qquad \frac{1}{2} + 4 = \frac{1}{2} + \frac{8}{2} = \frac{9}{2}$$

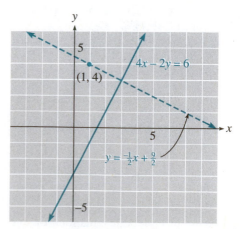

Figure 2.70

The given line and the perpendicular line are shown in Figure 2.70.

Exercise 2.5

In Exercises 1–6
a. graph each equation.
b. state the slope of each line.

1. $y = -3$ **2.** $x = -2$ **3.** $2x = 8$

4. $3y = 15$ **5.** $x = 0$ **6.** $y = 0$

In Exercises 7–8
a. determine whether the slope of each line is positive, negative, zero, or undefined.
b. list the lines in order of increasing slope.

7.

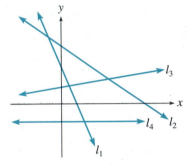

Figure 2.71

8.

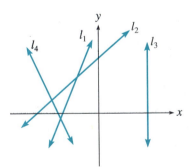

Figure 2.72

9. Determine whether the lines are parallel, perpendicular, or neither.

  a. $y = \dfrac{3}{5}x - 7; \quad 3x - 5y = 2$

  b. $y = 4x + 3; \quad y = \dfrac{1}{4}x - 3$

  c. $6x + 2y = 1; \quad x = 1 - 3y$

  d. $2y = 5; \quad 5y = -2$

10. Determine whether the lines are parallel, perpendicular, or neither.

  a. $2x - 7y = 14; \quad 7x - 2y = 14$

  b. $x + y = 6; \quad x - y = 6$

  c. $x = -3; \quad 3y = 5$

  d. $\dfrac{1}{4}x - \dfrac{3}{4}y = \dfrac{2}{3}; \quad \dfrac{1}{6}x = \dfrac{1}{2}y + \dfrac{1}{3}$

11. a. Sketch the triangle with vertices A (2, 5), B (5, 2), and C (10, 7).

  b. Show that the triangle is a right triangle.

12. a. Sketch the triangle with vertices P (−1, 3), Q (−3, 8), and R (4, 5).

  b. Show that the triangle is a right triangle.

13. a. Sketch the quadrilateral with vertices P (2, 4), Q (3, 8), R (5, 1), and S (4, −3).

  b. Show that the quadrilateral is a parallelogram.

14. a. Sketch the quadrilateral that has vertices A (−5, 4), B (7, −11), C (12, 25), and D (0, 40).

  b. Show that the quadrilateral is a parallelogram.

15. a. Write an equation for the line that is parallel to the graph of $x - 2y = 5$ and passes through the point (2, −1).

  b. Sketch the graphs of both equations on the same axes.

16. a. Write an equation for the line that is parallel to the graph of $2y - 3x = 5$ and passes through the point (−3, 2).

  b. Sketch the graphs of both equations on the same axes.

17. a. Write an equation for the line that is perpendicular to the graph of $2y - 3x = 5$ and passes through the point (1, 4).

  b. Sketch the graphs of both equations on the same axes.

18. a. Write an equation for the line that is perpendicular to the graph of $x - 2y = 5$ and passes through the point $(4, -3)$.

b. Sketch the graphs of both equations on the same axes.

19. Two of the vertices of rectangle $ABCD$ are $A(-5, 2)$ and $B(-2, -4)$.

a. Find an equation for the line that includes side $\overline{AB}$.

b. Find an equation for the line that includes side $\overline{BC}$.

20. Two of the vertices of rectangle $PQRS$ are $P(-2, -6)$ and $Q(4, -4)$.

a. Find an equation for the line that includes side $\overline{PQ}$.

b. Find an equation for the line that includes side $\overline{QR}$.

In Exercises 21 and 22 recall from geometry that the tangent line to a circle is perpendicular to the radius to the point of tangency.

21. The center of a circle is the point $C(2, 4)$, and $P(-1, 6)$ is a point on the circle. Find the equation of the line tangent to the circle at the point P. (See Figure 2.73.)

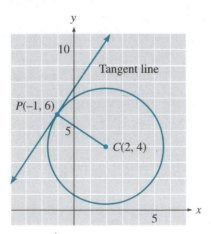

Figure 2.73

22. The center of a circle is the point $D(-2, 1)$, and $Q(1, -4)$ is a point on the circle. Find the

equation of the line tangent to the circle at the point Q. (See Figure 2.74.)

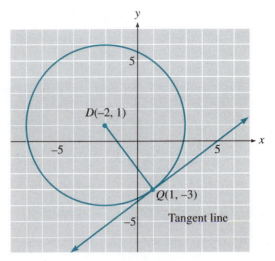

Figure 2.74

23. In this exercise we will show that parallel lines have the same slope. In Figure 2.75, l_1 and l_2 are two parallel lines that are neither horizontal nor vertical. Their y-intercepts are A and B. The segments $\overline{AC}$ and $\overline{CD}$ are constructed parallel to the x- and y-axes, respectively. Explain why each of the following statements is true.

a. Angle ACD equals angle CAB.

b. Angle DAC equals angle ACB.

c. Triangle ACD is similar to triangle CAB.

d. $m_1 = \dfrac{CD}{AC}; \quad m_2 = \dfrac{AB}{AC}.$

e. $m_1 = m_2.$

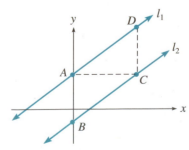

Figure 2.75

24. In this exercise we will show that if two lines with slopes m_1 and m_2 (where neither line is vertical) are perpendicular, then m_2 is the negative reciprocal of m_1. In Figure 2.76, l_1 and l_2 are perpendicular lines. Their y-intercepts are B and C. The segment $\overline{AP}$ is constructed through the point of intersection of l_1 and l_2 parallel to the x-axis. Explain why each of the following statements is true.

a. Angle ABC and angle ACB are complementary.

b. Angle ABC and angle BAP are complementary.

c. Angle BAP equals angle ACB.

d. Angle CAP and angle ACB are complementary.

e. Angle CAP equals angle ABC.

f. Triangle ABP is similar to triangle CAP.

g. $m_1 = \dfrac{BP}{AP}$; $m_2 = -\dfrac{CP}{AP}$ **h.** $m_2 = \dfrac{-1}{m_1}$

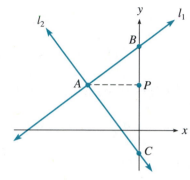

Figure 2.76

GRAPHICAL SOLUTION OF EQUATIONS AND INEQUALITIES

Using Graphs to Solve Equations

We can use graphs to find solutions of equations that are difficult or impossible to solve algebraically. Recall that the graph of an equation in two variables is just a picture of its solutions. When we read the coordinates of a point on the graph, we are reading a pair of x- and y-values that make the equation true.

Example 1

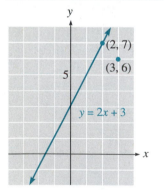

Figure 2.77

a. Verify both algebraically and graphically that the ordered pair $(2, 7)$ is a solution of the equation $y = 2x + 3$.

b. Verify that the ordered pair $(3, 6)$ is not a solution of $y = 2x + 3$.

Solutions a. In Figure 2.77 the point $(2, 7)$ lies on the graph of $y = 2x + 3$.   This confirms graphically that $(2, 7)$ is a solution. To verify algebraically that $(2, 7)$ is a solution, substitute $x = 2$ and $y = 7$ into the equation:

$$7 \overset{?}{=} 2\,(2) + 3 \qquad \text{True}$$

b. The point $(3, 6)$ does not lie on the graph of $y = 2x + 3$,   so $(3, 6)$ cannot be a solution of the equation. We can verify algebraically that $(3, 6)$ is not a solution:

$$6 \overset{?}{=} 2\,(3) + 3 \qquad \text{False}$$

We can also use the graph of an equation to find specific solutions. Consider the graph of

$$y = 285 - 15x \qquad\qquad (3)$$

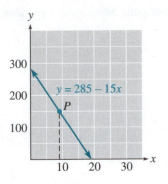

Figure 2.78

shown in Figure 2.78. Suppose we would like to find the solution to Equation (3) that has $y = 150$. We begin by locating the point on the graph for which $y = 150$. This point is labeled P in Figure 2.78. Now find the x-coordinate of point P by drawing an imaginary line straight down from P to the x-axis. The x-coordinate of P is $x = 9$. Thus P is the point $(9, 150)$, and the ordered pair $(9, 150)$ must be the solution we want.

We can verify algebraically that $(9, 150)$ is a solution by substituting $x = 9$ and $y = 150$ into Equation (3):

Does $\quad$ **150** $= 285 - 15\,(\mathbf{9})$?

$285 - 15(9) = 285 - 135 = 150$ $\qquad$ **Yes.**

The graphical method can also be used to solve equations in one variable. Compare the two statements below.

a. Solve the equation $\quad y = 285 - 15x$ $\quad$ when $\quad y = 150$.

b. Solve the equation $\quad 150 = 285 - 15x$.

The same value of x will solve the equations in both (a) and (b); they are really two different ways of stating the same problem. Using Figure 2.78, we solved the equation in (a) and found that when $\quad y = 150$ $\quad$ the corresponding value of x is $\quad x = 9$. Thus $\quad x = 9$ $\quad$ is also a solution for the equation in (b), $\quad 150 = 285 - 15x$.

We summarize by noting that the following three statements are equivalent.

1. The point $(9, 150)$ lies on the graph of $\quad y = 285 - 15x$.

2. The ordered pair $(9, 150)$ is a solution of the equation $\quad y = 285 - 15x$.

3. $x = 9$ $\quad$ is a solution of the equation $\quad 150 = 285 - 15x$.

Example 2

a. Use the graph of the equation $\quad y = 2.5x + 13$ $\quad$ to find the solution with $\quad y = 5.5$.

b. Use a graph to solve the equation $\quad 2.5x + 13 = 5.5$.

c. Verify your solutions algebraically.

Solutions $\quad$ **a.** Locate the point on the graph that has y-coordinate 5.5. This point is labeled Q in Figure 2.79 on page 120. Find the x-coordinate of Q by reading the value directly below Q on the x-axis. The coordinates of point Q are $(-3, 5.5)$, so the solution we seek is the ordered pair $(-3, 5.5)$.

b. Notice that y has been replaced by 5.5. Thus we are looking for a value of x that corresponds to $\quad y = 5.5$ $\quad$ on the graph. But this problem is the same one we solved in part (a): the x-value we need is $\quad x = -3$. Thus the solution to the equation $\quad 2.5x + 13 = 5.5$ $\quad$ is $\quad x = -3$.

c. We can verify both solutions by substituting $\quad x = -3$ $\quad$ into the equation:

$$2.5\,(\mathbf{-3}) + 13 = -7.5 + 13 = 5.5$$

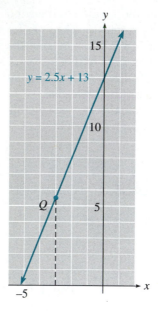

Figure 2.79

This graphical technique also works on nonlinear equations. Consider the following examples.

a. $\dfrac{1}{4} x^3 + \dfrac{1}{4} x^2 - 3x = 8$

b. $\dfrac{1}{4} x^3 + \dfrac{1}{4} x^2 - 3x = 0$

c. $\dfrac{1}{4} x^3 + \dfrac{1}{4} x^2 - 3x = 3$

We can solve all three of these equations using the graph of

$$y = \dfrac{1}{4} x^3 + \dfrac{1}{4} x^2 - 3x$$

shown in Figure 2.80. Each equation has a different value for *y*. To solve equation (a), we look for points on the graph with *y*-coordinate 8. The point *A* on Figure 2.80 has coordinates (4, 8), so $x = 4$ is a solution to (a). We can verify this algebraically:

$$\dfrac{1}{4}(4)^3 + \dfrac{1}{4}(4)^2 - 3\,(4) = \dfrac{1}{4}\,(64) + \dfrac{1}{4}\,(16) - 3\,(4)$$

$$= 16 + 4 - 12 = 8$$

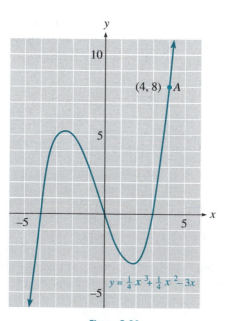

Figure 2.80

To solve (b), we look for points on the graph with y-coordinate 0. These points are the x-intercepts of the graph. We see in Figure 2.80 that there are three of them, at $x = -4$, $x = 0$, and $x = 3$. This means that the equation in (b) has three solutions, $-4, 0$, and 3. [You should verify algebraically that each of these x-values satisfies (b).]

To solve the equation in (c), we look for points on the graph with y-coordinate 3. It may help to draw in a horizontal line at $y = 3$ as shown in Figure 2.81. There are three points with y-coordinate 3 on the graph, and we need to find the x-coordinate of each of these points. The middle point has x-coordinate -1, and it is easy to verify that $x = -1$ is a solution of (c):

$$\frac{1}{4}(-1)^3 + \frac{1}{4}(-1)^2 - 3(-1)$$

$$= -\frac{1}{4} + \frac{1}{4} + 3 = 3$$

The other two points do not have integer-valued x-coordinates, so we have to estimate these values. This means that we will not find exact solutions for the equation in (c), but only approximations. For many applications an approximate value may be adequate. Later we will learn how to make our approximations more precise. For now it is sufficient to observe that the

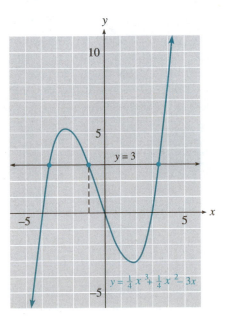

Figure 2.81

second solution of the equation in (c) is about -3.5 and the third solution is about 3.5.

If we try to verify these approximate solutions by evaluating the left side of (c), we will not get exactly 3 for the y-value, but it should be close. For example,

$$\frac{1}{4}(\mathbf{3.5})^3 + \frac{1}{4}(\mathbf{3.5})^2 - 3(\mathbf{3.5}) = 3.28125$$

which is not bad for a rough estimate.

Example 3

Solve each equation graphically.
a. $10 - 3x - x^2 = 6.5$ **b.** $10 - 3x - x^2 = 15$

Solutions a. Consider the graph of $y = 10 - 3x - x^2$ shown in Figure 2.82. To solve the equation in part (a), we need to find those points on the graph whose y-coordinate is 6.5. The line $y = 6.5$ crosses the graph in two points, whose x-coordinates are approximately -3.9 and 0.9. Thus the equation $10 - 3x - x^2 = 6.5$ has two solutions, which we estimate as $x \approx -3.9$ and $x \approx 0.9$.
b. From Figure 2.82 we see that there are no points on the graph of $y = 10 - 3x - x^2$ with y-coordinate 15, which means that there are no real solutions to the equation $10 - 3x - x^2 = 15$.

Graphs to Solve Inequalities

We can also use graphs to solve inequalities. Consider the inequality

$$9 - 2x \le 5$$

Remember that to "solve" this inequality means to find all values of x that make the expression $9 - 2x$ less than 5 or equal to 5. One direct (but inefficient) method for doing this would be to start evaluating $9 - 2x$ for various x-values and see which ones come out less than 5. Table 2.1 records the results of such a search.

x	-2	-1	0	1	2	3	4	5
$9 - 2x$	13	11	9	7	5	3	1	-1

Table 2.1

From Table 2.1 it appears that values of x greater than or equal to 2 give the desired result, which means that x-values of 2 or greater are solutions to the inequality. We can get a more complete picture by plotting our data on a graph. In Figure 2.83 we have plotted x-values on the horizontal axis and the corresponding values of $9 - 2x$ on the vertical axis. The coordinates of each point on the graph represent a value of x and the resulting value of 9

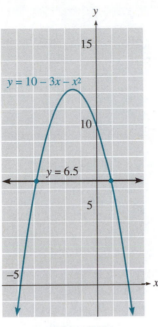

Figure 2.82

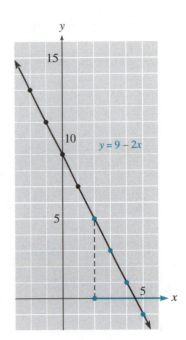

Figure 2.83

$- 2x$. We are interested in points on the graph whose vertical coordinate is less than or equal to 5. Those points are shown in color.

Now which x-values resulted in points with vertical coordinates less than or equal to 5? We can read the x-coordinates of these points by dropping straight down to the x-axis on the graph, as shown by the arrows. The x-value corresponding to $9 - 2x = 5$ is $x = 2$. For smaller values of $9 - 2x$, we must choose x-values greater than or equal to 2. This confirms our answer obtained from Table 2.1: the solutions are all values of x greater than or equal to 2, as shown in color on the x-axis in Figure 2.83. (You may also want to verify the solution by solving the inequality algebraically.)

In many cases a graphical solution may be the fastest way to solve an inequality.

Example 4

a. Use the graph of $y = x^3 - 6x^2 + 12x$ to solve the inequality

$$x^3 - 6x^2 + 12x > 7$$

b. Verify your solution for $x = 2$ and $x = -1$.

Solutions **a.** The graph is shown in Figure 2.84 on page 124. Keep in mind that the y-coordinate of each point gives the value of $x^3 - 6x^2 + 12x$. We are interested in x-values that correspond to y-values greater than 7. Points

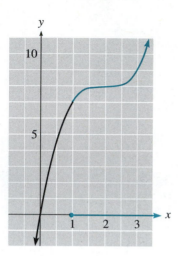

Figure 2.84

on the graph with y-value greater than 7 are shown in color. The x-coordinates of those points are the x-values we want. From the graph we see that the solution is all x-values greater than 1.

b.   To verify the solution, substitute the given x-values into the inequality. For $x = 2$,

$$(2)^3 - 6\,(2)^2 + 12\,(2) = 8 - 24 + 24 = 8 > 7$$

so $x = 2$ is a solution. For $x = -1$,

$$(-1)^3 - 6\,(-1)^2 + 12\,(-1) = -1 - 6 - 24 = -31 < 7$$

so $x = -1$ is not a solution.   ▭

Using a Graphing Calculator to Solve Equations and Inequalities

With the aid of a graphing calculator, we can quickly find (at least approximate) solutions to many different equations and inequalities.

Example 5

Use a graph to solve the equation $1.3x + 2.4 = 8.51$.   On a TI-82 set the WINDOW to

$$\text{Xmin} = -4.6 \qquad \text{Xmax} = 4.8$$
$$\text{Ymin} = -10 \qquad \text{Ymax} = 10$$

Solution   After setting the WINDOW values, graph $y = 1.3x + 2.4$. We want to find the x-coordinate that corresponds to $y = 8.51$. Activate the TRACE so that you can read coordinates of points on the graph. Press the right arrow key until the bug reaches the point with y-coordinate 8.51. Read the corresponding value of x to find $x = 4.7$.   (See Figure 2.85.)

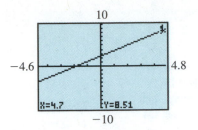

Figure 2.85

We can also check our solution using the calculator. First press $\boxed{\text{2nd}}$ $\boxed{\text{QUIT}}$ to get back to the home screen. Substitute 4.7 for x in the equation by entering

$$1.3 \boxed{\times} 4.7 \boxed{+} 2.4 \boxed{\text{ENTER}}$$

and you will see that $y = 8.51$, as expected.

Because the graphing calculator does not label the scales on the axes, it is often helpful to plot a horizontal line corresponding to the y-value of interest.

Example 6

Use a graph to solve the inequality $572 - 23x \geq 181$. Set the $\boxed{\text{WINDOW}}$ to

$$\begin{array}{ll} \text{Xmin} = -40 & \text{Xmax} = 54 \\ \text{Ymin} = -100 & \text{Ymax} = 900 \end{array}$$

Solution After setting the $\boxed{\text{WINDOW}}$, enter $Y_1 = 572 - 23X$ and $Y_2 = 181$. Press $\boxed{\text{GRAPH}}$ and you should see the graphs shown in Figure 2.86. Activate the $\boxed{\text{TRACE}}$ and use the arrow keys until the bug rests on the intersection of the two lines, that is, when the y-coordinate is 181. The corresponding value of x is $x = 17$. Now, to solve $572 - 23x \geq 181$, we want points with y-values greater than or equal to 181. Experiment by pressing the left and right arrow keys to move the bug to either side of the intersection point. Notice that points with y-coordinates *greater* than 181 have x-coordinates *less* than 17. Thus the solution to the inequality is $x \leq 17$.

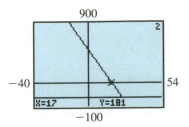

Figure 2.86

Exercise 2.6

In Exercises 1–6
a. use the graph to solve the equations and inequalities. (You will have to approximate some of the solutions.)
b. solve the equations and inequalities algebraically, and compare your answers with the solutions you read from the graph.

1. Figure 2.87 shows the graph of $y = 1.4x - 0.64$. Solve:

   a. $1.4x - 0.64 = 0.2$

   b. $1.4x - 0.64 = -1.2$

   c. $1.4x - 0.64 > 0.2$

   d. $-1.2 > 1.4x - 0.64$

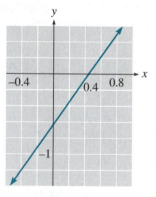

Figure 2.87

2. Figure 2.88 shows the graph of $y = 28x + 14$. Solve:

   a. $70 = 28x + 14$

   b. $28x + 14 = 210$

   c. $28x + 14 \leq 70$

   d. $210 \leq 28x + 14$

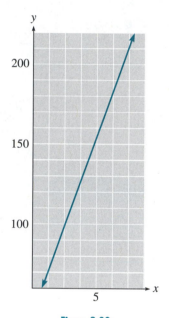

Figure 2.88

3. Figure 2.89 shows the graph of
$y = -36x + 226$.   Solve:

a. $-36x + 226 = 10$

b. $190 = -36x + 226$

c. $-36x + 226 \leq 10$

d. $190 \leq -36x + 226$

4. Figure 2.90 shows the graph of
$y = -2.4x + 2.32$.   Solve:

a. $1.6 = -2.4x + 2.32$

b. $-2.4x + 2.32 = 0.4$

c. $-2.4x + 2.32 \geq 1.6$

d. $0.4 \geq -2.4x + 2.32$

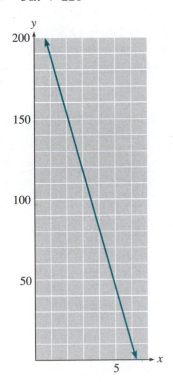

Figure 2.89

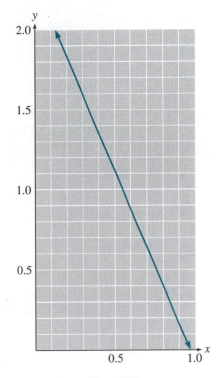

Figure 2.90

5. Figure 2.91 shows a graph of
$y = \sqrt{x} - 2$, for $x > 0$. Solve:
a. $\sqrt{x} - 2 = 1.5$
b. $\sqrt{x} - 2 = 2.25$
c. $\sqrt{x} - 2 < 1$
d. $\sqrt{x} - 2 > -0.25$

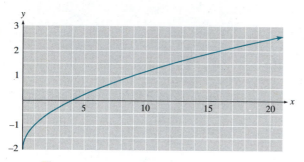

Figure 2.91

6. Figure 2.92 shows a graph of $y = \dfrac{4}{x + 2}$,

for $x > -2$. Solve:

a. $\dfrac{4}{x + 2} = 4$

b. $\dfrac{4}{x + 2} = 0.8$

c. $\dfrac{4}{x + 2} > 1$

d. $\dfrac{4}{x + 2} < 3$

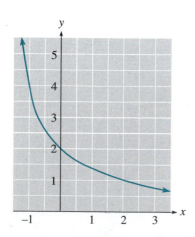

Figure 2.92

In Exercises 7–8 use the graph to solve the equations and inequalities and to answer the questions. Then check your answers algebraically.

7. Figure 2.93 shows the graph of
$w = -10(t + 1)^3 + 10$. Solve:
a. $-10(t + 1)^3 + 10 = 100$
b. $-10(t + 1)^3 + 10 = -140$
c. $-10(t + 1)^3 + 10 > -50$
d. $-20 < -10(t + 1)^3 + 10 < 40$
e. For what values of t is w decreasing?

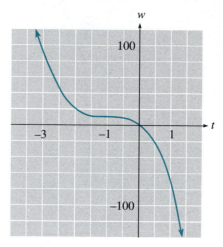

Figure 2.93

8. Figure 2.94 shows the graph of
$B = \frac{1}{3}p^3 - 3p + 2$. Solve:
a. $\frac{1}{3}p^3 - 3p + 2 = 5.5$
b. $\frac{1}{3}p^3 - 3p + 2 = 5$
c. $\frac{1}{3}p^3 - 3p + 2 < 1$
d. What range of values does B have for p between -2.5 and 0.5?
e. For what values of p is B increasing?

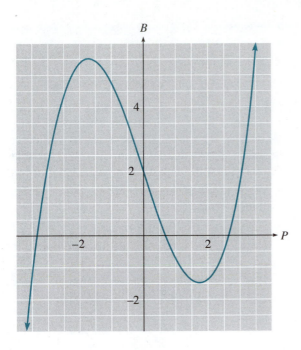

Figure 2.94

In Exercises 9–16 use the graph to answer the questions.

9. Figure 2.95 shows a graph of M in terms of q. Find all values of q for which

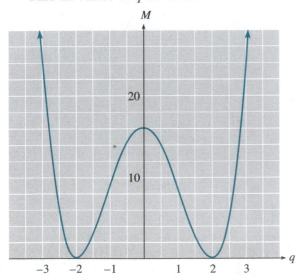

Figure 2.95

a. $M = 0$

b. $M = 16$

c. $M < 6$

d. For what values of q is M increasing?

e. What range of values does M have for q between -3 and -1?

10. Figure 2.96 shows a graph of P in terms of t. Find all values of t for which

a. $P = 3$

b. $P > 4.5$

c. For what values of t is P decreasing?

d. What is the maximum value of P? For what value of t does this maximum occur?

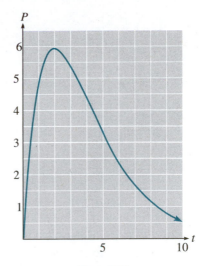

Figure 2.96

11. Figure 2.97 shows a graph of C in terms of u. Find all values of u for which

a. $C = -10$

b. $C = 0$

c. $C < 10$

d. $0 > C > -8$

a. $S = 0$

b. $S = -0.5$

c. $S < -0.75$

d. $-1 < S < 1$

13. Figure 2.99 shows the graph of C in terms of t. C stands for the number of students at State University who regard themselves as computer literate. C is measured in thousands. The variable t represents time, measured in years since 1990.

a. When did 2000 students consider themselves computer literate?

b. How long did it take that number to double?

c. How long did it take for the number to double again?

d. How many students became computer literate between January 1992 and June 1993?

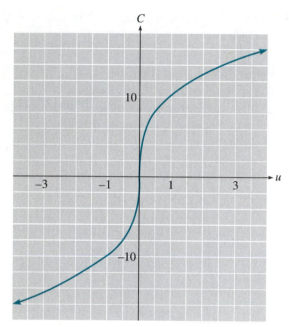

Figure 2.97

12. Figure 2.98 shows a graph of S in terms of v. Find all values of v for which

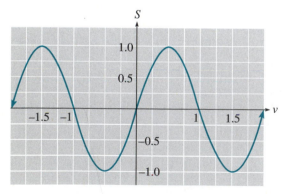

Figure 2.98

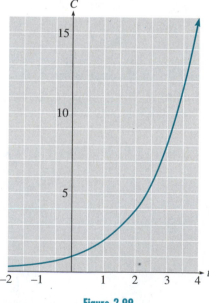

Figure 2.99

14. The graph in Figure 2.100 shows the number of people, P, in Cedar Grove who owned compact disc (CD) players t years after 1980.

a. When did 3500 people own CD players?

b. How many people owned CD players in 1986?

c. The number of owners of CD players seems to be leveling off at what number?

d. How many people acquired CD players between 1981 and 1984?

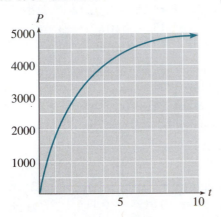

Figure 2.100

15. The graph in Figure 2.101 shows the revenue, *R*, a movie house takes in when it charges *d* dollars for a ticket.

a. What is the revenue if the movie house charges $6.50 for a ticket?

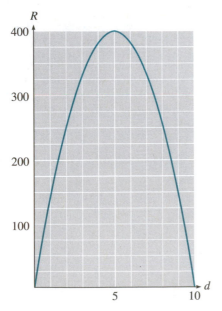

Figure 2.101

b. What should the movie house charge for a ticket to make $250 in revenue?

c. Solve $R > 350$.

16. The graph in Figure 2.102 shows the sales, *S*, of a best-selling book, in thousands of dollars, *w* weeks after it is released.

a. In what weeks were sales over $7000?

b. In what week did sales fall below $5000 on their way down?

c. Solve $S > 3.4$.

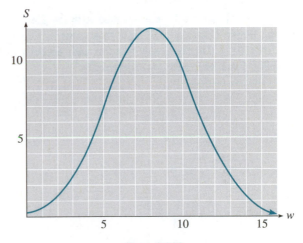

Figure 2.102

In Exercises 17–20 graph each equation with the ZInteger setting. (Press ZOOM 6 and then press ZOOM 8 ENTER.) Use the graph to answer each question. Use the equation to verify your answers.

17. Graph $y = 2x - 3$.

a. For what value of *x* is $y = 5$?

b. For what value of *x* is $y = -13$?

c. For what values of *x* is $y > -1$?

d. For what values of *x* is $y < 25$?

18. Graph $y = 4 - 2x$.

a. For what value of *x* is $y = 6$?

b. For what value of *x* is $y = -4$?

c. For what values of *x* is $y > -12$?

d. For what values of *x* is $y < 18$?

19. Graph $y = 6.5 - 1.8x$.
   a. For what value of x is $y = -13.3$?
   b. For what value of x is $y = 24.5$?
   c. For what values of x is $y \leq 15.5$?
   d. For what values of x is $y \geq -7.9$?

20. Graph $y = 0.2x + 1.4$.
   a. For what value of x is $y = -5.2$?
   b. For what value of x is $y = 2.8$?
   c. For what values of x is $y \leq -3.2$?
   d. For what values of x is $y \geq 4.4$?

In Exercises 21–30 graph each equation with the ZInteger setting. (Press $\boxed{\text{ZOOM}}$ $\boxed{6}$ and then press $\boxed{\text{ZOOM}}$ $\boxed{8}$ $\boxed{\text{ENTER}}$.) Use the graph to solve each equation or inequality. Check your solutions algebraically.

21. Graph $y = \dfrac{2}{3}x - 24$.

   a. Solve $\dfrac{2}{3}x - 24 = -10\dfrac{2}{3}$.

   b. Solve $\dfrac{2}{3}x - 24 \leq -19\dfrac{1}{3}$.

22. Graph $y = \dfrac{80 - 3x}{5}$.

   a. Solve $\dfrac{80 - 3x}{5} = 22\dfrac{3}{5}$.

   b. Solve $\dfrac{80 - 3x}{5} \leq 9\dfrac{2}{5}$.

23. Graph $y = -0.4x + 3.7$.
   a. Solve $-0.4x + 3.7 = 2.1$.
   b. Solve $-0.4x + 3.7 > -5.1$.

24. Graph $y = 0.4 (x - 1.5)$.
   a. Solve $0.4 (x - 1.5) = -8.6$.
   b. Solve $0.4 (x - 1.5) < 8.6$.

25. Graph $y = 4\sqrt{x} - 25$.
   a. Solve $4\sqrt{x} - 25 = 16$.
   b. Solve $8 < 4\sqrt{x} - 25 \leq 24$.

26. Graph $y = 15 - 0.01 (x - 2)^3$.
   a. Solve $15 - 0.01(x - 2)^3 = -18.75$.
   b. Solve $15 - 0.01(x - 2)^3 \leq 25$.

27. Graph $y = 24 - 0.25 (x - 6)^2$.
   a. Solve $24 - 0.25 (x - 6)^2 = -6.25$.
   b. Solve $24 - 0.25 (x - 6)^2 > 11.75$.

28. Graph $y = 0.1 (x + 12)^2 - 18$.
   a. Solve $0.1 (x + 12)^2 - 18 = 14.4$.
   b. Solve $0.1 (x + 12)^2 - 18 < 4.5$.

29. Graph $y = 0.01x^3 - 0.1x^2 - 2.75x + 15$.
   a. Solve $0.01x^3 - 0.1x^2 - 2.75x + 15 = 0$.
   b. Press $\boxed{\text{Y} =}$ and enter $Y_2 = 10$. Press $\boxed{\text{GRAPH}}$ and you should see the horizontal line $y = 10$ superimposed on your previous graph. How many solutions does the equation

$$0.01x^3 - 0.1x^2 - 2.75x + 15 = 10$$

have? Estimate each solution to the nearest whole number.

30. Graph $y = 2.5x - 0.025x^2 - 0.005x^3$.
   a. Solve $2.5x - 0.025x^2 - 0.005x^3 = 0$.
   b. Press $\boxed{\text{Y} =}$ and enter $Y_2 = -5$. Press $\boxed{\text{GRAPH}}$ and you should see the horizontal line $y = -5$ superimposed on your previous graph. How many solutions does the equation

$$2.5x - 0.025x^2 - 0.005x^3 = -5$$

have? Estimate each solution to the nearest whole number.

Chapter 2 Review

Write and graph a linear equation for each situation in Exercises 1–6. Then answer the questions.

1. Last year Pinwheel Industries introduced a new-model calculator. It cost $2000 to develop the calculator and $20 to manufacture each one.

   a. Write an equation that expresses the total costs, C, in terms of the number, n, of calculators produced.

   b. Graph the equation.

   c. What is the cost of producing 1000 calculators? Illustrate this as a point on your graph.

   d. How many calculators can be produced for $10,000? Illustrate this as a point on your graph.

2. Megan weighed 5 pounds at birth and gained 18 ounces per month during her first year.

   a. Write an equation that expresses Megan's weight, W, in terms of her age, m, in months.

   b. Graph the equation.

   c. How much did Megan weigh at 9 months? Illustrate this as a point on your graph.

   d. When did Megan weigh 9 pounds? Illustrate this as a point on your graph.

3. The world's oil reserves were 1660 billion barrels in 1976; total annual consumption is 20 billion barrels.

   a. Write an equation that expresses the remaining oil reserves, R, in terms of time, t (in years since 1976).

   b. Find the intercepts and graph the equation.

   c. What is the significance of the intercepts to the world's oil supply?

4. The world's copper reserves were 500 million tons in 1976; total annual consumption is 8 million tons.

   a. Write an equation that expresses the remaining copper reserves, R, in terms of time, t (in years since 1976).

   b. Find the intercepts and graph the equation.

   c. What is the significance of the intercepts to the world's copper supply?

5. The owner of a movie theater needs to bring in $1000 at each screening in order to stay in business. He sells adult tickets at $5 apiece and children's tickets at $2 each.

   a. Write an equation that relates the number of adult tickets, A, he must sell and the number of children's tickets, C.

   b. Find the intercepts and graph the equation.

   c. If the owner sells 120 adult tickets, how many children's tickets must he sell?

   d. What is the significance of the intercepts to the sale of tickets?

6. Alida plans to spend part of her vacation in Atlantic City and part in Saint-Tropez. She estimates that after airfare her vacation will cost $60 per day in Atlantic City and $100 per day in Saint-Tropez. She has $1200 to spend after airfare.

   a. Write an equation that relates the number of days, C, Alida can spend in Atlantic City and the number of days, T, in Saint-Tropez.

   b. Find the intercepts and graph the equation.

   c. If Alida spends 10 days in Atlantic City, how long can she spend in Saint-Tropez?

   d. What is the significance of the intercepts to Alida's vacation?

In Exercises 7–14 graph each equation. Use the most convenient method for each problem.

7. $4x - 3y = 12$ 8. $\dfrac{x}{6} - \dfrac{y}{12} = 1$

9. $50x = 40y - 20,000$

10. $1.4x + 2.1y = 8.4$

11. $3x - 4y = 0$

12. $x = -4y$

13. $4x = -12$

14. $2y - 6 = 0$

15. The table shows the amount of oil, B (in thousands of barrels), left in a tanker t minutes after it hits an iceberg and springs a leak.

t	0	10	20	30
B	800	750	700	650

a. Write a linear equation expressing B in terms of t.

b. Graph your equation.

c. Give the slope of the graph, including units, and explain the meaning of the slope in terms of the oil leak.

16. A traditional first experiment for chemistry students is to make 98 observations about a burning candle. Delbert records the height, h, of the candle in inches at various times t minutes after he lit it, as shown in the table.

t	0	10	30	45
h	12	11.5	10.5	9.75

a. Write a linear equation expressing h in terms of t.

b. Graph your equation.

c. Give the slope of the graph, including units, and explain the meaning of the slope in terms of the candle.

17. An interior decorator bases her fee on the cost of a remodeling job. The table shows her fee, F, for jobs of various costs, C, both given in dollars.

C	5000	10,000	20,000	50,000
F	1000	1500	2500	5500

a. Write a linear equation for F in terms of C.

b. Graph your equation.

c. Give the slope of the graph, and explain the meaning of the slope in terms of the fee.

18. Auto registration fees in Connie's home state depend on the value of the automobile. The table shows the registration fee, R, for a car whose value is V, both given in dollars.

V	5000	10,000	15,000	20,000
R	135	235	335	435

a. Write a linear equation for R in terms of V.

b. Graph your equation.

c. Give the slope of the graph, and explain the meaning of the slope in terms of the fee.

In Exercises 19–22 find the slope of the line segment joining each pair of points.

19. $(-1, 4)$, $(3, -2)$ **20.** $(5, 0)$, $(2, -6)$

21. $(6.2, 1.4)$, $(-2.1, 4.8)$

22. $(0, -6.4)$, $(-5.6, 3.2)$

The tables in Exercises 23 and 24 give values for a linear equation in two variables. Fill in the missing values.

23.

d	v
-5	-4.8
-2	-3
	-1.2
6	1.8
10	

24.

q	s
-8	-8
-4	56
3	
	200
9	264

25. The planners at AquaWorld want the small water slide to have a slope of 25%. If the slide is 20 feet tall, how far should the end of the slide be from the base of the ladder?

26. In areas with heavy snowfall, the pitch (or slope) of the roof of an A-frame house should be at least 1.2. If a small ski chalet is 40 feet wide at its base, how tall is the center of the roof?

In Exercises 27–30 find the slope and y-intercept of each line.

27. $2x - 4y = 5$ **28.** $\dfrac{1}{2}x + \dfrac{2}{3}y = \dfrac{5}{6}$

29. $8.4x + 2.1y = 6.3$ **30.** $y - 3 = 0$

In Exercise 31 and 32
a. graph the line that passes through the given point and has the given slope.
b. find an equation for the line.

31. $(-4, 6)$; $m = -\dfrac{2}{3}$

32. $(2, -5)$; $m = \dfrac{3}{2}$

In Exercises 33 and 34 find an equation for the line passing through the two given points.

33. $(3, -5), (-2, 4)$ **34.** $(0, 8), (4, -2)$

In Exercises 35 and 36
a. find a linear equation relating the variables.
b. state the slope of the line, including units, and explain its meaning in the context of the problem.

35. The population of Maple Rapids was 4800 in 1972 and had grown to 6780 by 1987. Assume that the population increases at a constant rate. Express the population, P, of Maple Rapids in terms of the number of years, t, since 1972.

36. Cicely's odometer read 112 miles when she filled up her 14-gallon gas tank and 308 when the gas gauge read half full. Express her odometer reading, M, in terms of the amounts of gas, g, she used.

In Exercises 37 and 38
a. find the slope and y-intercept of each line.
b. write an equation for the line.

37.

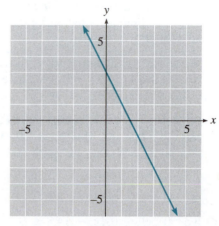

Figure 2.103

38.

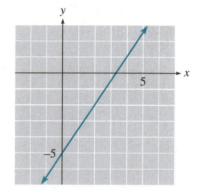

Figure 2.104

In Exercises 39 and 40 decide whether the lines are parallel, perpendicular, or neither.

39. $y = \dfrac{1}{2}x + 3$; $x - 2y = 8$

40. $4x - y = 6$; $x + 4y = -2$

41. Write an equation for the line that is parallel to the graph of $2x + 3y = 6$ and passes through the point $(1, 4)$.

42. Write an equation for the line that is perpendicular to the graph of $2x + 3y = 6$ and passes through the point $(1, 4)$.

43. Two vertices of the rectangle $ABCD$ are $A\,(3, 2)$ and $B\,(7, -4)$. Find an equation for the line that includes side $\overline{BC}$.

44. One leg of the right triangle PQR has vertices $P\,(-8, -1)$ and $Q\,(-2, -5)$. Find an equation for the line that includes the leg $\overline{QR}$.

In Exercises 45 and 46 graph each equation with the ZInteger setting. (Press ZOOM 6**, then press** ZOOM 8 ENTER**.) Use the graph to solve each equation or inequality.**

45. Graph $y = 0.2x^2 - 0.4x - 14.8$.

    a. Solve $0.2x^2 - 0.4x - 14.8 = -7.8$.

    b. Solve $0.2x^2 - 0.4x - 14.8 \leq 9.2$.

46. Graph $y = \dfrac{600}{(x - 20)^2} - 20$.

    a. Solve $\dfrac{600}{(x - 20)^2} - 20 = -14$.

    b. Solve $\dfrac{600}{(x - 20)^2} - 20 < 4$.

In Exercises 47 and 48 use the graph to solve the equations and inequalities.

47. Figure 2.105 is the graph of $y = \dfrac{12}{2 + x^2}$.

    a. Solve $\dfrac{12}{2 + x^2} = 4$.

b. For what values of x is $\dfrac{12}{2 + x^2}$ between 1 and 2?

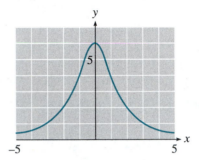

Figure 2.105

48. Figure 2.106 is the graph of $y = \dfrac{30\sqrt{x}}{1 + x}$.

    a. Solve $\dfrac{30\sqrt{x}}{1 + x} = 15$.

    b. For what values of x is $\dfrac{30\sqrt{x}}{1 + x}$ less than 12?

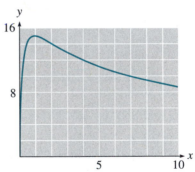

Figure 2.106

Chapter 3

Applications of Linear Models

*A*s we learned in Chapter 2, linear models describe variable quantities that increase or decrease at a constant rate. Such models are used routinely to solve problems in engineering, in business and economics, and in the life sciences. Even when a more complicated nonlinear model is more appropriate to describe a certain process, we can often use linear approximations to help us understand its behavior. In this chapter we consider some linear techniques used in applications, including linear systems, inequalities and linear programming, and curve-fitting by linear regression.

In Chapter 1 we practiced some basic techniques of problem solving: represent the unknown quantity by a variable, and write an equation that models information about the variable. If the problem has more than one unknown quantity, we can assign a different variable to each and try to write a system of several equations involving all the variables. In Sections 3.1 through 3.3 we will consider problems that can be modeled by systems of equations and learn methods for solving such systems.

3.1 SYSTEMS OF LINEAR EQUATIONS IN TWO VARIABLES

Solving Systems by Graphing

As an example, suppose a biologist wants to monitor the weights of two species of birds in a wildlife preserve. She sets up a feeder whose platform is actually a scale and mounts a camera to monitor the feeder. She waits until the feeder is occupied only by members of the two species she is studying, robins and thrushes. Then she takes a picture, which records the number of each species on the scale and the total weight registered.

From her two best pictures, she obtains the following information. The total weight of three thrushes and six robins is 48 ounces, and the total weight of five thrushes and two robins is 32 ounces. Using these data, the biologist would like to estimate the average weight of a thrush and of a robin. She begins by assigning variables to the two unknown quantities:

Average weight of a thrush: t

Average weight of a robin: r

Because there are two variables, the biologist must write two equations about the weights of the birds. Using her two photos, she writes the following equations:

$$3t + 6r = 48$$

$$5t + 2r = 32$$

This pair of equations is an example of a **linear system of two equations in two unknowns** (or, for short, a 2×2 linear system). A **solution** to the system is an ordered pair of numbers (t, r) that satisfies both equations in the system.

Recall that every point on the graph of an equation represents a solution to that equation. A solution to *both* equations thus corresponds to a point on *both* graphs. In other words, a solution to the system is a point at which the two graphs intersect. From Figure 3.1 it appears that the intersection point is $(4, 6)$, so we would expect that the values $t = 4$ and $r = 6$ are the solution to the system. We can check the solutions by verifying that these values satisfy both equations in the system.

$$3(\mathbf{4}) + 6(\mathbf{6}) \overset{?}{=} 48$$

$$5(\mathbf{4}) + 2(\mathbf{6}) \overset{?}{=} 32$$

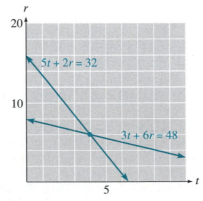

Figure 3.1

Both equations are true, so we conclude that the average weight of a thrush is 4 ounces and the average weight of a robin is 6 ounces.

We can obtain graphs for the equations in a system quickly and easily using a calculator.

Example 1

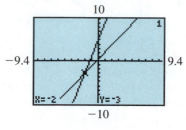

Figure 3.2

Use your calculator to solve the system

$$y = 1.7x + 0.4$$
$$y = 4.1x + 5.2$$

by graphing. Set the graphing window to

$$\text{Xmin} = -9.4, \quad \text{Xmax} = 9.4$$
$$\text{Ymin} = -10, \quad \text{Ymax} = 10$$

Solution   Set the window and enter the two equations. We can see from Figure 3.2 that the two lines intersect in the third quadrant. Now use the $\boxed{\text{TRACE}}$ key to find the coordinates of the intersection point. Check that the point $(-2, -3)$ lies on both graphs. (Remember that we use the up and down arrow keys to move the bug from one graph to the other.) The solution to the system is $x = -2, \quad y = -3$. $\square$

Since the values we obtain from the calculator may be only approximations, it is a good idea to check the solution algebraically. In Example 1 we find that both equations are true when we substitute $x = -2$ and $y = -3$.

$$-3 = 1.7(-2) + 0.4 \qquad \text{True.}$$
$$-3 = 4.1(-2) + 5.2 \qquad \text{True.}$$

Using the ZBox Feature   Because the $\boxed{\text{TRACE}}$ does not show every point on the graph, most often we can only approximate the solution of a system of equations with a graphing calculator. However, we can improve our estimate by "zooming in" on a small portion of the graphs around the intersection point. Consider the system

$$3x - 2.8y = 21.06$$
$$2x + 1.2y = 5.3$$

We can graph the system in the standard window by solving each equation for y. Enter

$$Y_1 = (21.06 - 3X)/-2.8$$
$$Y_2 = (5.3 - 2X)/1.2$$

and then press $\boxed{\text{ZOOM}}$ $\boxed{6}$.

Trace along the first line to find the intersection point. It appears to be at $x = 4.4680851, \quad y = -2.734195,$ as in Figure 3.3(a). However, if we press the up or down arrow to read coordinates off the second line, we see that for the same x-coordinate we obtain a different y-coordinate, -3.030142, as in Figure 3.3(b). This difference indicates that we have *not* found an intersection point, although we are close. We will zoom in to read the coordinates of the intersection point more accurately.

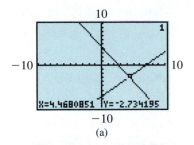

Figure 3.3

(a)   (b)

Press ZOOM ENTER to activate the **ZBox** command. We will create a box around the intersection point, using the arrow keys to mark two corners of the box. Start with the upper left corner: move the bug slightly up and to the left, perhaps three key presses each. Press the ENTER key to set that corner. Now use the down and right arrow keys to move to the lower right corner of the box, perhaps pressing six times on each key. Your display should resemble Figure 3.4(a). Press the ENTER key and the graph will be "blown up" in a new window showing just the contents of the zoom box we created.

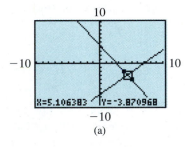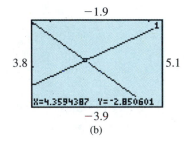

Figure 3.4

(a)   (b)

Activate the TRACE again to obtain a better estimate for the coordinates of the intersection point, as shown in Figure 3.4(b). At this stage we might estimate the solution to two decimal places as $x = 4.36$, $y = -2.85$. We can substitute these values into the original system to see whether they satisfy both equations.

$$3(\mathbf{4.36}) - 2.8(\mathbf{-2.85}) = 21.06$$

$$2(\mathbf{4.36}) + 1.2(\mathbf{-2.85}) = 5.3$$

Thus $(4.36, -2.85)$ is the exact solution to the system.

Inconsistent and Dependent Systems

Because two straight lines do not always intersect at a single point, a 2×2 system of linear equations does not always have a unique solution. In fact, there are three possibilities, as illustrated in Figure 3.5.

1. The graphs may be the same line, as shown in Figure 3.5(a).

2. The graphs may be parallel but distinct lines, as shown in Figure 3.5(b).

3. The graphs may intersect in one and only one point, as shown in Figure 3.5(c).

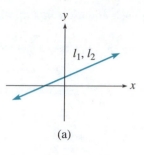

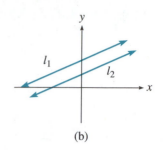

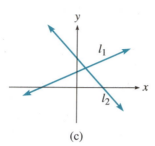

Figure 3.5

(a)   (b)   (c)

Example 2

Solve the system

$$y = -x + 5$$
$$2x + 2y = 3$$

Solution   We will use the calculator to graph both equations on the same axes, as shown in Figure 3.6. First rewrite the second equation in slope-intercept form by solving for *y*.

$2x + 2y = 3$	Subtract 2x from both sides.
$2y = -2x + 3$	Divide both sides by 2.
$y = -x + 1.5$	

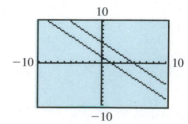

Figure 3.6

Now enter the equations as

$$Y_1 = -X + 5$$
$$Y_2 = -X + 1.5$$

The lines do not intersect within the viewing window; they appear to be parallel. If we look again at the equations of the lines, we recognize that both have slope -1, so they *are* parallel. Since parallel lines never meet, the system has no solution. ▭

A system with no solutions, such as the system in Example 2, is called **inconsistent**. A 2×2 system of linear equations is inconsistent when the two equations correspond to parallel lines. This situation occurs when the lines have the same slope.

Example 3

Solve the system

$$x = \frac{2}{3}y + 3$$
$$3x - 2y = 9$$

Solution   We begin by putting each equation in slope-intercept form.

$$x = \frac{2}{3}y + 3 \qquad \text{Subtract 3 from both sides.}$$

$$x - 3 = \frac{2}{3}y \qquad \text{Multiply both sides by } \frac{3}{2}.$$

$$\frac{3}{2}x - \frac{9}{2} = y$$

For the second equation

$$3x - 2y = 9 \qquad \text{Subtract } 3x \text{ form both sides.}$$

$$-2y = -3x + 9 \qquad \text{Divide both sides by } -2.$$

$$y = \frac{3}{2}x - \frac{9}{2}$$

The two equations are actually different forms of the same equation. Since they are equivalent, they share the same line as a graph, as shown in Figure 3.7. Because they share the same line, every solution of the first equation is also a solution of the second equation. Thus the system has infinitely many solutions.   ▭

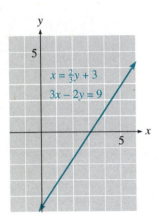

Figure 3.7

A linear system with infinitely many solutions, such as the system in Example 3, is called **dependent**. A 2 × 2 system is dependent when the two equations describe the same line. This situation occurs when the two lines have the same slope and the same *y*-intercept.

Here is a summary of the three cases for a 2 × 2 system of linear equations.

1. **Dependent system.** All the solutions of one equation are also solutions of the second equation and hence are solutions of the system. The graphs of the two equations are the same line. A dependent system has infinitely many solutions.

2. **Inconsistent system.** The graphs of the equations are parallel lines and hence do not intersect. An inconsistent system has no solutions.

3. **Consistent and independent system.** The system has exactly one solution.

Applications   In Chapter 1 we solved a variety of problems by writing equations in one variable to model the problem. This often required us to write several quantities in terms of a *single* variable. In most situations it is easier to assign *different* variables to represent different quantities, but we must then write two or more equations describing the conditions of the problem. This results in a system of equations.

Example 4

A cup of rolled oats provides 11 grams of protein. A cup of rolled wheat flakes provides 8.5 grams of protein. We want to combine oats and wheat to make a cereal with 10 grams of protein per cup. How much of each grain will we need in 1 cup of our mixture?

Solution

Step 1 Fraction of a cup of oats needed: x
Fraction of a cup of wheat needed: y

Step 2 Because we have two variables, we must find two equations that describe the problem. First, the wheat and oats together will make 1 cup of mixture, so

$$x + y = 1$$

Second, the 10 grams of protein must come from the protein in the oats plus the protein in the wheat, so

$$11x + 8.5y = 10$$

We now have a system of two equations.

Step 3 We will solve the system by graphing. First solve each equation for y in terms of x to get

$$y = -x + 1$$

$$y = (10 - 11x)/8.5$$

Although we could simplify the second equation, the calculator can graph both equations as they are. Since we know that x and y represent fractions of 1 cup, we can set the window (as shown in Figure 3.8) with

$$Xmin = 0, \quad Xmax = 0.94$$

$$Ymin = 0, \quad Ymax = 1$$

The lines intersect at (0.6, 0.4), which we can verify by substituting these values into the original two equations of our system.

Step 4 We need 0.6 cup of oats and 0.4 cup of wheat.

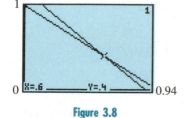

Figure 3.8

Being unable to read exact coordinates from a graph is not always a disadvantage. In many situations fractional values of the unknowns are not acceptable.

Example 5

The mathematics department has $40,000 to set up a new computer lab. They will need one printer for every four terminals they purchase. If a printer costs $560 and a terminal costs $1520, how many of each should they buy?

Solution

Step 1   Number of printers: p
            Number of terminals: t

Step 2   Since the math department needs four times as many terminals as printers,

$$t = 4p$$

The total cost for the printers will be $560p$ dollars and for the terminals $1520t$ dollars, so we have

$$560p + 1520t = 40,000$$

Step 3   Solve the second equation for t to get

$$t = \frac{40,000 - 560p}{1520}$$

Now we graph the equations

$$Y_1 = 4X$$

$$Y_2 = (40,000 - 560X)/1520$$

on the same set of axes. The second graph is not visible in the standard graphing window, but with a little experimentation we can find an appropriate window setting. The window values used for Figure 3.9 are

$$Xmin = 0, \quad Xmax = 9.4$$

$$Ymin = 0, \quad Ymax = 30$$

The lines intersect at approximately (6, 24). These values satisfy the first equation but not the second.

$$560(6) + 1520(24) \overset{?}{=} 40,000$$

$$39,840 \neq 40,000$$

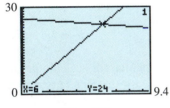

Figure 3.9

The exact solution to the system is $\left(\dfrac{500}{83}, \dfrac{2000}{83}\right)$. (We will explore methods for finding exact solutions in Section 3.2.) But this solution is not of practical use, since the math department cannot purchase fractions of printers or terminals. The department *can* purchase 6 printers and 24 terminals (with some money left over).   ⬜

An Application from Economics   In economics the number of an item that consumers will buy usually decreases as the price increases. On the other hand, the producer will be willing to supply more units of a product if the price increases. The **demand equation** gives the number of units of the product that consumers will buy in terms of the price per unit. The **supply equation** gives the number of units that the producer will supply in terms of the price per unit. The price at which the supply and demand are equal is called the **equilibrium price**. This is the price at which the consumer and the producer agree to do business.

Example 6

A woolens mill can produce 400x yards of fine suit fabric if they can charge x dollars per yard. Their clients in the garment industry will buy 6000 − 100x yards of wool fabric at a price of x dollars per yard. Find the equilibrium price and the amount of fabric that will change hands at that price.

Solution

Step 1   Price per yard: x
Number of yards: y

Step 2   The supply equation tells us how many yards of fabric the mill will produce for a price of x dollars per yard:

$$y = 400x$$

The demand equation tells us how many yards of fabric the garment industry will buy at a price of x dollars per yard:

$$y = 6000 - 100x$$

Step 3   Graph the two equations on the same set of axes, as shown in Figure 3.10. Set the window values to

$$Xmin = 0, \quad Xmax = 94$$

$$Ymin = 0, \quad Ymax = 6200$$

and use the $\boxed{\text{TRACE}}$ to locate the solution. The graphs intersect at the point (12, 4800).

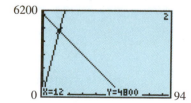

Figure 3.10

Step 4   The equilibrium price is $12 per yard, and the mill will sell 4800 yards of fabric at that price.

Exercise 3.1

In Exercises 1–4 solve each system of equations using the graphs given. Verify algebraically that your solution satisfies both equations.

1. $2.3x - 3.7y = 6.9$
$1.1x + 3.7y = 3.3$

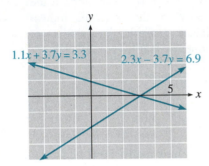

Figure 3.11

2. $-2.3x + 5.9y = 38.7$
$9.3x + 7.4y = -0.2$

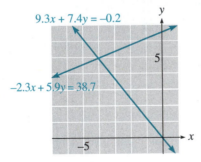

Figure 3.12

3. $35s - 17t = 560$
$24s + 15t = 2250$

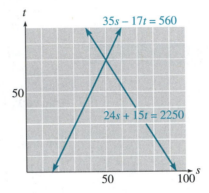

Figure 3.13

4. $56a + 32b = -880$
$23a - 7b = 1250$

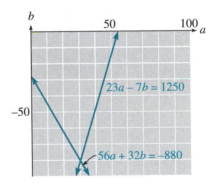

Figure 3.14

In Exercises 5–10 solve each system of equations by graphing. Use the "friendly" window

$$\text{Xmin} = -9.4, \quad \text{Xmax} = 9.4$$

$$\text{Ymin} = -10, \quad \text{Ymax} = 10$$

Verify algebraically that your solution satisfies both equations.

5. $y = -0.7x + 6.9$
$y = 1.2x - 6.4$

6. $y = 3.7x - 6.1$
$y = -1.6x + 9.8$

7. $y = 2.6x + 8.2$
$y = 1.8 - 0.6x$

8. $y = 5.8x - 9.8$
$y = 0.7 - 4.7x$

9. $y = 7.2 - 2.1x$
$-2.8x + 3.7y = 5.5$

10. $y = -2.3x - 5.5$
$3.1x + 2.4y = -1.1$

In Exercises 11–16 solve each system of equations by graphing. Choose a suitable window and use the ZOOM Box feature to estimate your answers to two decimal places.

11. $n - 32m = 2630$
$n = -21m - 1610$

12. $q = 47p - 1930$
$q + 19p = 710$

13. $38a + 2.3b = -55.2$
$b = 15a + 121$

14. $25u - 1.7v = 10.5$
$v + 5u = 49$

15. $64x + 58y = 707$
$82x - 21y = 496$

16. $35x - 76y = 293$
$15x + 44y = -353$

In Exercises 17–22 use graphs to identify each system as dependent, inconsistent, or consistent and independent.

17. $y = 3x + 6$
$6x - 2y = 14$

18. $3a - b = 6$
$6a - 2b = 12$

19. $2m = n + 1$
$8m - 4n = 4$

20. $6p = 8 - 2q$
$12p + 4q = 2$

21. $r - 3s = 6$
$2r + s = 8$

22. $2u + v = 5$
$u - 3v = 3$

23. $2L - 5W = 8$
$\dfrac{15W}{2} + 9 = 3L$

24. $-3A = 4B + 12$
$\dfrac{1}{2}A + 2 = \dfrac{-2}{3}B$

Solve Exercises 25–34 by graphing a system of equations.

25. The manager of Books for Cooks plans to spend $300 stocking a new diet cookbook. The paperback version costs her $4.95, and the hardback costs $9.95. She finds that she will sell three times as many paperbacks as hardbacks. How many of each should she buy?

26. Etienne plans to open a coffee house, and he has $7520 to spend on furniture. A table costs $460 and a chair costs $120. Etienne will buy four chairs for each table. How many tables can he buy?

27. Yasuo can afford to produce $50x$ bushels of wheat if he can sell them at x cents per bushel, and the market will buy $2100 - 20x$ bushels at x cents per bushel. Find the equilibrium price and the number of bushels of wheat Yasuo can sell at that price.

28. Mel's Pool Service can clean $1.5x$ pools per week if it charges x dollars per pool, and the public will book $120 - 2.5x$ pool cleanings at x dollars per pool. Find the equilibrium price and the number of pools that will be cleaned at that price.

29. The Aquarius Jewelry Company determines that each production run to manufacture a pendant involves an initial set-up cost of $200 and $4.00 for each pendant produced. The pendants sell for $12 each.

a. Express the cost, C, of production in terms of the number, x, of pendants produced.

b. Express the revenue, R, in terms of the number, x, of pendants sold.

c. Graph the revenue and cost on the same set of axes.

d. How many pendants must Aquarius sell to break even on a particular production run?

30. The Bread Alone Bakery has a daily overhead of $90. It costs $0.60 to bake each loaf of bread, and the bread sells for $1.50 per loaf.

a. Express the cost, C, in terms of the number, x, of loaves baked.

b. Express the revenue, R, in terms of the number, x, of loaves sold.

c. Graph the revenue and cost on the same set of axes.

d. How many loaves must the bakery sell to break even on a given day?

31. Dash Phone Company charges a monthly fee of $10 plus $0.09 per minute for long-distance calls. Friendly Phone Company charges $15 per month plus $0.05 per minute for long-distance calls.

a. On the same set of axes, graph the monthly charges for each company in terms of minutes of long-distance calls.

b. How many minutes of long-distance calls would result in equal bills from the two companies?

32. The Olympus Health Club charges an initial fee of $230 and $13 monthly dues. The Valhalla Health Spa charges $140 initially and $16 per month.

a. On the same set of axes, graph the cost of belonging to each club in terms of time.

b. After how many months of membership would the costs of belonging to the two clubs be equal?

33. The admission at a Bengals baseball game was $1.50 for adults and $0.85 for students. The ticket office took in $93.10 for 82 paid admissions. How many adults and how many students attended the game?

34. Forty-two passengers flew on an airplane flight for which first-class fare was $80 and tourist fare was $64. If the receipts for the flight totaled $2880, how many first-class and how many tourist passengers paid for the flight?

3.2   SOLUTION OF SYSTEMS BY ALGEBRAIC METHODS

Although graphing calculators and other technologies make it fairly easy to solve systems by graphing, it may be difficult to find an exact answer, rather than an approximation. If an exact answer is required, solutions to systems of equations are usually sought by algebraic methods. In this section we will review two algebraic methods: substitution and elimination.

Solving Systems by Substitution

You are probably familiar with the method of **substitution.** The basic strategy can be described as follows.

> **STEPS FOR SOLVING A 2 × 2 SYSTEM BY SUBSTITUTION**
>
> **1.** Solve one of the equations for one of the variables in terms of the other.
> **2.** Substitute this expression into the second equation, which will yield an equation in one variable.
> **3.** Solve the new equation.
> **4.** Use the result of Step 1 to find the other variable.

Example 1

Staci stocks two kinds of sleeping bags in her sporting goods store: a standard model and a down-filled model for colder temperatures. From past experience she estimates that she will sell twice as many of the standard variety as of the down-filled. She has room to stock 60 sleeping bags at a time. How many of each variety should Staci order?

Solution

Step 1 Number of standard sleeping bags: x
Number of down-filled sleeping bags: y

Step 2 Write two equations about the variables. First, Staci needs twice as many standard model as down-filled, so

$$x = 2y \qquad (1)$$

Also, the total number of sleeping bags is 60, so

$$x + y = 60 \qquad (2)$$

These two equations give us a system.

Step 3 We will solve this system using substitution. Notice that Equation (1) is already solved for x in terms of y: $x = 2y$. Substitute $2y$ for x in Equation (2) to obtain

$$2y + y = 60$$

$$3y = 60$$

Solving for y we find $y = 20$. Finally, substitute this value into Equation (1) to find

$$x = 2(20) = 40$$

The solution to the system is $x = 40, \quad y = 20$.

Step 4 Staci should order 40 standard sleeping bags and 20 down-filled ones.

The method of substitution is especially convenient when one of the variables has a coefficient of 1 or -1 in one of the equations, because then it is easy to solve for that variable. If none of the coefficients are 1 or -1, then a second method, called **elimination**, is usually more efficient for solving a system.

Solving Systems by Elimination The method of elimination is based on the following properties of linear equations.

PROPERTIES OF LINEAR SYSTEMS

1. Multiplying a linear equation by a (nonzero) constant does not change its solutions; that is, any solution of the equation

$$ax + by = c$$

is also a solution of the equation

$$kax + kby = kc$$

2. Adding (or subtracting) two linear equations does not change their common solutions; that is, any solution of the system

$$a_1x + b_1y = c_1$$
$$a_2x + b_2y = c_2$$

is also a solution of the equation

$$(a_1 + a_2)x + (b_1 + b_2)y = c_1 + c_2$$

Example 2

Rani kayaks downstream for 45 minutes and travels a distance of 6000 meters. On the return journey upstream, she covers only 4800 meters in 45 minutes. How fast is the current in the stream, and how fast would Rani kayak in still water? (Give your answers in meters per minute.)

Solution

Step 1   Rani's speed in still water: r
Speed of the stream current: s

Step 2   We must write two equations using the variables r and s. First organize the information into a table similar to Table 3.1.

	Rate	Time	Distance
Downstream	$r + s$	45	6000
Upstream	$r - s$	45	4800

Table 3.1

Using the formula   *Rate $\times$ Time = Distance,*   we can write one equation describing Rani's journey downstream and a second equation for the journey upstream.

$$(r + s) \cdot 45 = 6000$$
$$(r - s) \cdot 45 = 4800$$

Apply the distributive law to write each equation in standard form.

$$45r + 45s = 6000 \tag{3}$$
$$45r - 45s = 4800 \tag{4}$$

Step 3   To solve the system, we will eliminate the variable s by adding the two equations together vertically. (This operation is justified by property 2 of linear systems.)

$$45r + 45s = 6000$$

$$\underline{+\ 45r - 45s = 4800}$$

$$90r \qquad\quad = 10{,}800$$

Now we have an equation in one variable only, which we can solve for r.

$$90r = 10{,}800 \qquad \text{Divide both sides by 90.}$$

$$r = 120$$

To solve for s, we can substitute this value for r into any of our previous equations involving both r and s. We will use Equation (3).

$$45\,(\mathbf{120}) + 45s = 6000 \qquad \text{Simplify the left side.}$$

$$5400 + 45s = 6000 \qquad \text{Subtract 5400 from both sides.}$$

$$45s = 600 \qquad \text{Divide both sides by 45; reduce.}$$

$$s = \frac{40}{3}$$

Step 4   The speed of the current is 40/3 meters per minute, and Rani's speed in still water is 120 meters per minute. (Notice that since the stream's speed is not a whole number, we might not have been able to find the exact solution to the system by graphing.)   ☐

In Example 2 we were able to eliminate the variable s by adding the two equations together. This worked only because the coefficients of s in the two equations were opposites, namely 45 and -45. If the coefficients of the chosen variable are not opposites, we must multiply each equation by a suitable factor so that the new coefficients are opposites. We choose the "suitable factors" the same way we choose building factors to obtain a least common denominator when adding fractions.

Example 3

Solve the system

$$2x + 3y = 8 \tag{5}$$

$$3x - 4y = -5 \tag{6}$$

by the method of elimination.

Solution   We must first decide which variable to eliminate: x or y. We can choose whichever looks easiest. In this problem neither choice has a clear advantage, so we choose to eliminate x. We next look for the smallest number that both coefficients, 2 and 3, divide into evenly. This number is 6. We want the coefficients of x to become 6 and -6, so we multiply Equation (5) by 3 and Equation (6) by -2 to obtain

$$6x + 9y = 24 \tag{7}$$

$$-6x + 8y = 10 \tag{8}$$

Now add the corresponding terms of Equations (7) and (8). The x-terms are eliminated, yielding an equation in one variable.

$$
\begin{aligned}
6x + 9y &= 24 \\
-6x + 8y &= 10 \\
\hline
17y &= 34
\end{aligned}
\tag{9}
$$

Solve this equation for y to find $y = 2$.   We can substitute this value of y into any of our equations involving both x and y. If we choose Equation (5), then

$$2x + 3(2) = 8$$

and solving this equation yields $x = 1$.   The ordered pair $(1, 2)$ is a solution to the system. You should verify that these values satisfy both original equations.

We summarize the strategy for solving a linear system by elimination.

STEPS FOR SOLVING A 2 × 2 LINEAR SYSTEM BY ELIMINATION

1. Choose one of the variables to eliminate. Multiply each equation by a suitable factor so that the coefficients of that variable are opposites.

2. Add the two new equations termwise.

3. Solve the resulting equation for the remaining variable.

4. Substitute the value found in Step 3 into either of the original equations and solve for the other variable.

In Example 3 we added 3 times the first equation to -2 times the second equation. The result from adding a constant multiple of one equation to a constant multiple of another equation is called a **linear combination** of the two equations. The method of elimination is also called the method of linear combinations.

If either equation in a system has fractional coefficients, it is helpful to clear the fractions before applying the method of linear combinations.

Example 4

Solve the system by linear combinations:

$$\frac{2}{3}x - y = 2 \tag{10}$$

$$x + \frac{1}{2}y = 7 \tag{11}$$

Solution Multiply each side of Equation (10) by 3 and each side of Equation (11) by 2 to clear the fractions:

$$2x - 3y = 6 \tag{12}$$

$$2x + y = 14 \tag{13}$$

To eliminate the variable x, multiply Equation (13) by -1 and add the result to Equation (12) to get

$$-4y = -8$$

$$y = 2$$

Substitute 2 for y in one of the original equations and solve for x. Using Equation (11) we find

$$x + \frac{1}{2}(2) = 7$$

$$x = 6$$

Verify that $x = 6$ and $y = 2$ satisfy both Equations (10) and (11). The solution to the system is the ordered pair (6, 2). ▭

Systems of linear equations are some of the most useful mathematical tools for solving problems. Systems involving hundreds of variables and equations are not uncommon in applications such as scheduling airline flights or routing telephone calls. The method of substitution does not generalize easily to larger linear systems, but the method of elimination is the basis for a large number of sophisticated algorithms that can be implemented on computers.

Inconsistent and Dependent Systems

Recall that a system of equations may not have a unique solution. It is not always easy to tell from the equations themselves whether the system has one solution, no solution, or infinitely many solutions. However, the method of elimination will reveal which of the three cases applies.

Example 5

Solve each system.

a. $2x = 2 - 3y$ b. $3x - 4 = y$

   $6y = 7 - 4x$ $2y + 8 = 6x$

Solutions **a.** First rewrite the system in standard form as

$$2x + 3y = 2 \tag{14}$$

$$4x + 6y = 7 \tag{15}$$

Multiply Equation (14) by -2 and add the result to Equation (15) to obtain

$$
\begin{array}{r}
-4x - 6y = -4 \\
\underline{4x + 6y = 7} \\
0x + 0y = 3
\end{array}
$$

This equation has no solutions. The system is inconsistent. (Notice that both lines have slope $\dfrac{-2}{3}$, so their graphs are parallel.)

b. Rewrite the system in standard form as

$$3x - y = 4 \tag{16}$$

$$-6x + 2y = -8 \tag{17}$$

Multiply Equation (16) by 2 and add the result to Equation (17) to obtain

$$
\begin{array}{r}
6x - 2y = 8 \\
\underline{-6x + 2y = -8} \\
0x + 0y = 0
\end{array}
$$

This equation has infinitely many solutions. The system is dependent. (Notice that both original equations have the same slope-intercept form, $y = 3x - 4$, so they have the same graph.)

We can generalize the results from Example 5 as follows.

1. If an equation of the form

$$0x + 0y = k \qquad (k \neq 0)$$

is obtained as a linear combination of the equations in a system, then the system is **inconsistent.**

2. If an equation of the form

$$0x + 0y = 0$$

is obtained as a linear combination of the equations in a system, then the system is **dependent.**

Exercise 3.2

In Exercises 1–18 solve each system by substitution or by linear combinations.

1. $2x - 3y = 6$
$x + 3y = 3$

2. $a + 2b = -6$
$2a - 3b = 16$

3. $3m + n = 7$
$2m = 5n - 1$

4. $2r = s + 7$
$2s = 14 - 3r$

5. $2u - 3v = -4$
$5u + 2v = 9$

6. $3x + 5y = 1$
$2x - 3y = 7$

7. $3y = 2x - 8$
$4y + 11 = 3x$

8. $4L - 3 = 3W$
$25 + 5L = -2W$

9. $\dfrac{2}{3}A - B = 4$

$A - \dfrac{3}{4}B = 6$

10. $\dfrac{1}{8}w - \dfrac{3}{8}z = 1$

$\dfrac{1}{2}w - \dfrac{1}{4}z = -1$

11. $\dfrac{M}{4} = \dfrac{N}{3} - \dfrac{5}{12}$

$\dfrac{N}{5} = \dfrac{1}{2} - \dfrac{M}{10}$

12. $\dfrac{R}{3} = \dfrac{2S}{3} + 2$

$\dfrac{S}{3} = \dfrac{R}{6} - 1$

13. $\dfrac{s}{2} = \dfrac{7}{6} - \dfrac{t}{3}$

$\dfrac{s}{4} = \dfrac{3}{4} - \dfrac{t}{4}$

14. $\dfrac{2p}{3} + \dfrac{8q}{9} = \dfrac{4}{3}$

$\dfrac{p}{3} = 2 + \dfrac{q}{2}$

15. $4.8x - 3.5y = 5.44$
$2.7x + 1.3y = 8.29$

16. $6.4x + 2.3y = -14.09$
$-5.2x - 3.7y = -25.37$

17. $0.9x = 25.78 + 1.03y$
$0.25x + 0.3y = 85.7$

18. $0.02x = 0.6y - 78.72$
$1.1y = -0.4x + 108.3$

Identify each system in Exercises 19–24 as dependent, inconsistent, or consistent and independent.

19. $x + 3y = 6$
$2x + 6y = 12$

20. $3a - 2b = 6$
$6a - 4b = 8$

21. $2m = n + 1$
$8m - 4n = 3$

22. $6p = 1 - 2q$
$12p + 4q = 2$

23. $r - 3s = 4$
$2r + s = 6$

24. $2u + v = 4$
$u - 3v = 2$

25. $2L - 5W = 6$
$\dfrac{15W}{2} + 9 = 3L$

26. $-3A = 4B + 8$
$\dfrac{1}{2}A + \dfrac{4}{3} = \dfrac{-2}{3}B$

In Exercises 27–40 write a system of equations for each problem, and solve algebraically.

27. In a recent election 7179 votes were cast for the two candidates. If six votes had been switched from the winner to the loser, the loser would have won by one vote. How many votes were cast for each candidate?

28. Delbert answered 13 true-false and 9 fill-in questions correctly on his last test and got a score of 71. If he had answered 9 true-false and 13 fill-ins correctly, he would have made an 83. How many points was each type of problem worth?

29. Francine has $2000, part in bonds paying 10% and the rest in a certificate account at 8%. Find the amount Francine invested at each rate if her yearly income from the two investments is $184.

30. Carmella has $1200 invested in two stocks; one returns 8% per year and the other returns 12% per year. How much did she invest in each stock if the income from the 8% stock is $3 more than the income from the 12% stock?

31. Paul needs 40 pounds of a 48% silver alloy to finish a collection of jewelry. How many pounds of 45% silver alloy should he melt with a 60% silver alloy to obtain the alloy he needs?

32. Amal plans to make 10 liters of a 17% acid solution by mixing a 20% acid solution with a 15% acid solution. How much of each solution should she use?

33. Leon flies 1260 miles in the same time that Marlene drives 420 miles. If Leon flies 120 miles per hour faster than Marlene drives, find the speed of each.

34. Thelma and Louise start together and drive in the same direction, Thelma driving twice as fast as Louise. At the end of 3 hours they are 96 miles apart. How fast is each traveling?

35. Because of prevailing winds a flight from Detroit to Denver, a distance of 1120 miles, takes 4 hours on Econoflite, whereas the return trip takes 3.5 hours. Find the speed of the airplane and the speed of the wind.

36. On a breezy day Bonnie propelled her human-powered aircraft 100 meters in 15 seconds going into the wind and made the return trip in 10 seconds with the wind. Find the speed of the wind and Bonnie's speed in still air.

37. A cup of rolled oats provides 310 calories. A cup of rolled wheat flakes provides 290 calories. A new breakfast cereal combines wheat and oats to provide 302 calories. How much of each grain does 1 cup of the cereal include?

38. Acme Motor Company is opening a new plant to produce chassis for two of its models, a sports coupe and a wagon. Each sports coupe requires a riveter for 3 hours and a welder for 4 hours; each wagon needs a riveter for 4 hours and a welder for 5 hours. If the plant has available 120 hours of riveting and 155 hours of welding per day, how many of each model of chassis can it produce in a day?

39. Sanaz can afford to produce $35x$ pairs of sunglasses if she can sell them at x dollars per pair, and the market will buy $1700 - 15x$ at x dollars a pair. Find the equilibrium price and the number of sunglasses Sanaz will produce and sell at that price.

40. Benham will service $2.5x$ copy machines per week if he can charge x dollars per machine, and the public will pay for $350 - 4.5x$ jobs at x dollars per machine. Find the equilibrium price and the number of copy machines Benham will service at that price.

3.3   SYSTEMS OF LINEAR EQUATIONS IN THREE VARIABLES

Some problems involve three (or more) unknown quantities. The notion of a system of equations can be extended to any number of variables, and efficient techniques for solving linear systems in several variables are available. In this section we introduce some of these techniques for systems of three linear equations in three variables.

3 × 3 Linear Systems

A solution to an equation in three variables, such as

$$x + 2y - 3z = -4 \qquad (1)$$

is an **ordered triple** of numbers that satisfies the equation. For example, $(0, -2, 0)$ and $(-1, 0, 1)$ are solutions to Equation (1) but $(1, 1, 1)$ is not. You can verify this by substituting the coordinates into the equation to see if a true statement results.

For $(0, -2, 0)$: $0 + 2(-2) - 3(0) = -4$ True.

For $(-1, 0, 1)$: $-1 + 2(0) - 3(1) = -4$ True.

For $(1, 1, 1)$: $1 + 2(1) - 3(1) = -4$ Not true.

As with the two-variable case, a single linear equation in three variables has infinitely many solutions.

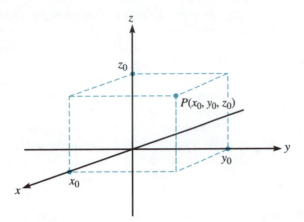

Figure 3.15

An ordered triple (x, y, z) can be represented geometrically as a point in space using a three-dimensional Cartesian coordinate system. (See Figure 3.15.) In this coordinate system the graph of a linear equation in three variables such as Equation (1) is a plane, and the fact that the equation has infinitely many solutions simply tells us that the corresponding plane has infinitely many points.

A solution to a *system* of three linear equations in three variables is an ordered triple that satisfies each equation in the system. That triple represents a point that must lie on all three graphs. Figure 3.16 shows the different ways in which three planes may intersect in space.

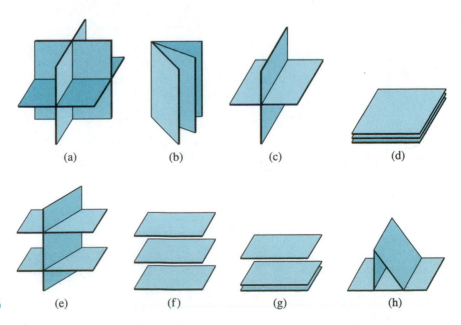

Figure 3.16

In Figure 3.16(a) the three planes intersect in a single point, so the corresponding system of three equations has a unique solution. In Figure 3.16(b), (c), and (d) the intersection is either a line or an entire plane, so the corresponding system has infinitely many solutions. As in the two-variable case, such a system is called **dependent**. In Figure 3.16(e), (f), (g), and (h) the three planes have no common intersection, so the corresponding system has no solution. In this case the system is said to be **inconsistent**.

It is impractical to solve 3×3 systems by graphing. Even when technology for producing three-dimensional graphs is available, we cannot read coordinates on such graphs with any confidence. Thus we will restrict our attention to algebraic methods of solving such systems.

Back-Substitution   When we review our algebraic methods for 2×2 systems—substitution and linear combinations—we find features common to both. In both methods we obtain an equation in one variable. And in both methods, once we have solved for that variable, we substitute the value into an earlier equation to find the other variable.

One strategy for solving a 3×3 system extends this idea to include another variable and another equation. We would like to get an equation that involves a single variable. Once we have found the value for that unknown, we can substitute the value back into the other equations to help us find the remaining unknowns. The following special case illustrates the substitution part of the procedure.

Example 1   Solve the system

$$x + 2y + 3z = 2$$
$$-2y - 4z = -2$$
$$3z = -3$$

Solution   The third equation involves only the variable z, so we solve that equation to find $z = -1$. Then we substitute -1 for z in the second equation and solve for y.

$$-2y - 4(-1) = -2$$
$$-2y + 4 = -2$$
$$-2y = -6$$
$$y = 3$$

Finally, we substitute -1 for z and 3 for y into the first equation to find x.

$$x + 2(3) + 3(-1) = 2$$
$$x + 6 - 3 = 2$$
$$x = -1$$

The solution is the ordered triple $(-1, 3, -1)$. You should verify that this triple satisfies all three equations of the system. ⬜

The technique used in Example 1 is called **back-substitution**. It works in the special case in which one of the equations involves exactly one variable and a second equation involves that same variable and just one other variable. (A 3×3 linear system with these properties is said to be in **triangular** form.) If we can transform a system into triangular form, then we can use back-substitution to complete the solution.

Gaussian Reduction

On pages 149–150 we stated two properties of linear systems. These properties allow us to form linear combinations of the equations in a system without changing the system's solution. They apply to linear systems with any number of variables. We can use linear combinations to reduce a 3×3 system to triangular form and then use back-substitution to find the solutions. Our strategy will be to eliminate one of the variables from each of the three equations by considering them in pairs. This results in a 2×2 system that we can solve using elimination, as we did in Section 3.2. As an example consider the system

$$x + 2y - 3z = -4 \tag{1}$$

$$2x - y + z = 3 \tag{2}$$

$$3x + 2y + z = 10 \tag{3}$$

We can choose any one of the three variables to eliminate first. For this example we will eliminate x. We then choose two of the equations, say (1) and (2), and use a linear combination: we multiply Equation (1) by -2 and add the result to Equation (2) to produce Equation (4).

$$-2x - 4y + 6z = 8 \tag{1a}$$

$$\underline{2x - y + z = 3} \tag{2}$$

$$-5y + 7z = 11 \tag{4}$$

Now we have an equation involving only two variables. But we need *two* equations in two unknowns to specify a unique solution. So we choose a different pair of equations, say (1) and (3), and eliminate x again. We multiply Equation (1) by -3 and add the result to Equation (3) to obtain Equation (5).

$$-3x - 6y + 9z = 12 \tag{1b}$$

$$\underline{3x + 2y + z = 10} \tag{3}$$

$$-4y + 10z = 22 \tag{5}$$

We now form a 2 × 2 system with Equations (4) and (5).

$$-5y + 7z = 11 \tag{4}$$

$$-4y + 10z = 22 \tag{5}$$

Finally, we eliminate either y or z to obtain an equation in a single variable. If we choose to eliminate y, we add 4 times Equation (4) to -5 times Equation (5) to obtain Equation (6).

$$-20y + 28z = 44 \tag{4a}$$

$$\underline{20y + 50z = -110} \tag{5a}$$

$$-22z = -66 \tag{6}$$

At this stage we are ready to start solving for the variables. To keep things organized, we will choose one of our original equations in three variables, one of the equations from our 2 × 2 system, and our final equation in one variable. We choose Equations (1), (4), and (6).

$$x + 2y - 3z = -4 \tag{1}$$

$$-5y + 7z = 11 \tag{4}$$

$$-22z = -66 \tag{6}$$

This new system is in triangular form, and it has the same solutions as the original system. We can complete its solution by back-substitution. We first solve Equation (6) to find $z = 3$. Substituting 3 for z in Equation (4), we find

$$-5y + 7(\mathbf{3}) = 11$$

$$-5y + 21 = 11$$

$$-5y = -10$$

$$y = 2$$

Next we substitute 3 for z and 2 for y into Equation (1) to find

$$x + 2(\mathbf{2}) - 3(\mathbf{3}) = -4$$

$$x + 4 - 9 = -4$$

$$x = 1$$

The solution to the system is the ordered triple (1, 2, 3). You should verify that this triple satisfies all three of the original equations.

The method described above for putting a linear system into triangular form is called **Gaussian reduction**, after the German mathematician Carl Gauss. We can summarize our method for solving a 3 × 3 linear system as follows.

STEPS FOR SOLVING A 3 × 3 LINEAR SYSTEM

1. Clear each equation of fractions and put it in standard form.

2. Choose two of the equations and eliminate one of the variables by forming a linear combination.

3. Choose a different pair of equations and eliminate the *same* variable.

4. Form a 2 × 2 system with the equations found in Steps 2 and 3. Eliminate one of the variables from this 2 × 2 system by using a linear combination.

5. Form a triangular system by choosing among the previous equations. Use back-substitution to solve the triangular system.

Example 2

Solve the system

$$x + 2y - z = -3 \tag{1}$$

$$\frac{1}{3}x - y + \frac{1}{3}z = 2 \tag{2}$$

$$x + \frac{1}{2}y + z = \frac{5}{2} \tag{3}$$

Solution Follow the steps outlined above.

Step 1 Multiply each side of Equation (2) by 3 and each side of Equation (3) by 2 to obtain the equivalent system

$$x + 2y - z = -3 \tag{1}$$

$$x - 3y + z = 6 \tag{2a}$$

$$2x + y + 2z = 5 \tag{3a}$$

Step 2 Eliminate z from Equations (1) and (2a) by adding.

$$x + 2y - z = -3 \tag{1}$$

$$\underline{x - 3y + z = 6} \tag{2a}$$

$$2x - y = 3 \tag{4}$$

Step 3 Eliminate z from Equations (1) and 3(a): multiply Equation (1) by 2 and add the result to Equation (3a).

$$2x + 4y - 2z = -6 \tag{1a}$$

$$\underline{2x + y + 2z = 5} \tag{3a}$$

$$x + 5y = -1 \tag{5}$$

Step 4 Form the 2×2 system consisting of Equations (4) and (5).

$$2x - y = 3 \tag{4}$$

$$4x + 5y = -1 \tag{5}$$

Eliminate x by adding -2 times Equation (4) to Equation (5).

$$-4x + 2y = -6 \tag{4a}$$

$$\underline{4x + 5y = -1} \tag{5}$$

$$7y = -7 \tag{6}$$

Step 5 Form a triangular system using Equations (1), (4), and (6).

$$x + 2y - z = -3 \tag{1}$$

$$2x - y = 3 \tag{4}$$

$$7y = -7 \tag{6}$$

Use back-substitution to find the solution. Divide both sides of Equation (6) by 7 to find $y = -1$. Substitute -1 for y in Equation (4) and solve for x.

$$2x - (\mathbf{-1}) = 3$$

$$2x = 2$$

$$x = 1$$

Finally, substitute 1 for x and -1 for y in Equation (1) and solve for z.

$$1 + 2(\mathbf{-1}) - z = -3$$

$$-1 - z = -3$$

$$-z = -2$$

$$z = 2$$

The solution is the ordered triple $(1, -1, 2)$. You should verify that this triple satisfies all three of the original equations of the system.

Inconsistent and Dependent Systems   The results on page 154 for identifying dependent and inconsistent systems can be extended to 3×3 linear systems. If at any step in forming linear combinations we obtain an equation of the form

$$0x + 0y + 0z = k \qquad (k \neq 0)$$

then the system is inconsistent and has no solution. If we obtain an equation of the form

$$0x + 0y + 0z = 0$$

then the system is dependent and has an infinite number of solutions.

Example 3

Solve the system

$$3x + y - 2z = 1 \qquad (1)$$
$$6x + 2y - 4z = 5 \qquad (2)$$
$$-2x - y + 3z = -1 \qquad (3)$$

Solution To eliminate y from Equations (2) and (3), multiply Equation (1) by -2, and add the result to Equation (2).

$$-6x - 2y + 4z = -2$$
$$\underline{6x + 2y - 4z = 5}$$
$$0x + 0y + 0z = 3$$

Since the resulting equation has no solution, the system is inconsistent.

Example 4

Solve the system

$$-x + 3y - z = -2 \qquad (1)$$
$$2x + y - 4z = 6 \qquad (2)$$
$$2x - 6y + 2z = 4 \qquad (3)$$

Solution To eliminate x from Equations (1) and (3), multiply Equation (1) by 2 and add Equation (3).

$$-2x + 6y - 2z = -4$$
$$\underline{2x - 6y + 2z = 4}$$
$$0x + 0y + 0z = 0$$

Since the resulting equation vanishes, the system is dependent.

Applications Here are some examples of problems that can be modeled by a system of three linear equations. When writing such systems we must be careful that we find three *independent* equations describing the conditions of the problem.

Example 5

One angle of a triangle measures 4° less than twice the second angle, and the third angle is 20° greater than the sum of the first two. Find the measure of each angle.

Solution

Step 1 Represent the measure of each angle by a separate variable.

First angle: x

Second angle: y

Third angle: z

Step 2   Write the conditions stated in the problem as three equations.

$$x = 2y - 4$$

$$z = x + y + 20$$

$$x + y + z = 180$$

(The third equation states the fact that the sum of the angles of a triangle is 180°.)

Step 3   We follow the steps outlined for solving a 3×3 linear system.

1. Write the three equations in standard form.

$$x - 2y = -4 \qquad (1)$$

$$x + y - z = -20 \qquad (2)$$

$$x + y + z = 180 \qquad (3)$$

2–3. Since Equation (1) has no z term, it will be most efficient to eliminate the variable z from Equations (2) and (3). Add these two equations.

$$x + y - z = -20 \qquad (2)$$

$$\underline{x + y + z = 180} \qquad (3)$$

$$2x + 2y \quad = 160 \qquad (4)$$

4. Form a 2×2 system from Equations (1) and (4). Add the two equations to eliminate the variable y, yielding

$$x - 2y = -4 \qquad (1)$$

$$\underline{2x + 2y = 160} \qquad (4)$$

$$3x \qquad = 156 \qquad (5)$$

5. Form a triangular system using Equations (3), (1), and (5). Use back-substitution to complete the solution.

$$x + y + z = 180 \qquad (3)$$

$$x - 2y = -4 \qquad (1)$$

$$3x = 156 \qquad (5)$$

Divide both sides of Equation (5) by 3 to find $x = 52$.   Substitute 52 for x in Equation (1) and solve for y to find

$$52 + 2y = 160$$

$$y = 28$$

Substitute 52 for x and 28 for y in Equation (3) to find

$$52 + 28 + z = 180$$

$$z = 100$$

Step 4   Thus the angles measure 52°, 28°, and 100°.

Example 6

A manufacturer of office supplies makes three types of file cabinets: two-drawer, four-drawer, and horizontal. The manufacturing process is divided into three phases: assembly, painting, and finishing. A two-drawer cabinet requires 3 hours to assemble, 1 hour to paint, and 1 hour to finish. The four-drawer model takes 5 hours to assemble, 90 minutes to paint, and 2 hours to finish. The horizontal cabinet takes 4 hours to assemble, 1 hour to paint, and 3 hours to finish. The manufacturer employs enough workers for 500 hours of assembly time, 150 hours of painting, and 230 hours of finishing per week. How many of each type of file cabinet should he make in order to use all the hours available?

Solution

Step 1 Represent the number of each model of file cabinet by a different variable.

Number of two-drawer cabinets: x

Number of four-drawer cabinets: y

Number of horizontal cabinets: z

Step 2 Write three equations describing the time constraints in each of the three manufacturing phases. It is usually helpful to organize the information into a table such as Table 3.2. All times are listed in hours.

	Two Drawers	Four Drawers	Horizontal	Total Available
Assembly	3	5	4	500
Painting	1	3/2	1	150
Finishing	1	2	3	230

Table 3.2

(For example, the assembly phase requires $3x$ hours for the two-drawer cabinets, $5y$ hours for the four-drawer cabinets, and $4z$ hours for the horizontal cabinets; the sum of these times should be the time available, 500 hours.)

$$3x + 5y + 4z = 500 \quad \text{Assembly time}$$

$$x + \frac{3}{2}y + z = 150 \quad \text{Painting time}$$

$$x + 2y + 3z = 230 \quad \text{Finishing time}$$

Step 3 Solve the system. Follow the steps outlined above.

1. Clear the fractions from the second equation to obtain the system

$$3x + 5y + 4z = 500 \quad (1)$$
$$2x + 3y + 2z = 300 \quad (2)$$
$$x + 2y + 3z = 230 \quad (3)$$

2. Subtract Equation (1) from 3 times Equation (3) to obtain

$$y + 5z = 190 \qquad (4)$$

3. Subtract Equation (2) from twice Equation (3) to obtain

$$y + 4z = 160 \qquad (5)$$

4. Equations (4) and (5) form a 2×2 system in y and z. Subtract Equation (5) from Equation (4) to find

$$z = 30 \qquad (6)$$

5. Form a triangular system with Equations (3), (4), and (6).

$$x + 2y + 3z = 230 \qquad (3)$$
$$y + 5z = 190 \qquad (4)$$
$$z = 30 \qquad (6)$$

Use back-substitution to complete the solution. Substitute 30 for z in Equation (4) to get

$$y + 5(30) = 190$$
$$y = 40$$

Substitute 40 for y and 30 for z in Equation (3) to obtain

$$x + 2(40) + 3(30) = 230$$
$$x = 60$$

Step 4 The manufacturer should make 60 two-drawer cabinets, 40 four-drawer cabinets, and 30 horizontal cabinets.   ▭

Exercise 3.3

Use back-substitution to solve Exercises 1–6.

1. $x + y + z = 2$
   $3y + z = 5$
   $-4y = -8$

2. $2x + 3y - z = -7$
   $y - 2z = -6$
   $5z = 15$

3. $2x - y - z = 6$
   $5y + 3z = -8$
   $13y = -13$

4. $x + y + z = 1$
   $x + 4y = 1$
   $3x = 3$

5. $2x + z = 5$
   $3y + 2z = 6$
   $5x = 20$

6. $3x - y = 6$
   $x - 2z = -7$
   $13x = 13$

Use Gaussian reduction to solve Exercises 7–22.

7. $x + y + z = 0$
   $2x - 2y + z = 8$
   $3x + 2y + z = 2$

8. $x - 2y + 4z = -3$
   $3x + y - 2z = 12$
   $2x + y - 3z = 11$

9. $x - 2y + z = -1$
   $2x + y - 3z = 3$
   $3x + 3y - 2z = 10$

10. $x - 5y - z = 2$
$3x - 9y + 3z = 6$
$x - 3y - z = -6$

11. $4x + z = 3$
$2x - y = 2$
$3y + 2z = 0$

12. $3y + z = 3$
$-2x + 3y = 7$
$3x + 2z = -6$

13. $2x + 3y - 2z = 5$
$3x - 2y - 5z = 5$
$5x + 2y + 3z = -9$

14. $3x - 4y + 2z = 20$
$4x + 3y - 3z = -4$
$2x - 5y + 5z = 24$

15. $4x + 6y + 3z = -3$
$2x - 3y - 2z = 5$
$-6x + 6y + 2z = -5$

16. $3x + 4y + 6z = 2$
$-2x + 2y - 3z = 1$
$4x - 10y + 9z = 0$

17. $x - \dfrac{1}{2}y - \dfrac{1}{2}z = 4$

$x - \dfrac{3}{2}y - 2z = 3$

$\dfrac{1}{4}x + \dfrac{1}{4}y - \dfrac{1}{4}z = 0$

18. $x + 2y + \dfrac{1}{2}z = 0$

$x + \dfrac{3}{5}y - \dfrac{2}{5}z = \dfrac{1}{5}$

$4x - 7y - 7z = 6$

19. $x + y - z = 2$

$\dfrac{1}{2}x - y + \dfrac{1}{2}z = -\dfrac{1}{2}$

$x + \dfrac{1}{3}y - \dfrac{2}{3}z = \dfrac{4}{3}$

20. $x + y - 2z = 3$

$x - \dfrac{1}{3}y + \dfrac{1}{3}z = \dfrac{5}{3}$

$\dfrac{1}{2}x - \dfrac{1}{2}y - z = \dfrac{3}{2}$

21. $x = -y$

$x + z = \dfrac{5}{6}$

$y - 2z = -\dfrac{7}{6}$

22. $x = y + \dfrac{1}{2}$

$y = z + \dfrac{5}{4}$

$2z = x - \dfrac{7}{4}$

Solve Exercises 23–34. If the system is inconsistent or dependent, say so.

23. $3x - 2y + z = 6$
$2x + y - z = 2$
$4x + 2y - 2z = 3$

24. $x + 3y - z = 4$
$-2x - 6y + 2z = 1$
$x + 2y - z = 3$

25. $2x + 3y - z = -2$

$x - y + \dfrac{1}{2}z = 2$

$4x - \dfrac{1}{3}y + 2z = 8$

26. $3x + 6y + 2z = -2$

$\dfrac{1}{2}x - 3y - z = 1$

$4x + y + \dfrac{1}{3}z = -\dfrac{1}{3}$

27. $2x + y = 6$
$x - z = 4$
$3x + y - z = 10$

28. $x - 2y + z = 5$
$-x + y = -2$
$y - z = -3$

29. $x = 2y - 7$
$y = 4z + 3$
$z = 3x + y$

30. $x = y + z$
$y = 2x - z$
$z = 3x - y$

31. $\dfrac{1}{2}x + y = \dfrac{1}{2}z$

$x - y = -z - 2$

$-x - 2y = -z + \dfrac{4}{3}$

32. $x = \dfrac{1}{2}y - \dfrac{1}{2}z + 1$

$x = 2y + z - 1$

$x = \dfrac{1}{2}y - \dfrac{1}{2}z + \dfrac{1}{4}$

33. $x - y = 0$
$2x + 2y + z = 5$
$2x + y - \dfrac{1}{2}z = 0$

34. $x + y = 1$
$2x - y + z = -1$
$x - 3y - z = -\dfrac{2}{3}$

Solve Exercises 35–44 using a system of equations.

35. A box contains $6.25 in nickels, dimes, and quarters. There are 85 coins in all, with three times as many nickels as dimes. How many coins of each kind are in the box?

36. Vanita has $446 in 10-dollar, 5-dollar, and 1-dollar bills. There are 94 bills in all and 10 more 5-dollar bills than 10-dollar bills. How many bills of each kind does she have?

37. The perimeter of a triangle is 155 inches. Side x is 20 inches shorter than side y, and side y is 5 inches longer than side z. Find the lengths of the triangle's sides.

38. One angle of a triangle measures 10° more than a second angle, and the third angle is 10° more than six times the measure of the smallest angle. Find the measure of each angle.

39. Vegetable Medley is made of carrots, green beans, and cauliflower. The package says that 1 cup of Vegetable Medley provides 29.4 milligrams of vitamin C and 47.4 milligrams of calcium. One cup of carrots contains 9 milligrams of vitamin C and 48 milligrams of calcium. One cup of green beans contains 15 milligrams of vitamin C and 63 milligrams of calcium. One cup of cauliflower contains 69 milligrams of vitamin C and 26 milligrams of calcium. How much of each vegetable is in 1 cup of Vegetable Medley?

40. The Java Shoppe sells a house brand of coffee that is only 2.25% caffeine for $6.60 per pound. The house brand is a mixture of Colombian coffee that sells for $6 per pound and is 2% caffeine, French Roast that sells for $7.60 per pound and

is 4% caffeine, and Sumatran at $6.80 per pound and 1% caffeine. How much of each variety is in a pound of house brand?

41. The ABC Psychological Testing Service offers three types of reports on test results: score only, evaluation, and narrative report. Each score-only test takes 3 minutes to score using an optical scanner and 1 minute to print the interpretation. Each evaluation takes 3 minutes to score, 4 minutes to analyze, and 2 minutes to print. Each narrative report takes 3 minutes to score, 5 minutes to analyze, and 8 minutes to print. If ABC uses its optical scanner 7 hours per day, has 8 hours in which to analyze results, and has 12 hours of printer time available per day, how many of each type of report can it complete each day when it is using all its resources?

42. Reliable Auto Company wants to ship 1700 Status Sedans to three major dealers in Los Angeles, Chicago, and Miami. From past experience Reliable figures that it will sell twice as many sedans in Los Angeles as in Chicago. It costs $230 to ship a sedan to Los Angeles, $70 to Chicago, and $160 to Miami. If Reliable has $292,000 to pay for shipping costs, how many sedans should it ship to each city?

43. Ace, Inc. produces three kinds of wooden rackets: tennis rackets, Ping-Pong paddles, and squash rackets. After the pieces are cut, each racket goes through three phases of production: gluing, sanding, and finishing. A tennis racket takes 3 hours to glue, 2 hours to sand, and 3 hours to finish. A Ping-Pong paddle takes 1 hour to glue, 1 hour to sand, and 1 hour to finish. A squash racket takes 2 hours to glue, 2 hours to sand, and $2\frac{1}{2}$ hours to finish. Ace has available 95 work-hours in its gluing department, 75 work-hours in sanding, and 100 work-hours in finishing per day. How many of each racket should it make in order to use all its available workforce?

44. A farmer has 1300 acres on which to plant wheat, corn, and soybeans. The seed costs $6 for an acre of wheat, $4 for an acre of corn, and $5 for an acre of soybeans. An acre of wheat requires

5 acre-feet of water during the growing season, an acre of corn requires 2 acre-feet, and an acre of soybeans requires 3 acre-feet. If the farmer has \$6150 to spend on seed and can count on 3800 acre-feet of water, how many acres of each crop should he plant in order to use all his resources?

3.4 LINEAR APPROXIMATIONS

Linear Interpolation and Extrapolation

We have examined numerous examples in which we use an equation or formula to find values of one variable when we know a second variable. In many situations we may have only a collection of data relating two variables, and we would like to find an equation that relates them. In this section we consider how to fit a linear equation to a set of data points.

Example 1

The temperature in Northridge was 49° at 8:00 A.M. and 81° by noon. Approximate the temperature at 9:00 A.M., 10:00 A.M., and 11:00 A.M.

Solution   We will assume that the relationship between temperature and time of day is linear. This assumption means that a graph of temperature versus time would be a straight line. The time variable goes on the horizontal axis and temperature goes on the vertical axis, and we know the coordinates of two points on the line: (8, 49) and (12, 81). From these two points we can compute the slope of the line:

$$m = \frac{\Delta y}{\Delta x} = \frac{81 - 49}{12 - 8} = \frac{32}{4} = 8$$

Next we can use the point-slope formula to get an equation of the line:

$$y - y_1 = m\,(x - x_1) \qquad \text{Substitute} \quad m = 8, \quad x_1 = 8, \quad \text{and} \quad y_1 = 49.$$

$$y - 49 = 8\,(x - 8) \qquad \text{Simplify.}$$

$$y = 8x - 15$$

We can use this equation to estimate temperatures at other times. At 9:00 A.M., $x = 9$,   so the temperature is given by

$$y = 8(\mathbf{9}) - 15 = 57$$

or 57°. At 10:00 A.M., $x = 10$,   so

$$y = 8(\mathbf{10}) - 15 = 65$$

At 11:00 A.M., $x = 11$,   so

$$y = 8(\mathbf{11}) - 15 = 73$$

Notice that the slope of the line in Example 1 is measured in degrees per hour. Thus the slope tells us that the temperature rises by 8° per hour, so we could also solve this problem by starting with 49° at 8:00 A.M. and

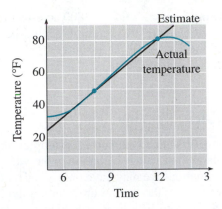

Figure 3.17

simply adding 8° (to get 57°) for the temperature at 9:00 A.M., then adding another 8° (to get 65°) for the temperature at 10:00 A.M., and another 8° (to get 73°) for the temperature at 11:00 A.M.

The process of estimating between known data points is called **interpolation**. In Example 1 we used the most common form of interpolation, **linear interpolation**. Our estimates depended on the assumption that the graph of temperature versus time is a straight line. This assumption might be fairly close to correct between 8:00 A.M. and noon, but it probably will not continue to hold throughout the night, since the temperatures will eventually start to fall again. The graph in Figure 3.17 shows the actual temperatures recorded each hour and the temperatures predicted by our linear model. As you can see, the model is fairly accurate close to the given data, but it is not so useful as time goes on. Making predictions beyond the range of known data is called **extrapolation**.

Example 2

Emily was 82 centimeters tall at age 36 months and 88 centimeters tall at age 48 months. Use linear interpolation to estimate her height when she was 38 months old, and extrapolate to estimate her height at age 50 months.

Solution   Let y represent Emily's height at age x months. We will look for a linear equation that relates x and y. Using the points $(36, 82)$ and $(48, 88)$, we find a slope of

$$m = \frac{88 - 82}{48 - 36} = \frac{6}{12} = \frac{1}{2}$$

Use the point-slope formula to find the equation of the line:

$$y - 82 = \frac{1}{2}(x - 36) \qquad \text{Simplify and solve for } y.$$

$$y = \frac{x}{2} + 64$$

We use this equation to approximate Emily's height at other times. When Emily was 38 months old, $x = 38$ and

$$y = \frac{38}{2} + 64 = 83$$

When $x = 50$,

$$y = \frac{50}{2} + 64 = 89$$

We estimate that Emily was 83 centimeters tall at age 38 months, and we predict that she will be approximately 89 centimeters tall at age 50 months.

If we try to extrapolate too far, we get unreasonable results. For example, if we use our model to predict Emily's height at 300 months (25 years old), we get

$$y = \frac{300}{2} + 64 = 214$$

It is unlikely that Emily will be 214 centimeters, or over 7 feet, tall when she is 25 years old.

Linear Regression   Figure 3.18 is called a **scatterplot**. Each point on a scatterplot exhibits a pair of measurements about a single event. In Figure 3.18 the measurements are the outside temperature and the number of cups of hot chocolate sold at a snack bar on 13 consecutive nights. Although we cannot tell from the scatterplot which dot corresponds to which night, we can tell, for instance, that when the temperature was 2°C, 45 cups of hot chocolate were sold, and

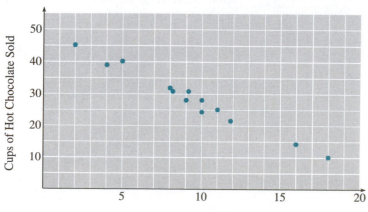

Figure 3.18

Temperature (°C)

Cups of Hot Chocolate Sold

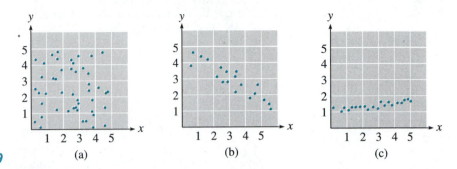

Figure 3.19

    (a)   (b)   (c)

on the two evenings it was 10°C, 24 cups and 28 cups of hot chocolate were sold.

    The points on a scatterplot may or may not show some sort of pattern. Consider the three plots in Figure 3.19. In Figure 3.19(a) the data points resemble a cloud of gnats; there is no discernible pattern to their locations. In Figure 3.19(b) the data follow a generally decreasing trend; as the values of x increase, most of the corresponding y-values decrease. The points in Figure 3.19(c) appear even more organized. Even though the points do not all lie on a straight line, they seem to be clustered around some imaginary line.

    If the data in a scatterplot are roughly linear, we can estimate the location of this imaginary line in order to make predictions about the data.

Example 3

Based on Figure 3.18, predict the number of cups of hot chocolate the snack bar will sell if the temperature is 7°C.

Solution We draw a line that "fits" the data points as best we can, as shown in Figure 3.20. On this line we see that when $x = 7$, the y-value is approximately 34. We therefore predict that about 34 cups of hot chocolate will be sold when the temperature is 7°C.

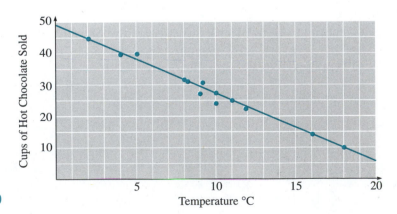

Figure 3.20

The process of predicting a value of y based on a straight line that fits the data is called **linear regression**, and the line itself is called the **regression line**. The equation of the regression line is usually used (instead of a graph) to predict values.

Example 4

Find the equation of the line in Example 3. Use the equation to predict the number of cups of hot chocolate that will be sold when the temperature is 10°C and when the temperature is 24°C.

Solution We choose two points on the line we drew in Figure 3.20. The line appears to pass through the points (2, 45) and (18, 10). The slope of the line is then

$$m = \frac{10 - 45}{18 - 2} = -\frac{35}{16} \approx -2.2$$

Thus the equation of the line is approximately

$$y - 45 = -2.2\,(x - 2)$$

or $y = -2.2x + 49.4$. We will use this equation to make our predictions. When the temperature is 10°C,

$$y = -2.2(\mathbf{10}) + 49.4 = 27.4$$

We predict that about 27 cups of hot chocolate will be sold if the temperature is 10°C. When $x = 24$,

$$y = -2.2(\mathbf{24}) + 49.4 = -3.4$$

This prediction is nonsensical, however, since a negative number of cups of hot chocolate cannot be sold. In this instance extrapolation leads to an invalid prediction.

Using a Calculator

Finding a line that appears to fit the data is a subjective process. Rather than base their estimates on the equation of a subjectively determined line, statisticians use an algorithm, or recipe, that depends only on the data (not on the appearance of a graph) to determine the equation of a regression line. The most commonly used regression line is called the **least squares regression line**. This regression line minimizes the sum of the squares of all the vertical distances between the data points and their corresponding points on the line. (See Figure 3.21.)

Graphing calculators and many scientific calculators can find the equation of the least squares regression line. We will use the statistics mode on the TI-82 calculator, so press [STAT]. You will see a display that looks like Figure 3.22(a) on page 174. Choose [1] to **Edit** (enter or alter) data.

If data are in column L_1 or L_2, clear them out: Use the [△] key to select L_1 and press [CLEAR] [ENTER]; then do the same for L_2. Now follow the instructions in Example 5 for using your calculator's statistical features.

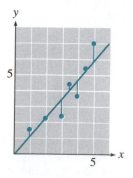

Figure 3.21

Figure 3.22   (a)   (b)

Example 5

a. Find the equation of the least squares regression line for the following data.

$$(10, 12), \quad (11, 14), \quad (12, 14), \quad (12, 16), \quad (14, 20)$$

b. Use the least squares regression line to predict the value of y when $x = 9$.

c. Plot the data points and the least squares regression line on the same axes.

Solutions   **a.** We must first enter the data. Press $\boxed{\text{STAT}}$, then press $\boxed{1}$ or $\boxed{\text{ENTER}}$ to select **Edit**. Enter the x-coordinates of the data points in the L_1 column and the y-coordinates in the L_2 column.

Now we are ready to find the regression equation for our data. Press $\boxed{\text{STAT}}$ $\boxed{\triangleright}$ to get the CALC (calculation) menu, then press $\boxed{5}$ for linear regression or **LinReg($ax + b$)**. Finally, press $\boxed{\text{ENTER}}$. The calculator will display the equation $y = ax + b$ and values for a, b, and r. You should find that your regression line is approximately

$$y = 1.95x - 7.86$$

Notice that the slope of the line is a and its y-intercept is b—if you do not have a TI-82, your calculator may reverse the roles of a and b. (The third value, r, is the correlation coefficient, which we will not discuss.)

b. First store the value $x = 9$ by pressing 9 $\boxed{\text{STO}\triangleright}$ $\boxed{\text{X, T, }\theta}$ $\boxed{\text{ENTER}}$. To access the regression equation, press $\boxed{\text{VARS}}$, then $\boxed{5}$ for **Statistics**, $\boxed{\triangleright}\boxed{\triangleright}$ to select **EQ** (equations), and $\boxed{7}$ for **RegEQ**. Finally, press $\boxed{\text{ENTER}}$ to evaluate the equation for $x = 9$. We find that $y \approx 9.7$ when $x = 9$.

c. First clear out all old definitions in the $\boxed{\text{Y} =}$ list. Then choose an appropriate graphing window for the data:

$$\text{Xmin} = 8, \quad \text{Xmax} = 15$$

$$\text{Ymin} = 10, \quad \text{Ymax} = 22$$

will work. To draw the scatterplot, we first press $\boxed{\substack{\text{STAT}\\\text{PLOT}}}$ (or $\boxed{\text{2nd}}$ $\boxed{\text{Y} =}$) to access the statistics plot options. Press $\boxed{1}$ to choose Plot 1 and $\boxed{\text{ENTER}}$ to activate that plot. Then choose the settings for Plot 1, as shown in Figure 3.23(b). Press $\boxed{\text{GRAPH}}$ to see the scatterplot.

Figure 3.23 (a) (b)

To graph the regression line over the scatterplot, press $\boxed{Y =}$, then access the regression equation **RegEQ** as before: press $\boxed{VARS}$, then $\boxed{5}$ for **Statistics**, $\boxed{\triangleright}\boxed{\triangleright}$ to select **EQ** (equations), and $\boxed{7}$ for **RegEQ** (regression equation). Finally, press $\boxed{GRAPH}$. How closely does the regression line "fit" the data points?

When you are through with the scatterplot, press $\boxed{\substack{STAT \\ PLOT}}$ $\boxed{4}$ $\boxed{ENTER}$ for "Plots Off." If you neglect to do this, the calculator will continue to show the scatterplot even after you ask it to plot a new equation.

Exercise 3.4

Use linear interpolation to give approximate answers to Exercises 1–8.

1. The temperature in Encino dropped from 81° at 1:00 A.M. to 73° at 5:00 A.M. Estimate the temperature at 3:00 A.M. and at 4:00 A.M.

2. The temperature in a room was 35°C at 1:00 P.M. when the air conditioning was turned on and dropped to 29°C by 5:00 P.M. Estimate the temperature at 2:00 P.M. and at 3:00 P.M.

3. A bean shoot that was 1 centimeter tall on the first day of the month was 8 centimeters tall on the sixth day of the month. Approximate its height on the third and fifth days of the month.

4. Bernice fasted from August 1 to August 10. She weighed 187 pounds on the first and weighed 159 pounds on the tenth. Approximate her weight on August 5 and August 7.

5. A car starts from a standstill and accelerates to a speed of 60 miles per hour in 6 seconds. Estimate the car's speed 2 seconds after it began to accelerate and 4 seconds after it began to accelerate.

6. A truck on a slippery road is moving at 24 feet per second when the driver hits the brakes. The truck needs 3 seconds to come to a stop. Estimate the truck's speed at 1 second and 2 seconds after the brakes were applied.

7. The temperature in Alto was 23°C at 10:00 P.M. and 13°C at 6:00 A.M. the following morning. Estimate the temperature at midnight and at 2:00 A.M. (*Hint:* Let t represent the number of hours since 10:00 P.M. Then 10:00 P.M. corresponds to $t = 0$, and 6:00 A.M. corresponds to $t = 8$. What is t at midnight?)

8. A plant that is 20 centimeters tall on March 25 was 50 centimeters tall on April 14. Estimate the plant's height on April 6 and on April 10. (*Hint:* Let t represent the number of days since March 25. Then March 25 corresponds to $t = 0$, and April 14 corresponds to $t = 20$. What is t on April 6?)

Use linear interpolation or extrapolation to solve Exercises 9–12.

9. The temperature of an automobile engine is 9° Celsius when the engine is started and is 51° seven minutes later. Use a linear model to predict the engine temperature for both two minutes and two hours after it started. Are your predictions reasonable?

10. The temperature in Death Valley is 95° at 5:00 A.M. and rises to 110° by noon. Use a linear model to predict the temperature at 2:00 P.M. and at midnight. Are your predictions reasonable?

11. Ben weighed 8 pounds at birth and 20 pounds at age 1 year. What weight will he be at age 10 if his weight increases at a constant rate?

12. The elephant at the City Zoo becomes ill and loses weight. She was 10,012 pounds when healthy and only 9,641 pounds a week later. Predict her weight after 10 days of illness.

Use information from the graphs to answer Exercises 13–24.

13. The scatterplot in Figure 3.24 shows scores that various computer science students attained on a video game versus the number of glasses of beer they had drunk immediately before the game.

a. Use a straight-edge to draw in a line that "fits" the data.

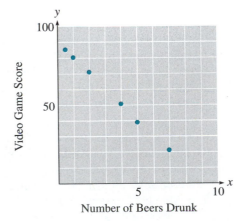

Figure 3.24

b. Use your line to predict the score of the next student, who has drunk 3 glasses of beer, and the score of a student who has drunk 8 glasses of beer.

c. Use your answers from part (b) to find the equation of your regression line.

d. Use your answer to part (c) to predict the score of a student who drinks 4.5 glasses of beer.

14. Good Food magazine commissioned several well-known bakers to rate the muffins produced by six shops. The scatterplot in Figure 3.25 shows the price of the muffins versus the score the panel of bakers gave the muffin.

a. Use a straight-edge to draw in a line that "fits" the data.

b. Use your line to predict the price of a muffin that the bakers rate with a score of 2, and the price of a muffin that the bakers rate with a score of 7.

c. Use your answers from part (b) to find the equation of your regression line.

d. Use your answer to part (c) to predict the price of a muffin rated 9.

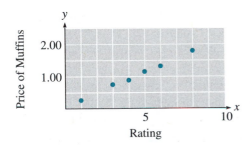

Figure 3.25

15. The scatterplot in Figure 3.26 shows weights (in pounds) and heights (in inches) of a team of distance runners.

a. Use a straight-edge to draw in a line that "fits" the data.

b. Use your line to predict the weight of a 65-inch-tall distance runner and the weight of a 71-inch-tall distance runner.

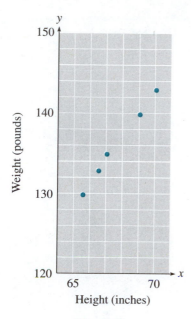

Weight (pounds) / Height (inches)

Figure 3.26

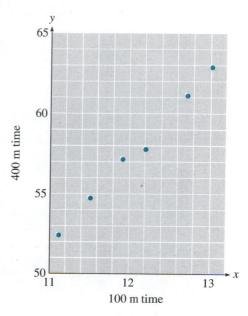

400 m time / 100 m time

Figure 3.27

c. Use your answers from part (b) to approximate the equation of a regression line.

d. Use your answer to part (c) to predict the weight of a 68-inch-tall distance runner.

16. The scatterplot in Figure 3.27 shows best times for various women running 400 meters and 100 meters.

a. Use a straight-edge to draw in a line that "fits" the data.

b. Use your line to predict the 400-meter time of a woman who runs the 100-meter dash in 11.2 seconds, and the 400-meter time of a woman who runs the 100-meter dash in 13.2 seconds.

c. Use your answers from part (b) to approximate the equation of a regression line.

d. Use your answer to part (c) to predict the 400-meter time of a woman who runs the 100-meter dash in 12.1 seconds.

17. The scatterplot in Figure 3.28 on page 178 shows the number of lawyers (per 100,000 residents) versus the number of physicians (per 100,000 residents) for various states.

a. Use a straight-edge to draw in a line that "fits" the data.

b. Use your line to predict the number of lawyers (per 100,000 residents) in a state with 210 physicians (per 100,000), and the number of lawyers (per 100,000) in a state with 310 physicians (per 100,000).

c. Use your answers from part (b) to approximate the equation of a regression line.

d. Use your answer to part (c) to predict the number of lawyers per 100,000 residents in a state with 220 physicians per 100,000 residents.

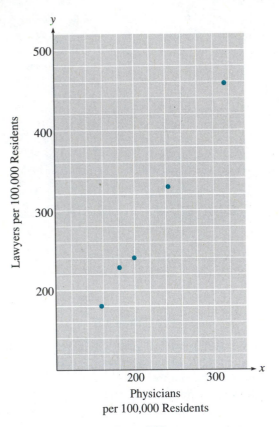

Figure 3.28

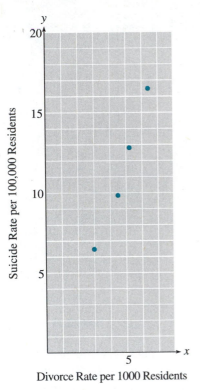

Figure 3.29

18. The scatterplot in Figure 3.29 shows the death rate by suicide (per 100,000 residents) versus the divorce rate (per 1000 residents) for various states.

a. Use a straight-edge to draw in a line that "fits" the data.

b. Use your line to predict the suicide rate in a state with a divorce rate of 4, and the suicide rate in a state with a divorce rate of 7.

c. Use your answers from part (b) to approximate the equation of a regression line.

d. Use your answer to part (c) to predict the suicide rate in a state with a divorce rate of 4.7.

19. The points on the scatterplot in Figure 3.24 are (0.5, 86), (1, 79), (2, 71), (4, 51), (5, 39), (7, 21). Use your calculator and the given points to find the least squares regression line. Compare the score this equation gives for part (d) of Exercise 13 with what you predicted earlier.

20. The points on the scatterplot in Figure 3.25 are (1, 0.29), (3, 0.72), (4, 0.91), (5, 1.12), (6, 1.29), (8, 1.79). Use your calculator and the given points to find the least squares regression line. Compare the price this equation gives for part (d) of Exercise 14 with what you predicted earlier.

21. The points on the scatterplot in Figure 3.26 are (65.5, 130), (66.5, 133), (67, 135), (69, 140), (70, 143). Use your calculator and the given points to find the least squares regression line. Compare the weight this equation gives for part (d) of Exercise 15 with what you predicted earlier.

22. The points on the scatterplot in Figure 3.27 are (11.1, 52.4), (11.5, 54.7), (11.9, 57.4), (12.2, 57.9), (12.7, 61.3), (13.0, 63.0). Use your calculator and the given points to find the least squares regression line. Compare the time this equation gives for part (d) of Exercise 16 with what you predicted earlier.

23. The points on the scatterplot in Figure 3.28 are Indiana (157, 181), New Mexico (183, 232), New Hampshire (200, 241), California (244, 329), New York (315, 457). Use your calculator and the given points to find the least squares regression line. Compare the number of lawyers this equation gives for part (d) of Exercise 17 with what you predicted earlier.

24. The points on the scatterplot in Figure 3.29 are New Jersey (3.0, 6.5), Hawaii (4.6, 9.8), California (5.2, 12.7), Colorado (6.4, 16.5). Use your calculator and the given points to find the least squares regression line. Compare the suicide rate this equation gives for part (d) of Exercise 18 with what you predicted earlier.

3.5   LINEAR INEQUALITIES IN TWO VARIABLES

Many relationships between variables are more naturally described using inequalities instead of equations. In this section we will study linear inequalities in two variables and how they arise in applications.

Graphs of Inequalities in Two Variables

Consider the following situation. Suppose Ivana is starting to make investments in the hotel business. She has bought two hotels and will decide to expand her investments if after 1 year her total profit from the two hotels exceeds $10,000. If we let x represent the profit from one hotel and let y represent the profit from the other, then Ivana will expand her investments when

$$x + y \geq 10,000 \tag{1}$$

Notice that the equation $x + y = 10,000$ is not appropriate to model our situation, since Ivana will be delighted if her profits are not exactly equal to $10,000 but exceed that amount.

A **solution** to an inequality in two variables is an ordered pair of numbers that satisfies the inequality. The graph of the inequality must show all the points whose coordinates are solutions. As an example we will graph inequality (1).

We will first rewrite the inequality by subtracting x from both sides to get

$$y \geq -x + 10,000 \tag{2}$$

Inequality (2) says that for each x-value, we must choose points with y-values greater than or equal to $-x + 10,000$. For example, when $x = 2000$ we must choose points with y-values greater than or equal to 8000. Solutions for several choices of x are shown in Figure 3.30(a).

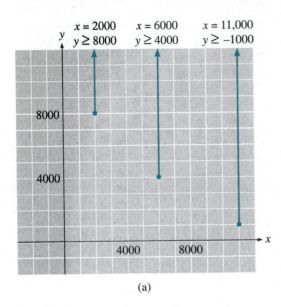

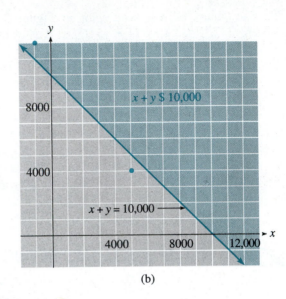

(a)

(b)

Figure 3.30

A more efficient way to find all the solutions of Inequality (2) is to start with the graph of the corresponding equation $y = -x + 10,000$. This graph is simply a straight line, as illustrated in Figure 3.30(b). To graph Inequality (2), we observe that any point *above* this line has a y-coordinate greater than $-x + 10,000$ and hence satisfies (2). Thus the graph of Inequality (2) includes all the points on or above the graph of $y = -x + 10,000$, as shown by the shaded region in Figure 3.30(b).

You can check that the shaded points are also solutions to Inequality (1). Consider the point $(-1000, 12,000)$, which lies in the shaded region above the line. This pair does satisfy (1) since $-1000 + 12,000 \geq 10,000$. (Ivana will expand her investments if her first hotel loses $1000 and her second has a profit of $12,000.) On the other hand, the point $(5000, 4000)$ does not lie in the graph of (1) because the coordinates do not satisfy (1).

Linear Inequalities   More generally, a **linear inequality** can be written in the form

$$ax + by + c \leq 0 \qquad \text{or} \qquad ax + by + c \geq 0$$

The solutions consist of the line $ax + by + c = 0$ and a **half-plane** on one side of that line. We shade the half-plane to show that all its points are included in the solution set. If the inequality is strict, then the graph includes only the half-plane and not the line. In that case we use a dashed line for the graph of the equation $ax + by + c = 0$ to show that it is not part of the solution.

One way to decide which side of the line to shade is to solve the inequality for y in terms of x. If we obtain

$$y \geq mx + b \qquad (\text{or} \quad y > mx + b)$$

then we shade the half-plane *above* the line. If the inequality is equivalent to

$$y \le mx + b \qquad (\text{or} \quad y < mx + b)$$

then we shade the half-plane *below* the line. Be careful when isolating y: we must remember to reverse the direction of the inequality whenever we multiply or divide by a negative number.

Example 1

Graph $4x - 3y \ge 12$.

Solution   Solve the inequality for y.

$$4x - 3y \ge 12 \qquad \text{Subtract } 4x \text{ from both sides.}$$

$$-3y \ge -4x + 12 \qquad \text{Divide both sides by } -3.$$

$$y \le \frac{4}{3}x - 4$$

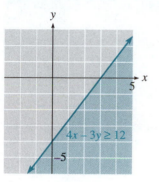

Figure 3.31

Graph the corresponding line $y = \frac{4}{3}x - 4$.   (Note that the y-intercept is -4 and the slope is $\frac{4}{3}$). Finally, shade the half-plane below the line. The completed graph is shown in Figure 3.31.

Using a Graphing Calculator

We can also use a graphing calculator to graph the inequality in Example 1. First clear any equations that are stored in the $\boxed{Y =}$ screen. Next choose a graphing window. (For this example we will use the standard window, $\boxed{ZOOM}$ $\boxed{6}$.)

On the TI calculators we use the **Shade** command to graph an inequality. First press the $\boxed{DRAW}$ button ($\boxed{2ND}$ $\boxed{PRGM}$), then press $\boxed{7}$ to select **Shade** from the option list. The display will show

Shade(

on the home screen. You must now enter two expressions separated by a comma. The first will represent the lower boundary of the shading, and the second will represent the upper boundary. Since our inequality is equivalent to $y \le \frac{4}{3}x - 4$,   we want the calculator to shade in the region *below* the line $y = \frac{4}{3}x - 4$,   all the way to the bottom of the screen. Thus for the lower boundary we can use any number below the bottom of the screen; we will use -11 since the bottom of the standard window is $y = -10$.

Next enter a comma by pressing the $\boxed{,}$ key. Finally, enter the upper boundary, 4X/3 $-$ 4, so that your display looks like this:

Shade(-11, 4X/3 $-$ 4)

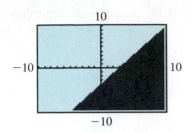

Figure 3.32

(You can also enter a close parenthesis, but it is not necessary.) Press [ENTER] to see the graph of the inequality, as shown in Figure 3.32.

Here is a summary of the steps for graphing an inequality.

TO GRAPH AN INEQUALITY WITH A CALCULATOR

1. Clear the [Y =] screen and select a graphing window.

2. Open the [DRAW] menu by pressing [2ND] [PRGM]. Press [7] to select **Shade(**.

3. Enter the lower boundary of the shaded region. (It can be a number or an algebraic expression.)

4. Enter a comma by pressing the [,] key, then enter the upper boundary of the shaded region.

5. Press [ENTER].

To remove the graph of an inequality, open the [DRAW] menu, select **ClrDraw**, and press [ENTER] again. The calculator should return **Done**, and you are ready to begin a new graph.

Using a Test Point   A second method for graphing inequalities does not require us to solve for *y*. Once we have graphed the boundary line, we can determine which half-plane to shade by using a "test point." The test point can be any point that is not on the boundary line itself.

Example 2

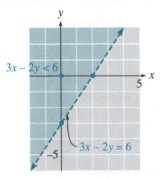

Figure 3.33

Graph $3x - 2y < 6$.

Solution   As shown in Figure 3.33, first graph the line $3x - 2y = 6$.   We will use the intercept method. The intercepts are $(2, 0)$ and $(0, -3)$, so we sketch the boundary line through those points. Next choose a test point. Since $(0, 0)$ does not lie on the line, we choose it as our test point. Substitute the coordinates of the test point into the inequality to obtain

$$3(\mathbf{0}) - 2(\mathbf{0}) < 6$$

Since this statement is true, $(0, 0)$ is a solution of the inequality. Since *all* the solutions lie on the same side of the boundary line, we shade the half-plane that contains the test point. In this example the boundary line is a dashed line because the original inequality was strict.   ▭

We can choose *any* point for the test point, as long as it does not lie on the boundary line. We chose $(0, 0)$ in Example 2 because the coordinates are easy to substitute into the inequality. If the test point *is* a solution to the inequality, then the half-plane including that point should be shaded. If the test point is *not* a solution to the inequality, then the *other* half-plane should be shaded. For example, suppose we had chosen (5, 0) as the test point in Example 2. When we substitute its coordinates into the inequality, we find

$$3(5) - 2(0) < 6$$

which is a false statement. Thus (5, 0) is not a solution to the inequality, so the solutions must lie on the other side of the boundary line. Using (5, 0) as the test point gives us the same solutions we found in Example 2.

Here is a summary of our test-point method for graphing inequalities.

TO GRAPH AN INEQUALITY USING A TEST POINT

1. Graph the corresponding equation to obtain the boundary line.

2. Choose a test point that does not lie on the boundary line.

3. Substitute the coordinates of the test point into the inequality.

  a. If the resulting statement is true, shade the half-plane that includes the test point.

  b. If the resulting statement is false, shade the half-plane that does not include the test point.

4. If the inequality is strict, make the boundary line a dashed line.

Recall that the equation of a vertical line has the form

$$x = k$$

where k is a constant, and a horizontal line has an equation of the form

$$y = k$$

Similarly, the inequality $x \geq k$ may represent the inequality in two variables

$$x + 0y \geq k$$

Its graph is then a region in the plane.

Example 3

Graph $x \geq 2$ in the plane.

Solution First graph the equation $x = 2$; its graph is a vertical line. Since the origin does not lie on this line, we can use it as a test point. Substitute 0 for x (there is no y) into the inequality to obtain

$$0 \geq 2$$

Since this statement is false, shade the half-plane that does not contain the origin. We see in Figure 3.34 (page 184) that the graph of the inequality contains all points whose x-coordinates are greater than or equal to 2.

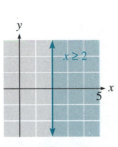

Figure 3.34

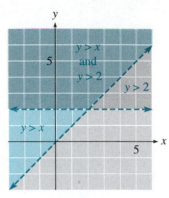

Figure 3.35

Systems of Inequalities   Some applications are best described by a system of two or more inequalities. The solutions to a system of inequalities include all points that are solutions to each inequality in the system. The graph of the system is the intersection of the shaded regions for each inequality in the system. For example, Figure 3.35 shows the solutions of the system

$$y > x \quad \text{and} \quad y > 2$$

Example 4   Laura is a finicky eater and coincidently dislikes most foods that are high in calcium. Her morning cereal satisfies some of her calcium requirements, but she needs an additional 500 milligrams of calcium, which she will get from a combination of broccoli, at 160 milligrams per serving, and zucchini, at 30 milligrams per serving. Draw a graph representing the possible combinations of broccoli and zucchini that will fulfill Laura's calcium requirements.

Solution

Step 1   Number of servings of broccoli: x
Number of servings of zucchini: y

Step 2   To take in at least 500 milligrams of calcium, Laura must choose x and y so that

$$160x + 30y \geq 500$$

It makes no sense to consider negative values of x or y, since Laura cannot consume a negative number of servings. Thus we have two more inequalities to satisfy:

$$x \geq 0 \quad \text{and} \quad y \geq 0$$

Step 3   Graph all three inequalities on the same axes. Note that the inequalities $x \geq 0$ and $y \geq 0$ restrict the solutions to lie in the first

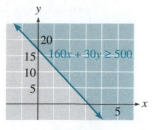

Figure 3.36

quadrant. The solutions common to all three inequalities are shown in Figure 3.36.

Step 4 Laura can choose any combination of broccoli and zucchini represented by points in the shaded region. For example, the point (3, 1) is a solution to the system of inequalities, so Laura could choose to eat 3 servings of broccoli and 1 serving of zucchini.

To describe the solutions of a system of inequalities, it is useful to locate the vertices of the boundary.

Example 5

Graph the solution set of the system below and find the coordinates of its vertices.

$$x - y - 2 \leq 0$$

$$x + 2y - 6 \leq 0$$

$$x \geq 0, \quad y \geq 0$$

Solution First notice that the last two inequalities, $x \geq 0$ and $y \geq 0$, restrict the solutions to the first quadrant. Next graph the line $x - y - 2 = 0$, and use the test point (0, 0) to decide to shade the half-plane including the origin. Finally, graph the line $x + 2y - 6 = 0$, and again use the test point (0, 0) to shade the half-plane below the line. The intersection of the shaded regions is shown in Figure 3.37.

To determine the coordinates of the vertices *A, B, C,* and *D,* solve simultaneously the equations of the two lines that intersect at the vertex. Thus

for *A* solve the system $x = 0$ to find (0, 0)
                                  $y = 0$

for *B* solve the system $x = 0$ to find (0, 3)
                              $x + 2y = 6$

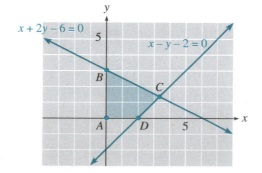

Figure 3.37

for C solve the system
$$x + 2y = 6$$
$$x - y = 2$$
to find $\left(\dfrac{10}{3}, \dfrac{4}{3}\right)$

for D solve the system
$$y = 0$$
$$x - y = 2$$
to find $(2, 0)$

The vertices are the points $(0, 0)$, $(0, 3)$, $\left(\dfrac{10}{3}, \dfrac{4}{3}\right)$, and $(2, 0)$.

We can use the graphing calculator to draw the graph in Example 5 as follows. Set the graphing window to

$$\text{Xmin} = 0, \quad \text{Xmax} = 9, \quad \text{Ymin} = 0, \quad \text{Ymax} = 6$$

This window shows only the first quadrant and thus automatically takes care of the inequalities $x \geq 0$ and $y \geq 0$. The inequality $x - y - 2 \leq 0$ is equivalent to $y \geq x - 2$, and the inequality $x + 2y - 6 \leq 0$ is equivalent to $y \leq \dfrac{-x}{2} + 3$. We want a region that is *above* $y = x - 2$ and *below* $y = \dfrac{-x}{2} + 3$. After choosing $\boxed{7}$ from the $\boxed{\text{DRAW}}$ menu, enter the following command.

$$\text{Shade}(X - 2, -X/2 + 3)$$

Press $\boxed{\text{ENTER}}$ to see the graph, which should resemble Figure 3.37 on page 185.

Exercise 3.5

Graph each inequality in Exercises 1–16.

1. $y > 2x + 4$

2. $y < 9 - 3x$

3. $3x - 2y \leq 12$

4. $2x + 5y \geq 10$

5. $x + 4y \geq -6$

6. $3x - y \leq -2$

7. $x > -3y + 1$

8. $x > 2y - 5$

9. $x \geq -3$

10. $y < 4$

11. $y > \dfrac{1}{2}$

12. $y > \dfrac{4}{3}x$

13. $0 \geq x - y$

14. $0 \geq x + 3y$

15. $-1 < y \leq 4$

16. $-2 \leq y < 0$

Graph each system of inequalities in Exercises 17–26.

17. $y > 2$
$\quad x \geq -2$

18. $y \leq -1$
$\quad x > 2$

19. $y < x$
$\quad y \geq -3$

20. $y \geq -x$
$\quad y < 2$

21. $x + y \leq 6$
$\quad x + y \geq 4$

22. $x - y < 3$
$\quad x - y > -2$

23. $2x - y \leq 4$
$\quad x + 2y > 6$

24. $2y - x < 2$
$\quad x + y \leq 4$

25. $3y - 2x < 2$
$\quad y > x - 1$

26. $2x + y < 4$
$\quad y > 1 - x$

In Exercises 27–36 graph each system of inequalities and find the coordinates of the vertices.

27. $2x + 3y - 6 < 0$
$\quad x \geq 0, \quad y \geq 0$

28. $3x + 2y < 6$
$\quad x \geq 0, \quad y \geq 0$

29. $5y - 3x \leq 15$
$\quad x + y \leq 11$
$\quad x \geq 0, \quad y \geq 0$

30. $y - 2x \geq -4$
$\quad x + y \leq 5$
$\quad x \geq 0, \quad y \geq 0$

31. $2y \leq x$
$2x \leq y + 12$
$x \geq 0, \quad y \geq 0$

32. $y \geq 3x$
$2y + x \leq 14$
$x \geq 0, \quad y \geq 0$

33. $x + y \geq 3$
$2y \leq x + 8$
$2y + 3x \leq 24$
$x \geq 0, \quad y \geq 0$

34. $2y + 3x \geq 6$
$2y + x \leq 10$
$y \geq 3x - 9$
$x \geq 0, \quad y \geq 0$

35. $3y - x \geq 3$
$y - 4x \geq -10$
$y - 2 \leq x$
$x \geq 0, \quad y \geq 0$

36. $2y + x \leq 12$
$4y \leq 2x + 8$
$x \leq 4y + 4$
$x \geq 0, \quad y \geq 0$

Graph the set of solutions to each problem in Exercises 37–42.

37. The math club is selling tickets for a show by a "mathemagician." Student tickets will cost $1 and faculty tickets will cost $2. The ticket receipts must be at least $250 to cover the performer's fee. Write a system of inequalities for the number of student tickets and the number of faculty tickets that must be sold and graph the solutions.

38. The math department is having a book sale of old textbooks to raise at least $300 for scholarships. Paperback textbooks will cost $2 and hardcover textbooks will cost $5. Write a system of inequalities for the number of paperback and hardback textbooks that must be sold and graph the solutions.

39. Vassilis plans to invest at most $10,000 in two banks. One bank pays 6% annual interest and the other pays 5% annual interest. Vassilis wants at least $540 total annual interest from his two investments. Write a system of inequalities for the amount Vassilis can invest in the two accounts and graph the system.

40. Jeannette has 180 acres of farmland for growing wheat or soybeans. She can get a profit of $36 per acre for wheat and $24 per acre for soybeans. She wants to have a profit of at least $5400 from her crops. Write a system of inequalities for the number of acres she can use for each crop and graph the solutions.

41. Gary's pancake recipe includes corn meal and whole wheat flour. Corn meal has 2.4 grams of linoleic acid and 2.5 milligrams of niacin per cup. Whole wheat flour has 0.8 gram of linoleic acid and 5 milligrams of niacin per cup. These two ingredients should not exceed 3 cups total. The mixture should provide at least 3.2 grams of linoleic acid and at least 10 milligrams of niacin. Write a system of inequalities for the amount of corn meal and the amount of whole wheat flour Gary can use and graph the solutions.

42. Cho and his brother go into business making comic book costumes. They need 1 hour of cutting and 2 hours of sewing to make a Batman costume. They need 2 hours of cutting and 1 hour of sewing to make a Wonder Woman costume. They have available at most 10 hours per day for cutting and at most 8 hours per day for sewing. They must make at least one costume each day to stay in business. Write a system of inequalities for the number of each type of costume they can make and graph the solutions.

3.6 LINEAR PROGRAMMING

The term **linear programming** was coined in the late 1940s. It describes a relatively young branch of mathematics compared with other subjects such as Euclidean geometry, in which the major ideas were already well understood twenty-three centuries ago. (The Greek mathematician Euclid wrote what can be considered the first geometry textbook about 300 B.C.) Since business managers must routinely solve linear programming problems for purchasing and marketing strategies, it is possible that linear programming affects your daily life as much as any other branch of mathematics.

**Objective Function
and Constraints**

The goal of a linear programming problem is to maximize or minimize some objective function, subject to one or more constraints. For example, suppose TrailGear produces two kinds of hiking boots, a Weekender model, on which it makes $8 profit per pair, and a Sierra model, on which it makes $10 profit per pair. TrailGear would like to know how many of each model to produce each week to maximize its profit. If we let x represent the number of Weekender boots and y the number of Sierra boots TrailGear produces, then the total weekly profit is given by

$$P = 8x + 10y$$

This expression is called the **objective function**.

Now if TrailGear had infinite resources and an infinite market, there would be no limit to the profit it could earn by producing more and more hiking boots. However, every business has to consider many factors, including its supplies of labor and materials, overhead and shipping costs, and the size of the market for its product. To keep things simple, we will concentrate on just two of these factors.

Each pair of Weekender boots requires 3 work-hours of labor to produce, and each pair of Sierra boots requires 6 work-hours. TrailGear has available 2400 work-hours of labor per week. Thus x and y must satisfy the inequality

$$3x + 6y \leq 2400$$

In addition, suppose that retail outlets can accept at most 1000 pairs of hiking boots each week, with twice as many pairs of Weekenders as Sierra models. This limitation means that

$$2x + y \leq 1000$$

Of course, we will also require that $x \geq 0$ and $y \geq 0$. These inequalities are called the **constraints** of the problem.

Solution by Graphing

We have formulated the original problem into an objective function,

$$P = 8x + 10y$$

and a system of inequalities called the constraints:

$$3x + 6y \leq 2400$$

$$2x + y \leq 1000$$

$$x \geq 0, \quad y \geq 0$$

Our goal is to find values for x and y that satisfy the constraints and produce the maximum value for P.

We begin by graphing the solutions to the constraint inequalities. These solutions are shown in the shaded region in Figure 3.38. The points in this region are called **feasible solutions** because they are the only values we can consider while looking for the maximum value of the objective function, P.

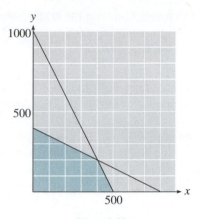

Figure 3.38   **Figure 3.39**

Example 1

a. Verify that the points (300, 100) and (200, 300) represent feasible solutions for the TrailGear problem. Show that (300, 400) is not a feasible solution.
b. Find the values of the objective function $P = 8x + 10y$ at the two feasible solutions in part (a).

Solutions   a. The two points (300, 100) and (200, 300) lie within the shaded region in Figure 3.38, but (300, 400) does not. We can also verify that the coordinates of (300, 100) and (200, 300) satisfy each of the constraint inequalities.

   b. For (300, 100) we have

$$P = 8(\textbf{300}) + 10(\textbf{100}) = 3400$$

For (200, 300) we have

$$P = 8(\textbf{200}) + 10(\textbf{300}) = 4600$$

Although the constraints restrict the values of x and y we can consider, there are still a lot of points that represent feasible solutions. We cannot check them all to see which one results in the largest profit. How can we find the optimal solution? This question has a simple answer, but it is not easy to understand at first glance.

Let's look at the objective function, $P = 8x + 10y$. If TrailGear would like to make $2000 on hiking boots, it could produce 200 pairs of Sierra boots or 250 pairs of Weekenders. Or it could produce some of each; for example, 50 pairs of Weekenders and 160 pairs of Sierras. In fact, every point on the line $8x + 10y = 2000$ represents a combination of Weekenders and Sierras that will yield a profit of $2000. This line is labeled $P = 2000$ in Figure 3.39.

If TrailGear would like to make $4000 on boots, it should choose a point on the line labeled $P = 4000$. Similarly, all the points on the line labeled $P = 6000$ will yield a profit of $6000, and so on. Different values of P correspond to parallel lines on the graph. Smaller values of P correspond to lines near the origin, and the values of P get larger for lines farther from the origin. Here is another example.

Figure 3.40 shows the feasible solutions for a linear programming problem. The objective function is $C = 3x + 5y$.   Answer the following questions.

a.   Find the value of C at the point $(0, 3)$. Do any other feasible solutions give the same value of C?

b.   Find all feasible solutions that result in an objective value of 30.

c.   How many feasible solutions result in an objective value of 39?

d.   Is it possible for a feasible solution to result in an objective value of 45?

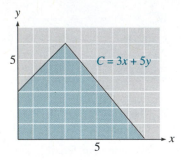

Figure 3.40

Solutions   **a.**   The objective value at the point $(0, 3)$ is

$$C = 3(0) + 5(3) = 15$$

Another point with the same objective value is $(5, 0)$. In fact, all points on the line $3x + 5y = 15$ have an objective value of 15. This line intersects the set of feasible solutions in a line segment, as shown in Figure 3.41. Thus there are infinitely many feasible solutions with objective value 15.

b.   Points that give an objective value of $C = 30$ lie on the line $3x + 5y = 30$,   as shown in Figure 3.41. Infinitely many feasible solutions lie on this line; one such point is $(5, 3)$.

c.   The line $3x + 5y = 39$ intersects the set of feasible solutions in only one point, $(3, 6)$. This point is the only feasible solution that yields an objective value of 39.

d.   The line $3x + 5y = 45$ includes all points for which $C = 45$. This line does not intersect the set of feasible solutions, as we see in Figure 3.41. Thus there are no feasible solutions that result in an objective value of 45.

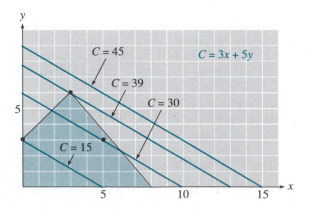

Figure 3.41

Now recall that we are allowed to choose points only from the set of feasible solutions. Imagine the parallel lines representing different values of the objective function sweeping across the graph of the feasible solutions. The objective values increase as the lines sweep across the graph. What is the last feasible solution the lines intersect before leaving the shaded region? If you study the examples above, perhaps you can see that the largest (and smallest) values of the objective function will occur at corner points of the set of feasible solutions. We have not proved this fact, but it is true.

> The maximum and minimum values of the objective function always occur at vertices of the graph of feasible solutions.

Depending on the exact formula for the objective function, the maximum and minimum values may occur at *any* of the vertices of the shaded region.

Example 3

Figure 3.42 shows the feasible solutions for a linear programming problem. The objective function is $R = x + 5y$. Answer the following questions.
a. Sketch lines for objective values of $R = 5$, $R = 15$, $R = 25$, and $R = 35$.
b. Evaluate the objective function at each vertex of the shaded region.
c. Which vertex corresponds to the maximum value of the objective function? What is the maximum value?
d. Which vertex corresponds to the minimum value of the objective function? What is the minimum value?

Solutions **a.** Sketch the lines

$$x + 5y = 5, x + 5y = 15,$$
$$x + 5y = 25, \text{ and } x + 5y = 35$$

on the set of feasible solutions, as shown in Figure 3.43.

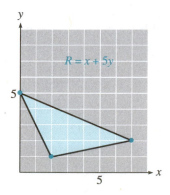

Figure 3.42

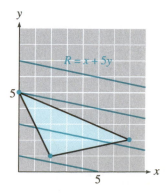

Figure 3.43

b. At $(2, 1)$, $R = 2 + 5(1) = 7$

    At $(0, 5)$, $R = 0 + 5(5) = 25$

    At $(7, 2)$, $R = 7 + 5(2) = 17$

c. The maximum value of R occurs at the point $(0, 5)$. The maximum value is 25.

d. The minimum value of R occurs at the point $(2, 1)$. The minimum value is 7.

We can now formulate a strategy for solving problems by linear programming.

> **TO SOLVE A LINEAR PROGRAMMING PROBLEM**
>
> **1.** Represent the unknown quantities by variables. Write the objective function and the constraints in terms of the variables.
>
> **2.** Graph the solutions to the constraint inequalities.
>
> **3.** Find the coordinates of each vertex of the solution set.
>
> **4.** Evaluate the objective function at each vertex.
>
> **5.** The maximum and minimum values of the objective function occur at vertices of the set of feasible solutions.

In Example 4 the set of feasible solutions is an unbounded region.

Example 4

Each week the Healthy Food Store buys both granola and muesli in bulk from two cereal companies. The store requires at least 12 kilograms of granola and 9 kilograms of muesli. Company A charges $15 for a package that contains 2 kilograms of granola and 1 kilogram of muesli. Company B charges $25 for a package of 3 kilograms of granola and 3 kilograms of muesli. How much should the Healthy Food Store purchase from each company in order to minimize its costs and still meet its needs for granola and muesli? What is the minimum cost?

Solution

Step 1   Number of packages purchased from Company A: x
             Number of packages purchased from Company B: y

First write the objective function. The store would like to minimize its cost, so

$$C = 15x + 25y$$

Next write the constraints. These will be a system of inequalities. It may help to organize the information into a table such as Table 3.3.

	Company A	Company B	Required
Granola	$2x$	$3y$	12
Muesli	x	$3y$	9

Table 3.3

The Healthy Food Store will have $2x$ kilograms of granola and x kilograms of muesli from Company A, and $3y$ kilograms of granola and $3y$ kilograms of muesli from Company B. The store's requirements are that

$$2x + 3y \geq 12$$
$$x + 3y \geq 9$$

Because the store cannot purchase negative quantities, we also have

$$x \geq 0, \quad y \geq 0$$

Step 2 Graph the solutions to the constraint system. The feasible solutions form the shaded region in Figure 3.44. Any ordered point on this graph corresponds to a way to purchase granola and muesli that meets the store's needs, but some of these choices cost more than others.

Step 3 We know that the minimum cost will occur at one of the vertex points, which are labeled in Figure 3.44. The coordinates of P and R are easy to see. To find the coordinates of Q, we notice that it is the intersection of the lines $2x + 3y = 12$ and $x + 3y = 9$. Thus we must solve the system

$$2x + 3y = 12$$
$$x + 3y = 9$$

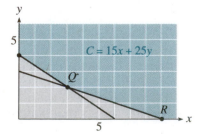

Figure 3.44

Subtracting the second equation from the first, we find that $x = 3$. Substituting this value into either of the original two equations we can find that $y = 2$. Thus the point Q has coordinates $(3, 2)$.

Step 4   Now we evaluate the objective function at each of the three vertices.

At P (0, 4), $C = 15 (0) + 25 (4) = 100$

At Q (3, 2), $C = 15 (3) + 25 (2) = 95$ Minimum cost.

At R (9, 0), $C = 15 (9) + 25 (0) = 135$

The minimum cost occurs at point Q.

Step 5   The Healthy Food Store should buy 3 packages from Company A and 2 packages from Company B. It will pay $95 for its stock of granola and muesli.   ☐

Using a Graphing Calculator

You can use your graphing calculator to solve the problem in Example 4. Set your ⟨WINDOW⟩ values at

$$\text{Xmin} = 0, \quad \text{Xmax} = 9.4$$

$$\text{Ymin} = 0, \quad \text{Ymax} = 6.2$$

(These values ensure that you will have a "nice" viewing window.) Next graph the set of feasible solutions. We have already taken care of the constraints $x \geq 0$ and $y \geq 0$ by setting Xmin and Ymin to zero. Solve each of the other constraints for y to get

$$y \geq \frac{12 - 2x}{3} \qquad \text{and} \qquad y \geq \frac{9 - x}{3}$$

We will graph just the boundary lines for the feasible solutions, so press ⟨Y =⟩ and enter

$$Y_1 = (12 - 2X)/3$$

$$Y_2 = (9 - X)/3$$

and then ⟨GRAPH⟩. Your display should look like Figure 3.45.

The set of feasible solutions lies *above* each of the boundary lines because in each constraint y is greater than the expression in x. Use the ⟨TRACE⟩ to find the coordinates of one of the vertices, say (0, 4). We can use the calculator to evaluate the objective function at that vertex.

First ⟨QUIT⟩ the graphing screen to get back to the home screen. We will enter the formula for the objective function by keying in

$$15X + 25Y$$

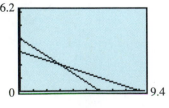

Figure 3.45

(Note that we enter Y by pressing [ALPHA] [1].) Your calculator has stored the values $x = 0$ and $y = 4$ from the [TRACE] key, so all you have to do now is press [ENTER], and the calculator returns 100 for the value of C. Thus when $x = 0$ and $y = 4$, $C = 100$.

Now we will evaluate the objective function at the other vertices. Press [TRACE] to get the graph back, and move the bug to another vertex point, say (9, 0). Then [QUIT] to get back to the home screen. Press [ENTER], and the calculator evaluates the objective function at (9, 0) to get 135. Thus when $x = 9$ and $y = 0$, $C = 135$. Repeat the process to evaluate the objective function at the last vertex: press [TRACE] and position the bug at the intersection of the two boundary lines, (3, 2). Then [QUIT] and press [ENTER] to see that when $x = 3$ and $y = 2$, $C = 95$.

As before, we find that the minimum cost of $95 occurs when $x = 3$ and $y = 2$.

Exercise 3.6

For Exercises 1–4 find the minimum value of the cost $C = 3x + 4y$ subject to the following constraints:

$$x + y \geq 10, \quad x \leq 8, \quad y \leq 7, \quad x \geq 0, \quad y \geq 0$$

The graph of the feasible solutions is shown in Figure 3.46.

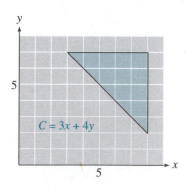

Figure 3.46

1. Use a graph to explain why it is impossible in this situation to have a cost as low as $12. (*Hint:* Draw the graph of $12 = 3x + 4y$ together with the graph of the feasible solutions.)

2. Use a graph to explain why the cost will not be as great as $60. (*Hint:* Draw the graph of $60 = 3x + 4y$ together with the graph of the feasible solutions.)

3. Use a graph to determine which vertex of the shaded region will correspond to the minimum cost. What is the minimum cost?

4. Use a graph to determine which vertex of the shaded region will correspond to the maximum cost. What is the maximum cost?

For Exercises 5–8 find the minimum value of the profit

$$P = 4x - 2y$$

subject to the following constraints:

$$5x - y \geq -2, \quad x + y \leq 8, \quad x \geq 0, \quad y \geq 0$$

The graph of the feasible solutions is shown in Figure 3.47.

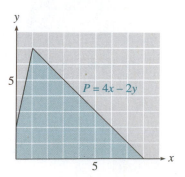

Figure 3.47

5. Graph the line that corresponds to a profit of $8. Find the coordinates of at least one feasible solution that gives a profit of $8.

6. Graph the line that corresponds to a profit of $22. Find the coordinates of at least one feasible solution that gives a profit of $22.

7. a. Which line is farther from the origin: the line for a profit of $8 or the line for a profit of $22?

b. Use a graph to determine which vertex corresponds to a maximum profit.

c. Find the maximum profit.

8. a. Use a graph to determine which vertex corresponds to a minimum profit.

b. Find the minimum profit.

For Exercises 9–12 objective functions and the graphs of the feasible solutions are given.
a.   Use the graph to find the vertex that yields the minimum value of the objective function.
b.   Find the minimum value.
c.   Use the graph to find the vertex that yields the maximum value of the objective function.
d.   Find the maximum value.

9. $C = 3x + y$

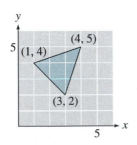

Figure 3.48

10. $C = x + 4y$

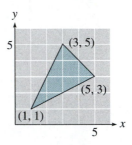

Figure 3.49

11. $C = 5x - 2y$

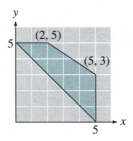

Figure 3.50

12. $C = 2x - y$

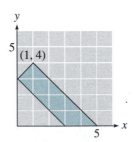

Figure 3.51

For Exercises 13–20
a. graph the set of feasible solutions.
b. find the vertex that gives the minimum of the objective function, and find the minimum value.
c. find the vertex that gives the maximum of the objective function, and find the maximum value.

13. Objective function $C = 3x + 2y$ with constraints $x \geq 0$, $y \geq 0$, $2x + y \leq 8$, $4x + 6y \leq 24$.

14. Objective function $C = -2x + y$ with constraints $x \geq 0$, $y \geq 0$, $x - 2y \geq -10$, $2x + y \leq 10$.

15. Objective function $C = 3x - y$ with constraints $x \geq 0$, $y \geq 0$, $x + y \leq 14$, $5x + y \leq 50$.

16. Objective function $C = 5x + 4$ with constraints $x \geq 0$, $y \geq 0$, $2x + y \leq 10$, $x - 3y \geq -3$.

17. Objective function $C = 200x - 20y$ with constraints $x \geq 0$, $y \geq 0$, $3x + 2y \leq 24$, $x + y \leq 9$, $x + 2y \leq 16$.

18. Objective function $C = 54x + 24y$ with constraints $x \geq 0$, $y \geq 0$, $3x + 2y \leq 24$, $3x - y \leq 15$, $3x - 4y \geq -12$.

19. Objective function $C = 18x + 48y$ with constraints $x \geq 0$, $y \geq 0$, $3x + y \geq 3$, $2x + y \leq 12$, $x + 5y \leq 15$.

20. Objective function $C = 10x - 8y$ with constraints $x \geq 0$, $y \geq 0$, $5x - y \geq 2$, $x + 2y \leq 18$, $x - y \leq 3$.

For Exercises 21–26 solve each linear programming problem by graphing.

21. The math club is selling tickets for a show by a "mathemagician." Student tickets will cost $1 and faculty tickets will cost $2. The ticket receipts must be at least $250 to cover the performer's fee. An alumnus promises to donate one calculator for each student ticket sold and three calculators for each faculty ticket sold. What is the minimum number of calculators that the alumnus will donate?

22. The math department is having a book sale of old textbooks to raise at least $300 for scholarships. Paperback textbooks will cost $2 and hardcover textbooks will cost $5. If the paperback textbooks weigh 2 pounds each and the hardcover textbooks weigh 3 pounds each, find the minimum weight of textbooks the department must sell in order to raise its required funds.

23. Jeannette has 180 acres of farmland for growing wheat or soybeans. Each acre of wheat requires 2 hours of labor at harvest time, and each acre of soybeans requires 1 hour of labor. She will have 240 hours of labor available at harvest time. Find the maximum profit Jeannette can make from her two crops if she can get a profit of $36 per acre for wheat and $24 per acre for soybeans.

24. Vassilis has at most $10,000 to invest in two banks. Alpha Bank will pay 6% annual interest and Bank Beta pays 5% annual interest. Alpha Bank will only insure up to $6000, so Vassilis will invest no more than that with Alpha. What is the maximum amount of interest Vassilis can earn in 1 year?

25. Gary's pancake recipe includes corn meal and whole wheat flour. Corn meal has 2.4 grams of linoleic acid and 2.5 milligrams of niacin per cup. Whole wheat flour has 0.8 gram of linoleic acid and 5 milligrams of niacin per cup. These two dry ingredients do not exceed 3 cups total. They combine for at least 3.2 grams of linoleic acid and at least 10 milligrams of niacin. Minimize the number of calories possible in the recipe if corn meal has 433 calories per cup and whole wheat flour has 400 calories per cup.

26. Cho requires 1 hour of cutting and 2 hours of sewing to make a Batman costume. He requires 2 hours of cutting and 1 hour of sewing to make a Wonder Woman costume. At most, 10 hours per day are available for cutting and 8 hours per day are available for sewing. At least one costume must be made each day to stay in business. Find Cho's maximum income from selling 1 day's costumes if a Batman costume costs $68 and a Wonder Woman costume costs $76.

For Exercises 27–32 use a graphing calculator to find approximate values for the maximum and minimum of the objective function.

27. Objective function $C = 8.7x - 4.2y$ with constraints $x \geq 0,\ y \geq 0,\ 1.7x - 4.5y \geq -9,$ $14.3x + 10.9y \leq 28.6.$

28. Objective function $C = -142x + 83y$ with constraints $x \geq 0,\ y \geq 0,\ 21x - 49y \geq -147,$ $19x + 21y \leq 171.$

29. Objective function $C = 5.3x + 4.2y$ with constraints $x \geq 0,\ y \geq 0,\ 18x + 17y \leq 284,$ $51x + 11y \leq 656.$

30. Objective function $C = 5.3x + 4.2$ with constraints $x \geq 0,\ y \geq 0,\ 2.5x + 1.7y \leq 20.1,$ $0.09x - 0.31y \geq -0.39.$

31. Objective function $C = 202x + 220y$ with constraints $x \geq 0,\ y \geq 0,\ 38x + 24y \leq 294,$ $35x + 34y \leq 310,\ 13x + 29y \leq 197.$

32. Objective function $C = 54x + 24y$ with constraints $x \geq 0,\ y \geq 0,\ 43x + 32y \leq 333,$ $23x - 9y \leq 152,\ 73x - 94y \geq -296.$

Chapter 3 Review

Solve each system in Exercises 1–2 by graphing. Use the ZDecimal window.

1. $y = -2.9x - 0.9$
$y = 1.4 - 0.6x$

2. $y = 0.6x - 1.94$
$y = -1.1x + 1.29$

Solve each system in Exercises 3–6 using elimination.

3. $x + 5y = 18$
$x - y = -3$

4. $x + 5y = 11$
$2x + 3y = 8$

5. $\frac{2}{3}x - 3y = 8$

$x + \frac{3}{4}y = 12$

6. $3x = 5y - 6$
$3y = 10 - 11x$

Decide whether each system in Exercises 7–10 is inconsistent, dependent, or consistent and independent.

7. $2x - 3y = 4$
$x + 2y = 7$

8. $2x - 3y = 4$
$6x - 9y = 4$

9. $2x - 3y = 4$
$6x - 9y = 12$

10. $x - y = 6$
$x + y = 6$

Solve each system in Exercises 11–16 using Gaussian reduction.

11. $x + 3y - z = 3$
$2x - y + 3z = 1$
$3x + 2y + z = 5$

12. $x + y + z = 2$
$3x - y + z = 4$
$2x + y + 2z = 3$

13. $x + z = 5$
$y - z = -8$
$2x + z = 7$

14. $x + 4y + 4z = 0$
$3x - 2y + z = -10$
$2x - 4y + z = -11$

15. $\frac{1}{2}x + y + z = 3$

$x - 2y - \frac{1}{3}z = -5$

$\frac{1}{2}x - 3y - \frac{2}{3}z = -6$

16. $\frac{3}{4}x - \frac{1}{2}y + 6z = 2$

$\frac{1}{2}x + y - \frac{3}{4}z = 0$

$\frac{1}{4}x + \frac{1}{2}y - \frac{1}{2}z = 0$

Solve Exercises 17–22 using two or three variables.

17. A math contest exam has 40 questions. A contestant scores 5 points for each correct answer but loses 2 points for each wrong answer. Lupe answered all the questions and her score was 102. How many questions did she answer correctly?

18. A game show contestant wins $25 for each correct answer he gives but loses $10 for each incorrect response. Roger answered 24 questions and won $355. How many answers did he get right?

19. Barbara wants to earn $500 a year by investing $5000 in two accounts, a savings plan that pays 8% annual interest and a high-risk option that pays 13.5% interest. How much should she invest in each account?

20. An investment broker promises his client a 12% return on her funds. If the broker invests $3000 in bonds paying 8% interest, how much must he invest in stocks paying 15% interest to keep his promise?

21. The perimeter of a triangle is 30 centimeters. The length of one side is 7 centimeters shorter than the second side, and the third side is 1 centimeter longer than the second side. Find the length of each side.

22. A company ships its product to three cities: Boston, Chicago, and Los Angeles. The cost of shipping is $10 per crate to Boston, $5 per crate to Chicago, and $12 per crate to Los Angeles. The company's shipping budget for April is $445. It has 55 crates to ship, and demand for its product is twice as high in Boston as in Los Angeles. How many crates should the company ship to each destination?

In Exercises 23–26 graph each inequality.

23. $3x - 4y < 12$

24. $x > 3y - 6$

25. $y < -\dfrac{1}{2}$

26. $-4 \le x < 2$

In Exercises 27–30 graph the solutions to each system of inequalities.

27. $y > 3, \quad x \le 2$

28. $y \ge x, \quad x > 2$

29. $3x - y < 6, \quad x + 2y > 6$

30. $x - 3y > 3, \quad y < x + 2$

In Exercises 31–34 graph the solutions to the system of inequalities and find the coordinates of the vertices.

31. $3x - 4y \le 12$
$\quad x \ge 0, \quad y \le 0$

32. $x - 2y \le 6$
$\quad y \le x$
$\quad x \ge 0, \quad y \ge 0$

33. $x + y \le 5$
$\quad y \ge x$
$\quad y \ge 2, \quad x \ge 0$

34. $x - y \le -3$
$\quad x + y \le 6$
$\quad x \le 4$
$\quad x \ge 0, \quad y \ge 0$

35. Ruth wants to provide cookies for the customers at her video rental store. Each batch of peanut butter cookies takes 20 minutes of mixing the ingredients and 10 minutes of baking time. Each batch of granola cookies takes 8 minutes of mixing and 10 minutes of baking. Because of energy considerations, Ruth cannot use the oven more than 2 hours a day, and she will not spend more than 2 hours a day mixing ingredients. Write a system of inequalities for the number of batches of peanut butter and of granola cookies that Ruth can make in 1 day, and graph the solutions.

36. A vegetarian recipe calls for a combination of tofu and brown rice. The tofu has 2 grams of protein per ounce and the rice has 1.6 grams of protein per ounce. Suppose the total of tofu and rice should be no more than 32 ounces and the protein from them must be at least 56 grams. Write a system of inequalities for the amounts of tofu and rice for the recipe, and graph the solutions.

37. Suppose Ruth (from Exercise 35) decides to sell the cookies. She can charge 25¢ per peanut butter cookie and 20¢ per granola cookie and sell all that she bakes. How many batches of each type of cookie should she bake to maximize her income from the cookies if each batch contains 50 cookies?

38. Suppose tofu costs 12¢ per ounce and brown rice costs 16¢ per ounce. What would be the least expensive combination of tofu and brown rice for the vegetarian recipe of Exercise 36?

39. There had been one incident of space shuttle "O-ring" thermal distress when the temperature was 70°F and three incidents when the temperature was 54°F. Use linear extrapolation to estimate the number of incidents of O-ring thermal distress on a morning with a temperature of 30°F. (The space shuttle *Challenger* exploded because of O-ring failures on a morning when the temperature was about 30°F.)

40. Thelma typed a 19-page technical report in 40 minutes. She required only 18 minutes for an 8-page technical report. Use linear interpolation to estimate how long Thelma would need to type a 12-page technical report.

41. The scatterplot in Figure 3.52 on page 201 shows the metabolic rate of various members of the women's swim team compared with the athlete's body mass in kilograms.

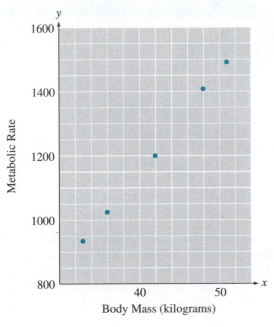

Figure 3.52

a. Predict the metabolic rate of a woman swimmer whose body mass is 30 kilograms.

b. Predict the metabolic rate of a woman swimmer whose body mass is 50 kilograms.

c. Use your answers from parts (a) and (b) to approximate the equation of a regression line.

d. Use your answer to part (c) to predict the metabolic rate of a woman swimmer with a body mass of 45 kilograms.

e. Use your calculator and the points given on Figure 3.52 to find the least squares regression line. Compare the score this equation gives for part (d) with what you predicted earlier. The ordered pairs defining the data are (33, 935), (36, 1027), (42, 1202), (48, 1409), (51, 1489).

42. An archaeopteryx is an extinct beast with characteristics of both birds and reptiles. Only six fossil specimens are known, and only five of those include both a femur (leg bone) and a humerus (forearm bone) The scatterplot in Figure 3.53 shows the lengths of femur and humerus for the five archaeopteryx specimens.

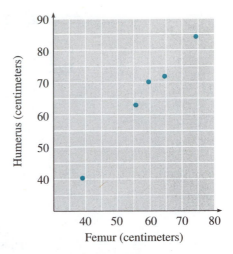

Figure 3.53

a. Predict the humerus length of an archaeopteryx whose femur is 40 centimeters.

b. Predict the humerus length of an archaeopteryx whose femur is 75 centimeters.

c. Use your answers from parts (a) and (b) to approximate the equation of a regression line.

d. Use your answer to part (c) to predict the humerus length of an archaeopteryx whose femur is 60 centimeters.

e. Use your calculator and the points given on Figure 3.53 to find the least squares regression line. Compare the score this equation gives for part (d) with what you predicted earlier. The ordered pairs defining the data are (38, 41), (56, 63), (59, 70), (64, 72), (74, 84).

Quadratic Models

$\mathcal{I}$n Chapters 2 and 3 we considered problems that can be modeled by linear equations or systems of linear equations. However, not all situations can be described by linear models. In this chapter we will investigate some problems that are best described by **quadratic** equations. These are equations that involve the square of the variable; that is, they have the form

$$ax^2 + bx + c = 0$$

As part of our study, we will learn how to solve quadratic equations and how to graph quadratic equations in two variables.

4.0 SKILLS REVIEW

In order to analyze the models in this chapter, you must be able to find products of monomials and binomials and to factor certain quadratic expressions. Check your skills by completing Exercise 4.0. If you need to review any of these operations, consult Appendix A.2, Review of Products and Factoring.

Exercise 4.0

In Exercises 1–12 write each product as a polynomial in simplest form.

1. $3x(x - 5)$

2. $-5a(2a + 3)$

3. $(b + 6)(2b - 3)$

4. $(3z - 8)(4z - 1)$

5. $(4w - 3)^2$

6. $(2d + 8)^2$

7. $3p(2p - 5)(p - 3)$

8. $2v(v + 4)(3v - 4)$

9. $-50(1 + r)^2$

10. $12(1 - t)^2$

11. $3q^2(2q - 3)^2$

12. $-5m^2(3m + 4)^2$

Factor Exercises 13–30 completely.

13. $x^2 - 7x + 10$

14. $x^2 - 7x + 12$

15. $x^2 - 225$

16. $x^2 - 121$

17. $w^2 - 4w - 32$

18. $w^2 + 5w - 150$

19. $2z^2 + 11z - 40$

20. $5z^2 - 28z - 12$

21. $9n^2 + 24n + 16$ **22.** $4n^2 - 28n + 49$ **27.** $-10u^2 - 100u + 390$

23. $3a^4 + 6a^3 + 3a^2$ **24.** $2a^3 - 12a^2 + 18a$ **28.** $270 - 15u + 5u^2$

25. $4h^4 - 36h^2$ **26.** $80h - 5h^3$ **29.** $24t^4 + 6t^2$ **30.** $27t^3 + 75t$

4.1   SOME EXAMPLES OF QUADRATIC MODELS

We will start by investigating several situations that can be modeled by quadratic equations.

Do all rectangles with the same perimeter, say 36 inches, have the same area? Two different rectangles with perimeter 36 inches are shown in Figure 4.1. Notice that the rectangle with base 10 inches and height 8 inches has an area of 80 square inches, and the rectangle with base 12 inches and height 6 inches has an area of 72 square inches.

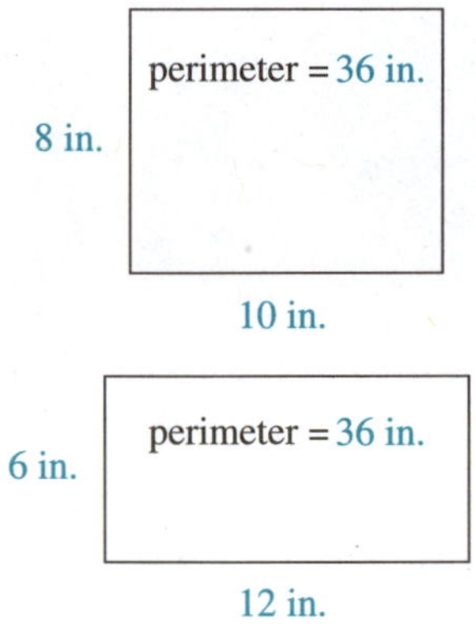

1. Fill in the values for Table 4.1 on page 204. All the rectangles have perimeters of 36 inches.

2. We are interested in what happens to the area of the rectangle when we change its base. On the grid in Figure 4.2, plot the points with coordinates (Base, Area). The first two points, (1, 17) and (2, 32), are shown. Connect your data points with a smooth curve.

3. What are the coordinates of the highest point on your graph?

4. Notice that each point on your graph represents a particular rectangle with perimeter 36 inches. The first coordinate of the point gives the base of the rectangle, and the second coordinate gives the area of the rectangle. What is the largest area you found among rectangles with perimeter 36 inches? What is the base for that rectangle? What is its height?

5. Describe the rectangle corresponding to the point (13, 65).

6. Find two points on your graph with y-coordinate 80.

7. If the rectangle has area 80 square inches, what is its base? Why are two different answers appropriate here? Describe the rectangle corresponding to each answer.

8. Let x represent the base of the rectangle. Express the height of the rectangle in terms of x. (*Hint:* If the perimeter of the rectangle is 36 inches, what is the sum of the base and the height?) Now write an expression for the area of the rectangle in terms of x.

9. Use your formula from part (8) to compute the area of the rectangle when the base is 5 inches. Does your answer agree with the values in your table and the point on your graph?

perimeter = 36 in.

8 in.

10 in.

perimeter = 36 in.

6 in.

12 in.

Figure 4.1

Base	Height	Area
1	17	17
2	16	32
3		
4		
5		
6		
7		
8		
9		
10		
11		
12		
13		
14		
15		
16		
17		

Table 4.1

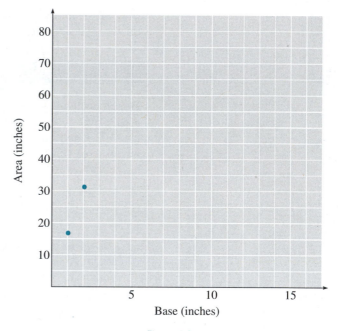

Figure 4.2

10. Use your formula to compute the area of the rectangle when $x = 0$ and when $x = 18$. Describe the "rectangles" that correspond to these data points.

11. Continue your graph to include the points corresponding to $x = 0$ and $x = 18$.

Investigation #2

Suppose a baseball player "pops up," that is, hits the baseball straight up into the air. The height, h, of the baseball t seconds after it leaves the bat can be calculated using a formula from physics that takes into account the initial speed of the ball and the height at which it was hit. The formula for the height of the ball (in feet) is

$$h = -16t^2 + 64t + 4$$

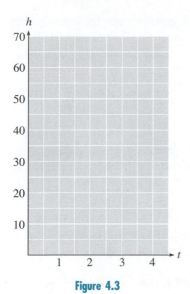

Figure 4.3

1. On the grid in Figure 4.3, graph the height of the baseball versus time. Plot data points for $t = 0, 1, 2, 3,$ and 4 seconds, and connect the points with a smooth curve.

2. What are the coordinates of the highest point on the graph? When does the baseball reach its maximum height? What is that height?

3. Use the formula to find the height of the baseball after 1/2 second.

4. Check that your answer to part (3) corresponds to a point on your graph. Approximate from your graph another time at which the baseball is at the same height as your answer to part (3).

5. Use your graph to find two times when the baseball is at a height of 64 feet.

6. Use your graph to approximate two times when the baseball is at a height of 20 feet. Then use the formula to find the actual heights at those times.

7. Suppose the catcher catches the baseball at a height of 4 feet, before it strikes the ground. At what time was the ball caught?

 Investigation #3

The local theater group sold tickets to its opening night performance for $5 and drew an audience of 100 people. The next night they reduced the ticket price by $0.25 and 10 more people attended; that is, 110 people bought tickets at $4.75 apiece. In fact, for each $0.25 reduction in ticket price, 10 additional tickets were sold.

1. Make a table with the following column headings. The first two rows are filled in for you.

Number of Price Reductions	Price of Ticket	Number of Tickets Sold	Total Revenue
0	5.00	100	500.00
1	4.75	110	522.50

Table 4.2

Continue your table until the price of a ticket drops to $2.50.

2. Use your table to make a graph. Plot "Total Revenue" on the vertical axis and "Number of Price Reductions" on the horizontal axis.

3. Let x represent the "Number of Price Reductions," as in the first column of Table 4.2. Write algebraic expressions in terms of x for the price of a ticket after x price reductions, the number of tickets sold at that price, and the total revenue from ticket sales.

4. Verify that your algebraic expressions agree with the entries in your table.

5. Use your graphing calculator to graph an equation for total revenue in terms of x, using your expression from part (3). Experiment with the WINDOW settings until you get a good window for the graph, showing the high point of the graph and both x-intercepts.

6. What is the maximum revenue possible from ticket sales? What price should the theater group charge for a ticket to generate that revenue? How many tickets will they sell at that price?

The equations we considered in the three preceding investigations are all examples of quadratic equations. A **quadratic equation** is one that can be written in the standard form

$$y = ax^2 + bx + c$$

where a, b, and c are constants, and a is not equal to zero. (If a is zero, then there is no x-squared term. It is the x-squared term that makes the equation quadratic.)

A graphing calculator can be very helpful for studying quadratic equations.

 *Investigation #4*

An acrobat is catapulted into the air from a springboard at ground level. His height, h, in meters is given by the formula

$$h = -4.9t^2 + 14.7t \tag{1}$$

where t is the time in seconds from launch. We will use the calculator to graph the acrobat's height versus time. Set the WINDOW values on your calculator to

$$Xmin = 0, \quad Xmax = 4.7$$

$$Ymin = 0, \quad Ymax = 12$$

The graph of Equation (1) is shown in Figure 4.4.

Figure 4.4
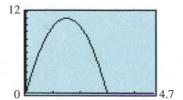

1. Use the TRACE key to find the coordinates of the highest point on the graph. When does the acrobat reach his maximum height? What is that height?

2. Use the formula to find the height of the acrobat after 2.4 seconds.

3. Use the TRACE key to verify your answer to part (2). Find another time when the acrobat is at the same height.

4. Use the TRACE key to find two times when the acrobat is at a height of 6.125 meters.

5. What are the coordinates of the horizontal intercepts of your graph? What do these points have to do with the acrobat?

6. Use the ZOOM feature to estimate the time when the acrobat is at a height of 10 meters on the way up. Repeat the zooming in process until you can estimate the time to two decimal places. (To review the procedure for zooming in, see page 139.)

Using the Table Feature on a Graphing Calculator

We can use the graphing calculator to make a table of values for an equation in two variables. Let's consider again the quadratic model from Investigation #2:

$$h = -16t^2 + 64t + 4$$

Begin by entering the equation: press the Y = key, then define $Y_1 = -16X^2 + 64X + 4$, and clear out any other definitions.

Next we must choose the settings we want for the table. We will make a table starting with $x = 0$ and with increments of one unit in the x-values. Press 2nd WINDOW to access the **TblSet** (table set-up) menu and set it to look like Figure 4.5(a). This setting will give us an initial x-value of 0 (that is, **TblMin = 0**) and an x-increment of 1 (**ΔTbl = 1**). It also fills in values of both the independent and dependent variables automatically. Now press 2nd GRAPH to see the table of values, as shown in Figure 4.5(b). From this table we can check the heights we found in parts (1) and (2) of Investigation #2.

To answer the rest of the questions in Investigation #2, we would need a more detailed table of values. Go back to the table set-up (2nd WINDOW) and change the value of **ΔTbl** to 0.5. Now when we return to the table

Figure 4.5

(a)

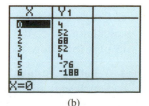
(b)

([2nd] [GRAPH]), we get the baseball's height at every half second. Use the [△] and [▽] keys to "scroll" up and down this table and verify your answers to parts (3) through (7).

Exercise 4.1

1. Suppose you want to enclose a rectangular area with a 36-inch tape measure and you can use a wall of the room for one side of the rectangle, as shown in Figure 4.6.

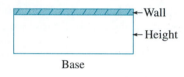

Figure 4.6

a. Make a table with these columns:

Height	Base	Perimeter	Area

Fill it in for all possible integer heights, starting with 1 inch.

b. Make a graph with "Height" on the horizontal axis and "Area" on the vertical axis. Draw a smooth curve through your data points. (Use your table to help you decide on appropriate scales for the axes.)

c. Let x stand for the height of a rectangle, and write algebraic expressions for the rectangle's base, perimeter, and area.

d. Verify that your algebraic expressions agree with your table entries.

e. Graph your formula for area on your graphing calculator. Does the graph agree with your graph in part (b)? How is it different?

f. What is the area of the largest rectangle you can enclose in this way? What are its dimensions? Label the point on your graph that corresponds to this rectangle with the letter M.

g. What is the height of the rectangle whose area is 149.5 square inches? Label the point on your graph that corresponds to this rectangle with the letter B.

2. We are going to make an open box from a square piece of cardboard by cutting 3-inch squares from

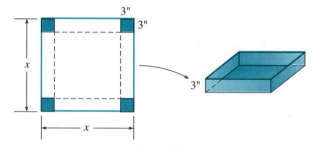

Figure 4.7

each corner and then turning up the edges, as shown in Figure 4.7.

a. Make a table showing the side of the original sheet of cardboard, the dimensions of the box created from it, and the volume of the box. (What is the side of the smallest cardboard sheet you can use?) Your table should have the following columns: Side, Length of Box, Width of Box, Height of Box, Volume of Box.

b. Make a graph with "Side" on the horizontal axis and "Volume of Box" on the vertical axis.

c. Let x represent the side of the original sheet of cardboard. Write algebraic expressions for the box's dimensions and for its volume.

d. Verify that your expression for the volume agrees with the values in your table.

e. Graph your formula for volume on your graphing calculator. Does the graph agree with your graph in part (b)? How is it different?

f. Use your graph to find out how large a square of cardboard you need to make a box with volume 126.75 cubic inches.

g. Does your graph have a highest point? What happens to the volume of the box as you increase x?

3. Delbert stands at the top of a 300-foot cliff and throws his algebra book directly upward with a velocity of 20 feet per second. The height of his book above the ground t seconds later is given by the formula $h = -16t^2 + 20t + 300$, where h is in feet.

a. Use your graphing calculator to graph the formula. Experiment with the range settings until you get a good picture.

Use the $\boxed{\text{TRACE}}$ **to answer the following questions. Use the** $\boxed{\text{ZOOM}}$ **if necessary.**

b. What is the highest altitude Delbert's book reaches? When does it reach that height?

c. When does Delbert's book pass him on its way down? (Delbert is standing at a height of 300 feet.)

d. How long will it take Delbert's book to hit the ground at the bottom of the cliff?

4. James Bond stands on top of a 240-foot building and throws a film canister upward to a fellow agent in a helicopter 16 feet above the building. The height of the film above the ground t seconds later is given by $h = -16t^2 + 32t + 240$, where h is in feet.

a. Use your graphing calculator to graph the formula. Experiment with the range settings until you get a good picture.

Use the $\boxed{\text{TRACE}}$ **to answer the following questions. Use the** $\boxed{\text{ZOOM}}$ **if necessary.**

b. How long will it take the film canister to reach the agent in the helicopter? (What is the agent's altitude?)

c. If the agent misses the canister, when will it pass James Bond on the way down?

d. How long will it take to hit the ground?

5. The owner of a motel has 60 rooms to rent. She finds that if she charges $20 per room per night, all the rooms will be rented. For every $2 that she increases the price of a room, three rooms will stand vacant.

a. Make a table with the following columns. The first two rows are filled in for you.

Number of Price Increases	Price of Room	Number of Rooms Rented	Total Revenue
0	20	60	1200
1	22	57	1254

Table 4.3

b. Let x stand for the number of $2 price increases the owner makes. Write algebraic expressions for the price of a room, the number of rooms that will be rented, and the total revenue earned at that price.

c. Use your calculator to make a table of values for your algebraic expressions. Let Y_1 stand for the price of a room, Y_2 for the number of rooms rented, and Y_3 for the total revenue. Verify the values you calculated in part (a).

d. Use your table to find a value of x that causes the total revenue to be zero.

e. Use your graphing calculator to graph your formula for total revenue.

f. What is the lowest price the owner can charge for a room if she wants her revenue to exceed $1296 per night? What is the highest price she can charge to obtain this revenue?

g. What is the maximum revenue the owner can earn in one night? How much should she charge for a room to maximize her revenue? How many rooms will she rent at that price?

6. The owner of a video store sells 96 blank tapes per week if he charges $6 per tape. For every $0.50 he increases the price, he sells four fewer tapes per week.

a. Make a table with the following columns. The first two rows are filled in for you.

Number of Price Increases	Price of Tape	Number of Tapes Sold	Total Revenue
0	6.00	96	576
1	6.50	92	598

Table 4.4

b. Let x stand for the number of $0.50 price increases the owner makes. Write algebraic expressions for the price of a tape, the number of tapes sold, and the total revenue.

c. Use your calculator to make a table of values for your algebraic expressions. Let Y_1 stand for the price of a tape, Y_2 for the number of tapes sold, and Y_3 for the total revenue. Verify the values you calculated in part (a).

d. Use your table to find a value of x for which the total revenue is zero.

e. Use your graphing calculator to graph your formula for total revenue.

f. How much should the owner charge for a tape in order to bring in $630 per week from tapes? (You should have two answers.)

g. What is the maximum revenue the owner can earn from tapes in 1 week? How much should he charge for a tape to maximize his revenue? How many tapes will he sell at that price?

4.2   SOLVING QUADRATIC EQUATIONS

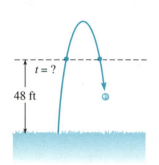

Figure 4.8

Suppose a ball thrown into the air reaches a height, h, in feet given by the formula

$$h = 64t - 16t^2 \qquad (2)$$

where t is the time in seconds after the throw. How long will it take the ball to reach a height of 48 feet on its way up?

In Section 4.1 we answered such questions by reading points from the graph of the equation. For this example we would look for a point whose h-coordinate is 48 and then read the corresponding t-coordinate, as shown in Figure 4.8. Is there a way to answer the same question algebraically?

We are looking for the value of t that results in a value for h of 48 feet. In other words, if we substitute $h = 48$ into Equation (2), we would like to solve the equation

$$48 = 64t - 16t^2$$

for t. This equation is an example of a quadratic equation in one variable. The standard form for such equations is

$$ax^2 + bx + c = 0$$

In this section and the next we discuss how to solve quadratic equations.

Solution by Factoring   If the left side of a quadratic equation in standard form can be factored, we can solve the equation by applying the following principle, called the **zero-factor principle**.

ZERO-FACTOR PRINCIPLE
The product of two factors equals zero if and only if one or both of the factors equals zero. In symbols,

$$ab = 0 \quad \text{if and only if} \quad a = 0 \quad \text{or} \quad b = 0$$

To apply the zero-factor principle, we must first write the quadratic equation in standard form; that is, one side of the equation must be zero. We then factor the other side and set each factor equal to zero separately.

Example 1

Solve $3x\,(x + 1) = 2x + 2$.

Solution   First write the equation in standard form.

$$3x\,(x + 1) = 2x + 2 \qquad \text{Apply the distributive law to the left side.}$$

$$3x^2 + 3x = 2x + 2 \qquad \text{Subtract } 2x + 2 \text{ from both sides.}$$

$$3x^2 + x - 2 = 0$$

Next factor the left side to obtain

$$(3x - 2)\,(x + 1) = 0$$

Apply the zero-factor principle: set each factor equal to zero.

$$3x - 2 = 0 \quad \text{or} \quad x + 1 = 0$$

Finally, solve each equation to find

$$x = \frac{2}{3} \quad \text{or} \quad x = -1$$

The solutions are $\frac{2}{3}$ and -1.

 COMMON ERROR

In order for us to apply the zero-factor principle, one side of the equation *must* be zero. For example, to solve the equation

$$(x - 2)\,(x - 4) = 15$$

it is *incorrect* to set each factor equal to 15! (There are many ways in which the product of two numbers can equal 15; it is not necessary that one of the numbers be 15.) We must first simplify the left side and write the equation in standard form. (The correct solutions are 7 and -1; check that you can find these solutions.)

We can use the solution-by-factoring technique to answer the question that opened this section.

Example 2

A ball thrown into the air reaches a height, h, in feet given by the formula $h = 64t - 16t^2$, where t is the time in seconds after the throw. How long will it take the ball to reach a height of 48 feet on its way up?

Solution Substitute 48 for h in the formula and solve for t.

$$48 = 64t - 16t^2 \qquad \text{Write the equation in standard form.}$$

$$16t^2 - 64t + 48 = 0 \qquad \text{Factor 16 from the left side.}$$

$$16\,(t^2 - 4t + 3) = 0 \qquad \text{Factor the quadratic expression.}$$

$$16\,(t - 3)\,(t - 1) = 0 \qquad \text{Set each variable factor equal to zero.}$$

$$t - 3 = 0 \quad \text{or} \quad t - 1 = 0 \qquad \text{Solve each equation.}$$
$$t = 3 \quad \text{or} \quad t = 1$$

The quadratic equation has two solutions, but only one of them answers the question asked: it takes 1 second for the ball to reach a height of 48 feet *on the way up*. (When $t = 3$ seconds the ball is also at a height of 48 feet, but on the way down. See Figure 4.8.)

In the solution of Example 2, note that the factor of 16 does not affect the solutions of the equation at all. You can understand why this is true by looking at some graphs. Use your graphing calculator to graph the equation

$$Y_1 = X^2 - 4X + 3$$

Notice that when $y = 0$, $x = 1$ or $x = 3$. (These two points are the x-intercepts of the graph.) Now on the same window graph

$$Y_2 = 16(X^2 - 4X + 3)$$

(See Figure 4.9.) This graph has the same x-values when $y = 0$. The factor of 16 makes the graph "skinnier" but does not change the location of the x-intercepts.

Here is another example of how quadratic equations arise in applications.

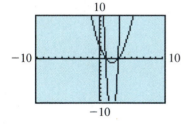

Figure 4.9

Example 3

The size of a rectangular computer monitor screen is taken to be the length of its diagonal. (See Figure 4.10.) If the length of the screen should be 3 inches greater than its width, what are the dimensions of a 15-inch monitor?

Solution Express the two dimensions of the screen in terms of a single variable:

Width of screen: w

Length of screen: $w + 3$

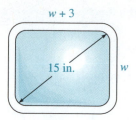

Figure 4.10

Use the Pythagorean theorem to write an equation:

$$w^2 + (w + 3)^2 = 15^2$$

Solve the equation. Begin by simplifying the left side.

$$w^2 + w^2 + 6w + 9 = 225 \qquad \text{Write the equation in standard form.}$$

$$2w^2 + 6w - 216 = 0 \qquad \text{Factor 2 from the left side.}$$

$$2\,(w^2 + 3w - 108) = 0 \qquad \text{Factor the quadratic expression.}$$

$$2\,(w - 9)\,(w + 12) = 0 \qquad \text{Set each factor equal to zero.}$$

$$w - 9 = 0 \qquad \text{or} \qquad w + 12 = 0 \qquad \text{Solve each equation.}$$

$$w = 9 \qquad \text{or} \qquad w = -12$$

Since the width of the screen cannot be a negative number, the width is 9 inches and the length is 12 inches.

Solutions of Quadratic Equations

A quadratic equation in one variable always has two solutions. However, if the left side is the square of a binomial, both solutions are the same. For example, the equation

$$x^2 - 2x + 1 = 0$$

can be solved by factoring as follows:

$$(x - 1)\,(x - 1) = 0 \qquad \text{Apply the zero-factor principle.}$$

$$x - 1 = 0 \qquad \text{or} \qquad x - 1 = 0$$

Both these equations have solution 1. We say that 1 is a solution of *multiplicity two,* meaning that it occurs twice as a solution of the quadratic equation.

Notice that the solutions of the quadratic equation

$$(x - r_1)\,(x - r_2) = 0 \tag{3}$$

are r_1 and r_2. Thus if we know the two solutions of a quadratic equation, we can work backwards and reconstruct the equation starting from its factored form as in Equation (3). We can then write the equation in standard form by multiplying the two factors.

Example 4

Find a quadratic equation whose solutions are $\frac{1}{2}$ and -3.

Solution The quadratic equation is

$$\left(x - \frac{1}{2}\right) [x - (-3)] = 0$$

or

$$\left(x - \frac{1}{2}\right) (x + 3) = 0$$

To write the equation in standard form, multiply the factors:

$$x^2 + \frac{5}{2}x - \frac{3}{2} = 0$$

We can also find an equation with integer coefficients if we clear the equation of fractions; multiply both sides by 2:

$$2\left(x^2 + \frac{5}{2}x - \frac{3}{2}\right) = 2\,(0)$$

or

$$2x^2 + 5x - 3 = 0$$

You should check that the solutions of this last equation are in fact $\frac{1}{2}$ and -3. Multiplying by a constant factor does not change the equation's solutions.

Extraction of Roots   In Section 1.4 we solved simple quadratic equations by extraction of roots; you may want to review that material now. Recall that we solved quadratic equations of the form

$$ax^2 + c = 0$$

where the linear term bx is missing, by isolating x^2 on one side of the equation and then taking the square root of each side. For example, to solve the equation

$$2x^2 - 6 = 0$$

we first solve for x^2 to get

$$x^2 = 3$$

and then take square roots to find

$$x = \pm\sqrt{3}$$

(Do not forget that every positive number has *two* square roots.)
Equations of the form

$$(x - p)^2 = q$$

can also be solved by extraction of roots.

Example 5   Solve the equation $3\,(x - 2)^2 = 48$.

Solution   First isolate the perfect square, $(x - 2)^2$.

$$3\,(x - 2)^2 = 48 \qquad \text{Divide both sides by 3.}$$
$$(x - 2)^2 = 16 \qquad \text{Take the square root of each side.}$$
$$x - 2 = \pm\sqrt{16}$$

We now have two equations for x:

$$x - 2 = 4 \quad \text{or} \quad x - 2 = -4 \qquad \text{Solve each equation.}$$
$$x = 6 \quad \text{or} \quad x = -2$$

The solutions are 6 and -2.

Exercise 4.2

In Exercises 1–10 graph each equation on a graphing calculator and then use your graph to solve the equation $y = 0$. Check your answers with the zero-factor principle.

1. $y = (2x + 5)(x - 2)$

2. $y = (x + 1)(4x - 1)$

3. $y = x(3x + 10)$ 4. $y = x(3x - 7)$

5. $y = (x - 3)(2x + 3)$

6. $y = (2x - 7)(x + 1)$

7. $y = (4x + 3)(x + 8)$

8. $y = (x - 2)(x - 9)$

9. $y = (x - 4)^2$ 10. $y = (x + 6)^2$

Solve Exercises 11–24 by factoring.

11. $2a^2 + 5a - 3 = 0$

12. $3b^2 - 4b - 4 = 0$

13. $2x^2 = 6x$ 14. $5z^2 = 5z$

15. $3y^2 - 6y = -3$ 16. $4y^2 + 4y = 8$

17. $x(2x - 3) = -1$

18. $2x(x - 2) = x + 3$

19. $t(t - 3) = 2(t - 3)$

20. $5(t + 2) = t(t + 2)$

21. $z(3z + 2) = (z + 2)^2$

22. $(z - 1)^2 = 2z^2 + 3z - 5$

23. $(v + 2)(v - 5) = 8$

24. $(w + 1)(2w - 3) = 3$

Use a graphing calculator to graph each set of equations in Exercises 25–28. Use the standard graphing window. What do you notice about the x-intercepts? Try to generalize your observation, and test your idea with a few examples.

25. a. $y = x^2 - x - 20$
    b. $y = 2(x^2 - x - 20)$
    c. $y = 0.5(x^2 - x - 20)$

26. a. $y = x^2 + 2x + 15$
    b. $y = 3(x^2 + 2x + 15)$
    c. $y = 0.2(x^2 + 2x + 15)$

27. a. $y = x^2 + 6x - 16$
    b. $y = -2(x^2 + 6x - 16)$
    c. $y = -0.1(x^2 + 6x - 16)$

28. a. $y = x^2 - 16$
    b. $y = -1.5(x^2 - 16)$
    c. $y = -0.4(x^2 - 16)$

In Exercises 29–36 write a quadratic equation in standard form, with integer coefficients whose solutions are given.

29. -2 and 1   30. -4 and 3

31. 0 and -5 32. 0 and 5

33. -3 and $\dfrac{1}{2}$ 34. $\dfrac{-2}{3}$ and 4

35. $\dfrac{-1}{4}$ and $\dfrac{3}{2}$ 36. $\dfrac{-1}{3}$ and $\dfrac{-1}{2}$

Graph each equation in Exercises 37–40 on a graphing calculator. (Use the *ZInteger* setting.) Locate the x-intercepts of the graph. Use the x-intercepts to write the quadratic expression in factored form.

37. $y = 0.1(x^2 - 3x - 270)$

38. $y = 0.1(x^2 + 9x - 360)$

39. $y = -0.08(x^2 + 14x - 576)$

40. $y = -0.06(x^2 - 22x - 504)$

Solve Exercises 41–52 by extraction of roots.

41. $(x - 2)^2 = 9$

42. $(x + 3)^2 = 4$

43. $(2x - 1)^2 = 16$

44. $(3x + 1)^2 = 25$

45. $(x + 2)^2 = 3$

46. $(x - 5)^2 = 7$

47. $\left(x - \dfrac{1}{2}\right)^2 = \dfrac{3}{4}$

48. $\left(x - \dfrac{2}{3}\right)^2 = \dfrac{5}{9}$

49. $\left(x + \dfrac{1}{3}\right)^2 = \dfrac{1}{81}$

50. $\left(x + \dfrac{1}{2}\right)^2 = \dfrac{1}{16}$

51. $(8x - 7)^2 = 8$

52. $(5x - 12)^2 = 24$

Solve Exercises 53–56 by algebraic and by graphical methods.

53. The area of an equilateral triangle is given by the formula $A = \dfrac{\sqrt{3}}{4} s^2$, where s is the length of the side.

a. Graph the equation and explain what the coordinates of a point (s, A) on the graph represent.

b. How long is the side of an equilateral triangle whose area is 12 square centimeters?

54. The area of the ring in Figure 4.11 is given by the formula $A = \pi R^2 - \pi r^2$, where R is the radius of the outer circle and r is the radius of the inner circle.

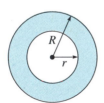

Figure 4.11

a. Graph the equation when $r = 4$. What do the coordinates of a point (R, A) on the graph represent?

b. If the area of the ring is 11π centimeters, what is the radius of the outer circle?

55. One end of a ladder is 10 feet from the base of a wall, and the other end reaches a window in the wall. The ladder is 2 feet longer than the height of the window. (See Figure 4.12.)

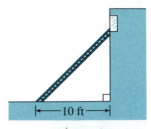

Figure 4.12

a. Write an equation about the height of the window.

b. Use your equation to find the height of the window.

56. The diagonal of a rectangle is 20 inches. One side of the rectangle is 4 inches shorter than the other side.

a. Write an equation about the length of the rectangle.

b. Use your equation to find the dimensions of the rectangle.

If an object is thrown into the air from a height s_0 above the ground with an initial velocity v_0, then its height t seconds later is given by the formula

$$h = -\frac{1}{2} gt^2 + v_0 t + s_0$$

where g is a constant that measures the force of gravity. Use this formula to answer Exercises 57 and 58.

57. a. Write a quadratic equation that gives the height of a tennis ball thrown into the air with an initial velocity of 16 feet per second from a height of 8 feet. The value of g is 32.

b. Find the height of the tennis ball at $t = \frac{1}{2}$ second and at $t = 1$ second.

c. At what time is the tennis ball 11 feet high?

d. Graph your equation and verify your answers to parts (b) and (c).

58. a. A mountain climber stands on a ledge 80 feet above the ground and tosses a rope down to a companion clinging to the rock face at a height of 17 feet above the ground. The initial velocity of the rope is -8 feet per second, and the value of g is 32. Write a quadratic equation that gives the height of the rope at time t.

b. What is the height of the rope after $\frac{1}{2}$ second? After 1 second?

c. How long does it take the rope to reach the second climber?

d. Graph your equation and verify your answers to parts (b) and (c).

59. A rancher has 360 yards of fence to enclose a rectangular pasture. If the pasture should be 8000 square yards in area, what should its dimensions be? (*Hint:* See Investigation #1 in Section 4.1.) Choose one of the following methods to solve the problem.

a. Make a table with column headings

Width	Length	Area

What is the sum of the length plus the width if there are 360 yards of fence? Use the table to find the pasture whose area is 8000 square yards.

b. Write an expression for the length of the pasture if its width is x. Next write an expression for the area A of the pasture if its width is x. Graph the equation for A and use the graph to find the pasture of area 8000 square yards.

c. Write an equation for the area A of the pasture in terms of its width, x. Solve your equation when $A = 8000$.

60. If the rancher in Exercise 59 uses a riverbank to border one side of the pasture, he can enclose 16,000 square yards with 360 yards of fence. What will the dimensions of the pasture be then? (*Hint:* See Exercise 1 in Exercise 4.1.) Choose one of the following methods to solve the problem.

a. Make a table with column headings

Width	Length	Area

(Be careful computing the length: remember that one side of the pasture does not need any fence!) Use the table to find the pasture whose area is 16,000 square yards.

b. Write an expression for the length of the pasture if its width is x. Next write an expression for the area, A, of the pasture if its width is x. Graph the equation for A and use the graph to find the pasture of area 16,000 square yards.

c. Write an equation for the area, A, of the pasture in terms of its width, x. Solve your equation when $A = 16,000$.

61. A box is made from a square piece of cardboard by cutting 2-inch squares from each corner and turning up the edges.

a. Write an expression for the volume, V, of the box if the piece of cardboard is x inches square. (See Figure 4.13.)

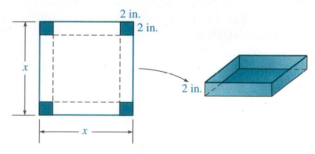

Figure 4.13

b. Graph your expression for the volume. What happens to V as x increases?

c. If the volume of the box must be 50 cubic inches, how large should the piece of cardboard be?

62. A length of rain gutter is made from a piece of aluminum 6 feet long and 1 foot wide.

a. Write an expression for the volume, V, of the gutter if a strip of width x is turned up along each long edge. (See Figure 4.14 on page 218.)

b. Graph your expression for the volume. What happens to V as x increases?

c. How much should be turned up along each long edge so that the gutter has a capacity of $\frac{3}{4}$ cubic foot of rainwater?

63. A travel agency offers a group rate of $600 per person for a weekend in Lake Tahoe if 20 people

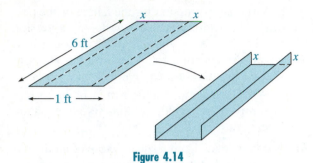

Figure 4.14

sign up. For each additional person who signs up, the price for all participants is reduced by $10 per person. (See Investigation #3 in Section 4.1.)

a. Write algebraic expressions for the size of the group and the price per person if x additional people sign up.

b. Write a quadratic equation for the travel agency's total income if x additional people sign up for the trip.

c. If 25 members of a ski club sign up for the weekend, what is the travel agency's income? What if 30 people sign up?

d. How many people must sign up in order for the agency to bring in $15,750?

e. Graph your equation on a graphing calculator and use the graph to verify your answers.

64. A farmer inherits an apple orchard on which 60 trees are planted per acre. Each tree yields 12 bushels of apples. Experimentation has shown that for each additional tree planted per acre, the yield per tree decreases by $\frac{1}{2}$ bushel. (See Investigation #3 in Section 4.1.)

a. Write algebraic expressions for the number of trees per acre and for the yield per tree if x additional trees are planted per acre.

b. Write a quadratic equation for the total yield per acre if x additional trees are planted per acre.

c. What is the yield per acre if 10 additional trees are planted per acre? Using your equation, predict the yield per acre if 10 trees per acre are removed.

d. How many trees should be planted per acre in order to harvest 880 bushels per acre?

e. Graph your equation on a graphing calculator and use your graph to verify your answers.

4.3   COMPLETING THE SQUARE; QUADRATIC FORMULA

Not every quadratic equation can be solved by factoring. For example, $x^2 + x - 1$ cannot be factored, so the equation $x^2 + x - 1 = 0$ cannot be solved by factoring. For some equations that can be solved by factoring, the correct factorization may be difficult to find because the coefficients are large numbers. In this section we will learn two methods that can be used to solve any quadratic equation.

Completing the Square   Completing the square is really an algebraic technique that allows us to apply extraction of roots to any quadratic equation. In Section 4.2 we used extraction of roots to solve equations of the form

$$(x - p)^2 = q \tag{4}$$

where the left side of the equation is the square of a binomial, or a "perfect square." We can write any quadratic equation in the form of Equation (4) by completing the square.

Consider the following squares of binomials:

1. $(x + 5)^2 = x^2 + 10x + 25$ $\qquad \frac{1}{2}(10) = 5;\quad 5^2 = 25$

2. $(x - 3)^2 = x^2 - 6x + 9$ $\qquad \frac{1}{2}(-6) = -3;\quad (-3)^2 = 9$

3. $(x - 12)^2 = x^2 - 24x + 144$ $\qquad \frac{1}{2}(-24) = -12;\quad (-12)^2 = 144$

Notice that in each resulting trinomial the constant term is equal to *the square of one-half the coefficient of x*. In other words, we can find the constant term by taking one-half the coefficient of x and then squaring the result. Obtaining the constant term in this way is called **completing the square**.

Example 1

Complete the square by adding an appropriate constant, and write the result as the square of a binomial.

a. $x^2 - 12x +$ _____ **b.** $x^2 + 5x +$ _____

Solutions **a.** One-half of -12 is -6, so the constant term is $(-6)^2$, or 36. Add 36 to obtain

$$x^2 - 12x + \mathbf{36} = (x - 6)^2$$

b. One-half of 5 is $\frac{5}{2}$, so the constant term is $\left(\frac{5}{2}\right)^2$, or $\frac{25}{4}$. Add $\frac{25}{4}$ to obtain

$$x^2 + 5x + \frac{25}{4} = \left(x + \frac{5}{2}\right)^2 \qquad \square$$

You may have noticed that one-half the coefficient of x also appears in the squared binomial. If we think of the coefficient of x as $2p$, the constant term is then p^2, and the resulting trinomial can be factored as $(x + p)^2$. In general,

$$x^2 + 2px + p^2 = (x + p)^2$$

Now let's see how to apply the technique of completing the square to solving quadratic equations. Consider the equation

$$x^2 - 6x - 7 = 0 \qquad (5)$$

Our goal is to write the left side of the equation as a perfect square. To do this, we must find the correct constant term, so we move the present constant term to the other side of the equals sign:

$$x^2 - 6x \qquad = 7$$

Now we use completing the square to fill in the blank with the correct constant term. The coefficient of x is -6, so $p = \frac{1}{2}(-6) = -3$, and $p^2 = (-3)^2 = 9$. Thus we add 9 to *both* sides of our equation to get

$$x^2 - 6x + 9 = 7 + 9$$

The left side of the equation is now the square of a binomial, namely $(x - 3)^2$, so we have

$$(x - 3)^2 = 16$$

You can check that this equation is equivalent to the original one; if you expand the left side and collect like terms, you will return to the original form of the equation. However, with this new form we can use extraction of roots to find the solutions. Taking square roots of both sides we get

$$x - 3 = 4 \qquad \text{or} \qquad x - 3 = -4$$

and finally

$$x = 7 \qquad \text{or} \qquad x = -1$$

The solutions are 7 and -1.

We can also solve Equation (5) by factoring instead of completing the square. Of course, we would obtain the same solutions by either method. However, completing the square can be used to solve equations that cannot be solved by factoring. These equations have solutions that are irrational numbers (or complex numbers).

Example 2   Solve $x^2 - 4x - 3 = 0$.

Solution   First write the equation with the constant term on the right side:

$$x^2 - 4x = 3$$

Now complete the square on the left side. The coefficient of x is -4, so $p = \dfrac{1}{2}(-4) = -2$, and $p^2 = (-2)^2 = 4$. We add 4 to both sides of the equation.

$$x^2 - 4x + 4 = 3 + 4$$

Write the left side as the square of a binomial, and combine terms on the right side:

$$(x - 2)^2 = 7$$

Finally, use extraction of roots to obtain

$$x - 2 = \sqrt{7} \qquad \text{or} \qquad x - 2 = -\sqrt{7}$$

and solve each equation for x:

$$x = 2 + \sqrt{7} \qquad \text{or} \qquad x = 2 - \sqrt{7}$$

We can use a calculator to find decimal approximations for each solution: $2 + \sqrt{7} \approx 4.646$ and $2 - \sqrt{7} \approx -0.646$.   ▭

The method we have just learned for completing the square works only if the coefficient of x^2 is 1. If we want to solve a quadratic equation whose lead coefficient is not 1, we must first divide each term of the equation by the lead coefficient.

Example 3

Solve $2x^2 - 6x - 5 = 0$.

Solution   Since the coefficient of x^2 is 2, we must divide each term of the equation by 2 to get

$$x^2 - 3x - \frac{5}{2} = 0$$

Now proceed as before. Rewrite the equation with the constant on the right side.

$$x^2 - 3x = \frac{5}{2}$$

Complete the square:

$$p = \frac{1}{2}(-3) = \frac{-3}{2} \quad \text{and} \quad p^2 = \left(\frac{-3}{2}\right)^2 = \frac{9}{4}$$

Add $\frac{9}{4}$ to both sides of the equation:

$$x^2 - 3x + \frac{9}{4} = \frac{5}{2} + \frac{9}{4}$$

Rewrite the left side as the square of a binomial and simplify the right side to get

$$\left(x - \frac{3}{2}\right)^2 = \frac{19}{4}$$

Finally, extract roots and solve each equation for x.

$$x - \frac{3}{2} = \sqrt{\frac{19}{4}} \quad \text{or} \quad x - \frac{3}{2} = -\sqrt{\frac{19}{4}}$$

The solutions are $\frac{3}{2} + \sqrt{\frac{19}{4}}$ and $\frac{3}{2} - \sqrt{\frac{19}{4}}$. Using a calculator, we can find decimal approximations for the solutions: 3.679 and -0.679.

COMMON ERROR

In Example 3 it is essential that we first divide each term of the equation by 2, the coefficient of x^2. The following attempt at a solution is *incorrect*.

$$2x^2 - 6x = 5$$

$$2x^2 - 6x + 9 = 5 + 9$$

$$(2x - 3)^2 = 14$$

You can check that $(2x - 3)^2$ is *not* equal to $2x^2 - 6x + 9$. We have not written the left side of the equation as a perfect square, so the solutions we obtain by extracting roots will not be correct.

Here is a summary of the steps for solving quadratic equations by completing the square.

TO SOLVE A QUADRATIC EQUATION BY COMPLETING THE SQUARE

1. Write the equation in standard form.
2. Divide both sides of the equation by the coefficient of the quadratic term, and subtract the constant term from both sides.
3. Complete the square on the left side:
   a. Multiply the coefficient of the first-degree term by $\frac{1}{2}$, then square the result.
   b. Add the value obtained in (a) to both sides of the equation.
4. Write the left side of the equation as the square of a binomial. Simplify the right side.
5. Use extraction of roots to finish the solution.

Quadratic Formula　Instead of following the procedure for completing the square every time we solve a new quadratic equation, we can complete the square on the general quadratic equation,

$$ax^2 + bx + c = 0$$

and obtain the following formula for the solutions.

QUADRATIC FORMULA

The solutions of the equation $ax^2 + bx + c = 0$ are

$$x = \frac{-b \pm \sqrt{b^2 - 4ac}}{2a}$$

This formula expresses the solutions of a quadratic equation in terms of its coefficients. (The proof of the formula is left to Exercises 41 and 42 on page 227.) The symbol $\pm$, read "plus or minus," is used to combine the two equations

$$x = \frac{-b + \sqrt{b^2 - 4ac}}{2a} \quad \text{and} \quad x = \frac{-b - \sqrt{b^2 - 4ac}}{2a}$$

into a single equation. All we have to do is substitute the coefficients a, b, and c into the formula to find the solutions of the equation.

Example 4

Solve $2x^2 + 1 = 4x$.

Solution Write the equation in standard form as

$$2x^2 - 4x + 1 = 0$$

Substitute **2** for a, **−4** for b, and **1** for c into the quadratic formula and simplify.

$$x = \frac{-(-4) \pm \sqrt{(-4)^2 - 4(2)(1)}}{2(2)} = \frac{4 \pm \sqrt{8}}{4}$$

Using a calculator, we find that the solutions are approximately 1.707 and 0.293. ☐

Not all quadratic equations have solutions that are real numbers.

Example 5

Solve $x^2 - \dfrac{x}{2} + 1 = 0$.

Solution Notice that for this equation, b is a fraction, $-\frac{1}{2}$. It is easier to apply the quadratic formula if the coefficients are integers, so we will first multiply both sides of the equation by 2 to clear the fraction. (Recall that multiplying both sides of an equation by the same number does not change the solutions of the equation.) This gives us

$$2x^2 - x + 2 = 0$$

For this new equation $a = \mathbf{2}$, $b = \mathbf{-1}$, and $c = \mathbf{2}$. Substitute these values into the quadratic formula to obtain

$$x = \frac{-(-1) \pm \sqrt{(-1)^2 - 4(2)(2)}}{2(2)} = \frac{1 \pm \sqrt{-15}}{4}$$

Since $\sqrt{-15}$ is not a real number, this equation does not have real-number solutions. (The solutions are called **complex numbers**. Complex numbers are considered in Appendix A.3.) ☐

We have now examined four different algebraic methods for solving quadratic equations:

1. Factoring

2. Extraction of roots

3. Completing the square

4. Quadratic formula

Factoring and extraction of roots are relatively fast and simple, but they do not work on all quadratic equations. We used completing the square to derive the quadratic formula, and either of these techniques will work on any quadratic equation.

Applications   Most often quadratic equations that arise in applications cannot be solved by factoring. In such cases we use the quadratic formula or (if we are lucky) extraction of roots.

Example 6

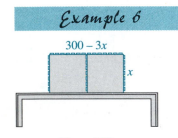

Figure 4.15

The owners of a day-care center plan to enclose a divided play area against the back wall of their building, as shown in Figure 4.15. They have 300 feet of picket fence and would like the total area of the playground to be 6000 square feet. Can they enclose the playground with the fence they have? If so, what should the dimensions of the playground be?

Solution   Suppose the width of the play area is x feet. Since three sections of fence run across the width of the play area, that leaves $300 - 3x$ feet of fence for its length, as shown in Figure 4.15. The area of the play area should be 6000 square feet, so we have the equation

$$x\,(300 - 3x) = 6000$$

This is a quadratic equation. In standard form

$$3x^2 - 300x + 6000 = 0 \qquad \text{Divide each term by 3.}$$

or

$$x^2 - 100x + 2000 = 0$$

The left side cannot be factored, so we use the quadratic formula with $a = 1$, $b = -100$, and $c = 2000$.

$$x = \frac{-(-100) \pm \sqrt{(-100)^2 - 4(1)(2000)}}{2(1)}$$

$$= \frac{100 \pm \sqrt{2000}}{2}$$

$$\approx \frac{100 \pm 44.7}{2}$$

Thus $x \approx 72.35$ or $x \approx 27.65$. Both values give feasible solutions to the problem. If the width of the play area is 72.35 feet, then the length is $300 - 3(72.35)$, or 82.95 feet. If the width is 27.65 feet, the length is $300 - 3(27.65)$, or 217.05 feet.

Compound Interest

Many savings institutions offer accounts on which the interest is compounded annually; that is, at the end of each year the interest earned that year is added to the principal, and the interest for the next year is computed on this larger sum of money. After n years the amount of money in the account is given by the formula

$$A = P(1 + r)^n$$

where P is the original principal and r is the interest rate.

Example 7

Carmella invests $3000 in an account that pays an interest rate, r, compounded annually.
a. Write an expression for the amount in Carmella's account after 2 years.
b. What interest rate would be necessary for Carmella's account to grow to $3500 in 2 years?

Solutions a. Use the formula above with $P = 3000$ and $n = 2$. Carmella's account balance will be

$$A = 3000(1 + r)^2$$

b. Substitute 3500 for A in the equation.

$$3500 = 3000(1 + r)^2$$

This is a quadratic equation in the variable r, which we can solve by extraction of roots. First isolate the perfect square.

$3500 = 3000(1 + r)^2$	Divide both sides by 3000.
$1.1\overline{6} = (1 + r)^2$	Take the square root of both sides.
$\pm 1.0801 \approx 1 + r$	Subtract 1 from both sides.

$$r \approx 0.0801 \text{ or } r \approx -2.0801$$

Since the interest rate must be a positive number, we discard the negative solution. Carmella needs an account with interest rate $r \approx 0.0801$, or over 8%, in order to have an account balance of $3500 in 2 years.

The formula we used for compound interest also applies to calculating the effects of inflation. For instance, if there is a steady inflation rate of 4% per year, then an item that costs $100 now will cost

$$A = 100(1 + 0.04)^2 = \$108.16$$

2 years from now.

Exercise 4.3

In Exercises 1–8 complete the square and write the result as the square of a binomial.

1. $x^2 + 8x$ **2.** $x^2 - 14x$ **3.** $x^2 - 7x$

4. $x^2 + 3x$ **5.** $x^2 + \dfrac{3}{2}x$ **6.** $x^2 - \dfrac{5}{2}x$

7. $x^2 - \dfrac{4}{5}x$ **8.** $x^2 + \dfrac{2}{3}x$

Solve Exercises 9–20 by completing the square.

9. $x^2 - 2x + 1 = 0$ **10.** $x^2 + 4x + 4 = 0$

11. $x^2 + 9x + 20 = 0$ **12.** $x^2 - x - 20 = 0$

13. $x^2 = 3 - 3x$ **14.** $x^2 = 5 - 5x$

15. $2x^2 + 4x - 3 = 0$ **16.** $3x^2 + x - 4 = 0$

17. $4x^2 - 3 = 2x$ **18.** $2x^2 - 5 = 3x$

19. $3x^2 - x - 4 = 0$ **20.** $2x^2 - x - 3 = 0$

Solve Exercises 21–30 using the quadratic formula. Round your answers to three decimal places.

21. $x^2 - x - 1 = 0$ **22.** $x^2 + x - 1 = 0$

23. $y^2 + 2y = 5$ **24.** $y^2 - 4y = 4$

25. $3z^2 = 4.2z + 1.5$ **26.** $2z^2 = 7.5z - 6.3$

27. $0 = x^2 - \dfrac{5}{3}x + \dfrac{1}{3}$

28. $0 = -x^2 + \dfrac{5}{2}x - \dfrac{1}{2}$

29. $-5.2z^2 + 176z + 1218 = 0$

30. $15z^2 - 18z - 2750 = 0$

Solve Exercises 31–40 using whichever technique is most convenient.

31. A car traveling at s miles per hour on a dry road surface will require approximately d feet to stop, where d is given by

$$d = \frac{s^2}{24} + \frac{s}{2}$$

a. Graph the equation for d in terms of s.

b. If a car must be able to stop in 50 feet, what is the maximum safe speed it can travel?

32. A car traveling at s miles per hour on a wet road surface will require approximately d feet to stop, where d is given by

$$d = \frac{s^2}{12} + \frac{s}{2}$$

a. Graph the equation for d in terms of s.

b. Insurance investigators at the scene of an accident find skid marks 100 feet long leading up to the point of impact. How fast was the car traveling when it put on the brakes?

33. A skydiver jumps out of an airplane at 11,000 feet. While she is in free-fall, her altitude in feet t seconds after jumping is given by

$$h = -16t^2 - 16t + 11{,}000$$

a. If she must open her parachute at 1000 feet, how long can she free-fall?

b. If the skydiver drops a marker just before she opens her parachute, how long will it take the marker to hit the ground?

c. Graph the equation for h in terms of t. Find the points on the graph corresponding to your answers to parts (a) and (b).

34. A high diver jumps from the 10-meter springboard. His height in meters above the water t seconds after leaving the board is given by

$$h = -4.9t^2 + 8t + 10$$

a. How long is it before the diver passes the board on the way down?

b. How long is it before the diver hits the water?

c. Graph the equation for h in terms of t. Find the points on the graph corresponding to your answers to parts (a) and (b).

35. Cyril plans to invest $5000 in a money market account paying interest compounded annually. If he would like to have $6250 in 2 years, what interest rate must the account pay?

36. Two years ago Carol's living expenses were $1200 per month. This year the same items cost Carol $1400 per month. What has the annual inflation rate been for the past 2 years?

37. A dog trainer has 100 meters of chain-link fence. She wants to enclose 250 square meters in three pens of equal size, as shown in Figure 4.16. Find the dimensions of each pen.

Figure 4.16

38. An architect is planning to include a rectangular window topped by a semicircle in his plans for a new house, as shown in Figure 4.17. To allow for enough light, the window should have an area of 120 square feet. The architect wants the rectangular portion of the window to be 2 feet wider than it is tall. Find the dimensions of the window.

Figure 4.17

39. When you look down from a height, say a tall building or a mountain peak, your line of sight is tangent to the earth at the horizon. (See Figure 4.18.)

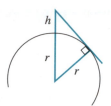

Figure 4.18

a. Suppose you are standing on top of the World Trade Center in New York, 1350 feet high. How far can you see on a clear day? You will need to use the Pythagorean theorem and the fact that the earth's radius is 3960 miles.

b. How tall a building would you need in order to see 100 miles?

40. a. If the earth's radius is 6370 kilometers, how far can you see from an airplane at an altitude of 10,000 meters? (See Exercise 39.)

b. How high would an airplane have to be in order for you to see a distance of 10 kilometers?

In Exercises 41 and 42 we will derive the quadratic formula.

41. Complete the square to find the solutions of the equation $x^2 + bx + c = 0$. (Your answers will be expressions in b and c.)

42. Complete the square to find the solutions of the equation $ax^2 + bx + c = 0$. (Your answers will be expressions in a, b, and c.)

43. What is the sum of the two solutions of a quadratic equation? (*Hint:* The two solutions are given by the quadratic formula.)

44. What is the product of the two solutions of a quadratic equation? (*Hint:* Do *not* try to multiply the two solutions given by the quadratic formula! Think about the factored form of the equation.)

4.4 GRAPHING PARABOLAS

In Section 4.1 we graphed several quadratic equations of the form

$$y = ax^2 + bx + c$$

The graphs of such equations are called **parabolas**. The constants a, b, and c determine the relative size and position of the graph. Some examples of parabolas are shown in Figure 4.19.

All parabolas have certain features in common. The graph has either a highest point (if the parabola opens downward, as in Figure 4.19(a)) or a lowest point (if the parabola opens upward, as in Figure 4.19(b)). This high or low point is called the **vertex** of the graph. The parabola is symmetric about a vertical line, called the **axis of symmetry**, that runs through the vertex. The **y-intercept** is the point at which the parabola intersects the y-axis. We shall see that a parabola may intersect the x-axis in zero, one, or two points, called the **x-intercepts**. If there are two x-intercepts, they are equidistant from the axis of symmetry.

If we can locate the vertex of a parabola and a few other points, perhaps the x- and y-intercepts, we can sketch a fairly accurate graph. We can find these points by analyzing the coefficients a, b, and c. We will begin by considering some examples.

Graph of $y = ax^2$ The simplest quadratic equation in two variables is

$$y = x^2$$

We can sketch its graph by plotting a few points, as shown in Figure 4.20. We will refer to the graph of $y = x^2$ as the *standard* parabola.

First we will investigate the effect of the constant a on the shape of the

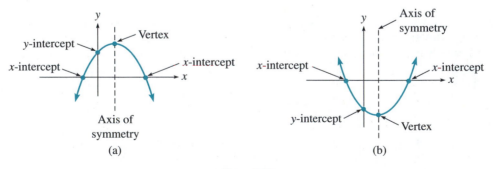

Figure 4.19

x	y
−2	4
−1	1
0	0
1	1
2	4

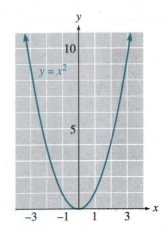

Figure 4.20

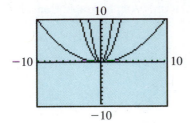

Figure 4.21

graph. Use your calculator to graph the following three equations on the same screen:

$$y = x^2$$

$$y = 3x^2$$

$$y = 0.1x^2$$

The graphs are shown in Figure 4.21. You can see that compared to the standard parabola the graph of $y = 3x^2$ is narrower, and the graph of $y = 0.1x^2$ is wider. We can also describe the graphs by saying that as x increases, the graph of $y = 3x^2$ increases faster than the standard parabola, and the graph of $y = 0.1x^2$ increases more slowly. This terminology is clearer if we compare the corresponding y-values for the three graphs, as in Table 4.5.

x	$y = x^2$	$y = 3x^2$	$y = 0.1x^2$
−2	4	12	0.4
1	1	3	0.1
3	9	27	0.9

Table 4.5

For each x-value the points on the graph of $y = 3x^2$ are higher than the points on the standard parabola, whereas the points on the graph of $y = 0.1x^2$ are lower. Multiplying x^2 by a positive constant greater than 1 stretches the graph vertically, and multiplying by a positive constant less than 1 squashes the graph vertically.

What about negative values for a? Consider the graphs of

$$y = x^2$$
$$y = -2x^2$$
$$y = -0.5x^2$$

shown in Figure 4.22. We see that multiplying x^2 by a negative constant reflects the graph about the x-axis. These parabolas open downward. In general, the graph of $y = ax^2$ opens upward if $a > 0$ and opens downward if $a < 0$. The magnitude of a determines how "wide" or "narrow" the parabola is. For equations of the form $y = ax^2$, the vertex, the x-intercepts, and the y-intercept all coincide at the origin.

Graph of $y = x^2 + c$ Next we will consider the effect of the constant term, c, on the graph. Consider the graphs of

$$y = x^2$$
$$y = x^2 + 4$$
$$y = x^2 - 4$$

shown in Figure 4.23. The graph of $y = x^2 + 4$ is shifted upward four units compared to the standard parabola, and the graph of $y = x^2 - 4$ is shifted downward four units. Look at Table 4.6, the y-values for the three graphs.

x	$y = x^2$	$y = x^2 + 4$	$y = x^2 - 4$
-1	1	5	-3
0	0	4	-4
2	4	8	0

Table 4.6

Each point on the graph of $y = x^2 + 4$ is four units higher than the corresponding point on the standard parabola, and each point on the graph of $y = x^2 - 4$ is four units lower. Therefore the vertex of the graph of $y = x^2 + 4$ is the point $(0, 4)$, and the vertex of the graph of $y = x^2 - 4$ is the point $(0, -4)$. In general, adding a positive constant c shifts the graph upward by c units, and adding a negative constant c shifts the graph downward.

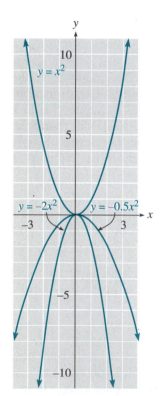

Figure 4.22

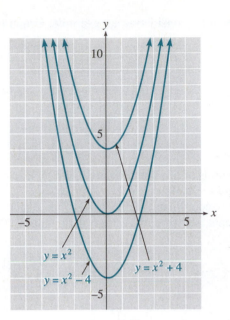

Figure 4.23

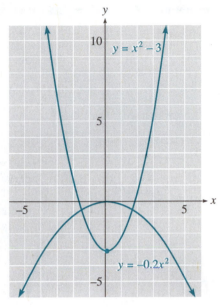

Figure 4.24

The x-intercepts of the graph of $y = x^2 - 4$ can be found by setting y equal to zero and solving for x:

$$0 = x^2 - 4$$
$$= (x - 2)(x + 2)$$

so the x-intercepts are the points $(2, 0)$ and $(-2, 0)$. The graph of $y = x^2 + 4$ has no x-intercepts, since the equation

$$0 = x^2 + 4$$

(obtained by setting $y = 0$) has no real-number solutions.

Example 1

Sketch graphs for the following quadratic equations.
a. $y = x^2 - 3$ **b.** $y = -0.2x^2$

Solutions   a.   The graph of $y = x^2 - 3$ is shifted downward by three units compared to the standard parabola. The vertex is the point $(0, -3)$, and the x-intercepts are the solutions of the equation

$$0 = x^2 - 3$$

or $\sqrt{3}$ and $-\sqrt{3}$. (See Figure 4.24.)
b.   The graph of $y = -0.2x^2$ opens downward and is wider than the standard parabola. Its vertex is the point $(0, 0)$. (See Figure 4.24 above; you can also verify both graphs with your graphing calculator.)

Graph of $y = ax^2 + bx$ Let's begin by considering an example. Graph the equation

$$y = 2x^2 + 8x$$

on your calculator. The graph is shown in Figure 4.25. Note that since $a = 2$ and $2 > 0$, the parabola opens upward. We can find the x-intercepts of the graph by setting y equal to zero:

$$0 = 2x^2 + 8x$$
$$= 2x\,(x + 4)$$

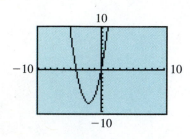

Figure 4.25

The solutions of this equation are 0 and -4, so the x-intercepts are the points $(0, 0)$ and $(-4, 0)$.

The x-coordinate of the vertex lies exactly halfway between the two x-intercepts, so we can average their values. The x-coordinate of the vertex is

$$x = \frac{1}{2}\,[0 + (-4)] = -2$$

To find the y-coordinate of the vertex, substitute $x = -2$ into the equation for the parabola:

$$y = 2\,(-2)^2 + 8(-2)$$
$$= 8 - 16 = -8$$

Thus the vertex is the point $(-2, -8)$.

We can use the same method to find a formula for the vertex of any parabola. All we need to do is consider the general quadratic equation instead of a specific example. First consider the equation

$$y = ax^2 + bx$$

We can find the x-intercepts of its graph by setting y equal to zero and solving for x:

$$0 = ax^2 + bx$$
$$= x\,(ax + b)$$

Thus

$$x = 0 \qquad \text{or} \qquad ax + b = 0$$

so

$$x = 0 \qquad \text{or} \qquad x = -\frac{b}{a}$$

The x-intercepts are the points $(0, 0)$ and $(-\frac{b}{a}, 0)$. The axis of symmetry of the parabola is located halfway between the x-intercepts, so we can find its

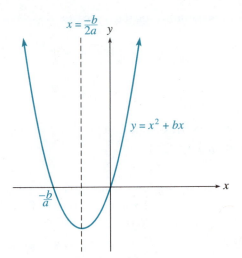

Figure 4.26

equation by taking the average of the two x-values found above.

$$x = \frac{1}{2} \left[0 + (-\frac{b}{a}) \right] = \frac{-b}{2a}$$

The axis of symmetry is the vertical line $x = \frac{-b}{2a}$. (See Figure 4.26.) Since the vertex lies on the axis of symmetry, its x-coordinate must be $\frac{-b}{2a}$. We can find the y-coordinate of the vertex by substituting its x-coordinate into the equation for the parabola.

Example 2

Sketch a graph of the quadratic equation $y = -x^2 + 5x$.

Solution Since $a = -1$ and $-1 < 0$, the parabola opens downward. Set $y = 0$ to find the x-intercepts:

$$0 = x^2 + 5x$$
$$= -x(x - 5)$$

Thus the x-intercepts are 0 and 5.

The vertex has x-coordinate

$$x = \frac{-b}{2a} = \frac{-5}{2(-1)}$$
$$= \frac{-5}{-2} = \frac{5}{2}$$

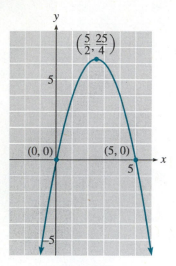

Figure 4.27

The y-coordinate of the vertex is

$$y = -\left(\frac{5}{2}\right)^2 + 5\left(\frac{5}{2}\right)$$

$$= -\frac{25}{4} + \frac{25}{2} = \frac{25}{4}$$

Thus the vertex is the point $(\frac{5}{2}, \frac{25}{4})$. The graph of $y = -x^2 + 5x$ is shown in Figure 4.27.

You can use your calculator to find the y-value that corresponds to a given x-value. In Example 2 we can find the y-coordinate of the vertex as follows. We first "store" the x-value in the calculator's memory by entering

2.5 $\boxed{\text{STO}\triangleright}$ $\boxed{\text{X,T,}\theta}$

and then pressing $\boxed{\text{ENTER}}$. (The $\boxed{\text{STO}\triangleright}$ button is located just above the $\boxed{\text{ON}}$ button.) Next we enter the expression we want the calculator to evaluate, $-x^2 + 5x$, by keying in

$\boxed{(\,\_\,)}$ $\boxed{\text{X,T,}\theta}$ $\boxed{x^2}$ $\boxed{+}$ 5 $\boxed{\text{X,T,}\theta}$

and pressing $\boxed{\text{ENTER}}$. The calculator returns the result, 6.25.

General Case:
$y = ax^2 + bx + c$

As an example of the general case $y = ax^2 + bx + c$, consider the equation

$$y = 2x^2 + 8x + 6$$

On page 232 we graphed the equation $y = 2x^2 + 8x$, and the graph is shown in Figure 4.28. Adding 6 to $2x^2 + 8x$ will have the effect of shifting the graph six units upward, as shown in Figure 4.28. Notice that the x-coordinate of the vertex will not be affected by an upward shift. Thus the formula

$$x_v = \frac{-b}{2a}$$

for the x-coordinate of the vertex still holds. We have

$$x_v = \frac{-8}{2(2)} = -2$$

and

$$y_v = 2(-2)^2 + 8(-2) + 6$$

$$= 8 - 16 + 6 = -2$$

so the vertex is the point $(-2, -2)$. (Notice that this point is shifted six units upward from the vertex of $y = 2x^2 + 8x$.)

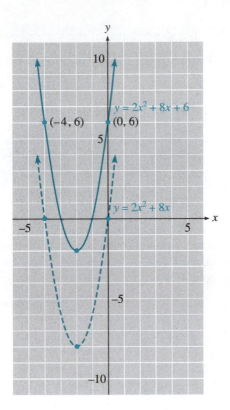

Figure 4.28

We find the x-intercepts of the graph by setting y equal to zero:

$$0 = 2x^2 + 8x + 6$$
$$= 2(x + 1)(x + 3)$$

Thus

$$x + 1 = 0 \qquad \text{or} \qquad x + 3 = 0$$
$$x = -1 \qquad\qquad\qquad x = -3$$

The x-intercepts are the points $(-1, 0)$ and $(-3, 0)$.

The y-intercept of the graph is found by setting x equal to zero:

$$y = 2(0)^2 + 8(0) + 6 = 6$$

Note that the y-intercept, 6, is just the constant term of the quadratic equation. The completed graph is shown in Figure 4.28.

Example 3

Use a calculator to locate the vertex of the graph of $y = -2x^2 + x + 1$.

Solution For this equation $a = -2$, $b = 1$, and $c = 1$. The x-coordinate of the vertex is given by

$$x_v = \frac{-b}{2a} = \frac{-1}{2(-2)} = \frac{1}{4}$$

To find the y-coordinate of the vertex, we substitute $x = \dfrac{1}{4} = 0.25$ into the equation. First store the x-value in the calculator's memory by entering

$$0.25 \; \boxed{\text{STO} \triangleright} \; \boxed{\text{X,T,}\theta}$$

and pressing $\boxed{\text{ENTER}}$. Then evaluate $-2x^2 + x + 1$ for $x = 0.25$. Enter

$$\boxed{(-)} \; 2 \; \boxed{\text{X,T,}\theta} \; \boxed{x^2} \; \boxed{+} \; \boxed{\text{X,T,}\theta} \; \boxed{+} \; 1$$

and then press $\boxed{\text{ENTER}}$. The calculator returns the y-value, 1.125. Thus the vertex is the point (0.25, 1.125).

Number of x-Intercepts

The graph of the quadratic equation $y = ax^2 + bx + c$ may have two, one, or no x-intercepts, according to the number of real-valued solutions of the equation $ax^2 + bx + c = 0$. Consider the three equations graphed in Figure 4.29.

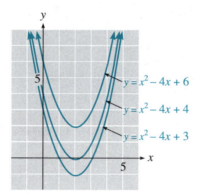

Figure 4.29

The graph of

$$y = x^2 - 4x + 3$$

has two x-intercepts, since the equation

$$x^2 - 4x + 3 = 0$$

has two real-valued solutions. The graph of

$$y = x^2 - 4x + 4$$

has only one x-intercept, since the equation

$$x^2 - 4x + 4 = 0$$

has only one real-valued solution. Because the solutions of the equation

$$x^2 - 4x + 6 = 0$$

are complex numbers, they do not appear on the graph. Thus the graph of

$$y = x^2 - 4x + 6$$

has no x-intercepts.

Once we have located the vertex of the parabola, the x-intercepts, and the y-intercept, we can sketch a reasonably accurate graph. Recall that the graph should be symmetric about a vertical line through the vertex. We summarize the procedure as follows.

TO GRAPH THE QUADRATIC EQUATION $y = ax^2 + bx + c$

1. Determine whether the parabola opens upward (if $a > 0$) or downward (if $a < 0$).

2. Locate the vertex of the parabola.

a. The x-coordinate of the vertex is $x_v = \dfrac{-b}{2a}$.

b. Find the y-coordinate of the vertex by substituting x_v into the equation of the parabola.

3. Locate the x-intercepts (if any) by setting $y = 0$ and solving for x.

4. Locate the y-intercept by evaluating y for $x = 0$.

5. If necessary, locate one or two additional points on the graph by evaluation.

Example 4

Graph the equation $y = x^2 + 3x + 1$.

Solution   We follow the steps outlined above.

Step 1   Since $a = 1$ and $1 > 0$, the parabola opens upward.

Step 2   Compute the coordinates of the vertex:

$$x_v = \frac{-b}{2a} = \frac{-3}{2(1)} = -1.5$$

$$y_v = (-1.5)^2 + 3(-1.5) + 1 = -1.25$$

The vertex is the point $(-1.5, -1.25)$.

Step 3   Set y equal to zero to find the x-intercepts:

$$0 = x^2 + 3x + 1 \qquad \text{Use the quadratic formula.}$$

$$x = \frac{-3 \pm \sqrt{3^2 - 4(1)(1)}}{2(1)}$$

$$= \frac{-3 \pm \sqrt{5}}{2}$$

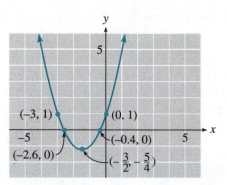

Figure 4.30

Rounding to the nearest tenth, we find that the x-intercepts are approximately $(-2.6, 0)$ and $(-0.4, 0)$.

Step 4   Substitute $x = 0$ to find the y-intercept, $(0, 1)$. Finally, plot the x- and y-intercepts and the vertex and draw a parabola through them. The finished graph is shown in Figure 4.30.

Exercise 4.4

In Exercises 1–8 describe what each graph will look like compared to the standard parabola. Then sketch the graph and label the coordinates of three points on the graph.

1. $y = 2x^2$ **2.** $y = 4x^2$

3. $y = \dfrac{1}{2}x^2$ **4.** $y = 0.6x^2$

5. $y = -x^2$ **6.** $y = -3x^2$

7. $y = -0.2\,x^2$ **8.** $y = \dfrac{3}{4}x^2$

In Exercises 9–16 describe what each graph will look like compared to the standard parabola. Then sketch each graph. Label the vertex and the x-intercepts (if there are any) with their coordinates.

9. $y = x^2 + 2$ **10.** $y = x^2 + 5$

11. $y = x^2 - 1$ **12.** $y = x^2 - 9$

13. $y = x^2 - 5$ **14.** $y = x^2 - 3$

15. $y = 100 - x^2$ **16.** $y = 225 - x^2$

Find the x-intercepts and the vertex of each graph in Exercises 17–24. Then sketch the graph.

17. $y = x^2 - 4x$ **18.** $y = x^2 - 2x$

19. $y = x^2 + 2x$ **20.** $y = x^2 + 6x$

21. $y = 3x^2 + 6x$ **22.** $y = 2x^2 - 6x$

23. $y = -2x^2 + 5x$ **24.** $y = -3x^2 - 8x$

In Exercises 25–32 find the coordinates of the vertex.

25. $y = 3x^2 - 6x + 4$

26. $y = -2x^2 + 5x - 1$

27. $y = 2 + 3x - x^2$ **28.** $y = 3 - 5x + x^2$

29. $y = \dfrac{1}{2}x^2 - \dfrac{2}{3}x + \dfrac{1}{3}$

30. $y = \dfrac{-3}{4}x^2 + \dfrac{1}{2}x - \dfrac{1}{4}$

31. $y = 2.3 - 7.2x - 0.8x^2$

32. $y = 5.1 - 0.2x + 4.6x^2$

Sketch each parabola in Exercises 33–44. Give the coordinates of the vertex and the intercepts. Use your calculator to verify your graphs.

33. $y = x^2 - 5x + 4$ **34.** $y = x^2 + x - 6$

35. $y = -2x^2 + 7x + 4$

36. $y = -3x^2 + 2x + 8$

37. $y = 0.6x^2 + 0.6x - 1.2$

38. $y = 0.5x^2 - 0.25x - 0.75$

39. $y = x^2 + 4x + 7$　　**40.** $y = x^2 - 6x + 10$

41. $y = x^2 + 2x - 1$　　**42.** $y = x^2 - 6x + 2$

43. $y = -2x^2 + 6x - 3$

44. $y = -2x^2 - 8x - 5$

45. a. Write an equation for a parabola that has x-intercepts at $(2, 0)$ and $(-3, 0)$.

b. Write an equation for another parabola that has the same x-intercepts.

46. a. Write an equation for a parabola that has x-intercepts $(-1, 0)$ and $(4, 0)$ and opens upward.

b. Write an equation for a parabola that has x-intercepts $(-1, 0)$ and $(4, 0)$ and opens downward.

47. Write equations for two different parabolas with y-intercept $(0, 2)$.

48. Write equations for two different parabolas with vertex $(0, 2)$.

In Exercises 49–52
a. find the vertex of each parabola. What do you notice about the coordinates of the vertex compared to the form of the equation as written? Explain your answer.

b. write each equation in standard form.

49. $y = 2(x - 3)^2 + 4$

50. $y = -3(x + 1)^2 - 2$

51. $y = -\dfrac{1}{2}(x + 4)^2 - 3$

52. $y = 4(x - 2)^2 - 6$

53. Explain why the equation

$$y = a(x - x_v)^2 + y_v$$

is useful when we know the vertex of a parabola.

54. Explain how to obtain the vertex form for the equation of a parabola from the standard form.

55. a. Write an equation for a parabola whose vertex is the point $(-2, 6)$. (Many answers are possible.)

b. Find the value of a if the y-intercept of the parabola in part (a) is 18.

56. a. Write an equation for a parabola whose vertex is the point $(5, -10)$. (Many answers are possible.)

b. Find the value of a if the y-intercept of the parabola in part (a) is -5.

57. Write an equation for a parabola whose vertex and y-intercept coincide at $(0, -3)$.

58. Write an equation for a parabola whose vertex is $(4, 0)$. (*Hint:* What are the x-intercepts of the parabola?)

4.5 QUADRATIC INEQUALITIES

Solving Quadratic Inequalities Graphically

In Chapter 2 we used graphs to help us solve equations and inequalities in one variable. For instance, to solve the inequality

$$1.3x - 4.2 \geq 2.3$$

we can consider the graph of the equation

$$y = 1.3x - 4.2$$

as shown in Figure 4.31 on page 240. We are looking for x-values that make the expression $1.3x - 4.2$ greater than or equal to 2.3. This search is the same as looking for points on the graph whose y-coordinates are greater than or equal to 2.3. These points are shown in color on the graph in Figure 4.31.

The *x*-coordinates of these points are the solutions of the inequality. From the graph we see that all these points have *x*-coordinates greater than or equal to 5, so the solutions to the inequality are $x \geq 5$.

The graphing technique is also helpful for solving quadratic inequalities.

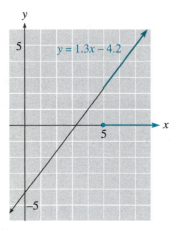

Figure 4.31

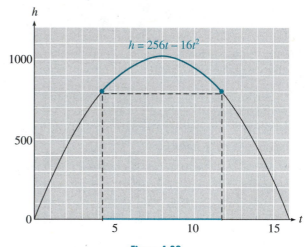

Figure 4.32

Example 1

The Chamber of Commerce in River City wants to put on a Fourth of July fireworks display. City ordinances require that fireworks at public gatherings explode at least 800 feet above the ground. The mayor particularly wants to include the Freedom Starburst model, which is launched from the ground so that its height after *t* seconds is given by

$$h = 256t - 16t^2$$

When should the Freedom Starburst explode in order to satisfy the safety ordinance?

Solution   We can find an approximate answer to this question by looking at the graph of the rocket's height shown in Figure 4.32. We would like to know when the rocket's height is at least 800 feet, or in mathematical terms, for what values of *t* is $h \geq 800$? Answering this question is equivalent to solving the inequality

$$256t - 16t^2 \geq 800$$

Points on the graph with $h \geq 800$ are shown in color, and the *t*-coordinates of those points are marked on the horizontal axis. If the Freedom Starburst explodes at any of these times, it will satisfy the safety regulation. From the graph the safe time interval runs from approximately 4.25 seconds to 11.75 seconds after launch, or $4.25 \leq t \leq 11.75$.

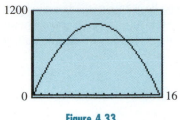

Figure 4.33

You can use your graphing calculator to solve the problem in Example 1. Graph the two equations

$$Y_1 = 256X - 16X^2$$

$$Y_2 = 800$$

on the same screen. (See Figure 4.33.) Experiment with the $\boxed{\text{WINDOW}}$ settings until you get a good picture. Then use the $\boxed{\text{TRACE}}$ feature to find the coordinates of the points (there are two) where the two graphs intersect. These points will have y-coordinates of 800. You can use the $\boxed{\text{ZOOM}}$ Box to increase the accuracy of your estimate. To two decimal places, you can show that $4.26 \leq t \leq 11.74$.

In Example 1 we solved the inequality $256t - 16t^2 \geq 800$ by graphing the equation $h = 256t - 16t^2$ and locating points on the graph for which the h-coordinate is at least 800. In many cases it may be easier to rewrite the inequality so that zero is on one side of the inequality sign.

Example 2

Solve the inequality $x^2 - 2x - 9 \geq 6$ by using a graph.

Solution First rewrite the inequality so that the right side is zero.

$$x^2 - 2x - 9 \geq 6 \qquad \text{Subtract 6 from both sides.}$$

$$x^2 - 2x - 15 \geq 0$$

Now consider the graph of the equation $y = x^2 - 2x - 15$ shown in Figure 4.34. We are interested in points whose y-coordinates are greater than

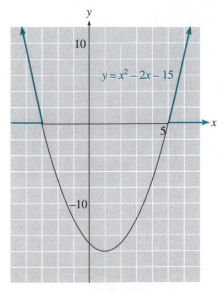

Figure 4.34

or equal to zero. We see that *two* parts of the graph have positive y-coordinates: points with x-coordinates greater than or equal to 5, and points with x-coordinates less than or equal to -3. Thus the solutions of the inequality are x-values satisfying $x \geq 5$ or $x \leq -3$.   ▱

Example 3

Consider the quadratic expression $x^2 - 4$. Find the solutions of (a)–(c).

a. $x^2 - 4 = 0$ **b.** $x^2 - 4 < 0$ **c.** $x^2 - 4 > 0$

Solutions   Look at the graph of the equation $y = x^2 - 4$ shown in Figure 4.35. For each value of x that we substitute into the expression $x^2 - 4$, the result is either positive, negative, or zero. You can see these results more clearly if you compute a few values yourself to complete the table below; your values should agree with the coordinates of points on the graph.

x	-3	-2	-1	0	1	2	3
y							

a. First locate the two points on the graph where $y = 0$. These points are $(-2, 0)$ and $(2, 0)$. Their x-coordinates, -2 and 2, are the solutions of the equation $x^2 - 4 = 0$. Note that the two points, -2 and 2, divide the x-axis into three sections, which are labeled on the graph in Figure 4.35. On each of these sections, the value of $x^2 - 4$ is either always positive or always negative.

b. Next find the points on the graph where $y < 0$. These are points with x-coordinates between -2 and 2 (labeled II in Figure 4.35). Thus the solution to the inequality $x^2 - 4 < 0$ is $-2 < x < 2$.

c. Finally, locate the points on the graph where $y > 0$. There are two sections of the x-axis that correspond to positive y-values, those labeled I and III on Figure 4.35. In section I, $x < -2$, and in section III, $x > 2$. Thus the solution to the inequality $x^2 - 4 > 0$ consists of two sets of numbers: $x < -2$ or $x > 2$.   ▱

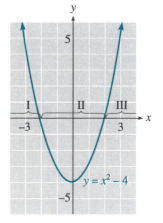

Figure 4.35

($y = x^2 - 4$)

Interval Notation

The solution to the inequality $x^2 - 4 < 0$ in Example 3(b) is called an interval. An **interval** is a set that consists of all the real numbers between two numbers a and b. If the set includes both of the endpoints a and b, so that $a \leq x \leq b$, then the set is called a **closed interval** and is denoted with square brackets: $[a, b]$. If the set does not include its endpoints, so that $a < x < b$, then it is called an **open interval** and is denoted with parentheses: (a, b).

The solution to Example 3(b) is an open interval, and we denote it by $(-2, 2)$. (See Figure 4.36(a). Do not confuse the open interval $(-2, 2)$ with the point $(-2, 2)$! The notation is the same, so you must decide from the context whether an interval or a point is being discussed.)

The solution to Example 3(c) consists of two **infinite intervals**, $x < -2$ or $x > 2$, shown in Figure 4.36(b). We denote the interval

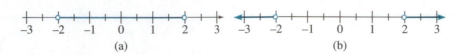

Figure 4.36

(a)   (b)

$x < -2$ by $(-\infty, -2)$, and the interval $x > 2$ by $(2, \infty)$. The symbol ∞, for "infinity," does not represent a specific real number; it indicates that the interval continues forever along the real line. The set consisting of both these intervals is called the **union** of the two intervals and is denoted by $(-\infty, -2) \cup (2, \infty)$.

Many solutions of inequalities are intervals or unions of intervals.

Example 4

Write each of the solution sets with interval notation, and graph the solution set on a number line.

a. $3 \leq x < 6$ **b.** $x \geq -9$

c. $x \leq 1$ or $x > 4$ **d.** $-8 < x \leq -5$ or $-1 \leq x < 3$

Solutions **a.** [3, 6). This interval is referred to as **half-open** or **half-closed**. (See Figure 4.37.)

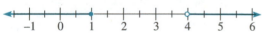

Figure 4.37   **Figure 4.38**

b. $[-9, \infty)$. We always use parentheses next to the symbol ∞ because ∞ is not a specific number and is not included in the set. (See Figure 4.38.)

c. $(-\infty, 1] \cup (4, \infty)$. The word *or* describes the union of two sets. (See Figure 4.39.)

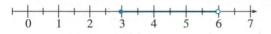

Figure 4.39   **Figure 4.40**

d. $(-8, -5] \cup [-1, 3)$. (See Figure 4.40.)   ⬚

Solving Quadratic Inequalities Algebraically

Although the graph of a quadratic relationship is very helpful in solving inequalities, it is not completely necessary. Every quadratic inequality can be put into one of the forms

$$ax^2 + bx + c < 0, \quad ax^2 + bx + c > 0$$

$$ax^2 + bx + c \leq 0, \quad ax^2 + bx + c \geq 0$$

What we really need to know is whether the corresponding parabola $y = ax^2 + bx + c$ opens upward or downward. Consider the parabolas shown in Figure 4.41 on page 244.

The parabola in Figure 4.41(a) opens upward. It crosses the x-axis at two points: $x = r_1$ and $x = r_2$. At these points $y = 0$. The solutions

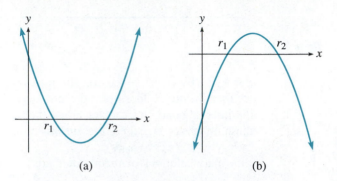

Figure 4.41
(a)
(b)

to the inequality $y < 0$ lie between r_1 and r_2, because y is negative on that portion of the graph. The solutions to the inequality $y > 0$ are the x-values less than r_1 or greater than r_2, because y is positive in those regions.

If the parabola opens downward, as in Figure 4.41(b), the situation is reversed. The solutions to the inequality $y > 0$ lie between the x-intercepts, and the solutions to $y < 0$ lie outside the x-intercepts.

From the graphs in Figure 4.41, we see that the x-intercepts are the "boundary points" between the portions of the graph with positive y-coordinates and the portions with negative y-coordinates. To solve a quadratic inequality, we need only locate the x-intercepts of the corresponding graph and then decide which intervals of the x-axis result in the correct sign for y.

> **TO SOLVE A QUADRATIC INEQUALITY ALGEBRAICALLY**
>
> **1.** Write the inequality in standard form: one side is zero, and the other has the form $ax^2 + bx + c$.
>
> **2.** Make a rough sketch of the graph of $y = ax^2 + bx + c$, using the sign of a to determine whether the parabola opens upward or downward.
>
> **3.** Find the x-intercepts of the graph by setting $y = 0$ and solving for x.
>
> **4.** Decide which intervals on the x-axis give the correct sign for y.

Example 5

Solve the inequality $36 + 6x - x^2 \leq 20$ algebraically.

Solution First subtract 20 from both sides of the inequality so that zero is on the right side.

$$16 + 6x - x^2 \leq 0$$

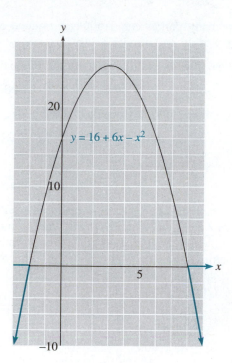

Figure 4.42

Next consider the equation $y = 16 + 6x - x^2$. We are interested in points on the graph for which $y \leq 0$. Now the graph is a parabola that opens downward, so the points with negative y-coordinates lie outside the x-intercepts of the graph. To locate the x-intercepts, set $y = 0$ and solve for x.

$$16 + 6x - x^2 = 0 \qquad \text{Multiply each term by } -1.$$
$$x^2 - 6x - 16 = 0 \qquad \text{Factor the left side.}$$
$$(x - 8)(x + 2) = 0 \qquad \text{Apply the zero-factor principle.}$$
$$x - 8 = 0 \quad \text{ or } \quad x + 2 = 0$$
$$x = 8 \quad \text{ or } \quad x = -2$$

The x-intercepts are $x = -2$ and $x = 8$. The points on the graph with $y < 0$ lie outside the x-intercepts, so the solution of the inequality is $(-\infty, -2] \cup [8, \infty)$.

Example 6

TrailGear manufactures camping equipment. They find that the cost of producing x alpine parkas per week is given in dollars by

$$C = 0.1x^2 - 40x + 5200$$

How many parkas can TrailGear produce each week if they have to keep their weekly costs under $3000?

Solution   We would like to solve the inequality

$$0.1x^2 - 40x + 5200 < 3000$$

or, subtracting 3000 from both sides,

$$0.1x^2 - 40x + 2200 < 0$$

Consider the equation

$$y = 0.1x^2 - 40x + 2200$$

The graph of this equation is a parabola that opens upward, since the coefficient of x^2 is positive. (You may want to make a rough sketch of the graph, as shown in Figure 4.43.) We first locate the x-intercepts of the graph by setting $y = 0$ and solving for x. We will use the quadratic formula to solve the equation $0.1x^2 - 40x + 2200$, so $a = \mathbf{0.1}$, $b = \mathbf{-40}$, and $c = \mathbf{2200}$.

$$x = \frac{-(\mathbf{-40}) \pm \sqrt{(\mathbf{-40})^2 - 4(\mathbf{0.1})(\mathbf{2200})}}{2(\mathbf{0.1})}$$

$$= \frac{40 \pm \sqrt{1600 - 880}}{0.2} = \frac{40 \pm \sqrt{720}}{0.2}$$

To two decimal places, the solutions to the equation are 65.84 and 334.16. Negative values of y correspond to x-values between the two x-intercepts, that is, for $65.84 < x < 334.16$. Because we cannot produce a fraction of a parka, we restrict the interval to the closest whole-number x-values included, namely 66 and 334. (See Figure 4.44.) Thus TrailGear can produce as few as 66 parkas or as many as 334 parkas to keep their costs under $3000 per week.

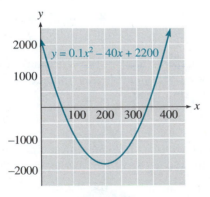

Figure 4.43

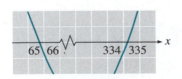

Figure 4.44

Exercise 4.5

In Exercises 1–4 use the graphs provided to estimate the solutions to each inequality.

1. $x^2 - 3x - 180 > 0$

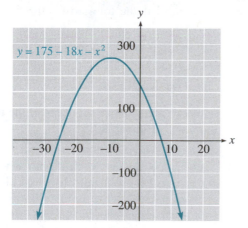

Figure 4.45

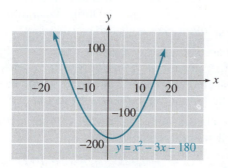

Figure 4.46

2. $175 - 18x - x^2 < 0$

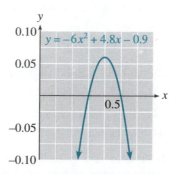

Figure 4.47

3. $-6x^2 + 4.8x - 0.9 \geq 0$

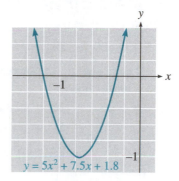

Figure 4.48

4. $5x^2 + 7.5x + 1.8 \leq 0$

In Exercises 5–10 solve each inequality by graphing. Use the following WINDOW settings on your calculator:

$$Xmin = -9.4, \quad Xmax = 9.4$$
$$Ymin = -12.8, \quad Ymax = 12.4$$

5. $(x - 3)(x + 2) > 0$

6. $(x + 1)(x - 5) > 0$

7. $k(4 - k) \geq 0$ **8.** $-m(7 + m) \geq 0$

9. $6 + 5p - p^2 < 0$ **10.** $q^2 + 9q + 18 < 0$

In Exercises 11–16 solve each inequality by graphing. Use the following WINDOW settings on your calculator:

$$Xmin = -9.4, \quad Xmax = 9.4$$
$$Ymin = -64, \quad Ymax = 62$$

11. $x^2 - 1.4x - 20 < 9.76$

12. $-x^2 + 3.2x + 20 > 6.56$

13. $5x^2 + 39x + 27 \geq 5.4$

14. $-6x^2 - 36x - 20 \leq 25.3$

15. $-8x^2 + 112x - 360 < 6.08$

16. $10x^2 + 96x + 180 > 17.8$

In Exercises 17–22 solve each inequality by graphing. Use the ZOOM feature to estimate your solutions accurate to one decimal place.

17. $x^2 > 12.2$ **18.** $x^2 \leq 45$

19. $-3x^2 + 7x - 25 \leq 0$

20. $2.4x^2 - 5.6x + 18 \leq 0$

21. $0.4x^2 - 54x < 620$ **22.** $-0.05x^2 - 3x > 76$

In Exercises 23–32 write each set with interval notation, and graph the set on a number line.

23. $-5 < x \leq 3$ **24.** $0 \leq x < 4$

25. $0 \geq x \geq -4$ **26.** $8 > x > 5$

27. $x > -6$ **28.** $x \leq 1$

29. $x < -3$ or $x \geq -1$

30. $x \geq 3$ or $x \leq -3$

31. $-6 \leq x < -4$ or $-2 < x \leq 0$

32. $x < 2$ or $2 < x < 3$

In Exercises 33–38 solve each inequality algebraically. Write your answers in interval notation, and round to two decimal places if necessary.

33. $2z^2 - 7z > 4$ **34.** $6h^2 + 13h < 15$

35. $v^2 < 5$ **36.** $t^2 \geq 7$

37. $5a^2 - 32a + 12 \geq 0$

38. $6b^2 + 16b - 9 < 0$

In Exercises 39–44
a. solve by writing and solving an inequality.
b. graph the given equation and verify your solution on the graph.

39. A fireworks rocket is fired from ground level. Its height in feet t seconds after launch is given by $h = 320t - 16t^2$. During what time interval is the rocket higher than 1024 feet?

40. A baseball thrown vertically reaches a height, h, in feet given by $h = 56t - 16t^2$, where t is measured in seconds. During what time intervals is the ball between 40 and 48 feet high?

41. The cost in dollars of manufacturing x pairs of garden shears is given by

$$C = -0.02x^2 + 14x + 1600$$

for $0 \leq x \leq 700$. How many pairs of shears can be produced if the total cost must be kept under $2800?

42. The cost in dollars of producing x cashmere sweaters is given by

$$C = x^2 + 4x + 90.$$

How many sweaters can be produced if the total cost must be kept under $1850?

43. The Locker Room finds that it sells $1200 - 30p$ sweatshirts each month when it charges p dollars per sweatshirt. It would like its revenue from sweatshirts to be over $9000 per month. In what range should it keep the price of a sweatshirt? *Hint:* Recall that *Revenue = (Number of items sold) (Price per item).*

44. Green Valley Nursery sells $120 - 10p$ boxes of rose food per month at a price of p dollars per box. It would like to keep its monthly revenue from rose food over $350. In what range should it price a box of rose food?

4.6 PROBLEM SOLVING

Maximum or Minimum Values

Finding the maximum or minimum value for a variable expression is a common problem in applications. For example, if you own a company that manufactures blue jeans, you might like to know how much to charge for your jeans in order to maximize your revenue. Recall that

> *Revenue = (Price of one item)(Number of items sold)*

As you increase the price of your jeans, your revenue may increase for a while. But if you charge too much for your jeans, consumers will not buy as many pairs and your revenue may start to decrease. Is there some optimum price you should charge for a pair of jeans in order to achieve the greatest revenue?

Example 1

Late Nite Blues finds that it can sell $600 - 15x$ pairs of jeans per week if it charges x dollars per pair. (Notice that as the price increases, the number of pairs of jeans sold decreases.)
a. Write an equation for the revenue in terms of the price of a pair of jeans.
b. Graph the equation.
c. How much should Late Nite Blues charge for a pair of jeans in order to maximize its revenue?

Solutions   a. Using the formula for revenue stated above, we find

$$\text{Revenue} = (\text{Price of one item}) (\text{Number of items sold})$$
$$R = x (600 - 15x)$$
$$R = 600x - 15x^2$$

b. We recognize the equation as quadratic, so the graph is a parabola, as shown in Figure 4.49.

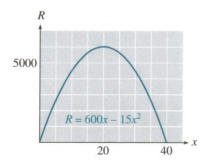

Figure 4.49

c. The maximum value of R occurs at the vertex of the parabola. Thus

$$x_v = \frac{-b}{2a} = \frac{-600}{2(-15)} = 20$$

$$R_v = 600\,(20) - 15\,(20)^2 = 6000$$

The revenue takes on its maximum value when $x = 20$, and the maximum value is $R = 6000$. Late Nite Blues should charge \$20 for a pair of jeans in order to maximize its revenue at \$6000. (Of course, we can also graph the equation on a calculator and use the $\boxed{\text{TRACE}}$ to estimate the coordinates of the vertex.)

If the equation relating two variables is quadratic, then the maximum or minimum value is easy to find: it is the value at the vertex. Notice that if the parabola opens downward, as in Example 1, the maximum value is at the vertex. If the parabola opens upward, there is a minimum value at the vertex.

Example 2

TrailGear can manufacture x alpine parkas per week at a cost of

$$C = 0.1x^2 - 40x + 5200$$

dollars. How many parkas should they produce each week in order to minimize their weekly costs?

Solution   The cost equation is quadratic and the coefficient of x^2 is positive, so the graph of the equation is a parabola that opens upward. (See Figure 4.50.) Hence the minimum value of C occurs at the vertex of the parabola. The coordinates of the vertex are

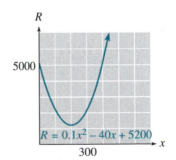

$$x_v = \frac{-b}{2a} = \frac{40}{2(0.1)} = 200$$

and

$$y_v = 0.1(\mathbf{200})^2 - 40(\mathbf{200}) + 5200$$

$$= 1200$$

Thus TrailGear should produce 200 parkas per week to minimize their costs at \$1200.

Figure 4.50

Curve Fitting

We have used linear and quadratic equations to model a variety of variable relationships. In Section 2.4 we used the point-slope formula to find the equation of a line when we knew the coordinates of two points on the line. In Section 3.4 we extended this idea to "fit" a line through a collection of data points that lie roughly on a line.

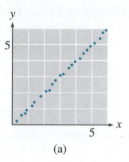

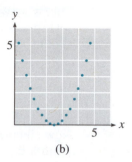

Figure 4.51   (a)   (b)

If the data points do not lie roughly on a line, it does not make sense to describe them by a linear equation. For example, compare the data points in Figures 4.51(a) and (b). The points in Figure 4.51(a) are roughly linear in appearance, but the points in Figure 4.51(b) are not. We can, however, visualize a parabola that would approximate the data.

To find a linear approximation by hand, we chose two data points and computed the equation of the line through those two points. Every linear equation can be written in the form

$$y = mx + b$$

and to find a specific line we must find values for the two parameters (constants) m and b. We need two data points in order to find those two parameters. A quadratic equation, however, has the form

$$y = ax^2 + bx + c$$

and we must find *three* parameters: a, b, and c. To do this we need *three* data points. Here is an example of how to find the parabola that passes through three given points.

Example 3

Find values for a, b, and c so that the points (1, 3), (3, 5), and (4, 9) lie on the graph of $y = ax^2 + bx + c$.

Solution Substitute the coordinates of each of the three points into the equation of the parabola to obtain three equations:

$$3 = a(1)^2 + b(1) + c$$
$$5 = a(3)^2 + b(3) + c$$
$$9 = a(4)^2 + b(4) + c$$

or, equivalently,

$$a + b + c = 3 \qquad\qquad (1)$$
$$9a + 3b + c = 5 \qquad\qquad (2)$$
$$16a + 4b + c = 9 \qquad\qquad (3)$$

This is a system of three equations in the three unknowns a, b, and c. To solve the system, we will first eliminate c. Subtract Equation (1) from Equation (2) to obtain

$$8a + 2b = 2 \qquad\qquad (4)$$

and subtract Equation (1) from Equation (3) to get

$$15a + 3b = 6 \qquad\qquad (5)$$

Now eliminate b from Equations (4) and (5): add -3 times Equation (4) to 2 times Equation (5) to get

$$
\begin{array}{ll}
-24a - 6b = -6 & \text{−3 times Equation (4).} \\
\underline{30a + 6b = 12} & \text{2 times Equation (5).} \\
6a = 6 &
\end{array}
$$

or $a = 1$.   Substitute 1 for a in Equation (4) to find

$$8(\mathbf{1}) + 2b = 2 \qquad \text{Solve for } b.$$
$$b = -3$$

Finally, substitute -3 for b and 1 for a in Equation (1) to find

$$\mathbf{1} + (\mathbf{-3}) + c = 3 \qquad \text{Solve for } c.$$
$$c = 5$$

Thus the equation of the parabola is

$$y = x^2 - 3x + 5$$

The parabola and the three points are shown in Figure 4.52.

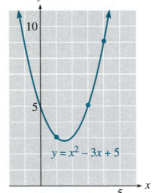

Figure 4.52

The simplest way to fit a parabola to a set of data points is to pick three of the points and find the equation of the parabola that passes through those three points. More accurate methods for locating a quadratic regression curve are available, but we will not discuss them here.

Example 4

Major Motors Corporation is testing a new car designed for in-town driving. They have done tests to measure the cost of driving the car at different speeds and have gathered the data shown in Table 4.7. The speeds, v, are given in miles per hour, and the cost, C, includes fuel and maintenance for driving the car 100 miles at that speed. Find a possible quadratic model

$$C = av^2 + bv + c$$

that expresses C in terms of v.

Solution   When the data are plotted, it is clear that the relationship between v and C is not linear (see Figure 4.53), but it may be quadratic. We will use the last three data points, $(50, 6.20)$, $(60, 7.80)$, and $(70, 10.60)$, to fit a

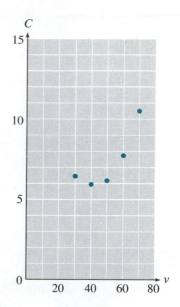

v	C
30	6.50
40	6.00
50	6.20
60	7.80
70	10.60

Table 4.7

Figure 4.53

parabola to the data. We would like to find the coefficients a, b, and c of a parabola $C = av^2 + bv + c$ that includes the three data points. We substitute the coordinates of each data point (v, C) into the equation $C = av^2 + bv + c$ to obtain

$$6.20 = a(50)^2 + b(50) + c$$

$$7.80 = a(60)^2 + b(60) + c$$

$$10.60 = a(70)^2 + b(70) + c$$

Simplifying these equations gives us a system of equations:

$$2500a + 50b + c = 6.20 \qquad (1)$$

$$3600a + 60b + c = 7.80 \qquad (2)$$

$$4900a + 70b + c = 10.60 \qquad (3)$$

Eliminating c from Equations (1) and (2) yields Equation (4), and eliminating c from Equations (2) and (3) yields Equation (5).

$$1100a + 10b = 1.60 \qquad (4)$$

$$1300a + 10b = 2.80 \qquad (5)$$

Eliminating b from Equations (4) and (5) gives us

$$200a = 1.20$$

$$a = 0.006$$

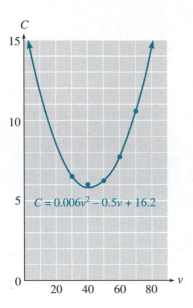

Figure 4.54

$C = 0.006v^2 - 0.5v + 16.2$

Substitute this value into Equation (4) to find $b = -0.5$, then substitute both values into Equation (1) to find $c = 16.2$. Thus our quadratic model is

$$C = 0.006v^2 - 0.5v + 16.2$$

The graph of this equation, along with the data points, is shown in Figure 4.54.

Systems Involving Quadratic Equations

Recall that the solution to a 2×2 system of linear equations is the intersection point of the graphs of the equations. The solution of a system in which one of the equations is quadratic also occurs at the intersection point of their graphs. However, such a system may have either one solution, two solutions, or no solutions, as you can see in the graphs in Figure 4.55.

In Example 5 we will use both graphical and algebraic techniques to solve such a system.

Figure 4.55

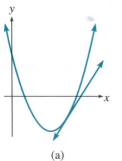

(a)

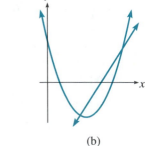

(b)

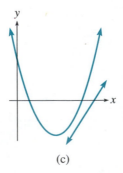

(c)

Example 5

The Pizza Connection calculates that the cost in dollars of producing x pizzas per day is

$$C = 240 + 2x$$

Its revenue from selling x pizzas is

$$R = -0.05x^2 + 10x$$

How many pizzas per day must the Pizza Connection sell in order to break even?

Solution Recall that to "break even" means to make zero profit. Since

$$Profit = Revenue - Cost$$

the break-even point occurs when the revenue equals the cost. In mathematical terms we would like to find any values of x for which $R = C$. If we graph the revenue and cost equations on the same axes, these values correspond to points at which the two graphs intersect. Use the WINDOW settings

$$Xmin = 0, \quad Xmax = 188$$
$$Ymin = 0, \quad Ymax = 500$$

on your calculator to obtain the graph shown in Figure 4.56. With the TRACE keys you can verify that the two intersection points are (40, 320) and (120, 480).

Thus the Pizza Connection must sell either 40 pizzas or 120 pizzas to break even. Notice on the graph that for x-values between 40 and 120, the revenue is greater than the cost, so that the Pizza Connection will make a profit if it sells between 40 and 120 pizzas. We can also solve algebraically for the break-even points. The intersection points of the two graphs correspond to the solutions of the system of equations

$$y = -0.05x^2 + 10x$$
$$y = 240 + 2x$$

To solve this system, we equate the two expressions for y and then solve for x.

$$-0.05x^2 + 10x = 240 + 2x \qquad \text{Subtract } 240 + 2x \text{ from both sides.}$$

$$-0.05x^2 + 8x - 240 = 0 \qquad \text{Use the quadratic formula.}$$

$$x = \frac{-8 \pm \sqrt{8^2 - 4(-0.05)(-240)}}{2(-0.05)} \qquad \text{Simplify.}$$

$$= \frac{-8 \pm \sqrt{64 - 48}}{-0.1}$$

The solutions are 40 and 120, as we found from the graph in Figure 4.56.

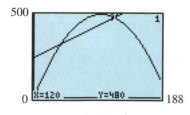

Figure 4.56

Solving Systems with the Graphing Calculator

When the TRACE does not locate the intersection point of two curves exactly, we can use the ZOOM to get a more accurate approximation. However, zooming in can be time-comsuming, so we will take advantage of another feature of the graphing calculator.

Consider the system of two quadratic equations

$$y = (x + 1.1)^2$$

$$y = 7.825 - 2x - 2.5x^2$$

We will graph these two equations in the standard window. The two intersection points are visible in the window, but we do not find their exact coordinates when we trace the graphs.

Instead of zooming in, we will use the CALC, or **Calculate**, feature of the calculator. Press 2nd TRACE to get the CALC menu, shown in Figure 4.57(a), and choose 5 to select **intersect**. Then press ENTER twice in response to the questions on the screen. (This tells the calculator that you want the intersection of graphs of Y_1 and Y_2.)

In response to the query **Guess?** move the bug close to one of the intersection points and press ENTER. After a short pause the calculator will display the x- and y-coordinates of that intersection point. The result is shown in Figure 4.57(b). You can check that the point (0.9, 4) is an exact solution to the system by substituting $x = 0.9$ and $y = 4$ into each equation of the system. (The calculator is not always able to find the exact coordinates, but it usually gives a very good approximation.)

You can find the other solution of the system by following the same steps and moving the bug close to the other intersection point. You should verify that the other solution is the point $(-2.1, 1)$.

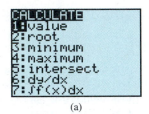

(a)

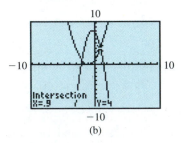

(b)

Figure 4.57

Exercise 4.6

For Exercises 1–12

a. find the maximum or minimum value algebraically.

b. obtain a good graph on your calculator, and verify your answer with the TRACE key.

1. The equation $d = 96t - 16t^2$ gives the distance, d, in feet above the ground of a toy water rocket t seconds after it is launched. When will the rocket reach its greatest height? What will that height be?

2. The equation $h = -12 + 32t - 16t^2$ gives the height, h, in feet of a wrench t seconds after it is tossed into the air from the bottom of a trench 12 feet deep. When will the wrench reach its greatest height? What will that height be?

3. As part of a collage for her art class, Sheila wants to enclose a rectangle with 100 inches of yarn.

a. Letting w represent the width of the rectangle, write an expression for the rectangle's length. Then write an expression in terms of w for the area A of the rectangle.

b. What is the area of the largest rectangle Sheila can enclose with 100 inches of yarn?

4. Gavin has rented space for a booth at the county fair. As part of his display, he wants to rope off a rectangular area with 80 yards of rope.

   a. Letting w represent the width of the roped-off rectangle, write an expression for the rectangle's length. Then write an expression in terms of w for the area A of the roped-off space.

   b. What is the largest area Gavin can rope off? What will the rectangle's dimensions be?

5. A farmer plans to fence a rectangular grazing area along a river with 300 yards of fence, as shown in Figure 4.58.

   a. Write an expression for the area A of the grazing land in terms of the width w of the rectangle.

   b. What is the largest area he can enclose?

Figure 4.58

6. A breeder of horses wants to fence two rectangular grazing areas along a river with 600 meters of fence, as shown in Figure 4.59.

Figure 4.59

   a. Write an expression for the total area A of the grazing land in terms of the width w of the rectangles.

   b. What is the largest area she can enclose?

7. A travel agent offers a group rate of $2400 per person for a week in London if 16 people sign up for the tour. For each additional person who signs up, the price per person is reduced by $100.

   a. Let x represent the number of additional people who sign up. Write expressions for the total number of people signed up, the price per person, and the total revenue.

   b. How many people must sign up for the tour in order for the travel agent to maximize her revenue?

8. An entrepreneur buys an apartment building with 40 units. The previous owner charged $240 per month for a single apartment and on the average rented 32 apartments at that price. The entrepreneur discovers that for every $20 he raises the price, another apartment stands vacant.

   a. Let x represent the number of $20 price increases. Write expressions for the new price, the number of rented apartments, and the total revenue.

   b. What price should the entrepreneur charge for an apartment in order to maximize his revenue?

9. The owners of a small fruit orchard decide to produce gift baskets. Their costs for producing x baskets are

$$C = 0.1x^2 - 20x + 1800$$

   How many baskets should they produce in order to minimize their costs? How much will each basket cost to produce?

10. A new electronics firm is considering marketing a line of telephones. Its costs for producing x telephones are

$$C = 0.2x^2 - 60x + 9000$$

    How many telephones should it produce in order to minimize its costs? How much will each telephone cost to produce?

11. During a statistical survey, a public-interest group obtains two estimates for the average monthly income of young adults aged 18 to 25. The first estimate is $860 and the second estimate is $918. To refine their estimate, they will take a weighted average of these two figures:

$$I = 860a + (1 - a)\,918 \qquad \text{where} \quad 0 \le a \le 1$$

To get the best estimate, they must choose a to minimize the expression

$$V = 576a^2 + 5184(1 - a)^2$$

(The numbers that appear in this expression reflect the **variance** of the data, which measures how closely the data cluster around the mean, or average.) Find the value of a that minimizes V, and use this value to get a refined estimate for the average income.

12. The rate at which an antigen precipitates during an antigen-antibody reaction depends upon the amount of antigen present. For a fixed quantity of antibody, the time required for a particular antigen to precipitate is given in minutes by

$$t = 2w^2 - 20w + 54$$

where w is the quantity of antigen present, in grams. For what quantity of antigen will the reaction proceed most rapidly? How long will the precipitation take?

In Exercises 13–20 find a quadratic equation that fits the given data.

13. Find values for a, b, and c so that the graph of the parabola $y = ax^2 + bx + c$ includes the points $(-1, 0)$, $(2, 12)$, and $(-2, 8)$.

14. Find values for a, b, and c so that the graph of the parabola $y = ax^2 + bx + c$ includes the points $(-1, 2)$, $(1, 6)$, and $(2, 11)$.

15. The following data were collected in a survey to determine what percent of different age groups regularly use marijuana.

Age	15	20	25	30
Percent	4	13	11	7

a. Use the percents for ages 15, 20, and 30 to fit a quadratic equation to the data,

$$P = ax^2 + bx + c$$

where x represents age.
b. What does your equation predict for the percent of 25-year-olds who use marijuana?
c. Sketch the graph of your quadratic equation and the given data on the same axes.

16. The following data show the number of people of certain ages who were victims of homicide in a large city last year.

Age	10	20	30	40
Number of Victims	12	62	72	40

a. Use the first three data points to fit a quadratic equation to the data, $N = ax^2 + bx + c$, where x represents age.
b. What does your equation predict for the number of 40-year-olds who were victims of homicide?
c. Sketch the graph of your quadratic equation and the given data on the same axes.

17. The following data show Americans' annual per capita consumption of chicken for the years 1986 to 1990.

Year	1986	1987	1988	1989	1990
Pounds of Chicken	51.3	55.5	57.4	60.8	63.6

a. Use the values for 1987 through 1989 to fit a quadratic equation to the data,

$$C = at^2 + bt + c$$

where t is measured in years since 1985.
b. What does your equation predict for per capita chicken consumption in 1990?
c. Sketch the graph of your equation and the given data on the same axes.

18. The following data show sales of in-line skates at a sporting goods store at the beach.

Year	1991	1992	1993	1994	1995
Skates Sold	54	82	194	446	726

a. Use the values for 1992 through 1994 to fit a quadratic equation to the data,

$$S = at^2 + bt + c$$

where t is measured in years since 1991.

b. What does your equation predict for the number of pairs of skates sold in 1995?

c. Sketch the graph of your equation and the given data on the same axes.

19. Find a quadratic formula for the number of diagonals that can be drawn in a polygon of n sides. Some data are provided.

Sides	4	5	6	7
Diagonals	2	5	9	14

20. Find a quadratic formula for the distance in feet that your car travels in t seconds after you step on the brakes while traveling at 60 miles per hour. Some data are provided.

Seconds	1	2	3	4
Feet	81	148	210	240

21. The cables on a suspension bridge hang in the shape of parabolas. Imagine a coordinate system superimposed on a diagram of a suspension bridge, as shown in Figure 4.60. Each tower is 500 feet high, and the span between the towers is 4000 feet long. At its lowest point the cable hangs 20 feet above the roadway. Find the coordinates of three points on the cable, and use them to find an equation for the shape of the cable.

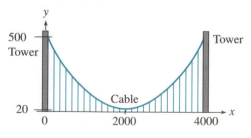

Figure 4.60

22. Some comets move about the sun in parabolic orbits. In 1973 the comet Kohoutek passed within 0.14 AU (astronomical units), or 21 million kilo-

meters, of the sun. Imagine a coordinate system superimposed on a diagram of the comet's orbit, as shown in Figure 4.61. In that system the comet's coordinates at perihelion (its closest approach to the sun) were (0, 0.14). When the comet was first discovered, its coordinates were (1.68, −4.9). A third point on its orbit had coordinates (1, −1.66). Find an equation for comet Kohoutek's orbit.

In Exercises 23–34 solve each system algebraically, and verify your solutions with a graph.

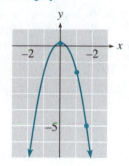

Figure 4.61

23. $y = x^2 - 4x + 7$
$y = 11 - x$

24. $y = x^2 + 6x + 4$
$y = 3x + 8$

25. $y = -x^2 - 2x + 7$
$y = 2x + 11$

26. $y = x^2 - 8x + 17$
$y + 4x = 13$

27. $y = x^2 + 8x + 8$
$3y + 2x = -36$

28. $y = -x^2 + 4x + 2$
$4y - 3x = 24$

29. $y = x^2 - 9$
$y = -2x^2 + 9x + 21$

30. $y = 4 - x^2$
$y = 3x^2 - 12x - 12$

31. $y = x^2 - 0.5x + 3.5$
$y = -x^2 + 3.5x + 1.5$

32. $y = x^2 + 10x + 22$
$y = -0.5x^2 - 8x - 32$

33. $y = x^2 - 4x + 4$
$y = x^2 - 8x + 16$

34. $y = 0.5x^2 + 3x + 5.5$
$y = 2x^2 + 12x + 4$

For Exercises 35–38 use the fact that

Profit = Revenue − Cost

a. Graph the equations for revenue and cost in the same window.
b. Use a system of equations to find the break-even point(s), and verify your solutions on the graph.

35. Writewell makes fountain pens. It costs Writewell $C = 8x + 4000$ dollars to manufacture x pens, and the company receives

$$R = -0.02(x - 800)^2 + 12,800$$

dollars in revenue from the sale of the pens. Use the equation for R to explain why the maximum revenue is earned when 800 pens are sold, and find the maximum revenue. Use these values to help you choose an appropriate window for your graphs.

36. It costs The Sweetshop $C = 3x + 3000$ dollars to produce x pounds of chocolate creams. The company brings in

$$R = -0.03(x - 600)^2 + 10,800$$

dollars in revenue from the sale of the chocolates. Use the equation for R to explain why the maximum revenue is earned when 600 pounds of chocolates are sold, and find the maximum revenue. Use these values to help you choose an appropriate window for your graphs.

37. It costs an appliance manufacturer $C = 175x + 21,875$ dollars to produce x top-loading washing machines, which will then bring in revenues of

$$R = -1.75(x - 200)^2 + 70,000$$

dollars. Use the equation for R to explain why the maximum revenue is earned when 200 washing machines are sold, and find the maximum revenue. Use these values to help you choose an appropriate window for your graphs.

38. A company can produce x lawn mowers for a cost of $C = 80x + 12,800$ dollars. The sale of the lawn mowers will generate

$$R = -2(x - 120)^2 + 28,800$$

dollars in revenue. Use the equation for R to explain why the maximum revenue is earned when 120 lawn mowers are sold, and find the maximum revenue. Use these values to help you choose an appropriate window for your graphs.

Chapter 4 Review

In Exercises 1–4 solve by factoring.

1. $x(3x + 2) = (x + 2)^2$

2. $6y = (y + 1)^2 + 3$

3. $4x - (x + 1)(x + 2) = -8$

4. $3(x + 2)^2 = 15 + 12x$

In Exercises 5 and 6 write a quadratic equation with integer coefficients and with the given solutions.

5. $\dfrac{-3}{4}$ and 8

6. $\dfrac{5}{3}$ and $\dfrac{5}{3}$

In Exercises 7 and 8 graph each equation using the *ZDecimal* setting. Locate the *x*-intercepts and use them to write the quadratic expression in factored form.

7. $y = x^2 - 0.6x - 7.2$

8. $y = -x^2 + 0.7x + 2.6$

Solve Exercises 9 and 10 by extraction of roots.

9. $(2x - 5)^2 = 9$ **10.** $(7x - 1)^2 = 15$

Solve Exercises 11–14 by completing the square.

11. $x^2 - 4x - 6 = 0$ **12.** $x^2 + 3x = 3$

13. $2x^2 + 3 = 6x$ **14.** $3x^2 = 2x + 3$

Solve Exercises 15–18 by using the quadratic formula. Give your answers as decimal approximations accurate to two decimal places.

15. $\dfrac{1}{2}x^2 + 1 = \dfrac{3}{2}x$ **16.** $x^2 - 3x + 1 = 0$

17. $x^2 - 4x + 2 = 0$ **18.** $2x^2 + 2x = 3$

19. In a tennis tournament among n competitors, $\dfrac{n(n-1)}{2}$ matches must be played. If the organizers can schedule 36 matches, how many players should they invite?

20. The formula $S = \dfrac{n(n+1)}{2}$ gives the sum of the first n positive integers. How many consecutive integers must be added to make a sum of 91?

21. Lewis invested $2000 in an account that compounds interest annually. He made no deposits or withdrawals after that. Two years later he closed the account, withdrawing $2464.20. What interest rate did Lewis earn?

22. Earl borrowed $5500 from his uncle for 2 years with interest compounded annually. At the end of 2 years he owed his uncle $6474.74. What was the interest rate on the loan?

23. The credit union divides $12,600 equally among its members each year as a dividend. This year there are six fewer members than last year, and each member receives $5 more. How many members are in the credit union this year?

24. Irene wants to enclose two adjacent chicken coops of equal size against the henhouse wall. She has 66 feet of chicken-wire fencing and would like the total area of the two coops to be 360 square feet. What should the dimensions of the chicken coops be?

For Problems 25–32 sketch the graph. Find the coordinates of the vertex and the intercepts algebraically. Verify your answers using your calculator's TRACE feature.

25. $y = \dfrac{1}{2}x^2$ **26.** $y = x^2 - 4$

27. $y = x^2 - 9x$ **28.** $y = -2x^2 - 4x$

29. $y = x^2 - x - 12$

30. $y = -2x^2 + x - 4$

31. $y = -x^2 + 2x + 4$

32. $y = x^2 - 3x + 4$

In Exercises 33–38 solve each inequality algebraically. Verify your solutions by graphing.

33. $(x - 3)(x + 2) > 0$ **34.** $y^2 - y - 12 \leq 0$

35. $2y^2 - y \leq 3$ **36.** $3z^2 - 5z > 2$

37. $s^2 \leq 4$ **38.** $4t^2 > 12$

39. The Sub Station sells $120 - \frac{1}{4}p$ submarine sandwiches at lunchtime if it sells them at p cents each. What range of prices can the Sub Station charge if it wants to keep its daily revenue from subs over \$140?

40. When it charges p dollars for an electric screwdriver, Handy Hardware will sell $30 - \frac{1}{2}p$ screwdrivers per month. How much should Handy Hardware charge per screwdriver if it wants the monthly revenue from the screwdrivers to be over \$400?

41. A farmer inherits an apple orchard on which 60 trees are planted. Each tree yields 12 bushels of apples. Experimentation has shown that for each tree removed, the yield per tree increases by $\frac{1}{2}$ bushel. How many trees should be removed in order to maximize the total apple harvest?

42. A small company manufactures radios. When it charges \$20 for a radio, it sells 500 radios per month. For each dollar the price is increased, 10 fewer radios are sold per month. What should

the company charge for a radio in order to maximize its monthly revenue?

In Exercises 43–48 solve each system algebraically, and verify your solution with a graph.

43. $y + x^2 = 4$
$y = 3$

44. $y = 3 - x^2$
$5x + y = 7$

45. $y = x^2 - 5$
$y = 4x$

46. $y = x^2 - 2x + 1$
$y = 3 - x$

47. $y = x^2 - 6x + 20$
$y = 2x^2 - 2x - 25$

48. $y = x^2 - 5x - 28$
$y = -x^2 + 4x + 28$

49. Find values of a, b, and c so that the graph of the parabola $y = ax^2 + bx + c$ contains the points $(-1, -4)$, $(0, -6)$, and $(4, 6)$.

50. Find a parabola that fits the following data points.

x	−8	−4	2	4
y	10	18	0	−14

Functions and Their Graphs

$\mathcal{M}$any applied problems involve relationships among two or more variables. We often want to predict the values of one variable from the values of a related variable. For example, when a physician prescribes a drug in a certain dosage, she needs to know how long the dose will remain in the bloodstream. A pilot needs to know how the speed of the prevailing winds affects flight time. Such a relationship between variables is called a function. In this chapter we study some properties of functions and their graphs and examine several common functions.

5.1 DEFINITIONS AND NOTATION

We have already encountered some examples of functions in Chapters 2 and 4. For example, suppose it costs $800 for flying lessons plus $30 per hour to rent a plane. If we let C represent the total cost for t hours of flying lessons, then the equation

$$C = 30t + 800 \qquad (t \geq 0) \tag{1}$$

relates the variables t and C. In other words, with each value of $t \geq 0$ we associate the value of C given by Equation (1). Thus, for example,

$$\begin{aligned}
\text{when} \quad t = 0, \qquad &C = 30(0) + 800 = 800 \\
\text{when} \quad t = 4, \qquad &C = 30(4) + 800 = 920 \\
\text{when} \quad t = 10, \qquad &C = 30(10) + 800 = 1100
\end{aligned}$$

The variable t in Equation (1) is called the **independent variable**; C is the **dependent variable** because its values are determined by the value of t. We can display an association between the values of two variables by a table (such as Table 5.1) or by ordered pairs. The independent variable is the first component of the ordered pair, and the dependent variable is the second component. For the example above we have

t	C	(t, C)
0	800	(0, 800)
4	920	(4, 920)
10	1100	(10, 1100)

Table 5.1

Notice that for this relationship, defined by Equation (1), we can always determine the value of C associated with any given value of t. All we have to do is substitute the value of t into Equation (1) and solve for C. Notice also that the result has no ambiguity: only *one* value for C corresponds to each value of t. This type of relationship between variables is called a **function**. In general, we make the following definition.

> A **function** is a relationship between two variables for which a *unique* value of the **dependent variable** can be determined from a value of the **independent variable**.

What distinguishes functions from other variable relationships? The definition of a function calls for a *unique value,* that is, *exactly one value,* of the dependent variable corresponding to each value of the independent variable. This property makes functions more useful in applications than other types of variable relationships. In fact, the linear relationships we studied in Chapter 2 and the quadratic relationships in Chapter 4 were all examples of functions.

Example 1

a.   The distance, d, traveled by a car in 2 hours is a function of its speed, r. If we know the speed, r, of the car, we can determine the distance it travels by the formula $d = r \cdot 2$.
b.   The price of a fill-up with unleaded gasoline is a function of the number of gallons purchased. The gas pump represents the function by displaying the corresponding values of the independent variable (number of gallons) and the dependent variable (total cost).
c.   Score on the Scholastic Aptitude Test (SAT) is *not* a function of score on an IQ test, since two people with the same score on an IQ test may score differently on the SAT; that is, a person's score on the SAT is not uniquely determined by his or her score on an IQ test.

A function can be described in several different ways. In the following examples we consider functions defined by tables and by equations. In Section 5.2 we will consider functions defined by graphs.

Example 2

Table 5.2 shows data on sales compiled over several years by the accounting office for Eau Claire Auto Parts, a division of Major Motors.

Year (t)	Total Sales (S)
1990	$612,000
1991	$663,000
1992	$692,000
1993	$749,000
1994	$804,000

Table 5.2

In this example the year is considered the independent variable, and we say that the total sales, *S, is a function of t.* ▭

Example 3

Table 5.3 gives the cost of sending printed material by first-class mail.

Weight in Ounces (w)	Postage (p)
$0 < w \le 1$	$0.32
$1 < w \le 2$	$0.55
$2 < w \le 3$	$0.78
$3 < w \le 4$	$1.01
$4 < w \le 5$	$1.24
$5 < w \le 6$	$1.47
$6 < w \le 7$	$1.70

Table 5.3

If we know the weight of the article being shipped, we can determine the required postage from Table 5.3. For instance, a catalog weighing 4.5 ounces would require $1.24 in postage. In this example *w* is the independent variable and *p* is the dependent variable. We say that *p is a function of w.* ▭

Example 4

Table 5.4 on page 266 records the age and cholesterol count for 20 patients tested in a hospital survey. According to these data, cholesterol count is *not* a function of age, since several patients who are the same age have different cholesterol levels. For example, patients 316, 332, and 340 are all 51 years old but have cholesterol counts of 227, 209, and 216, respectively. Thus we

Patient Number	Age	Cholesterol Count	Patient Number	Age	Cholesterol Count
301	53	217	332	51	209
308	48	232	336	53	241
312	55	198	339	49	186
313	56	238	340	51	216
316	51	227	343	57	208
320	52	264	347	52	248
322	53	195	356	50	214
324	47	203	359	56	271
325	48	212	362	53	193
328	50	234	370	48	172

Table 5.4

cannot determine a *unique* value of the dependent variable (cholesterol count) from the value of the independent variable (age). Other factors besides age must influence a patient's cholesterol count.   ▭

Example 5 illustrates a function defined by an equation.

Example 5

The Sears Tower in Chicago is the world's tallest building at 1454 feet. If an algebra book is dropped from the top of the Sears Tower, its height above the ground after t seconds is given by the equation

$$h = 1454 - 16t^2 \tag{2}$$

Thus after 1 second the book's height is

$$h = 1454 - 16(1)^2 = 1438 \text{ feet}$$

and after 2 seconds its height is

$$h = 1454 - 16(2)^2 = 1390 \text{ feet}$$

For this function, t is the independent variable and h is the dependent variable. Notice that for any value of t the corresponding unique value of h can be determined from Equation (2). We say that *h is a function of t.*   ▭

Function Notation

There is a convenient notation we can use when discussing functions. First we choose a letter, such as f, g, or h, or F, G, or H, to name a particular function. (We can use any letter, but these are the most common choices.)

For instance, in Example 5 we expressed the height, h, of an algebra book falling from the top of the Sears Tower as a function of the time, t, that it has been falling. We will call this function f; that is, f is the name of the relationship between the variables h and t. We can then write

$$h = f(t)$$

which means "h is a function of t, and f is the name of the function."

The new symbol $f(t)$, read "f of t," is another name for the height, h. Note that the parentheses in the symbol $f(t)$ do *not* indicate multiplication. (It would not make sense to multiply the name of a function by a variable.) Think of the symbol $f(t)$ as a single entity that represents the dependent variable of the function.

With this new notation we may write

$$h = f(t) = 1454 - 16t^2$$

or just

$$f(t) = 1454 - 16t^2$$

instead of

$$h = 1454 - 16t^2$$

to describe the function.

Perhaps it seems we are complicating things needlessly by introducing a new symbol for h, but the notation $f(t)$ is very useful for showing the correspondence between specific values of the variables h and t. In Example 5 we saw that

$$\text{when } t = 1, \quad h = 1438$$
$$\text{when } t = 2, \quad h = 1390$$

Using function notation, these relationships can be expressed more concisely as

$$f(1) = 1438 \quad \text{and} \quad f(2) = 1390$$

which we read as "f of 1 equals 1438" and "f of 2 equals 1390." Notice that the values for the independent variable, t, appear *inside* the parentheses, and the values for the dependent variable, h, appear on the other side of the equation.

$$f(t) = h$$

Independent variable ⎯⎯⎯⎯⎯ ⎯⎯ Dependent variable

Example 6 Let g be the name of the postage function defined by Table 5.3 in Example 3. Find $g(1)$, $g(3)$, and $g(6.75)$.

Solution According to Table 5.3,

$$\text{when} \quad w = 1, \qquad p = 0.32 \qquad \text{so} \qquad g(1) = 0.32$$

$$\text{when} \quad w = 3, \qquad p = 0.78 \qquad \text{so} \qquad g(3) = 0.78$$

$$\text{when} \quad w = 6.75, \quad p = 1.70 \qquad \text{so} \qquad g(6.75) = 1.70$$

Thus a letter weighing 1 ounce costs \$0.32 to mail, a letter weighing 3 ounces costs \$0.78, and a letter weighing 6.75 ounces costs \$1.70.

Finding the value of the dependent variable that corresponds to a particular value of the independent variable is called **evaluating the function**. If a function is described by an equation, we simply substitute the given value into the equation to find the corresponding function value.

Example 7 The function H is defined by $H(s) = \dfrac{\sqrt{s + 3}}{s}$.   Evaluate the function at the given values.

a. $s = 6$ **b.** $s = -1$

Solutions Substitute the given values for s into the equation defining H.

a. $H(6) = \dfrac{\sqrt{6 + 3}}{6} = \dfrac{\sqrt{9}}{6} = \dfrac{3}{6} = \dfrac{1}{2}$

b. $H(-1) = \dfrac{\sqrt{-1 + 3}}{-1} = \dfrac{\sqrt{2}}{-1} = -\sqrt{2}$

Example 8 Make a table displaying four ordered pairs for the function $f(x) = 5 - x^3$.

Solution Choose several values for x and evaluate the function to find the corresponding $f(x)$-value for each. (See Table 5.5.)

x	f(x)
−2	13
0	5
1	4
3	−22

$f(-2) = 5 - (-2)^3 = 5 - (-8) = 13$

$f(0) = 5 - 0^3 = 5 - 0 = 5$

$f(1) = 5 - 1^3 = 5 - 1 = 4$

$f(3) = 5 - 3^3 = 5 - 27 = -22$

Table 5.5

To simplify the notation, we sometimes use the same letter for the dependent variable and for the name of the function. In the next example C is used in this way.

Example 9

TrailGear decides to market a line of backpacks. The cost, C, of manufacturing backpacks is a function of the number, x, of backpacks produced, given by the equation

$$C(x) = 3000 + 20x$$

where $C(x)$ is measured in dollars. Find the cost of producing 500 backpacks.

Solution   To find the value of C that corresponds to $x = 500$, evaluate $C(500)$:

$$C(500) = 3000 + 20(500) = 13,000$$

The cost of producing 500 backpacks is $13,000.

Operations with Function Notation

Sometimes we need to evaluate a function at an algebraic expression rather than a specific number.

Example 10

TrailGear finds that the monthly demand for backpacks increases by 50% during the summer. Their backpacks are produced at several small co-ops in different states. If each co-op usually produces b backpacks per month, how many should they produce during the summer months? What costs for producing backpacks should they expect during the summer? (See Example 9.)

Solution   A co-op that usually produces b backpacks should produce $1.5b$ backpacks during the summer months. The cost of production will be

$$C(1.5b) = 3000 + 20(1.5b) = 3000 + 30b$$

Example 11

Evaluate the function $f(x) = 4x^2 - x + 5$ for the following expressions.
a. $x = 2h$ **b.** $x = a + 3$

Solutions
a. $f(2h) = 4(2h)^2 - (2h) + 5$
$$= 4(4h^2) - 2h + 5 = 16h^2 - 2h + 5$$

b. $f(a + 3) = 4(a + 3)^2 - (a + 3) + 5$
$$= 4(a^2 + 6a + 9) - a - 3 + 5$$
$$= 4a^2 + 24a + 36 - a + 2$$
$$= 4a^2 + 23a + 38$$

COMMON ERRORS

In Example 11 notice that

$$f(2h) \neq 2f(h)$$

and

$$f(a + 3) \neq f(a) + f(3)$$

To compute $f(a) + f(3)$,   we must first compute $f(a)$ and $f(3)$ and then add them:

$$f(a) + f(3) = (4a^2 - a + 5) + (4 \cdot 3^2 - 3 + 5)$$
$$= 4a^2 - a + 43$$

In general, it is *not* true that $f(a + b) = f(a) + f(b)$.   Remember that the parentheses in the expression $f(x)$ do not indicate multiplication, so the distributive law does not apply to the expression $f(a + b)$. However, keep in mind that $f(x)$ represents a *single* value, so we can perform operations on it just as we do on any other variable.

Example 12

Define $f(x) = x^3 - 1$,   and evaluate each expression.
a. $f(2) + f(3)$ **b.** $f(2 + 3)$ **c.** $2f(x) + 3$

Solutions

a. $f(2) + f(3) = (2^3 - 1) + (3^3 - 1)$
$$= 7 + 26 = 33$$

b. $f(2 + 3) = f(5) = 5^3 - 1 = 124$

c. $2f(x) + 3 = 2(x^3 - 1) + 3$
$$= 2x^3 - 2 + 3 = 2x^3 + 1$$

Using the Table Feature to Evaluate Functions

The TI-82 table feature gives us a convenient and simple tool for evaluating functions. We will demonstrate this feature using the function of Example 8, $f(x) = 5 - x^3$.

Press $\boxed{Y =}$, clear any old functions, and enter

$$Y_1 = 5 - X \boxed{\wedge} 3$$

Then press $\boxed{\text{TblSet}}$ ($\boxed{\text{2nd}}$ $\boxed{\text{WINDOW}}$) and choose **Ask** after **Indpnt**, as shown in Figure 5.1(a) on page 271. This setting allows you to enter any x-values you like. Next press $\boxed{\text{TABLE}}$ (using $\boxed{\text{2nd}}$ $\boxed{\text{GRAPH}}$).

To follow Example 8, key in $\boxed{(-)}$ 2 $\boxed{\text{ENTER}}$ for the x-value, and the calculator will fill in the y-value. Continue by entering 0, 1, 3, or any other x-values you choose. One such table is shown in Figure 5.1(b).

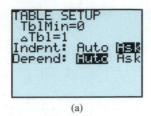

Figure 5.1   (a)   (b)

If you would like to evaluate a new function, you do not have to return to the $\boxed{Y =}$ screen. Use the $\boxed{\triangleright}$ and $\boxed{\triangle}$ keys to highlight Y_1 at the top of the second column. The definition of Y_1 will appear at the bottom of the display, as shown in Figure 5.1(b). You can key in a new definition here, and the second column will be updated automatically to show the y-values of the new function.

Exercise 5.1

For which of the pairs in Exercises 1–6 is the second quantity a function of the first? Explain your answers.

1. Price of an item; sales tax on the item at 4%

2. Time traveled at constant speed; distance traveled

3. Number of years of education; annual income

4. Distance flown in an airplane; price of the ticket

5. Volume of a container of water; its weight

6. Amount of a paycheck; amount of Social Security tax withheld

Each object in Exercises 7–14 establishes a correspondence between two variables. Suggest appropriate independent and dependent variables and decide whether the relationship is a function.

7. An itemized grocery receipt

8. An inventory list

9. An index

10. A will

11. An instructor's grade book

12. An address book

13. A bathroom scale

14. A radio dial

Which of the tables in Exercises 15–26 describe functions? Why or why not?

15.

x	t
−1	2
0	9
1	−2
0	−3
−1	5

16.

y	w
0	8
1	12
3	7
5	−3
7	4

17.

x	y
−3	8
−2	3
−1	0
0	−1
1	0
2	3
3	8

18.

s	t
2	5
4	10
6	15
8	20
6	25
4	30
2	35

19.

r	−4	−2	0	2	4
v	6	6	3	6	8

20.

p	−5	−4	−3	−2	−1
d	−5	−4	−3	−2	−1

21.

Pressure (p)	Volume (v)
15	100.0
20	75.0
25	60.0
30	50.0
35	42.8
40	37.5
45	33.3
50	30.0

22.

Frequency (f)	Wavelength (w)
5	60.0
10	30.0
20	15.0
30	10.0
40	7.5
50	6.0
60	5.0
70	4.3

23.

Temperature (T)	Humidity (h)
Jan. 1 34°F	42%
Jan. 2 36°F	44%
Jan. 3 35°F	47%
Jan. 4 29°F	50%
Jan. 5 31°F	52%
Jan. 6 35°F	51%
Jan. 7 34°F	49%

24.

Inflation Rate (I)	Unemployment Rate (U)
1972 5.6%	5.1%
1973 6.2%	4.5%
1974 10.1%	4.9%
1975 9.2%	7.4%
1976 5.8%	6.7%
1977 5.6%	6.8%
1978 6.7%	7.4%

25.

Adjusted Gross Income (*I*)	Tax Bracket (*T*)
0–2479	0%
2480–3669	11%
3670–4749	12%
4750–7009	14%
7010–9169	15%
9170–11,649	16%
11,650–13,919	18%

26.

Cost of Merchandise (*M*)	Shipping Charge (*C*)
$0.01–10.00	$2.50
10.01–20.00	3.75
20.01–30.00	4.85
30.01–50.00	5.95
50.01–75.00	6.95
75.01–100.00	7.95
Over 100.00	8.95

Evaluate each function for the given values in Exercises 27–34.

27. $f(x) = 6 - 2x$

  a. $f(3)$ **b.** $f(-2)$ **c.** $f(12.7)$ **d.** $f\left(\dfrac{2}{3}\right)$

28. $g(t) = 5t - 3$

  a. $g(1)$ **b.** $g(-4)$

  c. $g(14.1)$ **d.** $g\left(\dfrac{3}{4}\right)$

29. $h(v) = 2v^2 - 3v + 1$

  a. $h(0)$ **b.** $h(-1)$

  c. $h\left(\dfrac{1}{4}\right)$ **d.** $h(-6.2)$

30. $r(s) = 2s - s^2$

  a. $r(2)$ **b.** $r(-4)$

  c. $r\left(\dfrac{1}{3}\right)$ **d.** $r(-1.3)$

31. $H(z) = \dfrac{2z - 3}{z + 2}$

  a. $H(4)$ **b.** $H(-3)$ **c.** $H\left(\dfrac{4}{3}\right)$ **d.** $H(4.5)$

32. $F(x) = \dfrac{1 - x}{2x - 3}$

  a. $F(0)$ **b.** $F(-3)$ **c.** $F\left(\dfrac{5}{2}\right)$ **d.** $F(9.8)$

33. $E(t) = \sqrt{t - 4}$

  a. $E(16)$ **b.** $E(4)$ **c.** $E(7)$ **d.** $E(4.2)$

34. $D(r) = \sqrt{5 - r}$

  a. $D(4)$ **b.** $D(-3)$ **c.** $D(-9)$ **d.** $D(4.6)$

35. The function described in Exercise 21 is called *g*, so that $v = g(p)$. Find the following.
  a. $g(25)$ **b.** $g(40)$
  c. *x* so that $g(x) = 50$

36. The function described in Exercise 22 is called *h*, so that $w = h(f)$. Find the following.
  a. $h(20)$ **b.** $h(60)$
  c. *x* so that $h(x) = 10$

37. The function described in Exercise 25 is called *T*, so that $T = T(I)$. Find the following.
  a. $T(8750)$ **b.** $T(6249)$
  c. *x* so that $T(x) = 15\%$

38. The function described in Exercise 26 is called *C*, so that $C = C(M)$. Find the following.
  a. $C(11.50)$ **b.** $C(47.24)$
  c. *x* so that $C(x) = 7.95$

For each function described in Exercises 39–46,
a. write short phrases describing the independent and dependent variables.
b. make a table of ordered pairs for at least five choices of the independent variable.
c. answer the question in the exercise.

39. A computer costs $28,000 and depreciates according to the formula

$$V(t) = 28{,}000(1 - 0.06t)$$

where *V* is the value of the computer after *t* years. Evaluate $V(10)$ and explain what it means.

40. In a profit-sharing plan an employee receives a salary of

$$S(x) = 20{,}000 + 0.01x$$

where x represents the company's profit for the year. Evaluate $S(850{,}000)$ and explain what it means.

41. An advertising agency accrues revenue in thousands of dollars according to the formula

$$R(x) = 50x^2 - 200x + 800$$

where x represents the number of its clients. Evaluate $R(40)$ and explain what it means.

42. A manufacturer of machine parts finds that his profit is given by the formula

$$P(x) = 0.02x^2 - 800x - 500{,}000 \ (x \geq 10{,}000)$$

where x is the number of parts manufactured. Evaluate $P(20{,}000)$ and explain what it means.

43. The number of compact cars that a large dealership can sell at price p is given by

$$N(p) = \frac{12{,}000{,}000}{p}$$

Evaluate $N(6000)$ and explain what it means.

44. A department store finds that the market value of its Christmas-related merchandise is given by

$$M(t) = \frac{600{,}000}{t} \qquad (1 \leq t \leq 30)$$

where t is the number of weeks after Christmas. Evaluate $M(12)$ and explain what it means.

45. The velocity of a car that brakes suddenly can be determined from the length of its skid marks, d, by

$$v(d) = \sqrt{12d}$$

where d is in feet and v is in miles per hour. Evaluate $v(250)$ and explain what it means.

46. The distance, d, in miles that a person can see on a clear day from a height h in feet is given by

$$d(h) = 1.22\sqrt{h}$$

Evaluate $d(20{,}320)$ and explain what it means.

Evaluate the function at the given algebraic expressions in Exercises 47–52.

47. $G(s) = 3s^2 - 6s$
    a. $G(3a)$ **b.** $G(a + 2)$
    c. $G(a) + 2$ **d.** $G(-a)$

48. $h(x) = 2x^2 + 6x - 3$
    a. $h(2a)$ **b.** $h(a + 3)$
    c. $h(a) + 3$ **d.** $h(-a)$

49. $g(x) = 8$
    a. $g(2)$ **b.** $g(8)$
    c. $g(a + 1)$ **d.** $g(-x)$

50. $f(t) = -3$
    a. $f(4)$ **b.** $f(-3)$
    c. $f(b - 2)$ **d.** $f(-t)$

51. $P(x) = x^3 - 1$
    a. $P(2x)$ **b.** $2P(x)$ **c.** $P(x^2)$ **d.** $[P(x)]^2$

52. $Q(t) = 5t^3$
    a. $Q(2t)$ **b.** $2Q(t)$ **c.** $Q(t^2)$ **d.** $[Q(t)]^2$

For each function in Exercises 53–60, compute

  a. $f(2) + f(3)$ **b.** $f(2 + 3)$
  c. $f(a) + f(b)$ **d.** $f(a + b)$

For which functions does $f(a + b) = f(a) + f(b)$?

53. $f(x) = 3x - 2$ **54.** $f(x) = 1 - 4x$

55. $f(x) = x^2 + 3$ **56.** $f(x) = x^2 - 1$

57. $f(x) = \sqrt{x + 1}$ **58.** $f(x) = \sqrt{6 - x}$

59. $f(x) = \dfrac{-2}{x}$ **60.** $f(x) = \dfrac{3}{x}$

For each pair of functions in Exercises 61–64,
  a. compute $f(0)$ and $g(0)$.
  b. find all values of x for which $f(x) = 0$.
  c. find all values of x for which $g(x) = 0$.
  d. find all values of x for which $f(x) = g(x)$.

61. $f(x) = 2x^2 + 3x, \quad g(x) = 5 - 6x$

62. $f(x) = 3x^2 - 6x, \quad g(x) = 8 + 4x$

63. $f(x) = \sqrt{x + 2}, \quad g(x) = 3x - 4$

64. $f(x) = \sqrt{x - 3}, \quad g(x) = 2x - 7$

Find an equation that describes each function in Exercises 65–68.

65.

x	f(x)
1	1
2	4
3	7
4	10
5	13

66.

x	g(x)
1	16
2	12
3	8
4	4
5	0

67.

t	G(t)
1	2
2	5
3	10
4	17
5	26

68.

t	H(t)
1	2
2	8
3	18
4	32
5	50

5.2   GRAPHS OF FUNCTIONS

In Section 5.1 we considered functions defined by tables or by equations. Functions can also be described by graphs. For example, the graph in Figure 5.2 gives the Dow-Jones Industrial Average (the average value of the stock prices of 500 major companies) for the stock market "correction" of October 1987.

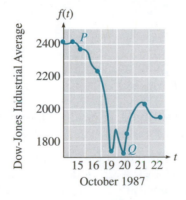

Figure 5.2
October 1987

In this example the Dow-Jones Industrial Average (DJIA) is given as a function of time during the 8 days from October 15 to October 22; that is, $f(t)$ is the DJIA recorded at time t. The values of the independent variable, time, are displayed on the horizontal axis, and the values of the dependent variable, DJIA, are displayed on the vertical axis. The coordinates of each point on the graph of the function represent a pair of corresponding values of the two variables. For example, point P shows that the DJIA was 2412 at noon on October 15, and point Q shows that the DJIA was 1726 at noon on October 20. In general, we can make the following statement.

> The point (a, b) lies on the graph of the function f if and only if $f(a) = b$.

Another way of saying this is

Each point on the graph has coordinates $(x, f(x))$ for some value of x.

Example 1   Consider the graph of the function g shown in Figure 5.3.

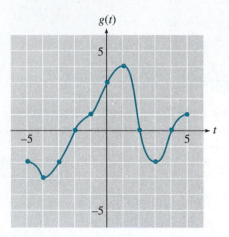

Figure 5.3

a. Find $g(-2)$, $g(0)$, and $g(5)$.

b. For what value(s) of t is $g(t) = -2$? For what value(s) of t is $g(t) = 0$?

c. What is the largest, or maximum, value of $g(t)$? For what value of t does the function take on its maximum value?

Solutions **a.** The points $(-2, 0)$, $(0, 3)$, and $(5, 1)$ lie on the graph of g. Therefore, $g(-2) = 0$, $g(0) = 3$, and $g(5) = 1$.

b. Since the points $(-3, -2)$ and $(3, -2)$ lie on the graph, $g(-3) = -2$ and $g(3) = -2$. (The t-values are -3 and 3.) Since the points $(-2, 0)$, $(2, 0)$, and $(4, 0)$ lie on the graph, $g(-2) = 0$, $g(2) = 0$, and $g(4) = 0$. (The t-values are -2, 2, and 4.)

c. The maximum value of $g(t)$ is the second coordinate of the highest point on the graph, $(1, 4)$. Thus the maximum value of $g(t)$ is 4, and it occurs when $t = 1$. $\Box$

Although we *define* some functions by their graphs, we can also construct graphs for functions described by tables or equations. We make these graphs in the same way we graph equations in two variables: substitute values for the independent variable into the equation to find the corresponding values of the dependent variable, and then plot the resulting ordered pairs.

| *Example 2* | Graph the function $f(x) = \sqrt{x + 4}$. |

Solution Choose several convenient values for x and evaluate the function to find the corresponding $f(x)$ values. (See Table 5.6.) Notice that we cannot choose x-values less than -4, because we cannot take the square root of a negative number.

x	f(x)
-4	0
-3	1
0	2
2	$\sqrt{6}$
5	3

Table 5.6

$f(-4) = \sqrt{-4 + 4} = \sqrt{0} = 0$

$f(-3) = \sqrt{-3 + 4} = \sqrt{1} = 1$

$f(0) = \sqrt{0 + 4} = \sqrt{4} = 2$

$f(2) = \sqrt{2 + 4} = \sqrt{6} \approx 2.45$

$f(5) = \sqrt{5 + 4} = \sqrt{9} = 3$

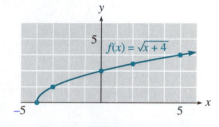

Figure 5.4

Plot the points and connect them with a smooth curve to obtain the graph in Figure 5.4. Notice that no points on the graph have x-coordinates less than -4.

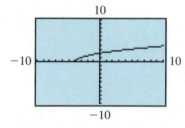

Figure 5.5

We can also use a graphing calculator to obtain the table and graph in Example 2. Choose the standard window by pressing ZOOM 6; then enter the equation that describes the function. Press Y = and type in

(Your calculator does not use the $f(x)$ notation for graphs, so we will continue to use Y_1, Y_2, and so on for the dependent variable.) Do not forget to enclose $x + 4$ in parentheses, since it appears under a radical. The calculator's picture of the graph is shown in Figure 5.5.

Domain and Range   In Example 2 we graphed the function $f(x) = \sqrt{x + 4}$ and observed that $f(x)$ is undefined for x-values less than -4. When we studied linear and quadratic functions, we could find y-values for *any* x-value, but for this function we must choose x-values in the interval $[-4, \infty)$. The set of all permissible values of the independent variable is called the **domain** of the function f. All the points on the graph have x-coordinates greater than or equal to -4.

We can also see from the graph that no points have negative $f(x)$-values: all the points have $f(x)$-values greater than or equal to zero. The set of all function values corresponding to the x-values in the domain is called the **range** of the function. Thus the domain of the function $f(x) = \sqrt{x + 4}$

is the set $[-4, \infty)$, and its range is the set $[0, \infty)$. In general, we make the following definitions.

> The **domain** of a function is the set of permissible values for the independent variable. The **range** is the set of function values (i.e., values of the dependent variable) that correspond to the domain values.

We often think of the elements of the domain as the "input" values for a function and the elements of the range as the corresponding "output" values. Thus for the function $f(x) = \sqrt{x + 4}$, if we choose $x = 0$ as an input value, then the output value is $f(0) = \sqrt{0 + 4} = 2$.

Example 3

a. Determine the domain and range of the function h graphed in Figure 5.6.
b. For the indicated points show the domain values and their corresponding range values in the form of ordered pairs.

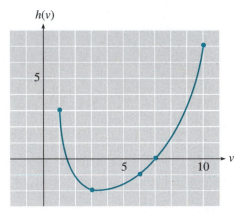

Figure 5.6

Solutions   a. All the points on the graph have v-coordinates between 1 and 10, so the domain of the function h is the interval $[1, 10]$. The $h(v)$-coordinates have values between -2 and 7, inclusive, so the range of the function is the interval $[-2, 7]$.
b. Recall that the points on the graph of a function have coordinates $(v, h(v))$. In other words, the coordinates of each point are made up of a domain value and its corresponding range value. Read the coordinates of the indicated points to obtain the ordered pairs $(1, 3)$, $(3, -2)$, $(6, -1)$, $(7, 0)$, and $(10, 7)$.   ▭

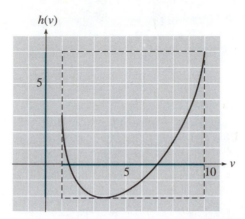

Figure 5.7

Figure 5.7 shows the graph of the function h in Example 3 with the domain values marked on the horizontal axis and the range values marked on the vertical axis. Imagine a rectangle whose length and width are determined by those segments, as shown in Figure 5.7. All the points $(v, h(v))$ on the graph of the function will lie within this rectangle. This rectangle is a convenient "window" in the plane for viewing the function. Of course, if the domain or range of the function is an infinite interval, we can never include the whole graph within a viewing rectangle and must be satisfied with studying only the "important" parts of the graph.

In many applications we may restrict the domain and range of a function to suit the situation at hand. Consider the graph of the function

$$h = f(t) = 1454 - 16t^2$$

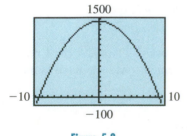

Figure 5.8

This function appeared in Example 5 of Section 5.1. It gives the height of an algebra book dropped from the top of the Sears Tower as a function of time. The graph of f is a parabola that opens downward. Its vertex occurs at the point $(0, 1454)$. You can use the window

$$\text{Xmin} = -10, \quad \text{Xmax} = 10$$

$$\text{Ymin} = -100, \quad \text{Ymax} = 1500$$

to obtain the graph shown in Figure 5.8.

Because t represents the time in seconds after the book was dropped, only positive t-values make sense for the problem. Also notice that the book stops falling when it hits the ground, at $h = 0$. Using the $\boxed{\text{TRACE}}$ key and the $\boxed{\text{ZOOM}}$ feature, you can verify that $h = 0$ at approximately $t = 9.5$ seconds. Thus only t-values between 0 and 9.5 are realistic in this application. During that time period, the height, h, of the book decreases from 1454 feet to 0 feet. The conditions of the problem prompt us to restrict the domain of the function f to the interval $[0, 9.5]$ and its range to $[0, 1454]$. The graph of this function is shown in Figure 5.9.

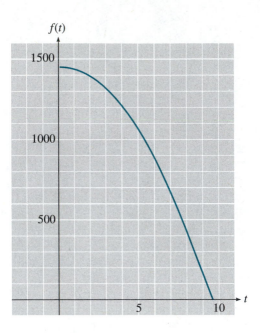

Figure 5.9

Sometimes the domain is given as part of a function's definition. In that case it is easiest to determine the range of the function from its graph.

Example 4

Graph the function $f(x) = x^2 - 6$ on the domain $0 \leq x \leq 4$ and give its range.

Solution   The graph is a parabola that opens upward. Obtain several points on the graph by evaluating the function at convenient x-values in the domain. (See Table 5.7.).

x	f(x)	
0	-6	since $f(0) = 0^2 - 6 = -6$
1	-5	since $f(1) = 1^2 - 6 = -5$
2	-2	since $f(2) = 2^2 - 6 = -2$
3	3	since $f(3) = 3^2 - 6 = 3$
4	10	since $f(4) = 4^2 - 6 = 10$

Table 5.7

The range of the function is the set of all $f(x)$-values that appear on the graph. We can see in Figure 5.10 on page 281 that the lowest point on the

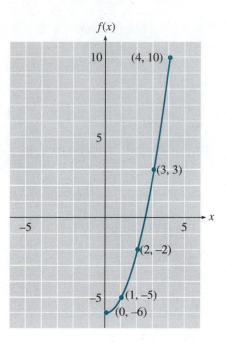

Figure 5.10

graph is $(0, -6)$, so the smallest $f(x)$-value is -6. The highest point on the graph is $(4, 10)$, so the largest $f(x)$-value is 10. Thus the range of the function f is the interval $[-6, 10]$.

Not all functions have domains and ranges that are intervals.

Example 5

a. Graph the postage function given in Example 3 of Section 5.1.
b. Determine the domain and range of the function.

Solutions   a.   From Table 5.3 (see page 265) note that articles of any weight up to 1 ounce require \$0.32 postage. This means that for all w-values greater than 0 but less than or equal to 1, the p-value is 0.32. Thus the graph of $p = g(w)$ between $w = 0$ and $w = 1$ looks like a small piece of the horizontal line $p = 0.32$. Similarly, for all w-values greater than 1 but less than or equal to 2, the p-value is 0.55, so the graph on this interval looks like a small piece of the line $p = 0.55$. Continue in this way to obtain the graph shown in Figure 5.11.

The open circles at the left endpoint of each horizontal segment indicate that that point is not included in the graph; the closed circles are points on the graph. For instance, if $w = 3$ the postage p is \$0.78, not \$1.01. Consequently, the point $(3, 0.78)$ is part of the graph of g, but the point $(3, 1.01)$ is not.

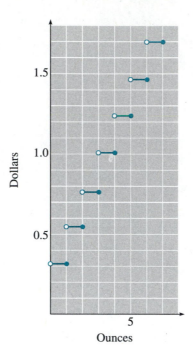

Figure 5.11   Ounces

b. Postage rates are given for all weights greater than 0 ounces up to and including 7 ounces, so the domain of the function is the half-open interval (0, 7]. (Note that there is a point on the graph for *every* w-value from 0 to 7.) The range of the function is *not* an interval, however, since the possible values for p do not include *all* the real numbers between 0.32 and 1.70. The range is the set of discrete values 0.32, 0.55, 0.78, 1.01, 1.24, 1.47, and 1.70.

Vertical Line Test   Using the notions of domain and range, we can restate the definition of a function as follows.

> A relationship between two variables is a function if each element of the domain is paired with only one element of the range.

If a variable relationship is a function, two different ordered pairs cannot have the same first component. What does this restriction mean in terms of the graph of the function?

Consider the graph shown in Figure 5.12(a). Every vertical line intersects the graph in at most one point; only one point on the graph corresponds to any given element of the domain. This graph represents a function. In Figure 5.12(b) the line $x = 2$ intersects the graph at two points, (2, 1) and

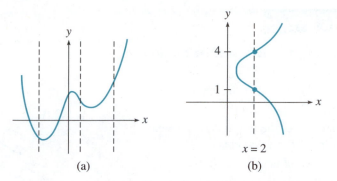

Figure 5.12   (a)   (b)

(2, 4). The domain value 2 is associated with two different range values, 1 and 4. This graph cannot be the graph of a function.

We summarize these observations as follows.

> **VERTICAL LINE TEST**
>
> A graph represents a function if every vertical line intersects the graph in at most one point.

Example 6

Use the vertical line test to determine which of the following graphs represent functions.

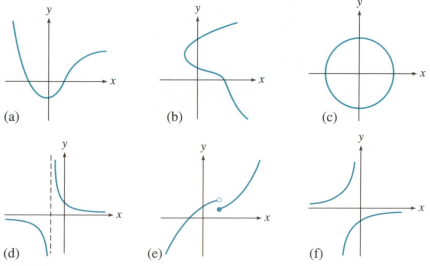

(a)   (b)   (c)

(d)   (e)   (f)

Solutions   Graphs (a), (d), and (e) represent functions, since no vertical line intersects the graph in more than one point. Graphs (b), (c), and (f) do not represent functions.

Exercise 5.2

In Exercises 1–8 use the graphs to answer the questions about the functions.

1. a. Find $h(-3)$, $h(1)$, and $h(3)$ in Figure 5.13.

  b. For what value(s) of z is $h(z) = 3$?

  c. Find the intercepts of the graph. List the function values given by the intercepts.

  d. What is the maximum value of $h(z)$?

  e. For what value(s) of z does h take on its maximum value?

  f. Give the domain and range of the function.

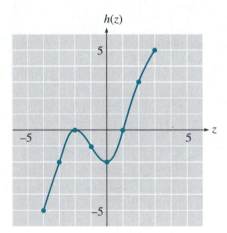

Figure 5.13

2. a. Find $G(-3)$, $G(-1)$, and $G(2)$ in Figure 5.14.

  b. For what value(s) of s is $G(s) = 3$?

  c. Find the intercepts of the graph. List the function values given by the intercepts.

  d. What is the minimum value of $G(s)$?

  e. For what value(s) of s does G take on its minimum value?

  f. Give the domain and range of the function.

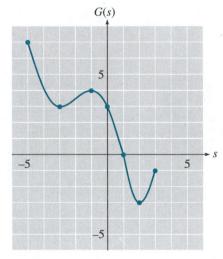

Figure 5.14

3. a. Find $R(1)$ and $R(3)$ in Figure 5.15.

  b. For what value(s) of p is $R(p) = 2$?

  c. Find the intercepts of the graph. List the function values given by the intercepts.

  d. Find the maximum and minimum values of $R(p)$.

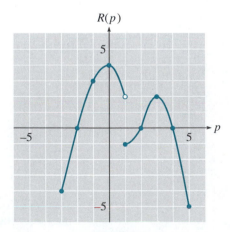

Figure 5.15

e. For what value(s) of p does R take on its maximum and minimum values?

f. Give the domain and range of the function.

4. a. Find $f(-1)$ and $f(3)$ in Figure 5.16.

   b. For what value(s) of t is $f(t) = 5$?

   c. Find the intercepts of the graph. List the function values given by the intercepts.

   d. Find the maximum and minimum values of $f(t)$.

   e. For what value(s) of t does f take on its maximum and minimum values?

   f. Give the domain and range of the function.

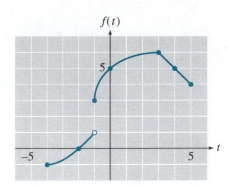

Figure 5.16

5. a. Find $S(0)$, $S\left(\dfrac{1}{6}\right)$, and $S(-1)$ in Figure 5.17.

   b. Estimate the value of $S\left(\dfrac{1}{3}\right)$ from the graph.

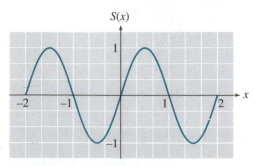

Figure 5.17

c. For what value(s) of x is $S(x) = -\dfrac{1}{2}$?

d. Find the maximum and minimum values of $S(x)$.

e. For what value(s) of x does S take on its maximum and minimum values?

f. Give the domain and range of the function.

6. a. Find $C(0)$, $C\left(-\dfrac{1}{3}\right)$, and $C(1)$ in Figure 5.18.

   b. Estimate the value of $C\left(\dfrac{1}{6}\right)$ from the graph.

   c. For what value(s) of x is $C(x) = \dfrac{1}{2}$?

   d. Find the maximum and minimum values of $C(x)$.

   e. For what value(s) of x does C take on its maximum and minimum values?

   f. Give the domain and range of the function.

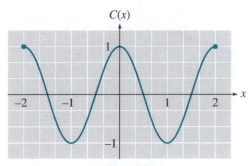

Figure 5.18

7. a. Find $F(-3)$, $F(-2)$, and $F(2)$ in Figure 5.19 on page 286.

   b. For what value(s) of s is $F(s) = -1$?

   c. Find the maximum and minimum values of $F(s)$.

   d. For what value(s) of s does F take on its maximum and minimum values?

   e. Give the domain and range of the function.

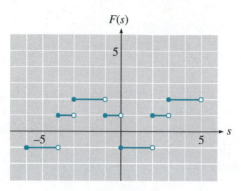

Figure 5.19

8. a. Find $P(-3)$, $P(-2)$, and $P(1)$ in Figure 5.20.

b. For what value(s) of n is $P(n) = 0$?

c. Find the maximum and minimum values of $P(n)$.

d. For what value(s) of n does P take on its maximum and minimum values?

e. Give the domain and range of the function.

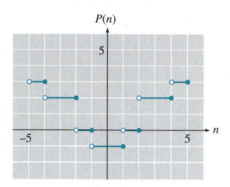

Figure 5.20

In Exercises 9–20 use a graphing calculator to graph each function on the given domain. Using the $\boxed{\text{TRACE}}$ key, adjust Ymin and Ymax until you can determine the range of the function. Then verify the range algebraically by evaluating the function. State the domain and range in interval notation.

9. $f(x) = x^2 - 4x; \quad -2 \le x \le 6$

10. $g(x) = 6x - x^2; \quad -1 \le x \le 7$

11. $g(t) = -t^2 - 2t; \quad -5 \le t \le 3$

12. $f(t) = -t^2 - 4t; \quad -6 \le t \le 2$

13. $h(x) = x^3 - 1; \quad -2 \le x \le 2$

14. $q(x) = x^3 + 4; \quad -3 \le x \le 2$

15. $F(t) = \sqrt{8 - t}; \quad -1 \le t \le 8$

16. $G(t) = \sqrt{t + 6}; \quad -6 \le t \le 3$

17. $G(x) = \dfrac{1}{3 - x}; \quad -1.25 \le x \le 2.75$

18. $H(x) = \dfrac{1}{x - 1}; \quad -3.25 \le x \le -1.25$

19. $G(x) = \dfrac{1}{3 - x}; \quad 3 < x \le 6$

20. $H(x) = \dfrac{1}{x - 1}; \quad 1 < x \le 4$

Which of the graphs in Exercises 21–30 represent functions?

21.

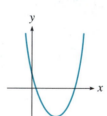

22.

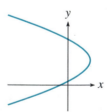

23.

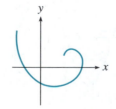

24.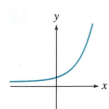

25.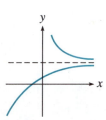

26.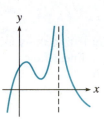

27.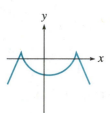

28.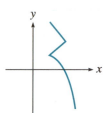

29.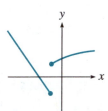

30.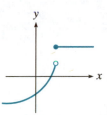

5.3   SOME BASIC GRAPHS

In this section we will study the graphs of some important basic functions. Many functions fall into families, or classes, of similar functions, and recognizing the appropriate family for a given situation can be an important first step in modeling the situation. We have already encountered two such families of functions: linear functions, whose graphs are straight lines, and quadratic functions, whose graphs are parabolas.

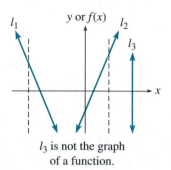

Figure 5.21

l_3 is not the graph of a function.

Linear Functions

The graphs of all lines except vertical lines pass the vertical line test and hence are the graphs of functions. (See Figure 5.21.)

Functions of the form

$$f(x) = mx + b$$

are therefore called **linear functions**. By letting the vertical axis, or y-axis, represent the values of $f(x)$, we can graph linear functions the same way we graphed linear equations, $y = mx + b$.

Example 1

Graph the function $f(x) = -2x + 5$.

Solution   The graph is a line of slope $m = -2$ and y-intercept $b = 5$. Plot the point $(0, 5)$ and use the definition of slope

$$m = \frac{\Delta y}{\Delta x} = \frac{-2}{1}$$

to find a second point on the line. Draw the line through the two points. (See Figure 5.22.)

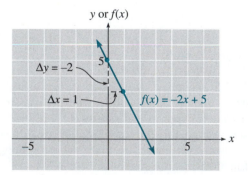

Figure 5.22

If the domain of a function is not given as part of its definition, we assume that the domain is as large as possible. This assumption means that we include in the domain all x-values that "make sense" when substituted into the equation that defines the function. A linear function $f(x) = mx + b$ can be evaluated at any real-number value of x, so its domain is the set of all real numbers. This set is represented in interval notation as $(-\infty, \infty)$.

The range of the linear function $f(x) = mx + b$ (if $m \neq 0$) is also the set of all real numbers, since any number can be obtained as the output value of the function. In other words, the graph of a linear function continues infinitely at both ends. (See Figure 5.23a.) If $m = 0$, then $f(x) = b$, and the graph of f is a horizontal line. In this case the range consists of a single number, b.

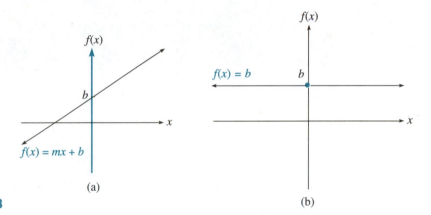

Figure 5.23

(a)

(b)

Most of the graphs in the rest of this section will be new to you. In the exercises you will be asked to verify each graph by making a table of values and plotting points and to check the result using your graphing calculator. Because these graphs are fundamental to further study of mathematics and its applications, you should become familiar enough with the properties of each that you can sketch them easily from memory.

Some Powers and Roots

The functions $f(x) = x^2$ and $g(x) = x^3$ are graphed in Figure 5.24 on page 290. Each function has as its domain the set of all real numbers, and the graphs extend infinitely, as the arrows indicate.

Recall that the graph of the function $f(x) = x^2$ is called a parabola. If we think of moving along the x-axis from left to right, we see that the parabola is "falling" for x-values less than zero and "rising" for x-values greater than zero. The graph of the cubic function $g(x) = x^3$, however, is rising for all values of x. From their graphs we see that the range of the

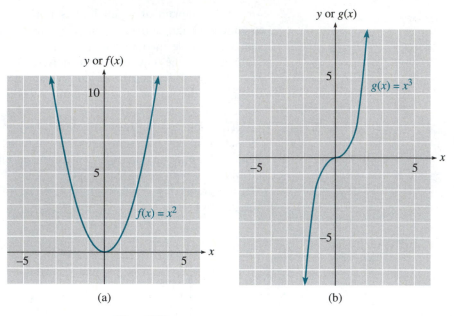

(a)   (b)

Figure 5.24

function $f(x) = x^2$ is the interval $[0, \infty)$ (all the function values are non-negative) and the range of $g(x) = x^3$ is the interval $(-\infty, \infty)$.

The two graphs in Figure 5.25 are examples of **radical** functions, or **roots**. You are already familiar with square roots, and the graph of the square root function $f(x) = \sqrt{x}$ is shown in Figure 5.25(a). In this book we will use several other kinds of roots, one of which is called the **cube root**.

b is the **cube root** of *a* if *b* cubed equals *a*.

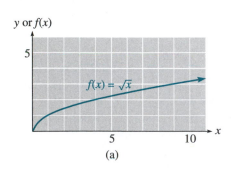

(a)

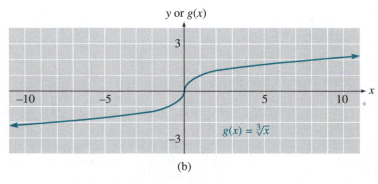

(b)

Figure 5.25

We use the symbol $\sqrt[3]{a}$ for the cube root of a, so in symbols we can write

$$b = \sqrt[3]{a} \quad \text{if} \quad b^3 = a$$

For example,

$$4 = \sqrt[3]{64} \quad \text{since} \quad 4^3 = 64$$

and

$$-3 = \sqrt[3]{-27} \quad \text{since} \quad (-3)^3 = -27$$

We can use the calculator to find cube roots as follows. Press the MATH key to get a menu of options. Option 4 is labeled $\sqrt[3]{}$, which is the cube root key. If we want to find the cube root of, say, 15.625, we key in

MATH 4 15.625 ENTER

and the calculator returns the result 2.5. Thus $\sqrt[3]{15.625} = 2.5$. You can check this result by verifying that $2.5^3 = 15.625$.

Notice that although we cannot take the square root of a negative number, we can take the *cube* root of *any* real number. The graph of the function $g(x) = \sqrt[3]{x}$ is shown in Figure 5.25(b).

The domain of the function $f(x) = \sqrt{x}$ is the interval $[0, \infty)$. (Do you see why?) As x increases the graph rises, and although the function values grow slowly, every non-negative real number will eventually be attained as a function value. Thus the range of f is the interval $[0, \infty)$.

Because we can take the cube root of every real number, the domain of $g(x) = \sqrt[3]{x}$ is the set of all real numbers. The graph rises for all values of x, and the range of g is the set of all real numbers.

Asymptotes   The functions $f(x) = \dfrac{1}{x}$ and $g(x) = \dfrac{1}{x^2}$ graphed in Figure 5.26 on page 292 are examples of **rational functions**. Neither function can be evaluated at $x = 0$, since division by zero is undefined. The domain of each function is the set of all real numbers *except* zero.

Because zero is not in the domain of f, there is no point on the graph of $f(x) = \dfrac{1}{x}$ that has x-coordinate 0. To understand the shape of the graph near $x = 0$, we evaluate the function for several x-values close to zero.

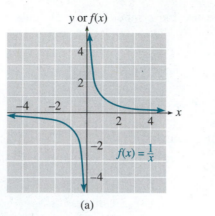

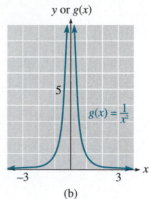

Figure 5.26   (a)   (b)

We can use the table feature of the calculator to generate the tables shown in Figure 5.27.

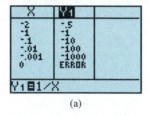   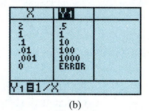

Figure 5.27   (a)   (b)

As we choose x-values closer and closer to zero but still greater than zero, the function values increase. As we choose x-values less than zero but approaching zero, the function values decrease. We say that f has a **vertical asymptote** at $x = 0$; that is, the graph approaches but never touches the vertical line $x = 0$.

Now consider the rest of the graph. For positive x, as the value of x increases, the value of its reciprocal, $\frac{1}{x}$, decreases (but remains positive). (You can verify this fact by evaluating the function for some large values of x.) Thus the value of $f(x)$ gets closer to zero and the graph approaches the x-axis. For negative x-values, as x decreases, the function values increase toward zero through negative values, and the graph approaches the x-axis from below. Since $\frac{1}{x}$ never *equals* zero for any x-value, the graph never actually touches the x-axis. (See Figure 5.26a.) We say that the x-axis is a **horizontal asymptote** for the graph. The range of f is the set of all real numbers except zero.

The graph of $g(x) = \frac{1}{x^2}$ is similar to the graph of $f(x) = \frac{1}{x}$, except that the function values of g are always positive. (See Figure 5.26b.) Thus the range of g is the interval $(0, \infty)$.

Functions Defined Piecewise   A function for which different equations are used to determine the function values on different portions of the domain is said to be defined "piecewise." To graph a function defined piecewise, we consider each piece of the domain separately.

Example 2   Graph the function defined by

$$f(x) = \begin{cases} x + 1 & \text{if } x \le 1 \\ 3 & \text{if } x > 1 \end{cases}$$

Solution   Think of the plane as divided into two regions by the vertical line $x = 1$. In the left-hand region $(x \le 1)$, graph the line $y = x + 1$. Notice that the value $x = 1$ is included in the first region, so $f(1) = 2$, and the point $(1, 2)$ is included on the graph. We indicate this with a solid dot at the point $(1, 2)$. In the right-hand region $(x > 1)$, graph the line $y = 3$. The value $x = 1$ is *not* included in the second region, so the point $(1, 3)$ is *not* part of the graph. We indicate this with an open circle at the point $(1, 3)$. (See Figure 5.28.)

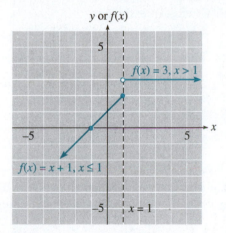

Figure 5.28

Absolute Value   The absolute value function is used to discuss problems involving distance. For example, consider the number line in Figure 5.29. Starting at the origin,

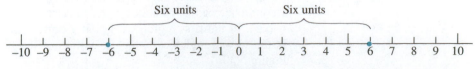

Figure 5.29

we travel in opposite *directions* to reach the two numbers 6 and -6, but the *distance* we travel in each case is the same.

The distance from a number c to the origin is called the **absolute value** of c, denoted by $|c|$. Because distance is always positive, the absolute value of a number is always positive. Thus $|6| = 6$ and $|-6| = 6$. In general, we define the absolute value of a number x as

$$|x| = \begin{cases} x & \text{if } x \geq 0 \\ -x & \text{if } x < 0 \end{cases}$$

This definition says that the absolute value of a positive number (or zero) is the same as the number. To find the absolute value of a negative number, we take the opposite of the number, which is then positive. For instance,

$$|-6| = -(-6) = 6$$

Absolute value bars serve as grouping devices in the order of operations: you should complete any operations that appear inside absolute value bars before you compute the absolute value.

Example 3

Simplify each expression.
a. $|3 - 8|$ **b.** $|3| - |8|$

Solutions   **a.**   Simplify the expression inside the absolute value bars first.

$$|3 - 8| = |-5| = 5$$

b.   Simplify each absolute value, and then subtract.

$$|3| - |8| = 3 - 8 = -5$$

The absolute value function $f(x) = |x|$ is an example of a function that is defined piecewise. From the definition of absolute value, we have

$$f(x) = |x| = \begin{cases} x & \text{if } x \geq 0 \\ -x & \text{if } x < 0 \end{cases}$$

To graph the absolute value function, think of dividing the plane into two regions, the first region including all points with x-coordinates less than zero (to the left of the y-axis) and the second region including all points with x-coordinates greater than or equal to zero (to the right of and including the y-axis). In the first region we graph a portion of the line $y = -x$, and in the second region we graph a portion of the line $y = x$, as shown in Figure 5.30.

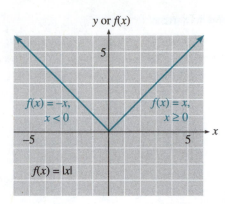

Figure 5.30

Notice that in Figure 5.30 the two "pieces" of the graph of the absolute value function meet at the origin, but the two pieces of the graph in Figure 5.28 did not meet.

Exercise 5.3

Sketch graphs of the linear and quadratic functions in Exercises 1–12.

1. $f(x) = 3x - 4$

2. $g(x) = -2x + 5$

3. $G(s) = -\dfrac{5}{3}s + 50$

4. $F(s) = -\dfrac{3}{4}s + 60$

5. $R(u) = 2.6u - 120$

6. $H(v) = 1.8v - 240$

7. $h(t) = t^2 - 8$

8. $p(t) = 3 - t^2$

9. $g(w) = (w + 2)^2$

10. $f(w) = (w - 1)^2$

11. $F(z) = -2z^2$

12. $G(z) = 0.5z^2$

In Exercises 13–20 compute each cube root, approximating to three decimal places if necessary. Verify your answers by cubing them.

13. $\sqrt[3]{512}$

14. $\sqrt[3]{-125}$

15. $\sqrt[3]{-0.064}$

16. $\sqrt[3]{1.728}$

17. $\sqrt[3]{9}$

18. $\sqrt[3]{258}$

19. $\sqrt[3]{-0.002}$

20. $\sqrt[3]{-3.1}$

In Exercises 21–30 simplify each expression according to the order of operations.

21. a. $-|-9|$

   b. $-(-9)$

22. a. $2 - (-6)$

   b. $2 - |-6|$

23. a. $|-8| - |12|$

   b. $|-8 - 12|$

24. a. $|-3| + |-5|$

   b. $|-3 + (-5)|$

25. $4 - 9|2 - 8|$

26. $2 - 5|-6 - 3|$

27. $|-4 - 5||1 - 3(-5)|$

28. $|-3 + 7||-2(6 - 10)|$

29. $||-5| - |-6||$

30. $||4| - |-6||$

For each function in Exercises 31–36,
a.   construct a table of values and sketch the graph.
b.   state the domain and range of the function.
c.   use a graphing calculator to verify your table and your graph.

31. $f(x) = x^3$

32. $f(x) = |x|$

33. $f(x) = \sqrt{x}$

34. $f(x) = \sqrt[3]{x}$

35. $f(x) = \dfrac{1}{x}$

36. $f(x) = \dfrac{1}{x^2}$

Use graphs to solve the equations and inequalities in Exercises 37–40.

37. Use the graph of $y = x^2$ in Figure 5.31 to answer the questions. (You will have to estimate your answers.)

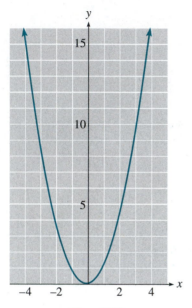

Figure 5.31

a. Estimate the values of 1.8^2 and $(-2.5)^2$.

b. Find all numbers whose square is 7, and all numbers whose square is 12.

c. Estimate the values of $\sqrt{10.5}$ and $-\sqrt{6.5}$.

d. Find all solutions of the equation $m^2 = 15$, and all solutions of the equation $p^2 = 3.5$.

e. Find all solutions of the equation $\sqrt{w} = 3.2$, and all solutions of the equation $\sqrt{z} = 2.3$.

f. Find all solutions of the inequality $4 \leq x^2 \leq 10$.

38. Use the graph of $y = x^3$ in Figure 5.32 to answer the questions. (You will have to estimate your answers.)

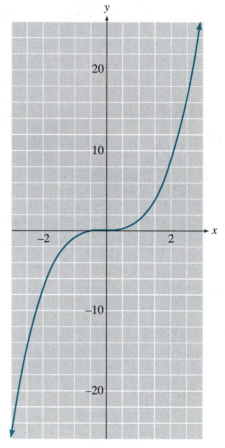

Figure 5.32

a. Estimate the values of $(1.4)^3$ and $(-2.2)^3$.

b. Find all numbers whose cube is 13, and all numbers whose cube is -20.

c. Estimate the values of $\sqrt[3]{24}$ and $\sqrt[3]{-7}$.

d. Find all solutions of the equation $q^3 = 6$, and all solutions of the equation $n^3 = -25$.

e. Find all solutions of the equation $\sqrt[3]{b} = 2.8$, and all solutions of the equation $\sqrt[3]{h} = 1.5$.

f. Find all solutions of the inequality $-12 \leq x^3 \leq 15$.

39. Use the graph of $y = \dfrac{1}{x}$ in Figure 5.33 to answer the questions. (You will have to estimate your answers.)

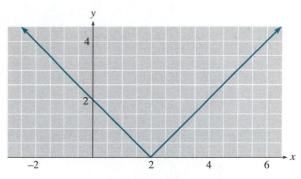

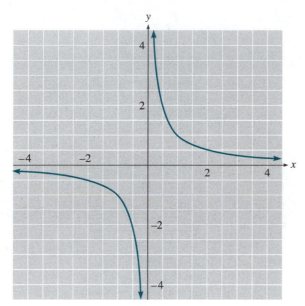

Figure 5.33

a. Estimate the values of $\dfrac{1}{3.4}$ and $\dfrac{1}{-0.8}$.

b. Find all numbers whose reciprocal is -2.5, and all numbers whose reciprocal is 1.1.

c. Find all solutions of the equation $\dfrac{1}{v} = 4.8$, and all solutions of the equation $\dfrac{1}{n} = -0.2$.

d. Find all solutions of the inequality $0.3 \le \dfrac{1}{h} \le 4.5$.

40. Use the graph of $y = |x - 2|$ in Figure 5.34 to answer the questions. (You will have to estimate your answers.)

a. Estimate the values of $|-1.4 - 2|$ and $|1.6 - 2|$.

Figure 5.34

b. Find all values of x for which $|x - 2| = 3$, and all values of x for which $|x - 2| = -2$.

c. Find all solutions of the equation $|x - 2| = 3.6$, and all solutions of the equation $|x - 2| = 0.4$.

d. Find all solutions of the inequality $|x - 2| < 4$, and all solutions of $|x - 2| > 1$.

Graph the piecewise-defined functions in Exercises 41–52.

41. $f(x) = \begin{cases} -2 & \text{if } x \le 1 \\ x - 3 & \text{if } x > 1 \end{cases}$

42. $h(x) = \begin{cases} -x + 2 & \text{if } x \le -1 \\ 3 & \text{if } x > -1 \end{cases}$

43. $G(t) = \begin{cases} 3t + 9 & \text{if } t < -2 \\ -3 - \dfrac{1}{2}t & \text{if } t \ge -2 \end{cases}$

44. $F(s) = \begin{cases} \dfrac{1}{3}s + 3 & \text{if } s < 3 \\ 2s - 3 & \text{if } s \ge 3 \end{cases}$

45. $H(t) = \begin{cases} t^2 & \text{if } t \le 1 \\ \dfrac{1}{2}t + \dfrac{1}{2} & \text{if } t > 1 \end{cases}$

46. $g(t) = \begin{cases} \dfrac{3}{2}t + 7 & \text{if } t \le -2 \\ t^2 & \text{if } t > -2 \end{cases}$

47. $k(x) = \begin{cases} |x| & \text{if } x \le 2 \\ \sqrt{x} & \text{if } x > 2 \end{cases}$

48. $S(x) = \begin{cases} \dfrac{1}{x} & \text{if } x < 1 \\ |x| & \text{if } x \ge 1 \end{cases}$

49. $D(x) = \begin{cases} |x| & \text{if } x < -1 \\ x^3 & \text{if } x \ge -1 \end{cases}$

50. $m(x) = \begin{cases} x^2 & \text{if } x \le \dfrac{1}{2} \\ |x| & \text{if } x > \dfrac{1}{2} \end{cases}$

51. $P(t) = \begin{cases} t^3 & \text{if } t \le 1 \\ \dfrac{1}{t^2} & \text{if } t > 1 \end{cases}$

52. $Q(t) = \begin{cases} t^2 & \text{if } t \le -1 \\ \sqrt[3]{t} & \text{if } t > -1 \end{cases}$

Recall that the graphs of all quadratic functions are parabolas. They all have the same basic shape as the graph of $y = x^2$, but they may have a different position or orientation in the plane. For example, the following four graphs are graphs of quadratic functions.

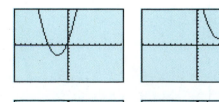

Each graph in Exercises 53–64 represents a variation of one of the basic functions listed here. Identify the basic graph for each exercise.

$$y = x^3 \qquad y = |x| \qquad y = \sqrt{x} \qquad y = \sqrt[3]{x}$$

$$y = x \qquad y = x^2 \qquad y = \frac{1}{x} \qquad y = \frac{1}{x^2}$$

53.

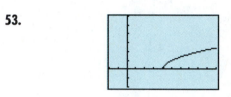

54.

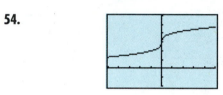

55.

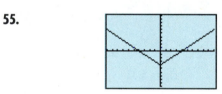

56.

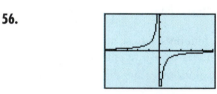

57.

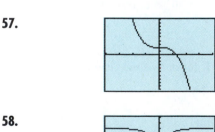

58.

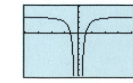

59.

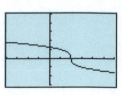

60.

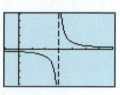

61.

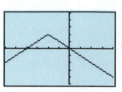

62.

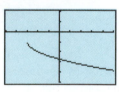

63.

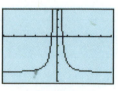

64.

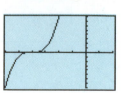

In Exercises 65–70 use a graphing calculator to explore some properties of the basic functions. Choose any *x*-value in the

given interval, and decide which graph has the larger *y*-value corresponding to that *x*-value.

65. Graph $f(x) = x^2$ and $g(x) = x^3$ on the domain $[0, 1]$ and the range $[0, 1]$. On the interval $(0, 1)$, which is greater, $f(x)$ or $g(x)$?

66. Graph $f(x) = x^2$ and $g(x) = x^3$ on the domain $[1, 10]$ and the range $[1, 100]$. On the interval $(1, 10)$, which is greater, $f(x)$ or $g(x)$?

67. Graph $f(x) = \sqrt{x}$ and $g(x) = \sqrt[3]{x}$ on the domain $[0, 1]$ and the range $[0, 1]$. On the interval $(0, 1)$, which is greater, $f(x)$ or $g(x)$?

68. Graph $f(x) = \sqrt{x}$ and $g(x) = \sqrt[3]{x}$ on the domain $[1, 100]$ and the range $[1, 10]$. On the interval $(0, 100)$, which is greater, $f(x)$ or $g(x)$?

69. Graph $f(x) = \dfrac{1}{x}$ and $g(x) = \dfrac{1}{x^2}$ on the domain $[0.01, 1]$ and the range $(0, 10]$. On the interval $(0, \infty)$, which is greater, $f(x)$ or $g(x)$?

70. Graph $f(x) = \dfrac{1}{x}$ and $g(x) = \dfrac{1}{x^2}$ on the domain $[1, 10]$ and the range $(0, 1]$. On the interval $(1, \infty)$, which is greater, $f(x)$ or $g(x)$?

For Exercises 71–74 refer to the list of eight basic functions that precedes Exercise 53.

71. Which of the eight basic functions are increasing on their entire domain? Which are decreasing on their entire domain?

72. Which of the eight basic functions can be evaluated at any real number? Which can take on any real number as a function value?

73. Which of the eight basic functions can be graphed in one piece, without lifting the pencil from the paper?

74. Which of the eight basic functions have no negative numbers in their range?

DIRECT AND INVERSE VARIATION

Two types of functional variation are widely used in the sciences and through their applications have come to be known by special names: **direct variation** and **inverse variation**.

Direct Variation

Two variables are said to be **directly proportional** (or just **proportional**) if the ratios of their corresponding values are always equal. Consider the functions described in Table 5.8.

The first example involves the price of gasoline as a function of the number of gallons purchased.

Notice that the ratio $\dfrac{\text{total price}}{\text{number of gallons}}$, or price per gallon, is the same for each pair of values in Table 5.8. This agrees with everyday experience: the price per gallon of gasoline is the same no matter how many gallons you buy. Thus the total cost of a gasoline purchase is directly proportional to the number of gallons purchased.

The next example considers the population of a small town as a function of the town's age. (See Table 5.9.)

The ratios we computed, $\dfrac{\text{number of people}}{\text{number of years}}$, give the average growth rate of the town's population in people per year. You can see from Table 5.9 that this ratio is *not* constant; in fact, it increases as time goes on. Thus the population of the town is *not* proportional to its age.

Gallons of Gasoline	Total Price	$\dfrac{\text{Price}}{\text{Gallons}}$
4	$ 4.60	$\dfrac{4.60}{4} = 1.15$
6	$ 6.90	$\dfrac{6.90}{6} = 1.15$
8	$ 9.20	$\dfrac{9.20}{8} = 1.15$
12	$13.80	$\dfrac{13.80}{12} = 1.15$
15	$17.25	$\dfrac{17.25}{15} = 1.15$

Table 5.8

Years	Population	$\dfrac{\text{People}}{\text{Years}}$
10	432	$\dfrac{432}{10} \approx 43$
20	932	$\dfrac{932}{20} \approx 47$
30	2013	$\dfrac{2013}{30} \approx 67$
40	4345	$\dfrac{4345}{40} \approx 109$
50	9380	$\dfrac{9380}{50} \approx 188$
60	20,251	$\dfrac{20,251}{60} \approx 338$

Table 5.9

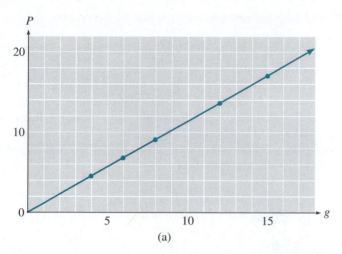 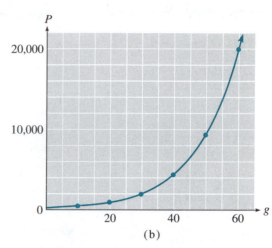

(a)   (b)

Figure 5.35

The graphs of these two functions are shown in Figure 5.35. We see that for the gasoline example, the price, P, of a fill-up is a linear function of the number of gallons, g, purchased. This should not be surprising if we consider an equation relating the variables g and P. Because the ratio of their values is a constant, we can write

$$\frac{P}{g} = k$$

where k is a constant. In this example the constant, k, is 1.15, the price of gasoline per gallon. Solving for P in terms of g, we have

$$P = kg = 1.15g$$

which we recognize as the equation of a line through the origin.

In general, we say that the variable y **varies directly** with x if

$$y = kx$$

where k is a positive constant. From the discussion above we see that *direct variation* means exactly the same thing as *directly proportional*. The two terms are interchangeable.

Example 1

a.   The circumference, C, of a circle varies directly with its radius, r, since

$$C = 2\pi r$$

b.   The amount of interest, I, earned in 1 year on an account paying 7% simple interest varies directly with the principal, P, invested, since

$$I = 0.07P$$

Thus direct variation defines a linear function of the form

$$y = f(x) = kx$$

The fact that the constant term is zero is significant: if we double the value of x, then the value of y will double also. In fact, increasing x by any factor will result in y increasing by the same factor. For example, in Table 5.8 doubling the number of gallons of gas purchased, say, from 4 gallons to 8 gallons or from 6 gallons to 12 gallons, causes the total price to double also. As another example, consider investing \$800 for 1 year at 7% simple interest, as in Example 1(b). The interest earned is

$$I = 0.07(800) = \$56$$

If we increase the investment by a factor of 1.6 to 1.6(800), or \$1280, the interest will be

$$I = 0.07(1280) = \$89.60$$

You can check that multiplying the original interest of \$56 by a factor of 1.6 does give the same figure for the new interest, \$89.60.

Example 2

a.   Tuition at Woodrow University is \$400 plus \$30 per unit. Is the tuition proportional to the number of units you take?
b.   Imogen makes a 15% commission on her sales of environmentally friendly products marketed by her co-op. Do her earnings vary directly with her sales?

Solutions   a.   Let u represent the number of units you take, and let T represent your tuition. Then

$$T = 400 + 30u$$

Thus T is a linear function of u, but the T-intercept is 400, not zero. Your tuition is *not* proportional to the number of units you take, so this is *not* an example of direct variation. You can check that doubling the number of units does not double the tuition. For example,

$$T(6) = 400 + 30(6) = 580$$

and

$$T(12) = 400 + 30(12) = 760$$

Doubling \$580 tuition does not result in \$760 tuition. (See Figure 5.36(a).)
  b.   Let S represent Imogen's sales, and let C represent her commission. Then

$$C = 0.15S$$

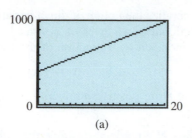

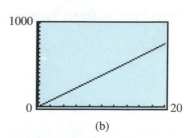

Figure 5.36 (a) (b)

Thus C is a linear function of S with C-intercept zero, so Imogen's earnings do vary directly with her sales. This *is* an example of direct variation. (See Figure 5.36(b).)

The positive constant k in the equation $y = kx$ is called the **constant of variation**. It is just the slope of the graph, so it tells us how rapidly the graph increases. If we know any one pair of associated values for the variables, we can find the constant of variation. We can then use the constant to express one of the variables as a function of the other.

Example 3

If an object is dropped from a great height, say, off the rim of the Grand Canyon, its speed, v, varies directly with the time, t, the object has been falling. A rock kicked off the edge of the canyon is falling at a speed of 39.2 meters per second when it passes a lizard on a ledge 4 seconds later.

a. Express v as a function of t.
b. What is the speed of the rock after it has fallen for 6 seconds?
c. Sketch a graph of v versus t.

Solutions **a.** Since v varies directly with t, there is a positive constant k for which

$$v = kt$$

Substitute $v = 39.2$ when $t = 4$ and solve for k to find

$$39.2 = k(4)$$

or

$$k = 9.8$$

Thus $v = 9.8t$.
b. Substitute $t = 6$ into the equation you found in part (a).

$$v = 9.8(6) = 58.8$$

The rock is falling at a speed of 58.8 meters per second.
c. The graph of the equation $v = 9.8t$ is shown in Figure 5.37.

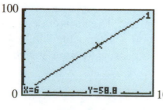

Figure 5.37

Other Types of Direct Variation   We can generalize the notion of direct variation to include situations in which y is proportional to a power of x instead of x itself. This type of variable relation is modeled by a function of the form

$$y = f(x) = kx^n$$

where k and n are positive numbers.

Example 4

a. The area of a circle varies directly with the *square* of the radius, since

$$A = \pi r^2$$

b. The volume of a sphere varies directly with the *cube* of the radius, since

$$V = \frac{4}{3}\pi r^3$$

In any example of direct variation, as the independent variable increases through positive values, the dependent variable increases also. Thus a direct variation is an example of an increasing function, as we can see when we consider the graphs of some typical direct variations in Figure 5.38.

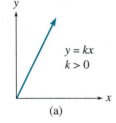
(a)

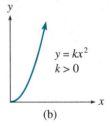
(b)

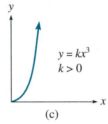
(c)

Figure 5.38

Notice that the graph of a direct variation always passes through the origin, so when the independent variable is zero, the dependent variable is zero also. This requirement means that the functions $y = 3x + 2$ and $y = 0.4x^2 - 2.3$, for example, are *not* direct variations even though they are increasing functions for positive x.

Inverse Variation   How long does it take to travel a distance of 600 miles? The answer depends upon the average speed at which you travel. If you are on a bicycle, your average speed might be 15 miles per hour. In that case your traveling time would be

$$T = \frac{D}{R} = \frac{600}{15} = 40 \text{ hours}$$

(Of course, you will have to add time for rest stops; the 40 hours are *just* your travel time.) If you are driving a car, you might average 50 miles per hour. Your travel time would then be

$$T = \frac{D}{R} = \frac{600}{50} = 12 \text{ hours}$$

If you fly on a commercial airline, the plane's speed might be 400 miles per hour, and your flight time would be

$$T = \frac{D}{R} = \frac{600}{400} = 1.5 \text{ hours}$$

You can see that for higher average speeds the travel time is shorter. In other words, the time required for a 600-mile journey is a decreasing function of average speed. A formula for this function is

$$T = \frac{600}{R}$$

This function is an example of **inverse variation**. A graph of the function is shown in Figure 5.39.

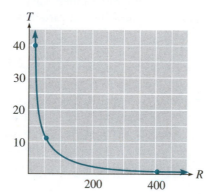

Figure 5.39

In general, we say that y **varies inversely** with x^n if

$$y = \frac{k}{x^n}$$

where k is a positive constant and $n > 0$. (Notice that x cannot be equal to zero.) We may also say that y is **inversely proportional** to x^n.

Example 5

a. The weight, w, of an object varies inversely with the square of its distance, d, from the center of the earth. Thus

$$w = \frac{k}{d^2}$$

b. The amount of force, F (in pounds), needed to lift a heavy object with the help of a lever is inversely proportional to the length, l, of the lever. Thus

$$F = \frac{k}{l}$$

In each case in Example 5, as the independent variable increases through positive values, the dependent variable decreases. An inverse variation is an example of a decreasing function. The graphs of some typical inverse variations are shown in Figure 5.40.

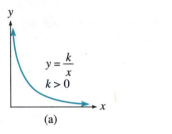
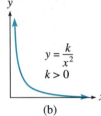

Figure 5.40
      (a)   (b)

If we know that two variables vary inversely and we can find one pair of corresponding values for the variables, we can determine k, the constant of variation.

Example 6

The intensity of electromagnetic radiation, such as light or radio waves, varies inversely with the square of the distance from their source. Radio station KPCC broadcasts a signal that is measured at 0.016 watt per square meter by a receiver 1 kilometer away. If you live 5 kilometers from the station, what is the strength of the signal you will receive?

Solution   Let I stand for the intensity of the signal in watts per square meter, and d for the distance from the station in kilometers. Then

$$I = \frac{k}{d^2}$$

To find the constant k, we substitute 0.016 for I and 1 for d. Solving for k gives us

$$0.016 = \frac{k}{1^2}$$

$$k = 0.016\,(1^2) = 0.016$$

Thus $I = \dfrac{0.016}{d^2}$.   Now we can substitute 5 for d and solve for I.

$$I = \frac{0.016}{5^2} = 0.00064$$

At a distance of 5 kilometers from the station, the signal strength is 0.00064 watt per square meter.

Exercise 5.4

For each function described in Exercises 1–6,
a. use the values in the table to find the constant of variation, *k*, and write *y* as a function of *x*.
b. fill in the rest of the table with the correct values.
c. graph the function.
d. What happens to *y* when you double the value of *x*?

1. *y* varies directly with *x*.

x	y
2	
5	1.5
	2.4
12	
	4.5

2. *y* varies directly with *x*.

x	y
0.8	
1.5	54
	108
	126
6.0	

3. *y* varies directly with the square of *x*.

x	y
3	
6	24
	54
12	
	150

4. *y* varies directly with the cube of *x*.

x	y
2	120
3	
	1875
6	
	15,000

5. *y* varies inversely with *x*.

x	y
4	
	15
20	6
30	
	3

6. *y* varies inversely with the square of *x*.

x	y
0.2	
	80.00
2.0	
4.0	1.25
	0.80

Which of the graphs in Exercises 7 and 8 could describe direct variation?

7.

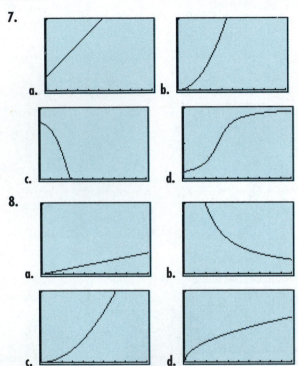

8.

Which of the graphs in Exercises 9 and 10 could describe inverse variation?

9.

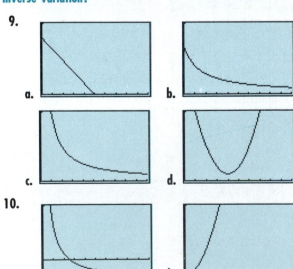

10.

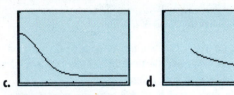

c.   d.

For each table in Exercises 11–14, decide whether
a.   *y* varies directly with *x*.
b.   *y* varies directly with x^2.
c.   *y* does not vary directly with a power of *x*.
If your choice is (a) or (b), find the constant of variation, *k*.

11.

x	y
2	2.0
3	4.5
5	12.5
8	32.0

12.

x	y
2	12
4	28
6	44
9	68

13.

x	y
1.5	3.0
2.4	7.2
5.5	33.0
8.2	73.8

14.

x	y
1.2	7.20
2.5	31.25
6.4	204.80
12.0	720.00

For each table in Exercises 15–18, decide whether
a. *y* varies inversely with *x*.
b. *y* varies inversely with *x²*.
c. *y* does not vary inversely with a power of *x*.
If your choice is (a) or (b), find the constant of variation, *k*.

15.

x	y
0.5	288
2.0	18
3.0	8
6.0	2

16.

x	y
0.5	100.0
2.0	25.0
4.0	12.5
5.0	10.0

17.

x	y
1.0	4.0
1.3	3.7
3.0	2.0
4.0	1.0

18.

x	y
0.5	180.00
2.0	11.25
3.0	5.00
5.0	1.80

19. The length of a rectangle is 10 inches, and its width is 8 inches. Suppose we increase the length of the rectangle while holding the width constant.

a. Fill in Table 5.10.
b. Does the perimeter vary directly with the length?
c. Write a formula for the perimeter of the rectangle in terms of its length.
d. Does the area vary directly with the length?
e. Write a formula for the area of the rectangle in terms of its length.

Length	Width	Perimeter	Area
10	8		
12	8		
15	8		
20	8		

Table 5.10

20. The base of an isosceles triangle is 12 centimeters, and the equal sides have length 15 centimeters. Suppose we increase the base of the triangle while holding the sides constant.

Base	Sides	Height	Perimeter	Area
12	15			
15	15			
18	15			
20	15			

Table 5.11

a. Fill in Table 5.11.
b. Does the perimeter vary directly with the base?
c. Write a formula for the perimeter of the triangle in terms of its base.
d. Write a formula for the area of the triangle in terms of its base.
e. Does the area vary directly with the base?

21. The weight of an object on the moon varies directly with its weight on earth. A person who weighs 150 pounds on earth would weigh only 24.75 pounds on the moon.

a. Find a function that gives the weight, *m*, of an object on the moon in terms of its weight, *w*, on earth. Graph your function.

b. How much would a person weigh on the moon if she weighs 120 pounds on earth?

c. A piece of rock weighs 50 pounds on the moon. How much will it weigh on earth?

d. Locate the points on your graph that correspond to your answers for (b) and (c).

22. Hubble's law says that distant galaxies are receding from us at a rate that varies directly with their distance. (The speeds of the galaxies are measured using a phenomenon called "redshifting.") A galaxy in the constellation Ursa Major is 980 million light-years away and is receding at a speed of 15,000 kilometers per second.

a. Find a function that gives the speed, *v*, of a galaxy in terms of its distance, *d*, from earth. Graph your function.

b. How far away is a galaxy in the constellation Hydra that is receding at 61,000 kilometers per second?

c. A galaxy in Leo is 1240 million light-years away. How fast is it receding?

d. Locate the points on your graph that correspond to your answers for (b) and (c).

23. The length, *L*, of a pendulum in feet varies directly with the square of its period, *T*, the time in seconds required for the pendulum to make one complete swing back and forth. The pendulum on a grandfather clock is 3.25 feet long and has a period of 2 seconds.

a. Express *L* as a function of *T*. Graph your function.

b. How long is the Foucault pendulum in the Pantheon in Paris, which has a period of 17 seconds?

c. A hypnotist uses a gold pendant as a pendulum to mesmerize his clients. If the chain on the pendant is 9 inches long, what is the period of its swing?

d. Locate the points on your graph that correspond to your answers for (b) and (c).

24. The load, *L*, that a beam can support varies directly with the square of its vertical thickness, *h*. A beam that is 4 inches thick can support a load of 2000 pounds.

a. Express *L* as a function of *h*. Graph your function.

b. What size load can be supported by a beam that is 6 inches thick?

c. How thick a beam is needed to support a load of 100 pounds?

d. Locate the points on your graph that correspond to your answers for (b) and (c).

25. Computer monitors produce a magnetic field. The effect of the field on the user varies inversely with his or her distance from the screen. The field from a certain 13-inch color monitor was measured at 22 milligauss 4 inches from the screen.

a. Express the field strength as a function of distance from the screen. Graph your function.

b. What is the field strength 10 inches from the screen?

c. An elevated risk of cancer can result from exposure to field strengths of 4.3 milligauss. How far from the screen should the computer user sit to keep the field level below 4.3 milligauss?

d. Locate the points on your graph that correspond to your answers for (b) and (c).

26. The amount of current that flows through a circuit varies inversely with the resistance on the circuit. An iron with a resistance of 12 ohms draws 10 amps of current.

a. Express the current as a function of the resistance. Graph your function.

b. How much current is drawn by a compact fluorescent light bulb with a resistance of 533.3 ohms?

c. What is the resistance of a toaster that draws 12.5 amps of current?

d. Locate the points on your graph that correspond to your answers for (b) and (c).

27. The amount of power generated by a windmill varies directly with the cube of the wind speed. A windmill on Oahu, Hawaii, produces 7300 kilowatts of power when the wind speed is 32 miles per hour.

a. Express the power as a function of wind speed. Graph your function.

b. How much power would the windmill produce in a light breeze of 15 miles per hour?

c. What wind speed is needed to produce 10,000 kilowatts of power?

d. Locate the points on your graph that correspond to your answers for (b) and (c).

28. A crystal form of pyrite (a compound of iron and sulfur) has the shape of a regular solid with 12 faces. Each face is a regular pentagon. This compound is called "pyritohedron," and its mass varies directly with the cube of the length of one edge. If each edge is 1.1 centimeters, then the mass is 51 grams.

a. Express the mass of pyritohedron as a function of the length of one edge. Graph your function.

b. What is the weight of a chunk of pyritohedron if each edge is 2.2 centimeters?

c. How long would each edge be for a 3264-gram piece of pyritohedron?

d. Locate the points on your graph that correspond to your answers for (b) and (c).

Each of the functions described in Exercises 29–36 by a table of data or by a graph is an example of direct or inverse variation. (The dependent variable varies with some power of the independent variable.) For each exercise
a.   find an algebraic formula for the second variable in terms of the first variable, including the constant of variation, *k*.
b.   answer the question asked.

29. The faster a car is moving, the more difficult it is to stop it. The graph in Figure 5.41 shows the relationship between the velocity (in kilometers per hour), *v*, of a car before the brakes are applied and the distance (in meters), *d*, required to stop the car. How far would it take to stop a car that was moving at 100 kilometers per hour?

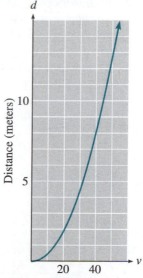

Figure 5.41

30. A wide pipe can handle a greater water flow than a narrow pipe. The graph in Figure 5.42 shows the relationship between the radius (in inches), *r*, of the pipe and the water flow (in gallons per second), *w*. How great is the water flow when the pipe has a radius of 10 inches?

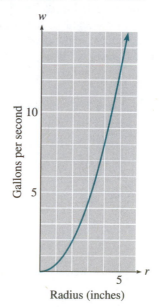

Figure 5.42

31. If the price of mushrooms goes up, then the amount that consumers are willing to buy goes down. The graph in Figure 5.43 relates the price, p, of shiitake mushrooms (in dollars per pound) and the weekly amount, m, of the mushrooms (in tons) sold in California. What amount of mushrooms do you predict will be sold if the price of shiitake mushrooms rises to $10 per pound?

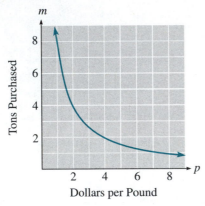

Figure 5.43

32. When an adult plays with a small child on a seesaw, the adult must sit closer to the pivot point for the seesaw to balance. The graph in Figure 5.44 relates the weight of the adult and the distance he must sit from the pivot. How far from the pivot must Kareem sit if he weighs 280 pounds?

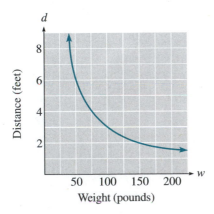

Figure 5.44

33. Ocean temperatures are generally colder at greater depths. Table 5.12 shows how the depth in kilometers is related to the temperature of the water in degrees Celsius. What would be the temperature at a depth of 6 kilometers?

Depth (km)	Temperature (°C)
0.5	12
1.0	6
2.0	3
3.0	2

Table 5.12

34. The shorter the length of a vibrating guitar string, the higher the frequency of the vibrations will be. The fifth string is 65 centimeters long and is tuned to A (with a frequency of 220 vibrations per second). The placement of the fret relative to the bridge changes the effective length of the guitar string. Table 5.13 gives effective lengths and corresponding frequencies. How far from the bridge should the fret for the note C (256 vibrations per second) be placed?

Length (cm)	Frequency
55.0	260
57.2	250
65.0	220
71.5	200

Table 5.13

35. The strength of a cylindrical rod depends on its diameter. The greater the diameter of a 1-meter rod made of a particular metal alloy, the more weight it can support before collapsing. Table 5.14 shows the maximum weight that can be supported by a rod of different diameters. How much weight could a 1.2-centimeter rod support before collapsing?

Diameter (cm)	Weight (newtons)
0.5	150
1.0	600
1.5	1350
2.0	2400

Table 5.14

36. The maximum height attainable by a cannonball depends on the speed at which it was shot. Table 5.15 shows various initial speeds and the associated maximum heights attainable by a cannonball shot with those initial speeds. What is the maximum height a cannonball shot with an initial upward speed of 100 feet per second can attain?

Speed (ft/sec)	Height (ft)
40	200.0
50	312.5
60	450.0
70	612.5

Table 5.15

37. The intensity of illumination, I, from a lamp varies inversely with the square of your distance, d, from the lamp. If you double your distance from a reading lamp, what happens to the illumination?

38. The resistance, R, of an electrical wire varies inversely with the square of its diameter, d. If you replace an old wire with a new one whose diameter is two-thirds of the old one, what happens to the resistance?

39. The wind resistance, W, experienced by a vehicle on the freeway varies directly with the square of its speed, v. If you decrease your speed by 10%, what happens to the wind resistance?

40. The weight of a bronze statue varies directly with the cube of its height. If you increase the height of a statue by 50%, what happens to its weight?

41. y varies directly with x. Show that if you multiply x by any constant c, then y will be multiplied by the same constant.

42. y varies inversely with x. Show that if you multiply x by any constant c, then y will be divided by the same constant.

43. Explain why the ratio $\dfrac{y}{x^2}$ is a constant when y varies directly with x^2.

44. Explain why the product yx^2 is a constant when y varies inversely with x^2.

45. If x varies directly with y and y varies directly with z, does x vary directly with z?

46. If x varies inversely with y and y varies inversely with z, does x vary inversely with z?

♪.♪ FUNCTIONS AS MATHEMATICAL MODELS

Shape of the Graph Constructing a good model for a situation often begins with deciding what kind of function to use. Your decision can depend on qualitative considerations, such as the general shape of the graph. What sort of function has the right shape to describe the process we want to model? Should it be increasing or decreasing, or some combination? Is the slope constant or is it changing? In Examples 1 and 2 we investigate how the shape of a graph illustrates the nature of the process it models.

Example 1

Forrest leaves his house to go to school. Sketch a possible graph of Forrest's distance from the house versus the time since he left, for each of the following situations.

a. Forrest walks at a constant speed until he reaches the bus stop.

b. Forrest walks at a constant speed until he reaches the bus stop, then waits there until the bus arrives.

c. Forrest walks at a constant speed until he reaches the bus stop and waits there until the bus arrives. Then the bus drives him to school at a constant speed.

Solutions   a.   The graph could look like the straight-line segment in Figure 5.45(a). It begins at the origin because at the instant Forrest leaves the house, his distance from home is 0. In other words, when $t = 0$, $y = 0$. The graph is a straight line because Forrest has a constant speed. The slope of the line is equal to Forrest's walking speed.

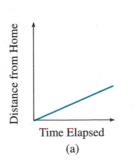

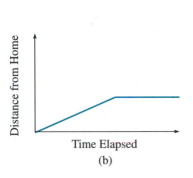

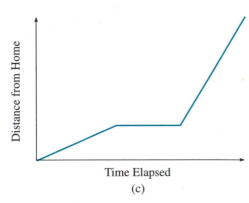

Figure 5.45

b.   The graph begins exactly as the graph in Figure 5.45(a) does. But while Forrest waits for the bus his distance from home remains constant, so the graph at that time must be a horizontal line. The fact that this line has slope 0 corresponds to the fact that during the time Forrest is waiting for the bus his speed is 0.

c.   The graph begins just as the graph in Figure 5.45(b) does. The section of the graph that represents the bus ride should have a constant slope because the bus is moving at a constant speed. Because the bus (probably) moves much faster than Forrest walks, the slope of the line should be larger for the bus section of the graph than it was for the walking section. The graph is shown in Figure 5.45(c).

Example 2

The two functions described here are both examples of increasing functions, but they increase in different ways. Match each function to its graph and to the appropriate table of values.

a. The number of flu cases reported at an urban medical center during an epidemic is an increasing function of time, and it is growing at a faster and faster rate.

b. The temperature of a potato placed in a hot oven increases rapidly at first, and then slows down as it approaches the temperature of the oven.

(1)

x	0	2	5	10	15
y	70	89	123	217	383

(2)

x	0	2	5	10	15
y	70	219	341	419	444

(A)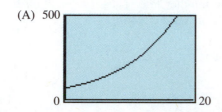

(B)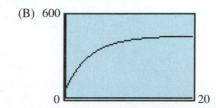

Solutions **a.** The number of flu cases is described by graph (A) and table (1). The function values in table (1) increase at an increasing rate. We can see this by computing the slope over successive time intervals:

$$x = 0 \quad \text{to} \quad x = 5: \qquad m = \frac{\Delta y}{\Delta x} = \frac{123 - 70}{5 - 0} = 10.6$$

$$x = 5 \quad \text{to} \quad x = 10: \qquad m = \frac{\Delta y}{\Delta x} = \frac{217 - 123}{10 - 5} = 18.8$$

$$x = 10 \quad \text{to} \quad x = 15: \qquad m = \frac{\Delta y}{\Delta x} = \frac{383 - 217}{15 - 10} = 33.2$$

The increasing slopes are reflected in graph (A); the graph bends upward as the slopes increase.

b. The temperature of the potato is described by graph (B) and table (2). The function values in table (2) increase, but at a decreasing rate:

$$x = 0 \quad \text{to} \quad x = 5: \qquad m = \frac{\Delta y}{\Delta x} = \frac{341 - 70}{5 - 0} = 54.2$$

$$x = 5 \quad \text{to} \quad x = 10: \qquad m = \frac{\Delta y}{\Delta x} = \frac{419 - 341}{10 - 5} = 15.6$$

$$x = 10 \quad \text{to} \quad x = 15: \qquad m = \frac{\Delta y}{\Delta x} = \frac{441 - 419}{15 - 10} = 4.4$$

The decreasing slopes are illustrated by graph (B); the graph is increasing but bends downward.

Using the Basic Functions as Models

In Chapters 2 and 3 we examined a number of situations that can be modeled by linear functions, and in Chapter 4 we examined some applications of quadratic functions. Direct and inverse variation, which we studied in

Section 5.4, use functions of the form $y = kx^n$ and $y = \dfrac{k}{x^n}$. In this section we will look at a few of the other basic functions.

First we will consider the absolute value function. Recall that the absolute value of a number gives the distance from the origin to that number on the real-number line. More generally, the distance between two points x and a can be denoted by $|x - a|$. For example, the equation $|x - 2| = 6$ means that "the distance between x and 2 is six units." The absolute value ensures that the distance will be a positive number, regardless of whether x lies to the left or the right of a on the number line. Thus $|x - 2| = 6$ has two solutions, 8 and -4, as shown in Figure 5.46.

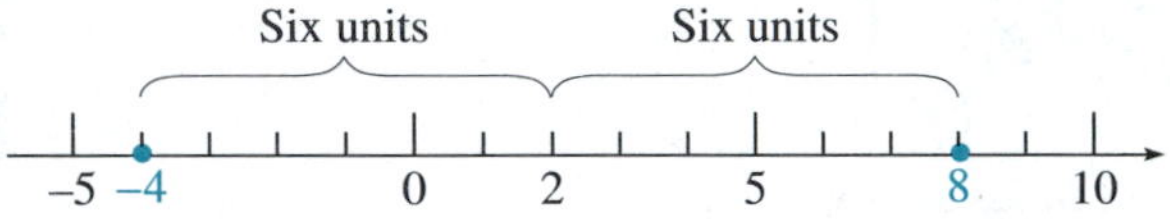

Figure 5.46

Example 3

Write each statement using absolute value notation.
a. x is three units from the origin.
b. p is two units from 5.
c. a is within four units of -2.

Solutions   First restate each sentence in terms of distance.
a.   The distance between x and the origin is three units, or $|x| = 3$.
b.   The distance between p and 5 is two units, or $|p - 5| = 2$.
c.   The distance between a and -2 is less than four units, or
$$|a - (-2)| < 4, \quad \text{or} \quad |a + 2| < 4$$

In Example 4 we use a graph to solve equations and inequalities involving absolute values.

Example 4

Marlene is driving to a new outlet mall on Highway 17. A gas station is by Marlene's on-ramp, and she buys gas there and resets her odometer to zero before getting on the highway. The mall is only 15 miles from Marlene's on-ramp. Unfortunately, she mistakenly drives past the mall and continues down the highway. Marlene's distance from the mall is a function of how far she has driven on Highway 17. (See Figure 5.47.)

Figure 5.47   Gas Station   Mall

a. Make a table of values showing how far Marlene has driven on Highway 17 and how far she is from the mall.

b. Make a graph of Marlene's distance from the mall versus the number of miles she has driven on the highway. Which of the basic graphs from Section 5.3 does your graph most resemble?

c. Find a piecewise-defined formula that describes Marlene's distance from the mall as a function of the distance she has driven on the highway.

d. Use your graph to determine how far Marlene has driven when she is within 5 miles of the mall.

e. Determine how far Marlene has driven when she is at least 10 miles from the mall.

Solutions　a. Marlene gets closer to the mall for each mile that she has driven on the highway until she has driven 15 miles, and after that she gets farther from the mall.

b. Plot the points in Table 5.16 to obtain the graph shown in Figure 5.48. This graph looks like the absolute value function defined in Section 5.3, except that the vertex has moved from the origin to (15, 0).

Miles on highway	0	5	10	15	20	25	30
Miles from mall	15	10	5	0	5	10	15

Table 5.16

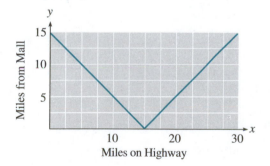

Figure 5.48

c. Let x represent the number of miles on the highway and $f(x)$ the number of miles from the mall. For x-values less than 15, the graph is a straight line with slope -1 and y-intercept at (0, 15), so its equation is $y = -x + 15$. Thus

$$f(x) = -x + 15 \qquad \text{when} \qquad 0 \le x < 15$$

When $x \ge 15$ however, the graph of f is a straight line with slope 1 that passes through the point (15, 0). The point-slope form of this line is $y - 0 = 1(x - 15)$, so $y = x - 15$. Thus

$$f(x) = x - 15 \qquad \text{when} \qquad x \ge 15$$

Combining our two pieces together, we obtain

$$f(x) = \begin{cases} -x + 15 & \text{when} & 0 \le x < 15 \\ x - 15 & \text{when} & x \ge 15 \end{cases}$$

The graph of $f(x)$ is a part of the graph of $y = |x - 15|$. If we think of the highway as a portion of the real line, with Marlene's on-ramp located at the origin, then the outlet mall is located at 15. (See Figure 5.48.) Marlene's coordinate as she drives along the highway is x, and the distance from Marlene to the mall is given by $|x - 15|$.

d.   Marlene is within 5 miles of the mall when she has driven between 10 and 20 miles, or when $10 < x < 20$. (See Figure 5.49.) Note that we can restate this problem as "find Marlene's position when the distance between Marlene and the mall is less than 5," or in mathematical terms, solve the inequality

$$|x - 15| < 5$$

The solution is $10 < x < 20$.

e.   Marlene is at least 10 miles from the mall when she has driven less than 5 miles or more than 25 miles. (See Figure 5.50.) In mathematical terms we are solving the inequality

$$|x - 15| \ge 10$$

The solution is $x \le 5$ or $x \ge 25$.

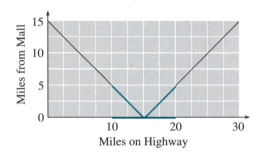

Miles from Mall / Miles on Highway

Figure 5.49

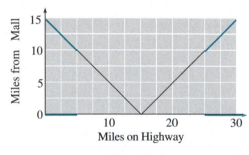

Miles from Mall / Miles on Highway

Figure 5.50

Example 5 illustrates an application of the function $f(x) = \sqrt{x}$.

Example 5

The speed of sound is a function of the temperature of the air in degrees Kelvin. (The temperature, T, in degrees Kelvin is given by $T = C + 273$, where C is the temperature in degrees Celsius.) Table 5.17 shows some data giving the speed of sound, s, in meters per second, at various temperatures, T.

T	0	20	50	100	200	400
s	0	89.7	141.8	200.6	283.7	401.2

Table 5.17

a.   Plot the data to obtain a graph. Which of the basic functions does your graph most resemble?
b.   This function is an example of direct variation. Find a value of k that fits the data.
c.   On a summer night when the temperature is 20°C, you see a flash of lightning, and 6 seconds later you hear a thunderclap. Use your function to estimate your distance from the thunderstorm.

Solutions   a.   The graph of the data is shown in Figure 5.51. The shape of the graph reminds us of the square root function, $y = \sqrt{x}$.
b.   We are looking for a value of k so that the function $f(T) = k\sqrt{T}$ fits the data. Recall that we solved this type of problem in Section 5.4 on variation: we will substitute one of the data points into the formula and solve for k. If

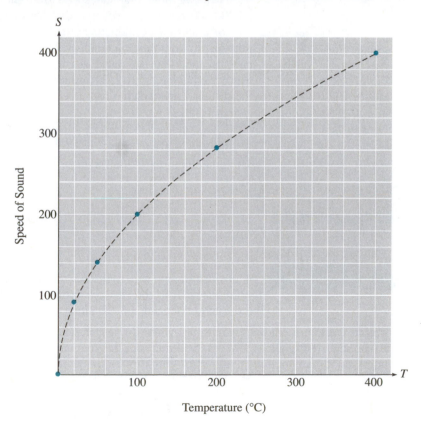

Figure 5.51

we choose the point (100, 200.6), we obtain

$$200.6 = k\sqrt{100}$$

and solving for k yields $k = 20.06$. We can check that the formula $s = 20.06\sqrt{T}$ is a good fit for the rest of the data points as well. Thus we suggest the function

$$f(T) = 20.06\sqrt{T}$$

as a model for the speed of sound.

c. First use the model to calculate the speed of sound at a temperature of 20°C. The Kelvin temperature is

$$T = 20 + 273 = 293$$

so we evaluate $s = f(T)$ for $T = 293$.

$$f(293) = 20.06\sqrt{293} \approx 343.4$$

so s is approximately 343.4 meters per second.

Next we note that the speed of light is fast enough (approximately 30,000,000 meters per second) that for distances on earth we can consider the transmission of light to be instantaneous. The lightning flash and the thunderclap occur simultaneously, so if we see the lightning immediately we can conclude that the sound of the thunderclap takes 6 seconds to reach us. Since we have already calculated the speed of sound, it is now easy to find the distance it traveled.

$$\begin{aligned} \text{Distance} &= \text{speed} \times \text{time} \\ &= (343.4 \text{ m/sec})(6 \text{ sec}) = 2060.4 \text{ meters} \end{aligned}$$

The thunderstorm is 2060 meters, or about 1.3 miles, away.

Some Geometric Models

We often need to use knowledge of geometry to find a formula for a function.

Example 6

Consider several different lines that pass through the point (2, 1) and form right triangles in the first quadrant, as shown in Figure 5.52. The shape of the triangle depends on the slope of the line.

a. Find the area of the triangle as a function of the slope, m, of the line through (2,1).
b. What is the domain of this function?
c. Graph your function for values of m between -3 and 0.
d. For what value of m does the triangle have the smallest area?

Solutions **a.** First draw several different triangles that satisfy the conditions of the problem. Let a and b stand for the x- and y-intercepts of the line

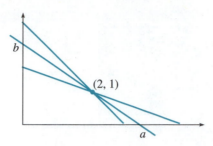

Figure 5.52

through (2, 1). Note that the values of a and b depend upon the slope of the line. Now the area of the triangle is $A = \frac{1}{2} \cdot \text{base} \cdot \text{height}$, or

$$A = \frac{1}{2} \cdot a \cdot b$$

Thus we will be able to express the area, A, as a function of slope if we can express both a and b in terms of the slope.

The point-slope equation of the line is

$$y - 1 = m(x - 2)$$

To find the x-intercept, we substitute the point $(a, 0)$ into the equation of the line and solve for a.

$$0 - 1 = m(a - 2) \qquad \text{Divide by } m.$$

$$\frac{-1}{m} = a - 2 \qquad \text{Add 2.}$$

$$2 - \frac{1}{m} = a$$

This calculation tells us that the x-intercept is $a = 2 - \frac{1}{m}$.

To find the y-intercept, we substitute the point $(0, b)$ into the equation of the line and solve for b.

$$b - 1 = m(0 - 2) \qquad \text{Add 1.}$$

$$b = 1 - 2m$$

The y-intercept is $b = 1 - 2m$.

Since the area of the triangle is $A = \frac{1}{2} \cdot a \cdot b$, we obtain

$$A(m) = \frac{1}{2}\left(2 - \frac{1}{m}\right)(1 - 2m)$$

b. The domain of the function is the set of all values of the slope that will give a triangle in the first quadrant. We see that we have such a triangle

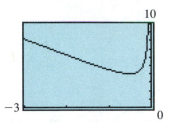

Figure 5.53

exactly when the slope is negative, so the domain is the set of all negative numbers.

c. Use your calculator to graph the function $A(m)$ on the domain $[-3, 0)$. The graph is shown in Figure 5.53. Each point on the graph represents a different triangle. The x-coordinate gives the slope of the line that forms the triangle, and the y-coordinate gives the area of the triangle. For example, the point $(-1.5, 5.\overline{3})$ represents the triangle formed by the line with slope -1.5. The area of that triangle is $5\frac{1}{3}$ square units.

d. Use the $\boxed{\text{TRACE}}$ feature to find the coordinates of the lowest point on the graph. This point represents the triangle with the smallest area. Its x-coordinate is approximately -0.5. By zooming in we can verify that the coordinates of the minimum point are $(-0.5, 4)$, so the smallest triangle is formed by a line of slope -0.5, and its area is 4.

🔍 *Investigation*

Periodic Functions

In this investigation we consider a very simple example of a periodic function. A **periodic function** is one that repeats its range values at evenly spaced intervals, or periods, of the domain. Periodic functions are used to model a great variety of phenomena that exhibit cyclical behavior, from growth patterns in plants and animals to radio waves and planetary motion.

Delbert is standing at the point $(5,0)$ on the square shown in Figure 5.54. If he walks around the square in the counterclockwise direction, his position at any time depends on the distance, d, he has walked. In other words, Delbert's y-coordinate is a function of the distance he has walked.

1. What is the y-coordinate of Delbert's position after he has walked 0 units? 2 units? 5 units? (Start filling in Table 5.18 in part (6).)

2. What do you conclude about the y-coordinate of Delbert's position after he has walked between 0 and 5 units?

3. After Delbert reaches the upper right corner of the square, he will turn left and start walking along the top of the square. What is his y-coordinate along the top of the square?

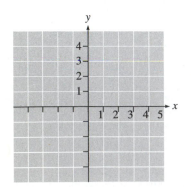

Figure 5.54

4. What do you conclude about the *y*-coordinate of Delbert's position after he has walked between 5 units and 15 units from the start?

5. Delbert will turn left again after he reaches the upper left corner and start walking down the left side of the square. What will be the *y*-coordinate of his position after he has walked a total of 18 units? 20 units? 22 units?

6. Fill in the rest of Table 5.18 with the *y*-coordinates of Delbert's position after walking a distance *d*. (Assume that he continues to go around the square more than once.)

d	0	2	5	8	10	12	15	18	20	22	25	28	30	32	35	38	40	42	45	48	50
y																					

Table 5.18

7. On the grid in Figure 5.55, plot a connected graph of Delbert's *y*-coordinate versus distance.

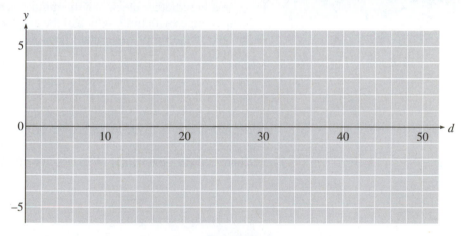

Figure 5.55

8. How far does Delbert walk before he starts his second time around the square? Before he starts his third time around?

9. Suppose you know where Delbert is on the square at some given time. Can you predict where he will be after he has walked another 40 units?

10. Let *f* be the name of the function relating Delbert's *y*-coordinate with the distance he has walked, that is, $y = f(d)$. What does your answer to part (9) say about $f(d)$ and $f(d + 40)$ for any positive value of *d?*

11. Look at the portion of your graph that corresponds to $0 \le d \le 5$. This portion of the graph should be part of a straight line. Find a formula for $f(d)$ that is valid when $0 \le d \le 5$. (*Hint:* First find the equation of the line. Since $f(d)$ is the y-coordinate, we can simply replace y with $f(d)$ in the equation of the line.)

12. Find a formula for $f(d)$ that is valid when $5 \leq d \leq 15$. (*Hint:* Find the equation of the corresponding line.)

13. Find a formula for $f(d)$ that is valid when $15 \leq d \leq 25$. (*Hint:* Find the equation of the corresponding line.)

14. How does the graph of f for values of d with $0 \leq d \leq 40$ compare with the graph for values of d with $40 \leq d \leq 80$?

15. Describe how the graph of f would continue if we considered all the d-values from 0 to 400.

16. Choose any positive value for d. How does the value of $f(d)$ compare with the value of $f(d + 20)$?

Here is another example about periodic functions.

Example 7

Imagine a grandfather clock. As the minute hand sweeps around, the height of its tip changes with time. Which of the graphs in Figure 5.56 best represents the height of the tip of the minute hand as a function of time?

Solution   Figure 5.56(a) is not the graph of a function at all: some values of t, such as $t = 0$, correspond to more than one value of h, which is not possible in the graph of a function.

   Figure 5.56(b) shows the height of the minute hand varying between a maximum and minimum value. However, the height decreases at a constant rate (the graph is straight and the slope is constant) until the minimum is reached, and then the height increases at a constant rate (the graph is straight

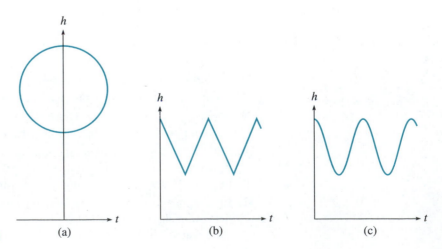

Figure 5.56   (a)   (b)   (c)

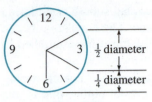

Figure 5.57

and the slope is constant). But notice in Figure 5.57 that during the 10 minutes from 12:10 to 12:20, the height of the minute hand decreases about half the diameter of the clock, whereas its decrease in height from 12:20 to 12:30 is only about a quarter of the diameter of the clock. Thus the height does *not* decrease at a constant rate.

Figure 5.56(c) is the best choice. The graph is curved because the slopes are not constant. The graph is steep when the height is changing rapidly, and the graph is nearly horizontal when the height is changing slowly. The height changes slowly near the hour and the half-hour, and more rapidly near the quarter-hours.

Exercise 5.5

Which graph best illustrates each of the situations in Exercises 1–4?

1. Your pulse rate during an aerobics class

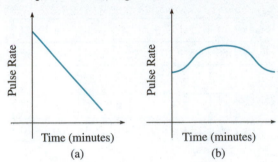

2. The stopping distances for cars traveling at various speeds

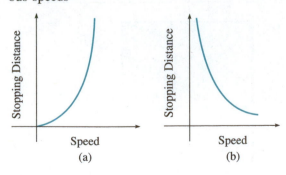

3. Your income in terms of the number of hours you worked

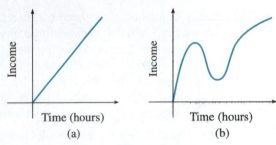

4. Your temperature during an illness

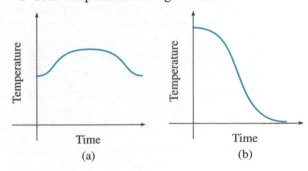

Sketch graphs to illustrate the situations in Exercises 5–10.

5. The height of your head above the ground during a ride on a Ferris wheel.

6. The height above the ground of a rubber ball dropped from the top of a 10-foot ladder.

7. Halfway from your English class to your math class, you realize that you left your math book in the classroom. You retrieve the book, then walk to your math class. Graph the distance between you and your English classroom as a function of time, from the moment you originally leave the English classroom until you reach the math classroom.

8. After you leave your math class, you start toward your music class. Halfway there you meet an old friend, so you stop and chat awhile. Then you continue to the music class. Graph the distance between you and your math classroom as a function of time, from the moment you originally leave the math classroom until you reach the music classroom.

9. Toni drives from home to meet her friend at the gym, which is halfway between their homes. They work out together at the gym, then they both go to the friend's home for a snack. Finally, Toni drives home. Graph the distance between Toni and her home as a function of time, from the moment she leaves home until she returns.

10. While bicycling from home to school, Greg gets a flat tire. He repairs the tire in just a few minutes but decides to backtrack a few miles to a service station where he cleans up. Finally, he bicycles the rest of the way to school. Graph the distance between Greg and his home as a function of time, from the moment he leaves home until he arrives at school.

11. Four different functions are described here. Match each description with the appropriate table of values and with its graph.

a. As a chemical pollutant pours into a lake, its concentration is a function of time. The concentration of the pollutant initially increases quite rapidly, but due to the natural mixing and self-cleansing action of the lake, the concentration levels off and stabilizes at some saturation level.

b. An overnight express train travels at a constant speed across the Great Plains. The train's distance from its point of origin is a function of time.

c. The population of a small suburb of a Florida city is a function of time. The population began increasing rather slowly, but it has continued to grow at a faster and faster rate.

d. The level of production at a manufacturing plant is a function of capital outlay, that is, the amount of money invested in the plant. At first, small increases in capital outlay result in large increases in production, but eventually the investors begin to experience diminishing returns on their money so that although production continues to increase, it is at a disappointingly slow rate.

(1)

x	1	2	3	4	5	6	7	8
y	60	72	86	104	124	149	179	215

(2)

x	1	2	3	4	5	6	7	8
y	60	85	103	120	134	147	159	169

(3)

x	1	2	3	4	5	6	7	8
y	60	120	180	240	300	360	420	480

(4)

x	1	2	3	4	5	6	7	8
y	60	96	118	131	138	143	146	147

(A)

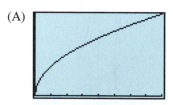

(B)

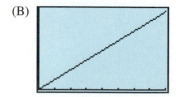

(C)

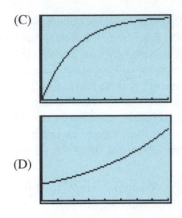

(D)

(3)

x	0	1	2	3	4
y	480	340	240	160	120

(4)

x	0	1	2	3	4
y	250	180	170	150	80

(A)

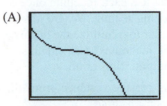

(B)

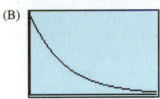

(C)

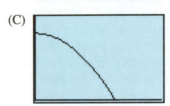

(D)

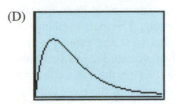

12. Four different functions are described here. Match each description with the appropriate table of values and with its graph.

a. After a chemical spill that greatly increased the phosphate concentration of Crystal Lake last year, fresh water flowing through the lake has gradually reduced the phosphate concentration to its natural level. The concentration of phosphate since the spill is a function of time.

b. The number of bacteria in a person during the course of an illness is a function of time. It increases rapidly at first and then decreases slowly as the patient recovers.

c. A squirrel drops a pine cone from the top of a California redwood. The height of the pine cone is a function of time, decreasing ever more rapidly as gravity accelerates its descent.

d. Enrollment in Ginny's Weight Reduction program is a function of time. It began declining last fall. After the holidays enrollment stabilized for a while but soon began to fall off again.

(1)

x	0	1	2	3	4
y	160	144	96	16	0

(2)

x	0	1	2	3	4
y	20	560	230	90	30

Use absolute value notation to write each expression in Exercises 13–20 as an equation or an inequality. (It may be helpful to restate each sentence using the word *distance*.)

13. x is six units from the origin.

14. a is seven units from the origin.

15. The distance from p to -3 is five units.

16. The distance from q to -7 is two units.

17. t is within three units of 6.

18. w is no more than one unit from -5.

19. b is at least 0.5 unit from -1.

20. m is more than 0.1 unit from 8.

21. A small pottery is setting up a workshop to produce mugs. The potter will work at a long table that holds three machines, as shown in Figure 5.58. The potter must use each machine once in the course of producing a mug. Where should the potter stand in order to minimize the distance she must walk to the machines? Let x represent the coordinate of the potter's station.

  a. Write expressions for the distance from the potter's station to each machine.

  b. Write a function that gives the sum of the distances from the potter's station to the three machines.

  c. Graph your function on the domain $[-20, 30]$, and use the graph to answer the question.

22. Suppose the potter in Exercise 21 adds a fourth machine to the procedure for producing a mug. This machine is located at $x = 16$ in Figure 5.58. Where should the potter stand now to minimize the distance she has to walk while producing each mug?

23. Richard and Marian are moving to Parkville after they graduate to take jobs. The main road through Parkville runs east and west and crosses a river in the center of town. Richard's job is located 10 miles east of the river on the main road, and Marian's job is 6 miles west of the river. A health club they both like is located 2 miles east of the river. If they plan to visit the health club every workday, where should Richard and Marian look for an apartment to minimize their total daily driving distance?

24. Romina's Bakery has just signed contracts to provide baked goods for three new restaurants located on Route 28 outside of town. The Coffee Stop is 2 miles north of town center, Sneaky Pete's is 8 miles north, and the Sea Shell is 12 miles south. Romina wants to open a branch bakery on Route 28 to handle the new business. Where should she locate the bakery in order to minimize the distance she must drive for deliveries?

25. Graph $y = |x + 3|$. Use your graph to solve the following equations and inequalities.

  a. $|x + 3| = 2$

  b. $|x + 3| \leq 4$

  c. $|x + 3| > 5$

26. Graph $y = |x - 2|$. Use your graph to solve the following equations and inequalities.

  a. $|x - 2| = 5$

  b. $|x - 2| < 8$

  c. $|x - 2| \geq 4$

Use graphs to solve the equations and inequalities in Exercises 27–32.

27. $|2x - 1| = 4$ **28.** $|3x - 1| = 5$

29. $|2x + 6| < 3$ **30.** $|5 - 3x| \leq 15$

31. $|3 - 2x| \geq 7$ **32.** $|3x + 2| > 10$

In each of Exercises 33–36, one quantity varies directly with the square root of the other; that is, $y = k\sqrt{x}$.
a. Find the value of k and write a function relating the variables.
b. Use your function to answer the question.
c. Graph your function and verify your answer to part (b) on the graph.

33. The stream speed necessary to move a granite particle is a function of the diameter of the parti-

Figure 5.58

cle; faster river currents can move larger particles. Table 5.19 shows the stream speed necessary
to move particles of different sizes. What speed
is needed to carry a particle with diameter 0.36
centimeter?

Diameter (cm)	Speed (cm/second)
0.01	5
0.04	10
0.09	15
0.16	20

Table 5.19

34. The speed at which water comes out of the spigot
at the bottom of a water jug is a function of the
water level in the jug; it slows down as the water
level drops. Table 5.20 shows different water
levels and the resulting exit speeds. What is the
exit speed when the water level is at 16 inches?

Level (in.)	Speed (gal/min)
9.00	1.50
6.25	1.25
4.00	1.00
2.25	0.75

Table 5.20

35. Table 5.21 gives the distance, d, in miles that
you can see from various heights, h, given in
feet. How far can you see from an airplane flying
at 20,000 feet?

h	100	500	1000	1500
d	12.25	27.39	38.74	47.44

Table 5.21

36. When a layer of ice forms on a pond, the thickness, d, of the ice in centimeters is a function of

time, t, in minutes. (See Table 5.22.) How thick
is the ice after 3 hours?

t	10	30	40	60
d	0.50	0.87	1.01	1.24

Table 5.22

Find appropriate formulas for each function in Exercises 37–42.

37. Sketch a rectangle with its base on the x-axis
and its upper vertices both on the graph of the
semicircle $y = \sqrt{16 - x^2}$. (See Figure
5.59.) Note that you can draw many different
rectangles, depending on where you choose the
vertex (a, b).

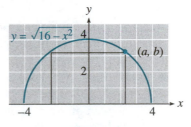

Figure 5.59

a. Express the length of the rectangle's base in
terms of a.

b. Express the height of the rectangle in terms
of a. (*Hint:* Do not forget that the point (a, b)
lies on the graph of the semicircle.)

c. Find the area of the rectangle as a function of
the x-coordinate, a, of its upper right corner.

d. What is the domain of your function? (What
values of a result in a rectangle?)

e. Graph your function on its domain. What is the
area of the rectangle formed if $a = 2$? What
value(s) of a result in a rectangle with area
10 square units?

f. What value of a results in the rectangle of
largest area?

38. Sketch a rectangle with its base on the x-axis and its upper vertices both on the graph of the parabola $y = 25 - x^2$. (See Figure 5.60.) Note that you can draw many different rectangles, depending on where you choose the vertex (a, b).

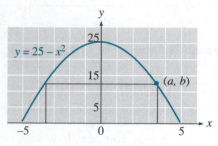

Figure 5.60

a. Express the length of the rectangle's base in terms of a.

b. Express the height of the rectangle in terms of a. *(Hint:* Do not forget that the point (a, b) lies on the graph of the parabola.)

c. Find the area of the rectangle as a function of the x-coordinate, a, of its upper right corner.

d. What is the domain of your function? (What values of a result in a rectangle?)

e. Graph your function on its domain. What is the area of the rectangle formed if $a = 4$? What value(s) of a result in a rectangle with area 75 square units?

f. What value of a results in the rectangle of largest area?

39. A triangle is formed in the first quadrant as follows: Two of the sides are formed by the x- and y-axes, and the third side is on a line passing through the point $(2, 1)$. Note that you can draw many different triangles in this way, depending on the slope of the line through $(2, 1)$.

a. Let $(a, 0)$ be the x-intercept of the third side of the triangle. Express the slope of the third side in terms of a.

b. Let $(0, b)$ be the y-intercept of the third side. Express b in terms of a. (Use your expression for the slope.)

c. Express the area of the triangle as a function of a.

d. What is the domain of your function? (What values of a result in a triangle?)

e. Graph your function on the domain $(2, 20.8]$. What is the area of the triangle if $a = 3$? What value(s) of a result in a triangle of area 7.2 square units?

f. What value of a results in the triangle with the smallest area? Describe what happens to the area of the triangle as a increases from 2 to larger values.

40. A triangle is formed in the first quadrant as follows: Two of the sides are formed by the x- and y-axes, and the third side is on a line passing through the point $(4, 2)$.

a. Let $(0, b)$ be the y-intercept of the third side of the triangle. Express the slope of the third side in terms of b.

b. Let $(a, 0)$ be the x-intercept of the third side. Express a in terms of b. (Use your expression for the slope.)

c. Express the area of the triangle as a function of b.

d. What is the domain of your function? (What values of b result in a triangle?)

e. Graph your function on the domain $(2, 11.4]$. What is the area of the triangle if $b = 3$? What value(s) of b result in a triangle of area 25 square units?

f. What value of b results in the triangle with the smallest area? Describe what happens to the area of the triangle as the values of b increase.

41. Suppose you want to measure the radius of a circular pond in the park. It will be easier to measure the circumference of the pond and then calculate the radius.

a. Express the radius of a circle in terms of its circumference.

b. Express the area of a circle as a function of its circumference.

c. What type of function is your answer to part (b)? What is the domain of your function?

d. What is the area of a pond whose circumference is 100 yards?

42. The hypotenuse of a right triangle is 12 centimeters long.

a. If the length of one leg of the triangle is x, what is the length of the other leg?

b. Express the area of the triangle as a function of x.

c. What is the domain of your function? Graph your function on its domain.

d. What is the area of the triangle if the shortest side is $\sqrt{6} \approx 2.45$ inches long?

43. Refer to the investigation on page 322. Sketch a graph of the x-coordinate of Delbert's position as a function of the distance he has walked.

44. Refer to Example 7 on page 324. Sketch a graph of the x-coordinate of the hand of the grandfather clock as a function of time.

Chapter 5 Review

Which of the tables in Exercises 1–4 describe functions? Why or why not?

1.

x	-2	-1	0	1	2	3
y	6	0	1	2	6	8

2.

p	3	-3	2	-2	-2	0
q	2	-1	4	-4	3	0

3.

Student	Score on IQ Test	Score on SAT Test
(A)	118	649
(B)	98	450
(C)	110	590
(D)	105	520
(E)	98	490
(F)	122	680

4.

Student	Correct Answers on Math Quiz	Quiz Grade
(A)	13	85
(B)	15	89
(C)	10	79
(D)	12	82
(E)	16	91
(F)	18	95

5. The total number of barrels of oil pumped by the AQ oil company is given by the formula

$$N(t) = 2000 + 500t$$

where N is the number of barrels of oil t days after a new well is opened. Evaluate $N(10)$ and explain what it means.

6. The number of hours required for a boat to travel upstream between two cities is given by the formula

$$H(v) = \frac{24}{v - 8} \qquad (v > 8)$$

where v represents the boat's top speed in miles per hour. Evaluate $H(16)$ and explain what it means.

Evaluate each function in Exercises 7–10 for the values given.

7. $F(t) = \sqrt{1 + 4t^2}$, $F(0)$ and $F(-3)$

8. $H(t) = t^2 + 2t$, $H(2a)$ and $H(a + 1)$

9. $f(x) = 2 - 3x$, $f(2) + f(3)$ and $f(2 + 3)$

10. $f(x) = 2x^2 - 4$, $f(a) + f(b)$ and $f(a + b)$

11. $P(x) = x^2 - 6x + 5$

   a. Compute $P(0)$.

   b. Find all values of x for which $P(x) = 0$.

12. $R(x) = \sqrt{4 - x^2}$

   a. Compute $R(0)$.

   b. Find all values of x for which $R(x) = 0$.

Consider the graphs shown for Exercises 13 and 14.

13.

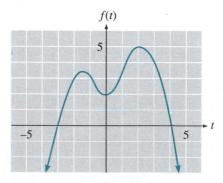

Figure 5.61

a. Find $f(-2)$ and $f(2)$ in Figure 5.61.

b. For what value(s) of t is $f(t) = 4$?

c. Find the t- and $f(t)$-intercepts of the graph.

d. What is the maximum value of f? For what value(s) of t does f take on its maximum value?

14.

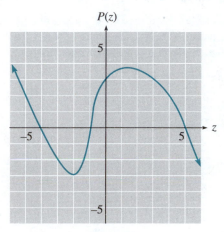

Figure 5.62

a. Find $P(-3)$ and $P(3)$ in Figure 5.62.

b. For what value(s) of z is $P(z) = 2$?

c. Find the z- and $P(z)$-intercepts of the graph.

d. What is the minimum value of P? For what value(s) of z does P take on its minimum value?

In Exercises 15–18 use a graphing calculator to graph each function on the given domain. Adjust Ymin and Ymax until you can determine the range of the function using the [TRACE] key. Then verify your answer algebraically by evaluating the function. State the domain and corresponding range in interval notation.

15. $f(t) = -t^2 + 3t; \quad -2 \le t \le 4$

16. $g(s) = \sqrt{s - 2}; \quad 2 \le s \le 6$

17. $F(x) = \dfrac{1}{x + 2}; \quad -2 < x \le 4$

18. $H(x) = \dfrac{1}{2 - x}; \quad -4 \le x < 2$

Which of the graphs in Exercises 19–22 represent functions?

19.

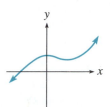

20.

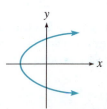

21.

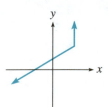

22.

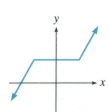

Graph each function in Exercises 23–32.

23. $f(t) = -2t + 4$

24. $g(s) = -\dfrac{2}{3}s - 2$

25. $p(x) = 9 - x^2$

26. $q(x) = x^2 - 16$

27. $f(x) = \begin{cases} x + 1 & \text{if } x \le 0 \\ x^2 & \text{if } x > 0 \end{cases}$

28. $g(x) = \begin{cases} x - 1 & \text{if } x \le 1 \\ x^3 & \text{if } x > 1 \end{cases}$

29. $H(x) = \begin{cases} x^2 & \text{if } x \le 0 \\ \sqrt{x} & \text{if } x > 0 \end{cases}$

30. $F(x) = \begin{cases} |x| & \text{if } x \le 0 \\ \dfrac{1}{x} & \text{if } x > 0 \end{cases}$

31. $S(x) = \begin{cases} x^3 & \text{if } x \le 1 \\ |x| & \text{if } x > 1 \end{cases}$

32. $T(x) = \begin{cases} \dfrac{1}{x^2} & \text{if } x < 0 \\ \sqrt{x} & \text{if } x \ge 0 \end{cases}$

Graph the functions given in Exercises 33–36 on a graphing calculator. Then use the graph to solve the equations and inequalities. Round your answers to one decimal place if necessary.

33. $y = \sqrt[3]{x}$
  a. Solve $\sqrt[3]{x} = 0.8$
  b. Solve $\sqrt[3]{x} = 1.5$
  c. Solve $\sqrt[3]{x} > 1.7$
  d. Solve $\sqrt[3]{x} \le 1.26$

34. $y = \dfrac{1}{x}$

  a. Solve $\dfrac{1}{x} = 2.5$

  b. Solve $\dfrac{1}{x} = 0.3125$

  c. Solve $\dfrac{1}{x} \ge 0.\overline{2}$

  d. Solve $\dfrac{1}{x} < 5$

35. $y = \dfrac{1}{x^2}$

  a. Solve $\dfrac{1}{x^2} = 0.03$

  b. Solve $\dfrac{1}{x^2} = 6.25$

  c. Solve $\dfrac{1}{x^2} > 0.16$

  d. Solve $\dfrac{1}{x^2} \le 4$

36. $y = \sqrt{x}$
  a. Solve $\sqrt{x} = 0.707$
  b. Solve $\sqrt{x} = 1.7$
  c. Solve $\sqrt{x} < 1.5$
  d. Solve $\sqrt{x} \ge 1.3$

For each table in Exercises 37–40, y varies directly or inversely with a power of x. Find the power of x and the constant of variation, k. Write a formula for each function of the form

$$y = kx^n \quad \text{or} \quad y = \dfrac{k}{x^n}.$$

37.

x	y
2	4.8
5	30.0
8	76.8
11	145.2

38.

x	y
1.4	75.6
2.3	124.2
5.9	318.6
8.3	448.2

39.

x	y
0.5	40.0
2.0	10.0
4.0	5.0
8.0	2.5

40.

x	y
1.5	320.0
2.5	115.2
4.0	45.0
6.0	20.0

41. The distance, s, a pebble falls through a thick liquid varies directly with the square of the length of time, t, it falls.

a. If the pebble falls 28 centimeters in 4 seconds, express the distance it will fall as a function of time.

b. Find the distance the pebble will fall in 6 seconds.

42. The volume, V, of a gas varies directly with the temperature, T, and inversely with the pressure, P, of the gas.

a. If $V = 40$ when $T = 300$ and $P = 30$, express the volume of the gas as a function of the temperature and pressure of the gas.

b. Find the volume when $T = 320$ and $P = 40$.

43. The demand for bottled water is inversely proportional to the price per bottle. If Droplets can sell 600 bottles at $8 each, how many bottles can it sell at $10 each?

44. The intensity of illumination from a light source varies inversely with the square of the distance from the source. If a reading lamp has an intensity of 100 lumens at a distance of 3 feet, what is its intensity 8 feet away?

45. A person's weight, w, varies inversely with the square of his or her distance, r, from the center of the earth.

a. Express w as a function of r. Let k stand for the constant of variation.

b. Make a rough graph of your function.

c. How far from the center of the earth must Neil be in order to weigh one-third of his weight on the surface? The radius of the earth is about 3960 miles.

46. The period, T, of a pendulum varies directly with the square root of its length, L.

a. Express T as a function of L. Let k stand for the constant of variation.

b. Make a rough graph of your function.

c. If a certain pendulum is replaced by a new one four-fifths as long as the old one, what happens to the period?

Sketch graphs to illustrate the situations in Exercises 47–48.

47. Inga runs hot water into the bathtub until it is about half full. Because the water is too hot, she lets it sit awhile before getting into the tub. After several minutes of bathing, she gets out and drains the tub. Graph the water level in the bathtub as a function of time, from the moment Inga starts filling the tub until it is drained.

48. David turns on the oven and it heats up steadily until the proper baking temperature is reached. The oven maintains that temperature during the time David bakes a pot roast. David leaves the oven door open for a few minutes, and the oven temperature drops fairly rapidly during that time. After David closes the door, the temperature continues to drop, but at a much slower rate. Graph the temperature of the oven as a function of time, from the moment David first turns on the oven until shortly after he closes the door when the oven is cooling.

For Exercises 49 and 50
a. plot the points and sketch a smooth curve through them.
b. use your graph to help you discover the equation that describes the function.

49.

x	$g(x)$
2	12
3	8
4	6
6	4
8	3
12	2

50.

x	$F(x)$
-2	9
-1	1
0	0
1	-1
2	-8
3	-27

Use absolute value notation to write each expression in Exercises 51–54 as an equation or inequality.

51. x is four units from the origin.

52. The distance from y to -5 is three units.

53. p is within four units of 7.

54. q is at least three-tenths of a unit from -4.

Use a graph to solve each inequality in Exercises 55–58.

55. $|3x - 2| < 4$ **56.** $|2x + 0.3| \le 0.5$

57. $|3y + 1.2| \ge 1.5$ **58.** $|3z + \frac{1}{2}| > \frac{1}{3}$

59. Sketch the equilateral triangle whose sides have length s.

 a. Use the Pythagorean theorem to find the altitude of the triangle in terms of s.

 b. Express the area of the triangle as a function of s.

 c. Sketch the graph of your function.

 d. Find the area of an equilateral triangle whose side is 4 centimeters long.

 e. Find the side of an equilateral triangle whose area is 2.7 square feet.

60. A rectangle is formed in the first quadrant as follows: The base and left side lie on the x- and y-axis, respectively. The upper right vertex (a, b) lies on the line $y = 10 - \frac{1}{2}x$.

 a. Express the length and height of the rectangle in terms of a.

 b. Express the area of the rectangle as a function of a.

 c. What is the domain of your function?

 d. Graph your function on its domain.

 e. Express the perimeter of the rectangle as a function of a.

Powers and Roots

*I*n this chapter we will learn more about the operation of raising to a power and its inverse operation, extracting a root. We will also learn how to handle both these operations using exponents. A working knowledge of exponents is essential for many applications, including scientific notation, the distance formula, and the exponential functions we will study in Chapter 7.

6.1 LAWS OF EXPONENTS

In this section we will briefly review the basic rules for performing operations on powers with positive integer exponents. In the sections that follow, we will study negative and fractional exponents and discuss how they are related to roots and radicals.

Products of Powers   Recall that an exponent tells us how many times its base occurs as a factor in an expression. For example,

$$4a^3b^2 \qquad \text{means} \qquad 4aaabb$$

where a is used as a factor three times and b is used as a factor twice. Now consider a product of two powers with the same base, $(a^3)(a^2)$, which can be written as

$$(a^3)(a^2) = aaa \cdot aa = a^5$$

since a occurs as a factor five times. We see that the number of a's in the product is the *sum* of the number of a's in each factor. In general, we have the following rule.

> **FIRST LAW OF EXPONENTS**
> To multiply two powers with the same base, add the exponents and leave the base unchanged.
> $$a^m \cdot a^n = a^{m+n}$$

Example 1

a.
$$5^3 \cdot 5^4 = 5^{3+4} = 5^7$$
Same base ———— Add exponents.

b.
$$x^4 \cdot x^2 = x^{4+2} = x^6$$
Same base ———— Add exponents.

In applying the laws of exponents, many opportunities for error arise. We must learn how to recognize when the laws do *not* apply as well as how to apply them correctly.

  COMMON ERRORS

1. We do not *multiply* the exponents when simplifying a product. For example,

$$b^4 \cdot b^2 \neq b^8$$

You can verify this with your calculator by choosing a value for b, for instance, $b = 3$:

$$3^4 \cdot 3^2 \neq 3^8$$

2. To apply the first law of exponents, the bases must be the same. For example,

$$2^3 \cdot 3^5 \neq 6^8$$

(Check this on your calculator.)

3. We do not multiply the bases when simplifying a product. In Example 1(a) note that

$$5^3 \cdot 5^4 \neq 25^7$$

4. Although we can simplify the product $x^2 x^3$ as x^5, we cannot simplify the *sum* $x^2 + x^3$ because x^2 and x^3 are not like terms.

We can use the first law of exponents, along with the commutative and associative laws of multiplication, to find the product of two monomials.

Example 2

Multiply $(-3x^4z^2)(5x^3z)$.

Solution   Rearrange the factors to group together the numerical coefficients and the powers of each base. Then multiply the coefficients and use the first law of exponents to find the powers of the variable factors.

$$(-3x^4z^2)(5x^3z) = (-3)(5)x^4x^3z^2z$$

$$= -15x^7z^3$$

Quotients of Powers

Recall that we can reduce fractions by "canceling," that is, by dividing both the numerator and denominator by any common factors. We can apply the same technique to quotients of powers with the same base. Consider the following example:

$$\frac{x^7}{x^4} = \frac{\cancel{x}\cancel{x}\cancel{x}\cancel{x}xxx}{\cancel{x}\cancel{x}\cancel{x}\cancel{x}} = \frac{x^3}{1} = x^3$$

We can obtain the exponent of the quotient by *subtracting* the exponent of the denominator from the exponent of the numerator. In other words,

$$\frac{x^7}{x^4} = x^{7-4} = x^3$$

What if the larger power occurs in the denominator of the fraction? Consider the example

$$\frac{x^4}{x^7} = \frac{\cancel{x}\cancel{x}\cancel{x}\cancel{x}}{\cancel{x}\cancel{x}\cancel{x}\cancel{x}xxx} = \frac{1}{x^3}$$

We can obtain the quotient by subtracting the exponent of the numerator from the exponent of the denominator, that is,

$$\frac{x^4}{x^7} = \frac{1}{x^{7-4}} = \frac{1}{x^3}$$

These examples suggest the following law.

SECOND LAW OF EXPONENTS

To divide two powers with the same base, subtract the smaller exponent from the larger one and keep the same base.

a. If the larger exponent occurs in the numerator, put the power in the numerator.

$$\text{If} \quad m > n, \quad \text{then} \quad \frac{a^m}{a^n} = a^{m-n} \quad (a \neq 0)$$

b. If the larger exponent occurs in the denominator, put the power in the denominator.

$$\text{If} \quad m < n, \quad \text{then} \quad \frac{a^m}{a^n} = \frac{1}{a^{n-m}} \quad (a \neq 0)$$

Example 3

a. $\dfrac{3^8}{3^2} = 3^{8-2} = 3^6$ Subtract exponents: 8 > 2.

b. $\dfrac{w^3}{w^6} = \dfrac{1}{w^{6-3}} = \dfrac{1}{w^3}$ Subtract exponents: 3 < 6.

To divide one monomial by another, we can apply the second law of exponents to the powers of each variable.

Example 4

Divide $\dfrac{3x^2y^4}{6x^3y}$.

Solution   Consider the numerical coefficients and the powers of each variable separately. Use the second law of exponents to simplify each quotient of powers.

$$\frac{3x^2y^4}{6x^3y} = \frac{3}{6} \cdot \frac{x^2}{x^3} \cdot \frac{y^4}{y}$$

$$= \frac{1}{2} \cdot \frac{1}{x^{3-2}} \cdot y^{4-1} \qquad \text{Subtract exponents.}$$

$$= \frac{1}{2} \cdot \frac{1}{x} \cdot \frac{y^3}{1} = \frac{y^3}{2x}$$

Powers of a Power   Consider the expression $(a^4)^3$, the third power of a^4. Because raising to a power is actually repeated multiplication, we can apply the first law of exponents repeatedly to simplify the expression.

$$(a^4)^3 = (a^4)(a^4)(a^4) = a^{4+4+4} = a^{12} \qquad \text{Add exponents.}$$

Of course, repeated addition is actually multiplication, so we can obtain the same result by multiplying the exponents together.

> **THIRD LAW OF EXPONENTS**
>
> To raise a power to a power, keep the same base and multiply the exponents.
>
> $$(a^m)^n = a^{mn}$$

Example 5

a. $(4^3)^5 = 4^{3 \cdot 5} = 4^{15}$ Multiply exponents.

b. $(y^5)^2 = y^{5 \cdot 2} = y^{10}$ Multiply exponents.

 COMMON ERROR

Notice the difference between the expressions

$$(x^3)(x^4) = x^{3+4} = x^7$$

and

$$(x^3)^4 = x^{3 \cdot 4} = x^{12}$$

The first expression is a product, so we add the exponents. The second expression raises a power to a power, so we multiply the exponents.

Powers of a Product In addition to the three basic laws governing products, quotients, and powers of powers, we can formulate two additional rules for the power of a product or a quotient.

To simply the expression $(5a)^3$, we can use the associative and commutative laws to regroup the factors as follows:

$$(5a)^3 = (5a)(5a)(5a)$$
$$= 5 \cdot 5 \cdot 5 \cdot a \cdot a \cdot a$$
$$= 5^3 a^3$$

Thus to raise a product to a power, we can simply raise each factor to the power.

FOURTH LAW OF EXPONENTS

A power of a product is equal to the product of the powers of each of its factors.

$$(ab)^n = a^n b^n$$

Example 6

a. $(5a)^3 = 5^3 a^3 = 125a^3$ Cube each factor.

b. $(-xy^2)^4 = (-x)^4 (y^2)^4$ Raise each factor to the fourth power.
$\qquad\quad = x^4 y^8$ Apply the third law of exponents.

 COMMON ERRORS

1. Compare the two expressions $3a^2$ and $(3a)^2$; they are not the same. In the expression $3a^2$, only the factor a is squared, whereas in $(3a)^2$ both 3 and a are squared. Thus

$$3a^2 \text{ cannot be simplified}$$

but

$$(3a)^2 = 3^2 a^2 = 9a^2$$

2. Compare the two expressions $(3a)^2$ and $(3 + a)^2$. The fourth law of exponents applies to the *product* $3a$ but not to the *sum* $3 + a$. Thus

$$(3 + a)^2 \neq 3^2 + a^2$$

In order to simplify $(3 + a)^2$, we must expand the binomial product:

$$(3 + a)^2 = (3 + a)(3 + a) = 9 + 6a + a^2$$

Powers of a Quotient   To simplify the expression $\left(\dfrac{x}{3}\right)^4$, we multiply together four copies of the fraction $\dfrac{x}{3}$; that is,

$$\left(\frac{x}{3}\right)^4 = \frac{x}{3} \cdot \frac{x}{3} \cdot \frac{x}{3} \cdot \frac{x}{3} = \frac{x \cdot x \cdot x \cdot x}{3 \cdot 3 \cdot 3 \cdot 3}$$

$$= \frac{x^4}{3^4} = \frac{x^4}{81}$$

In general, we have the following rule.

FIFTH LAW OF EXPONENTS

To raise a quotient to a power, raise both the numerator and the denominator to the power.

$$\left(\frac{a}{b}\right)^n = \frac{a^n}{b^n} \qquad (b \neq 0)$$

For easy reference we state all the laws of exponents together. All the laws are valid when a and b are not equal to zero and when the exponents m and n are positive integers.

I. $a^m \cdot a^n = a^{m+n}$

II. a. $\left(\dfrac{a^m}{a^n}\right) = a^{m-n} \qquad (m > n)$

   b. $\left(\dfrac{a^m}{a^n}\right) = \dfrac{1}{a^{n-m}} \qquad (m < n)$

III. $(a^m)^n = a^{mn}$

IV. $(ab)^n = a^n b^n$

V. $\left(\dfrac{a}{b}\right)^n = \dfrac{a^n}{b^n}$

We often need to use two or more laws of exponents to simplify an expression involving powers.

Example 7

Simplify $5x^2 y^3 (2xy^2)^4$.

Solution   According to the order of operations, we should perform any powers before multiplications. Thus we begin by simplifying $(2xy^2)^4$.   We apply the fourth law.

$$5x^2 y^3 (2xy^2)^4 = 5x^2 y^3 \cdot 2^4 x^4 (y^2)^4 \qquad \text{Apply the third law.}$$

$$= 5x^2 y^3 \cdot 2^4 x^4 y^8$$

Finally, multiply powers with the same base. Apply the first law.

$$5x^2 y^3 \cdot 2^4 x^4 y^8 = 5 \cdot 2^4 x^2 x^4 y^3 y^8 = 80 x^6 y^{11}$$

Example 8

Simplify $\left(\dfrac{2x}{z^2}\right)^3$.

Solution   Begin by applying the fifth law.

$$\left(\frac{2x}{z^2}\right)^3 = \frac{(2x)^3}{(z^2)^3}$$ Apply the fourth law to the numerator and the third law to the denominator.

$$= \frac{2^3 x^3}{z^6} = \frac{8x^3}{z^6}$$

Exercise 6.1

Find the products in Exercises 1–6 by applying the first law of exponents.

1. $b^4 \cdot b^5$

2. $b^2 \cdot b^8$

3. $6^5 \cdot 6^3$

4. $2^7 \cdot 2^2$

5. $(q^3)(q)(q^5)$

6. $(p^2)(p^4)(p^4)$

Multiply in Exercises 7–18.

7. $(4y)(-6y)$

8. $(-4z)(-8z)$

9. $(2wz^3)(-8z)$

10. $(4wz)(-9w^2z^2)$

11. $-4x(3xy)(xy^3)$

12. $-5x^2(2xy)(5x^2)$

13. $-7ab^2(-3ab^3)$

14. $-4a^2b(-3a^3b^2)$

15. $-2x^3(x^2y)(-4y^2)$

16. $3xy^3(-x^4)(-2y^2)$

17. $y^2z(-3x^2z^3)(-y^4z)$

18. $-3xy(2xz^4)(3x^3y^2z)$

Find the quotients in Exercises 19–24 by applying the second law of exponents.

19. $\dfrac{w^6}{w^3}$

20. $\dfrac{c^{12}}{c^4}$

21. $\dfrac{2^9}{2^4}$

22. $\dfrac{8^6}{8^2}$

23. $\dfrac{z^6}{z^9}$

24. $\dfrac{b^4}{b^8}$

Divide in Exercises 25–32.

25. $\dfrac{2a^3b}{8a^4b^5}$

26. $\dfrac{8a^2b}{12a^5b^3}$

27. $\dfrac{-12qw^4}{8qw^2}$

28. $\dfrac{-12rz^6}{20rz}$

29. $\dfrac{14x^4yz^6}{10x^6y^2z^2}$

30. $\dfrac{26xy^2z^2}{6x^4y^3z^7}$

31. $\dfrac{-15b^3c^2}{-3b^3c^4}$

32. $\dfrac{-25c^3d^2}{-5c^8d^2}$

Compute the powers in Exercises 33–38 by applying the third law of exponents.

33. $(d^3)^5$

34. $(d^4)^2$

35. $(5^4)^3$

36. $(4^3)^3$

37. $(t^8)^4$

38. $(t^2)^9$

Compute the powers in Exercises 39–44 by applying the fourth law of exponents.

39. $(6x)^3$

40. $(3y)^4$

41. $(2t^3)^5$

42. $(6s^2)^2$

43. $(-4a^2b^4)^4$

44. $(-5ab^8)^3$

Simplify in Exercises 45–56 by applying the laws of exponents.

45. $b^3(b^2)^5$

46. $b(b^4)^6$

47. $(p^2q)^3(pq^3)$

48. $(p^3)^4(p^3q^4)$

49. $(2x^3y)^2(xy^3)^4$

50. $(3xy^2)^3(2x^2y^2)^2$

51. $(yz^2)^3(zw^3)(-2yzw)$

52. $(3w^2z^2)(w^2y^4)^2(y^3z^2)$

53. $-a^2(-a)^2$

54. $-a^3(-a)^3$

55. $-(-xy^2)^3(-x^3)^3$

56. $-(-xy)^2(-xy^2)$

Compute the powers in Exercises 57–62 by applying the fifth law of exponents.

57. $\left(\dfrac{w}{2}\right)^6$

58. $\left(\dfrac{5}{u}\right)^4$

59. $\left(\dfrac{-4}{p^5}\right)^3$

60. $\left(\dfrac{-3}{q^4}\right)^5$

61. $\left(\dfrac{h^2}{m^3}\right)^4$

62. $\left(\dfrac{n^3}{k^4}\right)^8$

Simplify in Exercises 63–70 by applying the laws of exponents.

63. $\left(\dfrac{-2x}{3y^2}\right)^3$ **64.** $\left(\dfrac{-x^2}{2y}\right)^4$

65. $\dfrac{(4x)^3}{(-2x^2)^2}$ **66.** $\dfrac{(5x)^2}{(-3x^2)^3}$

67. $\dfrac{(xy)^2(-x^2y)^3}{(x^2y^2)^2}$ **68.** $\dfrac{(-x)^2(-x^2)^4}{(x^2)^3}$

69. $\left(\dfrac{-2x}{y^2}\right)^3\left(\dfrac{y^2}{3x}\right)^2$ **70.** $\left(\dfrac{x^2z}{2}\right)^2\left(\dfrac{-2}{x^2z}\right)^3$

If possible, simplify in Exercises 71–78.

71. a. $w + w$
　　b. $w(w)$

72. a. $m^2 - m^2$
　　b. $m^2(-m^2)$

73. a. $4z^2 - 6z^2$
　　b. $4z^2(-6z^2)$

74. a. $t^3 + 3t^3$
　　b. $t^3(3t^3)$

75. a. $4p^2 + 3p^3$
　　b. $4p^2(3p^3)$

76. a. $2w^2 - 5w^4$
　　b. $(2w^2)(-5w^4)$

77. a. $3^9 \cdot 3^8$
　　b. $3^9 + 3^8$

78. a. $(-2)^7(-2)^5$
　　b. $-2^7 - 2^5$

Evaluate each function for the expressions given in Exercises 79–82 and then simplify.

79. $f(x) = x^3$
　　a. $f(a^2)$
　　b. $a^3 \cdot f(a^3)$
　　c. $f(ab)$
　　d. $f(a + b)$

80. $g(x) = x^4$
　　a. $g(a^3)$
　　b. $a^4 \cdot g(a^4)$
　　c. $g(ab)$
　　d. $g(a + b)$

81. $F(x) = 3x^5$
　　a. $F(2a)$
　　b. $2F(a)$
　　c. $F(a^2)$
　　d. $[F(a)]^2$

82. $G(x) = 4x^3$
　　a. $G(3a)$
　　b. $3G(a)$
　　c. $G(a^4)$
　　d. $[G(a)]^4$

6.2　INTEGER EXPONENTS

Negative Exponents　We can simplify the second law of exponents if we allow negative numbers as exponents. Consider the quotient $\dfrac{x^2}{x^5}$. Since $2 < 5$, we can use law II(b) to write

$$\frac{x^2}{x^5} = \frac{1}{x^{5-2}} = \frac{1}{x^3}$$

If we allow negative exponents, however, we can apply law II(a) to find

$$\frac{x^2}{x^5} = x^{2-5} = x^{-3}$$

These two results for the same quotient motivate us to define x^{-3} to mean $\dfrac{1}{x^3}$. In general, for $a \neq 0$, we define negative powers as follows.

$$a^{-n} = \frac{1}{a^n} \qquad (a \neq 0)$$

Example 1

a. $2^{-3} = \dfrac{1}{2^3} = \dfrac{1}{8}$ **b.** $9x^{-2} = 9 \cdot \dfrac{1}{x^2} = \dfrac{9}{x^2}$

⚠ **COMMON ERROR**

In Example 1(b), note that

$$9x^{-2} \neq \dfrac{1}{9x^2}$$

The exponent, -2, applies *only* to the base, x, not to 9.

We see that a negative power is the *reciprocal* of a positive power. In particular, to simplify a negative power of a fraction, we need only compute the positive power of its reciprocal.

Example 2

a. $\left(\dfrac{3}{5}\right)^{-2} = \left(\dfrac{5}{3}\right)^{2} = \dfrac{25}{9}$ **b.** $\left(\dfrac{x^3}{y^2}\right)^{-3} = \left(\dfrac{y^2}{x^3}\right)^{3} = \dfrac{(y^2)^3}{(x^3)^3} = \dfrac{y^6}{x^9}$

In general, we have

$$\left(\dfrac{a}{b}\right)^{-n} = \left(\dfrac{b}{a}\right)^{n} \qquad a \neq 0, \;\; b \neq 0$$

Also, the reciprocal of a^{-n} is just a^n, since

$$\dfrac{1}{a^{-n}} = \dfrac{1}{\dfrac{1}{a^n}} = a^n$$

Example 3

a. $\dfrac{1}{5^{-3}} = 5^3 = 125$ **b.** $\dfrac{k^2}{m^{-4}} = k^2 m^4$

In general,

$$\dfrac{1}{a^{-n}} = a^n \qquad \text{and} \qquad \dfrac{b}{a^{-n}} = b \cdot a^n \qquad a \neq 0$$

 COMMON ERROR

A negative exponent does *not* mean that the power is negative! For example,

$$3^{-4} \neq -3^4$$

Laws of Exponents   With the definition given above for a^{-n}, the laws of exponents discussed in Section 6.1 apply to *all* integer powers. In particular, we can replace the two cases of the second law of exponents by a single rule.

$$\text{II.} \qquad \frac{a^m}{a^n} = a^{m-n} \qquad (a \neq 0)$$

As an immediate consequence of this new form of the second law of exponents, we have that

$$a^0 = 1, \qquad a \neq 0$$

You can convince yourself that this makes sense by observing that the quotient of any (nonzero) number divided by itself is 1, and thus

$$1 = \frac{a^m}{a^m} = a^{m-m} = a^0$$

For example,

$$3^0 = 1, \qquad (-528)^0 = 1, \qquad \text{and} \qquad (6xy)^0 = 1$$

We can use the laws of exponents to simplify expressions involving negative exponents.

Example 4

a. $x^3 \cdot x^{-5} = x^{3-5} = x^{-2}$ Apply the first law: add exponents.

b. $\dfrac{8x^{-2}}{4x^{-6}} = \dfrac{8}{4}x^{-2-(-6)} = 2x^4$ Apply the second law: subtract exponents.

We often need to apply more than one law to simplify an expression. For easy reference we restate the laws of exponents below. The laws are valid for all integer exponents m and n, and for $a \neq 0$, $b \neq 0$.

LAWS OF EXPONENTS

I. $a^m \cdot a^n = a^{m+n}$ IV. $(ab)^n = a^n b^n$

II. $\left(\dfrac{a^m}{a^n}\right) = a^{m-n}$ V. $\left(\dfrac{a}{b}\right)^n = \dfrac{a^n}{b^n}$

III. $(a^m)^n = a^{mn}$

Example 5

Simplify $\dfrac{(3ab^{-4})^{-2}}{2b^{-3}}$. Write your answer without negative exponents.

Solution Begin by applying the fourth law to the numerator.

$$\frac{(3ab^{-4})^{-2}}{2b^{-3}} = \frac{3^{-2}a^{-2}(b^{-4})^{-2}}{2b^{-3}} \qquad \text{Apply the third law: } (b^{-4})^{-2} = b^{8}.$$

$$= \frac{3^{-2}a^{-2}b^{8}}{2b^{-3}} \qquad \text{Apply the second law: } \frac{b^{8}}{b^{-3}} = b^{8-(-3)} = b^{11}.$$

$$= \frac{3^{-2}a^{-2}b^{11}}{2} \qquad \text{Rewrite without negative exponents.}$$

$$= \frac{b^{11}}{2 \cdot 3^{2}a^{2}} = \frac{b^{11}}{18a^{2}}$$

Sums of Powers and the Distributive Law

The laws of exponents apply only to products and quotients of powers, not to sums or differences of powers. In many cases it is helpful to write all powers with positive exponents first and then treat the expressions as algebraic fractions.

Example 6

Write each expression with positive exponents only.
a. $x^{-1} - y^{-1}$ **b.** $(x^{-2} + y^{-2})^{-1}$

Solutions **a.** $x^{-1} - y^{-1} = \dfrac{1}{x} - \dfrac{1}{y},$ or $\dfrac{y - x}{xy}$

b. $(x^{-2} + y^{-2})^{-1} = \left(\dfrac{1}{x^{2}} + \dfrac{1}{y^{2}}\right)^{-1} = \left(\dfrac{y^{2} + x^{2}}{x^{2}y^{2}}\right)^{-1} = \left(\dfrac{x^{2}y^{2}}{y^{2} + x^{2}}\right)$

If you have forgotten how to add and subtract algebraic fractions, we will review those skills in Chapter 8. Meanwhile, you should concentrate on avoiding some tempting but *incorrect* algebraic operations.

 COMMON ERRORS

In Example 6(a) note that

$$\frac{1}{x} - \frac{1}{y} \ne \frac{1}{x - y}$$

For example, you can check that for $x = 2$ and $y = 3$,

$$\frac{1}{2} - \frac{1}{3} \ne \frac{1}{2 - 3} = -1$$

In Example 6(b) note that

$$(x^{-2} + y^{-2})^{-1} \neq x^2 + y^2$$

In general, the fourth law of exponents does *not* apply to sums and differences; that is,

$$(a + b)^n \neq a^n + b^n$$

We can use the distributive law to simplify products or to factor out negative powers.

Example 7

Multiply $x^{-2}(2x^3 + x^2 - 4x)$.

Solution   Apply the distributive law: multiply each term inside the parentheses by x^{-2}.

$$x^{-2}(2x^3 + x^2 - 4x) = x^{-2}(2x^3) + x^{-2}(x^2) + x^{-2}(-4x)$$

$$= 2x + 1 - 4x^{-1}$$

A useful technique for simplifying expressions with negative exponents is to factor out the power with the smallest exponent and then write the factored form as a fraction.

Example 8

Factor $6x^{-2} - 3x^{-1}$ and write the factored form with positive exponents only.

Solution   The smallest exponent in the expression is -2, so we factor out $3x^{-2}$. Recall that "factoring out" means *dividing out,* so we divide each term of the given expression by $3x^{-2}$.

$$\frac{6x^{-2}}{3x^{-2}} = 2x^{-2-(-2)} = 2x^0 = 2$$

$$\frac{-3x^{-1}}{3x^{-2}} = -x^{-1-(-2)} = -x^1 = -x$$

Thus

$$6x^{-2} - 3x^{-1} = 3x^{-2}(2 - x) = \frac{3(2 - x)}{x^2}$$

Scientific Notation   Scientists and engineers regularly encounter very large numbers, such as

$$5{,}980{,}000{,}000{,}000{,}000{,}000{,}000{,}000$$

(the mass of the earth in kilograms) and very small numbers, such as

$$0.000\ 000\ 000\ 000\ 000\ 000\ 000\ 001\ 67$$

(the mass of a hydrogen atom in grams) in their work. These numbers can be written in a more compact and useful form by using powers of 10.

Recall that multiplying a number by a positive power of 10 has the effect of moving the decimal point k places to the right, where k is the exponent in the power of 10. For example,

$$3.529 \times 10^2 = 352.9 \qquad \text{and} \qquad 25 \times 10^4 = 250,000$$

Multiplying a number by a negative power of 10 moves the decimal point to the left. For example,

$$1728 \times 10^{-3} = 1.728 \qquad \text{and} \qquad 4.6 \times 10^{-5} = 0.000046$$

A number is said to be written in **scientific notation** if it is expressed as the product of a number between 1 and 10 (including 1) and a power of 10. Thus the mass of the earth and the mass of a hydrogen atom can be expressed in scientific notation as

$$5.98 \times 10^{24} \text{ kilograms} \qquad \text{and} \qquad 1.67 \times 10 \text{ gram}$$

respectively.

TO WRITE A NUMBER IN SCIENTIFIC NOTATION

1. Position the decimal point so that exactly one nonzero digit is to its left.

2. Count the number of places you moved the decimal point: this number determines the power of 10.
   a. If the original number is greater than 10, the exponent is positive.
   b. If the original number is less than 1, the exponent is negative.

Example 9

Write each number in scientific notation.

a. $478,000 = 4.78000 \times 10^5$ **b.** $0.00032 = 00003.2 \times 10^{-4}$

     Five places   Four places

     $= 4.78 \times 10^5$ $= 3.2 \times 10^{-4}$

Your calculator displays numbers in scientific notation if they are too large or too small to fit in the display screen. For example, try squaring the number   123, 456, 789   on your calculator. Enter

$$123456789 \;\boxed{x^2}$$

The calculator will display the result as

$$1.524157875 \text{ E } 16$$

which is how the calculator displays the number $1.524157875 \times 10^{16}$.   Notice that the calculator displays the power 10^{16} as "E 16."

To enter a number in scientific form, use the key labeled $\boxed{\text{EE}}$. For example, to enter 3.26×10^{-18},   use the keying sequence

$$3.26 \quad \boxed{\text{EE}} \quad \boxed{(-)} \quad 18$$

Example 10

The average American eats 110 kilograms (kg) of meat per year. It takes about 16 kg of grain to produce 1 kg of meat, and advanced farming techniques can produce about 6000 kg of grain on each hectare of arable land. (The "hectare" is a unit of area equivalent to just under $2\frac{1}{2}$ acres.) The total land area of the earth is about 13 billion hectares, but only about 11% of that land is arable. Is it possible for each of the 5.5 billion people on earth to eat as much meat as Americans do?

Solution   First we will compute the amount of meat necessary to feed every person on earth 110 kg per year. There are 5.5×10^9 people on earth.

$$(5.5 \times 10^9 \text{ people}) \times (110 \text{ kg/person}) = 6.05 \times 10^{11} \text{ kg of meat}$$

Next we will compute the amount of grain needed to produce that much meat.

$$(16 \text{ kg of grain/kg of meat}) \times (6.05 \times 10^{11} \text{ kg of meat})$$
$$= 9.68 \times 10^{12} \text{ kg of grain}$$

Next we will see how many hectares of land are needed to produce that much grain.

$$(9.68 \times 10^{12} \text{ kg of grain}) \div (6000 \text{ kg/hectare}) = 1.61\overline{3} \times 10^9 \text{ hectares}$$

Finally, we will compute the amount of arable land available for grain production.

$$0.11 \times (13 \times 10^9 \text{ hectares}) = 1.43 \times 10^9 \text{ hectares}$$

Thus if we used every hectare of arable land to produce grain for livestock, we would not have enough to provide every person on earth with 110 kilograms of meat per year. $\boxed{}$

Exercise 6.2

Write the expressions in Exercises 1–20 without negative exponents and simplify.

1. 2^{-1}

2. 3^{-2}

3. $(-5)^{-2}$

4. $(-2)^{-3}$

5. $\left(\dfrac{1}{3}\right)^{-3}$

6. $\left(\dfrac{3}{5}\right)^{-2}$

7. $\dfrac{1}{(-2)^{-4}}$

8. $\dfrac{1}{(-3)^{-3}}$

9. $\dfrac{5}{4^{-3}}$

10. $\dfrac{3}{2^{-6}}$

11. $(2q)^{-5}$

12. $(4k)^{-3}$

13. $-4x^{-2}$

14. $-7x^{-4}$

15. $y^{-2} + y^{-3}$

16. $z^{-1} - z^{-2}$ **17.** $(m - n)^{-2}$ **18.** $(p + q)^{-3}$

19. $\dfrac{-5y^{-2}}{x^{-5}}$ **20.** $\dfrac{-6y^{-3}}{x^{-3}}$

Write the expressions in Exercises 21–28 using negative exponents.

21. $\dfrac{1}{5^{13}}$ **22.** $\dfrac{1}{2^{36}}$ **23.** $\dfrac{7}{3^{24}}$ **24.** $\dfrac{3}{5^{12}}$

25. $\dfrac{3}{r^4}$ **26.** $\dfrac{9}{s^3}$ **27.** $\dfrac{2}{5w^3}$ **28.** $\dfrac{3}{8v^6}$

Compute each power in Exercises 29–32.

29. a. 2^3 **b.** $(-2)^3$ **c.** 2^{-3} **d.** $(-2)^{-3}$

30. a. 4^2 **b.** $(-4)^2$
   c. 4^{-2} **d.** $(-4)^{-2}$

31. a. $\left(\dfrac{1}{2}\right)^3$ **b.** $\left(-\dfrac{1}{2}\right)^3$

   c. $\left(\dfrac{1}{2}\right)^{-3}$ **d.** $\left(-\dfrac{1}{2}\right)^{-3}$

32. a. $\left(\dfrac{1}{4}\right)^2$ **b.** $\left(-\dfrac{1}{4}\right)^2$

   c. $\left(\dfrac{1}{4}\right)^{-2}$ **d.** $\left(-\dfrac{1}{4}\right)^{-2}$

Use your calculator to fill in the tables in Exercises 33–36. Round your answers to two decimal places.

33.

x	1.2	5.4	12.1	63.4	128
f(x)					

a. $f(x) = x^{-2}$
b. What happens to the values of $f(x)$ as the values of x increase? Explain why.

34.

x	0.9	2.6	10.1	98.7	132.5
x^{-3}					

a. $g(x) = x^{-3}$
b. What happens to the values of $g(x)$ as the values of x increase? Explain why.

35.

x	2.5	1.3	0.8	0.02	0.004
x^{-2}					

a. $f(x) = x^{-2}$
b. What happens to the values of $f(x)$ as the values of x decrease toward 0? Explain why.

36.

x	1.5	0.6	0.1	0.03	0.002
x^{-3}					

a. $g(x) = x^{-3}$
b. What happens to the values of $g(x)$ as the values of x decrease toward 0? Explain why.

37. Use your calculator to graph each of the following functions.
   a. $f(x) = x^2$ **b.** $f(x) = x^{-2}$
   c. $f(x) = \dfrac{1}{x^2}$ **d.** $f(x) = \left(\dfrac{1}{x}\right)^2$
   e. Which functions yield the same graphs? Explain your results.

38. Use your calculator to graph each of the following functions.
   a. $f(x) = x^3$ **b.** $f(x) = x^{-3}$
   c. $f(x) = \dfrac{1}{x^3}$ **d.** $f(x) = \left(\dfrac{1}{x}\right)^3$
   e. Which functions yield the same graphs? Explain your results.

39. Use an appropriate graph from Exercise 37 to find the following powers. Use the window setting

$$\text{Xmin} = -4.7, \quad \text{Xmax} = 4.7$$
$$\text{Ymin} = -5, \quad \text{Ymax} = 25$$

   a. $(0.7)^{-2}$ **b.** $(3.8)^{-2}$
   c. $(-0.4)^{-2}$ **d.** $(-2.4)^{-2}$

40. Use an appropriate graph from Exercise 38 to find the following powers. Use the window setting

$$\text{Xmin} = -4.7, \quad \text{Xmax} = 4.7$$
$$\text{Ymin} = -125, \quad \text{Ymax} = 125$$

a. $(0.4)^{-3}$ **b.** $(4.3)^{-3}$

c. $(-0.2)^{-3}$ **d.** $(-1.9)^{-3}$

Use the laws of exponents to simplify and write without negative exponents the expressions in Exercises 41–60.

41. $a^{-3} \cdot a^8$ **42.** $b^2 \cdot b^{-6}$

43. $5^{-4} \cdot 5^{-3}$ **44.** $4^{-2} \cdot 4^{-6}$

45. $(4x^{-5})(5x^2)$ **46.** $(3y^{-8})(2y^4)$

47. $\dfrac{p^{-7}}{p^{-4}}$ **48.** $\dfrac{w^{-9}}{w^2}$ **49.** $\dfrac{3u^{-3}}{9u^9}$

50. $\dfrac{4c^{-4}}{8c^{-8}}$ **51.** $\dfrac{5^6 t^0}{5^{-2} t^{-1}}$ **52.** $\dfrac{3^{10} s^{-1}}{3^{-5} s^0}$

53. $(7^{-2})^5$ **54.** $(9^{-4})^3$

55. $(3x^{-2}y^3)^{-2}$ **56.** $(2x^3 y^{-4})^{-3}$

57. $\left(\dfrac{6a^{-3}}{b^2}\right)^{-2}$ **58.** $\left(\dfrac{a^4}{4b^{-5}}\right)^{-3}$

59. $\dfrac{5h^{-3}(h^4)^{-2}}{6h^{-5}}$ **60.** $\dfrac{4v^{-5}(v^{-2})^{-4}}{3v^{-8}}$

Use the distributive law to write each product in Exercises 61–66 as a sum of powers.

61. $x^{-1}(x^2 - 3x + 2)$ **62.** $3x^{-2}(2x^4 + x^2 - 4)$

63. $-3t^{-2}(t^2 - 2 - 4t^{-2})$

64. $-t^{-3}(3t^2 - 1 - t^{-2})$

65. $2u^{-3}(-2u^3 - u^2 + 3u)$

66. $2u^{-1}(-1 - u - 2u^2)$

In Exercises 67–72, factor as indicated, and write the factored form with positive exponents only.

67. $4x^2 + 16x^{-2} = 4x^{-2}(?)$

68. $20y - 15y^{-1} = 5y^{-1}(?)$

69. $6p^3 - 3 - 12p^{-1} = p^{-1}(?)$

70. $2 - 4q^{-2} - 8q^{-4} = 2q^{-4}(?)$

71. $\dfrac{3}{2}a^{-3} - 3a + a^3 = \dfrac{1}{2}a^{-3}(?)$

72. $\dfrac{2}{3}b^{-2} - b^{-1} + 3 = \dfrac{1}{3}b^{-2}(?)$

Write each expression in Exercises 73–84 with positive exponents only.

73. $x^{-2} + y^{-2}$ **74.** $x^{-2} - y^{-2}$

75. $2w^{-1} - (2w)^{-2}$ **76.** $3w^{-3} + (3w)^{-1}$

77. $a^{-1}b - ab^{-1}$ **78.** $a - b^{-1}a - b^{-1}$

79. $(x^{-1} + y^{-1})^{-1}$ **80.** $(1 - xy^{-1})^{-1}$

81. $\dfrac{x + x^{-2}}{x}$ **82.** $\dfrac{x^{-1} - y}{x^{-1}}$

83. $\dfrac{a^{-1} + b^{-1}}{(ab)^{-1}}$ **84.** $\dfrac{x}{x^{-2} - y^{-2}}$

Write each number in Exercises 85–92 in scientific notation.

85. 285   **86.** 68,742

87. 8,372,000   **88.** 481,000,000,000

89. 0.024   **90.** 0.421

91. 0.000523   **92.** 0.000004

Write each number in Exercises 93–100 in standard notation.

93. 2.4×10^2 **94.** 4.8×10^3

95. 6.87×10^{15} **96.** 8.31×10^{12}

97. 5.0×10^{-3} **98.** 8.0×10^{-1}

99. 2.02×10^{-4} **100.** 4.31×10^{-5}

Compute Exercises 101–104 with the aid of a calculator. Write your answers in standard notation.

101. $\dfrac{(2.4 \times 10^{-8})(6.5 \times 10^{32})}{5.2 \times 10^{18}}$

102. $\dfrac{(8.4 \times 10^{-22})(1.6 \times 10^{15})}{3.2 \times 10^{-11}}$

103. $\dfrac{(7.5 \times 10^{-13})(3.6 \times 10^{-9})}{(1.5 \times 10^{-15})(1.6 \times 10^{-11})}$

104. $\dfrac{(9.4 \times 10^{24})(7.2 \times 10^{-18})}{(4.5 \times 10^{26})(6.4 \times 10^{-16})}$

105. In 1993 the public debt of the United States was $4,351,200,000,000.

 a. Express this number in scientific notation.

 b. If the population of the United States in 1993 was 257,908,000, what was the per capita debt for that year?

106. The speed of light is approximately 300,000,000 meters per second.

 a. Express this number in scientific notation.

 b. Express the speed of light in inches per second. (One inch equals 2.54 centimeters, and 1 meter equals 100 centimeters.)

107. One light-year is the number of miles traveled by light in 1 year (365 days). The speed of light is approximately 186,000 miles per second.

 a. Compute the number of miles in 1 light-year, and express your answer in scientific notation.

 b. The star nearest to the sun is Proxima Centauri, at a distance of 4.3 light-years. How long would it take Pioneer 10 (the first space vehicle to achieve escape velocity from the solar system), traveling at 32,114 miles per hour, to reach Proxima Centauri?

108. Lake Superior has an area of 31,700 square miles and an average depth of 483 feet.

 a. Find the approximate volume of Lake Superior in cubic feet.

 b. If 1 cubic foot of water is equivalent to 7.48 gallons, how many gallons of water are in Lake Superior?

109. On November 6, 1923, the number of Reichsbank marks in circulation in Germany was over 400,338,326,350,000,000,000.

 a. Express this number in scientific notation.

 b. Assume that each note is approximately 15 square inches in area. How large an area, in square feet, would that many 1-mark notes cover?

 c. The total surface area of the earth is 196,937,400 square miles. How many times could you paper the earth with that many 1-mark notes?

110. The mass of the earth is 6,585,600,000,000,000,000,000 tons; its volume is 259,875,300,000 cubic miles.

 a. Express these numbers in scientific notation.

 b. Find the average density of the earth in tons per cubic mile. (Density is defined as weight per unit volume.)

 c. Find the density of the earth in pounds per cubic foot.

6.3 n^{th} ROOTS AND IRRATIONAL NUMBERS

n^{th} Roots Recall that s is a square root of b if $s^2 = b$, and s is a cube root of b if $s^3 = b$. In a similar way we can look for the fourth, fifth, or sixth root of a number. For instance, the fourth root of b is a number, s, whose fourth power is b. In general, we make the following definition.

> s is called the nth **root of** b if $s^n = b$.

 We use the symbol $\sqrt[n]{b}$ to denote the nth root of b. An expression of the form $\sqrt[n]{b}$ is called a **radical**, b is called the **radicand**, and n is called the **index** of the radical.

Example 1

   a. $\sqrt[4]{81} = 3$ because $3^4 = 81$.   **b.** $\sqrt[5]{32} = 2$ because $2^5 = 32$.

   c. $\sqrt[6]{64} = 2$ because $2^6 = 64$.   **d.** $\sqrt[4]{1} = 1$ because $1^4 = 1$.

   e. $\sqrt[5]{100,000} = 10$ because $10^5 = 100,000$.   ▭

     We cannot find an even root of a negative number. For example, $\sqrt{-9}$ is not a real number, since no real number has a square of -9. Similarly, $\sqrt[4]{-16}$ is not a real number, because there is no real number r for which $r^4 = -16$. (Both of these radicals exist in the complex numbers, but we will not study them here.)

     Every positive number, however, has *two* even roots. You know that both 3 and -3 are square roots of 9, but we use the symbol $\sqrt{9}$ to refer only to the positive root, or **principal root**, of 9. If we want to refer to the negative square root of 9, we must write $-\sqrt{9} = -3$. Similarly, both 2 and -2 are fourth roots of 16, since $2^4 = 16$ and $(-2)^4 = 16$. However, the symbol $\sqrt[4]{16}$ refers to the principal, or positive, fourth root only. Thus

$$\sqrt[4]{16} = 2 \qquad \text{and} \qquad -\sqrt[4]{16} = -2$$

     Things are simpler for odd roots of numbers. We *can* take an odd root of a negative number. For instance, $\sqrt[3]{-8} = -2$, since $(-2)^3 = -8$. In fact, every real number, whether positive, negative, or zero, has exactly one real-valued odd root. For example,

$$\sqrt[5]{32} = 2 \qquad \text{and} \qquad \sqrt[5]{-32} = -2$$

Example 2

   a. $\sqrt[4]{-625}$ is not a real number.   **b.** $-\sqrt[4]{625} = -5$

   c. $\sqrt[5]{-1} = -1$ **d.** $\sqrt[4]{-1}$ is not a real number. ▭

Exponential Notation

A convenient notation for radicals uses fractional exponents. Consider the expression $9^{1/2}$. What meaning can we attach to an exponent that is a fraction? Recall the third law of exponents, which says that when we raise a power to a power we multiply the exponents together. Thus

$$(x^a)^b = x^{ab}$$

If we square the number $9^{1/2}$, we can apply the third law to get

$$(9^{1/2})^2 = 9^{(1/2)(2)} = 9^1 = 9$$

We see that $9^{1/2}$ is a number whose square is 9, which means that $9^{1/2}$ is a square root of 9, or

$$9^{1/2} = \sqrt{9} = 3$$

     In general, any non-negative number raised to the $\frac{1}{2}$ power is equal to the square root of the number, or

$$a^{1/2} = \sqrt{a}$$

Example 3

a. $25^{1/2} = 5$ **b.** $-25^{1/2} = -5$
c. $(-25)^{1/2}$ is not a real number.   **d.** $0^{1/2} = 0$

COMMON ERROR

Note that $25^{1/2} \neq \frac{1}{2}(25)$. An *exponent* of $\frac{1}{2}$ denotes the *square root* of its base.

The same reasoning works for roots of any index. For example, $8^{1/3}$ is the cube root of 8, because

$$(8^{1/3})^3 = 8^{(1/3)(3)} = 8^1 = 8$$

In general, we make the following definition for fractional exponents.

> For any natural number $n \geq 2$,
> $$a^{1/n} = \sqrt[n]{a}$$

Example 4

a. $81^{1/4} = 3$ **b.** $125^{1/3} = 5$
c. $(-32)^{1/5} = -2$ **d.** $-64^{1/6} = -2$

COMMON ERROR

Remember that we cannot find an even root of a negative number. For example,

$$(-64)^{1/6} \text{ is not a real number.}$$

Compare this statement with Example 4(d): there the exponent 1/6 applies only to 64, and the negative sign is applied after the root is computed.

Fractional exponents can greatly simplify many calculations involving radicals. You should learn to convert easily between exponential and radical notation.

Example 5

Convert each radical to exponential notation.

a. $\sqrt[3]{12} = 12^{1/3}$

b. $\sqrt[4]{2y} = (2y)^{1/4}$

c. $\dfrac{1}{\sqrt[5]{ab}} = (ab)^{-1/5}$ **d.** $3\sqrt[6]{w} = 3w^{1/6}$

Example 6

Convert each power to radical notation.
a. $5^{1/2} = \sqrt{5}$ **b.** $x^{1/5} = \sqrt[5]{x}$

c. $2x^{1/3} = 2\sqrt[3]{x}$ **d.** $8a^{-1/4} = \dfrac{8}{\sqrt[4]{a}}$

Irrational Numbers

A number that can be expressed as the quotient of two integers $\dfrac{a}{b}$, where $b \neq 0$, is called a **rational number**. All integers are rational numbers, and so are common fractions. Some examples of rational numbers are 5, -2, 0, $\dfrac{2}{9}$, $\sqrt{16}$, and $\dfrac{-4}{17}$.

Every rational number has a decimal form that either terminates or repeats a pattern of digits.

Example 7

Find decimal representations for $\dfrac{3}{4}$ and $\dfrac{9}{37}$.

Solutions   For each fraction divide the denominator into the numerator to find

$$\frac{3}{4} = 3 \div 4 = 0.75, \text{ a terminating decimal,}$$

and

$$\frac{9}{37} = 9 \div 37 = 0.243243243 \ldots$$

where the pattern of digits 243 is repeated endlessly. We use the "repeater bar" notation to write a repeating decimal fraction:

$$\frac{9}{37} = 0.\overline{243}$$

Some real numbers *cannot* be written as a quotient of integers. For example, the number $\sqrt{2}$ is not equal to any common fraction. Such numbers are called **irrational numbers**. The decimal form of an irrational number never terminates and its digits do not follow a repeating pattern, which means that it is impossible to write an exact decimal equivalent for an irrational number. However, we can obtain decimal *approximations* correct to any desired degree of accuracy by rounding off. For example, your calculator gives the decimal representation of π as 3.141592654. This number is not the *exact* value of π, but for most calculations it is quite adequate.

Some n^{th} roots are rational numbers and some are irrational numbers. For example,

$$\sqrt{49}, \qquad \sqrt[3]{\frac{-27}{8}}, \qquad \text{and} \qquad 81^{1/4}$$

are rational numbers because they are equal to 7, $\dfrac{-3}{2}$, and 3, respectively.

On the other hand,

$$\sqrt{2}, \qquad \sqrt[3]{56}, \qquad \text{and} \qquad 7^{1/5}$$

are irrational numbers.

We can use a calculator to obtain decimal approximations to irrational radicals.

Example 8

Obtain decimal approximations to three decimal places for $\sqrt{7}$ and $\sqrt[3]{56}$.

Solution   To approximate $\sqrt{7}$, we can use the square root key. Press

$$\boxed{\sqrt{}}\ 7\ \boxed{\text{ENTER}}$$

and the calculator returns   2.645751311.   Round the result to three decimal places to obtain

$$\sqrt{7} \approx 2.646$$

To approximate $\sqrt[3]{56}$, we can use the cube root option from the $\boxed{\text{MATH}}$ menu. Press

$$\boxed{\text{MATH}}\ \boxed{4}\ 56\ \boxed{\text{ENTER}}$$

and the calculator returns   3.825862366.   Round the result to three decimal places to obtain

$$\sqrt[3]{56} \approx 3.826$$ $\boxed{\qquad}$

We can also use fractional exponents to evaluate n^{th} roots.

Example 9

Obtain a decimal approximation to three decimal places for $\sqrt[4]{285}$.

Solution   Since $\sqrt[4]{285} = 285^{1/4}$,   we enter

$$285\ \boxed{\wedge}\ \boxed{(}\ 1\ \boxed{\div}\ 4\ \boxed{)}\ \boxed{\text{ENTER}}$$

to obtain   4.108764171.   Round the result to three decimal places to obtain

$$\sqrt[4]{285} \approx 4.109$$

Because $\frac{1}{4} = 0.25$,   we can also enter $285\ \boxed{\wedge}\ .25\ \boxed{\text{ENTER}}$ to get the same answer. $\boxed{\qquad}$

  COMMON ERROR

In Example 9 the following keying sequence would be incorrect:

$$285\ \boxed{\wedge}\ 1\ \boxed{\div}\ 4\ \boxed{\text{ENTER}}$$

You can check that this sequence calculates $\dfrac{25^1}{4}$ instead of $25^{1/4}$. Recall that according to the order of operations, powers are computed before multiplications or divisions. Thus it would be incorrect to omit the parentheses around the exponent $\frac{1}{4}$.

Solving Equations   In Chapter 1 we learned that squaring and taking square roots are *inverse operations,* that is, each operation undoes the effects of the other. We use this fact when solving equations involving x^2 or $\sqrt{x}$. For example, to solve the equation

$$x^2 = 15$$

we take the square root of both sides of the equation to find

$$x = \pm\sqrt{15} \approx 3.873$$

To solve the equation

$$\sqrt{x} = 6.2$$

we square both sides to find

$$x = (6.2)^2 = 38.44$$

Similarly, the operations of cubing and taking cube roots are inverse operations, which enables us to solve equations involving x^3 or $\sqrt[3]{x}$.

Example 10

Solve $4\sqrt[3]{x - 9} = 12$.

Solution   First isolate the radical by dividing both sides of the equation by 4.

$$\sqrt[3]{x - 9} = 3$$

Next cube both sides of the equation.

$$(\sqrt[3]{x - 9})^3 = 3^3$$
$$x - 9 = 27$$
$$x = 36$$

The solution is 36.

In general, we can solve an equation in which one side is an nth root of x by raising both sides of the equation to the nth power. (If negative values of x are allowed, we must check for extraneous roots when n is even.) To solve an equation involving n^{th} powers of x, we first try to isolate the power, then take the nth root of both sides.

Example 11

A spherical fish tank in the lobby of the Atlantis Hotel holds 904.78 cubic feet of water. What is the radius of the fish tank?

Solution   The volume of a sphere is given by the formula $V = \dfrac{4}{3}\pi r^3$,   where r is the radius. Substitute 904.78 cubic feet for the volume of the sphere, and solve for r.

$$\frac{4}{3}\pi r^3 = 904.78$$

First isolate the variable. Divide both sides by $\frac{4}{3}\pi$.

$$\frac{\frac{4}{3}\pi r^3}{\frac{4}{3}\pi} = \frac{904.78}{\frac{4}{3}\pi}$$

$$r^3 = \frac{904.78}{\pi} \cdot \frac{3}{4} \approx 216$$

Now take the cube root of both sides by raising each side to the power $\frac{1}{3}$.

$$(r^3)^{1/3} \approx 216^{1/3}$$

$$r \approx 6$$

The radius of the fish tank is approximately 6 feet.

Exercise 6.3

Find the indicated root in Exercises 1–12. If the radical is not a real number, say so.

1. $\sqrt{121}$ **2.** $-\sqrt{169}$ **3.** $\sqrt[3]{-27}$

4. $\sqrt[3]{64}$ **5.** $\sqrt[4]{-625}$ **6.** $-\sqrt[4]{81}$

7. $-\sqrt[5]{32}$ **8.** $\sqrt[5]{-100,000}$

9. $\sqrt[4]{16}$ **10.** $\sqrt[4]{-1296}$

11. $\sqrt[3]{729}$ **12.** $-\sqrt[3]{343}$

Find the indicated power in Exercises 13–24. If the power is not a real number, say so.

13. $9^{1/2}$ **14.** $25^{1/2}$ **15.** $(-81)^{1/4}$

16. $-81^{1/4}$ **17.** $-64^{1/6}$ **18.** $(-64)^{1/6}$

19. $(-32)^{1/5}$ **20.** $(-27)^{1/3}$

21. $(-8)^{-1/3}$ **22.** $(-243)^{-1/5}$

23. $64^{-0.5}$ **24.** $49^{-0.5}$

Write each expression in Exercises 25–36 in radical form.

25. $3^{1/2}$ **26.** $7^{1/2}$ **27.** $4x^{1/3}$

28. $3x^{1/4}$ **29.** $(4x)^{0.2}$ **30.** $(3x)^{0.25}$

31. $8^{-1/4}$ **32.** $6^{-1/3}$ **33.** $3(xy)^{-1/3}$

34. $y(5x)^{-1/2}$ **35.** $(x-2)^{1/4}$ **36.** $(y+2)^{1/3}$

Write each expression in Exercises 37–48 in exponential form.

37. $\sqrt{7}$ **38.** $\sqrt{5}$ **39.** $\sqrt[3]{2x}$

40. $\sqrt[3]{4y}$ **41.** $2\sqrt[5]{z}$ **42.** $-5\sqrt[3]{x}$

43. $\dfrac{-3}{\sqrt[4]{6}}$ **44.** $\dfrac{2}{\sqrt[5]{3}}$

45. $\sqrt[4]{x-3y}$ **46.** $\sqrt[3]{y+2x}$

47. $\dfrac{-1}{\sqrt[3]{1+3b}}$ **48.** $\dfrac{-1}{\sqrt[4]{3a-2b}}$

Write each fraction in Exercises 49–56 in decimal form. Does the decimal terminate or repeat a pattern?

49. $\dfrac{3}{8}$ **50.** $\dfrac{7}{16}$ **51.** $\dfrac{5}{6}$ **52.** $\dfrac{5}{12}$

53. $\dfrac{2}{7}$ **54.** $\dfrac{11}{13}$ **55.** $\dfrac{43}{11}$ **56.** $\dfrac{25}{6}$

In Exercises 57–68 use a calculator to approximate each irrational number to the nearest thousandth.

57. $2^{1/2}$ **58.** $3^{1/2}$ **59.** $\sqrt[3]{75}$

60. $\sqrt[4]{60}$ **61.** $\sqrt[5]{-43}$ **62.** $\sqrt[5]{-87}$

63. $\sqrt[4]{1.6}$ **64.** $\sqrt[3]{1.4}$ **65.** $365^{-1/3}$

66. $1058^{-1/4}$ **67.** $0.006^{-0.2}$ **68.** $1.05^{-0.1}$

69. When the Concorde lands at Heathrow airport near London, the width, w, of the sonic boom felt on the ground is given in kilometers by the following formula:

$$w = 4\left(\frac{Th}{m}\right)^{1/2}$$

T stands for the temperature on the ground in degrees Kelvin, h is the altitude of the Concorde when it breaks the sound barrier, and m is the drop in temperature for each gain in altitude of 1 kilometer. Find the width of the sonic boom if the ground temperature is 293° Kelvin, the altitude of the Concorde is 15 kilometers, and the temperature drop is 4° Kelvin per kilometer of altitude.

70. The manager of an office supply store must decide how many of each item in stock she should order. The Wilson lot-size formula gives the most cost-efficient quantity, Q, in terms of the cost, C, of placing an order, the number of items, N, sold per week, and the weekly inventory cost, I, per item (cost of storage, maintenance, and so on).

$$Q = \left(\frac{2CN}{I}\right)^{1/2}$$

How many reams of computer paper should she order if she sells on average 80 reams per week, the weekly inventory cost for a ream is $0.20, and the cost of ordering, including delivery charges, is $25?

71. Membership in the County Museum has been increasing slowly since it was built in 1950. The number of members is given by the function

$$M(t) = 72 + 100t^{1/3}$$

where t is the number of years since 1950.

a. How many members were there in 1960? In 1970?

b. In what year did the museum have 400 members? If the membership continues to grow according to the given function, when will the museum have 500 members?

c. Graph the function $M(t)$. How would you describe the growth of the membership over time?

72. Because of improvements in technology, the annual electricity cost of running most major appliances has decreased steadily since 1970. The average annual cost of running a refrigerator is given in dollars by the function

$$C(t) = 148 - 28t^{1/3}$$

where t is the number of years since 1970.

a. How much did it cost to run a refrigerator in 1980? In 1990?

b. When was the cost of running a refrigerator half of the cost in 1970? If the cost continues to decline according to the given function, when will it cost $50 per year to run a refrigerator?

c. Graph the function $C(t)$. Do you think that the cost will continue to decline indefinitely according to the given function? Why or why not?

73. a. Graph the following functions in the window

$$\text{Xmin} = 0, \quad \text{Xmax} = 100$$
$$\text{Ymin} = 0, \quad \text{Ymax} = 10$$

$$y_1 = \sqrt{x}, \quad y_2 = \sqrt[3]{x}, \quad y_3 = \sqrt[4]{x}, \quad y_4 = \sqrt[5]{x},$$

What do you observe?

b. Use your graphs to evaluate $100^{1/2}$, $100^{1/3}$, $100^{1/4}$, and $100^{1/5}$.

c. Use your calculator to evaluate $100^{1/n}$ for $n = 10$, $n = 100$, and $n = 1000$. What happens when n gets large?

74. a. Graph the following functions in the window

$$\text{Xmin} = 0, \quad \text{Xmax} = 1$$
$$\text{Ymin} = 0, \quad \text{Ymax} = 1$$

$$y_1 = \sqrt{x}, \quad y_2 = \sqrt[3]{x}, \quad y_3 = \sqrt[4]{x}, \quad y_4 = \sqrt[5]{x}$$

What do you observe?

b. Use your graphs to evaluate $0.5^{1/2}$, $0.5^{1/3}$, $0.5^{1/4}$, and $0.5^{1/5}$.

c. Use your calculator to evaluate $0.5^{1/n}$ for $n = 10$, $n = 100$, and $n = 1000$. What happens when n gets large?

In Exercises 75–78 graph each set of functions in the given window. What do you observe?

75. Xmin = 0, Xmax = 4
Ymin = 0, Ymax = 4

$$y_1 = \sqrt{x}, \quad y_2 = x^2, \quad y_3 = x$$

76. Xmin = −4, Xmax = 4
Ymin = −4, Ymax = 4

$$y_1 = \sqrt[3]{x}, \quad y_2 = x^3, \quad y_3 = x$$

77. Xmin = −2, Xmax = 2
Ymin = −2, Ymax = 2

$$y_1 = \sqrt[5]{x}, \quad y_2, = x^5, \quad y_3 = x$$

78. Xmin = 0, Xmax = 2
Ymin = 0, Ymax = 2

$$y_1 = \sqrt[4]{x}, \quad y_2 = x^4, \quad y_3 = x$$

79. Use the laws of exponents to show that

$$\sqrt[4]{x} = \sqrt{\sqrt{x}}$$

80. Use the laws of exponents to show that

$$\sqrt[6]{x} = \sqrt{\sqrt[3]{x}}$$

Simplify each expression in Exercises 81–88.

81. $(\sqrt[3]{125})^3$ **82.** $(\sqrt[4]{16})^4$ **83.** $(\sqrt[4]{2})^4$

84. $(\sqrt[3]{6})^3$ **85.** $(3\sqrt{7})^2$ **86.** $(2\sqrt[3]{12})^3$

87. $(-x^2\sqrt[3]{2x})^3$ **88.** $(-a^3\sqrt[4]{a^2})^4$

Solve the equations in Exercises 89–100.

89. $2\sqrt[3]{x} - 5 = -17$

90. $3\sqrt[3]{x} + 1 = -11$

91. $\sqrt[4]{x-1} = 2$ **92.** $\sqrt[4]{x+6} = 3$

93. $4(x+2)^{1/5} = 12$

94. $-9(x-3)^{1/5} = 18$

95. $(2x-3)^{-1/4} = \dfrac{1}{2}$ **96.** $(5x+2)^{-1/3} = \dfrac{1}{4}$

97. $\sqrt[3]{x^2 - 3} = 3$ **98.** $\sqrt[4]{x^3 - 7} = 2$

99. $\sqrt[3]{2x^2 - 15x} = 5$

100. $\sqrt[3]{2x^2 - 11x} = 6$

In Exercises 101 and 102 evaluate each function for the given values.

101. $M(h) = 2h - h^{1/3}$

   a. $M(1)$ **b.** $M(8)$

   c. $M(0.001)$ **d.** $M(24)$

102. $V(y) = 5y^{1/4} + y^2$

   a. $V(0)$ **b.** $V(16)$

   c. $V\left(\dfrac{1}{16}\right)$ **d.** $V(0.5)$

103. If the radius and height of a right circular cylinder are both equal to a, then its volume is given by the function

$$V = \pi a^3$$

Find the height of such a cylinder if its volume is 816.814 cubic meters.

104. The power generated by a windmill is given in watts by the function

$$P = 0.015v^3$$

where v is the velocity of the wind. Find the wind velocity needed to generate 500 watts of power.

105. The Stefan-Boltzmann law says that the temperature, T, of the sun can be computed, in degrees Kelvin, from the following formula:

$$sT^4 = \frac{L}{4\pi R^2}$$

where $L = 3.9 \times 10^{33}$ is the total luminosity of the sun, $R = 9.96 \times 10^{10}$ centimeters is the radius of the sun, and $s = 5.7 \times 10^{-5}$ is a constant governing radiation. Calculate the temperature of the sun.

106. Poiseuille's law for the flow of liquid through a tube can be used to describe blood flow through an artery. The rate of flow, F, in liters per minute is given by

$$F = \frac{kr^4}{L}$$

where r is the radius of the artery in centimeters, L is its length in centimeters, and $k = 7.8 \times 10^5$ is a constant determined by the blood pressure and viscosity. If a certain artery is 20 centimeters long, what should its radius be in order to allow a blood flow of 5 liters per minute?

Solve each formula in Exercises 107–110 for the specified variable.

107. $r = \sqrt[3]{\dfrac{3V}{4\pi}}$ for V

108. $d = \sqrt[3]{\dfrac{16Mr^2}{m}}$ for M

109. $R = \sqrt[4]{\dfrac{8Lvf}{\pi p}}$ for p

110. $T = \sqrt[4]{\dfrac{E}{SA}}$ for A

6.4 RATIONAL EXPONENTS

Powers of the Form $a^{m/n}$

When a sky diver jumps from an airplane, she experiences free fall until she opens her parachute. At first her downward velocity increases due to the force of gravity. The faster she falls, the greater the air resistance she feels, until the force of the air resistance balances her own weight and she reaches "terminal velocity." What speed is terminal velocity?

If the sky diver weighs 120 pounds, the air resistance can be described by the function

$$R = 0.057v^{3/2}$$

where the velocity is measured in feet per second and the air resistance is measured in pounds. We can use a calculator to find the air resistance at various speeds. For example, to calculate the air resistance at 25 feet per second, we enter

$$0.057 \boxed{\times} 25 \boxed{\wedge} 1.5 \boxed{\text{ENTER}}$$

to find that the air resistance is 7.125 pounds. To find out the terminal velocity, we can make a table of values for the air resistance, as shown in Table 6.1.

From Table 6.1 we see that the air resistance reaches 120 pounds at a velocity somewhere between 150 and 175 feet per second. To get a better estimate, we can graph R and see where it reaches 120. The graph in Figure 6.1 shows that R is approximately 120 when the velocity is 164 feet per second. Thus the sky diver will reach a terminal velocity of about 164 feet per second, or about 112 miles per hour.

How does the calculator compute the power $v^{3/2}$ in the application above? Notice that the exponent $\dfrac{3}{2} = 3\left(\dfrac{1}{2}\right)$, and by the third law of exponents

$$(v^{1/2})^3 = v^{(1/2)3} = v^{3/2}$$

v	R
50	20.153
75	37.023
100	57.000
125	79.660
150	104.716
175	131.957
200	161.220

Table 6.1

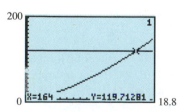

Figure 6.1

This means that we can compute $v^{3/2}$ by first taking the square root of v and then cubing the result. For example,

$$100^{3/2} = (100^{1/2})^3 \qquad \text{Take the square root of 100.}$$
$$= 10^3 = 1000 \qquad \text{Cube the result.}$$

In general, we can define a fractional power as follows. We will define fractional exponents only when the base is a positive number.

$$a^{m/n} = (a^{1/n})^m = (a^m)^{1/n} \qquad a > 0$$

To compute $a^{m/n}$, we can compute either the nth root or the mth power first, whichever is easiest. For example,

$$8^{2/3} = (8^{1/3})^2 = 2^2 = 4$$

or

$$8^{2/3} = (8^2)^{1/3} = 64^{1/3} = 4$$

Example 1

a. $81^{3/4} = (81^{1/4})^3$
$\qquad\;\; = 3^3 = 27$

b. $-27^{5/3} = -(27^{1/3})^5$
$\qquad\qquad\;\; = -3^5 = -243$

c. $27^{-2/3} = \dfrac{1}{(27^{1/3})^2}$
$\qquad\qquad = \dfrac{1}{3^2} = \dfrac{1}{9}$

d. $5^{3/2} = (5^{1/2})^3$
$\qquad\quad \approx (2.236)^3 \approx 11.180$

You can verify all the calculations in Example 1 on your calculator. For example, to evaluate $81^{3/4}$, you can key in

81 ^ [3 ÷ 4] ENTER

or simply

8 ^ 0.75 ENTER

Radical Notation Because $a^{1/n} = \sqrt[n]{a}$, we can write any power with a fractional exponent in radical form as follows.

$$a^{m/n} = \sqrt[n]{a^m} = (\sqrt[n]{a})^m$$

Example 2
a. $125^{4/3} = \sqrt[3]{125^4}$ or $(\sqrt[3]{125})^4$ b. $x^{2/5} = \sqrt[5]{x^2}$ or $(\sqrt[5]{x})^2$

c. $6w^{-3/4} = \dfrac{6}{\sqrt[4]{w^3}}$

Usually we will want to convert from radical notation to fractional exponents because exponential notation is easier to use.

Example 3
a. $\sqrt{x^5} = x^{5/2}$

b. $5\sqrt[4]{p^3} = 5p^{3/4}$

c. $\dfrac{3}{\sqrt[5]{t^2}} = 3t^{-2/5}$

d. $\sqrt[3]{2y^2} = (2y^2)^{1/3} = 2^{1/3}y^{2/3}$

Operations with Rational Exponents   Powers with rational exponents—positive, negative, or zero—obey the laws of exponents developed in Section 6.1. You may want to review those laws before studying the following examples.

Example 4
a. $\dfrac{7^{0.75}}{7^{0.5}} = 7^{0.75-0.5} = 7^{0.25}$ Apply the second law of exponents.

b. $v^{5/6}v^{-2/3} = v^{5/6+(-4/6)} = v^{1/6}$ Apply the first law of exponents.

c. $(x^8)^{0.5} = x^{8(0.5)} = x^4$ Apply the third law of exponents.

d. $\dfrac{(5^{1/2}y^2)^2}{(5^{2/3}y)^3} = \dfrac{5y^4}{5^2y^3}$ Apply the fourth law of exponents.

$= \dfrac{y^{4-3}}{5^{2-1}} = \dfrac{y}{5}$ Apply the second law of exponents.

We can use the distributive law to multiply factors that involve more than one term.

Example 5
a. $x^{1/3}(x + x^{2/3}) = x^{1/3+1} + x^{1/3+2/3}$

$= x^{4/3} + x$

b. $(a^{1/2} - 3a^{-1/2})^2 = (a^{1/2} - 3a^{-1/2})(a^{1/2} - 3a^{-1/2})$

$= a^{1/2+1/2} - 6a^{1/2-1/2} + 9a^{-1/2-1/2}$

$= a - 6 + 9a^{-1}$

c. $\dfrac{5n + n^{4/3}}{4n^{2/3}} = \dfrac{1}{4}n^{-2/3}(5n + n^{4/3})$

$= \dfrac{5}{4}n^{-2/3+1} + \dfrac{1}{4}n^{-2/3+4/3}$

$= \dfrac{5}{4}n^{1/3} + \dfrac{1}{4}n^{2/3}$

We can also factor expressions involving fractional exponents. It is often useful to factor out the power with the smallest exponent.

Example 6

Factor the smallest power from each binomial.
a. $3t + t^{1/2}$ **b.** $r^{-1/2} + 1$

Solutions a. The smallest power is $t^{1/2}$, so we factor out $t^{1/2}$. To do this, we divide $t^{1/2}$ into each term.

$$\frac{3t}{t^{1/2}} = 3t^{1/2} \quad \text{and} \quad \frac{t^{1/2}}{t^{1/2}} = 1$$

so

$$3t + t^{1/2} = t^{1/2}(3t^{1/2} + 1)$$

b. The smallest power is $r^{-1/2}$, so we divide $r^{-1/2}$ into each term.

$$\frac{r^{-1/2}}{r^{-1/2}} = 1 \quad \text{and} \quad \frac{1}{r^{-1/2}} = r^{1/2}$$

so

$$r^{-1/2} + 1 = r^{-1/2}(1 + r^{1/2})$$ ▢

Solving Equations

According to the third law of exponents, when we raise a power to a power, we multiply the exponents together. In particular, if the two exponents are reciprocals, then their product is 1. For example,

$$(x^{2/3})^{3/2} = x^{(2/3)(3/2)} = x^1 = x$$

This observation can help us to solve equations involving fractional exponents. For example, to solve the equation

$$x^{2/3} = 4$$

we can raise both sides of the equation to the reciprocal power, 3/2, which gives us

$$(x^{2/3})^{3/2} = 4^{3/2}$$

$$x = 8$$

Example 7

Solve $(2x + 1)^{3/4} = 27$.

Solution Raise both sides of the equation to the reciprocal power, 4/3.

$$[(2x + 1)^{3/4}]^{4/3} = 27^{4/3}$$

$$2x + 1 = 81$$

$$x = 40$$ ▢

We can also solve the problem about a sky diver's terminal velocity (see page 362) by solving an equation. Recall that we were looking for the velocity that resulted in air resistance of 120 pounds, where air resistance, *R,* was related to velocity, *v,* by the equation

$$R = 0.057v^{3/2}$$

We substitute 120 for *R* and solve for *v.*

$120 = 0.057v^{3/2}$	Divide both sides by 0.057.
$2105.26 \approx v^{3/2}$	Raise both sides to the 2/3 power.
$2105.26^{2/3} \approx (v^{3/2})^{2/3}$	Simplify.
$164.26 \approx v$	

Thus terminal velocity occurs at approximately 164 feet per second.

Exercise 6.4

For Exercise 6.4 we will assume that all variables represent positive numbers.

Evaluate each power in Exercises 1–12.

1. $81^{3/4}$ **2.** $125^{2/3}$ **3.** $-8^{2/3}$

4. $-64^{2/3}$ **5.** $16^{-3/2}$ **6.** $8^{-4/3}$

7. $-125^{-4/3}$ **8.** $-32^{-3/5}$ **9.** $625^{0.75}$

10. $243^{0.4}$ **11.** $32^{-1.6}$ **12.** $100^{-2.5}$

Write each power in Exercises 13–24 in radical form.

13. $x^{4/5}$ **14.** $y^{3/4}$ **15.** $3x^{2/5}$

16. $5y^{2/3}$ **17.** $b^{-5/6}$ **18.** $a^{-2/7}$

19. $(pq)^{-2/3}$ **20.** $(st)^{-3/5}$ **21.** $4z^{-2/3}$

22. $6w^{-3/2}$ **23.** $-2x^{1/4}y^{3/4}$ **24.** $-3x^{2/5}y^{3/5}$

Write each expression in Exercises 25–36 with fractional exponents.

25. $\sqrt[3]{x^2}$ **26.** $\sqrt{y^3}$ **27.** $\sqrt[3]{(ab)^2}$

28. $\sqrt[3]{ab^2}$ **29.** $2\sqrt[5]{ab^3}$ **30.** $6\sqrt[5]{(ab)^3}$

31. $\dfrac{8}{\sqrt[4]{x^3}}$ **32.** $\dfrac{5}{\sqrt[3]{y^2}}$ **33.** $\dfrac{-4m}{\sqrt[6]{p^7}}$

34. $\dfrac{-2n}{\sqrt[8]{q^{11}}}$ **35.** $\dfrac{R}{3\sqrt{TK^5}}$ **36.** $\dfrac{S}{4\sqrt{VH^3}}$

Evaluate each root in Exercises 37–48.

37. $\sqrt[5]{32^3}$ **38.** $\sqrt[4]{16^5}$

39. $-\sqrt[3]{27^4}$ **40.** $-\sqrt[3]{125^2}$

41. $\sqrt[4]{16y^{12}}$ **42.** $\sqrt[5]{243x^{10}}$

43. $-\sqrt{a^8b^{16}}$ **44.** $-\sqrt{a^{10}b^{36}}$

45. $\sqrt[3]{8x^9y^{27}}$ **46.** $\sqrt[3]{64x^6y^{18}}$

47. $-\sqrt[4]{81a^8b^{12}}$ **48.** $-\sqrt[5]{32x^{25}y^5}$

In Exercises 49–56 use a calculator to approximate each power or root to the nearest thousandth.

49. $12^{5/6}$ **50.** $20^{5/4}$ **51.** $\sqrt[3]{6^4}$

52. $\sqrt[5]{8^3}$ **53.** $37^{-2/3}$ **54.** $128^{-3/4}$

55. $4.7^{2.3}$ **56.** $16.1^{0.29}$

Simplify the expressions in Exercises 57–68 by applying the laws of exponents. Write your answers with positive exponents only.

57. $4a^{6/5}a^{4/5}$ **58.** $9b^{4/3}b^{1/3}$ **59.** $(-2m^{2/3})^4$

60. $(-5n^{3/4})^3$ **61.** $\dfrac{8w^{9/4}}{2w^{3/4}}$ **62.** $\dfrac{12z^{11/3}}{4z^{5/3}}$

63. $(-3u^{5/3})(5u^{-2/3})$ **64.** $(-2v^{7/8})(-3v^{-3/8})$

65. $\dfrac{k^{3/4}}{2k}$ **66.** $\dfrac{4h^{2/3}}{3h}$

67. $c^{-2/3}\left(\dfrac{2}{3}c^2\right)$

68. $\dfrac{r^3}{4}(r^{-5/2})$

Use the distributive law to find the product in Exercises 69–76.

69. $2x^{1/2}(x - x^{1/2})$

70. $x^{1/3}(2x^{2/3} - x^{1/3})$

71. $\dfrac{1}{2}y^{-1/3}(y^{2/3} + 3y^{-5/6})$

72. $3y^{-3/8}\left(\dfrac{1}{4}y^{-1/4} + y^{3/4}\right)$

73. $(2x^{1/4} + 1)(x^{1/4} - 1)$

74. $(2x^{1/3} - 1)(x^{1/3} + 1)$

75. $(a^{3/4} - 2)^2$

76. $(a^{2/3} + 3)^2$

Factor out the smallest power from each expression in Exercises 77–82. Write your answers with positive exponents only.

77. $x^{3/2} + x$

78. $y - y^{2/3}$

79. $y^{3/4} - y^{-1/4}$

80. $x^{-3/2} + x^{-1/2}$

81. $a^{1/3} + 3 - a^{-1/3}$

82. $3b - b^{3/4} + 4b^{-3/4}$

Evaluate each function in Exercises 83 and 84 for the given values.

83. $Q(x) = 4x^{5/2}$

  a. $Q(16)$ **b.** $Q\left(\dfrac{1}{4}\right)$

  c. $Q(3)$ **d.** $Q(100)$

84. $T(w) = -3w^{2/3}$

  a. $T(27)$ **b.** $T\left(\dfrac{1}{8}\right)$

  c. $T(20)$ **d.** $T(1000)$

Solve the equations in Exercises 85–90. Round your answers to the nearest thousandth if necessary.

85. $x^{2/3} - 1 = 15$

86. $x^{3/4} + 3 = 11$

87. $x^{-2/5} = 9$

88. $x^{-3/2} = 8$

89. $2(5.2 - x^{5/3}) = 1.4$

90. $3(8.6 - x^{5/2}) = 6.5$

91. If $f(x) = (3x - 4)^{3/2}$, find x so that $f(x) = 27$.

92. If $g(x) = (6x - 2)^{5/3}$, find x so that $g(x) = 32$.

93. If $S(x) = 12x^{-5/4}$, find x so that $S(x) = 20$.

94. If $T(x) = 9x^{-6/5}$, find x so that $T(x) = 15$.

95. During a flue epidemic in a small town, health officials estimate that the number of people infected t days after the first case was discovered is given by

$$I(t) = 50t^{3/5}$$

a. How many people are infected 5 days after the start of the epidemic? After 10 days? After 15 days?

b. How long will it be before 300 people are ill?

c. Graph the function $I(t)$ and verify your answers to parts (a) and (b) on your graph.

96. The research division of an advertising firm estimates that the number of people who have seen their ads t days after the campaign begins is given by the function

$$N(t) = 2000t^{5/4}$$

a. How many people have seen the ads after 6 days? After 10 days? After 2 weeks?

b. How long will it be before 75,000 people have seen the ads?

c. Graph the function $N(t)$ and verify your answers to parts (a) and (b) on your graph.

97. According to Kepler's law, the period, p, of a planet's revolution and its average distance, a, from the sun are related. The period is given in years by

$$p = K^{1/2}a^{3/2}$$

where $K = 1.243 \times 10^{-24}$ and a is measured in miles. Find the period of Mars if its average distance from the sun is 1.417×10^8 miles.

98. If the period, p, of a planet is known, we can calculate its distance, a, from the sun in miles by Kepler's law. The distance is given by

$$a = K^{-1/3}\, p^{2/3}$$

where $K = 1.243 \times 10^{-24}$ and the period is given in years. Find the distance from Venus to the sun if its period is 0.615 years.

99. A brewery wants to replace its old vats with new ones that hold 1.8 times as much as the old ones. To estimate the cost of the new equipment, the accountant uses the 0.6 rule for industrial costs, which states that the cost of a new container is approximately $Cr^{0.6}$, where C is the cost of the old container and r is the ratio of the capacity of the new container to the old one. If an old vat cost \$5000, how much should the accountant budget for a new one?

100. If a quantity of air expands without changing temperature, its pressure in pounds per square inch is given by $kV^{-1.4}$, where V is the volume of the air in cubic inches and $k = 2.79 \times 10^4$. Find the air pressure of an air sample when its volume is 50 cubic inches.

101. A population is in a period of "supergrowth" if its rate of growth, R, at any time is proportional to P^k, where P is the population at that time and k is a constant greater than 1. Suppose R is given by

$$R = 0.015P^{1.2}$$

where P is measured in thousands and R is measured in thousands per year.

a. Find R when $P = 20{,}000$, when $P = 40{,}000$, and when $P = 60{,}000$.

b. What will the population be when its growth rate is 5000 per year?

c. Graph R and use your graph to verify your answers to parts (a) and (b).

102. The length, L, of a stream channel is related to the area, A, that it drains according to the formula

$$L = 1.4A^{0.6}$$

where L is given in miles and A in square miles.

a. Find L when $A = 10$ square miles, when $A = 40$ square miles, and when $A = 100$ square miles.

b. What area can a stream of length 33 miles drain?

c. Graph L and use your graph to verify your answers to parts (a) and (b).

6.5   RADICALS

Although most applications involving powers and roots can be described using fractional exponents, in some situations radical notation is more convenient. In these cases we are often better off if we simplify any radical expressions algebraically before we use a calculator to obtain decimal approximations.

Simplifying Radicals   Because $\sqrt[n]{a} = a^{1/n}$ we can use the laws of exponents to derive two important properties that are useful in simplifying radicals.

Property 1. $\sqrt[n]{ab} = \sqrt[n]{a}\,\sqrt[n]{b}$ for $a, b \geq 0$

Property 2. $\sqrt[n]{\dfrac{a}{b}} = \dfrac{\sqrt[n]{a}}{\sqrt[n]{b}}$ for $a \geq 0, \quad b > 0$

We can use Property 1 to simplify radical expressions as follows. Look for factors of the radicand that are perfect n^{th} powers, where n is the index of the radical. For example, to simply $\sqrt[3]{108}$, we look for perfect cubes that divide evenly into 108. The easiest way to do this is to try the perfect cubes in order: 1, 8, 27, 64, 125, and so on, until we find one that is a factor. For this example we note that $108 = 27 \cdot 4$. Using Property 1, we can write

$$\sqrt[3]{108} = \sqrt[3]{27}\sqrt[3]{4}$$

Simplify the first factor to find

$$\sqrt[3]{108} = 3\sqrt[3]{4}$$

We can also simplify radicals containing variables. First note that if the exponent on the variable is a multiple of the index, we can extract the variable from the radical. For instance,

$$\sqrt[3]{x^{12}} = x^{12/3} = x^4$$

(You can verify this result by noting that $(x^4)^3 = x^{12}$.) If the exponent on the variable is not a multiple of the index, then we can use Property 1 to factor out the highest power that is. For example,

$$\sqrt[3]{x^{11}} = \sqrt[3]{x^9 \cdot x^2} \qquad \text{Apply Property 1.}$$
$$= \sqrt[3]{x^9} \cdot \sqrt[3]{x^2} \qquad \text{Simplify } \sqrt[3]{x^9} = x^{9/3}.$$
$$= x^3 \sqrt[3]{x^2}$$

Example 1

Simplify each radical.

a. $\sqrt{18x^5}$ **b.** $\sqrt[3]{24x^6y^8}$

Solutions **a.** The index of the radical is 2, so we look for perfect square factors of $18x^5$. We note that 9 is a perfect square, and x^4 has an exponent divisible by 2. Thus

$$\sqrt{18x^5} = \sqrt{9x^4 \cdot 2x} \qquad \text{Apply Property 1.}$$
$$= \sqrt{9x^4}\sqrt{2x} \qquad \text{Take square roots.}$$
$$= 3x^2\sqrt{2x}$$

b. The index of the radical is 3, so we look for perfect cube factors of $24x^6y^8$. We note that 8 is a perfect cube, and x^6 and y^6 have exponents divisible by 3. Thus

$$\sqrt[3]{24x^6y^8} = \sqrt[3]{8x^6y^6 \cdot 3y^2} \qquad \text{Apply Property 1.}$$
$$= \sqrt[3]{8x^6y^6}\sqrt[3]{3y^2} \qquad \text{Take cube roots.}$$
$$= 2x^2y^2\sqrt[3]{3y^2}$$

COMMON ERROR

Property 1 applies only to *products* under the radical, not to sums. Thus, for example,

$$\sqrt{4 \cdot 9} = \sqrt{4}\,\sqrt{9} = 2 \cdot 3 \qquad \text{but} \qquad \sqrt{4 + 9} \ne \sqrt{4} + \sqrt{9}$$

and

$$\sqrt[3]{x^3 y^6} = \sqrt[3]{x^3}\,\sqrt[3]{y^6} = xy^2 \qquad \text{but} \qquad \sqrt[3]{x^3 + y^6} \ne \sqrt[3]{x^3} + \sqrt[3]{y^6}$$

We can use Property 2 to simplify radicals of fractions.

Example 2

a. $\sqrt{\dfrac{3}{4}} = \dfrac{\sqrt{3}}{\sqrt{4}} = \dfrac{\sqrt{3}}{2}$ **b.** $\sqrt[3]{\dfrac{5}{8}} = \dfrac{\sqrt[3]{5}}{\sqrt[3]{8}} = \dfrac{\sqrt[3]{5}}{2}$

We can also use Properties 1 and 2 to simplify products and quotients of radicals.

Example 3

Simplify.

a. $\sqrt[4]{6x^2}\sqrt[4]{8x^3}$ **b.** $\dfrac{\sqrt[3]{16y^5}}{\sqrt[3]{y^2}}$

Solutions **a.** Apply Property 1.

$$\sqrt[4]{6x^2}\sqrt[4]{8x^3} = \sqrt[4]{48x^5} \qquad \text{\color{teal}Factor out perfect fourth powers.}$$

$$= \sqrt[4]{16x^4}\sqrt[4]{3x} \qquad \text{\color{teal}Simplify.}$$

$$= 2x\sqrt[4]{3x}$$

b. Apply Property 2.

$$\dfrac{\sqrt[3]{16y^5}}{\sqrt[3]{y^2}} = \sqrt[3]{\dfrac{16y^5}{y^2}} \qquad \text{\color{teal}Reduce.}$$

$$= \sqrt[3]{16y^3} \qquad \text{\color{teal}Factor out perfect cubes.}$$

$$= \sqrt[3]{8y^3}\sqrt[3]{2} \qquad \text{\color{teal}Simplify.}$$

$$= 2y\sqrt[3]{2}$$

Simplifying $\sqrt[n]{a^n}$ In Section 6.3 we saw that raising to a power is the *inverse* operation for extracting roots; that is,

$$(\sqrt[n]{a})^n = a$$

For example, if $a = 16$ and $n = 4$, then

$$(\sqrt[4]{16})^4 = 2^4 = 16$$

Now consider the power and root operations in the opposite order; that is, consider

$$\sqrt[n]{a^n}$$

If the index, n, is an odd number, then

$$\sqrt[n]{a^n} = a \qquad (n \text{ odd})$$

For example,

$$\sqrt[3]{2^3} = \sqrt[3]{8} = 2 \qquad \text{and} \qquad \sqrt[3]{(-2)^3} = \sqrt[3]{-8} = -2$$

However, if n is even, $\sqrt[n]{a^n}$ is always the principal or non-negative root of a^n, regardless of whether a itself is positive or negative. For example, if $a = 3$,

$$\sqrt{3^2} = \sqrt{9} = 3$$

so $\sqrt{a^2} = a$, but if a is negative, say $a = -3$

$$\sqrt{(-3)^2} = \sqrt{9} = 3$$

so $\sqrt{a^2} = -a$. Thus, because the principal root is always a non-negative number, we have the following special relationship for even n:

$$\sqrt[n]{a^n} = |a|$$

We summarize our results in the box below.

> **1.** If n is odd, $\sqrt[n]{a^n} = a$.
>
> **2.** If n is even, $\sqrt[n]{a^n} = |a|$.   In particular, $\sqrt{a^2} = |a|$.

Example 4

a. $\sqrt{16x^2} = 4|x|$ **b.** $\sqrt{(x - 1)^2} = |x - 1|$

c. $\sqrt[4]{(-5)^4} = |-5| = 5$

 COMMON ERROR

Note that $\sqrt[4]{(-5)^4} \neq -5$, because the symbol $\sqrt[4]{}$ represents the principal, or positive, fourth root.

Sums and Differences of Radicals

Recall that sums or differences of like terms can be combined by adding or subtracting their coefficients:

$$3xy + 5xy = (3 + 5)xy = 8xy$$

"Like radicals," that is, radicals of the same index and radicand, can be combined in the same way.

Example 5

a. $3\sqrt{3} + 4\sqrt{3} = (3 + 4)\sqrt{3}$
$$= 7\sqrt{3}$$

b. $4\sqrt[3]{2y} - 6\sqrt[3]{2y} = (4 - 6)\sqrt[3]{2y}$
$$= -2\sqrt[3]{2y}$$

 COMMON ERRORS

1. In Example 5(a), $3\sqrt{3} + 4\sqrt{3} \neq 7\sqrt{6}$. Only the coefficients are added; the radicals are not changed.

2. Sums of radicals with different radicands or different indices *cannot* be combined. Thus

$$\sqrt{11} + \sqrt{5} \neq \sqrt{16}$$
$$\sqrt[3]{10x} - \sqrt[3]{2x} \neq \sqrt[3]{8x}$$

and

$$\sqrt[3]{7} + \sqrt{7} \neq \sqrt[5]{7}$$

None of the expressions above can be simplified.

It may be possible to simplify the radicals in a sum or difference so that two or more terms contain like radicals. Those terms can then be combined into a single term.

Example 6

a. $\sqrt{32} + \sqrt{2} - \sqrt{18} = 4\sqrt{2} + \sqrt{2} - 3\sqrt{2}$ Simplify each radical.
$$= (4 + 1 - 3)\sqrt{2}$$ Combine like terms.
$$= 2\sqrt{2}$$

b. $5\sqrt[3]{24} - \sqrt[3]{81} + 3\sqrt[3]{32}$
$$= 5(2\sqrt[3]{3}) - 3\sqrt[3]{3} + 3(2\sqrt[3]{4})$$ Simplify each radical.
$$= (10 - 3)\sqrt[3]{3} + 6\sqrt[3]{4}$$ Combine like terms.
$$= 7\sqrt[3]{3} + 6\sqrt[3]{4}$$

Products and Factors

Radicals of the same index can be multiplied together according to Property 1 on page 368.

$$\sqrt[n]{a}\sqrt[n]{b} = \sqrt[n]{ab} \qquad (a, b \geq 0)$$

Example 7

a. $\sqrt{2}\sqrt{18} = \sqrt{36} = 6$

b. $\sqrt[3]{2x}\sqrt[3]{4x^2} = \sqrt[3]{8x^3} = 2x$

We can apply the distributive law to products involving binomials.

Example 8

a. $\sqrt{3}(\sqrt{2x} + \sqrt{6}) = \sqrt{6x} + \sqrt{18} = \sqrt{6x} + 3\sqrt{2}$

b. $\sqrt[3]{x}(\sqrt[3]{2x^2} - \sqrt[3]{x}) = \sqrt[3]{2x^3} - \sqrt[3]{x^2} = x\sqrt[3]{2} - \sqrt[3]{x^2}$

c. $(\sqrt{x} - \sqrt{y})(\sqrt{x} + \sqrt{y}) = \sqrt{x^2} - \sqrt{xy} + \sqrt{xy} - \sqrt{y^2} = x - y$ ☐

We can also use the distributive law to factor expressions that contain radicals.

Example 9

a. $3 + \sqrt{45} = 3 + 3\sqrt{5} = 3(1 + \sqrt{5})$

b. $\sqrt{a} + \sqrt{ab} = \sqrt{a} + \sqrt{a}\sqrt{b} = \sqrt{a}(1 + \sqrt{b})$

c. $3\sqrt[3]{2} - \sqrt[3]{10} = 3\sqrt[3]{2} - \sqrt[3]{2}\sqrt[3]{5} = \sqrt[3]{2}(3 + \sqrt[3]{5})$ ☐

Rationalizing the Denominator

Radicals are easier to work with if there are no roots in the denominators of fractions. We can use the fundamental principle of fractions to remove roots from the denominators of fractions. This process is called **rationalizing the denominator**. For square roots, we multiply the numerator and denominator of the fraction by the radical in the denominator.

Example 10

Rationalize the denominator of each fraction.

a. $\sqrt{\dfrac{1}{3}}$ b. $\dfrac{\sqrt{2}}{\sqrt{50x}}$

Solutions   a.   Apply Property 2 to write the radical as a quotient:

$$\sqrt{\frac{1}{3}} = \frac{\sqrt{1}}{\sqrt{3}} = \frac{1}{\sqrt{3}}$$ Multiply numerator and denominator by $\sqrt{3}$.

$$= \frac{1 \cdot \sqrt{3}}{\sqrt{3} \cdot \sqrt{3}} = \frac{\sqrt{3}}{3}$$

b.   Simplify the denominator:

$$\frac{\sqrt{2}}{\sqrt{50x}} = \frac{\sqrt{2}}{5\sqrt{2x}}$$ Multiply numerator and denominator by $\sqrt{2x}$.

$$= \frac{\sqrt{2} \cdot \sqrt{2x}}{5\sqrt{2x} \cdot \sqrt{2x}} = \frac{\sqrt{4x}}{5\,(2x)}$$ Simplify.

$$= \frac{2\sqrt{x}}{10x} = \frac{\sqrt{x}}{5x}$$ ☐

In Exercise 6.5 we consider how to rationalize denominators that involve radicals with higher indices.

If a the denominator of a fraction is a *binomial* in which one or both terms are radicals, we can use a special building factor to rationalize it. First recall that

$$(p - q)(p + q) = p^2 - q^2$$

Notice that the product consists of perfect squares only. Each of the two factors, $p - q$ and $p + q$, is said to be the **conjugate** of the other.

Now consider a fraction of the form

$$\frac{a}{b + \sqrt{c}}$$

If we multiply the numerator and denominator of this fraction by the conjugate of the denominator, we get

$$\frac{a(b - \sqrt{c})}{(b + \sqrt{c})(b - \sqrt{c})} = \frac{ab - a\sqrt{c}}{b^2 - (\sqrt{c})^2} = \frac{ab - a\sqrt{c}}{b^2 - c}$$

The denominator of the fraction no longer contains any radicals—it has been rationalized.

Multiplying the numerator and denominator by the conjugate of the denominator also works on fractions of the forms

$$\frac{a}{\sqrt{b} + c} \quad \text{and} \quad \frac{a}{\sqrt{b} + \sqrt{c}}$$

We leave the verification of these cases as exercises.

Example 11

Rationalize the denominator:

$$\frac{x}{\sqrt{2} + \sqrt{x}}$$

Solution   Multiply the numerator and denominator by the conjugate of the denominator, $\sqrt{2} - \sqrt{x}$.

$$\frac{x(\sqrt{2} - \sqrt{x})}{(\sqrt{2} + \sqrt{x})(\sqrt{2} - \sqrt{x})} = \frac{x(\sqrt{2} - \sqrt{x})}{2 - x}$$

Exercise 6.5

Simplify the radicals in Exercises 1–40. Assume that all variables represent positive numbers.

1. $\sqrt{18}$

2. $\sqrt{50}$

3. $\sqrt[3]{24}$

4. $\sqrt[3]{54}$

5. $-\sqrt[4]{64}$

6. $-\sqrt[4]{162}$

7. $\sqrt{60{,}000}$

8. $\sqrt{800{,}000}$

9. $\sqrt[3]{900{,}000}$

10. $\sqrt[3]{24{,}000}$

11. $\sqrt[3]{\dfrac{-40}{27}}$

12. $\sqrt[4]{\dfrac{80}{625}}$

13. $\sqrt[3]{x^{10}}$

14. $\sqrt[3]{y^{16}}$

15. $\sqrt{27z^3}$

16. $\sqrt{12t^5}$

17. $\sqrt[4]{48a^9b^{12}}$

18. $\sqrt[3]{81a^{12}b^8}$

19. $-\sqrt[5]{96p^7q^9}$

20. $-\sqrt[6]{256k^7u^{12}}$

21. $-\sqrt{18s}\sqrt{2s^3}$

22. $\sqrt{3w^3}\sqrt{27w^3}$

23. $\sqrt[3]{7h^2}\sqrt[3]{-49h}$

24. $-\sqrt[4]{2m^3}\sqrt[4]{8m}$

25. $\sqrt{16-4x^2}$

26. $\sqrt{9Y^2+18}$

27. $\sqrt[3]{8A^3+A^6}$

28. $\sqrt[3]{b^9-27b^{27}}$

29. $\sqrt{\dfrac{125p^{13}}{a^4}}$

30. $\sqrt{\dfrac{c^5}{169n^6}}$

31. $\sqrt[3]{\dfrac{56v^2}{w^6}}$

32. $\sqrt[4]{\dfrac{48k^3}{m^8}}$

33. $\dfrac{\sqrt{a^5b^3}}{\sqrt{ab}}$

34. $\dfrac{\sqrt{x}\sqrt{xy^3}}{\sqrt{y}}$

35. $\dfrac{\sqrt{98x^2y^3}}{\sqrt{xy}}$

36. $\dfrac{\sqrt{45x^3}\sqrt{y^3}}{\sqrt{5y}}$

37. $\dfrac{\sqrt[3]{8b^7}}{\sqrt[3]{a^6b^2}}$

38. $\dfrac{\sqrt[3]{16r^4}}{\sqrt[3]{4t^3}}$

39. $\dfrac{\sqrt[5]{a}\sqrt[5]{b^2}}{\sqrt[5]{ab}}$

40. $\dfrac{\sqrt[5]{x^2}\sqrt[5]{y^3}}{\sqrt[5]{xy^2}}$

For Exercises 41–46 do not assume that variables represent positive numbers. Use absolute value bars as necessary to simplify the radicals.

41. $\sqrt{4x^2}$

42. $\sqrt{9x^2y^4}$

43. $\sqrt{(x-5)^2}$

44. $\sqrt{(2x-1)^2}$

45. $\sqrt{x^2-6x+9}$

46. $\sqrt{9x^2-6x+1}$

In Exercises 47 and 48 use your calculator to graph each function, and explain your result.

47. a. $y=\sqrt{x^2}$

    b. $y=\sqrt[3]{x^3}$

48. a. $y=(x^4)^{1/4}$

    b. $y=(x^5)^{1/5}$

Simplify Exercises 49–60 by combining like terms.

49. $3\sqrt{7}+2\sqrt{7}$

50. $5\sqrt{2}-3\sqrt{2}$

51. $4\sqrt{3}-\sqrt{27}$

52. $\sqrt{75}+2\sqrt{3}$

53. $\sqrt{50x}+\sqrt{32x}$

54. $\sqrt{8y}-\sqrt{18y}$

55. $3\sqrt[3]{16}-\sqrt[3]{2}-2\sqrt[3]{54}$

56. $\sqrt[3]{81}+2\sqrt[3]{24}-3\sqrt[3]{3}$

57. $4\sqrt[3]{40}+6\sqrt{80}-5\sqrt{45}-\sqrt[3]{135}$

58. $6\sqrt[3]{32}-3\sqrt{32}+\sqrt[3]{128}-2\sqrt{128}$

59. $3\sqrt{4xy^2}-4\sqrt{9xy^2}+2\sqrt{4x^2y}$

60. $2\sqrt{8y^2z}-3\sqrt{9yz^2}+3\sqrt{32y^2z}$

Multiply in Exercises 61–78.

61. $2(3-\sqrt{5})$

62. $5(2-\sqrt{7})$

63. $\sqrt{2}(\sqrt{6}+\sqrt{10})$

64. $\sqrt{3}(\sqrt{12}-\sqrt{15})$

65. $\sqrt[3]{2}(\sqrt[3]{20}-2\sqrt[3]{12})$

66. $\sqrt[3]{3}(2\sqrt[3]{18}+\sqrt[3]{36})$

67. $2\sqrt{x}(\sqrt{24x}+\sqrt{12})$

68. $\sqrt{3y}(\sqrt{6y}-\sqrt{18})$

69. $(\sqrt{x}-3)(\sqrt{x}+3)$

70. $(2+\sqrt{x})(2-\sqrt{x})$

71. $(\sqrt{2}-\sqrt{3})(\sqrt{2}+2\sqrt{3})$

72. $(\sqrt{3}-\sqrt{5})(2\sqrt{3}+\sqrt{5})$

73. $(\sqrt{5}-\sqrt{2})^2$

74. $(\sqrt{2}-2\sqrt{3})^2$

75. $(3\sqrt{x}+\sqrt{2y})(2\sqrt{x}-3\sqrt{2y})$

76. $(\sqrt{5x}-2\sqrt{y})(2\sqrt{5x}-3\sqrt{y})$

77. $(\sqrt{a}-2\sqrt{b})^2$

78. $(\sqrt{2a}-2\sqrt{b})(\sqrt{2a}+2\sqrt{b})$

Factor the expressions in Exercises 79–90.

79. $2+2\sqrt{3}=2(?)$

80. $5+10\sqrt{2}=5(?)$

81. $2\sqrt{27}+6=6(?)$

82. $5\sqrt{5}-\sqrt{25}=5(?)$

83. $4+\sqrt{16y}=4(?)$

84. $3+\sqrt{18x}=3(?)$

85. $\sqrt{2}-\sqrt{6}=\sqrt{2}(?)$

86. $\sqrt{12}-2\sqrt{6}=2\sqrt{3}(?)$

87. $2y\sqrt{x} + 3\sqrt{xy} = \sqrt{x}(?)$

88. $a\sqrt{5b} - \sqrt{3ab} = \sqrt{b}(?)$

89. $4x - \sqrt{12x} = 2\sqrt{x}(?)$

90. $6y + \sqrt{18y} = 3\sqrt{y}(?)$

Rationalize the denominator in Exercises 91–110.

91. $\dfrac{6}{\sqrt{3}}$ **92.** $\dfrac{10}{\sqrt{5}}$

93. $\dfrac{-\sqrt{3}}{\sqrt{7}}$ **94.** $\dfrac{-\sqrt{5}}{\sqrt{6}}$

95. $\sqrt{\dfrac{7x}{18}}$ **96.** $\sqrt{\dfrac{27x}{20}}$

97. $\sqrt{\dfrac{2a}{b}}$ **98.** $\sqrt{\dfrac{5p}{q}}$

99. $\dfrac{2\sqrt{3}}{\sqrt{2k}}$ **100.** $\dfrac{6\sqrt{2}}{\sqrt{3v}}$

101. $\dfrac{-9x^2\sqrt{5x^3}}{2\sqrt{6x}}$ **102.** $\dfrac{-8y\sqrt{21y^5}}{3\sqrt{5y}}$

103. $\dfrac{4}{1 + \sqrt{3}}$ **104.** $\dfrac{3}{7 - \sqrt{2}}$

105. $\dfrac{x}{x - \sqrt{3}}$ **106.** $\dfrac{y}{\sqrt{5} - y}$

107. $\dfrac{\sqrt{6} - 3}{2 - \sqrt{6}}$

108. $\dfrac{\sqrt{x} + \sqrt{y}}{\sqrt{x} - \sqrt{y}}$

109. $\dfrac{\sqrt{5}}{5\sqrt{3} + 3\sqrt{5}}$

110. $\dfrac{\sqrt{3}}{2\sqrt{3} - 3\sqrt{2}}$

In Exercises 111–120 rationalize radical denominators with an index greater than 2. Follow the method described in Exercises 111 and 112.

111. Rationalize the denominator of $\dfrac{1}{\sqrt[3]{2x}}$.

We need a third power, $(2x)^3$, under the radical in order to extract the root. Therefore, we must multiply the denominator $\sqrt[3]{2x}$ by *two* additional factors of $\sqrt[3]{2x}$. Complete the solution:

$$\frac{1}{\sqrt[3]{2x}} = \frac{1 \cdot \sqrt[3]{2x}\,\sqrt[3]{2x}}{\sqrt[3]{2x} \cdot \sqrt[3]{2x}\,\sqrt[3]{2x}}$$

$$= ?$$

112. Rationalize the denominator of $\sqrt[5]{\dfrac{6}{16x^3}}$.

First note that

$$\sqrt[5]{\frac{6}{16x^3}} = \sqrt[5]{\frac{6}{2^4 x^3}}$$

Since we want a fifth power, $(2x)^5$, in the denominator, we multiply numerator and denominator by $\sqrt[5]{2x^2}$. Complete the solution:

$$\sqrt[5]{\frac{6}{16x^3}} = \frac{\sqrt[5]{6} \cdot \sqrt[5]{2x^2}}{\sqrt[5]{2^4 x^3} \cdot \sqrt[5]{2x^2}}$$

$$= ?$$

113. $\dfrac{1}{\sqrt[3]{x^2}}$ **114.** $\dfrac{1}{\sqrt[4]{y^3}}$

115. $\sqrt[3]{\dfrac{2}{3y}}$ **116.** $\sqrt[4]{\dfrac{2}{3x}}$

117. $\sqrt[4]{\dfrac{x}{8y^3}}$ **118.** $\sqrt[3]{\dfrac{x}{4y^2}}$

119. $\dfrac{9x^3}{\sqrt[4]{27x}}$ **120.** $\dfrac{15x^4}{\sqrt[3]{5x}}$

6.6 RADICAL EQUATIONS AND THE DISTANCE FORMULA

Solving Radical Equations   To solve equations containing radicals, we use the following property.

> If each side of an equation is raised to the same natural-number power, the solutions of the original equation are also solutions of the new equation.

We must be careful, however: the new equation *may* have additional solutions that are *not* solutions of the original equation. These are called **extraneous solutions**. For example, by squaring each side of the equation

$$x = 3$$

(which has solution 3), we obtain

$$x^2 = 9$$

The solutions of the new equation are 3 and -3. In this instance, -3 does not satisfy the original equation.

If each side of an equation is raised to an odd power, extraneous solutions will not be introduced. If an even power is used, however, each solution obtained *must* be checked in the original equation to verify its validity. The check is part of the solution process.

Example 1

Solve the equation $\sqrt{x + 2} + 4 = x$.

Solution   First isolate the radical expression on one side of the equation. (This will make it easier to square both sides.)

$$\sqrt{x + 2} = x - 4 \qquad \text{Square both sides of the equation.}$$
$$(\sqrt{x + 2})^2 = (x - 4)^2$$
$$x + 2 = x^2 - 8x + 16 \qquad \text{Subtract } x + 2 \text{ from both sides.}$$
$$x^2 - 9x + 14 = 0 \qquad \text{Factor the left side.}$$
$$(x - 2)(x - 7) = 0 \qquad \text{Set each factor equal to zero.}$$
$$x = 2 \qquad \text{or} \qquad x = 7$$

Check   Does $\sqrt{2 + 2} + 4 = 2$?   No; 2 is not a solution.
           Does $\sqrt{7 + 2} + 4 = 7$?   Yes; 7 is a solution.
Thus the solution to the original equation is 7.

Sometimes we need to square both sides of an equation more than once to eliminate all the radicals.

Example 2

Solve $\sqrt{x - 7} + \sqrt{x} = 7$.

Solution   First isolate the more complicated radicals on one side of the equation. (This will make it easier to square both sides.)

$$\sqrt{x - 7} = 7 - \sqrt{x}$$

Now square each side to remove one radical.

$$(\sqrt{x - 7})^2 = (7 - \sqrt{x})^2$$
$$x - 7 = 49 - 14\sqrt{x} + x$$

Simplify the result, and isolate the radical on one side of the equation.

$$-56 = -14\sqrt{x}$$
$$4 = \sqrt{x}$$

Now square again to obtain

$$(4)^2 = (\sqrt{x})^2$$
$$16 = x$$

Check   Does $\sqrt{16 - 7} + \sqrt{16} = 7$?   Yes. The solution is 16.

 COMMON ERROR

We cannot solve the equation in Example 2 by squaring each term separately. In other words, it is *incorrect* to begin by writing

$$(\sqrt{x - 7})^2 + (\sqrt{x})^2 = 7^2$$

We must square the *entire expression* on each side of the equals sign as one piece.

Applications   Equations involving radicals arise in a variety of applications.

Example 3

Two oil derricks are located the same distance offshore from a straight section of coast. A supply boat returns from the derrick at point *A* to the harbor at *C* (in Figure 6.2), reloads, and sails to the derrick at *B*, traveling a total distance of 102 miles. How far are the derricks from the coast?

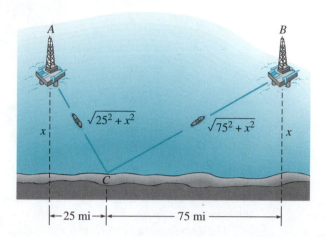

Figure 6.2

|←25 mi→|←————— 75 mi —————→|

Solution   Let x represent the distance from each derrick to the coast. Use the Pythagorean theorem to write an expression for the distance the supply boat travels on each leg of its journey. The sum of these distances is 102 miles.

$$\sqrt{x^2 + (25)^2} + \sqrt{x^2 + (75)^2} = 102$$

Solve the equation. First rewrite the equation with one radical on each side and square both sides.

$$\left(\sqrt{x^2 + (25)^2}\right)^2 = \left(102 - \sqrt{x^2 + (75)^2}\right)^2$$

$$x^2 + 625 = 10{,}404 - 204\sqrt{x^2 + 5625} + x^2 + 5625$$

Collect like terms and square again:

$$204\sqrt{x^2 + 5625} = 15{,}404$$

$$\sqrt{x^2 + 5625} \approx 75.5$$

$$x^2 + 5625 \approx 5700.25$$

$$x \approx \pm 8.7$$

Since x must be greater than 0, we find that $x \approx 8.7$.   The derricks are about 8.7 miles offshore.

Distance and Midpoint Formulas

If we know the coordinates of two points, P_1 and P_2, we can find the distance between them by using the Pythagorean theorem. We first create a right triangle by drawing a horizontal line through P_1 and a vertical line through P_2. These lines meet at a point P_3, forming a right triangle, as shown in Figure 6.3. Note that the x-coordinate of P_3 is the same as the x-coordinate

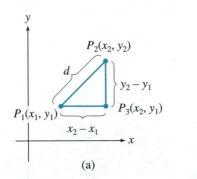

Figure 6.3   (a)   (b)

of P_2, and the y-coordinate of P_3 is the same as the y-coordinate of P_1. Thus the coordinates of P_3 are (x_2, y_1).

The distance between P_1 and P_3 is $|x_2 - x_1|$ and the distance between P_2 and P_3 is $|y_2 - y_1|$. These two numbers are the lengths of the legs of the right triangle. The length of the hypotenuse is the distance between P_1 and P_2, which we will call d. By the Pythagorean theorem,

$$d^2 = (x_2 - x_1)^2 + (y_2 - y_1)^2$$

Taking the (positive) square root of each side of this equation gives us the **distance formula**.

> The distance, d, between points $P_1(x_1, y_1)$ and $P_2(x_2, y_2)$ is
> $$d = \sqrt{(x_2 - x_1)^2 + (y_2 - y_1)^2}$$

Example 4

Find the distance between $(2, -1)$, and $(4, 3)$.

Solution   Substitute $(2, -1)$ for (x_1, y_1) and $(4, 3)$ for (x_2, y_2) in the distance formula to obtain

$$\begin{aligned}
d &= \sqrt{(x_2 - x_1)^2 + (y_2 - y_1)^2} \\
&= \sqrt{(4 - 2)^2 + [3 - (-1)]^2} \\
&= \sqrt{4 + 16} \\
&= \sqrt{20} = 2\sqrt{5}
\end{aligned}$$

Notice in Example 4 that we would obtain the same answer if we used $(4, 3)$ for P_1 and $(2, -1)$ for P_2:

$$\begin{aligned}
d &= \sqrt{(2 - 4)^2 + [(-1) - 3]^2} \\
&= \sqrt{4 + 16} = 2\sqrt{5}
\end{aligned}$$

Similar triangles can be used to derive the following formula for the **midpoint** of the line segment joining two points. The proof is left as an exercise. Note, however, that each coordinate of the midpoint is just the average of the corresponding coordinates of the two points.

The midpoint of the line segment joining the points $P_1(x_1, y_1)$ and $P_2(x_2, y_2)$ is the point $M\ (\bar{x}, \bar{y})$, where

$$\bar{x} = \frac{x_1 + x_2}{2} \qquad \text{and} \qquad \bar{y} = \frac{y_1 + y_2}{2}$$

Example 5

Find the midpoint of the line segment joining the points $(-2, 1)$ and $(4, 3)$.

Solution   Substitute $(-2, 1)$ for (x_1, y_1) and $(4, 3)$ for (x_2, y_2) in the midpoint formula to obtain

$$\bar{x} = \frac{x_1 + x_2}{2} = \frac{-2 + 4}{2} = 1$$

$$\bar{y} = \frac{y_1 + y_2}{2} = \frac{1 + 3}{2} = 2$$

The midpoint of the segment is the point $(\bar{x}, \bar{y}) = (1, 2)$.

Circles   A **circle** is the set of all points in a plane that lie at a given distance, called the **radius**, from a fixed point called the **center**. We can use the distance formula to find an equation for a circle. First consider the circle in Figure 6.4(a), whose center is the origin, $(0, 0)$.

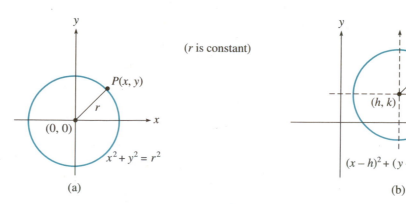

Figure 6.4

The distance from the origin to any point $P(x, y)$ on the circle is r. Therefore,

$$\sqrt{(x - 0)^2 + (y - 0)^2} = r$$

or, squaring both sides,

$$(x - 0)^2 + (y - 0)^2 = r^2$$

Thus the equation for a circle of radius r centered at the origin is

$$x^2 + y^2 = r^2$$

Now consider the circle in Figure 6.4(b), whose center is the point (h, k). Every point $P(x, y)$ on the circle lies a distance r from (h, k), so the equation of the circle is given by the following formula.

> The equation for a circle of radius r centered at the point (h, k) is
>
> $$(x - h)^2 + (y - k)^2 = r^2$$

This equation is the *standard form* for a circle of radius r with center at (h, k). It is easy to graph a circle if its equation is given in standard form.

Example 6

Graph the circles.
a. $(x - 2)^2 + (y + 3)^2 = 16$
b. $x^2 + (y - 4)^2 = 7$

Solutions a. The graph of $(x - 2)^2 + (y + 3)^2 = 16$ is a circle with radius 4 and center at $(2, -3)$. To sketch the graph, first locate the center of the circle. (Note that the center is not part of the graph of the circle.) From the center move a distance of four units (the radius of the circle) in each of four directions: up, down, left, and right. You have now located four points that lie on the circle: $(2, 1)$, $(2, -7)$, $(-2, -3)$, and $(6, -3)$. Sketch the circle through these four points. (See Figure 6.5.)

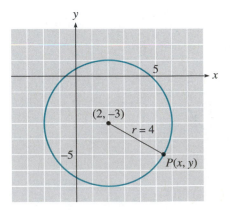

Figure 6.5

b.   The graph of $x^2 + (y - 4)^2 = 7$ is a circle with radius $\sqrt{7}$ and center at (0, 4). From the center move approximately $\sqrt{7}$, or 2.6, units in each of the four coordinate directions to obtain the points (0. 6.6), (0. 1.4), (−2.6, 4), and (2.6, 4). Sketch the circle through these four points. (See Figure 6.6.)

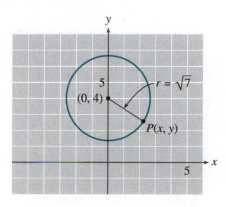

Figure 6.6

The equations of circles often appear in a general quadratic form rather than the standard form described above. For example, we can expand the squares of binomials in Example 6(a),

$$(x - 2)^2 + (y + 3)^2 = 16$$

to obtain

$$x^2 - 4x + 4 + y^2 + 6y + 9 = 16$$

or

$$x^2 + y^2 - 4x + 6y - 3 = 0$$

Conversely, an equation of the form $x^2 + y^2 + ax + by + c = 0$ can be converted to standard form by completing the square in both variables. Once this is done, the center and radius of the circle can be determined directly from the equation.

Example 7

Write the equation of the circle

$$x^2 + y^2 + 8x - 2y + 6 = 0$$

in standard form, and graph the equation.

Solution   Prepare to complete the square in both variables by writing the equation as

$$(x^2 + 8x + \underline{\quad}) + (y^2 - 2y + \underline{\quad}) = -6$$

Complete the square in x by adding 16 to each side of the equation, and complete the square in y by adding 1 to each side, to get

$$(x^2 + 8x + 16) + (y^2 - 2y + 1) = -6 + 16 + 1$$

from which we obtain the standard form,

$$(x + 4)^2 + (y - 1)^2 = 11$$

Thus the circle has its center at $(-4, 1)$, and its radius is $\sqrt{11}$, or approximately 3.3. The graph is shown in Figure 6.7.

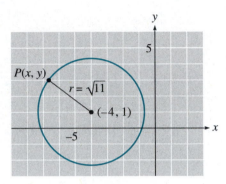

Figure 6.7

We can write an equation for any circle if we can find its center and radius.

Example 8

Find an equation for a circle whose diameter has endpoints $(7, 5)$ and $(1, -1)$.

Solution   The center of the circle is the midpoint of its diameter. (See Figure 6.8.) Use the midpoint formula to find the center:

$$h = \bar{x} = \frac{7 + 1}{2} = 4$$

$$k = \bar{y} = \frac{5 - 1}{2} = 2$$

Thus the center is the point $(h, k) = (4, 2)$.   The radius is the distance from the center to either endpoint of the diameter, say the point $(7, 5)$. Use the distance formula with the points $(7, 5)$ and $(4, 2)$ to find the radius:

$$r = \sqrt{(7 - 4)^2 + (5 - 2)^2}$$
$$= \sqrt{3^2 + 3^2} = \sqrt{18}$$

Finally, substitute 4 for h and 2 for k (the coordinates of the center) and $\sqrt{18}$ for r (the radius) into the standard form

$$(x - h)^2 + (y - k)^2 = r^2$$

to obtain

$$(x - 4)^2 + (y - 2)^2 = 18$$

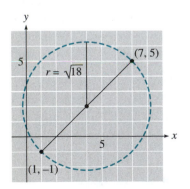

Figure 6.8

Exercise 6.6

Solve the equations in Exercises 1–16.

1. $3z + 4 = \sqrt{3z + 10}$

2. $2x - 3 = \sqrt{7x - 3}$

3. $2x + 1 = \sqrt{10x + 5}$

4. $4x + 5 = \sqrt{3x + 4}$

5. $\sqrt{y + 4} = y - 8$ **6.** $4\sqrt{x - 4} = x$

7. $\sqrt{2y - 1} = \sqrt{3y - 6}$

8. $\sqrt{4y + 1} = \sqrt{6y - 3}$

9. $\sqrt{x - 3}\sqrt{x} = 2$ **10.** $\sqrt{x}\sqrt{x - 5} = 6$

11. $\sqrt{y + 4} = \sqrt{y + 20} - 2$

12. $4\sqrt{y} + \sqrt{1 + 16y} = 5$

13. $\sqrt{x} + \sqrt{2} = \sqrt{x + 2}$

14. $\sqrt{4x + 17} = 4 - \sqrt{x + 1}$

15. $\sqrt{5 + x} + \sqrt{x} = 5$

16. $\sqrt{y + 7} + \sqrt{y + 4} = 3$

17. Grandview College is located 8 miles from a long straight stretch of the state highway. Station KGVC broadcasts from the college and has a signal range of 15 miles. Francine leaves the college and travels along the highway at a speed of 60 miles per hour (or 1 mile per minute). (See Figure 6.9.)

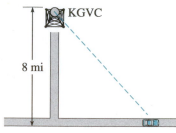

Figure 6.9

a. How far has Francine traveled from the junction t minutes after turning onto the highway?

b. Express Francine's distance from the college as a function of t.

c. Graph your function in the window

$$\text{Xmin} = 0, \qquad \text{Ymin} = 0$$
$$\text{Xmax} = 23.5, \quad \text{Ymax} = 25$$

d. How long will it be until Francine is out of radio range for station KGVC?

18. Brenda is flying at an altitude of 1500 feet on a heading that will take her directly above Van Nuys airport. Her speed is 70 knots (1 knot equals 1.15 miles per hour), and she passes over the airport at precisely 10:00 A.M. The airport has a radio range of 4 miles. *Hint:* Convert all units to miles. (See Figure 6.10.)

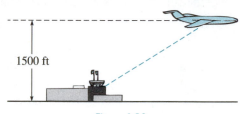

1500 ft

Figure 6.10

a. What is Brenda's horizontal distance from the airport at t minutes after 10:00 A.M.?

b. Express Brenda's distance from the airport as a function of t.

c. Graph your function in the window

$$\text{Xmin} = 0, \qquad \text{Ymin} = 0$$
$$\text{Xmax} = 0.1, \quad \text{Ymax} = 5$$

d. How long will it be before Brenda is out of radio range for Van Nuys airport?

19. A UFO is hovering directly over your head at an altitude of 700 feet. It begins descending at a rate of 10 feet per second. At the same time, you start to run at a rate of 15 feet per second.

a. What is the altitude of the UFO t seconds later?

b. How far have you run t seconds later?

c. Sketch a diagram of your position and the UFO at time t seconds. Label the sketch with your answers to parts (a) and (b).

d. Express the distance, D, between you and the UFO as a function of t. (Use your answers to parts (a) and (b).)

e. What is the distance between you and the UFO when $t = 10$?

f. Graph your function D in the window

$$Xmin = 0, \quad Ymin = 0$$

$$Xmax = 94, \quad Ymax = 1240$$

Verify your answer to part (e) on the graph. Use the graph to find the distance between you and the UFO when $t = 45$.

g. When is the distance between you and the UFO the smallest? (Use the graph to estimate to the nearest second.)

h. Actually, the window we used in part (f) is too big. Find the domain of the function D and change Xmax to a more realistic value. (*Hint:* When does the UFO reach the ground?)

20. A private plane flying north at 200 miles per hour leaves St. Louis at noon. A small jet plane flying east at 200 miles per hour is 600 miles west of St. Louis at noon.

a. How far is the jet from St. Louis t hours after noon (in hundreds of miles)?

b. How far is the private plane from St. Louis t hours after noon (in hundreds of miles)?

c. Sketch a diagram showing the positions of the two planes t hours after noon. Label the sketch with your answers to parts (a) and (b).

d. Express the distance, d, between the two planes as a function of t. (Use your answers to parts (a) and (b).)

e. What is the distance between the two planes at 1:00 P.M.?

f. Graph your function in the window

$$Xmin = 0, \quad Ymin = 0$$

$$Xmax = 4.7, \quad Ymax = 10$$

Verify your answer to part (e) on the graph. Use your graph to find the distance between the planes at 12:30 P.M.

g. When is the distance between the two planes the smallest? (Use the graph to estimate to the nearest tenth of an hour.)

h. Actually, the window we used in part (f) is too big. Find the domain of the function d and change Xmax to a more realistic value. (*Hint:* When does the jet reach St. Louis?)

21. Delbert drives past a service station on the highway and 5 miles later turns right onto a country road. After he has traveled for 1 mile, his car's engine throws a rod, and he decides to walk back to the service station. Delbert can walk 4 miles per hour along the road, but he estimates that he can go only 3 miles per hour through the fields. He would like to reach the service station in the shortest time possible. (See Figure 6.11.)

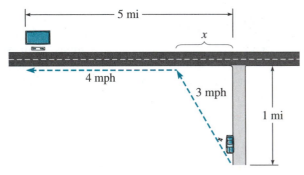

Figure 6.11

a. How long will it take Delbert if he retraces his route, walking along the road and then along the highway?

b. How long will it take him if he heads out across the fields, straight for the service station?

c. What if he walks through the fields to some intermediate point, P, on the highway, as shown in Figure 6.11, and then continues along the highway? Let the distance from the junction to point P be x miles, and express Delbert's walking time, t, as a function of x. Use the following steps:

i. Express the distance Delbert walks along the road and the distance he walks through the fields in terms of x.

ii. Express the time Delbert takes on each part of his walk in terms of x.

iii. Add the two times you found in part (ii).

d. Graph your function in the window

$$\text{Xmin} = -5, \quad \text{Ymin} = 0$$
$$\text{Xmax} = 10, \quad \text{Ymax} = 3$$

Locate the points corresponding to your answers to parts (a) and (b).

e. Choose an appropriate domain and range for this problem. (What are the smallest and largest values of x that make sense?)

f. Graph your function on the domain from part (e). Use your graph to find the value of x that results in the shortest walking time for Delbert to reach the service station. What is that time?

22. The telephone company is laying a cable from an island 12 miles offshore to a relay station 18 miles down the coast, as shown in Figure 6.12. It costs $8000 per mile to lay the cable underwater and $3000 per mile to lay the cable underground. The telephone company would like to find the cheapest route to run the cable.

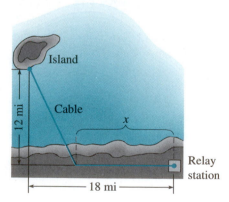

Figure 6.12

a. How much would it cost to run the cable directly to the shore (12 miles) and then 18 miles along the shore to the relay station?

b. How much would it cost to run the cable directly to the relay station, entirely underwater?

c. What if the phone company runs the cable from the island to some intermediate point Q on the shore and then continues laying the cable along the shore? Let the distance from the relay station to the point Q be $18 - x$ miles, and express the cost of laying the cable (in thousands of dollars) as a function of x. Use the following steps:

i. Express the length of the cable laid underwater and the length of the cable laid underground in terms of x.

ii. Express the cost of laying each portion of the cable in terms of x.

iii. Add the two costs you found in part (ii).

d. Graph your function in the window

$$\text{Xmin} = -10, \quad \text{Ymin} = 0$$
$$\text{Xmax} = 20, \quad \text{Ymax} = 200$$

Locate the points corresponding to your answers to parts (a) and (b).

e. Choose an appropriate domain and range for this problem. (What are the smallest and largest values of x that make sense?)

f. Graph your function on the domain from part (e). Use your graph to find the value of x that results in the least expensive route for the cable. What is the cost for that route?

Find the distance between each of the given pairs of points in Exercises 23–28, and find the midpoint of the segment joining them.

23. $(1, 1), (4, 5)$ **24.** $(-1, 1), (5, 9)$

25. $(2, -3), (-2, -1)$ **26.** $(5, -4), (-1, 1)$

27. $(3, 5), (-2, 5)$ **28.** $(-2, -5), (-2, 3)$

29. Find the perimeter of the triangle with vertices $(10, 1), (3, 1), (5, 9)$.

30. Find the perimeter of the triangle with vertices $(-1, 5), (8, -7), (4, 1)$.

31. Show that the rectangle with vertices $(-4, 1)$, $(2,6), (7, 0)$, and $(1, -5)$ is a square.

32. Show that the triangle with vertices $(0, 0)$, $(6, 0)$, and $(3, 3)$ is a right isosceles triangle—that is, a right triangle with two sides of the same length.

Graph each equation in Exercises 33–44.

33. $x^2 + y^2 = 25$ **34.** $x^2 + y^2 = 16$

35. $4x^2 + 4y^2 = 16$ **36.** $2x^2 + 2y^2 = 18$

37. $(x - 4)^2 + (y + 2)^2 = 9$

38. $(x - 1)^2 + (y - 3)^2 = 16$

39. $(x + 3)^2 + y^2 = 10$

40. $x^2 + (y + 4)^2 = 12$

41. $x^2 + y^2 + 2x - 4y - 6 = 0$

42. $x^2 + y^2 - 6x + 2y - 4 = 0$

43. $x^2 + y^2 + 8x = 4$

44. $x^2 + y^2 - 10y = 2$

In Exercises 45–52 write an equation for the circle with the given properties.

45. Center at $(-2, 5)$, radius $2\sqrt{3}$.

46. Center at $(4, -3)$, radius $2\sqrt{6}$.

47. Center at $\left(\dfrac{3}{2}, -4\right)$, one point on the circle $(4, -3)$.

48. Center at $\left(\dfrac{-3}{2}, \dfrac{-1}{2}\right)$, one point on the circle $(-4, -2)$.

49. Endpoints of a diameter at $(1, 5)$ and $(3, -1)$.

50. Endpoints of a diameter at $(3, 6)$ and $(-5, 2)$.

51. Center at $(-3, -1)$, x-axis tangent to the circle.

52. Center at $(1, 7)$, y-axis tangent to the circle.

53. Find an equation for the circle that passes through the points $(2, 3)$, $(3, 2)$, and $(-4, -5)$. (*Hint:* Find values a, b, and c so that the three points lie on the graph of $x^2 + y^2 + ax + by + c = 0$.)

54. Find an equation for the circle that passes through the points $(0, 0)$, $(6, 0)$, and $(0, 8)$. (See the hint for Exercise 53.)

55. In Figure 6.13, $M(\bar{x}, \bar{y})$ is the midpoint of the line segment $\overline{AB}$. Using similar triangles, show that

$$AC = 2AE \qquad \text{and} \qquad BC = 2BD$$

(Assume that $\angle ACB$, $\angle MDB$, and $\angle AEM$ are right angles.)

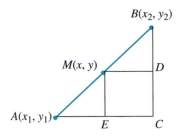

Figure 6.13

56. Use the result of Exercise 55 and the distance formula to show that

$$\bar{x} = \frac{x_1 + x_2}{2} \qquad \text{and} \qquad \bar{y} = \frac{y_1 + y_2}{2}$$

Chapter 6 Review

In Exercises 1–12 expand the products and simplify the quotients using the laws of exponents.

1. $(2x^3)(5x^4)$

2. $(3mn^5)(7m^8n)$

3. $(-a^2b^3)(4ab)$

4. $(5s^3t^2)(-2s^4t)$

5. $\dfrac{3u^4v}{6uv^6}$

6. $\dfrac{-24wz^5}{6wz^7}$

7. $\dfrac{54r^2s^7t^5}{-18r^3st^3}$

8. $\dfrac{48a^3bc^2}{36ab^4c^3}$

9. $(-3x^3)^2$

10. $(-4y^4)^3$

11. $\left(\dfrac{-3m^2n}{2m^3}\right)^3$

12. $\left(\dfrac{7vw^8}{21v^9w^3}\right)^4$

Write the expressions in Exercises 13–24 without negative exponents and simplify.

13. $(-3)^{-4}$

14. 4^{-3}

15. $\left(\dfrac{1}{3}\right)^{-2}$

16. $\dfrac{3}{5^{-2}}$

17. $(3m)^{-5}$

18. $-7y^{-8}$

19. $a^{-1} + a^{-2}$

20. $\dfrac{3q^{-9}}{r^{-2}}$

21. $6c^{-7} \cdot 3^{-1}c^4$

22. $\dfrac{11z^{-7}}{3^{-2}z^{-5}}$

23. $(2d^{-2}k^3)^{-4}$

24. $\dfrac{2w^3(w^{-2})^{-3}}{5w^{-5}}$

Factor the expressions in Exercises 25–34 as indicated, and write the factored form without negative exponents.

25. $\dfrac{5}{3}n^{-1} + 2n^{-2} - n^{-3} = \dfrac{1}{3}n^{-3}(?)$

26. $\dfrac{3}{7}p - 2 + \dfrac{2}{7}p^{-1} = \dfrac{1}{7}p^{-1}(?)$

27. $7 + 7(5n)^{1/3} = 7(?)$

28. $12 - 8y^{1/4} = 4(?)$

29. $3xy^{1/2} - 2(xy)^{1/2} = x^{1/2}(?)$

30. $3c(2d)^{1/2} + (3cd)^{1/2} = \sqrt{d}\,(?)$

31. $q^{-1/3} - q^{1/3} = q^{-1/3}(?)$

32. $5(x+2)^{3/4} - (x+2)^{-3/4} = (x+2)^{-3/4}(?)$

33. $9t + \sqrt{27t} = 3\sqrt{t}\,(?)$

34. $6w + \sqrt{8w} = 2\sqrt{w}\,(?)$

Write each power in Exercises 35–42 in radical form.

35. $25m^{1/2}$

36. $8n^{1/3}$

37. $(13d)^{2/3}$

38. $6x^{2/5}y^{3/5}$

39. $(3q)^{-3/4}$

40. $7(uv)^{3/2}$

41. $(a^2 + b^2)^{0.5}$

42. $(16 - x^2)^{0.25}$

Rationalize the denominator in Exercises 43–54.

43. $\dfrac{7}{\sqrt{5y}}$

44. $\dfrac{6d}{\sqrt{2d}}$

45. $\sqrt{\dfrac{3r}{11s}}$

46. $\sqrt{\dfrac{26}{2m}}$

47. $\dfrac{-3}{\sqrt{a}+2}$

48. $\dfrac{-3}{\sqrt{z}-4}$

49. $\dfrac{2x - \sqrt{3}}{x - \sqrt{3}}$

50. $\dfrac{m - \sqrt{3}}{5m + 2\sqrt{3}}$

51. $\dfrac{2}{\sqrt[3]{4}}$

52. $\dfrac{9}{\sqrt[4]{3}}$

53. $\sqrt[4]{\dfrac{q}{8w^3}}$

54. $\sqrt[3]{\dfrac{5}{49t}}$

Solve the equations in Exercises 55–62.

55. $x - 3\sqrt{x} + 2 = 0$

56. $\sqrt{x+1} + \sqrt{x+8} = 7$

57. $(x+7)^{1/2} + x^{1/2} = 7$

58. $(y-3)^{1/2} + (y+4)^{1/2} = 7$

59. $\sqrt[3]{x+1} = 2$

60. $x^{2/3} + 2 = 6$

61. $(x-1)^{-3/2} = \dfrac{1}{8}$

62. $(2x+1)^{-1/2} = \dfrac{1}{3}$

Solve each formula in Exercises 63–66 for the indicated variable.

63. $t = \sqrt{\dfrac{2v}{g}}$ for g

64. $q - 1 = 2\sqrt{\dfrac{r^2 - 1}{3}}$ for r

65. $R = \dfrac{1 + \sqrt{p^2 + 1}}{2}$ for p

66. $q = \sqrt[3]{\dfrac{1 + r^2}{2}}$ for r

67. The speed of light is approximately 186,000 miles per second. How long will it take light to travel a distance of 1 foot? (1 mile = 5280 feet) Express your answer in both scientific and standard notation.

68. The national debt is approaching 5×10^{12} dollars. How many hours would it take you to earn an amount equal to the national debt if you were paid $20 per hour? Express your answer in standard notation, in terms of both hours and years.

69. The land area of the earth is 57,267,400 square miles. In the year 2000 the population of the earth will be 6,100,000,000.

a. Express both of these numbers in scientific notation.

b. In the year 2000, how many people will there be for each square mile of the earth's surface?

70. The average distance from the sun to the earth is 92,956,000 miles. Sunlight travels at 186,000 miles per second.

a. Express each of these numbers in scientific notation.

b. How long does it take sunlight to reach the earth?

71. According to the theory of relativity, the mass of an object traveling at velocity v is given by the function

$$m = \dfrac{M}{\sqrt{1 - \dfrac{v^2}{c^2}}}$$

where M is the mass of the object at rest and c is the speed of light. Find the mass of a man traveling at a velocity of $0.7c$ if his rest mass is 80 kilograms (176 pounds).

72. The cylinder of smallest surface area for a given volume has a radius and height both equal to $\sqrt[3]{\dfrac{V}{\pi}}$. Find the dimensions of the tin can of smallest surface area with volume 60 cubic inches.

73. Membership in the Wildlife Society has grown according to the function

$$M(t) = 30t^{3/4}$$

where t is the number of years since the society's founding in 1970.

a. What was the society's membership in 1990?

b. In what year will the society's membership reach 810 people?

74. The heron population is Saltmarsh Refuge is estimated by conservationists at

$$P(t) = 360t^{-2/3}$$

where t is the number of years since the refuge was established in 1990.

a. How many heron were there in 1995?

b. In what year will there be only 40 heron left?

75. Two businesswomen start a small company to produce saddlebags for bicycles. The number, q, of saddlebags they can produce depends on the amount of money, m, they invest and the number of hours of labor, w, they employ, according to the formula

$$q = 0.6\, m^{1/4} w^{3/4}$$

where m is measured in thousands of dollars.

a. If the businesswomen invest $100,000 and employ 1600 hours of labor in their first month of production, how many bicycle saddlebags can they expect to produce?

b. With the same initial investment, how many hours of labor would they need in order to produce 200 bicycle saddlebags?

76. A child who weighs w pounds and is h inches tall has a surface area (in square inches) given approximately by

$$S = 8.5h^{0.35}w^{0.55}$$

a. What is the surface area of a child who weighs 60 pounds and is 40 inches tall?

b. What is the weight of a child who is 50 inches tall and whose surface area is 397 square inches?

77. Two highways intersect at right angles as shown in Figure 6.14. At the instant a car heading east at 50 miles per hour passes the intersection, a car traveling north at 40 miles per hour is already 5 miles north of the intersection. When will the cars be 200 miles apart?

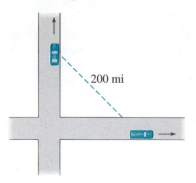

Figure 6.14

78. Two radio antennae are 75 feet apart. They are supported by a guy wire attached to the first antenna at a height of 20 feet, anchored to the ground between the antennae, and attached to the second antenna at a height of 25 feet. If the wire is 90 feet long, how far from the base of the second antenna should it be anchored to the ground? (See Figure 6.15.)

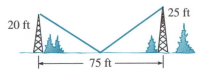

Figure 6.15

79. Find the perimeter of the triangle with vertices $A(-1, 2)$, $B(5, 4)$, $C(1, -4)$. Determine whether $\triangle ABC$ is a right triangle.

80. Find the perimeter of the triangle obtained by joining the midpoints of the sides of $\triangle ABC$ from Exercise 79.

Graph each equation in Exercises 81–84.

81. $x^2 + y^2 = 81$ **82.** $3x^2 + 3y^2 = 12$

83. $(x - 2)^2 + (y + 4)^2 = 9$

84. $x^2 + y^2 + 6y = 0$

In Exercises 85–88 write the equation for the circle with the given properties.

85. Center at $(5, -2)$, radius $4\sqrt{2}$.

86. Center at $(7, -1)$, one point of the circle $(2, 3)$.

87. Endpoints of diameter at $(-2, 3)$ and $(4, 5)$.

88. Passing through $(6, 2)$, $(-1, 1)$, and $(0, -6)$.

Exponential and Logarithmic Functions

*I*n previous chapters we considered several kinds of functions, including linear and quadratic functions and some simple examples of radical functions. We now turn to another class of functions, the exponential functions, which are used to model relationships in many fields, from biology to business.

7.1 EXPONENTIAL GROWTH AND DECAY

We first consider some examples of population growth.

Investigation #1

In a laboratory experiment a colony of 100 bacteria is established and the growth of the colony is monitored. The experimenters discover that the colony's population triples every day.

1. Fill in Table 7.1, showing the population, $P(t)$, of bacteria t days later.

t	$P(t)$
0	100
1	
2	
3	
4	
5	

$P(0) = 100$

$P(1) = 100 \cdot 3 =$

$P(2) = (100 \cdot 3) \cdot 3 =$

$P(3) =$

$P(4) =$

$P(5) =$

Table 7.1

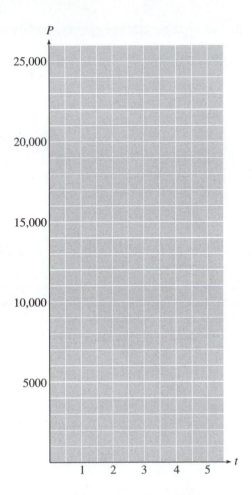

Figure 7.1

2. Plot the data points from Table 7.1 on Figure 7.1 and connect them with a smooth curve.

3. Write a function that gives the population of the colony at any time, t, in days. (Express the values you calculated in part (1) using powers of 3. Do you see a connection between the value of t and the exponent on 3?)

4. Graph your function from part (3) using a calculator. (Use Table 7.1 to choose an appropriate domain and range.) The graph should resemble your hand-drawn graph from part (2).

5. Evaluate the function to find the number of bacteria present after 8 days. How many bacteria are present after 36 hours?

 *Investigation #2*

Under ideal conditions the number of rabbits in a certain area can double every 3 months. A rancher estimates that 60 rabbits live on his land.

1. Fill in Table 7.2, showing the population, $P(t)$, of rabbits t months later.

t	$P(t)$
0	60
3	
6	
9	
12	
15	

$P(0)\ = 60$

$P(3)\ = 60 \cdot 2 =$

$P(6)\ = (60 \cdot 2) \cdot 2 =$

$P(9)\ =$

$P(12) =$

$P(15) =$

Table 7.2

2. Plot the data points from Table 7.2 on Figure 7.2 and connect them with a smooth curve.

3. Write a function that gives the population of rabbits at any time, t, in months. (Express the values you calculated in part (1) using powers of 2. Do you see a connection between the value of t and the exponent on 2?) Note that the population of rabbits is multiplied by 2 every 3 months.

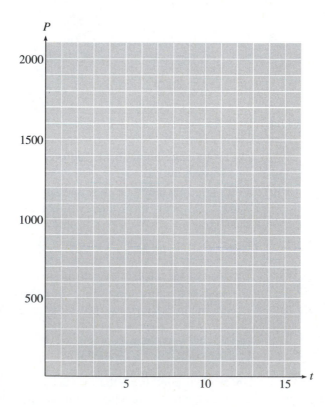

Figure 7.2

4. Graph your function from part (3) using a calculator. (Use Table 7.2 to choose an appropriate domain and range.) The graph should resemble your hand-drawn graph from part (2).

5. Evaluate your function to find the number of rabbits present after 2 years. How many rabbits are present after 8 months?

The functions we just examined describe **exponential growth**. During each time interval of a fixed length, the population is *multiplied* by a certain constant amount. In Investigation #1 the bacteria population grew by a factor of 3 every day. We call 3 the **growth factor** for the function. Functions that describe exponential growth can be put in the standard form

$$P(t) = P_0 a^t$$

where $P_0 = P(0)$ is the initial value of the function and a is the growth factor. In Investigation #1 we have

$$P(t) = 100 \cdot 3^t$$

so $P_0 = 100$ and $a = 3$.

In Investigation #2 the rabbit population grew by a factor of 2 every 3 months. We can write the function as

$$P(t) = 60 \cdot 2^{t/3}$$
$$= 60 \cdot (2^{1/3})^t$$

> Recall the third law of exponents:
> $(2^{1/3})^t = 2^{t(1/3)} = 2^{t/3}$

so that the initial value of the function is $P_0 = 60$ and the growth factor is $a = 2^{1/3}$.

If the units are the same, a population with a larger growth factor grows faster than one with a smaller growth factor.

Example 1

A lab technician compares the growth of two species of bacteria. She starts two colonies of 50 bacteria each. Species A doubles in population every 2 days, and species B triples every 3 days. Find the growth factor for each species.

Solution A function describing the growth of species A is

$$P(t) = 50 \cdot 2^{t/2} = 50 \cdot (2^{1/2})^t$$

so the growth factor for species A is $2^{1/2}$, or approximately 1.41. For species B,

$$P(t) = 50 \cdot 3^{t/3} = 50 \cdot (3^{1/3})^t$$

so the growth factor for species B is $3^{1/3}$, or approximately 1.44. Thus species B grows faster than species A.

Percent Increase

Other phenomena besides populations can exhibit exponential growth. For example, if the interest on a savings account is compounded annually, the amount of money in the account grows exponentially.

Consider a principal of $100 invested at 5% interest compounded annually. At the end of 1 year, the amount is

$$A = P + Prt$$
$$= P(1 + rt)$$
$$= 100[1 + 0.05(1)]$$
$$= 100(1.05) = \mathbf{105}$$

This amount becomes the new principal, so at the end of the second year,

$$A = P(1 + rt)$$
$$= \mathbf{105}[1 + 0.05(1)]$$
$$= 105(1.05) = 110.25$$

To find the amount at the end of the year, we multiply the principal by a factor of $1 + r = 1.05$. Thus we can express the amount at the end of the second year as

$$A = [100(1.05)](1.05)$$
$$= 100(1.05)^2$$

We organize our results into Table 7.3.

	Principal	Amount
First year	100	$100(1.05)$
Second year	$100(1.05)$	$[100(1.05)](1.05) = 100(1.05)^2$
Third year	$100(1.05)^2$	$[100(1.05)^2](1.05) = 100(1.05)^3$

Table 7.3

Continuing in this way, we find that the amount of money accumulated after t years is $100(1.05)^t$. In general, for an initial investment of P dollars at an interest rate $100r$ compounded annually, the amount accumulated after t years is

$$A(t) = P(1 + r)^t$$

This function describes exponential growth with an initial value of P and a growth factor of $a = 1 + r$. Note that the interest rate $100r$, which indicates the *percent increase* in the account each year, corresponds to a *growth factor* of $1 + r$. This notion of percent increase is often used to describe the growth factor for quantities that grow exponentially.

Example 2

During a period of rapid inflation, prices rose by 12% over 6 months. At the beginning of the inflationary period, a pound of butter cost $2.

a. Write a function that gives the price of a pound of butter t years after inflation began.

b. How much did a pound of butter cost after 3 years? After 15 months?

c. Graph the function found in part (a).

Solutions a. The percent increase in the cost of butter is 12% every 6 months. Therefore, the growth factor for the cost of butter is $1 + 0.12 = 1.12$ every half year. If P represents the price of butter after t years, then

$$\text{when } t = 0, \quad P = 2$$

$$\text{when } t = \frac{1}{2}, \quad P = 2(1.12)$$

$$\text{when } t = 1, \quad P = 2(1.12)^2$$

$$\text{when } t = \frac{3}{2}, \quad P = 2(1.12)^3$$

$$\text{when } t = 2, \quad P = 2(1.12)^4$$

In general, after t years of inflation, the original price of $2 has been multiplied $2t$ times by a factor of 1.12. Thus $P = 2(1.12)^{2t}$.

b. To find the price of a pound of butter at any time after inflation began, evaluate the function at the appropriate value of t. After 3 years the price was

$$P(3) = 2(1.12)^{2(3)}$$

$$= 2(1.12)^6 \approx 3.95$$

After 15 months, or 1.25 years, the price was

$$P(1.25) = 2(1.12)^{2(1.25)}$$

$$= 2(1.12)^{2.5} \approx 2.66$$

c. To graph the function $P(t) = 2(1.12)^{2t}$, evaluate it for several values. (See Table 7.4.) Then connect the points with a smooth curve to obtain the graph, as shown in Figure 7.3.

t	$P(t)$
0	2.00
1	2.51
2	3.15
3	3.95
4	4.95

Table 7.4

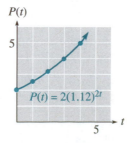

$P(t) = 2(1.12)^{2t}$

Figure 7.3

Exponential Decay   In the examples above, exponential growth was modeled by increasing functions of the form

$$P(t) = P_0 a^t$$

where $a > 1$. The function $P(t) = P_0 a^t$ is a *decreasing* function if we have $0 < a < 1$. In this case the function is said to describe exponential decay, and the constant a is called the **decay factor**.

 *Investigation #3*

A small coal-mining town has been losing population since 1940, when 5000 people lived there. At each census thereafter (taken at 10-year intervals), the population has been approximately 0.90 of its earlier figure.

1. Fill in Table 7.5, showing the population, $P(t)$, of the town t years after 1940.

t	$P(t)$
0	5000
10	
20	
30	
40	
50	

$P(0) = 5000$

$P(10) = 5000 \cdot 0.90 =$

$P(20) = (5000 \cdot 0.90) \cdot 0.90 =$

$P(30) =$

$P(40) =$

$P(50) =$

Table 7.5

2. Plot the data points from Table 7.5 on Figure 7.4 and connect them with a smooth curve.

3. Write a function that gives the population of the town at any time, t, in years after 1940. (Express the values you calculated in part (1) using powers of 0.90. Do you see a connection between the value of t and the exponent on 0.90?)

4. Graph your function from part (3) using a calculator. (Use Table 7.5 to choose an appropriate domain and range.) The graph should resemble your hand-drawn graph from part (2).

5. Evaluate your function to find the population of the town in 1990. What was the population in 1995?

On page 396 we noted that a percent increase of $100r$ corresponds to a growth factor of $a = 1 + r$. A percent decrease of $100r$ in a quantity corresponds to a decay factor of $a = 1 - r$.

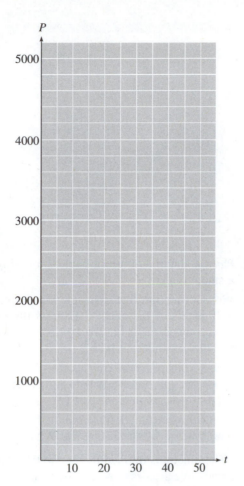

Figure 7.4

Investigation #4

A plastic window coating 1 millimeter thick decreases the light coming through the window by 25%. This means that 75% of the original amount of light comes through 1 millimeter of the plastic. Each additional millimeter reduces the light by another 25%.

1. Fill in Table 7.6, showing the percent of the light, $P(x)$, that shines through x millimeters of the window coating.

$P(0) = 100$

$P(1) = 100 \cdot 0.75 =$

$P(2) = (100 \cdot 0.75) \cdot 0.75 =$

$P(3) =$

$P(4) =$

$P(5) =$

x	P(x)
0	100
1	
2	
3	
4	
5	

Table 7.6

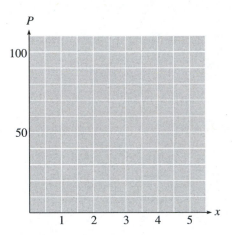

Figure 7.5

2. Plot the data points from Table 7.6 on Figure 7.5 and connect them with a smooth curve.

3. Write a function that gives the percent of the light that shines through x millimeters of the plastic. (Express the values you calculated in part (1) using powers of 0.75. Do you see a connection between the value of t and the exponent on 0.75?)

4. Graph your function from part (3) using a calculator. (Use Table 7.6 to choose an appropriate domain and range.) The graph should resemble your hand-drawn graph from part (2).

5. Evaluate your function to find the percent of the light that comes through 6 millimeters of plastic coating. What percent comes through $\frac{1}{2}$ millimeter?

    The functions defined by the models above are examples of exponential functions. We discuss these functions in more detail in the following sections.

Linear Growth and Exponential Growth

It may be helpful to compare the notions of linear growth and exponential growth. Consider the two functions

$$L(t) = 5 + 2t \quad \text{and} \quad E(t) = 5 \cdot 2^t \quad (t \geq 0)$$

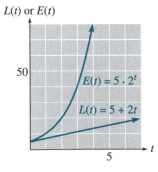

Figure 7.6

whose graphs are shown in Figure 7.6.

    L is a linear function with y-intercept 5 and slope 2; E is an exponential function with initial value 5 and growth factor 2. The growth factor of an exponential function is in a sense analogous to the slope of a linear function: each measures how quickly the function is increasing (or decreasing). However, for each unit that t increases, 2 units are *added* to the value of $L(t)$, whereas the value of $E(t)$ is *multiplied* by 2. An exponential function with growth factor 2 grows much more rapidly than a linear function with slope 2.

Example 3

A solar energy company sold $80,000 worth of solar collectors last year, its first year in operation. This year its sales rose to $88,000, an increase of 10%. The marketing department must estimate its projected sales for the next 3 years.
a. If the marketing department predicts that sales will grow linearly, what should it expect the sales total to be next year? Graph the projected sales figures over the next 3 years assuming that sales will grow linearly.
b. If the marketing department predicts that sales will grow exponentially, what should it expect the sales total to be next year? Graph the projected sales figures over the next 3 years assuming that sales will grow exponentially.

Solutions a. Let $L(t)$ represent the company's total sales t years after starting business, where $t = 0$ is considered the first year of operation, and assume that sales grow linearly. Then L is a linear function of the form $L(t) = mt + b$. Since $L(0) = 80,000$, the intercept b is 80,000. The slope, m, of the function is

$$\frac{\Delta S}{\Delta t} = \frac{8000 \text{ dollars}}{1 \text{ year}} = 8000$$

where $\Delta S = 8000$ is the increase in sales during the first year. Thus $L(t) = 8000t + 80,000$, and the expected sales total for the next year is

$$L(2) = 8000(2) + 80,000 = 96,000$$

b. Let $E(t)$ represent the company's sales under the assumption that sales will grow exponentially. Then E is a function of the form $E(t) = E_0 a^t$. The percent increase in sales over the first year was $r = 0.10$, so the growth factor is $a = 1 + r = 1.10$. The initial value E_0 of the function is 80,000. Thus $E(t) = 80,000(1.10)^t$, and the expected sales total for the next year is

$$E(2) = 80,000(1.10)^2 = 96,800$$

Evaluate each function at several points (as in Table 7.7) to obtain the graphs shown in Figure 7.7.

t	$L(t)$	$E(t)$
0	80,000	80,000
1	88,000	88,000
2	96,000	96,800
3	104,000	106,480
4	112,000	117,128

Table 7.7

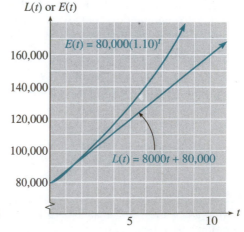

Figure 7.7

Exercises 7.1

For Exercises 1–10
a. make a table of values for the situation described.
b. write a function that describes exponential growth.
c. graph the function.
d. evaluate the function at the given values.

1. A colony of bacteria starts with 300 organisms and doubles every week. How many bacteria will there be after 8 weeks? After 5 days?

2. A population of 24 fruit flies triples every month. How many fruit flies will there be after 6 months? After 3 weeks? (Assume that a month equals 4 weeks.)

3. A typical beehive contains 20,000 insects. The population can increase in size by a factor of 2.5 every 6 weeks. How many bees will there be after 4 weeks? After 20 weeks?

4. A rancher who started with 800 head of cattle finds that his herd increases by a factor of 1.8 every 3 years. How many head of cattle will he have after 1 year? After 10 years?

5. A sum of $4000 is invested in an account that pays 8% interest compounded annually. How much is in the account after 2 years? After 10 years?

6. Otto invests $600 in an account that pays 7.3% interest compounded annually. How much is in Otto's account after 3 years? After 6 years?

7. Since 1963, housing prices have risen an average of 5% per year. Paul bought a house for $20,000 in 1963. How much was the house worth in 1975? In 1990?

8. Housing prices in Los Angeles rose an average of 10% per year from 1982 to 1990. If Marlene bought a house for $135,000 in 1982, how much was the house worth in 1985? In 1989?

9. Sales of Windsurfers have increased 12% per year since 1990. If Sunsails sold 1500 Windsurfers in 1990, how many did it sell in 1995? How many should it expect to sell in 2002?

10. Sales of personal computers have increased 2.3% per year since 1990. If Compucalc sold 500 personal computers in 1990, how may did it sell in 1995? How many should it expect to sell in 2005?

For Exercises 11–16
a. make a table of values for the situation described.
b. write a function that describes exponential decay.
c. graph the function.
d. evaluate the function at the given values.

11. During a vigorous spraying program, the mosquito population was reduced to three-fourths of its previous size every 2 weeks. If the mosquito population was originally estimated at 250,000, how many mosquitoes remained after 3 weeks of spraying? After 8 weeks?

12. The number of perch in Hidden Lake has declined to half of its previous value every 5 years since 1960, when the perch population was estimated at 8000. How many perch were there in 1970? In 1988?

13. Scuba divers find that the water in Emerald Lake filters out 15% of the sunlight for each 4 feet they descend. How much sunlight penetrates to a depth of 20 feet? Of 45 feet?

14. Arch's motorboat cost $15,000 in 1980 and has depreciated by 10% every 3 years. How much was the boat worth in 1989? In 1990?

15. Plutonium 238 is a radioactive element that decays over time into a less harmful element at a rate of 0.8% per year. A power plant has 50 pounds of plutonium 238 to dispose of. How much plutonium 238 will be left after 10 years? After 100 years?

16. Iodine 131 is a radioactive element that decays at a rate of 8.3% per day. How much of a 12-gram sample will be left after 1 week? After 15 days?

In Exercises 17 and 18 calculate the growth factor for each population to determine which one grows faster.

17. A researcher starts two populations of fruit flies of different species, each with 30 flies. Species A increases by 30% in 6 days, and species B increases by 20% in 4 days. Which species multiplies more rapidly?

18. A biologist isolates two strains of a particular virus and monitors the growth of each, starting with samples of 0.01 gram. Strain A increases by 10% in 8 hours, and strain B increases by 12% in 9 hours. Which strain grows more rapidly?

Compare linear and exponential growth in Exercises 19 and 20.

19. At a large university six students start a rumor that final exams have been canceled. After 2 hours nine students (including the first six) have heard the rumor.

a. Assuming that the rumor grows linearly, write a function that gives the number of students who have heard the rumor at time t. Graph the function.

b. Repeat part (a) assuming that the rumor grows exponentially.

20. Over the weekend the Midland Infirmary identifies five cases of Asian flu. Three days later it has treated a total of nine cases.

a. Assuming that the number of flu cases grows linearly, write at function that gives the number of people infected at time t. Graph the function.

b. Repeat part (a) assuming that the flu grows exponentially.

Answer the questions in Exercises 21 and 22 by writing exponential functions.

21. An eccentric millionaire offers you a summer job for the month of June. She will pay you 2 cents for your first day of work and will double your wages every day thereafter.

a. Make a table showing your wages on each day. Do you see a pattern?

b. Write a function that gives your wages in terms of the number of days you have worked.

c. How much will you make on June 15? On June 30?

22. The king of Persia offered his subject anything he desired in return for services rendered. The subject requested that the king give him an amount of grain calculated as follows: place one grain of wheat on the first square of a chessboard, two grains on the second square, four grains on the third square, and so on, until the entire chessboard is covered.

a. Make a table showing the number of grains of wheat on each square of the chessboard.

b. Write a function for the amount of wheat on each square.

c. How many grains of wheat should be placed on the last (64th) square?

Find the growth rate in Exercises 23–28.

23. The population of Texas was 9,579,700 in 1960. In 1970 the population was 11,196,700. What was the annual growth rate to the nearest hundredth of a percent?

24. The population of Florida was 4,951,600 in 1960. In 1970 the population was 6,789,400. What was the annual growth rate to the nearest hundredth of a percent?

25. a. The population of Rainville was 10,000 in 1970 and doubled in 20 years. What was the annual growth rate to the nearest hundredth of a percent?

b. The population of Elmira was 350,000 in 1970 and doubled in 20 years. What was the annual growth rate to the nearest hundredth of a percent?

c. If a population doubles in 20 years, does the growth rate depend on the size of the original population?

d. The population of Grayling doubled in 20 years. What was the annual growth rate to the nearest hundredth of a percent?

26. a. The population of Boomtown was 300 in 1908 and tripled in 7 years. What was the annual growth rate to the nearest hundredth of a percent?

b. The population of Fairview was 15,000 in 1962 and tripled in 7 years. What was the annual growth rate to the nearest hundredth of a percent?

c. If a population triples in 7 years, does the growth rate depend on the size of the original population?

d. The population of Pleasant Lake tripled in 7 years. What was the annual growth rate to the nearest hundredth of a percent?

27. You receive a chain letter with a list of six names. You are instructed to send $10 to the name at the top of the list, then cross it out and add your name to the bottom of the list. You should then send copies of the letter to six friends.

a. If the chain is not broken, how much money should you receive?

b. If you are one of the six people who receive the original letter, how many people will have received the letter before you start receiving money?

c. Why do you think chain letters are illegal?

28. A friend asks you to take part in a pyramid scheme selling pet supplies. Each salesperson must pay 30% of his monthly earnings, or at least $25, to his manager. If a salesperson can recruit six more salespeople, he becomes their manager and does not have to sell any more pet supplies himself. However, he must still pay 30% of his monthly earnings to *his* manager.

a. What is the minimum amount that a level-1 manager, with six salespeople under him, will make each month? What is the minimum amount that a level-2 manager, with six level-1 managers under him, will make each month? Find the minimum monthly earnings of managers up to level 6.

b. How many salespeople (including managers at lower levels) does a level-6 manager have working under her?

c. Why are pyramid schemes a risky venture?

7.2 EXPONENTIAL FUNCTIONS

In Section 7.1 we studied a number of functions that described exponential growth or decay. More formally, we define an **exponential function** to be one of the form

$$f(x) = b^x \quad \text{where} \quad b > 0 \quad \text{and} \quad b \neq 1$$

The positive constant b is called the **base** of the exponential function. We do not allow $b < 0$ as a base because if b is negative, then b^x is not a real number for some values of x. (For example, if $b = -4$, then $f(\frac{1}{2}) = (-4)^{1/2}$ is an imaginary number.) We also exclude $b = 1$ as a base because $1^x = 1$ for all values of x; hence the function $f(x) = 1^x$ is not exponential but is actually the constant function $f(x) = 1$.

If b is a positive number not equal to 1, then b^x is defined for all real values of x. In Chapter 6 we defined b^x for rational values of x by $b^{m/n} = \sqrt[n]{b^m}$. We now assume that powers such as $2^{\sqrt{3}}$ and $(\frac{1}{2})^\pi$, in which the exponent is an *irrational number*, are also defined. Your calculator gives decimal approximations for such powers.

Example 1

Find decimal approximations to three decimal places for the following powers.
a. $2^{\sqrt{3}}$ **b.** $(\frac{1}{2})^{\pi}$

Solutions **a.** A calculator keying sequence for evaluating $2^{\sqrt{3}}$ is

$$2 \; \boxed{\wedge} \; \boxed{\text{2nd}} \; \boxed{\sqrt{}} \; 3 \; \boxed{\text{ENTER}}$$

which yields 3.321997085. Thus to three decimal places, $2^{\sqrt{3}} \approx 3.322$.
b. A calculator keying sequence for evaluating $(\frac{1}{2})^{\pi}$ is

$$0.5 \; \boxed{\wedge} \; \boxed{\text{2nd}} \; \boxed{\pi} \; \boxed{\text{ENTER}}$$

which yields 0.1133147323. Thus to three decimal places, $(\frac{1}{2})^{\pi} \approx 0.113$.

Graphs of Exponential Functions

The graphs of exponential functions have two characteristic shapes, depending on whether the base, b, is greater than 1 or less than 1. As typical examples consider the graphs of $f(x) = 2^x$ and $g(x) = (\frac{1}{2})^x$ shown in Figure 7.8. (Some values for f and g are shown in Tables 7.8 and 7.9.)

x	$f(x)$
-3	1/8
-2	1/4
-1	1/2
0	1
1	2
2	4
3	8

Table 7.8

x	$g(x)$
-3	8
-2	4
-1	2
0	1
1	1/2
2	1/4
3	1/8

Table 7.9

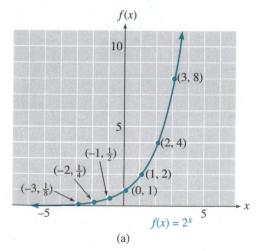

(a)

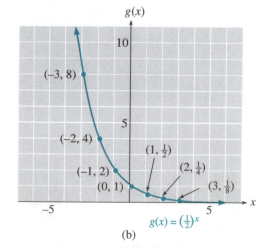

(b)

Figure 7.8

Notice that $f(x) = 2^x$ is an increasing function and $g(x) = (\frac{1}{2})^x$ is a decreasing function. In general, exponential functions have the following properties.

PROPERTIES OF EXPONENTIAL FUNCTIONS, $f(x) = b^x$

1. Domain: all real numbers.

2. Range: all positive numbers.

3. If $b > 1$, the function is increasing; if $0 < b < 1$, the function is decreasing.

Notice also that the negative x-axis is an asymptote for exponential functions with $b > 1$; for exponential functions with $0 < b < 1$, the positive x-axis is an asymptote.

Example 2

Compare the graphs of $f(x) = 3^x$ and $g(x) = 4^x$ in Figure 7.9.

Solution   Evaluate each function for several convenient values, as shown in Table 7.10.

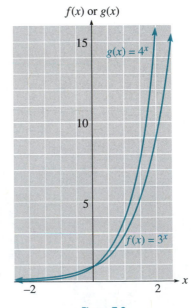

x	$f(x)$	$g(x)$
-2	1/9	1/16
-1	1/3	1/4
0	1	1
1	3	4
2	9	16

Table 7.10   **Figure 7.9**

Plot the points for each function and connect them with smooth curves. Notice that $g(x) = 4^x$ grows more rapidly than $f(x) = 3^x$. Both graphs cross the y-axis at $(0, 1)$.

Example 3

Use a graphing calculator to graph the following functions. Describe how these graphs compare with the graph of $y = 2^x$.

a. $y = 2^x + 3$ **b.** $y = 2^{x+3}$

Solutions   The graphs of $y = 2^x$, $y = 2^x + 3$, and $y = 2^{x+3}$ are shown in Figure 7.10 using a graphing window with Xmin $= -10$, Xmax $= 10$, Ymin $= 0$, Ymax $= 10$.

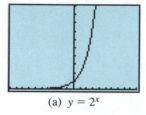
(a) $y = 2^x$

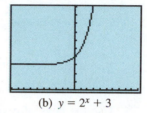
(b) $y = 2^x + 3$

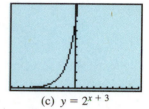
(c) $y = 2^{x+3}$

Figure 7.10

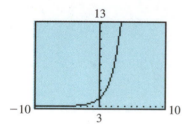

Figure 7.11

a. The graph of $y = 2^x + 3$ has the same basic shape as that of $y = 2^x$, but it has a horizontal asymptote at $y = 3$ instead of $y = 0$ (the x-axis). If every point on the graph of $y = 2^x$ were to move three units upward, the result would be the graph of $y = 2^x + 3$ shown in Figure 7.10(b). We can more easily see that the two graphs have the same shape if we graph $y = 2^x + 3$ using the window Xmin $= -10$, Xmax $= 10$, Ymin $= 3$, Ymax $= 13$, as shown in Figure 7.11.

b. The graph of $y = 2^{x+3}$ also has the same basic shape as that of $y = 2^x$, but this time the graph has been translated horizontally. If every point on the graph of $y = 2^x$ were to be moved three units to the left, the result would be the graph of $y = 2^{x+3}$ shown in Figure 7.10(c).

Exponential Equations   An **exponential equation** is one in which the variable is part of an exponent. For example, the equation

$$3^x = 81 \qquad (1)$$

is exponential. Many exponential equations can be solved by writing both sides of the equation as powers with the same base. To solve Equation (1), we would write

$$3^x = 3^4$$

which is true if and only if $x = 4$. In general, if two equivalent powers have the same positive base, then their exponents must be equal also.

Sometimes the laws of exponents can be used to express both sides of an equation as single powers of a common base.

Example 4

Solve the following equations.
a. $3x^{-2} = 9^3$ **b.** $27 \cdot 3^{-2x} = 9^{x+1}$

Solutions   **a.**   Using the fact that $9 = 3^2$,   write each side of the equation as a power of 3:

$$3^{x-2} = (3^2)^3$$
$$3^{x-2} = 3^6$$

Now equate the exponents to obtain

$$x - 2 = 6$$
$$x = 8$$

b.   Write each factor as a power of 3:

$$3^3 \cdot 3^{-2x} = (3^2)^{x+1}$$

Use the laws of exponents to simplify each side:

$$3^{3-2x} = 3^{2x+2}$$

Now equate the exponents to obtain

$$3 - 2x = 2x + 2$$
$$-4x = -1$$
$$x = \frac{1}{4}$$

Exponential equations arise frequently in the study of exponential growth.

Example 5

During the summer a population of fleas doubles every 5 days. If a population starts with 10 fleas, how long will it be before there are 10,240 fleas?

Solution   Let P represent the number of fleas present after t days. The original population of 10 is multiplied by a factor of 2 every 5 days, or

$$P(t) = 10 \cdot 2^{t/5}$$

Set $P = 10{,}240$ and solve for t:

$$10{,}240 = 10 \cdot 2^{t/5}$$

Divide both sides by 10 to obtain

$$1024 = 2^{t/5}$$
$$2^{10} = 2^{t/5}$$

Equate the exponents to get

$$10 = \frac{t}{5}$$

or $t = 50$.   The population will grow to 10,240 fleas in 50 days.

It is not always so easy to express both sides of an equation as powers of the same base. In the following sections we will develop more general methods for finding exact solutions to exponential equations. But we can use a graphing calculator to obtain approximate solutions.

Example 6

Find an approximate solution to the equation $2^x = 5$ that is accurate to the nearest hundredth using the graph of $y = 2^x$ with the ZOOM and TRACE keys.

Solution Set $Y_1 = 2$ ^ X and use the standard graphing window (ZOOM 6) to get a graph like the one shown in Figure 7.12(a). We are looking for a point on this graph that has a y-coordinate of 5. Using the TRACE button and the arrow keys, we see that the y-coordinates are too small when $x < 2.1$ and too large when $x > 2.3$. Thus the value we want lies somewhere between $x = 2.1$ and $x = 2.3$. But this approximation is not accurate enough.

Use the **Zoom Box** option to zoom into a region of the curve between $x = 2.1$ and $x = 2.4$. Depending on the box we choose, we find that the point we want lies between $x = 2.32$ and $x = 2.33$. With another zoom we can see that the point we want lies between $x = 2.321$ and $x = 2.323$, so that to the nearest hundredth, $x \approx 2.32$. (See Figure 17.12[b].)

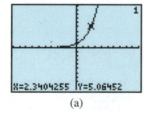
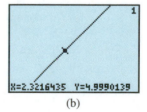

Figure 7.12 (a) (b)

We can verify that our estimate is reasonable by checking to see if it is a solution to the equation.

$$2^{2.32} \overset{?}{=} 5$$

We enter 2 ^ 2.32 ENTER on our calculator to get 4.993322196. This number is not equal to 5, but it is close, so we believe that $x = 2.32$ is a reasonable approximation to the solution of the equation $2^x = 5$.

Exponential and Polynomial Growth

We have seen that an exponential function with a growth factor 2 grows much more rapidly than a linear function with slope 2. Any exponential growth function will eventually grow faster than any linear function. In fact, exponential growth functions grow faster than *all* polynomial functions.

Example 7

a. Find a graphing window that shows that the function $y = 2^x$ eventually overtakes and continues to grow more rapidly than the function $y = 10x$.

b. Find a graphing window that shows that the function $y = 2^x$ eventually overtakes and continues to grow more rapidly than the function $y = x^3$.

Solutions a. Set $Y_1 = 10X$ and $Y_2 = 2\boxed{\wedge}X$ in the standard graphing window ($\boxed{\text{ZOOM}}$ 6). The linear function $y = 10x$ goes off the top of the display before the exponential function $y = 2^x$. (See Figure 7.13[a].) But if we set $Xmin = 0$, $Xmax = 10$, $Ymin = 0$, $Ymax = 100$, we get a graph like Figure 7.13(b), in which the curve is clearly above the straight line after about $x = 6$.

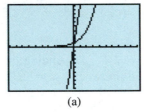 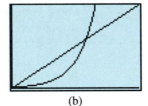

Figure 7.13   (a)   (b)

b. Set $Y_1 = X\boxed{\wedge}3$ and $Y_2 = 2\boxed{\wedge}X$. In the standard graphing window, the cubic function $y = x^3$ goes off the top of the display before the exponential function $y = 2^x$. (See Figure 7.14[a].) But if we set $Xmin = 0$, $Xmax = 20$, $Ymin = 0$, $Ymax = 5000$, we get a graph like Figure 7.14(b), in which the exponential curve is clearly above the graph of $y = x^3$ after about $x = 10$. (*Note:* The calculator normally graphs Y_1 before graphing Y_2, so you can tell which curve is which. If the two curves are graphed simultaneously, press the $\boxed{\text{MODE}}$ key, select **Sequential** and then press the $\boxed{\text{GRAPH}}$ key.)

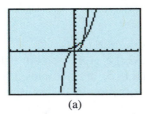 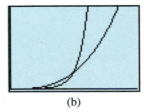

Figure 7.14   (a)   (b)

Exercise 7.2

Find decimal approximations for the expressions in Exercises 1–12.

1. $3^{\sqrt{2}}$

2. $8^{\sqrt{5}}$

3. $4^{\pi-1}$

4. $5^{\pi+2}$

5. $-0.6^{2\sqrt{3}}$

6. $-1.2^{\sqrt{13}/2}$

7. $6^{-\sqrt{5}}$

8. $10^{-\sqrt{2}}$

9. $2.8(9)^{\sqrt{7}}$

10. $0.3(11)^{\sqrt{3}}$

11. $8 - 4^{\sqrt{13}}$

12. $9 - 7^{\sqrt{8}}$

Graph the functions in Exercises 13–22 by making a table of values and plotting points by hand. Choose appropriate scales for the axes.

13. $f(x) = 5^x$

14. $g(x) = 10^x$

15. $h(t) = 3^{-t}$

16. $q(t) = 5^{-t}$

17. $G(z) = -4^z$

18. $F(z) = -3^z$

19. $p(x) = \left(\dfrac{1}{10}\right)^x$

20. $R(x) = \left(\dfrac{1}{4}\right)^x$

21. $y = \left(\dfrac{1}{2}\right)^{-x}$

22. $y = \left(\dfrac{1}{3}\right)^{-x}$

Use a graphing calculator to obtain the graphs in Exercises 23–26.

23. $g(t) = 1.3^t$

24. $h(t) = 2.4^t$

25. $N = 0.8^x$

26. $P = 0.7^x$

In Exercises 27–34 use a calculator to graph each pair of functions in the same window. Describe how the second graph differs from the first.

27. $y_1 = 3^x$
$y_2 = 3^x - 5$

28. $y_1 = 4^x$
$y_2 = 4^x + 2$

29. $y_1 = 4^t$
$y_2 = 4^{t-3}$

30. $y_1 = 3^t$
$y_2 = 3^{t+2}$

31. $f(h) = 2^{-h}$
$g(h) = 1 - 2^{-h}$

32. $g(p) = 5^{-p}$
$h(p) = 10 - 5^{-p}$

33. $N(t) = 10^t$
$M(t) = 20 + 10^{t+2}$

34. $N(t) = 10^t$
$M(t) = 10^{t-3} - 50$

Solve each equation in Exercises 35–46 algebraically.

35. $2^x = 32$

36. $5^x = 125$

37. $5^{x+2} = 25^{4/3}$

38. $3^{x-1} = 27^{1/2}$

39. $3^{2x-1} = \dfrac{\sqrt{3}}{9}$

40. $2^{3x-1} = \dfrac{\sqrt{2}}{16}$

41. $4 \cdot 2^{x-3} = 8^{-2x}$

42. $9 \cdot 3^{x+2} = 81^{-x}$

43. $27^{4x+2} = 81^{x-1}$

44. $16^{2-3x} = 64^{x+5}$

45. $10^{x^2-1} = 1000$

46. $5^{x^2-x-4} = 25$

47. During an introductory advertising campaign in a large city, the makers of Chip-O's corn chips estimate that the number of people who have heard of Chip-O's increases by a factor of 8 every 4 days.

   a. If 100 people are given trial bags of Chip-O's to start the campaign, write a function $N(t)$ for the number of people who have heard of Chip-O's after t days of advertising.

   b. Graph the function $N(t)$.

   c. How many days should the campaign run in order for Chip-O's to be familiar to 51,200 people? Use algebraic methods to find your answer, and verify it on your graph.

48. A nationwide association of cosmetologists finds that news of a new product will spread among its members so that the number of cosmetologists who have tried the product is increased by a factor of 9 every 5 weeks.

   a. If 20 cosmetologists try a new product at its launching, write a function $N(t)$ for the number of cosmetologists who have tried the product after t weeks.

   b. Graph the function $N(t)$.

   c. How long will it be before 14,580 cosmetologists have tried a new product? Use algebraic methods to find your answer, and verify it on your graph.

49. Before the advent of antibiotics an outbreak of cholera might spread through a city so that the number of cases doubled every 6 days.

a. Twenty-six cases were discovered on July 5. Write a function for the number of cases of cholera t days later.

b. Graph your function.

c. When should hospitals expect to be treating 106,496 cases? Use algebraic methods to find your answer, and verify it on your graph.

50. An outbreak of ungulate fever can sweep through a region's livestock so that the number of animals affected triples every 4 days.

a. A rancher discovers four cases of ungulate fever among his herd. Write a function for the number of cases of ungulate fever t days later.

b. Graph your function.

c. If the rancher does not act quickly, how long will it be until 324 head are affected? Use algebraic methods to find your answer, and verify it on your graph.

51. A color television set loses 30% of its value every 2 years.

a. Write a function for the value of a television set t years after it was purchased if it cost $700 originally.

b. Graph your function.

c. How long will it be before a $700 television set depreciates to $343? Use algebraic methods to find your answer, and verify it on your graph.

52. A mobile home loses 20% of its value every 3 years.

a. A certain mobile home costs $20,000. Write a function for its value after t years.

b. Graph your function.

c. How long will it be before a $20,000 mobile home depreciates to $12,800? Use algebraic methods to find your answer, and verify it on your graph.

In Exercises 53–60 use a graphing calculator to find an approximate solution accurate to the nearest hundredth.

53. $2^x = 3$ **54.** $3^x = 7$ **55.** $5^x = 2$
56. $2^x = 0.7$ **57.** $3^{x-1} = 4$ **58.** $2^{x+3} = 5$
59. $4^{-x} = 7$ **60.** $6^{-x} = 3$

Which of the tables in Exercises 61 and 62 could describe exponential functions?

61. a.

x	y
0	3
1	6
2	12
3	24
4	48

b.

t	P
0	6
1	7
2	10
3	15
4	22

c.

x	N
0	2
1	6
2	34
3	110
4	258

d.

p	R
0	405
1	135
2	45
3	15
4	5

62. a.

t	y
1	100
2	50
3	$33\frac{1}{3}$
4	25
5	20

b.

x	P
1	1/2
2	1
3	2
4	4
5	8

c.

h	a
0	70
1	7
2	0.7
3	0.07
4	0.007

d.

t	Q
0	0
1	1/4
2	1
3	9/4
4	4

63. The graph of $f(x) = P_0 a^x$ is shown in Figure 7.15.

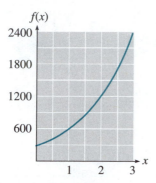

Figure 7.15

a. Read the value of P_0 from the graph.

b. Find the growth factor, a, by comparing two convenient points on the graph.

c. Using your answers to parts (a) and (b), write a formula for $f(x)$.

64. The graph of $g(x) = P_0 a^x$ is shown in Figure 7.16.

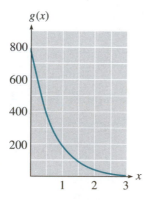

Figure 7.16

a. Read the value of P_0 from the graph.

b. Find the decay factor, a, by comparing two convenient points on the graph.

c. Using your answers to parts (a) and (b), write a formula for $g(x)$.

65. For several days after the Northridge earthquake on January 17, 1994, the area received a number of significant aftershocks. The black graph in Figure 7.17 shows that the number of aftershocks decreased exponentially over time. The graph of the function $S(d) = S_0 a^d$, in color, approximates the data.

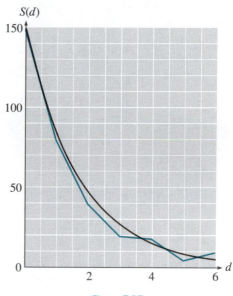

Figure 7.17

a. Read the value of S_0 from the graph.

b. Find the decay factor, a, by comparing two points on the graph. (Some of the points on the graph of $S(d)$ are approximately $(1, 82)$, $(2, 45)$, $(3, 25)$, and $(4, 14)$.)

c. Using your answers to parts (a) and (b), write a formula for $S(d)$.

66. The frequency of a musical note depends on its pitch. The graph in Figure 7.18 shows that the frequency increases exponentially. The function $F(p) = F_0 a^p$ gives the frequency as a function of the number of half tones above the starting point on the scale.

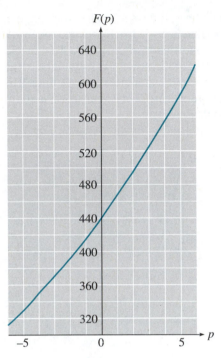

$F(p)$

Figure 7.18

a. Read the frequency F_0 from the graph. (This is the frequency of the note A above middle C.)

b. Find the growth factor, a, by comparing two points on the graph. (Some of the points on the graph of $F(p)$ are approximately $(1, 466)$, $(2, 494)$, $(3, 523)$, and $(4, 554)$.)

c. Using your answers to parts (a) and (b), write a formula for $F(p)$.

d. The frequency doubles when you raise a note by one octave, which is equivalent to 12 half tones. Can you use this information to find an exact value for a?

Fill in the tables in Exercises 67 and 68. Graph each pair of functions on the same set of axes.

67.

x	$f(x) = x^2$	$g(x) = 2^x$
-2		
-1		
0		
1		
2		
3		
4		
5		
6		

68.

x	$f(x) = x^3$	$g(x) = 3^x$
-2		
-1		
0		
1		
2		
3		
4		
5		
6		

7.3   LOGARITHMS

Suppose a colony of bacteria doubles in size every day. If the colony starts with 50 bacteria, how long will it be before there are 800 bacteria?

We solved problems of this type in Section 7.2 by writing and solving an appropriate exponential equation. The function

$$P(t) = 50 \cdot 2^t$$

gives the number of bacteria present on day t, so we must solve the equation

$$800 = 50 \cdot 2^t$$

Dividing both sides by 50 yields

$$16 = 2^t$$

The solution to this equation is the answer to the question "To what power must we raise 2 in order to get 16?" The value of t that solves the equation is called the base 2 **logarithm** of 16. Since $2^t = 16$,   the base 2 logarithm of 16 is 4. We write this as

$$\log_2 16 = 4$$

In general, we make the following definition.

> The **base b logarithm of x**, written **$\log_b x$**, is the exponent to which b must be raised in order to yield x.

Example 1

a. $\log_3 9 = 2$ because $3^2 = 9$

b. $\log_5 125 = 3$ because $5^3 = 125$

c. $\log_4 \dfrac{1}{16} = -2$ because $4^{-2} = \dfrac{1}{16}$

d. $\log_5 \sqrt{5} = \dfrac{1}{2}$ because $5^{1/2} = \sqrt{5}$

Note in particular that

$$\log_b b = 1 \qquad \text{since} \qquad b^1 = b$$
$$\log_b 1 = 0 \qquad \text{since} \qquad b^0 = 1$$

and

$$\log_b b^x = x \qquad \text{since} \qquad b^x = b^x$$

Example 2

a. $\log_2 2 = 1$ **b.** $\log_5 1 = 0$ **c.** $\log_3 3^4 = 4$ ☐

From the definition of a logarithm and the examples above, we see that the statements

$$y = \log_b x \quad \text{and} \quad x = b^y \qquad (2)$$

are equivalent. From Equation (2) we see that the logarithm, y, is the same as the *exponent* in $x = b^y$. Thus a logarithm is an exponent; it is the exponent to which b must be raised to yield x.

Every equation concerning logarithms can be rewritten in exponential form, and vice versa, by using Equation (2). Thus

$$3 = \log_2 8 \quad \text{and} \quad 8 = 2^3$$

are equivalent statement, as are

$$5 = \sqrt{25} \quad \text{and} \quad 25 = 5^2$$

The operation of taking a base b logarithm is the *inverse* of raising the base b to a power, just as extracting square roots is the inverse of squaring a number.

Example 3

Rewrite each equation in exponential form.

a. $\log_{10} 0.001 = z$ **b.** $\log_3 20 = t$
c. $\log_b (3x + 1) = 3$ **d.** $\log_q p = w$

Solutions First identity the base b and then the exponent or logarithm y. Rewrite the expression in the form $x = b^y$.

a. $10^z = 0.001$ **b.** $3^t = 20$
c. $b^3 = 3x + 1$ **d.** $q^w = p$ ☐

Example 4

Rewrite each equation in logarithmic form.

a. $2^{-1} = \dfrac{1}{2}$ **b.** $a^{1/5} = 2.8$

c. $6^{1.5} = T$ **d.** $M^v = 3K$

Solutions First identify the base b and then the exponent or logarithm y. Rewrite the expression in the form $y = \log_b x$.

a. $\log_2 \dfrac{1}{2} = -1$ **b.** $\log_a 2.8 = \dfrac{1}{5}$

c. $\log_6 T = 1.5$ **d.** $\log_M 3K = v$ ☐

Simple equations involving logarithms can sometimes be solved by rewriting them in exponential form.

Example 5

Solve for the unknown value in each equation.

a. $\log_2 x = 3$ **b.** $\log_b 2 = \dfrac{1}{2}$

c. $\log_3 (2x - 1) = 4$ **d.** $2(\log_3 x) - 1 = 4$

Solutions Write each equation in exponential form and solve for the variable.

a. $2^3 = x$ **b.** $b^{1/2} = 2$
      $x = 8$ $(b^{1/2})^2 = 2^2$
                        $b = 4$

c. $2x - 1 = 3^4$ **d.** $2(\log_3 x) = 5$
      $2x = 82$ $\log_3 x = \dfrac{5}{2}$
        $x = 41$ $x = 3^{5/2}$

Approximating Logarithms

Although we can easily evaluate $\log_3 81$ ($\log_3 81 = 4$ because $3^4 = 81$) or $\log_{10} 100$ ($\log_{10} 100 = 2$ because $10^2 = 100$), it is not easy to evaluate an expression like $\log_2 5$. We know that if $\log_2 5 = x$,   then $2^x = 5$.   This shows that $\log_2 5$ is the solution to the equation $2^x = 5$. When we approximated that solution in Example 6 of Section 7.2, we were actually approximating $\log_2 5$. Thus $\log_2 5 \approx 2.32$.

Example 6

Approximate the following logarithms to the nearest hundredth.

a. $\log_3 7$ **b.** $\log_{10} 386$

Solutions **a.** If $\log_3 7 = x$,   then $3^x = 7$.   We will use the graph of $y = 3^x$ to approximate a solution to this exponential equation. Set $Y_1 = 3 \;\boxed{\wedge}\; X$ and start with the standard graphing window ($\boxed{\text{ZOOM}}$ 6). Using the $\boxed{\text{TRACE}}$ feature, look for the point on the graph whose y-coordinate is 7. We see that the point we seek has x-value between 1.7 and 1.9. After more zooming and tracing, we find that x must lie between 1.77 and 1.772, so $\log_3 7 \approx 1.77$.   (See Figure 7.19[a]).

b. If $\log_{10} 386 = x$,   then $10^x = 386$.   We will use the graph of $y = 10^x$ to approximate the solution to this exponential equation. We start with the window

$$\text{Xmin} = 0, \quad \text{Xmax} = 3, \quad \text{Ymin} = 0, \quad \text{Ymax} = 500$$

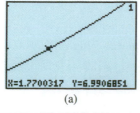
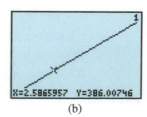

Figure 7.19   (a)   (b)

and graph $Y_1 = 10 \boxed{\wedge} X$. We see that the point on the graph with y-coordinate 386 has an x-value between 2.58 and 2.6. We can zoom in on this portion of the graph just as in part (a), but instead we will try a variation in strategy. This time we press the $\boxed{\text{WINDOW}}$ button and set Xmin = 2.58, Xmax = 2.6, Ymin = 380, and Ymax = 400. Using the $\boxed{\text{TRACE}}$ key, we find that x lies between 2.5865 and 2.5868, so $\log_{10} 386 \approx 2.59$. (See Figure 7.19[b].)

Base 10 Logarithms with a Graphing Calculator

Logarithms are used to solve exponential equations. For instance, to solve the equation

$$16 \cdot 10^t = 360$$

we first divide both sides by 16 to obtain

$$10^t = 22.5$$

and then use Equation (2) on page 416 to rewrite the equation in the form

$$t = \log_{10} 22.5$$

(Recall that $\log_{10} 22.5$ is the exponent to which 10 must be raised in order to yield 22.5.) Because 22.5 is not an integral power of 10, the value of $\log_{10} 22.5$ is not immediately apparent. However, approximate values of $\log_{10} 22.5$ can be found with a graphing calculator using the $\boxed{\text{LOG}}$ key. To find $\log_{10} 22.5$, we press

$$\boxed{\text{LOG}} \ 22.5 \ \boxed{\text{ENTER}}$$

and the calculator displays 1.352182518. The logarithms of many numbers are irrational, so their decimal representations are nonrepeating and nonterminating. The calculator gives a decimal approximation with as many digits as its display will allow. We can then round off the answer to whatever accuracy we need.

Notice that the calculator key is labeled $\boxed{\text{LOG}}$ rather than $\boxed{\log_{10}}$. Base 10 logarithms are used frequently in applications and consequently are called **common logarithms**. The subscript 10 is often omitted, so "$\log x$" is understood to mean "$\log_{10} x$."

Example 7

Approximate the following logarithms to two decimal places.

a. $\log_{10} 6.5$ **b.** $\log_{10} 256$

Solutions   **a.** The sequence $\boxed{\text{LOG}}$ 6.5 $\boxed{\text{ENTER}}$ yields 0.8129133566, so to two decimal places $\log_{10} 6.5 \approx 0.81$.

b. The sequence $\boxed{\text{LOG}}$ 256 $\boxed{\text{ENTER}}$ yields 2.408239965, so to two decimal places $\log_{10} 256 \approx 2.41$.

To solve exponential equations involving powers of 10, we can use the following steps.

STEPS FOR SOLVING EXPONENTIAL EQUATIONS

1. Isolate the power on one side of the equation.

2. Rewrite the equation in logarithmic form.

3. Use a calculator, if necessary, to evaluate the logarithm.

4. Solve for the variable.

Example 8

Solve the equation $38 = 95 - 15 \cdot 10^{0.4x}$.

Solution   First isolate the power of 10: subtract 95 from both sides of the equation and divide by -15 to obtain

$$-57 = -15 \cdot 10^{0.4x}$$

$$3.8 = 10^{0.4x}$$

Rewrite the equation in logarithmic form as

$$\log_{10} 3.8 = 0.4x$$

Solving for x yields

$$\frac{\log_{10} 3.8}{0.4} = x$$

We can evaluate this expression on the calculator by entering

$$\boxed{\text{LOG}}\ 3.8\ \boxed{\div}\ 0.4\ \boxed{\text{ENTER}}$$

which yields   1.449458992.   Thus to two decimal places, $x \approx 1.45$.

Example 9

The value of a large tractor originally worth \$30,000 declines exponentially according to the formula $V(t) = 30{,}000\,(10)^{-0.04t}$,   where t is in years. When will the tractor be worth half its original value?

Solution   We want to find the value of t for which $V(t) = 15{,}000$;   that is, we want to solve the equation

$$15{,}000 = 30{,}000(10)^{-0.04t}$$

Divide both sides by 30,000 to obtain

$$0.5 = 10^{-0.04t}$$

Rewrite the equation in logarithmic form as

$$\log_{10} 0.5 = -0.04t$$

and divide by -0.04 to obtain

$$\frac{\log_{10} 0.5}{-0.04} = t$$

To evaluate this expression, key in

$$\boxed{\text{LOG}} \ 0.5 \ \boxed{\div} \ \boxed{(-)} \ 0.04 \ \boxed{\text{ENTER}}$$

to find $t \approx 7.525749892$. The tractor will be worth \$15,000 in approximately $7\frac{1}{2}$ years.

At this stage you may think we will only be able to solve exponential equations in which the base is 10. In Section 7.5, however, we will see how the properties of logarithms enable us to solve exponential equations with any base, knowing only the values for base 10 logarithms.

Exercise 7.3

In Exercises 1–20 find each logarithm without using a calculator.

1. $\log_7 49$ **2.** $\log_2 32$ **3.** $\log_4 64$

4. $\log_3 27$ **5.** $\log_3 \sqrt{3}$ **6.** $\log_5 \sqrt{5}$

7. $\log_5 \dfrac{1}{5}$ **8.** $\log_3 \dfrac{1}{3}$ **9.** $\log_4 4$

10. $\log_{10} 10$ **11.** $\log_{10} 1$ **12.** $\log_6 1$

13. $\log_8 8^5$ **14.** $\log_7 7^6$

15. $\log_{10} 10^{-4}$ **16.** $\log_{10} 10^{-6}$

17. $\log_{10} 10{,}000$ **18.** $\log_{10} 1000$

19. $\log_{10} 0.1$ **20.** $\log_{10} 0.001$

Rewrite each equation in Exercises 21–30 in exponential form.

21. $\log_{16} 256 = w$ **22.** $\log_9 729 = y$

23. $\log_b 9 = -2$ **24.** $\log_b 8 = -3$

25. $\log_{10} A = -2.3$ **26.** $\log_{10} C = -4.5$

27. $\log_4 36 = 2q - 1$ **28.** $\log_5 3 = 6 - 2p$

29. $\log_u v = w$ **30.** $\log_m n = p$

Rewrite each equation in Exercises 31–40 in logarithmic form.

31. $8^{-1/3} = \dfrac{1}{2}$ **32.** $64^{-1/6} = \dfrac{1}{2}$

33. $t^{3/2} = 16$ **34.** $v^{5/3} = 12$

35. $0.8^{1.2} = M$ **36.** $3.7^{2.5} = Q$

37. $x^{5t} = W - 3$ **38.** $z^{-3t} = 2P + 5$

39. $3^{-0.2t} = 2N_0$ **40.** $10^{1.3t} = 3M_0$

In Exercises 41–52 solve for the unknown value.

41. $\log_b 8 = 3$ **42.** $\log_b 625 = 4$

43. $\log_4 x = 3$ **44.** $\log_{1/2} x = 5$

45. $\log_2 \dfrac{1}{2} = y$ **46.** $\log_5 \dfrac{1}{5} = y$

47. $\log_b 10 = \dfrac{1}{2}$ **48.** $\log_b 0.1 = -1$

49. $\log_2 (3x - 1) = 5$ **50.** $\log_5 (9 - 4x) = 3$

51. $3(\log_7 x) + 5 = 7$

52. $5(\log_2 x) + 6 = -14$

Use a graph to approximate each logarithm in Exercises 53–60 to the nearest hundredth.

53. $\log_2 25$ **54.** $\log_3 100$ **55.** $\log_{10} 50$

56. $\log_{10} 7$ **57.** $\log_8 5$ **58.** $\log_6 24$

59. $\log_3 67.9$ **60.** $\log_5 86.3$

Use a calculator to approximate each logarithm in Exercises 61–68 to four decimal places.

61. $\log_{10} 54.3$ **62.** $\log_{10} 27.9$

63. $\log_{10} 2344$ **64.** $\log_{10} 1476$

65. $\log_{10} 0.073$ **66.** $\log_{10} 0.00614$

67. $\log_{10} 0.6942$ **68.** $\log_{10} 0.0104$

Solve for x in Exercises 69–80.

69. $10^x = 200$ **70.** $10^x = 6$

71. $10^{-3x} = 5$ **72.** $10^{-5x} = 76$

73. $25 \cdot 10^{0.2x} = 80$ **74.** $8 \cdot 10^{1.6x} = 312$

75. $12.2 = 2(10^{1.4x}) - 11.6$

76. $163 = 3(10^{0.7x}) - 49.3$

77. $3(10^{-1.5x}) - 14.7 = 17.1$

78. $4(10^{-0.6x}) + 16.1 = 28.2$

79. $80(1 - 10^{-0.2x}) = 65$

80. $250(1 - 10^{-0.3x}) = 100$

The atmospheric pressure decreases with altitude above the surface of the earth. For Exercises 81–86 use the relationship

$$P(a) = 30(10)^{-0.09a}$$

between altitude, a, in miles and atmospheric pressure, P, in inches of mercury. Graph this function in the window

$$Xmin = 0, \quad Xmax = 9.4$$

$$Ymin = 0, \quad Ymax = 30$$

Solve the equations algebraically, and verify your answers with your graph.

81. The elevation of Mount Everest, the highest mountain in the world, is 29,028 feet. What is the atmospheric pressure at the top? (*Hint:* 1 mile = 5280 feet)

82. The elevation of Mount McKinley, the highest mountain in the United States, is 20,320 feet. What is the atmospheric pressure at the top?

83. How high above sea level is the atmospheric pressure 20.2 inches of mercury?

84. How high above sea level is the atmospheric pressure 16.1 inches of mercury?

85. Find the height above sea level at which the atmospheric pressure is equal to one-half the pressure at sea level.

86. Find the height above sea level at which the atmospheric pressure is equal to one-fourth the pressure at sea level.

Solve Exercises 87 and 88.

87. The population of California increased during the years 1960 to 1970 according to the formula $P(t) = 15,717,000(10)^{0.0104t}$, where t is measured in years since 1960.

a. What was the population in 1970?

b. Assuming the same growth rate, estimate the population of California in the years 1980, 1990, and 2000.

c. Estimate when the population of California will reach 20,000,000.

d. When will the population reach 30,000,000?

e. Graph the function P with a suitable domain and range, and verify your answers to parts (a) through (d).

88. The population of New York increased during the years 1960 to 1970 according to the formula $P(t) = 16,782,000(10)^{0.0036t}$, where t is measured in years since 1960.

a. What was the population in 1970?

b. Assuming the same growth rate, estimate the population of New York in the years 1980, 1990, and 2000.

c. Estimate when the population of New York will reach 20,000,000.

d. When will the population reach 30,000,000?

e. Graph the function P with a suitable domain and range, and verify your answers to parts (a) through (d).

91. $\log_{10} [\log_3 (\log_5 125)]$

92. $\log_{10} [\log_2 (\log_3 9)]$

93. $\log_2 [\log_2 (\log_3 81)]$ **94.** $\log_4 [\log_2 (\log_3 81)]$

95. $\log_b (\log_b b)$ **96.** $\log_b (\log_a a^b)$

Simplify each expression in Exercises 89–96.

89. $\log_2 (\log_4 16)$ **90.** $\log_5 (\log_5 5)$

7.4   LOGARITHMIC FUNCTIONS

In this section we will study some properties and applications of **logarithmic functions**. For example, if x is a positive number, we can define a function $f(x) = \log_2 x$. To understand logarithmic functions better, we first investigate how they are related to some more familiar functions, exponential functions.

Inverse of the Exponential Function

In Chapter 6 we saw that raising to the nth power and taking n^{th} roots are inverse operations. For example, if we first cube a number and then take the cube root of the result, we return to the original number.

$$x = 5 \xrightarrow{\quad\quad} x^3 = 125 \xrightarrow{\quad\quad} \sqrt[3]{x^3} = \sqrt[3]{125} = 5$$

Cube the number.   Take the cube root   Original
                     of the result.   number

If we graph the functions $f(x) = x^3$ and $g(x) = \sqrt[3]{x}$ on the same set of axes, we see that the graphs are related in an interesting way, as shown in Figure 7.20. We say that the graphs are "symmetric about the line $y = x$," which means that if we placed a mirror along the line $y = x$, each graph would be a reflection of the other.

We say that the two functions $f(x) = x^3$ and $g(x) = \sqrt[3]{x}$ are **inverse functions**. We will study inverse functions in more detail in Chapter 10, but for now we will just observe some basic facts about their relationship.

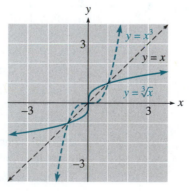

Figure 7.20

x	$f(x) = x^3$
-2	-8
-1	-1
$-1/2$	$-1/8$
0	0
$1/2$	$1/8$
1	1
2	8

Table 7.11

x	$g(x) = \sqrt[3]{x}$
-8	-2
-1	-1
$-1/8$	$-1/2$
0	0
$1/8$	$1/2$
1	1
8	2

Table 7.12

Notice that the table of values for $g(x) = \sqrt[3]{x}$ (Table 7.12) can be obtained from the table for $f(x) = x^3$ (Table 7.11) by interchanging the values of x and y in each ordered pair. This makes sense when we recall that each function undoes the effects of the other. In fact, we can state the following rule to define the cube root function:

$$y = \sqrt[3]{x} \qquad \text{if and only if} \qquad x = y^3$$

A similar rule relates the operations of raising a base b to a power and taking a base b logarithm:

$$y = \log_b x \qquad \text{if and only if} \qquad x = b^y$$

This relationship means that the function $g(x) = \log_b x$ is the inverse of the function $f(x) = b^x$. In particular, the function $g(x) = \log_2 x$ is the inverse of $f(x) = 2^x$. We can obtain a table of values for $g(x) = \log_2 x$ by making a table for $f(x) = 2^x$ and then interchanging the columns, as shown in Tables 7.13 and 7.14.

x	$f(x) = 2^x$
-2	$\frac{1}{4}$
-1	$\frac{1}{2}$
0	1
1	2
2	4

Table 7.13

x	$g(x) = \log_2 x$
$\frac{1}{4}$	-2
$\frac{1}{2}$	-1
1	0
2	1
4	2

Table 7.14

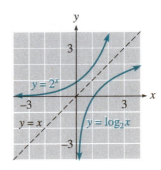

Figure 7.21

The graphs of $f(x) = 2^x$ and $g(x) = \log_2 x$ are shown in Figure 7.21. Notice that the graphs are symmetric about the line $y = x$. In general, if we want to find values for the function $y = \log_b x$, we can find values for the exponential function $y = b^x$ and then interchange the x- and y-values in each ordered pair.

Example 1

Graph the function $f(x) = 10^x$ and its inverse, $g(x) = \log_{10} x$, on the same axes.

Solution Make a table of values for the function $f(x) = 10^x$. A table of values for the inverse function, $g(x) = \log_{10} x$, can be obtained by interchanging the components of each ordered pair in the table for f. (See Tables 7.15 and 7.16.)

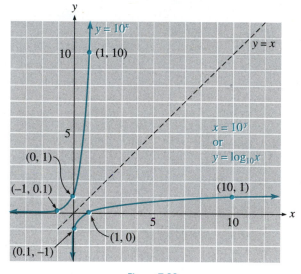

x	f(x)
−2	0.01
−1	0.1
0	1
1	10
2	100

Table 7.15

x	g(x)
0.01	−2
0.1	−1
1	0
10	1
100	2

Table 7.16

Figure 7.22

Plot each set of points and connect them with smooth curves to obtain the graphs shown in Figure 7.22.

In general, the logarithmic function $y = \log_b x$ has the following properties.

PROPERTIES OF LOGARITHMIC FUNCTIONS $y = \log_b x$
1. Domain: all positive real numbers.
2. Range: all real numbers.
3. The graphs of $y = \log_b x$ and $y = b^x$ are symmetric about the line $y = x$.

Notice that because the domain of a logarithmic function includes only the *positive* real numbers, *the logarithm of a negative number or zero is undefined.*

Because the exponential growth function increases very rapidly for positive values, its inverse, the logarithmic function, grows extremely slowly. Observe that for the common logarithmic function $y = \log_{10} x$,

$$\text{if} \quad 0 < x \leq 1, \qquad \text{then} \qquad \log_{10} x \leq 0$$

$$\text{if} \quad 1 < x \leq 10, \qquad \text{then} \qquad 0 \leq \log_{10} x \leq 1$$

$$\text{if} \quad 10 < x \leq 100, \qquad \text{then} \qquad 1 < \log_{10} x \leq 2$$

and so on. In fact,

$$\text{if} \quad 10^m < x \leq 10^n, \qquad \text{then} \qquad m < \log_{10} x \leq n$$

We can use a calculator to obtain more precise approximations for values of $\log_{10} x$.

Example 2

Let $f(x) = \log_{10} x$. Evaluate the following expressions.

a. $f(35)$ **b.** $f(-8)$ **c.** $2f(16) + 1$

Solutions **a.** $f(35) = \log_{10} 35 \approx 1.544$

b. Since -8 is not in the domain of f, $f(-8)$, or $\log_{10}(-8)$ is undefined.

c. $2f(16) + = 2(\log_{10} 16) + 1$

$$\approx 2(1.204) + 1 = 3.408$$

Example 3

Evaluate the expression $T = \dfrac{\log_{10}\left(\dfrac{M_f}{M_0} + 1\right)}{k}$ for $k = 0.028$,

$M_f = 1832$, and $M_0 = 15.3$.

Solution Follow the order of operations and calculate

$$T = \frac{\log_{10}\left(\dfrac{1832}{15.3} + 1\right)}{0.028} = \frac{\log_{10}(120.739)}{0.028}$$

$$\approx \frac{2.082}{0.028} \approx 74.35$$

A graphing calculator keying sequence for this computation is

$$\boxed{\text{LOG}}\ \boxed{(}\ 1832\ \boxed{\div}\ 15.3\ \boxed{+}\ 1\ \boxed{)}\ \boxed{\div}\ 0.028\ \boxed{\text{ENTER}}$$

Example 4

If $f(x) = \log_{10} x$, find x so that $f(x) = -3.2$.

Solution Solve the equation $\log_{10} x = -3.2$. Rewriting in exponential form yields

$$x = 10^{-3.2} \approx 0.00063$$

In Example 4 the expression $10^{-3.2}$ can be evaluated in two different ways with a calculator. One way is to use the $\boxed{\wedge}$ key and press

$$10 \;\boxed{\wedge}\; \boxed{(-)}\; 3.2 \;\boxed{\text{ENTER}}$$

which gives the answer 6.309573445E−4, which is approximately 0.00063. Alternatively, we can use the 10^x function of the calculator and press

$$\boxed{\text{2nd}}\; \boxed{\text{LOG}}\; \boxed{(-)}\; 3.2$$

which gives the same answer as before.

Logarithmic Models A number of phenomena can be modeled by logarithmic functions.

Example 5

The acidity of a substance is determined by the concentration of hydrogen ions, denoted by [H$^+$], in the substance. Chemists use a logarithmic scale called pH to measure acidity, where

$$\text{pH} = -\log_{10} [\text{H}^+] \tag{3}$$

The lower the pH value, the more acidic the substance.
a. Calculate the pH of a solution with a hydrogen ion concentration of 3.98×10^{-5}.
b. The water in a swimming pool should be maintained at a pH of 7.5. What is the hydrogen ion concentration of the water?

Solutions **a.** Evaluate Equation (3) with [H$^+$] = 3.95×10^{-5}:

$$\text{pH} = -\log_{10} (3.98 \times 10^{-5}) \approx 4.4$$

b. Solve the equation

$$7.5 = -\log_{10} [\text{H}^+]$$

for [H$^+$]. First write

$$-7.5 = \log_{10} [\text{H}^+]$$

Then rewrite the equation in exponential form to get

$$[\text{H}^+] = 10^{-7.5} \approx 3.2 \times 10^{-8}$$

The hydrogen ion concentration of the water is 3.2×10^{-8}.

Example 6

A sound's loudness is measured in decibels, D, by

$$D = 10 \log_{10} \left(\frac{I}{10^{-12}} \right) \qquad (4)$$

where I is the intensity of its sound waves (in watts per square meter).
a. A whisper generates about 10^{-10} watts per square meter at a distance of 3 feet. Find the number of decibels for a whisper 3 feet away.
b. Normal conversation registers at about 40 decibels. How many times more intense than a whisper is normal conversation?

Solutions **a.** Using Equation (4) with $I = 10^{-10}$,

$$D = 10 \log_{10} \left(\frac{10^{-10}}{10^{-12}} \right) = 10 \log_{10} 10^2$$

$$= 10(2) = 20 \text{ decibels}$$

b. Let I_w stand for the intensity of a whisper and let I_c stand for the intensity of normal conversation. We are looking for the ratio I_c/I_w. From part (a)

$$I_w = 10^{-10}$$

and from Equation (4) we have

$$40 = 10 \log_{10} \left(\frac{I_c}{10^{-12}} \right)$$

Dividing both members of the equation by 10 and rewriting in exponential form, we have

$$\frac{I_c}{10^{-12}} = 10^4$$

or

$$I_c = 10^4(10^{-12}) = 10^{-8}$$

so

$$\frac{I_c}{I_w} = \frac{10^{-8}}{10^{-10}} = 10^2$$

Normal conversation is 100 times more intense than a whisper.

Example 7

The magnitude of an earthquake is measured by comparing the amplitude, A, of its seismographic trace with the amplitude, A_0, of the smallest detectable earthquake. The log of their ratio is the Richter magnitude M. Thus

$$M = \log_{10} \left(\frac{A}{A_0} \right) \qquad (5)$$

a. The Northridge earthquake registering 6.9 on the Richter scale occurred in Los Angeles in January 1994. What would be the magnitude of an earthquake 100 times as powerful as the Northridge earthquake?

b. How many times more powerful than the Northridge earthquake was the San Francisco earthquake of 1989, which registered 7.1 on the Richter scale?

Solutions **a.** Let A_L represent the amplitude of the Northridge earthquake and let A_H represent the amplitude of an earthquake 100 times more powerful. From Equation (5) we have

$$6.9 = \log_{10}\left(\frac{A_L}{A_0}\right)$$

or, rewriting in exponential form,

$$\frac{A_L}{A_0} = 10^{6.9}$$

Because $A_H = 100A_L$,

$$\frac{A_H}{A_0} = \frac{100A_L}{A_0}$$

$$= 100\left(\frac{A_L}{A_0}\right) = 10^2(10^{6.9})$$

$$= 10^{8.9}$$

Thus from Equation (5) the magnitude of the more powerful earthquake is

$$\log_{10}\left(\frac{A_H}{A_0}\right) = \log_{10} 10^{8.9} = 8.9$$

b. Let A_S stand for the amplitude of the 1989 San Francisco earthquake. Then we are looking for the ratio A_S/A_L. From Equation (5) we have

$$6.9 = \log_{10}\left(\frac{A_L}{A_0}\right) \qquad \text{and} \qquad 7.1 = \log_{10}\left(\frac{A_S}{A_0}\right)$$

Rewriting each equation in exponential form, we have

$$\frac{A_L}{A_0} = 10^{6.9} \qquad \text{and} \qquad \frac{A_S}{A_0} = 10^{7.1}$$

or

$$A_L = 10^{6.9}A_0 \qquad \text{and} \qquad A_S = 10^{7.1}A_0$$

Thus

$$\frac{A_S}{A_L} = \frac{10^{7.1}A_0}{10^{6.9}A_0} = 10^{0.2}$$

The San Francisco earthquake was $10^{0.2}$, or approximately 1.58, times as powerful as the Northridge earthquake.

Notice in Example 7(a) that an earthquake 100, or 10^2, times as strong is only two units greater in magnitude on the Richter scale. In general, a *difference* of K units on the Richter scale corresponds to a *factor* of 10^K units in the intensity of the earthquake.

Exercise 7.4

In Exercises 1–4 graph each exponential function and its inverse logarithmic function on the same set of axes.

1. $f(x) = 2^x$

2. $f(x) = 3^x$

3. $f(x) = \left(\frac{1}{3}\right)^x$

4. $f(x) = \left(\frac{1}{2}\right)^x$

In Exercises 5–16 $f(x) = \log_{10} x$. **Evaluate each expression.**

5. $f(487)$ **6.** $f(93)$ **7.** $f(2.16)$

8. $f(6.95)$ **9.** $f(-7)$ **10.** $f(0)$

11. $6f(28)$ **12.** $3f(41)$

13. $18 - 5f(3)$ **14.** $15 - 4f(7)$

15. $\dfrac{2}{5 + f(0.6)}$ **16.** $\dfrac{3}{2 + f(0.2)}$

Evaluate each expression in Exercises 17–22 for the given values.

17. $t = \dfrac{1}{k} \log_{10} \dfrac{C_H}{C_L}$;   for $k = 0.05$, $C_H = 2$, and $C_L = 0.5$

18. $k = \dfrac{\log_{10} P - \log_{10} P_0}{t}$;   for $P = 35,000$, $P_0 = 18,000$, and $t = 26$

19. $R = \dfrac{1}{L} \log_{10} \left(\dfrac{P}{L - P}\right)$;   for $L = 8500$ and $P = 3600$

20. $T = \dfrac{H \log_{10} \dfrac{N}{N_0}}{\log_{10} \dfrac{1}{2}}$;   for $H = 5730$, $N = 180$, and $N_0 = 920$

21. $M = \sqrt{\dfrac{\log_{10} H}{k \log_{10} H_0}}$;   for $H = 0.93$, $H_0 = 0.02$, and $k = 0.006$

22. $h = a - \sqrt{\dfrac{\log_{10} B}{t}}$;   for $a = 56.2$, $B = 78$, and $t = 0.3$

In Exercises 23–28 $f(x) = \log_{10} x$. **Solve for x.**

23. $f(x) = 1.41$ **24.** $f(x) = 2.3$

25. $f(x) = 0.52$ **26.** $f(x) = 0.8$

27. $f(x) = -1.3$ **28.** $f(x) = -1.69$

Solve Exercises 29–42. (See the examples in Section 7.4 for the appropriate formulas.)

29. The hydrogen ion concentration of vinegar is about 6.3×10^{-4}. Calculate the pH of vinegar.

30. The hydrogen ion concentration of spinach is about 3.2×10^{-6}. Calculate the pH of spinach.

31. The pH of lime juice is 1.9. Calculate its hydrogen ion concentration.

32. The pH of ammonia is 9.8. Calculate its hydrogen ion concentration.

33. A lawn mower generates a noise of intensity 10^{-2} watts per square meter. Find the decibel level of the sound of a lawn mower.

34. A jet airplane generates 100 watts per square meter at a distance of 100 feet. Find the decibel level for a jet airplane.

35. The loudest sound emitted by any living source is made by the blue whale. Its whistles have been measured at 188 decibels and are detectable 500 miles away. Find the intensity of the blue whale's whistle in watts per square meter.

36. The loudest sound created in a laboratory registered at 210 decibels. The energy from such a sound is sufficient to bore holes in solid material. Find the intensity of a 210-decibel sound.

37. At a concert by The Who in 1976, the sound level 50 meters from the stage registered 120 decibels. How many times more intense was this level than a 90-decibel sound (the threshold of pain for the human ear)?

38. The loudest scientifically measured shouting by a human registered 123.2 decibels. How many times more intense was this shout than normal conversation at 40 decibels.?

39. In 1964 an earthquake in Alaska measured 8.4 on the Richter scale. An earthquake measuring 4.0 is considered small and causes little damage. How many times stronger was the Alaska earthquake than one measuring 4.0?

40. On April 30, 1986, an earthquake in Mexico City measured 7.0 on the Richter scale. On September 21 a second earthquake, measuring 8.1, hit Mexico City. How many times stronger was the September earthquake than the one in April?

41. A small earthquake measured 4.2 on the Richter scale. What would be the magnitude of an earthquake three times as strong?

42. Earthquakes measuring 3.0 on the Richter scale often go unnoticed. What would be the magnitude of a quake 200 times as strong as a 3.0 quake?

43. Let $f(x) = 3^x$ and $g(x) = \log_3 x$.
  a. Compute $f(4)$.
  b. Compute $g[f(4)]$.
  c. Explain why $\log_3 3^x = x$ for any x.
  d. Compute $\log_3 3^{1.8}$. Simplify $\log_3 3^a$.

44. Let $f(x) = \log_2 x$ and $g(x) = 2^x$
  a. Compute $f(32)$.
  b. Compute $g[f(32)]$.
  c. Explain why $2^{\log_2 x} = x$ for $x > 0$.
  d. Compute $2^{\log_2 6}$. Simplify $2^{\log_2 Q}$.

45. What is the inverse operation for "raise 6 to the power x"?

46. What is the inverse operation for "take the log base 5 of x"?

47. The log base 3 of a number is 5. What is the number?

48. Four raised to a certain power is 32. What is the exponent?

49. How large must x be before the graph of $y = \log_{10} x$ reaches a height of 5?

50. How large must x be before the graph of $y = \log_2 x$ reaches a height of 10?

51. For what values of x is $\log_b (x - 9)$ defined?

52. For what values of x is $\log_b (16 - 3x)$ defined?

53. a. Complete the table.

x	x^2	$\log_{10} x$	$\log_{10} x^2$
1			
2			
3			
4			
5			
6			

b. Do you notice a relationship between $\log_{10} x$ and $\log_{10} x^2$? State the relationship as an equation.

54. a. Complete the table.

x	$\dfrac{1}{x}$	$\log_{10} x$	$\log_{10} \dfrac{1}{x}$
1			
2			
3			
4			
5			
6			

b. Do you notice a relationship between $\log_{10} x$ and $\log_{10} \dfrac{1}{x}$? State the relationship as an equation.

Assuming that the relationships you found in Exercises 53 and 54 hold for logarithms to any base, complete the tables in Exercises 55 and 56 and use them to graph the given functions.

55.

x	$y = \log_e x$
1	0
2	0.693
4	
16	
$\frac{1}{2}$	
$\frac{1}{4}$	
$\frac{1}{16}$	

56.

x	$y = \log_4 x$
1	0
2	0.431
4	
16	
$\frac{1}{2}$	
$\frac{1}{4}$	
$\frac{1}{16}$	

7.5   PROPERTIES OF LOGARITHMS

Since logarithms are actually exponents, they have several properties that can be derived from the laws of exponents.

> **PROPERTIES OF LOGARITHMS**
>
> If $x, y > 0,$ then
>
> **Property 1.** $\log_b (xy) = \log_b x + \log_b y$
>
> **Property 2.** $\log_b \dfrac{x}{y} = \log_b x - \log_b y$
>
> **Property 3.** $\log_b x^m = m \log_b x$

As an example of Property 1, note that

$$\log_2 32 = \log_2 (4 \cdot 8) = \log_2 4 + \log_2 8$$
$$5 \qquad\qquad\qquad = \;\; 2 \;\; + \;\; 3$$

As an example of Property 2,

$$\log_2 8 = \log_2 \frac{16}{2} = \log_2 16 - \log_2 2$$
$$3 \qquad\qquad\quad = \;\; 4 \;\; - \;\; 1$$

As an example of Property 3,

$$\log_2 64 = \log_2 (4)^3 = 3\log_2 4$$
$$6 \qquad\qquad\quad = 3 \cdot 2$$

These three properties can be used to simplify expressions involving logarithms or to write them in more convenient forms for solving exponential and logarithmic equations.

Example 1

Simplify $\log_b \sqrt{\dfrac{xy}{z}}.$

Solution   First express $\sqrt{\dfrac{xy}{z}}$ using a fractional exponent:

$$\log_b \sqrt{\frac{xy}{z}} = \log_b \left(\frac{xy}{z}\right)^{1/2}$$

By Property 3,

$$\log_b \left(\frac{xy}{z}\right)^{1/2} = \frac{1}{2} \log_b \left(\frac{xy}{z}\right)$$

By Property 2,

$$\frac{1}{2}\log_b\left(\frac{xy}{z}\right) = \frac{1}{2}(\log_b xy - \log_b z)$$

and by Property 1,

$$\frac{1}{2}(\log_b xy - \log_b z) = \frac{1}{2}(\log_b x + \log_b y - \log_b z)$$

Thus

$$\log_b\sqrt{\frac{xy}{z}} = \frac{1}{2}(\log_b x + \log_b y - \log_b z)$$

Example 2

Given that $\log_b 2 \approx 0.6931$ and $\log_b 3 \approx 1.0986$, find the value of $\log_b 12$.

Solution   Using the properties of logarithms, express $\log_b 12$ in terms of $\log_b 2$ and $\log_b 3$. Since $12 = 2^2 \cdot 3$,

$$\begin{aligned}
\log_b 12 &= \log_b (2^2 \cdot 3) &&\text{Apply Property 1.}\\
&= \log_b 2^2 + \log_b 3 &&\text{Apply Property 3.}\\
&= 2\log_b 2 + \log_b 3 &&\text{Substitute the given values.}\\
&\approx 2(0.6931) + 1.0986\\
&= 2.4848
\end{aligned}$$

We can also use the three properties of logarithms to write sums and differences of logarithms as a single logarithm.

Example 3

Express $\frac{1}{2}(\log_b x - \log_b y)$ as a single logarithm with a coefficient of 1.

Solution   Use the properties of logarithms. Begin by applying Property 2.

$$\begin{aligned}
\frac{1}{2}(\log_b x - \log_b y) &= \frac{1}{2}\log_b\left(\frac{x}{y}\right) &&\text{Apply Property 3.}\\
&= \log_b\left(\frac{x}{y}\right)^{1/2}
\end{aligned}$$

Therefore,

$$\frac{1}{2}(\log_b x - \log_b y) = \log_b\left(\frac{x}{y}\right)^{1/2}$$

COMMON ERRORS

Note that

$$\log_b (x + y) \neq \log_b x + \log_b y$$

and

$$\log_b \frac{x}{y} \neq \frac{\log_b x}{\log_b y}$$

Solving Logarithmic Equations

The properties of logarithms are useful in solving equations in which the variable is part of a logarithmic expression.

Example 4

Solve $\log_{10} (x + 1) + \log_{10} (x - 2) = 1$.

Solution Use Property 1 of logarithms to rewrite the left-hand side as a single logarithm:

$$\log_{10} (x + 1)(x - 2) = 1$$

Once the left-hand side is expressed as a *single* logarithm, we can rewrite the equation in exponential form as

$$(x + 1)(x - 2) = 10^1$$

from which

$$x^2 - x - 2 = 10$$
$$x^2 - x - 12 = 0$$
$$(x - 4)(x + 3) = 0$$

Thus

$$x = 4 \quad \text{or} \quad x = -3$$

The number -3 is not a solution of the original equation because $\log_{10} (x + 1)$ and $\log_{10} (x - 2)$ are not defined for $x = -3$. The solution of the original equation is 4. ▭

Solving Exponential Equations

The properties of logarithms also enable us to solve exponential equations in which the base is not 10. For example, to solve the equation

$$5^x = 7 \tag{6}$$

we could rewrite the equation in logarithmic form to obtain the solution

$$x = \log_5 7$$

However, if we want a decimal approximation for the solution, we begin by taking the base 10 logarithm of both sides of Equation (6) to get

$$\log_{10} (5^x) = \log_{10} 7$$

Using Property 3, we rewrite the left-hand side as

$$x \log_{10}(5^x) = \log_{10} 7$$

and divide both sides by $\log_{10} 5$ to get

$$x = \frac{\log_{10} 7}{\log_{10} 5}$$

On a graphing calculator we can key in the sequence

$$\boxed{\text{LOG}}\ 7\ \boxed{\div}\ \boxed{\text{LOG}}\ 5\ \boxed{\text{ENTER}}$$

to find that

$$x \approx 1.2091$$

 COMMON ERROR

Note that

$$\frac{\log_{10} 7}{\log_{10} 5} \neq \log_{10} 7 - \log_{10} 5$$

Example 5

Solve $1640 = 80 \cdot 6^{0.03x}$

Solution   Divide both sides by 80 to obtain

$$20.5 = 6^{0.03x}$$

Take the base 10 logarithm of both sides of the equation and use Property 3 of logarithms to get

$$\log_{10} 20.5 = \log_{10} 6^{0.03x}$$
$$= 0.03x \log_{10} 6$$

Solve for x and use a calculator to evaluate the answer:

$$x = \frac{\log_{10} 20.5}{0.03 \log_{10} 6}$$
$$\approx 56.19$$

Example 6

The population of Silicon City was 6500 in 1970 and has been tripling every 12 years. When will the population reach 75,000?

Solution   The population of Silicon City grows according to the formula $P(t) = 6500 \cdot 3^{t/12}$,   where t is the number of years after 1970. We want to find the value of t for which $P(t) = 75,000;$ that is, we want to solve the equation

$$6500 \cdot 3^{t/12} = 75,000$$

Divide both sides by 6500 to get

$$3^{t/12} = \frac{150}{13}$$

Then take the base 10 logarithm of both sides and solve for t.

$$\log_{10}(3^{t/12}) = \log_{10}\left(\frac{150}{13}\right) \qquad \text{Apply Property 3.}$$

$$\frac{t}{12}\log_{10} 3 = \log_{10}\left(\frac{150}{13}\right) \qquad \text{Divide by } \log_{10} 3; \text{ multiply by 12.}$$

$$t = \frac{12\left(\log_{10}\dfrac{150}{13}\right)}{\log_{10} 3}$$

Use a calculator to evaluate the answer:

$$t \approx 26.71$$

The population of Silicon City will reach 75,000 about 27 years after 1970, or in 1997.

Solving Formulas

The techniques discussed above can also be used to solve formulas involving exponential or logarithmic expressions of one variable in terms of the others.

Example 7

Solve $P = Cb^{kt}$ for t $(C, k \neq 0)$.

Solution   First express the power b^{kt} in terms of the other variables:

$$b^{kt} = \frac{P}{C}$$

Write the exponential equation in logarithmic form:

$$kt = \log_b \frac{P}{C}$$

Multiply each side by $\dfrac{1}{k}$:

$$t = \frac{1}{k}\log_b \frac{P}{C}$$

Example 8

Solve $N = N_0 \log_b (ks)$ for s.

Solution   First express $\log_b (ks)$ in terms of the other variables:

$$\log_b (ks) = \frac{N}{N_0} \qquad (N_0 \neq 0)$$

Write the logarithmic equation in exponential form:

$$ks = b^{N/N_0}$$

or

$$s = \frac{1}{k} b^{N/N_0} \qquad (k \neq 0)$$

Exercise 7.5

Use Properties 1, 2, and 3 on page 432 to write each expression in Exercises 1–22 in terms of simpler logarithms. Assume that all variables denote positive numbers.

1. $\log_b 2x$

2. $\log_b xy$

3. $\log_b \dfrac{x}{y}$

4. $\log_b \dfrac{y}{x}$

5. $\log_b \dfrac{xy}{z}$

6. $\log_b \dfrac{x}{yz}$

7. $\log_b x^3$

8. $\log_b x^{1/3}$

9. $\log_b \sqrt{x}$

10. $\log_b \sqrt[5]{y}$

11. $\log_b \sqrt[3]{x^2}$

12. $\log_b \sqrt{x^3}$

13. $\log_b x^2 y^3$

14. $\log_b x^{1/3} y^2$

15. $\log_b \dfrac{x^{1/2} y}{z^2}$

16. $\log_b \dfrac{xy^2}{z^{1/2}}$

17. $\log_{10} \sqrt[3]{\dfrac{xy^2}{z}}$

18. $\log_{10} \sqrt{\dfrac{2L}{R^2}}$

19. $\log_{10} 2\pi \sqrt{\dfrac{l}{g}}$

20. $\log_{10} 2y \sqrt[3]{\dfrac{x}{y}}$

21. $\log_{10} \sqrt{(s-a)(s-b)}$

22. $\log_{10} \sqrt{s^2(s-a)^3}$

Given that $\log_b 2 = 0.6931$, $\log_b 3 = 1.0986$, and $\log_b 5 = 1.6094$, find the value of each expression in Exercises 23–34.

23. $\log_b 6$

24. $\log_b 10$

25. $\log_b \dfrac{2}{5}$

26. $\log_b \dfrac{3}{2}$

27. $\log_b 9$

28. $\log_b 25$

29. $\log_b \dfrac{15}{2}$

30. $\log_b \dfrac{6}{5}$

31. $\log_b (0.02)^3$

32. $\log_b \sqrt{50}$

33. $\log_b 75$

34. $\log_b \dfrac{0.08}{15}$

Express Exercises 35–44 as a single logarithm with a coefficient of 1.

35. $\log_b 8 - \log_b 2$

36. $\log_b 5 + \log_b 2$

37. $2 \log_b x + 3 \log_b y$

38. $\dfrac{1}{4} \log_b x - \dfrac{3}{4} \log_b y$

39. $-2 \log_b x$

40. $-\log_b x$

41. $\dfrac{1}{2} (\log_{10} y + \log_{10} x - 3 \log_{10} z)$

42. $\dfrac{1}{3} (\log_{10} x - 2 \log_{10} y - \log_{10} z)$

43. $\dfrac{1}{2} \log_b + 2(\log_b 2 - \log_b 8)$

44. $\dfrac{1}{2} (\log_b 6 + 2 \log_b 4) - \log_b 2$

Solve each logarithmic equation in Exercises 45–54.

45. $\log_{10} x + \log_{10} 2 = 3$

46. $\log_6 3 + \log_6 x = 2$

47. $\log_{10} x + \log_{10} (x + 21) = 2$

48. $\log_{10} (x + 3) + \log_{10} x = 1$

49. $\log_8 (x + 5) - \log_8 2 = 1$

50. $\log_{10} (x - 1) - \log_{10} 4 = 2$

51. $\log_{10}(x + 2) + \log_{10}(x - 1) = 1$

52. $\log_4(x + 8) + \log_4(x + 2) = 2$

53. $\log_3(x - 2) - \log_3(x + 1) = 3$

54. $\log_{10}(x + 3) - \log_{10}(x - 1) = 1$

Solve each equation in Exercises 55–66 by using logarithms in base 10.

55. $2^x = 7$

56. $3^x = 4$

57. $3^{x+1} = 8$

58. $2^{x-1} = 9$

59. $4^{x^2} = 15$

60. $3^{x^2} = 21$

61. $4.26^{-x} = 10.3$

62. $2.13^{-x} = 8.1$

63. $25 \cdot 3^{2.1x} = 47$

64. $12 \cdot 5^{1.5x} = 85$

65. $3600 = 20 \cdot 8^{-0.2x}$

66. $0.06 = 50 \cdot 4^{-0.6x}$

Solve Exercises 67–74 algebraically and check your answers with a graph.

67. A culture of *Salmonella* bacteria is started with 0.01 gram and triples in weight every 16 hours.

  a. Write a function for the weight of the culture as a function of time.

  b. How long will it take for the culture to weigh 0.5 gram?

68. In 1975 Summit City used 4.2×10^6 kilowatt-hours of electricity, and the demand for electricity has increased by a factor of 1.5 of every 10 years.

  a. Write a function for the demand for electricity as a function of time.

  b. When will Summit City need 10 million kilowatt-hours annually?

69. The annual growth rate of Hickory Corners is 3.7%.

  a. Write a function for the population of Hickory Corners as a function of time. Let the initial population be P_0.

  b. How long will it take for the population to double?

  c. Suppose $P_0 = 500$. Graph your function in the window Xmin $= 0$, Xmax $= 94$, Ymin $= 0$, Ymax $= 10,000$.

  d. Use the [TRACE] to verify that the population doubles from 500 to 1000, from 1000 to 2000, and from 2000 to 4000 in equal periods of time.

70. In 1986 the inflation rate in Bolivia was 8000% annually.

  a. Write a function for the price of an item as a function of time. Let P_0 be its initial price.

  b. How long did it take for prices to double?

  c. Suppose $P_0 = 5$. Graph your function in the window Xmin $= 0$, Xmax $= 0.94$, Ymin $= 0$, Ymax $= 100$.

  d. Use the [TRACE] to verify that the price doubles from 5 to 10, from 10 to 20, and from 20 to 40 in equal periods of time.

71. The concentration of a certain drug injected into the bloodstream decreases by 20% each hour as the drug is eliminated from the body. The initial dose creates a concentration of 0.7 milligram per milliliter.

  a. Write a function for the concentration of the drug in the bloodstream as a function of time.

  b. The minimum safe concentration of the drug is 0.4 milligram per milliliter. When should the second dose be administered?

  c. Verify your answer with a graph.

72. A small pond is tested for pollution and the concentration of toxic chemicals is found to be 80 parts per million. Clean water enters the pond from a stream, mixes with the polluted water, and then leaves the pond so that the pollution level is reduced by 10% each month.

  a. Write a function for the concentration of toxic chemicals as a function of time.

  b. How long it will be before the concentration of toxic chemicals reaches a safe level of 25 parts per million?

  c. Verify your answer with a graph.

73. Radioactive potassium 42, which is used by cardiologists as a tracer, decays at a rate of 5.4% per hour.

a. Find the half-life of potassium 42, that is, the time it takes for one-half of a sample of the isotope to decay.

b. How long will it take for three-fourths of the sample to decay? For seven-eighths of the sample?

74. Radium 226 decays at a rate of 0.4% per year.

a. Find the half-life of radium 226.

b. How long will it take for three-fourths of a sample of radium 226 to decay? For seven-eighths of the sample?

For Exercises 75–82 use the following information. If P dollars is invested at an annual interest rate r (expressed as a decimal) compounded n times yearly, the amount, A, after t years is given by

$$A = P\left(1 + \frac{r}{n}\right)^{nt}$$

75. a. Suppose that $5000 is invested at 12% annual interest. In the same window, graph the amount Y_1 in the account (as a function of t) if the interest is compounded annually, and the amount Y_2 in the account if the interest is compounded quarterly.

b. Find the compounded amount of $5000 invested at 12% for 10 years when compounded annually and when compounded quarterly. Verify your answers with your graphs.

76. a. Suppose that $800 is invested at 7% annual interest. In the same window, graph the amount Y_1 in the account (as a function of t) if the interest is compounded semiannually, and the amount Y_2 in the account if the interest is compounded monthly.

b. Find the compounded amount of $800 invested at 7% for 15 years when compounded semiannually and when compounded monthly. Verify your answers with your graphs.

77. a. Suppose that $1000 is invested at 6% annual interest. In the same window, graph the amount

Y_1 in the account (as a function of t) if the interest is compounded semiannually, and the amount Y_2 in the account if the interest is compounded quarterly.

b. How long would it take $1000 to grow to $2000 if compounded semiannually at 6%? If compounded quarterly at 6%? Verify your answers with your graphs.

78. a. Suppose that $500 is invested at 8% annual interest. In the same window, graph the amount Y_1 in the account (as a function of t) if the interest is compounded monthly, and the amount Y_2 in the account if the interest is compounded daily. (Use 365 days for a year.)

b. How long would it take $500 to grow to $1000 if compounded monthly at 8%? If compounded daily at 8%? Verify your answers with your graphs.

79. What interest rate is required so that $1000 will yield $1900 after 5 years if the interest is compounded monthly?

80. What interest rate is required so that $400 will yield $600 after 3 years if the interest is compounded quarterly?

81. How long will it take a sum of money to triple if it is invested at 10% compounded daily?

82. How long will it take a sum of money to increase fivefold if it is invested at 10% compounded quarterly?

Solve each formula in Exercises 83–88 for the specified variable.

83. $A = A_0 (10^{kt} - 1)$, for t

84. $B = B_0 (1 - 10^{-kt})$, for t

85. $w = pv^q$, for q **86.** $l = p^a q^b$, for b

87. $t = T \log_{10}\left(1 + \frac{A}{k}\right)$, for A

88. $\log_{10} R = \log_{10} R_0 + kt$, for R

Verify that each statement in Exercises 89–94 is true.

89. $\log_b 4 + \log_b 8 = \log_b 64 - \log_b 2$

90. $\log_b 24 - \log_b 2 = \log_b 3 + \log_b 4$

91. $2\log_b 6 - \log_b 9 = 2\log_b 2$

92. $4\log_b 3 - 2\log_b 3 = \log_b 9$

93. $\dfrac{1}{2}\log_b 12 - \dfrac{1}{2}\log_b 3 = \dfrac{1}{3}\log_b 8$

94. $\dfrac{1}{4}\log_b 8 + \dfrac{1}{4}\log_b 2 = \log_b 2$

95. Show by using a numerical example that $\log_{10}(x + y)$ is not equivalent to the expression $\log_{10} x + \log_{10} y$.

96. Show by a numerical example that $\log_{10}\dfrac{x}{y}$ is not equivalent to the expression $\dfrac{\log_{10} x}{\log_{10} y}$.

In Exercises 97–99 we will use the laws of exponents to prove the corresponding properties of logarithms.

97. We will use the first law of exponents, $a^p \cdot a^q = a^{p+q}$, to prove the first property of logarithms.

a. Let $m = \log_b x$ and $n = \log_b y$. Rewrite these relationships in exponential form:

$$x = \underline{\hspace{1cm}} \qquad \text{and} \qquad y = \underline{\hspace{1cm}}$$

b. Now consider the expression $\log_b(xy)$. Replace x and y by your answers to part (a).

c. Apply the first law of exponents to your expression in part (b).

d. Use the definition of logarithm to simplify your answer to part (c).

e. Refer to the definitions of m and n in part (a) to finish the proof.

98. We will use the second law of exponents, $\dfrac{a^p}{a^q} = a^{p-q}$, to prove the second property of exponents.

a. Let $m = \log_b x$ and $n = \log_b y$ Rewrite these relationships in exponential form:

$$x = \underline{\hspace{1cm}} \qquad \text{and} \qquad y = \underline{\hspace{1cm}}$$

b. Now consider the expression $\log_b\left(\dfrac{x}{y}\right)$. Replace x and y by your answers to part (a).

c. Apply the second law of exponents to your expression in part (b).

d. Use the definition of logarithm to simplify your answer to part (c).

e. Refer to the definitions of m and n in part (a) to finish the proof.

99. We will use the third law of exponents, $(a^p)^q = a^{pq}$, to prove the third property of logarithms.

a. Let $m = \log_b x$. Rewrite this relationship in exponential form:

$$x = \underline{\hspace{1cm}}$$

b. Now consider the expression $\log_b(x^k)$. Replace x by your answer to part (a).

c. Apply the third law of exponents to your expression in part (b).

d. Use the definition of logarithm to simplify your answer to part (c).

e. Refer to the definition of m in part (a) to finish the proof.

100. Use the fact that $b^0 = 1$ to explain why $\log_b 1 = 0$.

7.6   THE NATURAL BASE

Another base for logarithms and exponential functions is often used in applications. This base is an irrational number called e, where

$$e \approx 2.71828182845$$

The number e is essential for the study of numerous advanced topics and is often referred to as the **natural base**.

Natural Logarithmic Function   The base e logarithm of a number x, or $\log_e x$, is called the **natural logarithm** of x and is denoted by **ln x**. Thus

$$\boxed{\ln x = \log_e x}$$

The function $y = \ln x$ is called the natural logarithmic function. Recall that $\ln x$, or $\log_e x$, means "the exponent to which e must be raised to yield x." Because e is irrational, a calculator or a table of values is necessary for any calculations involving the natural base. To evaluate the natural log of a number, use the key labeled $\boxed{\text{LN}}$.

Example 1

Evaluate the following natural logarithms.
a.   ln 6.6   **b.**   ln 0.7

Solutions
a.   Use the keying sequence $\boxed{\text{LN}}$ 6.6 $\boxed{\text{ENTER}}$ to find $\ln x \approx 1.8871$.

b.   Use the keying sequence $\boxed{\text{LN}}$ 0.7 $\boxed{\text{ENTER}}$ to find $\ln x \approx -0.3567$.

Natural Exponential Function   The **natural exponential function** is the function $y = e^x$. Values for e^x can be obtained with a calculator using the $\boxed{\text{2nd}}$ and $\boxed{\text{LN}}$ keys.

Example 2

Evaluate the following powers.
a. $e^{2.4}$ **b.** $e^{-4.7}$

Solutions   **a.**   Use the keying sequence $\boxed{\text{2nd}}$ $\boxed{\text{LN}}$ 2.4 $\boxed{\text{ENTER}}$ to find $e^{2.4} \approx 11.023$.
b.   Use the keying sequence $\boxed{\text{2nd}}$ $\boxed{\text{LN}}$ $\boxed{(-)}$ 4.7 $\boxed{\text{ENTER}}$ to find $e^{-4.7} \approx 0.0090953$.

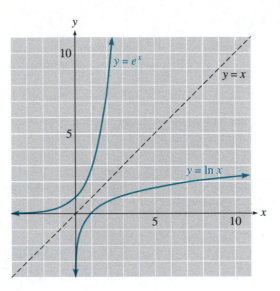

Figure 7.23

The graphs of the functions $y = e^x$ and $y = \ln x$ are shown in Figure 7.23.

Solving Equations   The techniques for solving logarithmic and exponential equations with other bases also apply to equations with base e.

Example 3

Solve each equation for x.
a. $e^x = 0.24$ **b.** $\ln x = 3.5$

Solutions   a.   Rewrite the equation in logarithmic form and evaluate:

$$x = \ln 0.24 \approx -1.427$$

b.   Rewrite the equation in exponential form and evaluate:

$$x = e^{3.5} \approx 33.115$$

In more complicated exponential equations, we isolate the power on one side of the equation before rewriting in logarithmic form.

Example 4

Solve $140 = 20\, e^{0.4x}$.

Solution   First divide each side by 20 to obtain

$$7 = e^{0.4x}$$

Then rewrite the equation in logarithmic form as

$$0.4x = \ln 7$$

Thus

$$x = \frac{\ln 7}{0.4}$$

which can be evaluated on a graphing calculator as

$$\boxed{\text{LN}}\ 7\ \boxed{\div}\ 0.4\ \boxed{\text{ENTER}}$$

to see that $x \approx 4.8648$.

Example 5

Solve $P = \dfrac{a}{1 + be^{-kt}}$ for t.

Solution Multiply both sides of the equation by the denominator, $1 + be^{-kt}$,   to get

$$P\,(1 + be^{-kt}) = a$$

Then isolate the power, e^{-kt}:

$$1 + be^{-kt} = \frac{a}{P}$$

$$be^{-kt} = \frac{a}{P} - 1 = \frac{a - P}{P}$$

$$e^{-kt} = \frac{a - P}{bP}$$

Take the natural logarithm of both sides to get

$$\ln e^{-kt} = \ln \frac{a - P}{bP}$$

or

$$-kt = \ln \frac{a - P}{bP}$$

Then solve for t:

$$t = \frac{-1}{k} \ln \frac{a - P}{bP}$$

Exponential Growth and Decay   In Section 7.1 we considered functions of the form

$$P(t) = P_0 \cdot a^t \tag{7}$$

which describe exponential growth when $a > 1$ and exponential decay when $0 < a < 1$.   Exponential growth and decay can also be modeled by functions of the form

$$P(t) = P_0 \cdot e^{kt} \tag{8}$$

where we have a substituted e^k for the growth factor a in Equation (7) so that

$$P(t) = P_0 \cdot a^t$$
$$= P_0 \cdot (e^k)^t = P_0 \cdot e^{kt}$$

The value of k can be found by solving the equation $a = e^k$ for k, which yields $k = \ln a$.

For instance, in Investigation 1 on page 392 we found that a colony of bacteria grew according to the formula

$$P(t) = 100 \cdot 3^t$$

We can express this function in the form $P(t) = 100 \cdot e^{kt}$ if we set

$$3 = e^k \quad \text{or} \quad k = \ln 3 \approx 1.0986$$

Thus the growth law for the colony of bacteria can be written

$$P(t) \approx 100 \cdot e^{1.0986t}$$

Example 6

Express the decay law $N(t) = 60(0.8)^t$ in the form $N(t) = N_0 e^{kt}$.

Solution Since $0.8 = e^k$, $k = \ln 0.8 \approx -0.2231$. $N_0 = 60$, so the decay law is

$$N(t) \approx 60\, e^{-0.2231t}$$

Example 7

Many savings institutions offer accounts on which the interest is compounded continuously. The amount accumulated in such an account after t years at interest rate r is given by the function

$$A(t) = P\, e^{rt}$$

where P is the principal invested.
a. Graph the function $A(t)$.
b. If \$500 is invested in an account offering 8% interest compounded continuously, how much will the account be worth after 5 years?
c. How long will it be before the account is worth \$1000?

Solutions **a.** The graph is shown in Figure 7.24.
b. Evaluate $A(t)$ for $t = 5$, with $P = 500$ and $r = 0.08$:

$$A(5) = 500\, e^{0.08(5)}$$

$$= 500\, e^{0.4}$$

$$\approx 500(1.4918) = 745.91$$

The account will be worth \$745.91 after 5 years.
c. Substitute 1000 for A and solve the equation:

$$1000 = 500\, e^{0.08t}$$

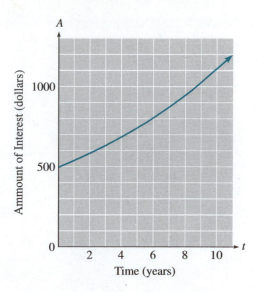

Figure 7.24

Divide both sides by 500 to get

$$2 = e^{0.08t}$$

Then rewrite the equation in logarithmic form as

$$0.08t = \ln 2 \approx 0.6931$$

Solve for t to find $t \approx 8.6643$. The account will be worth $1000 after approximately 8.7 years.

Example 8

Radioactive elements decay into more stable isotopes in such a way that the amount of radioactive material left at time t is given by an exponential decay law. A scientist starts with 25 grams of krypton 91, which decays according to the formula $N(t) = N_0 e^{-0.07t}$, where t is in seconds.

a. Graph the function $N(t)$.

b. How much krypton 91 (to the nearest hundredth of a gram) is left after 15 seconds?

c. How long does it take for 60% of the krypton 91 to decay?

Solutions   **a.**   The graph is shown in Figure 7.25 on page 446.

b.   Evaluate $N(t)$ for $t = 15$, with $N_0 = 25$:

$$N(15) = 25\, e^{-0.07(15)}$$

$$= 25\, e^{-1.05}$$

$$\approx 25(0.3499) = 8.7484$$

Approximately 8.75 grams of krypton 91 are left after 15 seconds.

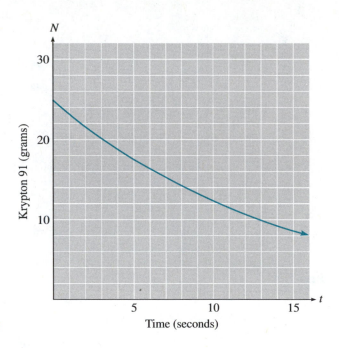

Figure 7.25

c. If 60% of the krypton 91 has decayed, then 40% of the original 25 grams, or 10 grams, is left. Substitute 10 for N and solve the equation:

$$10 = 25\, e^{-0.07t}$$

Divide both sides by 25 to get

$$0.4 = e^{-0.07t}$$

and rewrite the equation in logarithmic form as

$$-0.07t = \ln 0.4 \approx -0.9163$$

Solving for t yields $t \approx 13.0899$.   It takes approximately 13 seconds for 60% of the krypton 91 to decay.

Exercise 7.6

Evaluate each logarithm in Exercises 1–8.

1. $\ln 3.9$ **2.** $\ln 6.3$ **3.** $\ln 16$ **4.** $\ln 55$

5. $\ln 0.3$ **6.** $\ln 0.7$ **7.** $\ln 1$ **8.** $\ln e$

Find each power in Exercises 9–16.

9. $e^{0.4}$ **10.** $e^{0.73}$ **11.** $e^{2.34}$ **12.** $e^{3.16}$

13. $e^{-1.2}$ **14.** $e^{-2.3}$ **15.** $e^{-0.4}$ **16.** $e^{-0.62}$

In Exercises 17–38 solve for x.

17. $e^x = 1.9$ **18.** $e^x = 2.1$ **19.** $e^x = 45$

20. $e^x = 60$ **21.** $e^x = 0.3$ **22.** $e^x = 0.9$

23. $\ln x = 1.42$ **24.** $\ln x = 2.03$

25. $\ln x = 0.63$ **26.** $\ln x = 0.59$

27. $\ln x = -2.6$ **28.** $\ln x = -3.4$

29. $6.21 = 2.3 \, e^{1.2x}$ **30.** $22.26 = 5.3 \, e^{0.4x}$

31. $7.74 = 1.72 \, e^{0.2x}$ **32.** $14.15 = 4.03 \, e^{1.4x}$

33. $6.4 = 20 \, e^{0.3x} - 1.8$ **34.** $4.5 = 4 \, e^{2.1x} + 3.3$

35. $46.52 = 3.1 \, e^{1.2x} + 24.2$

36. $1.23 = 1.3 \, e^{2.1x} - 17.1$

37. $16.24 = 0.7 \, e^{-1.3x} - 21.7$

38. $55.68 = 0.6 \, e^{-0.7x} + 23.1$

Solve each equation in Exercises 39–44 for the specified variable.

39. $y = e^{kt}$,   for t **40.** $\dfrac{T}{R} = e^{t/2}$,   for t

41. $y = k(1 - e^{-t})$,   for t

42. $B - 2 = (A + 3) \, e^{-t/3}$,   for t

43. $T = T_0 \ln(k + 10)$,   for k

44. $P = P_0 + \ln 10k$,   for k

Express each growth or decay law in Exercises 45–50 in the form $N(t) = N_0 e^{kt}$.

45. $N(t) = 100 \cdot 2^t$ **46.** $N(t) = 50 \cdot 3^t$

47. $N(t) = 1200(0.6)^t$ **48.** $N(t) = 300(0.8)^t$

49. $N(t) = 10(1.15)^t$ **50.** $N(t) = 1000(1.04)^t$

51. The number of bacteria present in a culture is given by $N(t) = N_0 \, e^{0.04t}$, where N_0 is the number of bacteria present at time $t = 0$ and t is the time in hours.

a. If 6000 bacteria were present at $t = 0$, graph $N(t)$.

b. How many bacteria were present 10 hours later?

c. How much time must elapse (to the nearest tenth of an hour) for 6000 bacteria to increase to 10,000?

52. The amount of a radioactive element present at any time t is given by $y(t) = y_0 \, e^{-0.4t}$, where t is measured in seconds and y_0 is the amount present initially.

a. If 40 grams were present initially, graph $y(t)$.

b. How much of the element (to the nearest hundredth of a gram) remains after 3 seconds?

c. How much time must elapse (to the nearest hundredth of a second) for 40 grams to be reduced to 10 grams?

53. The intensity I (in lumens) of a light beam after passing through t centimeters of a filter having an absorption coefficient of 0.1 is given by $I = 1000 \, e^{-0.1t}$.

a. What is the intensity (to the nearest tenth of a lumen) of a light beam that has passed through 0.6 centimeter of the filter?

b. How many centimeters (to the nearest tenth) of the filter will reduce the illumination to 800 lumens?

54. The voltage, V, across a capacitor in a certain circuit is given by the formula

$$V = 100(1 - e^{-0.5t})$$

where t is the time in seconds.

a. What is the voltage (to the nearest tenth of a volt) after 10 seconds?

b. How much time must elapse (to the nearest hundredth of a second) for the voltage to reach 75 volts?

55. Hope invests $2000 at $9\frac{1}{2}\%$ interest compounded continuously.

a. Graph the function $A(t)$ that gives the amount of money in Hope's account after t years.

b. How much will Hope's account be worth after 7 years?

c. How long will it take for the account to grow to $5000?

56. D.G.'s savings account pays 6.25% interest compounded continuously.

a. Graph the function $A(t)$ that gives the amount of money in D.G.'s account after t years.

b. If D.G deposits $500 today, how much will be in his account 6 months from now?

c. How long will it take for the account to grow to $3000?

57. $600 compounded continuously for 4 years yields $809.92. What is the interest rate to the nearest tenth of a percent?

58. How long will it take $300 to yield $650 at 9% compounded continuously?

59. All living things contain a certain amount of the isotope carbon 14. When an organism dies the carbon 14 decays according to the formula $N = N_0 e^{-0.000124t}$. Scientists can estimate the age of an organic object by measuring the amount of carbon 14 remaining.

a. When the Dead Sea scrolls were discovered in 1947 they had 78.8% of their original carbon 14. How old were the Dead Sea scrolls then?

b. What is the half-life of carbon 14; that is, how long does it take for half of an object's carbon 14 to decay?

60. The absorption of X rays by a lead plate of thickness t inches is given by the formula

$$I = I_0 e^{-1.88t}$$

where I_0 is the X-ray count at the source and I is the X-ray count behind the lead plate.

a. What percent of an X-ray beam will penetrate a lead plate $\frac{1}{2}$ inch thick?

b. How thick should the lead plate be in order to screen out 70% of the X-rays?

61. The population of Citrus Valley was 20,000 in 1970. In 1980 it was 35,000.

a. What is P_0 if $t = 0$ in 1970?

b. Use the population in 1980 to find the growth factor e^k.

c. Write a growth law of the form $P(t) = P_0 e^{kt}$ for the population of Citrus Valley.

d. If it continues at the same growth rate, what will Citrus Valley's population be in 2000?

62. In 1981 a copy of *Time* magazine cost $1.50. In 1988 the cover price was $2.00.

a. What is P_0 if $t = 0$ in 1981?

b. Use the price in 1988 to find the growth factor e^k.

c. Find a growth law of the form $P(t) = P_0 e^{kt}$ for the price of *Time*.

d. If the price continues at the same rate of growth, what will *Time* cost in 2000?

63. The half-life (the time it takes for half of a sample of radioactive material to decay) of iodine 131 is approximately 8 days. Find a decay law of the form $N = N_0 e^{kt}$, where $k < 0$, for iodine 131.

64. The half-life of hydrogen 3 is 12.5 years. Find a decay law of the form $N = N_0 e^{kt}$, where $k < 0$, for hydrogen 3.

Exercises 65–70 deal with the "change of base" formula.

65. Follow the steps below to calculate $\log_8 20$.

a. Let $x = \log_8 20$. Write the equation in exponential form.

b. Take the logarithm base 10 of both sides of your new equation.

c. Simplify and solve for x.

66. Follow the steps in Exercise 65 to calculate $\log_8 5$.

67. Use Exercise 65 to find a formula for calculating $\log_8 Q$, where Q is any positive number.

68. Find a formula for calculating $\log_b Q$, where $b > 1$ and Q is any positive number.

69. Find a formula for calculating $\ln Q$ in terms of $\log_{10} Q$.

70. Find a formula for calculating $\log_{10} Q$ in terms of $\ln Q$.

Chapter 7 Review

For Exercises 1–4
a. make a table of values for the situation described.
b. write a function that describes the exponential growth or decay.
c. graph the function.
d. evaluate the function at the given values.

1. The number of computer science degrees awarded by Monroe College has increased by a factor of $\frac{3}{2}$ every 5 years since 1974. If the college granted 8 degrees in 1974, how many did it award in 1984? In 1995?

2. The price of public transportation has been rising by 10% per year since 1975. If it cost \$0.25 to ride the bus in 1975, how much did it cost in 1985? How much will it cost in the year 2000 if the current trend continues?

3. A certain medication is eliminated from the body at a rate of 15% per hour. If an initial dose of 100 milligrams is taken at 8:00 A.M., how much is left at 12:00 noon? At 6:00 P.M.?

4. After the World Series, sales of T-shirts and other memorabilia decline 30% per week. If \$200,000 worth of souvenirs were sold during the series, how much will be sold 4 weeks later? Six weeks after the series?

Find a decimal approximation for each expression in Exercises 5–8.

5. $7^{\sqrt{2}}$

6. $3^{\pi+1}$

7. $0.2(5)^{\sqrt{3}}$

8. $6 - 2^{\sqrt{7}}$

Graph each function in Exercises 9–12.

9. $f(t) = 1.2^t$

10. $g(t) = 0.6^{-t}$

11. $P(x) = 2^x - 3$

12. $R(x) = 2^{x+3}$

Solve each equation in Exericses 13–16.

13. $3^{x+2} = 9^{1/3}$

14. $2^{x-1} = 8^{-2x}$

15. $4^{2x+1} = 8^{x-3}$

16. $3^{x^2-4} = 27$

Find each logarithm in Exercises 17–22.

17. $\log_2 16$

18. $\log_4 2$

19. $\log_3 \frac{1}{3}$

20. $\log_7 7$

21. $\log_{10} 10^{-3}$

22. $\log_{10} 0.0001$

Write each equation in Exercises 23 and 24 in exponential form.

23. $\log_2 3 = x - 2$

24. $\log_n q = p - 1$

Write each equation in Exercises 25 and 26 in logarithmic form.

25. $0.3^{-2} = x + 1$

26. $4^{0.3t} = 3N_0$

In Exercises 27–34 solve for the unknown value.

27. $\log_3 \frac{1}{3} = y$

28. $\log_3 x = 4$

29. $\log_b 16 = 2$

30. $\log_2 (3x - 1) = 3$

31. $4 \cdot 10^{1.3x} = 20.4$

32. $127 = 2(10^{0.5x}) - 17.3$

33. $3(10^{-0.7x}) + 6.1 = 9$

34. $40(1 - 10^{-1.2x}) = 30$

Evaluate each expression in Exercises 35–38.

35. $k = \frac{1}{t}(\log_{10} N - \log_{10} N_0)$;
   for $t = 2.3$, $N = 12{,}000$, and $N_0 = 9000$

36. $P = \frac{1}{k}\sqrt{\dfrac{\log_{10} N}{t}}$;
   for $k = 0.4$, $N = 48$, and $t = 1.2$

37. $h = k\log_{10}\left(\dfrac{N}{N - N_0}\right)$;
   for $k = 1.2$, $N = 6400$, and $N_0 = 2000$

38. $Q = \frac{1}{t}\left(\dfrac{\log_{10} M}{\log_{10} N}\right)$;
   for $t = 0.3$, $M = 180$, and $N = 460$

Write each expression in Exercises 39–42 in terms of simpler logarithms. (Assume that all variables and variable expressions denote positive real numbers.)

39. $\log_b \left(\dfrac{xy^{1/3}}{z^2} \right)$

40. $\log_b \sqrt{\dfrac{L^2}{2R}}$

41. $\log_{10} \left(x \sqrt[3]{\dfrac{x}{y}} \right)$

42. $\log_{10} \sqrt{(s - a)(s - g)^2}$

Write each expression in Exercises 43–46 as a single logarithm with coefficient 1.

43. $\dfrac{1}{3} (\log_{10} x - 2 \log_{10} y)$

44. $\dfrac{1}{2} \log_{10} (3x) - \dfrac{2}{3} \log_{10} y$

45. $\dfrac{1}{3} \log_{10} 8 - 2(\log_{10} 8 - \log_{10} 2)$

46. $\dfrac{1}{2} (\log_{10} 9 + 2 \log_{10} 4) + 2 \log_{10} 5$

Solve each logarithmic equation in Exercises 47–50.

47. $\log_3 x + \log_3 4 = 2$

48. $\log_2 (x + 2) - \log_2 3 = 6$

49. $\log_{10} (x - 1) + \log_{10} (x + 2) = 3$

50. $\log_{10} (x + 2) - \log_{10} (x - 3) = 1$

Solve each equation in Exercises 51–54 by using base 10 logarithms.

51. $3^{x-2} = 7$

52. $4 \cdot 2^{1.2x} = 64$

53. $1200 = 24 \cdot 6^{-0.3x}$

54. $0.08 = 12 \cdot 3^{-1.5x}$

55. Solve $N = N_0 (10^{kt})$ for t.

56. Solve $Q = R_0 + R \log_{10} kt$ for t.

57. The population of Dry Gulch has been declining according to the function

$$P(t) = 3800 \cdot 2^{-t/20}$$

where t is the number of years since the town's heyday in 1910.

a. What was the population of Dry Gulch in 1990?

b. In what year will the population dip below 120 people?

58. The number of compact discs produced each year by Delta Discs is given by the function

$$N(t) = 8000 \cdot 3^{t/4}$$

where t is the number of years since discs were introduced in 1980.

a. How many discs did Delta produce in 1989?

b. In what year will Delta first produce over 2 million discs?

59. a. How much will a $90 camera cost 10 months from now if the inflation rate is 6% annually?

b. How long will it be before the camera costs $120?

60. a. How much will a $1200 sofa cost 20 months from now if the inflation rate is 8% annually?

b. How long will it be before the sofa costs $1500?

Solve each question in Exercises 61–66.

61. $e^x = 4.7$

62. $e^x = 0.5$

63. $\ln x = 6.02$

64. $\ln x = -1.4$

65. $4.73 = 1.2e^{0.6x}$

66. $1.75 = 0.3e^{-1.2x}$

67. Solve $y = 12e^{-kt} + 6$ for t.

68. Solve $N = N_0 + 4 \ln(k + 10)$ for k.

69. Express $N(t) = 600 (0.4)^t$ in the form

$$N(t) = N_0 e^{kt}$$

70. Express $N(t) = 100 (1.06)^t$ in the form

$$N(t) = N_0 e^{kt}$$

Chapter 8

Polynomial and Rational Functions

$\mathcal{T}$he quadratic functions we studied in Chapter 4 are examples of a larger class of functions called polynomials. The graphs of polynomials display great variety in their shapes, which makes them useful for modeling curves of many kinds. In addition, polynomial functions are relatively easy to evaluate and can be used to approximate the values of more difficult functions.

8.1 POLYNOMIAL FUNCTIONS

We have already encountered some examples of polynomial functions in our study of algebra. Linear functions,

$$f(x) = ax + b$$

and quadratic functions,

$$f(x) = ax^2 + bx + c$$

are special cases of polynomial functions. In general, a **polynomial function** has the form

$$f(x) = a_n x^n + a_{n-1} x^{n-1} + a_{n-2} x^{n-2} + \cdots + a_2 x^2 + a_1 x + a_0$$

where $a_0, a_1, a_2, \ldots, a_n$ are constants. Some additional examples of polynomials are

$$x^3 - 2x^2 + 1, \qquad 3x^5 - 2, \qquad \text{and} \qquad 2x^4 - x^3 + 3x^2 + 8x - 2$$

We can describe a polynomial as a sum of terms in which all the exponents on the variables are whole numbers and no variables appear in a denominator.

The following examples show some applications of polynomial functions.

Example 1

The power, P, in watts that a radar antenna must transmit to be able to detect an airplane at a range of R nautical miles is given by

$$P = 0.1R^4 \qquad (1)$$

451

a. Determine the power required to detect a plane at a range of 8 nautical miles.

b. Use a graph to approximate the range at which a plane can be detected when the transmitted power is 500 watts.

Solutions   a. Evaluate the function at $R = 8$ to obtain

$$P = 0.1(8)^4 = 409.6$$

The antenna must transmit 409.6 watts of power.

b. Graph the equations $Y_1 = 0.1X^4$ and $Y_2 = 500$ in the window

$$\text{Xmin} = 0, \quad \text{Xmax} = 10$$

$$\text{Ymin} = 0, \quad \text{Ymax} = 1000$$

We are looking for the value of X at which the two graphs intersect. Using the $\boxed{\text{TRACE}}$ key, we see (in Figure 8.1) that $Y \approx 500$ when $X \approx 8.4$. When the antenna transmits a 500-watt signal, the plane can be detected at a range of about 8.4 nautical miles.

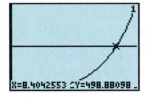

Figure 8.1

We can also answer part (b) of Example 1 by solving an equation algebraically. Use Equation (1) and replace P with 500 to obtain

$$500 = 0.1R^4 \qquad \text{Multiply both sides by 10.}$$

$$5000 = R^4 \qquad \text{Raise both sides to the power } \frac{1}{4}.$$

$$5000^{1/4} = R \qquad \text{Use a calculator for an approximation.}$$

$$8.409 \approx R$$

We see again that the range is approximately 8.4 nautical miles.

Example 2

An accounting firm is designing a large work space divided into square cubicles with 8-foot ceilings. The walls and ceiling of each cubicle will be covered with a soundproofing material, except for the door, which is 7 feet tall by 3 feet wide.

a. How much material will be needed for a cubicle if it measures 10 feet by 10 feet?

b. Write a function for the amount of soundproofing material needed for each cubicle in terms of the cubicle's size.

c. How big can the cubicles be if the soundproofing material costs $1.20 per square foot and the designers can budget at most $300 per cubicle for the material?

Solutions   a. To compute the amount of material needed for each cubicle, we add the areas of the ceiling and each wall and then subtract the area of the door.

Area of ceiling: $10 \times 10 = 100$ square feet

Area of each wall: $8 \times 10 = 80$ square feet

Area of door: $3 \times 7 = 21$ square feet

Because the cubicle has four walls, the total area to be covered is

$$100 + 4(80) - 21 = 399 \text{ square feet}$$

b.   Let d represent the dimensions (length and width) of the cubicle. Then compute the surface area as we did in part (a).

Area of ceiling: $d \times d = d^2$ square feet

Area of each wall: $8 \times d = 8d$ square feet

Area of door: $3 \times 7 = 21$ square feet

Finally, add the areas of the ceiling (d^2) and four walls   ($4 \times 8d = 32d$) and subtract the area of the door (21) to obtain a formula for the total soundproofed area.

$$A = d^2 + 32d - 21$$

c.   We can divide the amount budgeted by the cost per square foot to find the largest area the designers can afford to soundproof for each cubicle.

$$\frac{300 \text{ dollars}}{1.20 \text{ dollars per square foot}} = 250 \text{ square feet}$$

We substitute 250 for A and solve for d.

$$d^2 + 32d - 21 = 250$$

This is a quadratic equation, so we will write it in standard form and use the quadratic formula.

$$d^2 + 32d - 271 = 0$$

Thus

$$d = \frac{-32 \pm \sqrt{32^2 - 4(1)(-271)}}{2(1)}$$

$$d \approx 6.956 \quad \text{or} \quad d \approx -38.956$$

We discard the negative solution and find that the cubicle can have length and width 6.956 feet, or just under 7 feet.   ▭

Example 3

Leon is flying his plane to Au Gres, Michigan. He maintains a constant altitude until he passes over a marker just outside the neighboring town of Omer, when he begins his descent for landing. During the descent his altitude in feet is given by

$$A = 128x^3 - 960x^2 + 8000$$

where x is the number of miles Leon has traveled since passing over the marker in Omer.

a. What is Leon's altitude when he begins his descent?
b. What is Leon's altitude when he has gone 1.5 miles past the marker?
c. Use a graph to determine how far the plane travels past the marker before landing.

Solutions **a.** When Leon begins his descent, $x = 0$ and his altitude is

$$A = 128(0)^3 - 960(0)^2 + 8000 = 8000$$

Leon is at an altitude of 8000 feet when he begins his descent.

b. When $x = 1.5$,

$$A = 128(1.5)^3 - 960(1.5)^2 + 8000 = 6272$$

Leon is at an altitude of 6272 feet.

c. The altitude of the plane will be 0 when it lands, so we are trying to determine the value of x when $A = 0$.   A graph of

$$y = 128x^3 - 960x^2 + 8000$$

is shown in Figure 8.2, with

$$\text{Xmin} = 0, \quad \text{Xmax} = 5$$

$$\text{Ymin} = 0, \quad \text{Ymax} = 8000$$

Using the TRACE key, we can determine that the altitude is 0 when $x = 5$.

  The plane travels 5 miles from Omer until landing in Au Gres. (In other words, Au Gres is 5 miles from Omer.)

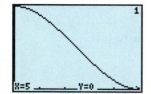

Figure 8.2

Polynomials in One Variable

The expressions $0.1 R^4$, $d^2 + 32d - 21$,   and $128x^3 - 960x^2 + 8000$ are all examples of polynomials in one variable. An algebraic expression consisting of one term of the form cx^n, where c is a constant and n is a whole number, is called a **monomial**. For example,

$$y^3, \qquad -3x^8, \qquad \text{and} \qquad 0.1R^4$$

are monomials. A polynomial is just a sum of one or more monomials.

  A polynomial with exactly two terms, like $\frac{1}{2}n^2 + \frac{1}{2}n$, is called a **binomial**. A polynomial with exactly three terms, like $d^2 + 32d - 21$ or $128x^3 - 960x^2 + 8000$, is called a **trinomial**. We have no special names for polynomials with more than three terms.

Example 4

Which of the following expressions are polynomials?

a. πr^2

b. $23.4s^6 - 47.9s^4$

c. $\frac{2}{3}w^3 - \frac{7}{3}w^2 + \frac{1}{3}w$

d. $7 + m^{-2}$

e. $\dfrac{x - 2}{x + 2}$

f. $\sqrt[3]{4y}$

Solutions The first three are polynomials: (a) is a monomial, (b) is a binomial, and (c) is a trinomial. The last three are not polynomials. The variable in (d) has a negative exponent, the variable in (e) occurs in the denominator, and the variable in (f) occurs under a radical, so its exponent is not a whole number. ▢

In a polynomial containing only one variable, the greatest exponent that appears on the variable is called the **degree** of the polynomial. If the polynomial has no variable, then it is called a **constant**, and the degree of a constant is zero.

Example 5

Give the degree of each polynomial.
a. $b^3 - 3b^2 + 3b - 1$
b. 10^{10}
c. $-4w^3$
d. $s^2 - s^6$

Solutions a. This is a polynomial in the variable b, and since the greatest exponent on b is 3, the degree of this polynomial is 3.
b. This is a constant polynomial, so its degree is 0. (The exponent on a *constant* does not affect the degree!)
c. This monomial has degree 3.
d. This is a binomial of degree 6. ▢

We can evaluate a polynomial just as we evaluate any other algebraic expression: we replace the variable with a number and simplify the result.

Example 6

Let $p(x) = -2x^2 + 3x - 1$. Find each of the following function values.
a. $p(2)$
b. $p(-1)$
c. $p(t)$
d. $p(t + 3)$

Solutions In each case we replace the x by the value given.
a. $p(2) = -2(2)^2 + 3(2) - 1 = -8 + 6 - 1 = -3$

b. $p(-1) = -2(-1)^2 + 3(-1) - 1 = -2 + (-3) - 1 = -6$

c. $p(t) = -2(t)^2 + 3(t) - 1 = -2t^2 + 3t - 1$

d. $p(t + 3) = -2(t + 3)^2 + 3(t + 3) - 1$
$$= -2(t^2 + 6t + 9) + 3(t + 3) - 1$$
$$= -2t^2 - 9t - 10$$ ▢

Products of Polynomials When we multiply two or more polynomials together, we get another polynomial of higher degree.

Example 7

a. $(x + 2)(5x^3 - 3x^2 + 4)$ Apply the distributive law.

$$= x(5x^3 - 3x^2 + 4) + 2(5x^3 - 3x^2 + 4)$$ Apply the distributive law again.

$$= 5x^4 - 3x^3 + 4x + 10x^3 - 6x^2 + 8$$ Combine like terms.

$$= 5x^4 + 7x^3 - 6x^2 + 4x + 8$$

b. $(x - 3)(x + 2)(x - 4)$ Multiply two of the factors first.

$$= (x - 3)(x^2 - 2x - 8)$$ Apply the distributive law.

$$= x(x^2 - 2x - 8) - 3(x^2 - 2x - 8)$$ Apply the distributive law again.

$$= x^3 - 2x^2 - 8x - 3x^2 + 6x + 24$$ Combine like terms.

$$= x^3 - 5x^2 - 2x + 24$$

In Example 7(a) we multiplied a polynomial of degree 1 by a polynomial of degree 3, and the product was a polynomial of degree 4. In Example 7(b) the product of three first-degree polynomials is a third-degree polynomial. In general,

> *The degree of a product of polynomials is the sum of the degrees of the factors.*

This fact will be useful to us in Section 8.2 when we graph polynomial functions.

There are a few special products resulting in cubic polynomials that are useful enough to merit specific mention. In Exercises 41 and 42 you will be asked to verify the following products.

CUBE OF A BINOMIAL

1. $(x + y)^3 = x^3 + 3x^2y + 3xy^2 + y^3$

2. $(x - y)^3 = x^3 - 3x^2y + 3xy^2 - y^3$

If you become familiar with these general forms, you can use them as patterns to find specific examples of such products.

Example 8

Write $(2w - 3)^3$ as a polynomial.

Solution   Use product 2, with x replaced by $2w$ and y replaced by 3.

$$(x - y)^3 = x^3 \quad - 3x^2y \quad + 3xy^2 \quad - y^3$$

$$(2w - 3)^3 = (2w)^3 - 3(2w)^2(3) + 3(2w)(3)^2 - 3^3 \qquad \text{Simplify.}$$

$$= 8w^3 \quad - 36w^2 \quad + 54w \quad - 27 \qquad \square$$

Of course, we can expand the product in Example 8 simply by performing the polynomial multiplication, which results in the same answer.

Another pair of products is useful for factoring cubic polynomials. In Exercises 43 and 44 you will be asked to verify the following products:

3. $(x + y)(x^2 - xy + y^2) = x^3 + y^3$

4. $(x - y)(x^2 + xy + y^2) = x^3 - y^3$

Viewing products (3) and (4) from right to left, we have the following special factorizations for the sum and difference of two cubes.

SUM OR DIFFERENCE OF TWO CUBES

3. $x^3 + y^3 = (x + y)(x^2 - xy + y^2)$

4. $x^3 - y^3 = (x - y)(x^2 + xy + y^2)$

In order to use these formulas as an aid in factoring, we must first recognize the polynomial to be factored as a sum or difference of two perfect cubes and then identify the expressions that are cubed.

Example 9

Factor each polynomial.

a. $8a^3 + b^3$ **b.** $1 - 27h^6$

Solutions   **a.** This polynomial is a sum of two cubes. The cubed expressions are $2a$ (because $(2a)^3 = 8a^3$) and b. Use formula (3) as a pattern, replacing x by $2a$ and y by b.

$$x^3 + y^3 = (x + y)(x^2 - xy + y^2)$$

$$(2a)^3 + b^3 = (2a + b)((2a)^2 - (2a)b + b^2) \qquad \text{Simplify.}$$

$$= (2a + b)(4a^2 - 2ab + b^2)$$

b. This polynomial is a difference of two cubes. The cubed expressions are 1 (because $1^3 = 1$) and $3h^2$ (because $(3h^2)^3 = 27h^6$). Use formula (4) as a pattern, replacing x by 1 and y by $3h^2$.

$$x^3 - y^3 = (x - y)(x^2 + xy + y^2)$$

$$1^3 - (3h^2)^3 = (1 - 3h^2)(1^2 + 1(3h^2) + (3h^2)^2) \qquad \text{Simplify.}$$

$$= (1 - 3h^2)(1 + 3h^2 + 9h^4) \qquad \square$$

Exercise 8.1

Identify each polynomial in Exercises 1–8 as a monomial, a binomial, or a trinomial. Give the degree of each polynomial.

1. $2x^3 - x^2$

2. $x - 2x + 1$

3. $5n^4$

4. $3n + 1$

5. $3r^2 - 4r + 2$

6. r^3

7. $y^3 - 2y^2 - y$

8. $3y^2 + 1$

Which of the expressions in Exercises 9–12 are not polynomials?

9. a. $1 - 0.04t^2$

b. $3x^2 - 4x + \dfrac{2}{x}$

c. $2\sqrt{z} - 7z^3 + 2$

d. $\sqrt{2}w^5 + \dfrac{3}{4}w^2 - w$

10. a. $\sqrt{3}p^2 - 7p + 2$

b. $2h^{4/3} + 6h^{1/3} - 2$

c. $\dfrac{2}{x^2 - 6x + 5}$

d. $\dfrac{1}{4}y^{-2} + 3y^{-1} + 4$

11. a. $\dfrac{1}{m^2 + 3}$

b. $v^2 - 16 + 2^v$

c. $\sqrt{x^3 - 4x}$

d. $\dfrac{m^4}{12}$

12. a. $3^t - 5t^3 + 2$

b. $\dfrac{q + 3}{q - 1}$

c. $c^{1/2} - c$

d. $\sqrt[3]{d} + 1$

In Exercises 13–20 evaluate each polynomial function for the given values of the variable.

13. $P(x) = x^3 - 3x^2 + x + 1$
a. $x = 2$ **b.** $x = -2$ **c.** $x = 2b$

14. $P(x) = 2x^3 + x^2 - 3x + 4$
a. $x = 3$ **b.** $x = -3$ **c.** $x = -a$

15. $Q(t) = t^2 + 3t + 1$
a. $t = \dfrac{1}{2}$ **b.** $t = -\dfrac{1}{3}$ **c.** $t = -w$

16. $Q(t) = 2t^2 - t + 1$
a. $t = \dfrac{1}{4}$ **b.** $t = -\dfrac{1}{2}$ **c.** $t = 3v$

17. $R(z) = 3z^4 - 2z^2 + 3$
a. $z = 1.8$ **b.** $z = -2.6$ **c.** $z = k - 1$

18. $R(z) = z^4 + 4z - 2$
a. $z = 2.1$ **b.** $z = -3.1$ **c.** $z = h + 2$

19. $N(a) = a^6 - a^5$
a. $a = -1$ **b.** $a = -2$ **c.** $a = \dfrac{m}{3}$

20. $N(a) = a^5 - a^4$
a. $a = -1$ **b.** $a = -2$ **c.** $a = \dfrac{q}{2}$

Multiply in Exercises 21–30.

21. $(y + 2)(y^2 - 2y + 3)$

22. $(t + 4)(t^2 - t - 1)$

23. $(3x - 2)(4x^2 + x - 2)$

24. $(2x + 3)(3x^2 - 4x + 2)$

25. $(x - 2)(x - 1)(x - 3)$

26. $(z - 5)(z + 6)(z - 1)$

27. $(2a^2 - 3a + 1)(3a^2 + 2a - 1)$

28. $(b^2 - 3b + 5)(2b^2 - b + 1)$

29. $(y - 2)(y + 2)(y + 4)(y + 1)$

30. $(z + 3)(z + 2)(z - 1)(z + 1)$

Without performing the polynomial multiplication, give the degree of each product in Exercises 31–36.

31. $(x^2 - 4)(3x^2 - 6x + 2)$

32. $(6x^2 - 1)(4x^2 - 9)$

33. $(x - 3)(2x - 5)(x^3 - x + 2)$

34. $(3x + 4)(3x + 1)(2x^3 + x^2 - 7)$

35. $(3x^2 + 2x)(x^3 + 1)(-2x^2 + 8)$

36. $(x^2 - 3)(2x^3 - 5x^2 + 2)(-x^3 - 5x)$

The polynomials in Exercises 37–40 will factor into a product of linear factors of the form $(x - a)$, where a can be either real or complex. How many linear factors are in the factored form?

37. $x^4 - 2x^3 + 4x^2 + 8x - 6$

38. $2x^5 - x^3 + 6x - 4$

39. $x^6 - 6x$

40. $x^3 + 3x^2 - 2x + 1$

Verify the products in Exercises 41–44, which were discussed in the text.

41. $(x + y)^3 = x^3 + 3x^2y + 3xy^2 + y^3$

42. $(x - y)^3 = x^3 - 3x^2y + 3xy^2 - y^3$

43. $(x + y)(x^2 - xy + y^2) = x^3 + y^3$

44. $(x - y)(x^2 + xy + y^2) = x^3 - y^3$

Write each product in Exercises 45–50 as a polynomial and simplify.

45. $(x - 1)(x^2 + x + 1)$

46. $(x + 2)(x^2 - 2x + 4)$

47. $(2x + 1)(4x^2 - 2x + 1)$

48. $(3x - 1)(9x^2 + 3x + 1)$

49. $(3a - 2b)(9a^2 + 6ab + 4b^2)$

50. $(2a + 3b)(4a^2 - 6ab + 9b^2)$

Factor completely in Exercises 51–62.

51. $x^3 + 27$ **52.** $y^3 - 1$ **53.** $(2x)^3 - y^3$

54. $y^3 + (3x)^3$ **55.** $a^3 - 8b^3$ **56.** $27a^3 + b^3$

57. $x^3y^3 - 1$ **58.** $8 + x^3y^3$

59. $27a^3 + 64b^3$ **60.** $8a^3 - 125b^3$

61. $125a^3b^3 - 1$ **62.** $64a^3b^3 + 1$

63. a. A closed box has a square base of length and width x inches and a height of 8 inches, as shown in Figure 8.3. Write a polynomial that gives the surface area of the box in terms of the dimensions of the base.

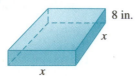

Figure 8.3

b. What is the surface area of a box of length and width 18 inches?

64. a. An empty reflecting pool is twice as long as it is wide and is 3 feet deep, as illustrated in Figure 8.4. Write a polynomial that gives the pool's surface area.

Figure 8.4

b. What is the surface area of the pool if it is 12 feet wide?

65. a. Write a polynomial that gives the area of the front face of the speaker frame (the color shaded region) in Figure 8.5.

b. If $x = 8$ inches, find the area of the front face of the frame.

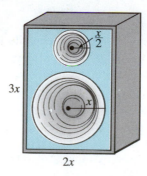

Figure 8.5

66. a. A Norman window is shaped like a rectangle whose length is twice its width, surmounted by a semicircle. (See Figure 8.6.) Write a polynomial that gives its area.

b. If $x = 3$ feet, find the area of the window.

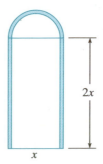

Figure 8.6

67. a. A grain silo is built in the shape of a cylinder with a hemisphere on top. (See Figure 8.7.) Write an expression for the volume of the silo in terms of the radius and height of the cylindrical portion of the silo.

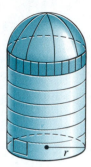

Figure 8.7

b. If the total height of the silo is five times its radius, write a polynomial in one variable for its volume.

68. a. A cold medication capsule is shaped like a cylinder with a hemispherical cap on each end. (See Figure 8.8.) Write an expression for the volume of the capsule in terms of the radius and length of the cylindrical portion.

Figure 8.8

b. If the radius of the capsule is one-fourth of its overall length, write a polynomial in one variable for its volume.

69. Jack invests $500 in an account bearing interest rate r, compounded annually; that is, each year his account balance increases by a factor of $1 + r$.

a. Write expressions for the amount of money in Jack's account after 2 years, after 3 years, and after 4 years.

b. Write the expressions you found in part (a) as polynomials.

c. How much money will be in Jack's account at the end of 2 years, 3 years, and 4 years if the interest rate is 8%? Do you get the same answers if you use the expressions from part (a) and the polynomials from part (b)?

70. A small company borrows $800 for start-up costs and agrees to repay the loan at interest rate r, compounded annually; that is, each year the debt increases by a factor of $1 + r$.

a. Write expressions for the amount of money the company will owe if it repays the loan after 2 years, after 3 years, or after 4 years.

b. Write the expressions you found in part (a) as polynomials.

c. How much money will the company owe after 2 years, 3 years, and 4 years at an interest rate of 12%? Do you get the same answers if you use the expressions from part (a) and the polynomials from part (b)?

71. A paper company plans to make boxes without tops from sheets of cardboard 12 inches wide and 16 inches long. They will cut out four squares of side x inches from the corners of the sheet and fold up the edges, as shown in Figure 8.9.

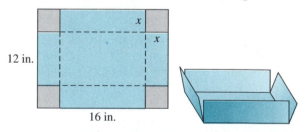

Figure 8.9

a. Write an expression in terms of x for the length, width, and height of the resulting box.

b. Write a formula for the volume, V, of the box as a function of x.

c. What is the domain of the function V? (What are the largest and smallest reasonable values for x?)

d. Evaluate your function for $x = 1$, $x = 2$, and $x = 3$.

e. Graph your function V in a suitable window.

f. Use your graph to find the value of x that will yield a box with maximum possible volume. What is the maximum possible volume?

72. The paper company also plans to make boxes with tops from 12-inch by 16-inch sheets of cardboard by cutting out the shaded areas shown in Figure 8.10 and folding along the dotted lines.

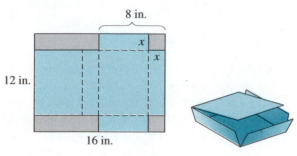

Figure 8.10

a. Write an expression in terms of x for the length, width, and height of the resulting box.

b. Write a formula for the volume, V, of the box as a function of x.

c. What is the domain of the function V? (What are the largest and smallest reasonable values for x?)

d. Evaluate your function for $x = 1$, $x = 2$, and $x = 3$.

e. Graph your function V in a suitable window.

f. Use your graph to find the value of x that will yield a box with maximum possible volume. What is the maximum possible volume?

Use the [TRACE] feature on the graphing calculator to help you answer the questions in Exercises 73–80. Verify your answers algebraically.

73. A doctor who is treating a heart patient wants to prescribe medication to lower the patient's blood pressure. The body's reaction to this medication is a function of the dose administered. If the patient takes x milliliters of the medication, his blood pressure should decrease by $R = f(x)$ points, where

$$f(x) = 3x^2 - \frac{1}{3}x^3$$

a. For what values of x is $R = 0$?

b. Graph the function f on a suitable domain.

c. How much should the patient's blood pressure drop if he takes 2 milliliters of medication?

d. What is the maximum drop in blood pressure that can be achieved with this medication?

e. There may be risks associated with a large change in blood pressure. How many milliliters of the medication should be administered to achieve half the maximum possible drop in blood pressure?

74. A soup bowl has the shape of a hemisphere of radius 6 centimeters. The volume, $V = f(x)$, of the soup in the bowl is a function of the depth, x, of the soup. (See Figure 8.11.)

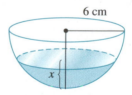

Figure 8.11

a. What is the domain of f?

b. The function f is given by

$$f(x) = 6\pi x^2 - \frac{\pi}{3}x^3$$

Graph the function on its domain.

c. What is the volume of the soup if it is 3 centimeters deep?

d. What is the maximum volume of soup that the bowl can hold?

e. Find the depth of the soup (to within two decimal places of accuracy) when the bowl is filled to half its capacity.

75. The population, $P(t)$, of Cyberville has been growing according to the formula

$$P(t) = t^3 - 63t^2 + 1403t + 900$$

where t is the number of years since 1960.

a. Graph $P(t)$ on the window

$$\begin{aligned}&\text{Xmin} = 0, \quad \text{Xmax} = 47\\&\text{Ymin} = 0, \quad \text{Ymax} = 20{,}000\end{aligned}$$

b. What was the population in 1960? 1975? 1994?

c. By how much did the population grow from 1960 to 1961? From 1975 to 1976? From 1994 to 1995?

d. Approximately when was the population growing the most slowly; that is, when is the graph the least steep?

76. The annual profit (in thousands of dollars), $P(t)$, of the Enviro Company is given by

$$P(t) = 2t^3 - 152t^2 + 3400t + 30$$

where t is the number of years since 1950, the first year the company showed a profit.

a. Graph $P(t)$ on the window

$$\begin{aligned}&\text{Xmin} = 0, \quad \text{Xmax} = 94\\&\text{Ymin} = 0, \quad \text{Ymax} = 50{,}000\end{aligned}$$

b. What were the profits in 1950? 1970? 1990?

c. How did the profits change from 1950 to 1951? From 1970 to 1971? From 1990 to 1991?

d. During which years did profits decrease from one year to the next?

77. During an earthquake Nordhoff Way split in two, and one section shifted up several centimeters. Engineers created a ramp from the lower section to the upper section. In the coordinate system shown in Figure 8.12, the ramp is part of the graph of

$$y = -0.00004x^3 - 0.006x^2 + 20$$

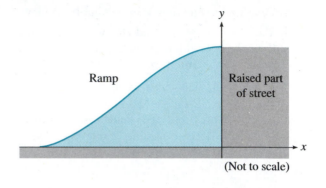

(Not to scale)

Figure 8.12

a. By how much did the upper section of the street shift during the earthquake?

b. What is the horizontal distance from the bottom of the ramp to the raised part of the street?

78. The off-ramp from a highway connects to a parallel one-way road. Figure 8.13 shows the highway, the off-ramp, and the road. The road lies on the x-axis, and the off-ramp begins at a point on the y-axis. The off-ramp is part of the graph of the polynomial

$$y = 0.00006x^3 - 0.009x^2 + 30$$

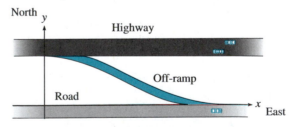

Figure 8.13

a. How far east of the exit does the off-ramp meet the one-way road?

b. How far apart are the highway and the road?

79. The number of minutes of daylight per day in Chicago is approximated by the polynomial

$$H(t) = 0.000000525t^4 - 0.0213t^2 + 864$$

where t is the number of days since the summer solstice. The approximation is valid for

$-74 < t < 74$. (A negative value of t corresponds to a number of days before the summer solstice.)

a. Graph the polynomial on its domain.

b. How many minutes of daylight are there on the summer solstice?

c. How much daylight is there 2 weeks before the solstice?

d. When are the days more than 14 hours long?

e. When are the days less than 13 hours long?

80. The water level (in feet) at a harbor is approximated by the polynomial

$$W(t) = 0.00733t^4 - 0.332t^2 + 9.1$$

where t is the number of hours since the high tide. The approximation is valid for $-4 \le t \le 4$. (A negative value of t corresponds to a number of hours before the high tide.)

a. Graph the polynomial on its domain.

b. What is the water level at high tide?

c. What is the water level 3 hours before high tide?

d. When is the water level below 8 feet?

e. When is the water level above 7 feet?

8.2 GRAPHING POLYNOMIAL FUNCTIONS

The graph of a polynomial function depends first of all on its degree. We have already studied the graphs of polynomials of degrees zero, one, and two. A polynomial of degree 0 is a constant, and its graph is a horizontal line. An example of such a polynomial function is $f(x) = 3$. (See Figure 8.14[a].)A polynomial of degree 1 is a linear function, and its graph is a straight line. The function $f(x) = 2x - 3$ is an example of a polynomial of degree 1. (See Figure 8.14[b].) Quadratic functions, such as $f(x) = -2x^2 + 6x + 8$, are polynomials of degree 2. The graph of every quadratic function is a parabola, with the same basic shape as the standard parabola, $y = x^2$. (See Figure 8.15.)

Do the graphs of all *cubic,* or third-degree, polynomials have a basic shape in common? We can graph a few examples and find out.

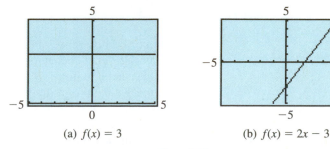

(a) $f(x) = 3$ (b) $f(x) = 2x - 3$

Figure 8.14

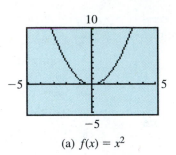

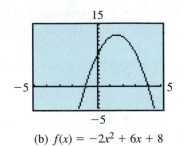

Figure 8.15   (a) $f(x) = x^2$ (b) $f(x) = -2x^2 + 6x + 8$

Example 1

Graph the cubic polynomial $P(x) = x^3 - 4x$, and compare its graph with that of the standard cubic, $y = x^3$.

Solution   The graph of the standard cubic is shown in Figure 8.16(a). To help us understand the graph of the polynomial $y = P(x)$, we will make a table of ordered pairs by evaluating the function. We can do this by hand (see Table 8.1) or by using the table feature on the graphing calculator.

x	−3	−2	−1	0	1	2	3
P(x)	−15	0	3	0	−3	0	15

Table 8.1

The graph of $y = P(x)$ is shown in Figure 8.16(b). It is not exactly the same shape as the basic cubic, but it is similar, especially at the edges of the graphs.

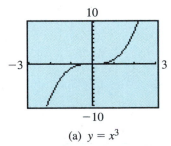

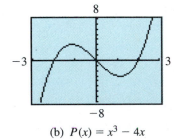

Figure 8.16   (a) $y = x^3$ (b) $P(x) = x^3 - 4x$

Despite the differences in the central portions of the two graphs, they exhibit similar "end" behavior (for very large and very small values of x). The y-values increase from $-\infty$ toward zero in the third quadrant and increase from zero toward $+\infty$ in the first quadrant. In simpler terms we might say that the graphs start at the lower left and extend to the upper right. All cubic polynomials display this behavior when their lead coefficients (the coefficient of the x^3 term) are positive, because for large values of $|x|$ the behavior of a polynomial is dominated by its term of highest degree.

    Notice that both of the graphs in Example 1 are smooth curves without any breaks or holes. This smoothness is a feature of the graphs of all polynomial functions. The domain of any polynomial function is the entire set of real numbers.

Now let's compare the end behavior of two quartic, or fourth-degree, polynomials.

Example 2

Graph the polynomials $f(x) = x^4 - 10x^2 + 9$ and $g(x) = x^4 + 2x^3$, and compare.

Solution   For each function we make a table of values, as shown in Tables 8.2 and 8.3.

x	-4	-3	-2	-1	0	1	2	3	4
$f(x)$	105	0	-15	0	9	0	-15	0	105

Table 8.2

x	-3	-2	-1	0	1	2	3
$g(x)$	27	0	-1	0	3	32	135

Table 8.3

The graphs are shown in Figure 8.17.

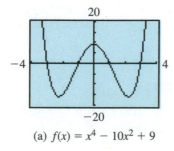

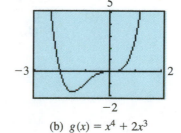

Figure 8.17   (a) $f(x) = x^4 - 10x^2 + 9$ (b) $g(x) = x^4 + 2x^3$

All the essential features of the graphs are shown in these viewing windows. In other words, the graphs continue forever in the directions indicated, without any additional twists or turns.

As in Example 1, both graphs have similar end behavior. The y-values decrease from $+\infty$ toward zero as x increases from $-\infty$, and the y-values increase toward $+\infty$ as x increases to $+\infty$. This behavior is similar to that of quadratic polynomials with positive lead coefficients. Their graphs start at the upper left and extend to the upper right, which supports our earlier observation that the end behavior of a polynomial is determined by its term of highest degree. (The other terms of the polynomial determine its behavior in the middle.)

In the exercises you will consider more graphs of polynomials of various degrees to help you verify the following observations:

---

1. A polynomial of *odd* degree (with a positive lead coefficient) has negative *y*-values for large negative *x*, and positive *y*-values for large positive *x*.

2. A polynomial of *even* degree (with a positive lead coefficient) has positive *y*-values for both large positive and large negative *x*.

---

x-Intercepts and the Factor Theorem

In Chapter 4 we saw that the *x*-intercepts of a quadratic polynomial *f* occur at values of *x* for which $f(x) = 0$, that is, at the real-valued solutions of the equation $ax^2 + bx + c = 0$. The same thing holds true for polynomials of higher degree.

Solutions of the equation $P(x) = 0$ are called **zeros** of the polynomial *P*. In Example 1 we graphed the cubic polynomial $P(x) = x^3 - 4x$. Its *x*-intercepts are the solutions of the equation $x^3 - 4x = 0$, which we can solve by factoring the polynomial $P(x)$:

$$x^3 - 4x = 0$$

$$x(x - 2)(x + 2) = 0$$

Thus the zeros of *P* are 0, 2, and -2. Note that each zero of *P* corresponds to a factor of $P(x)$. This result suggests the following theorem, which holds for any polynomial *P*.

FACTOR THEOREM

Let $P(x)$ be a polynomial with real-number coefficients. Then $(x - a)$ is a factor of $P(x)$ if and only if $P(a) = 0$.

Since a polynomial function of degree *n* can have at most *n* linear factors of the form $(x - a)$, it follows that *P* can have at most *n* distinct zeros. Another way of saying this is that if $P(x)$ is a polynomial of nth degree, the equation $P(x) = 0$ can have at most *n* distinct solutions, some of which may be complex. Because only real-valued solutions appear on the graph as *x*-intercepts, a polynomial of degree *n* can have at most *n* *x*-intercepts. If some of the zeros of *P* are complex numbers, they will not appear on the graph, so a polynomial of degree *n* may have fewer than *n* *x*-intercepts.

Example 3

Find the zeros of each polynomial, and list the *x*-intercepts of its graph.
a. $f(x) = x^3 + 6x^2 + 9x$ **b.** $g(x) = x^4 - 3x^2 - 4$

Solutions **a.** Factor the polynomial to obtain

$$f(x) = x(x^2 + 6x + 9)$$

$$= x(x + 3)(x + 3)$$

By the factor theorem, the zeros of f are 0, -3, and -3. (We say that f has a zero *of multiplicity two* at -3.) Since all of these are real numbers, all will appear as x-intercepts on the graph. Thus the x-intercepts occur at $(0, 0)$ and at $(-3, 0)$.

b. Factor the polynomial to obtain

$$g(x) = (x^2 - 4)(x^2 + 1)$$

$$= (x - 2)(x + 2)(x^2 + 1)$$

When we extract roots, the solutions of

$$x^2 + 1 = 0$$

are

$$x = \pm\sqrt{-1} = \pm i$$

Thus the zeros of g are -2, 2, $-i$, and i. Since only the first two of these are real numbers, the graph has only two x-intercepts, at $(-2, 0)$ and $(2, 0)$. The graphs of both polynomials are shown in Figure 8.18.

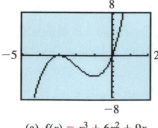
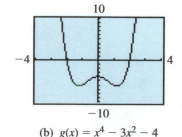

Figure 8.18   (a) $f(x) = x^3 + 6x^2 + 9x$ (b) $g(x) = x^4 - 3x^2 - 4$

Zeros of Multiplicity Two or Three

The appearance of the graph near an x-intercept is determined by the multiplicity of the zero there. Both real zeros of the polynomial

$$g(x) = x^4 - 3x^2 - 4$$

in Example 3(b) are of multiplicity one, and the graph *crosses* the x-axis at each intercept. However, the polynomial $f(x) = x^3 + 6x^2 + 9x$ in Example 3(a) has a zero of multiplicity two at $x = -3$. The graph of f just *touches* the x-axis and then reverses direction without crossing the axis.

To help us understand what happens in general, compare the graphs of the three polynomials in Figure 8.19. In Figure 8.19 $L(x) = x - 2$ has a zero of multiplicity one at $x = 2$, and its graph crosses the x-axis there. In Figure 8.19(b) $Q(x) = (x - 2)^2$ has a zero of multiplicity two at $x = 2$, and its graph touches the x-axis at $x = 2$ but changes direction

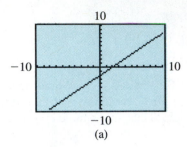

(a)

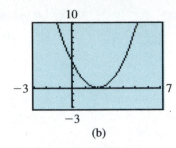

(b)

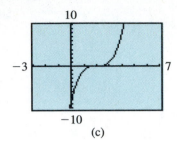

(c)

Figure 8.19

without crossing. In Figure 8.19(c) $C(x) = (x - 2)^3$ has a zero of multiplicity three at $x = 2,$ and its graphs makes an **S**-shaped curve at the intercept, like the graph of $y = x^3.$

Near its x-intercepts the graph of a polynomial takes one of the characteristic shapes illustrated in Figure 8.19. (Although we will not consider zeros of higher multiplicity, they correspond to similar behavior in the graph: at a zero of odd multiplicity the graph has an **S**-shaped curve at the intercept, and at a zero of even multiplicity the graph changes direction without crossing the x-axis.)

Example 4

Graph the polynomial

$$f(x) = (x + 2)^3 (x - 1)(x - 3)^2$$

Solution   The polynomial has degree 6, an even number, so its graph starts at the upper left and extends to the upper right. Its y-intercept is

$$f(0) = (2)^3 (-1)(-3)^2 = -72$$

f has a zero of multiplicity three at $x = -2,$ a zero of multiplicity one at $x = 1,$ and a zero of multiplicity two at $x = 3.$ The graph has an **S**-shaped curve at $x = -2,$ crosses the x-axis at $x = 1,$ touches the x-axis at $x = 3$ and then changes direction, as shown in Figure 8.20.

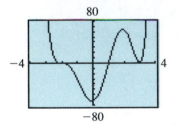

Figure 8.20

Exercise 8.2

Use your calculator to graph each cubic (third-degree) polynomial in Exercises 1–10. Write a sentence or two describing the end behavior of the graphs.

1. $y = x^3 + 4$
2. $y = x^3 - 8$
3. $y = -2 - 0.05x^3$
4. $y = 5 - 0.02x^3$
5. $y = x^3 - 3x$
6. $y = 9x - x^3$
7. $y = x^3 + 5x^2 - 4x - 20$

8. $y = -x^3 - 2x^2 + 5x + 6$
9. $y = x^3 + 2x^2 + 4x - 4$
10. $y = x^3 - 3x + 2$
11. Use a calculator to graph each cubic polynomial. Which graphs are the same?
   a. $y = x^3 - 2$ **b.** $y = (x - 2)^3$
   c. $y = x^3 - 6x^2 + 12x - 8$

12. Use a calculator to graph each cubic polynomial. Which graphs are the same?

a. $y = x^3 + 3$ **b.** $y = (x + 3)^3$

c. $y = x^3 + 9x^2 + 27x + 27$

Use a calculator to graph each quartic (fourth-degree) polynomial in Exercises 13–20. Write a sentence or two describing the end behavior of the graphs.

13. $y = 0.5x^4 - 4$ **14.** $y = 0.3x^4 + 1$

15. $y = -x^4 + 6x^2 - 10$

16. $y = x^4 - 8x^2 - 8$ **17.** $y = x^4 - 3x^3$

18. $y = -x^4 - 4x^3$

19. $y = -x^4 - x^3 - 2$

20. $y = x^4 + 2x^3 + 4x^2 + 10$

In Exercises 21–30 sketch a rough graph of each polynomial function by hand.

21. $f(x) = (x + 3)(x - 1)^2$

22. $g(x) = (x + 4)^2 (x - 2)$

23. $G(x) = (x - 2)^2 (x + 2)^2$

24. $F(x) = (x - 1)^2 (x - 3)^2$

25. $h(x) = x^3 (x + 2)(x - 2)$

26. $H(x) = (x + 1)^3 (x - 2)^2$

27. $P(x) = (x + 4)^2 (x + 1)^2 (x - 1)^2$

28. $Q(x) = x^2 (x - 5)(x - 1)^2 (x + 2)$

29. $q(x) = (x + 3)^2 (x + 1)^3$

30. $p(x) = (x + 2)^3 (x - 1)^3$

In Exercises 31–38
a. use a calculator to graph each polynomial and locate the x-intercepts.
b. write the polynomial in factored form.

31. $P(x) = x^3 - 7x - 6$

32. $Q(x) = x^3 + 3x^2 - x - 3$

33. $R(x) = x^4 - x^3 - 4x^2 + 4x$

34. $S(x) = x^4 + 3x^3 - x^2 - 3x$

35. $p(x) = x^3 - 3x^2 - 6x + 8$

36. $q(x) = x^3 + 6x^2 - x - 30$

37. $r(x) = x^4 - x^3 - 10x^2 + 4x + 24$

38. $s(x) = x^4 + x^3 - 8x^2 - 2x + 12$

In Exercises 39–50
a. find the zeros of each polynomial.
b. sketch a rough graph by hand.

39. $P(x) = x^4 + 4x^2$ **40.** $P(x) = x^3 + 3x$

41. $P(x) = x^3 - 8x$ **42.** $P(x) = 4x^4 - 9x^2$

43. $f(x) = x^4 + 4x^3 + 4x^2$

44. $g(x) = x^4 + 4x^3 + 3x^2$

45. $g(x) = 16x - x^3$ **46.** $f(x) = 12x - x^3$

47. $k(x) = x^4 - 10x^2 + 16$

48. $m(x) = x^4 - 15x^2 + 36$

49. $r(x) = (x^2 - 1)(x + 3)^2$

50. $s(x) = (x^2 - 9)(x - 1)^2$

Find the equation of each polynomial graphed in Exercises 51–56.

51.

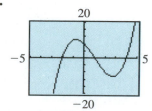

52.

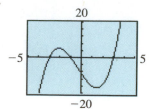

53.

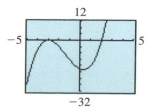

54.

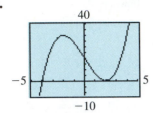

55.

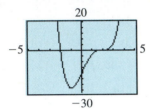

56.

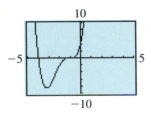

In Exercises 57–60 write the formula for each function and graph it with a calculator. Describe how the graph of each function differs from the graph of $y = f(x)$.

57. $f(x) = x^3 - 4x$

   a. $y = f(x) + 3$ **b.** $y = f(x) - 5$

   c. $y = f(x - 2)$ **d.** $y = f(x + 3)$

58. $f(x) = x^3 - x^2 + x - 1$

   a. $y = f(x) + 4$ **b.** $y = f(x) - 4$

   c. $y = f(x - 3)$ **d.** $y = f(x + 5)$

59. $f(x) = x^4 - 4x^2$

   a. $y = f(x) + 6$ **b.** $y = f(x) - 2$

   c. $y = f(x - 4)$ **d.** $y = f(x + 2)$

60. $f(x) = x^4 + 3x^3$

   a. $y = f(x) + 5$ **b.** $y = f(x) - 3$

   c. $y = f(x - 2)$ **d.** $y = f(x + 4)$

8.3   RATIONAL FUNCTIONS

A **rational function** is one of the form

$$f(x) = \frac{P(x)}{Q(x)}$$

where $P(x)$ and $Q(x)$ are polynomials. The graphs of rational functions can be quite different from the graphs of polynomials.

Example 1

Francine is planning a 60-mile training flight through the desert on her pedal-driven aircraft. If there is no wind, she can pedal at an average speed of 15 miles per hour, so she can complete the flight in 4 hours.

a. If there is a headwind of x miles per hour, it will take Francine longer to fly 60 miles. Express the time it will take for Francine to complete the training flight as a function of x.

b. Graph the function and explain what it tells you about the time Francine should allot for the ride.

Solutions   a. If there is a headwind of x miles per hour, Francine's ground speed will be $15 - x$ miles per hour. Since time $= \dfrac{\text{distance}}{\text{rate}}$,   we find that the time needed for the flight will be

$$t = \frac{60}{15 - x}$$

b.   The graph of the function is shown in Figure 8.21. You can use your calculator with the window

$$\text{Xmin} = -8.5, \quad \text{Xmax} = 15$$

$$\text{Ymin} = 0, \qquad \text{Ymax} = 30$$

to check the graph. Note that when $x = 0$, $t = 4$,   so if there is no wind Francine can fly the 60 miles in 4 hours, as we calculated earlier.

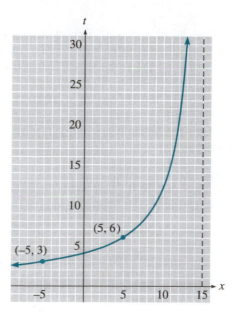

Figure 8.21

As x increases, so does the time it will take Francine to complete her flight. For example, if the headwind has a speed of 5 miles per hour, Francine's effective speed is only 10 miles per hour, and it will take her 6 hours to fly the 60 miles. You can use the $\boxed{\text{TRACE}}$ key to locate the point $(5, 6)$ on the graph. In fact, as the speed of the wind gets close to 15 miles per hour, Francine's flying time becomes extremely large. In theory, if the wind speed were exactly 15 miles per hour, Francine would never complete her flight. On the graph the time becomes infinite at $x = 15$.

What about negative values for x? If we interpret a negative headwind as a tailwind, Francine's flying time should decrease for negative x-values. For example, if $x = -5$ there is a tailwind of 5 miles per hour, so Francine's effective speed is 20 miles per hour and she can complete the flight in 3 hours. As the tailwind gets stronger (that is, as we move farther to the left in the x-direction), Francine's flying time continues to decrease and the graph approaches the x-axis.

The vertical dashed line at $x = 15$ on the graph of $t = \dfrac{60}{15 - x}$ is a *vertical asymptote* for the graph. We first encountered asymptotes in Section 5.3 when we studied the graph of $y = \dfrac{1}{x}$. Locating the vertical asymptotes of a rational function is an important part of determining the shape of the graph.

Example 2

EarthCare decides to sell T-shirts to raise money. They make an initial investment of $100 to pay for the design of the T-shirt and to set up the printing process for their design. After that the T-shirts cost $5 apiece for labor and materials.
a.   Express EarthCare's average cost per T-shirt as a function of the number of T-shirts they produce.
b.   Graph the function and explain what it tells you about the cost of the T-shirts.

Solutions   a.   If EarthCare produces x T-shirts, their total costs will be $100 + 5x$ dollars. To find the average cost per T-shirt, we divide the total cost by the number of T-shirts produced to get

$$C = \frac{100 + 5x}{x}$$

b.   The graph is shown in Figure 8.22. You can use your calculator, with the window

$$\text{Xmin} = 0, \quad \text{Xmax} = 470$$
$$\text{Ymin} = 0, \quad \text{Ymax} = 30$$

to check the graph.
   If EarthCare makes only one T-shirt, their cost is $105. But if they make more T-shirts, the cost of the original $100 investment is distributed among them. For example, the average cost per T-shirt for two T-shirts is

$$\frac{100 + 5(2)}{2} = 55$$

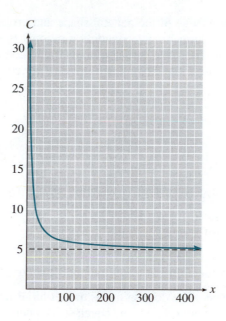

Figure 8.22

and the average cost for five T-shirts is

$$\frac{100 + 5(5)}{5} = 25$$

(You can locate this last point, (5, 25), with the [TRACE] feature on your graph.)

As the number of T-shirts increases, the average cost per shirt continues to decline, but not as rapidly as at first. Eventually the average cost levels off and approaches $5 per T-shirt. For example, if EarthCare produces 400 T-shirts, the average cost per shirt is

$$\frac{100 + 5(400)}{400} = 5.25$$

The horizontal line $C = 5$ on the graph of $C = \dfrac{100 + 5x}{x}$ is a *horizontal asymptote*. As x increases the graph approaches the line $C = 5$ but will never actually meet it. The average price per T-shirt will always be slightly more than $5. Horizontal asymptotes are also important in sketching the graphs of rational functions.

Graphing Rational Functions   Most applications of rational functions have restricted domains; that is, they make sense for only a subset of the real numbers on the x-axis. Consequently, only a portion of the graph is useful for analyzing the application. However, a knowledge of the general shape and properties of the whole graph can be very helpful in understanding a rational function.

As we stated earlier, a rational function is a quotient of two polynomials. Here are some examples of rational functions:

$$f(x) = \frac{2}{(x - 3)^2} \qquad g(x) = \frac{x}{x + 1}$$

$$h(x) = \frac{2x^2}{x^2 + 4} \qquad k(x) = \frac{x^2 - 1}{x^2 - 9}$$

Because we cannot divide by zero, a rational function $F(x) = \dfrac{P(x)}{Q(x)}$ is undefined for any value $x = a$ where $Q(a) = 0$. These x-values are not in the domain of the function.

Example 3 Find the domains of the rational functions f, g, h, and k defined above.

Solution The domain of f is the set of all real numbers except 3, because the denominator, $(x - 3)^2$, equals zero when $x = 3$. The domain of g is the set of all real numbers except -1, because $x + 1$ equals zero when $x = 1$. The denominator of the function h, $x^2 + 4$, is never equal to zero, so the domain of h is all real numbers. The domain of k is the set of all real numbers except 3 and -3, because $x^2 - 9$ equals zero when $x = 3$ or $x = -3$.   ▭

Note that we only need to exclude the zeros of the *denominator* from the domain of a rational function. We do not exclude the zeros of the numerator. In fact, the zeros of the numerator include the zeros of the rational function itself, since a fraction is equal to zero when its numerator is zero but its denominator is not zero.

Vertical Asymptotes As we saw in Section 8.2, a polynomial function is defined for all values of x, and its graph is a smooth curve without any breaks or holes. The graph of a rational function, on the other hand, will exhibit breaks or holes at those x-values where it is undefined.

For example, consider the function

$$f(x) = \frac{2}{(x - 3)^2}$$

This function is undefined for $x = 3$, so there is no point on the graph with x-coordinate 3. However, we can make a table of values for other values of x. Plotting the ordered pairs in Table 8.4 results in the points shown in Figure 8.23.

x	y
0	2/9
1	1/2
2	2
3	Undefined
4	2
5	1/2
6	2/9

Table 8.4

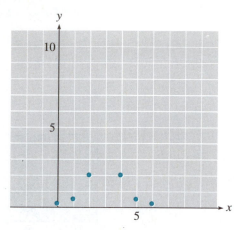

Figure 8.23

To fill in the rest of the graph, first use your calculator to make tables for x-values greater than 6 or less than 0, as shown in Figure 8.24. (Set the **Tbl Set** menu for **Ask** on the independent variable and **Auto** on the dependent variable.) These tables show that as x gets very large in absolute value, $\dfrac{2}{(x-3)^2}$ gets closer to zero. Consequently, the graph approaches the x-axis as we move away from the origin, as shown in Figure 8.25.

Next make tables by choosing x-values close to 3, as shown in Figure 8.26. As we choose x-values closer and closer to 3, $(x-3)^2$ gets closer

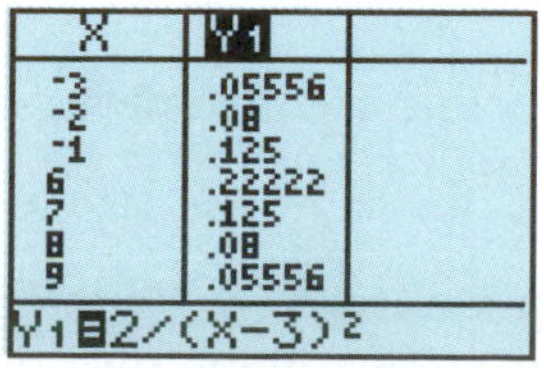

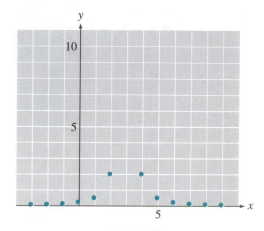

Figure 8.24

Figure 8.25

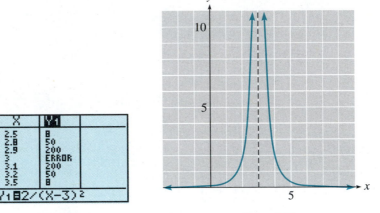

Figure 8.26   Figure 8.27

to zero, so the fraction $\dfrac{2}{(x-3)^2}$ gets very large in absolute value. The graph has a vertical asymptote at $x = 3$, which means that the graph approaches but never touches the vertical line $x = 3$. We indicate the vertical asymptote on the completed graph by a dashed line, as shown in Figure 8.27.

We see that the graph of $f(x) = \dfrac{2}{(x-3)^2}$ is a smooth curve except for a break at $x = 3$, where there is a vertical asymptote. The graph cannot be drawn in one piece like that of a polynomial; it is broken into two pieces separated at $x = 3$.

In general, we have the following result.

VERTICAL ASYMPTOTES

If $Q(a) = 0$ but $P(a) \neq 0$, then the graph of the rational function $f(x) = \dfrac{P(x)}{Q(x)}$ has a vertical asymptote at $x = a$.

If $P(a)$ and $Q(a)$ are both zero, then the graph of the rational function $\dfrac{P(x)}{Q(x)}$ may have a "hole" at $x = a$ rather than an asymptote. (This possibility is considered in Exercises 71–74 in Exercise 8.4.)

Near a vertical asymptote, the graph of a rational function has one of the four characteristic shapes illustrated in Figure 8.28. Locating the vertical asymptotes can help us make a quick sketch of a rational function.

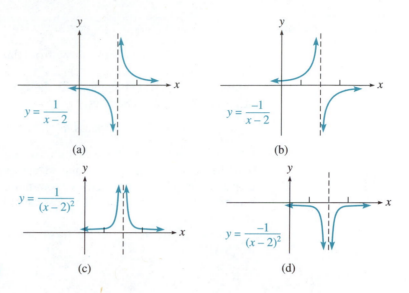

Figure 8.28

$$y = \frac{1}{x-2}$$ (a)

$$y = \frac{-1}{x-2}$$ (b)

$$y = \frac{1}{(x-2)^2}$$ (c)

$$y = \frac{-1}{(x-2)^2}$$ (d)

Example 4

Locate the vertical asymptotes and sketch the graph of $g(x) = \dfrac{x}{x+1}$.

Figure 8.29

X	Y1
-8	1.1429
-2	2
-1.2	6
-.9	-9
0	0
4	.8
8	.88889

Y1◻X/(X+1)

Solution The denominator, $x + 1$, equals zero when $x = -1$. Since the numerator does not equal zero when $x = -1$, there is a vertical asymptote at $x = -1$. The asymptote separates the graph into two pieces. Use the TABLE feature of your calculator to evaluate $g(x)$ for several values of x on either side of the asymptote, as shown in Figure 8.29. Plot the points found in this way and connect them on either side of the asymptote to obtain the graph shown in Figure 8.30.

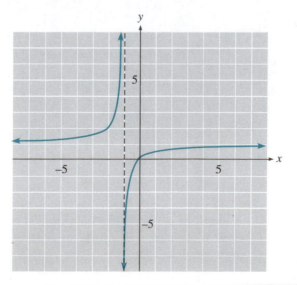

Figure 8.30

Horizontal Asymptotes   Look again at the graph of $g(x) = \dfrac{x}{x+1}$ in Example 4. Note that as $|x|$ gets large, that is, as we move away from the origin along the x-axis in either direction, the corresponding y-values get closer and closer to 1. The graph approaches but never coincides with the line $y = 1$.   We say that the graph has a horizontal asymptote at $y = 1$.

When does a rational function $f(x) = \dfrac{P(x)}{Q(x)}$ have a horizontal asymptote? It depends on the degrees of the two polynomials $P(x)$ and $Q(x)$. Notice that the degree of the polynomial in the numerator of $g(x)$ is equal to the degree of the polynomial in the denominator. Or we can say that the highest power of x in the numerator (1 in this case) is the same as the highest power in the denominator.

Consider the three rational functions whose graphs are shown in Figure 8.31.

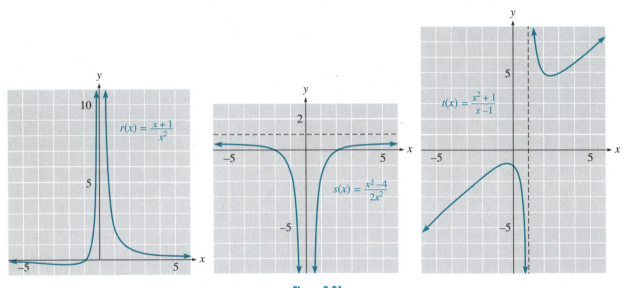

Figure 8.31

The graph of $r(x) = \dfrac{x+1}{x^2}$ in Figure 8.31(a) has a horizontal asymptote at $y = 0$,   the x-axis, because the degree of the denominator is larger than the degree of the numerator. Higher powers of x grow much more rapidly than smaller powers. Thus for large values of $|x|$, the denominator is much bigger in absolute value than the numerator of $r(x)$, and consequently the function values approach 0.

The graph of $s(x) = \dfrac{x^2 - 4}{2x^2}$ in Figure 8.31(b) has a horizontal asymptote at $y = \frac{1}{2}$, because the numerator and denominator of the fraction have the same degree. For large values of $|x|$, the terms of lower degree are negligible compared to the squared terms. As x increases, $s(x)$ is approximately equal to $\dfrac{x^2}{2x^2}$, or $\frac{1}{2}$. Thus the function values approach a constant value of $\frac{1}{2}$.

The graph of $t(x) = \dfrac{x^2 + 1}{x - 1}$ in Figure 8.31(c) does not have a horizontal asymptote, because the degree of the numerator is larger than the degree of the denominator. As $|x|$ increases, $x^2 + 1$ grows much faster than $x - 1$, so their ratio does not approach a constant value. The function values increase without bound.

We can summarize our discussion as follows.

HORIZONTAL ASYMPTOTES

Suppose $f(x) = \dfrac{P(x)}{Q(x)}$ is a rational function, where the degree of $P(x)$ is m and the degree of $Q(x)$ is n.

1. If $m < n$, the graph of f has a horizontal asymptote at $y = 0$.

2. If $m = n$, the graph of f has a horizontal asymptote at $y = \dfrac{a}{b}$, where a is the lead coefficient of $P(x)$ and b is the lead coefficient of $Q(x)$.

3. If $m > n$, the graph of f does not have a horizontal asymptote.

Example 5

Locate the horizontal asymptotes and sketch the graph of $h(x) = \dfrac{2x^2}{x^2 + 4}$.

Solution The numerator and denominator of the fraction are both second-degree polynomials, so the graph does have a horizontal asymptote. The asymptote is the horizontal line $y = \dfrac{a}{b}$, where a and b are the lead coefficients of $P(x)$ and $Q(x)$, respectively. For the function h the horizontal asymptote is $y = \dfrac{2}{1}$, or $y = 2$.

The function h does not have any vertical asymptotes because the denominator, $x^2 + 4$, is never equal to zero. The y-intercept of the graph is the point $(0, 0)$. We can plot several points by evaluating the function at convenient

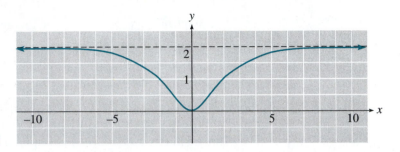

Figure 8.32

x-values and use the asymptote to help us sketch the graph, as shown in Figure 8.32.   ☐

Example 6   Locate the horizontal and vertical asymptotes and sketch the graph of

$$k(x) = \frac{x^2 - 1}{x^2 - 9}.$$

Solution   The denominator, $x^2 - 9$, equals zero if $x = 3$ or $x = -3$. Since the numerator does not equal zero at either of these *x*-values, there are vertical asymptotes at $x = 3$ and $x = -3$. Sketch in the vertical asymptotes as shown in Figure 8.33(a). The asymptotes divide the graph into three pieces.

The graph also has a horizontal asymptote because the degree of the numerator and the degree of the denominator are equal. The horizontal asymptote is the line $y = 1$.

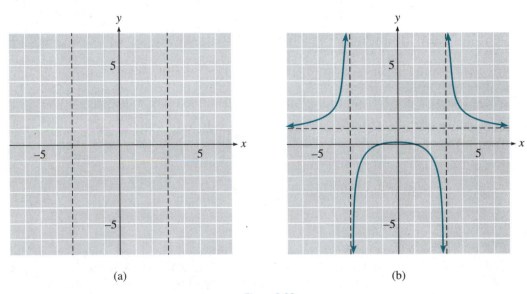

(a)   (b)

Figure 8.33

We can also find the x- and y-intercepts of the graph. The y-intercept is $k(0) = 1/9$, and the x-intercepts occur where the numerator of the fraction equals zero. Since $x^2 - 1 = 0$ at $x = 1$ and $x = -1$, these are the x-intercepts of the graph.

If we plot a few more points by evaluating the function, we can use the asymptotes to help us complete the sketch of the graph, which is shown in Figure 8.33(b).

Exercise 8.3

In Exercises 1–18
a. sketch the horizontal and vertical asymptotes for each function.
b. use the asymptotes to help you sketch the rest of the graph.

1. $y = \dfrac{1}{x + 3}$ **2.** $y = \dfrac{1}{x - 3}$

3. $y = \dfrac{2}{x^2 - 5x + 4}$ **4.** $y = \dfrac{4}{x^2 - x - 6}$

5. $y = \dfrac{x}{x + 3}$ **6.** $y = \dfrac{x}{x - 2}$

7. $y = \dfrac{x + 1}{x + 2}$ **8.** $y = \dfrac{x - 1}{x - 3}$

9. $y = \dfrac{2x}{x^2 - 4}$ **10.** $y = \dfrac{x}{x^2 - 9}$

11. $y = \dfrac{x - 2}{x^2 + 5x + 4}$ **12.** $y = \dfrac{x + 1}{x^2 - x - 6}$

13. $y = \dfrac{x^2 - 1}{x^2 - 4}$ **14.** $y = \dfrac{2x^2}{x^2 - 1}$

15. $y = \dfrac{x + 1}{(x - 1)^2}$ **16.** $y = \dfrac{2(x^2 - 1)}{x^2 + 4}$

17. $y = \dfrac{x}{x^2 + 3}$ **18.** $y = \dfrac{x^2 + 2}{x^2 + 4}$

19. The eider duck, one of the world's fastest flying birds, can exceed an airspeed of 65 miles per hour. A flock of eider ducks is migrating south at an average airspeed of 50 miles per hour against a moderate headwind. Its next feeding ground is 150 miles away.

a. Express the ducks' travel time, t, as a function of the wind speed, v.

b. How long will the ducks fly if the headwind is 5 miles per hour? What will happen to their travel time if the headwind increases?

c. Graph your function and give the equations of any horizontal or vertical asymptotes. What does the vertical asymptote signify in the context of this problem?

20. The fastest fish in the sea may be the bluefin tuna, which has been clocked at 43 miles per hour in short sprints. A school of tuna is migrating a distance of 200 miles at an average speed of 36 miles per hour in still water, but it has run into a current flowing against its direction of travel.

a. Express the tuna's travel time, t, as a function of the current speed, v.

b. How long will the tuna swim if the current is 10 miles per hour? What will happen to their travel time if the current decreases?

c. Graph your function and give the equations of any horizontal or vertical asymptotes. What does the vertical asymptote signify in the context of this problem?

21. The total cost in dollars of producing n calculators is approximately $20,000 + 8n$.

a. Express the average cost per calculator as a function of the number, n, of calculators produced. (*Hint:* Divide the total cost by the number of calculators.)

b. Suppose the average cost per calculator is $18. Use your answer to past (a) to write and solve an equation for n.

c. Graph the function from part (a). Use the graph to verify your answer to part (b). Use the window

$$\text{Xmin} = 0, \quad \text{Xmax} = 9400$$
$$\text{Ymin} = 0, \quad \text{Ymax} = 50$$

d. When is the average cost less than $12 per calculator?

e. Find the horizontal asymptote of the graph. What does it represent in this context?

22. The number of loaves of Mom's Bread sold in a month is approximated by the demand function

$$D(p) = \frac{100}{1 + (p - 1.10)^4}$$

where p is the price per loaf in dollars.

a. Graph the demand function in the window

$$\text{Xmin} = 0, \quad \text{Xmax} = 3.74$$
$$\text{Ymin} = 0, \quad \text{Ymax} = 170$$

What happens to the demand for Mom's Bread as the price increases?

b. Express the total monthly revenue, R, as a function of the price, p. (*Hint:* Multiply the number of loaves by the price per loaf.)

c. Graph the revenue function from part (b) in the same window. Estimate the maximum possible revenue.

d. Graph the functions D (the number of loaves sold) and R (the total revenue) on the same axes. Does the maximum for $D(p)$ occur at the same value of p as the maximum for $R(p)$?

e. Find the horizontal asymptote of the graphs. What does it represent in this context?

23. A computer store sells approximately 300 of its most popular model per year. The manager would like to determine how many computers to order at a time so as to minimize her annual inventory costs. The inventory cost is the sum of the storage costs and the reorder costs. If she orders x computers in each shipment, the cost of storage will

be $6x$ dollars, and the cost of reordering will be $\frac{300}{x}(15x + 10)$ dollars.

a. Use the distributive law to simplify the expression for the reordering costs. Then express the inventory costs, C, as a function of x.

b. Graph the function C in the window

$$\text{Xmin} = 0, \quad \text{Xmax} = 150$$
$$\text{Ymin} = 4500, \quad \text{Ymax} = 5500$$

Estimate the minimum possible value for C.

c. How many computers should the manager order in each shipment so as to minimize the inventory costs? How many orders will she make during the year?

d. Graph the function $y = 6x + 4500$ in the same window with C. What do you observe?

24. A chain of electronics stores sells approximately 500 portable phones every year. The owner would like to minimize his annual inventory costs by ordering the optimal number of phones, x, at regular intervals. The cost of storing the phones will then be $2x$ dollars, and the cost of reordering will be $\frac{500}{x}(4x + 10)$. The total annual inventory cost is the sum of the storage costs and the reordering costs.

a. Use the distributive law to simplify the expression for the reordering costs. Then express the inventory costs, C, as a function of x.

b. Graph your function C in the window

$$\text{Xmin} = 0, \quad \text{Xmax} = 150$$
$$\text{Ymin} = 2000, \quad \text{Ymax} = 2500$$

Estimate the minimum possible value for C.

c. How many portable phones should the owner order in each shipment so as to minimize the inventory costs? How many orders will he make during the year?

d. Graph the function $y = 2x + 2000$ in the same window with C. What do you observe?

25. Francine wants to make a rectangular box. To simplify construction and keep her costs down, she plans for the box to have a square base and a total surface area of 96 square centimeters. With these restrictions, if the square base has side x centimeters, then the height of the box is $h = \dfrac{24}{x} - \dfrac{x}{2}$ centimeters.

a. Write an expression for the box's volume in terms of the side x of its base.

b. Graph your expression for part (a) as a function of x. Approximate the maximum possible volume for Francine's box.

c. What value of x gives the maximum volume?

d. Graph the height, h, on the same axes. What are the dimensions of the box with the greatest volume? Find the height of the box with maximum volume directly from your graph and also by using the formula given for h.

26. Delbert wants a box with a square base and a volume of 64 cubic centimeters. With these restrictions, if the square base has side x centimeters, then the height of the box must be $h = \dfrac{64}{x^2}$ centimeters.

a. Write an expression for the surface area of the box in terms of the side x of its base. (*Hint:* The box's surface area is the sum of the areas of its six sides.)

b. Graph your expression for part (a) as a function of x. Approximate the minimum possible surface area for Delbert's box.

c. What value of x gives the minimum surface area?

d. Graph the height, h, on the same axes. What are the dimensions of the box with the smallest surface area? Find the height of the box with minimum surface directly from your graph and also by using the formula given for h.

27. The cost in thousands of dollars for immunizing $p\%$ of the residents of Emporia against a dangerous new disease is given by the function

$$C(p) = \frac{72p}{100 - p}$$

a. What is the domain of C?

b. What is the cost of immunizing 40% of the residents? Half of the residents? Three-fourths of the residents?

c. What percentage of the population can be immunized if the city is able to spend $108,000?

d. Graph the function C. Verify your answers to parts (b) and (c) graphically. (Use Xmin = 6, Xmax = 100 and appropriate values of Ymin and Ymax.) When is the total cost more than $1,728,000?

e. The graph has a vertical asymptote. What is it? What is its significance in the context of this problem?

28. The cost in thousands of dollars for extracting $p\%$ of a precious ore from a mine is given by the equation

$$C(p) = \frac{360p}{100 - p}$$

a. What is the domain of C?

b. What is the cost of extracting 40% of the ore? One-fourth of the ore? 90% of the ore?

c. What percent of the ore can be extracted if $540,000 can be spent?

d. Graph the function C. Verify your answers to parts (b) and (c) graphically. (Use Xmin = 6, Xmax = 100 and appropriate values of Ymin and Ymax.) When is the total cost less than $1,440,000?

e. The graph has a vertical asymptote. What is it? What is its significance in the context of this problem?

29. A train whistle sounds higher when the train is approaching you than when it is moving away. This phenomenon is known as the Doppler effect. If the actual pitch of the whistle is 440 hertz (this is the A note below middle C), then the note you hear will have the pitch

$$P(v) = \frac{440\,(332)}{332 - v}$$

where the velocity, v, in meters per second is positive as the train approaches and negative as the train moves away. (The number 332 that appears in this expression is the speed of sound in meters per second.)

a. What would be the pitch of the whistle if the train is approaching at a speed of 20 meters per second? Receding at a speed of 68 meters per second?

b. What is the velocity of the train if the note you hear has a pitch of 415 hertz (corresponding to the note A-flat)? A pitch of 553.3 hertz (C-sharp)?

c. Graph the function P and verify your answers to parts (a) and (b). (Use the window Xmin $= -94$, Xmax $= 94$ and appropriate values of Ymin and Ymax.) For what velocities will the pitch you hear be greater than 456.5 hertz?

d. The graph has a vertical asymptote (although it is not visible in the suggested window). Where is it? What is its significance in this context?

30. The maximum altitude (in meters) attained by a projectile shot from the surface of the earth is

$$h(v) = \frac{6.4 \times 10^6 v^2}{(19.6)(6.4 \times 10^6) - v^2}$$

where v is the speed (in meters per second) at which the projectile was launched. (The radius of the earth is 6.4×10^6 meters, and the constant 19.6 is related to the earth's gravitational constant.)

a. What is the projectile's maximum height if its initial speed is 80 meters per second? 140 meters per second?

b. Approximately what speed is needed to attain an altitude of 4000 meters? Of 16 kilometers?

c. Graph the function h and verify your answers to parts (a) and (b). (Use the window Xmin $= 0$, Xmax $= 940$ and appropriate values of Ymin and Ymax.) For what velocities will the projectile attain an altitude exceeding 32 kilometers?

d. The graph has a vertical asymptote (although it is not visible in the suggested window). Where is it? What is its significance in this context?

31. Graph the curve known as Newton's Serpentine:

$$y = \frac{4x}{x^2 + 1}$$

32. Graph the curve known as the Witch of Agnesi:

$$y = \frac{8}{x^2 + 4}$$

OPERATIONS ON ALGEBRAIC FRACTIONS

The algebraic expression that defines a rational function is called a **rational expression** or an **algebraic fraction**. Operations on algebraic fractions follow the same rules as operations on common fractions.

Reducing Fractions   When we reduce an ordinary fraction like $\frac{24}{36}$, we are using the fundamental principle of fractions.

> **FUNDAMENTAL PRINCIPLE OF FRACTIONS**
>
> If we multiply or divide the numerator and denominator of a fraction by the same (nonzero) number, the new fraction is equivalent to the old one. In symbols,
>
> $$\frac{ac}{bc} = \frac{a}{b} \qquad (b, c \neq 0)$$

Thus

$$\frac{24}{36} = \frac{2 \cdot 12}{3 \cdot 12} = \frac{2}{3}$$

We can use the same procedure to reduce algebraic fractions: we look for common factors in the numerator and denominator and then apply the fundamental principle.

Example 1

Reduce each algebraic fraction.

a. $\dfrac{8x^3y}{6x^2y^3}$ 　　　　　　　　　　**b.** $\dfrac{6x + 3}{3}$

Solutions Factor out any common factors from the numerator and denominator. Then divide the numerator and denominator by the common factors.

a. $\dfrac{8x^3y}{6x^2y^3} = \dfrac{4x \cdot 2x^2y}{3y^2 \cdot 2x^2y}$ 　　　　**b.** $\dfrac{6x + 3}{3} = \dfrac{3(2x + 1)}{3}$

$\qquad\quad = \dfrac{4x}{3y^2}$ 　　　　　　　　　　　　$= 2x + 1$

If the fraction's numerator or denominator contains more than one term, it is especially important to factor before we attempt to apply the fundamental principle. We can divide out common *factors* from the fraction's numerator and denominator, but the fundamental principle does not apply to common *terms*. For example,

$$\frac{2xy}{3y} = \frac{2x}{3}$$

because y is a common factor in the numerator and denominator. However,

$$\frac{2x + y}{3 + y} \neq \frac{2x}{3}$$

because y is a common term but *not a common factor* of the numerator and denominator. Furthermore,

$$\frac{5x + 3}{5y} \neq \frac{x + 3}{y}$$

because 5 is not a factor of the *entire* numerator.

Example 2

Reduce each fraction.

a. $\dfrac{4x + 2}{4}$ b. $\dfrac{9x^2 + 3}{6x + 3}$

Solutions   Factor the numerator and denominator. Then divide the numerator and denominator by the common factors.

a. $\dfrac{4x + 2}{4} = \dfrac{\cancel{2}(2x + 1)}{\cancel{2}(2)}$ b. $\dfrac{9x^2 + 3}{6x + 3} = \dfrac{\cancel{3}(3x^2 + 1)}{\cancel{3}(2x + 1)}$

$\qquad\quad = \dfrac{2x + 1}{2}$ $\qquad\quad = \dfrac{3x^2 + 1}{2x + 1}$

⚠ **COMMON ERRORS**

Note that in Example 2(a)

$$\frac{4x + 2}{4} \neq x + 2$$

and in Example 2(b)

$$\frac{9x^2 + 3}{6x + 3} \neq \frac{9x^2}{6x}$$

We summarize the procedure for reducing algebraic fractions as follows.

> **TO REDUCE AN ALGEBRAIC FRACTION**
> **1.** Factor the numerator and denominator.
> **2.** Divide the numerator and denominator by any common factors.

Example 3

Reduce each fraction.

a. $\dfrac{x^2 - 7x + 6}{36 - x^2}$ b. $\dfrac{27x^3 - 1}{9x^2 - 1}$

Solutions   **a.**   Factor the numerator and denominator to obtain

$$\frac{(x - 6)(x - 1)}{(6 - x)(6 + x)}$$

Notice that the factor $x - 6$ in the numerator is the opposite of the factor $6 - x$ in the denominator; that is, $x - 6 = -1(6 - x)$. Thus

$$\frac{-1(6 - x)(x - 1)}{(6 - x)(6 + x)} = \frac{-1(x - 1)}{6 + x} = \frac{1 - x}{6 + x}$$

b.   The numerator of the fraction is a difference of two cubes, and the denominator is a difference of two squares. Factor each to obtain

$$\frac{(3x - 1)(9x^2 + 3x + 1)}{(3x - 1)(3x + 1)} = \frac{9x^2 + 3x + 1}{3x + 1}$$

Products of Fractions   To multiply two or more common fractions together, we multiply their numerators together and multiply their denominators together. We follow the same procedure for a product of algebraic fractions. For example,

$$\frac{6x^2}{y} \cdot \frac{xy}{2} = \frac{6x^2 \cdot xy}{y \cdot 2} = \frac{6x^3y}{2y} \qquad \textcolor{blue}{\text{Reduce.}}$$

$$= \frac{3x^3 (2y)}{2y} = 3x^3$$

We can simplify the process by first factoring each numerator and denominator and dividing out any common factors.

$$\frac{6x^2}{y} \cdot \frac{xy}{2} = \frac{2 \cdot 3x^2}{y} \cdot \frac{xy}{2} = 3x^3$$

In general, we have the following procedure for finding the product of two algebraic fractions.

TO MULTIPLY ALGEBRAIC FRACTIONS

1. Factor each numerator and denominator.

2. Divide out any factors that appear in both a numerator and a denominator.

3. Multiply together the numerators; multiply together the denominators.

Example 4

Find each product.

a. $\dfrac{5}{x^2 - 1} \cdot \dfrac{x + 2}{x}$

b. $\dfrac{4y^2 - 1}{4 - y^2} \cdot \dfrac{y^2 - 2y}{4y + 2}$

Solutions a. The denominator of the first fraction factors into $(x + 1)(x - 1)$. There are no common factors to divide out, so we multiply the numerators together and multiply the denominators together.

$$\frac{5}{x^2 - 1} \cdot \frac{x + 2}{x} = \frac{5(x + 2)}{x(x^2 - 1)} = \frac{5x + 10}{x^3 - x}$$

b. Factor each numerator and each denominator. Look for common factors.

$$\frac{4y^2 - 1}{4 - y^2} \cdot \frac{y^2 - 2y}{4y + 2} = \frac{(2y - 1)(2y + 1)}{(2 - y)(2 + y)} \cdot \frac{\overset{-1}{y(y - 2)}}{2(2y + 1)} \qquad \text{Divide out common factors.}$$

$$= \frac{-y(2y - 1)}{2(y + 2)}$$

Quotients of Fractions   To divide two algebraic fractions, we multiply the first fraction by the reciprocal of the second fraction. For example,

$$\frac{2x^3}{3y} \div \frac{4x}{5y^2} = \frac{2x^3}{3y} \cdot \frac{5y^2}{4x}$$

$$= \frac{2x \cdot x^2}{3y} \cdot \frac{5y \cdot y}{2 \cdot 2x} = \frac{5x^2y}{6}$$

If the fractions involve polynomials of more than one term, we may need to factor each numerator and denominator in order to recognize any common factors. This requirement suggests the following procedures for dividing algebraic fractions.

TO DIVIDE ALGEBRAIC FRACTIONS

1. Multiply the first fraction by the reciprocal of the second fraction.

2. Factor each numerator and denominator.

3. Divide out any factors that appear in both a numerator and a denominator.

4. Multiply together the numerators; multiply together the denominators.

Example 5   Find each quotient.

a. $\dfrac{x^2 - 1}{x + 3} \div \dfrac{x^2 - x - 2}{x^2 + 5x + 6}$ **b.** $\dfrac{6ab}{2a + b} \div 4a^2b$

Solutions **a.** Multiply the first fraction by the reciprocal of the second fraction.

$$\frac{x^2 - 1}{x + 3} \div \frac{x^2 - x - 2}{x^2 + 5x + 6} = \frac{x^2 - 1}{x + 3} \cdot \frac{x^2 + 5x + 6}{x^2 - x - 2} \qquad \text{Factor.}$$

$$= \frac{(x - 1)(x + 1)}{x + 3} \cdot \frac{(x + 3)(x + 2)}{(x + 1)(x - 2)}$$

$$= \frac{(x - 1)(x + 2)}{x - 2}$$

b. Multiply the first fraction by the reciprocal of the second fraction.

$$\frac{6ab}{2a + b} \div 4a^2b = \frac{\overset{3}{\cancel{6ab}}}{2a + b} \cdot \frac{1}{\underset{2}{\cancel{4a^2b}}} \qquad \text{Divide out common factors.}$$

$$= \frac{3}{2a(2a + b)} = \frac{3}{4a^2 + 2ab}$$

Polynomial Division

If the quotient of two polynomials cannot be reduced, we can sometimes simplify the expression by treating it as a division. When the degree of the numerator is greater than the degree of the denominator, the quotient will be the sum of a polynomial and a simpler algebraic fraction.

If the divisor is a monomial, we can simply divide the monomial into each term of the numerator.

Example 6

Divide $\dfrac{9x^3 - 6x^2 + 4}{3x}$.

Solution Divide $3x$ into each term of the numerator.

$$\frac{9x^3 - 6x^2 + 4}{3x} = \frac{9x^3}{3x} - \frac{6x^2}{3x} + \frac{4}{3x}$$

$$= 3x^2 - 2x + \frac{4}{3x}$$

The quotient is the sum of a polynomial, $3x^2 - 2x$, and an algebraic fraction, $\dfrac{4}{3x}$. Note that the degree of the numerator in this remainder, 4, is less than the degree of the original divisor, $3x$.

If the denominator is not a monomial, we can use a method similar to the long-division algorithm used in arithmetic.

Example 7   Divide $\dfrac{2x^2 + x - 7}{x + 3}$.

Solution   First write

$$x + 3 \,\overline{)\,2x^2 + x - 7}$$

and divide $2x^2$ (the first term of the numerator) by x (the first term of the denominator) to obtain $2x$. (It may be helpful to write down the division: $\dfrac{2x^2}{x} = 2x$.)   Write $2x$ above the quotient bar as the first term of the quotient, as shown below.

Next multiply $x + 3$ by $2x$ to obtain $2x^2 + 6x$, and subtract this product from $2x^2 + x - 7$:

$$
\begin{array}{r}
2x \\
x + 3 \,\overline{)\,2x^2 + x - 7} \\
-\,(2x^2 + 6x) \\
\hline
-\,5x - 7
\end{array}
$$

Repeating the process, divide $-5x$ by x to obtain -5 as the second term of the quotient. Then multiply $x + 3$ by -5 to obtain $-5x - 15$, and subtract:

$$
\begin{array}{r}
2x - 5 \\
x + 3 \,\overline{)\,2x^2 + x - 7} \\
-\,(2x^2 + 6x) \\
\hline
-\,5x - 7 \\
-\,(-5x - 15) \\
\hline
8
\end{array}
$$

Since the degree of 8 is less than the degree of $x + 3$, the division is finished. The quotient is $2x - 5$, with a remainder of 8. We write the remainder as a fraction to obtain

$$\frac{2x^2 + x - 7}{x + 3} = 2x - 5 + \frac{8}{x + 3} \qquad \square$$

When using polynomial division it helps to write the polynomials in descending powers of the variable. If the numerator is "missing" any terms, we can insert terms with zero coefficients so that like powers will be aligned. For example, to perform the division

$$\frac{3x - 1 + 4x^3}{2x - 1}$$

we can first write the numerator in descending powers as $4x^3 + 3x - 1$. We then insert $0x^2$ between $4x^3$ and $3x$ and set up the quotient as

$$2x - 1 \overline{)4x^3 + 0x^2 + 3x - 1}$$

We then proceed as in Example 7. You can check that the quotient is

$$2x^2 + x + 2 + \frac{1}{2x - 1}$$

Exercise 8.4

Reduce each algebraic fraction in Exercises 1–24.

1. $\dfrac{14c^2d}{-7c^2d^3}$ **2.** $\dfrac{100mn}{-5m^2n^3}$ **3.** $\dfrac{-12r^2st}{-6rst^2}$

4. $\dfrac{-15xy^3z}{-3y^2z^4}$ **5.** $\dfrac{4x + 6}{6}$ **6.** $\dfrac{2y - 8}{8}$

7. $\dfrac{6a^3 - 4a^2}{4a}$ **8.** $\dfrac{3x^3 - 6x^2}{6x^2}$

9. $\dfrac{6 - 6t^2}{(t - 1)^2}$ **10.** $\dfrac{4 - 4x^2}{(x + 1)^2}$ **11.** $\dfrac{2y^2 - 8}{2y + 4}$

12. $\dfrac{5y^2 - 20}{2y - 4}$ **13.** $\dfrac{6 - 2v}{v^3 - 27}$ **14.** $\dfrac{4 - 2u}{u^3 - 8}$

15. $\dfrac{4x^3 - 36x}{6x^2 + 18x}$ **16.** $\dfrac{5x^2 + 10x}{5x^3 + 20x}$

17. $\dfrac{y^2 - 9x^2}{(3x - y)^2}$ **18.** $\dfrac{(2x - y)^2}{y^2 - 4x^2}$

19. $\dfrac{2x^2 + x - 6}{x^2 + x - 2}$ **20.** $\dfrac{6x^2 - x - 1}{2x^2 + 9x - 5}$

21. $\dfrac{x - 12 + 6x^2}{17x - 12 - 6x^2}$ **22.** $\dfrac{2x - 30 + 4x^2}{15 - 16x + 4x^2}$

23. $\dfrac{8z^3 - 27}{4z^2 - 9}$ **24.** $\dfrac{8z^3 - 1}{4z^2 - 1}$

25. Which of the following fractions are equivalent to $2a$ (on their common domain)?

a. $\dfrac{2a + 4}{4}$ b. $\dfrac{4a^2 - 2a}{2a - 1}$

c. $\dfrac{4a^2 - 2a}{2a}$ d. $\dfrac{a + 3}{2a^2 + 6a}$

26. Which of the following fractions are equivalent to $3b$ (on their common domain)?

a. $\dfrac{9b^2 - 3b}{3b}$ b. $\dfrac{b + 2}{3b^2 + 6b}$

c. $\dfrac{3b - 9}{9}$ d. $\dfrac{9b^2 - 3b}{3b - 1}$

27. Which of the following fractions are equivalent to -1 (where they are defined)?

a. $\dfrac{2a + b}{2a - b}$ b. $\dfrac{-(a + b)}{b - a}$

c. $\dfrac{2a^2 - 1}{2a^2}$ d. $\dfrac{-a^2 + 3}{a^2 + 3}$

28. Which of the following fractions are equivalent to -1 (where they are defined)?

a. $\dfrac{2a - b}{b - 2a}$ b. $\dfrac{-b^2 - 2}{b^2 + 2}$

c. $\dfrac{3b^2 - 1}{3b^2 + 1}$ d. $\dfrac{b - 1}{b}$

In Exercises 29–42 write each product as a single fraction in lowest terms.

29. $\dfrac{15n^2}{3p} \cdot \dfrac{5p^2}{n^3}$ **30.** $\dfrac{21t^2}{5s} \cdot \dfrac{15s^3}{7st}$

31. $\dfrac{-4}{3np} \cdot \dfrac{6n^2p^3}{16}$ **32.** $\dfrac{14a^3b}{3b} \cdot \dfrac{-6}{7a^2}$

33. $5a^2b^2 \cdot \dfrac{1}{a^3b^3}$ **34.** $15x^2y \cdot \dfrac{3}{35xy^2}$

35. $\dfrac{5x + 25}{2x} \cdot \dfrac{4x}{2x + 10}$ **36.** $\dfrac{3y}{4xy - 6y^2} \cdot \dfrac{2x - 3y}{12x}$

37. $\dfrac{4a^2 - 1}{a^2 - 16} \cdot \dfrac{a^2 - 4a}{2a + 1}$

38. $\dfrac{9x^2 - 25}{2x - 2} \cdot \dfrac{x^2 - 1}{6x - 10}$

39. $\dfrac{2x^2 - x - 6}{3x^2 + 4x + 1} \cdot \dfrac{3x^2 + 7x + 2}{2x^2 + 7x + 6}$

40. $\dfrac{3x^2 - 7x - 6}{2x^2 - x - 1} \cdot \dfrac{2x^2 - 9x - 5}{3x^2 - 13x - 10}$

41. $\dfrac{3x^4 - 48}{x^4 - 4x^2 - 32} \cdot \dfrac{4x^4 - 8x^3 + 4x^2}{2x^4 + 16x}$

42. $\dfrac{x^4 - 3x^3}{x^4 + 6x^2 - 27} \cdot \dfrac{x^4 - 81}{3x^4 - 81x}$

In Exercises 43–54 write each quotient as a single fraction in lowest terms.

43. $\dfrac{4x - 8}{3y} \div \dfrac{6x - 12}{y}$

44. $\dfrac{6y - 27}{5x} \div \dfrac{4y - 18}{x}$

45. $\dfrac{a^2 - a - 6}{a^2 + 2a - 15} \div \dfrac{a^2 - 4}{a^2 + 6a + 5}$

46. $\dfrac{a^2 + 2a - 15}{a^2 + 3a - 10} \div \dfrac{a^2 - 9}{a^2 - 9a + 14}$

47. $\dfrac{x^3 + y^3}{x} \div \dfrac{x + y}{3x}$

48. $\dfrac{8x^3 - y^3}{x + y} \div \dfrac{2x - y}{x^2 - y^2}$

49. $1 \div \dfrac{x^2 - 1}{x + 2}$ **50.** $1 \div \dfrac{x^2 + 3x + 1}{x - 2}$

51. $(x^2 - 5x + 4) \div \dfrac{x^2 - 1}{x^2}$

52. $(x^2 - 9) \div \dfrac{x^2 - 6x + 9}{3x}$

53. $\dfrac{x^2 + 3x}{2y} \div 3x$ **54.** $\dfrac{2y^2 + y}{3x} \div 2y$

Divide in Exercises 55–66.

55. $\dfrac{18r^2s^2 - 15rs + 6}{3rs}$

56. $\dfrac{8a^2x^2 - 4ax^2 + ax}{2ax}$

57. $\dfrac{15s^{10} - 21s^5 + 6}{-3s^2}$

58. $\dfrac{25m^6 - 15m^4 + 7}{-5m^3}$

59. $\dfrac{4y^2 + 12y + 7}{2y + 1}$ **60.** $\dfrac{4t^2 - 4t - 5}{2t - 1}$

61. $\dfrac{x^3 + 2x^2 + x + 1}{x - 2}$

62. $\dfrac{2x^3 - 3x^2 - 2x + 4}{x + 1}$

63. $\dfrac{4z^2 + 5z + 8z^4 + 3}{2z + 1}$

64. $\dfrac{7 - 3t^3 - 23t^2 + 10t^4}{2t + 3}$

65. $\dfrac{x^4 - 1}{x - 2}$ **66.** $\dfrac{y^5 + 1}{y - 1}$

In Exercises 67–70 verify that the value given is a zero of the polynomial. Find the other zeros. (*Hint:* Use polynomial division to write $P(x) = (x - a)Q(x)$, then factor $Q(x)$.)

67. $P(x) = x^3 - 2x^2 + 1$; $a = 1$

68. $P(x) = x^3 + 2x^2 - 1$; $a = -1$

69. $P(x) = x^4 - 3x^3 - 10x^2 + 24x$; $a = -3$

70. $P(x) = x^4 + 5x^3 - x^2 - 5x$; $a = -5$

Exercises 71–74 are examples of functions whose graphs have "holes."

a. Find the domain of the function.

b. Reduce the fraction to lowest terms.

c. Graph the function. (*Hint:* The graph of the original function is identical to the graph of the function in part (b) except that certain points are excluded from the domain.) Indicate a "hole" in the graph by an open circle.

71. $y = \dfrac{x^2 - 4}{x - 2}$ **72.** $y = \dfrac{x^2 - 1}{x + 1}$

73. $y = \dfrac{x + 1}{x^2 - 1}$ **74.** $y = \dfrac{x - 3}{x^2 - 9}$

8.5 MORE OPERATIONS ON ALGEBRAIC FRACTIONS

Sums and Differences of Like Fractions

Algebraic fractions with the same denominator are called **like fractions**. To add or subtract like fractions, we combine their numerators and keep the same denominator for the sum or difference. This method is actually an application of the distributive law and is valid regardless of how many terms appear in the numerator or denominator.

Example 1

Find each sum or difference.

a. $\dfrac{2x}{9z^2} + \dfrac{5x}{9z^2}$

b. $\dfrac{2x - 1}{x + 3} - \dfrac{5x - 3}{x + 3}$

Solutions   a.   Since these are like fractions, we add their numerators and keep the same denominator.

$$\frac{2x}{9z^2} + \frac{5x}{9z^2} = \frac{2x + 5x}{9z^2} = \frac{7x}{9z^2}$$

b.   Be careful to subtract the *entire* numerator of the second fraction: use parentheses to show that the subtraction applies to both terms of $5x - 3$.

$$\frac{2x - 1}{x + 3} - \frac{5x - 3}{x + 3} = \frac{2x - 1 - (5x - 3)}{x + 3}$$

$$= \frac{2x - 1 - 5x + 3}{x + 3} = \frac{-3x + 2}{x + 3} \quad \square$$

Least Common Denominator

If the fractions in a sum or difference have different denominators, we must first build the fractions to equivalent fractions with a common denominator. Recall that for arithmetic fractions we use as a common denominator the smallest natural number that is exactly divisible by each of the given denominators. For example, to add the fractions $\dfrac{1}{6}$ and $\dfrac{3}{8}$ we use 24 as the common denominator, because 24 is the smallest natural number that both 6 and 8 divide into evenly.

We define the **least common denominator (LCD)** of two or more algebraic fractions as the polynomial of least degree that is exactly divisible by each of the given denominators.

Example 2

Find the LCD for the fractions $\dfrac{3x}{x + 2}$ and $\dfrac{2x}{x - 3}$.

Solution   The LCD is the polynomial of smallest degree that both $x + 2$ and $x - 3$ will divide into evenly. Thus we need a polynomial that has as factors both $x + 2$ and $x - 3$. The simplest such polynomial is $(x + 2)(x - 3)$, or $x^2 - x + 6$. For our purposes it will be more convenient to leave the LCD in factored form: $(x + 2)(x - 3)$.   ▭

    The LCD in Example 2 was easy to find because each original denominator consisted of a single factor; that is, neither denominator could be factored. In that case the LCD is just the product of the original denominators. In fact, we can always find a *common* denominator simply by multiplying together all the denominators in the given fractions; this process may not, however, give us the *simplest* or *least* common denominator. Using anything other than the simplest common denominator will complicate our work needlessly.

    If any of the denominators in the given fractions can be factored, we must factor them before looking for the LCD.

> **TO FIND THE LCD OF ALGEBRAIC FRACTIONS**
>
> **1.** Factor each denominator completely.
>
> **2.** Include each different factor in the LCD as many times as it occurs in any *one* of the given denominators.

Example 3

Find the LCD for the fractions $\dfrac{2x}{x^2 - 1}$ and $\dfrac{x + 3}{x^2 + x}$.

Solution   Factor the denominators of each of the given fractions.

$$x^2 - 1 = (x - 1)(x + 1) \qquad \text{and} \qquad x^2 + x = x(x + 1)$$

The factor $(x - 1)$ occurs once in the first denominator, the factor x occurs once in the second denominator, and the factor $(x + 1)$ occurs once in each denominator. Therefore, we include in our LCD one copy of each of these factors. The LCD is $x(x + 1)(x - 1)$.   ▭

  COMMON ERROR

In Example 3 note that we do not include two factors of $(x + 1)$ in the LCD. We need only one factor of $(x + 1)$ because $(x + 1)$ occurs only once in either denominator. You should check that each original denominator divides evenly into the LCD, $x(x + 1)(x - 1)$.

Example 4

Find the LCD for the fractions $\dfrac{1}{2x^3}$, $\dfrac{1}{6x(x - 2)}$, and $\dfrac{1}{4(x - 2)^2}$.

Solution   Factor each of the given denominators.

$$2x^3 = 2 \cdot x \cdot x \cdot x$$

$$6x(x - 2) = 2 \cdot 3 \cdot x(x - 2)$$

$$4(x - 2)^2 = 2 \cdot 2(x - 2)(x - 2)$$

The factors that make up the given denominators are 2, 3, x, and $(x - 2)$. We will need some of each as factors in our LCD. To decide how many of each factor we should include, we ask which of the given denominators includes the most copies of that factor. For example, the factored form of $2x^3$ includes three copies of x, so we must include three copies of x in the LCD. Similarly, we must include two copies of 2, one copy of 3, and two copies of $(x - 2)$. Thus the LCD is $2 \cdot 2 \cdot 3 \cdot x \cdot x \cdot x(x - 2)(x - 2)$, or $12x^3(x - 2)^2$. $\quad\quad\quad\quad\square$

Building Fractions

After finding the LCD we must build each fraction to an equivalent one with the LCD as its denominator. The new fractions will be like fractions, and we can combine them as explained above.

Building fractions is the opposite of reducing fractions in the sense that we multiply, rather than divide, the numerator and denominator by an appropriate factor. To find the **building factor**, we compare the factors of the original denominator with those of the desired common denominator.

Example 5

Build each of the fractions $\dfrac{3x}{x + 2}$ and $\dfrac{2x}{x - 3}$ to equivalent fractions with the LCD $(x + 2)(x - 3)$ as denominator.

Solution   Compare the denominator of the given fraction to the LCD. We see that the fraction $\dfrac{3x}{x + 2}$ needs a factor of $(x - 3)$ in its denominator, so $(x - 3)$ is the building factor for the first fraction. By the fundamental principle of fractions, we can multiply the numerator and denominator of the first fraction by $(x - 3)$ and obtain an equivalent fraction:

$$\frac{3x}{x + 2} = \frac{3x(x - 3)}{(x + 2)(x - 3)} = \frac{3x^2 - 9x}{x^2 - x - 6}$$

The fraction $\dfrac{2x}{x - 3}$ needs a factor of $(x + 2)$ in the denominator, so we multiply the numerator and denominator by $(x + 2)$:

$$\frac{2x}{x - 3} = \frac{2x(x + 2)}{(x - 3)(x + 2)} = \frac{2x^2 + 4x}{x^2 - x - 6} \quad\quad \square$$

Notice that the two new fractions we obtained in Example 5 are like fractions; they have the same denominator.

Example 6   Build each of the fractions $\dfrac{2x}{x^2 - 1}$ and $\dfrac{x + 3}{x^2 + x}$ to equivalent fractions with the LCD $x(x + 1)(x - 1)$ as denominator.

Solution   Compare the factored form of the given denominator to the LCD. The first fraction, whose denominator is $(x - 1)(x + 1)$, needs a building factor of x, so

$$\frac{2x}{x^2 - 1} = \frac{2x \cdot x}{(x - 1)(x + 1) \cdot x} = \frac{2x^2}{x^3 - x}$$

The second fraction, whose denominator is $x(x + 1)$, needs a building factor of $(x - 1)$, so

$$\frac{x + 3}{x^2 + x} = \frac{(x + 3)(x - 1)}{(x^2 + x)(x - 1)} = \frac{x^2 + 2x - 3}{x^3 - x}$$

Sums and Differences of Unlike Fractions

We are now ready to add or subtract algebraic fractions with unlike denominators. We will do this in four steps.

TO ADD OR SUBTRACT FRACTIONS WITH UNLIKE DENOMINATORS

1. Find the LCD for the given fractions.

2. Build each fraction to an equivalent fraction with the LCD as its denominator.

3. Add or subtract the numerators of the resulting like fractions. Use the LCD as the denominator of the sum or difference.

4. Reduce the sum or difference, if possible.

Example 7   Subtract $\dfrac{3x}{x + 2} - \dfrac{2x}{x - 3}$.

Solution

Step 1   The LCD for these fractions is $(x + 2)(x - 3)$.

Step 2   In Example 5 we built each fraction to an equivalent one with the LCD:

$$\frac{3x}{x + 2} = \frac{3x^2 - 9x}{x^2 - x - 6} \quad \text{and} \quad \frac{2x}{x - 3} = \frac{2x^2 + 4x}{x^2 - x - 6}$$

Step 3   Combine the numerators over the same denominator.

$$\frac{3x}{x + 2} - \frac{2x}{x - 3} = \frac{3x^2 - 9x}{x^2 - x - 6} - \frac{2x^2 + 4x}{x^2 - x - 6}$$

$$= \frac{(3x^2 - 9x) - (2x^2 + 4x)}{x^2 - x - 6} = \frac{x^2 - 13x}{x^2 - x - 6}$$

Step 4 Reduce the result, if possible. If we factor both the numerator and the denominator, we find

$$\frac{x(x-13)}{(x-3)(x+2)}$$

The fraction cannot be reduced. ☐

Example 8

Simplify $\quad 1 - \dfrac{2}{a^2} + \dfrac{a-1}{a^2+a}$.

Solution

Step 1 To find the LCD, factor each denominator:

$$a^2 = a \cdot a$$

$$a^2 + a = a(a+1)$$

We need two factors of a and one factor of $(a+1)$. The LCD is $a^2(a+1)$.

Step 2 Build each term to an equivalent fraction with the LCD as denominator. (The building factors for each fraction are shown in color.)

$$1 = \frac{1 \cdot a^2(a+1)}{1 \cdot a^2(a+1)} = \frac{a^3+a}{a^2(a+1)}$$

$$\frac{2}{a^2} = \frac{2 \cdot (a+1)}{a^2 \cdot (a+1)} = \frac{2a+2}{a^2(a+1)}$$

$$\frac{a-1}{a^2+a} = \frac{(a-1) \cdot a}{a(a+1) \cdot a} = \frac{a^2-a}{a^2(a+1)}$$

Step 3 Combine the numerators over the LCD.

$$1 - \frac{2}{a^2} + \frac{a-1}{a^2+a} = \frac{a^3+a}{a^2(a+1)} - \frac{2a+2}{a^2(a+1)} + \frac{a^2-a}{a^2(a+1)}$$

$$= \frac{a^3 + a - (2a+2) + (a^2 - a)}{a^2(a+1)}$$

$$= \frac{a^3 + a^2 - 2a - 2}{a^2(a+1)}$$

Step 4 To see if this fraction can be reduced, we factor the numerator. Begin by factoring a^2 from the first two terms and -2 from the last two terms.

$$a^3 + a^2 - 2a - 2 = a^2(a+1) - 2(a+1) \qquad \text{Factor } (a+1) \text{ from each term.}$$

$$= (a+1)(a^2 - 2)$$

Thus

$$\frac{a^3 + a^2 - 2a - 2}{a^2(a + 1)} = \frac{(a + 1)(a^2 - 2)}{a^2(a + 1)} = \frac{a^2 - 2}{a^2} \qquad \square$$

Example 9

When estimating their travel time, pilots must take into account the prevailing winds. A tailwind adds to the plane's ground speed, whereas a headwind decreases it. Skyhigh Airlines is setting up a shuttle service from Dallas to Phoenix, a distance of 800 miles.

a. Express the time needed for a one-way trip, without wind, in terms of the plane's speed.

b. Suppose a prevailing wind of 30 miles per hour is blowing from the west. Write expressions for the flying time from Dallas to Phoenix and from Phoenix to Dallas.

c. Write an expression for the round-trip flying time, excluding stops, with a 30-mile-per-hour wind from the west.

Solutions   **a.**   Recall that $\text{time} = \dfrac{\text{distance}}{\text{rate}}$.   If we let r represent the speed of the plane in still air, then the time required for a one-way trip is

$$\frac{800}{r}$$

b.   On the trip from Dallas to Phoenix the plane encounters a headwind of 30 miles per hour, so its ground speed is $r - 30$. On the return trip the plane enjoys a tailwind of 30 miles per hour, so its ground speed is $r + 30$. Therefore, the flying times are

$$\text{Dallas to Phoenix:} \quad \frac{800}{r - 30}$$

and

$$\text{Phoenix to Dallas:} \quad \frac{800}{r + 30}$$

c.   The round-trip flying time from Dallas to Phoenix and back is

$$\frac{800}{r - 30} + \frac{800}{r + 30}$$

The LCD for these fractions is $(r - 30)(r + 30)$. Thus

$$\frac{800}{r - 30} + \frac{800}{r + 30} = \frac{800(r + 30)}{(r - 30)(r + 30)} + \frac{800(r - 30)}{(r + 30)(r - 30)}$$

$$= \frac{(800r + 24{,}000) + (800r - 24{,}000)}{(r + 30)(r - 30)} = \frac{1600r}{r^2 - 900}$$

$$\square$$

Complex Fractions A fraction that contains one or more fractions in either its numerator or its denominator or both is called a **complex fraction**. For example,

$$\frac{\frac{2}{3}}{\frac{5}{6}} \qquad \text{and} \qquad \frac{x + \frac{3}{4}}{x - \frac{1}{2}}$$

are complex fractions. Like simple fractions, complex fractions represent quotients. For the examples above,

$$\frac{\frac{2}{3}}{\frac{5}{6}} = \frac{2}{3} \div \frac{5}{6}$$

and

$$\frac{x + \frac{3}{4}}{x - \frac{1}{2}} = \left(x + \frac{3}{4}\right) \div \left(x - \frac{1}{2}\right)$$

Complex fractions may arise during calculations involving algebraic fractions. We can always simplify a complex fraction into a standard algebraic fraction as follows. If the denominator of the complex fraction is a single term, as in the first example above, we can treat the fraction as a division problem and multiply the numerator by the reciprocal of the denominator. Thus

$$\frac{\frac{2}{3}}{\frac{5}{6}} = \frac{2}{3} \div \frac{5}{6} = \frac{2}{3} \cdot \frac{6}{5} = \frac{4}{5}$$

If the numerator or denominator of the complex fraction contains more than one term, as in the second example above, it is easier to use the fundamental principle of fractions to simplify the expression.

Example 10

Simplify $\dfrac{x + \frac{3}{4}}{x - \frac{1}{2}}$.

Solution Consider all the simple fractions that appear in the complex fraction: in this example $\frac{1}{2}$ and $\frac{3}{4}$. The LCD of these fractions is 4. If we multiply the numerator and denominator of the complex fraction by 4, we will eliminate the fractions within the fraction. Be sure to multiply *each* term of the numerator and *each* term of the denominator by 4.

$$\frac{4\left(x + \dfrac{3}{4}\right)}{4\left(x - \dfrac{1}{2}\right)} = \frac{4(x) + 4\left(\dfrac{3}{4}\right)}{4(x) - 4\left(\dfrac{1}{2}\right)} = \frac{4x + 3}{4x - 2}$$

Thus the original complex fraction is equivalent to the simple fraction $\dfrac{4x + 3}{4x - 2}$.

We can verify that the answer to Example 10 is equivalent to the original complex fraction by graphing

$$Y_1 = \frac{x + \dfrac{3}{4}}{x - \dfrac{1}{2}}$$

$$Y_2 = \frac{4x + 3}{4x - 2}$$

on the same screen. The two graphs are identical, which means that for every *x*-value the two functions have the same *y*-value—exactly what it means for two algebraic expressions to be equivalent.

We summarize the method for simplifying complex fractions as follows.

TO SIMPLIFY A COMPLEX FRACTION

1. Find the LCD of all the fractions contained in the complex fraction.
2. Multiply the numerator and the denominator of the complex fraction by the LCD.
3. Reduce the resulting simple fraction, if possible.

Exercise 8.5

Write each sum or difference in Exercises 1–10 as a single fraction in lowest terms.

1. $\dfrac{x}{2} - \dfrac{3}{2}$

2. $\dfrac{y}{7} - \dfrac{5}{7}$

3. $\dfrac{1}{6}a + \dfrac{1}{6}b - \dfrac{5}{6}c$

4. $\dfrac{1}{3}x - \dfrac{2}{3}y + \dfrac{1}{3}z$

5. $\dfrac{x - 1}{2y} + \dfrac{x}{2y}$

6. $\dfrac{y + 1}{b} + \dfrac{y - 1}{b}$

7. $\dfrac{3}{x + 2y} - \dfrac{x - 3}{x + 2y} - \dfrac{x - 1}{x + 2y}$

8. $\dfrac{2}{a - 3b} - \dfrac{b - 2}{a - 3b} + \dfrac{b}{a - 3b}$

9. $\dfrac{a+1}{a^2-2a+1} - \dfrac{5-3a}{a^2-2a+1}$

10. $\dfrac{x+4}{x^2-x-2} - \dfrac{2x-3}{x^2-x-2}$

Find the LCD for each set of fractions in Exercises 11–18.

11. $\dfrac{5}{6(x+y)^2}, \dfrac{3}{4xy^2}$

12. $\dfrac{1}{8(a-b)^2}, \dfrac{5}{12a^2b^2}$

13. $\dfrac{2a}{a^2+5a+4}, \dfrac{2}{(a+1)^2}$

14. $\dfrac{3x}{x^2-3x+2}, \dfrac{3}{(x-1)^2}$

15. $\dfrac{x+2}{x^2-x}, \dfrac{x+1}{(x-1)^3}$

16. $\dfrac{y-1}{y^2+2y}, \dfrac{y-3}{(y+2)^2}$

17. $\dfrac{1}{6x^3}, \dfrac{x}{4x^2-4x}, \dfrac{x}{(x-1)^2}$

18. $\dfrac{1}{9y}, \dfrac{5y}{6y^3-6y}, \dfrac{y}{(y-1)^2}$

Build each fraction in Exercises 19–26 to an equivalent fraction with the given denominator.

19. $\dfrac{2}{6x}=\dfrac{?}{18x}$

20. $\dfrac{5}{3y}=\dfrac{?}{21y}$

21. $y=\dfrac{?}{xy}$

22. $x=\dfrac{?}{xy^3}$

23. $\dfrac{3y}{y+2}=\dfrac{?}{y^2-y-6}$

24. $\dfrac{5y}{y+3}=\dfrac{?}{y^2+y-6}$

25. $\dfrac{3}{a-b}=\dfrac{?}{a^2-b^2}$

26. $\dfrac{5}{2a+b}=\dfrac{?}{4a^2-b^2}$

Write each sum or difference in Exercises 27–52 as a single fraction in lowest terms.

27. $\dfrac{x}{2}+\dfrac{2x}{3}$

28. $\dfrac{3y}{4}+\dfrac{y}{3}$

29. $\dfrac{5}{6}y-\dfrac{3}{4}y$

30. $\dfrac{3}{4}x-\dfrac{1}{6}x$

31. $\dfrac{x+1}{2x}+\dfrac{2x-1}{3x}$

32. $\dfrac{y-2}{4y}+\dfrac{2y-3}{3y}$

33. $\dfrac{5}{x+1}+\dfrac{3}{x-1}$

34. $\dfrac{2}{y+2}+\dfrac{3}{y-2}$

35. $\dfrac{y}{2y-1}-\dfrac{2y}{y+1}$

36. $\dfrac{2x}{3x+1}-\dfrac{x}{x-2}$

37. $\dfrac{y-1}{y+1}-\dfrac{y-2}{2y-3}$

38. $\dfrac{x-2}{2x+1}-\dfrac{x+1}{x-1}$

39. $\dfrac{7}{5x-10}-\dfrac{5}{3x-6}$

40. $\dfrac{2}{3y+6}-\dfrac{3}{2y+4}$

41. $\dfrac{2}{x^2-x-2}+\dfrac{2}{x^2+2x+1}$

42. $\dfrac{1}{y^2-1}+\dfrac{1}{y^2+2y+1}$

43. $\dfrac{y-1}{y^2-3y}-\dfrac{y+1}{y^2+2y}$

44. $\dfrac{x+1}{x^2+2x}-\dfrac{x-1}{x^2-3x}$

45. $x-\dfrac{1}{x}$

46. $1+\dfrac{1}{y}$

47. $x+\dfrac{1}{x-1}-\dfrac{1}{(x-1)^2}$

48. $y-\dfrac{2}{y^2-1}+\dfrac{3}{y+1}$

49. $y-\dfrac{y^2}{y-1}+\dfrac{y^2}{y+1}$

50. $x+\dfrac{2x^2}{x+2}-\dfrac{3x^2}{x-1}$

51. $x-1+\dfrac{3}{x+2}$

52. $x+3+\dfrac{1}{x-1}$

Write each complex fraction in Exercises 53–72 as a simple fraction in lowest terms.

53. $\dfrac{1 - \dfrac{2}{3}}{3 + \dfrac{1}{3}}$

54. $\dfrac{\dfrac{1}{2} + \dfrac{3}{4}}{\dfrac{1}{2} - \dfrac{3}{4}}$

55. $\dfrac{\dfrac{2}{a} + \dfrac{3}{2a}}{5 + \dfrac{1}{a}}$

56. $\dfrac{\dfrac{2}{y} + \dfrac{1}{2y}}{y + \dfrac{y}{2}}$

57. $\dfrac{1 + \dfrac{2}{a}}{1 - \dfrac{4}{a^2}}$

58. $\dfrac{4 - \dfrac{1}{x^2}}{2 - \dfrac{1}{x}}$

59. $\dfrac{x + \dfrac{x}{y}}{1 + \dfrac{1}{y}}$

60. $\dfrac{1 + \dfrac{1}{x}}{1 - \dfrac{1}{x}}$

61. $\dfrac{1}{1 - \dfrac{1}{x}}$

62. $\dfrac{4}{\dfrac{2}{x} + 2}$

63. $\dfrac{y - 2}{y - \dfrac{4}{y}}$

64. $\dfrac{y + 3}{\dfrac{9}{y} - y}$

65. $\dfrac{x + y}{\dfrac{1}{x} + \dfrac{1}{y}}$

66. $\dfrac{x - y}{\dfrac{x}{y} - \dfrac{y}{x}}$

67. $\dfrac{x - \dfrac{x}{y}}{y + \dfrac{y}{x}}$

68. $\dfrac{y + \dfrac{x}{y}}{x - \dfrac{y}{x}}$

69. $\dfrac{\dfrac{4}{x^2} - \dfrac{4}{z^2}}{\dfrac{2}{z} - \dfrac{2}{x}}$

70. $\dfrac{\dfrac{6}{b} - \dfrac{6}{a}}{\dfrac{3}{a^2} - \dfrac{3}{b^2}}$

71. $\dfrac{\dfrac{1}{y + 1}}{1 - \dfrac{1}{y^2}}$

72. $\dfrac{\dfrac{1}{y - 1}}{\dfrac{1}{y^2} + 1}$

73. River Queen Tours offers a 50-mile round-trip excursion on the Mississippi River on a paddle wheeler. The current in the Mississippi is 8 miles per hour.

    a. Write an expression for the time required for the downstream journey in terms of the speed of the paddle wheel boat in still water.

    b. Write an expression for the time required for the return trip upstream.

    c. Write and simplify an expression for the time needed for the round-trip.

74. A rowing team can maintain a speed of 15 miles per hour in still water. The team's daily training session includes a 5-mile run up the Red Cedar River and the return downstream.

    a. Write an expression for the team's time on the upstream leg in terms of the speed of the current.

    b. Write an expression for the team's time on the downstream leg.

    c. Write and simplify an expression for the total time for the training run.

75. Two pilots for the Flying Express parcel service receive packages simultaneously. Orville leaves Boston for Chicago at the same time Wilbur leaves Chicago for Boston. Each selects an airspeed of 400 miles per hour for the 900-mile trip. The prevailing winds blow from east to west.

    a. Write an expression for Orville's flying time in terms of the wind speed.

    b. Write an expression for Wilbur's flying time.

    c. Who reaches his destination first? By how much time (in terms of airspeed)?

76. On New Year's Day a blimp leaves its berth in Carson, California, and heads north for the Rose Bowl, 23 miles away. A breeze is blowing from the north at 6 miles per hour.

    a. Write an expression for the time required for the trip, in terms of the blimp's airspeed.

    b. Write an expression for the time needed for the return trip.

    c. Which trip takes longer? By how much time (in terms of the blimp's airspeed)?

77. The focal length of a lens is given by the formula

$$\frac{1}{f} = \frac{1}{p} + \frac{1}{q}$$

where f stands for the focal length, p is the distance from the object viewed to the lens, and q is the distance from the image to the lens. Suppose you estimate that the distance from your cat (the object viewed) to your camera lens is 60 inches greater than the distance from the lens to the film inside the camera, where the image forms.

a. Write an expression for $1/f$ in terms of q.

b. Write and simplify an expression for f in terms of q.

78. If two resistors, R_1 and R_2, in an electrical circuit are connected in parallel, the total resistance, R, in the circuit is given by

$$\frac{1}{R} = \frac{1}{R_1} + \frac{1}{R_2}$$

a. Suppose the second resistor, R_2, is 10 ohms greater than the first. Write an expression for $1/R$ in terms of the first resistor.

b. Write and simplify an expression for R in terms of the first resistor.

79. Andy drives 300 miles to Lake Tahoe at 70 miles per hour and returns home at 50 miles per hour. What is his average speed for the round-trip? (It is not 60 miles per hour!)

a. Write expressions for the time it takes for each leg of the trip if Andy drives a distance d at speed r_1 and returns at speed r_2.

b. Write expressions for the total distance and total time for the trip.

c. Write an expression for the average speed for the entire trip.

d. Write your answer to part (c) as a simple fraction.

e. Use your formula to answer the question stated in the problem.

80. The owner of a print shop volunteers to produce fliers for his candidate's campaign. His large printing press can complete the job in 4 hours, and the smaller model can finish the fliers in 6 hours. How long will it take to print the fliers if he runs both presses simultaneously?

a. Suppose the large press can complete a job in t_1 hours and the smaller press takes t_2 hours. Write expressions for the fraction of a job that each press can complete in 1 hour.

b. Write an expression for the fraction of a job that can be completed in 1 hour with both presses running simultaneously.

c. Write an expression for the amount of time needed to complete the job with both presses running.

d. Write your answer to part (c) as a simple fraction.

e. Use your formula to answer the question stated in the problem.

8.6 EQUATIONS THAT INCLUDE ALGEBRAIC FRACTIONS

In Chapter 1 we solved simple equations involving fractions by multiplying each side of the equation by the denominator of the fraction. This process "cleared" the fraction and gave us an equivalent equation without fractions. For example, to solve the equation

$$\frac{20}{x - 2} = 5$$

we multiply both sides of the equation by $\;x - 2\;$ to obtain

$$(x - 2)\frac{20}{x - 2} = 5(x - 2)$$

or

$$20 = 5(x - 2)$$

From here we can proceed as usual to find the solution, $\;x = 6.$

If the equation contains more than one fraction, we can clear all the denominators at one time if we multiply both sides by the LCD of the fractions.

Example 1

Rani times herself as she kayaks 30 miles down the Derwent River with the help of the current. Returning upstream against the current, she manages only 18 miles in the same amount of time. Rani knows that she can kayak at a rate of 12 miles per hour in still water. What is the current's speed?

Solution   If we let x represent the current's speed, we can fill in Table 8.5.

	Distance	Rate	Time
Downstream	30	$12 + x$	$\dfrac{30}{12 + x}$
Upstream	18	$12 - x$	$\dfrac{18}{12 - x}$

Table 8.5

Because Rani paddled for equal amounts of time upstream and downstream, we have the equation

$$\frac{30}{12 + x} = \frac{18}{12 - x}$$

The LCD for the fractions in this equation is $(12 + x)(12 - x)$. We multiply both sides of the equation by the LCD to obtain

$$(12 + x)(12 - x)\frac{30}{12 + x} = \frac{18}{12 - x}(12 + x)(12 - x)$$

$$30(12 - x) = 18(12 + x)$$

Solving this equation, we find

$$360 - 30x = 216 + 18x$$

$$144 = 48x$$

$$3 = x$$

The current's speed is 3 miles per hour.

We can also solve the equation in Example 1 graphically. Graph the two functions

$$Y_1 = \frac{30}{12 + x}$$

$$Y_2 = \frac{18}{12 - x}$$

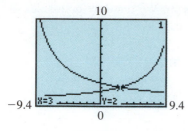

Figure 8.34

in the window

$$\text{Xmin} = -9.4, \quad \text{Xmax} = 9.4$$
$$\text{Ymin} = 0, \qquad \text{Ymax} = 10$$

to obtain the graph shown in Figure 8.34.

The function Y_1 gives the time it takes Rani to paddle 30 miles downstream, and Y_2 gives the time it takes her to paddle 18 miles upstream. Both of these times depend on the speed of the current, x. We are looking for a value of x that makes Y_1 and Y_2 equal. This value occurs at the intersection point of the two graphs. We see that the intersection is the point $(3, 2)$. Thus the current's speed is 3 miles per hour, as we found in Example 1. The y-coordinate of the intersection point gives the time Rani paddled on each part of her trip: 2 hours each way.

To solve an equation with fractions algebraically, we must be careful to multiply *each* term of the equation by the LCD, no matter whether each term involves fractions.

Example 2

Two trains travel from New Orleans to Birmingham, Alabama, a distance of 360 miles. The express train travels at twice the rate of the local and takes 4 hours less time for the trip. Find the rate of each train.

Solution Write algebraic expressions for the rate of each train.

$$\text{Rate of the local train:} \quad r$$

$$\text{Rate of the express train:} \quad 2r$$

Use the information in the problem to fill in Table 8.6.

	Distance	Rate	Time
Local train	360	r	$\dfrac{360}{r}$
Express train	360	$2r$	$\dfrac{360}{2r}$

Table 8.6

The travel time for the express train is 4 hours less than the travel time for the local. We now have an equation to solve.

$$\underset{\text{Express train's time}}{\frac{360}{r}} \underset{\text{is}}{=} \underset{\text{local train's time}}{\frac{360}{2r}} \underset{\text{less}}{-} \underset{4}{4}$$

The LCD of the fractions in this equation is $2r$. We multiply both sides of the equation by the LCD to clear the fractions:

$$2r\left(\frac{360}{2r}\right) = 2r\left(\frac{360}{r} - 4\right) \qquad \text{Apply the distributive law.}$$

$$360 = 720 - 8r \qquad \text{Solve.}$$

$$8r = 360$$

$$r = 45$$

The speed of the local train is 45 miles per hour, and the speed of the express train is $2(45) = 90$ miles per hour.

We can solve the equation in Example 2 graphically by graphing the two equations

$$Y_1 = \frac{360}{2r}$$

$$Y_2 = \frac{360}{r} - 4$$

in the window

$$\text{Xmin} = 0, \quad \text{Xmax} = 84.6$$

$$\text{Ymin} = 0, \quad \text{Ymax} = 16$$

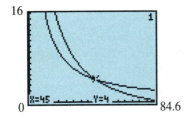

Figure 8.35

to obtain the graph shown in Figure 8.35.

The graphs intersect at the point $(45, 4)$, so the speed of the local train, r, is 45 miles per hour, and the speed of the express train is $2r = 90$ miles per hour. The y-coordinate of the intersection point, 4, is the value of Y_1, so it takes the express train 4 hours for the trip.

Extraneous Solutions

Recall that we can multiply both sides of an equation by any *nonzero* number to obtain a new equation with the same solutions as the old one. In the examples above we multiplied both sides by the LCD, an algebraic expression involving the variable in the problem. It is possible that the LCD will equal zero for some values of the variable, so that the new equation we obtain is not equivalent to the old one. When this happens, we may introduce new solutions that do not satisfy the original equation.

Consider the equation

$$\frac{x}{x - 3} = \frac{3}{x - 3} + 2 \tag{2}$$

When we multiply both sides by the LCD, $x - 3$, we obtain

$$(x - 3)\frac{x}{x - 3} = (x - 3)\frac{3}{x - 3} + (x - 3)2$$

or

$$x = 3 + 2x - 6$$

whose solution is

$$x = 3$$

However, $x = 3$ is *not* a solution of the original Equation (2). If we try to substitute $x = 3$ into Equation (2), we find that both sides of the equation are undefined at $x = 3$, which means that Equation (2) does not have a solution. If you graph the two functions

$$Y_1 = \frac{x}{x - 3}$$

$$Y_2 = \frac{3}{x - 3} + 2$$

you will find that the graphs never intersect, which means that Equation (2) has no solution. Both graphs have a vertical asymptote at $x = 3$.

An apparent solution that does not satisfy the original equation is called an **extraneous solution**. Notice that the extraneous solution in the example above causes the LCD, $x - 3$, to equal zero, so that we have really multiplied both sides of the original equation by zero. Whenever we multiply an equation by an expression containing the variable, we should check that the solution obtained does not cause any denominator to equal zero.

Example 3

Solve the equation $\dfrac{6}{x} + 1 = \dfrac{1}{x + 2}$ algebraically and graphically.

Solution To solve the equation algebraically, we multiply both sides by the LCD, $x(x + 2)$, to get

$$x(x + 2)\left(\frac{6}{x} + 1\right) = x(x + 2)\left(\frac{1}{x + 2}\right)$$

or

$$6(x + 2) + x(x + 2) = x$$

Use the distributive law to remove the parentheses and write the result in standard form:

$$6x + 12 + x^2 + 2x = x$$

$$x^2 + 7x + 12 = 0$$

This is a quadratic equation that we can solve by factoring.

$$(x + 3)(x + 4) = 0$$

so the solutions are $x = -3$ and $x = -4$. Notice that neither of these values causes the LCD to equal zero, so they are not extraneous solutions. To solve the equation graphically, graph the two functions

$$Y_1 = \frac{6}{x} + 1$$

$$Y_2 = \frac{1}{x + 2}$$

in the window

$$Xmin = -4.7, \quad Xmax = 4.7$$

$$Ymin = -10, \quad Ymax = 10$$

We see that the first graph has an asymptote at $x = 0$, and the second graph at $x = -2$. (See Figure 8.36[a].) The two graphs also appear to intersect around $x = -3$.

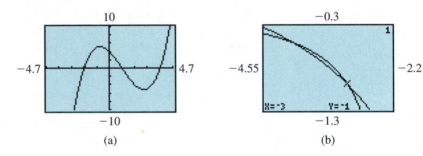

Figure 8.36   (a)   (b)

To investigate further, we change the window settings to

$$Xmin = -4.55, \quad Xmax = -2.2$$

$$Ymin = -1.3, \quad Ymax = -0.3$$

to obtain the close-up view shown in Figure 8.36(b). In this window we can see that the graphs intersect in two distinct points, and by using TRACE we can find that their x-coordinates are $x = -3$ and $x = -4$.

Systems   A system of equations may involve algebraic fractions.

Example 4   Find the intersection points of the graphs of

$$x^2 + y^2 = 5$$

$$xy = 2$$

shown in Figure 8.37.

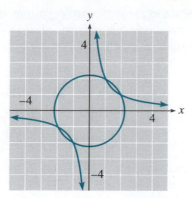

Figure 8.37

Solution   We will use substitution to solve the system. Solve the second equation for y to obtain

$$y = \frac{2}{x}$$

and substitute $\frac{2}{x}$ for y in the first equation to find

$$x^2 + \left(\frac{2}{x}\right)^2 = 5$$

or

$$x^2 + \frac{4}{x^2} = 5$$

Multiply both sides by x^2 to obtain

$$x^4 + 4 = 5x^2$$
$$x^4 - 5x^2 + 4 = 0$$

Factor the left side to get

$$(x^2 - 1)(x^2 - 4) = 0$$

and apply the zero-factor principle to find

$$x^2 - 1 = 0 \quad \text{or} \quad x^2 - 4 = 0$$

Solve each of these equations to find

$$x = 1, \quad x = -1, \quad x = 2, \quad \text{or} \quad x = -2$$

Finally, substitute each of these values into either of the original equations $\left(\text{or into} \quad y = \frac{2}{x}\right)$ to find the y-components of each solution. The intersection points of the two graphs are $(1, 2)$, $(-1, -2)$, $(2, 1)$, and $(-2, -1)$.

If we want to graph the two equations in Example 4 on a calculator, we solve each equation for y in terms of x. However, the circle $x^2 + y^2 = 5$ is not the graph of a function. When we solve for y we find

$$y = \pm\sqrt{5 - x^2}$$

We can enter this equation into the calculator as *two* functions:

$$Y_1 = \sqrt{5 - x^2}$$
$$Y_2 = -\sqrt{5 - x^2}$$

The first equation gives the top half of the circle, and the second equation gives the bottom half. Enter Y_1, Y_2, and

$$Y_3 = \frac{2}{x}$$

in the window

$$Xmin = -4.7, \quad Xmax = 4.7$$
$$Ymin = -3.1, \quad Ymax = 3.1$$

to obtain the graph in Figure 8.37.

Exercises 8.6

Solve each equation in Exercises 1–16.

1. $\dfrac{2}{x + 1} = \dfrac{x}{x + 1} + 1$

2. $\dfrac{5}{x - 3} = \dfrac{x + 2}{x - 3} + 3$

3. $\dfrac{3}{x - 2} = \dfrac{1}{2} + \dfrac{2x - 7}{2x - 4}$

4. $\dfrac{2}{x + 1} + \dfrac{1}{3x + 3} = \dfrac{1}{6}$

5. $\dfrac{4}{x + 2} - \dfrac{1}{x} = \dfrac{2x - 1}{x^2 + 2x}$

6. $\dfrac{1}{x - 1} + \dfrac{2}{x + 1} = \dfrac{x - 2}{x^2 - 1}$

7. $\dfrac{x}{x + 2} - \dfrac{3}{x - 2} = \dfrac{x^2 + 8}{x^2 - 4}$

8. $\dfrac{4}{2x - 3} + \dfrac{4x}{4x^2 - 9} = \dfrac{1}{2x + 3}$

9. $\dfrac{4}{3x} + \dfrac{3}{3x + 1} + 2 = 0$

10. $-3 = \dfrac{-10}{x + 2} + \dfrac{10}{x + 5}$

11. $\dfrac{2x}{x - 1} - \dfrac{x + 1}{2} = 0$

12. $\dfrac{3}{2x + 1} - \dfrac{2x - 3}{x} = 0$

13. $\dfrac{2x}{x - 1} - \dfrac{5}{x^2 - x} = \dfrac{x + 1}{x}$

14. $\dfrac{3x - 1}{3x + 1} + \dfrac{2x}{2x + 1} = 1$

15. $\dfrac{9}{x^2 + x - 2} + \dfrac{1}{x^2 - x} = \dfrac{4}{x - 1}$

16. $\dfrac{2}{x^2 - x} + \dfrac{1}{2x} = \dfrac{-1}{x^2 + 2x}$

17. The manager of Joe's Burgers discovers that he will sell $\dfrac{160}{x}$ burgers per day if the price of a burger is x dollars. On the other hand, he can afford to make $6x + 49$ burgers if he charges x dollars apiece for them.

a. Graph the demand function, $D(x) = \dfrac{160}{x}$, and the supply function, $S(x) = 6x + 49$, in the same window. At what price, x, does the demand for burgers equal the number that Joe can afford to supply? This value for x is called the equilibrium price.

b. Write and solve an equation to verify your equilibrium price.

18. A florist finds that she will sell $\dfrac{300}{x}$ dozen roses per week if she charges x dollars for a dozen. Her suppliers will sell her $5x - 55$ dozen roses if she sells them at x dollars per dozen.

a. Graph the demand function, $D(x) = \dfrac{300}{x}$, and the supply function, $S(x) = 5x - 55$, in the same window. At what equilibrium price, x, will the florist sell all the roses she purchases?

b. Write and solve an equation to verify your equilibrium price.

19. Francine wants to fence a rectangular area of 3200 square feet to grow vegetables for her family of three.

a. Express the length of the garden as a function of its width.

b. Express the perimeter, P, of the garden as a function of its width.

c. Graph your function for the perimeter and find the coordinates of the lowest point on the graph. Interpret those coordinates in the context of the problem.

d. Francine has 240 feet of chain-link to make a fence for the garden, and she would like to know how wide to make the garden. Write an equation that describes this situation.

e. Solve your equation and find the garden's dimensions.

20. The cost of wire fencing is $7.50 per foot. A rancher wants to enclose a rectangular pasture of 1000 square feet with this fencing.

a. Express the length of the pasture as a function of its width.

b. Express the cost of the fence as a function of its width.

c. Graph your function for the cost and find the coordinates of the lowest point on the graph. Interpret those coordinates in the context of the problem.

d. The rancher has $1050 to spend on the fence, and she would like to know how wide to make the pasture. Write an equation to describe this situation.

e. Solve your equation and find the pasture's dimensions.

21. Two student pilots leave the airport at the same time. They both fly at an airspeed of 180 miles per hour, but one flies with the wind and the other flies against it.

a. Express the time it takes the first pilot to travel 500 miles as a function of the wind speed.

b. Express the time it takes the second pilot to travel 400 miles as a function of the wind speed.

c. Graph the two functions in the same window, and find the coordinates of the intersection point. Interpret those coordinates in the context of the problem.

d. Both pilots check in with their instructors at the same time; the first pilot has traveled 500 miles, and the second pilot has gone 400 miles. Write an equation to describe this situation.

e. Solve your equation to find the speed of the wind.

22. Pam lives on the banks of the Cedar River and makes frequent trips in her outboard motorboat. The boat travels at 20 miles per hour in still water.

a. Express the time it takes Pam to travel 8 miles upstream to the gas station as a function of the speed of the current.

b. Express the time it takes Pam to travel 12 miles downstream to Marie's house as a function of the speed of the current.

c. Graph the two functions in the same window, and find the coordinates of the intersection point. Interpret those coordinates in the context of the problem.

d. Pam traveled to the gas station in the same time it took her to travel to Marie's house. Write an equation to describe this situation.

e. Solve your equation to find the speed of the current in the Cedar River.

23. A chartered flight over the Grand Canyon is scheduled to return to its departure point in 3 hours. The pilot would like to cover a distance of 144 miles before turning around, and he hears on the weather service that there will be a headwind of 20 miles per hour on the outward journey.

a. Express the time it takes for the outward journey as a function of the airspeed of the plane.

b. Express the time it takes for the return journey as a function of the speed of the plane.

c. Graph the sum of the two functions and find the point on the graph with y-coordinate 3. Interpret the coordinates of the point in the context of the problem.

d. The pilot would like to maintain an airspeed that allows her to complete the tour in 3 hours. Write an equation to describe this situation.

e. Solve your equation to find the appropriate airspeed.

24. The Explorer's Club is planning a canoe trip to go 90 miles up the Lazy River and return in 4 days. Club members plan to paddle for 6 hours each day, and they know that the current in the Lazy River is 2 miles per hour.

a. Express the time it will take for the upstream journey as a function of their paddling speed in still water.

b. Express the time it will take for the downstream journey as a function of their paddling speed in still water.

c. Graph the sum of the two functions and find the point on the graph with y-coordinate 24. Interpret the coordinates of the point in the context of the problem.

d. The Explorer's Club would like to know what average paddling speed they must maintain to complete their trip in 4 days. Write an equation to describe this situation.

e. Solve your equation to find the required paddling speed.

In Exercises 25–30 find the intersection points of the graphs by solving a system of equations. Verify your solutions by graphing.

25. $xy = 4$
$x^2 + y^2 = 8$

26. $x^2 - y^2 = 35$
$xy = 6$

27. $x^2 - 2y^2 = -4$
$xy = -4$

28. $x^2 + y^2 = 17$
$2xy = 23$

29. $x^2 - y^2 = 16$
$xy = 6$

30. $xy = -2$
$x^2 + y^2 = 5$

Chapter 8 Review

Reduce each fraction in Exercises 1–8 to lowest terms.

1. $\dfrac{2a^2(a-1)^2}{4a(a-1)^3}$

2. $\dfrac{a^2(2a-1)}{4a(1-2a)}$

3. $\dfrac{4y-6}{6}$

4. $\dfrac{2x^2y^3-4x^3y}{4x^2y}$

5. $\dfrac{2x^2+6x}{2(x+3)^2}$

6. $\dfrac{(x-2y)^2}{4y^2-x^2}$

7. $\dfrac{a^2-6a+9}{2a^2-18}$

8. $\dfrac{4x^2y^2+4xy+1}{4x^2y^2-1}$

Write each expression in Exercises 9–16 as a single fraction in lowest terms.

9. $\dfrac{2a^2}{3b}\cdot\dfrac{15b^2}{a}$

10. $-\dfrac{1}{3}ab^2\cdot\dfrac{3}{4}a^3b$

11. $\dfrac{4x+6}{2x}\cdot\dfrac{6x^2}{(2x+3)^2}$

12. $\dfrac{4x^2-9}{3x-3}\cdot\dfrac{x^2-1}{4x-6}$

13. $\dfrac{a^2-a-2}{a^2-4}\div\dfrac{a^2+2a+1}{a^2-2a}$

14. $\dfrac{a^3-8b^3}{a^2b}\div\dfrac{a^2-4ab+4b^2}{ab^2}$

15. $1\div\dfrac{4x^2-1}{2x+1}$

16. $\dfrac{y^2+2y}{3x}\div4y$

Divide in Exercises 17–22.

17. $\dfrac{12y^2z^2+6yz-3}{3yz}$

18. $\dfrac{36x^6-28x^4+16x^2-4}{4x^4}$

19. $\dfrac{y^3+3y^2-2y-4}{y+1}$

20. $\dfrac{x^3-4x^2+2x+3}{x-2}$

21. $\dfrac{x^2+2x^3-1}{2x-1}$

22. $\dfrac{4-y-3y^3}{3y+1}$

Write each expression in Exercises 23–30 as a single fraction in lowest terms.

23. $\dfrac{x+2}{3x}-\dfrac{x-4}{3x}$

24. $\dfrac{y-1}{y+3}-\dfrac{y+1}{y+3}+\dfrac{y}{y+3}$

25. $\dfrac{1}{2}a-\dfrac{2}{3}a$

26. $\dfrac{5}{6}b-\dfrac{1}{3}b+\dfrac{3}{4}b$

27. $\dfrac{3}{2x-6}-\dfrac{4}{x^2-9}$

28. $\dfrac{1}{y^2+4y+4}+\dfrac{3}{y^2-4}$

29. $\dfrac{2a+1}{a-3}-\dfrac{-2}{a^2-4a+3}$

30. $a-\dfrac{1}{a^2+2a+1}+\dfrac{3}{a^2-1}$

Write each complex fraction in Exercises 31–34 as a simple fraction in lowest terms.

31. $\dfrac{\dfrac{3}{4}-\dfrac{1}{2}}{\dfrac{3}{4}+\dfrac{1}{2}}$

32. $\dfrac{y-\dfrac{2y}{x}}{1+\dfrac{2}{x}}$

33. $\dfrac{x-4}{x-\dfrac{16}{x}}$

34. $\dfrac{\dfrac{1}{x-1}}{1-\dfrac{1}{x^2}}$

Solve the equations in Exercises 35–38.

35. $\dfrac{x}{x-2}=\dfrac{2}{x-2}+7$

36. $\dfrac{x}{x-3}+\dfrac{9}{x+3}=1$

37. $\dfrac{2}{x-1}-\dfrac{x+2}{x}=0$

38. $\dfrac{3x}{x+1}-\dfrac{2}{x^2+x}=\dfrac{4}{x}$

In Exercises 39 and 40 find the zeros of each polynomial function.

39. $Q(x) = x^5 - 4x^3$

40. $R(x) = 2x^3 + 3x^2 - 2x$

In Exercises 41–52
a. find the x-intercepts.
b. sketch the graph.

41. $f(x) = (x - 2)(x + 1)^2$

42. $g(x) = (x - 3)^2(x + 2)$

43. $G(x) = x^2\,(x - 1)(x + 3)$

44. $F(x) = (x + 1)^2(x - 2)^2$

45. $V(x) = x^3 - x^5$

46. $H(x) = x^4 - 9x^2$

47. $P(x) = x^3 + x^2 - x - 1$

48. $q(x) = x^3 - x^2 - 4x + 4$

49. $y = x^3 + x^2 - 2x$

50. $y = x^3 - 2x^2 + x - 2$

51. $y = x^4 - 7x^2 + 6$

52. $y = x^4 + x^3 - 3x^2 - 3x$

In Exercises 53–58
a. identify all asymptotes and intercepts.
b. sketch the graph.

53. $y = \dfrac{1}{x - 4}$

54. $y = \dfrac{2}{x^2 - 3x - 10}$

55. $y = \dfrac{x - 2}{x + 3}$

56. $y = \dfrac{x - 1}{x^2 - 2x - 3}$

57. $y = \dfrac{3x^2}{x^2 - 4}$

58. $y = \dfrac{2x^2 - 2}{x^2 - 9}$

Write each expression in Exercises 59–64 as a single fraction involving positive exponents.

59. $x^{-3} + y^{-1}$

60. $\dfrac{x^{-1}}{y} - \dfrac{x}{y - 1}$

61. $\dfrac{x^{-1} - y}{y - 1}$

62. $\dfrac{x^{-1} + y^{-1}}{x^{-1}}$

63. $\dfrac{x^{-1} - y^{-1}}{(x - y)^{-1}}$

64. $\dfrac{(xy)^{-1}}{x^{-1} - y^{-1}}$

65. The expression $\frac{1}{6}n(n - 1)(n - 2)$ gives the number of different three-item pizzas that can be created from a list of n toppings.

a. Write the expression as a trinomial.

b. If Mitch's Pizza offers 12 different toppings, how many three-item pizzas can be made?

c. Use a table or graph to determine how many different toppings are needed to be able to have more than a thousand possible three-item pizzas.

66. The expression $n(n - 1)(n - 2)$ gives the number of different triple-scoop ice cream cones that can be created from a list of n flavors.

a. Write the expression as a trinomial.

b. If Zanner's Ice Cream Parlor offers 21 different flavors, how many triple-scoop ice cream cones can be made?

c. Use a table or graph to determine how many different toppings are needed to be able to have more than ten thousand possible triple-scoop ice cream cones.

67. a. Write an expression for the area of the square in Figure 8.38.

b. Express the area as a polynomial.

c. Divide the square into four pieces whose areas are given by the terms of your answer to (b).

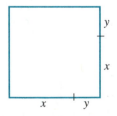

Figure 8.38

68. a. Write an expression for the area of the color shaded region in Figure 8.39.

b. Express the area in factored form.

c. By making one cut in the shaded region, rearrange the pieces into a rectangle whose area is given by your answer to (b).

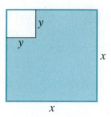

Figure 8.39

69. The sail pictured in Figure 8.40 is a right triangle of base and height x. It has a colored stripe along the hypotenuse and a white triangle of base and height y in the lower corner.

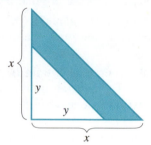

Figure 8.40

a. Write an expression for the area of the colored stripe.
b. Express the area of the stripe in factored form.
c. If the sail is $7\frac{1}{2}$ feet high and the white triangle is $4\frac{1}{2}$ feet high, use your answer to (b) to calculate mentally the area of the stripe.

70. An hors d'oeuvres tray has radius x, and the dip container has radius y, as shown in Figure 8.41.

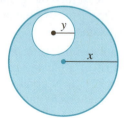

Figure 8.41

a. Write an expression for the area for the chips (the shaded region).

b. Express the area in factored form.

c. If the tray has radius $8\frac{1}{2}$ inches and the space for the dip has radius $2\frac{1}{2}$ inches, use your answer to (b) to calculate mentally the area for chips. (Express your answer as a multiple of π.)

71. The radius of a cylindrical can should be one-half its height.

a. Express the volume, V, of such a can as a function of its height.

b. What is the volume of such a can with a height of 2 centimeters? 4 centimeters?

c. Graph the volume as a function of the height and verify graphically your results of part (b). What is the approximate height of such a can if its volume is 100 cubic centimeters?

72. The Twisty-Freez machine dispenses soft ice cream in a cone-shaped peak with a height three times the radius of its base.

a. Express the volume, V, of Twisty-Freez that comes in a round dish with diameter d.

b. How much Twisty-Freez comes in a 3-inch-diameter dish? 4-inch?

c. Graph the volume as a function of the diameter and verify graphically your results of part (b). What is the approximate diameter of a Twisty-Freez if its volume is 5 cubic inches?

For Exercises 73–76 use the fact that

Profit = Revenue − Cost

to help you
a. write a polynomial to represent the profit.
b. graph the revenue, cost, and profit functions on the same set of axes.
c. determine when a net loss occurs, that is, determine when the profit is negative.
d. answer the questions.

73. Writewell, Inc. makes fountain pens. It costs Writewell $C = 8x + 4000$ dollars to manufacture x pens, and the company receives $R = -0.02(x - 800)^2 + 12{,}800$ dollars in revenue from the sale of the pens. How many pens should be manufactured for maximum revenue, and what is that maximum? How many pens should be manufactured for maximum profit, and what is that maximum?

74. It costs The Sweetshop $C = 3x + 3000$ dollars to produce x pounds of chocolate creams. The company brings in

$$R = -0.03 (x - 600)^2 + 10,800$$

dollars revenue from the sale of the chocolates. How many chocolate creams should be manufactured for maximum revenue, and what is that maximum? How many chocolate creams should be manufactured for maximum profit, and what is that maximum?

75. It costs an appliance manufacturer

$$C = 175x + 21,875$$

dollars to produce x top-loading washing machines, which will bring in revenues of

$$R = -1.75 (x - 200)^2 + 70,000$$

dollars. How many washing machines should be manufactured for maximum revenue, and what is that maximum? How many washing machines should be manufactured for maximum profit, and what is that maximum?

76. A company can produce x lawn mowers for a cost of $C = 80x + 12,800$ dollars. The sale of the lawn mowers will generate

$$R = -2 (x - 120)^2 + 28,800$$

dollars in revenue. How many lawn mowers should be manufactured for maximum revenue, and what is that maximum? How many lawn mowers should be manufactured for maximum profit, and what is that maximum?

Use an equation or a system of equations to solve Exercises 77–86.

77. Leon flies his plane 840 miles in the same time that Marlene drives her automobile 210 miles. Suppose Leon flies 180 miles per hour faster than Marlene drives. Find the rate of each.

78. Moia drives 180 miles in the same time that Fran drives 200 miles. Find the speed of each if Fran drives 5 miles per hour faster than Moia.

79. The perimeter of a rectangle is 26 inches and the area is 12 square inches. Find the dimensions of the rectangle.

80. The perimeter of a rectangle is 34 centimeters and the area is 70 square centimeters. Find the dimensions of the rectangle.

81. A rectangle has a perimeter of 18 feet. If the length is decreased by 5 feet and the width is increased by 12 feet, the area is doubled. Find the dimensions of the original rectangle.

82. The area of a rectangle is 216 square feet. If the perimeter is 60 feet, find the dimensions of the rectangle.

83. Norm takes a commuter train 10 miles to his job in the city. The evening train returns him home at a rate 10 miles per hour faster than the morning train takes him to work. If Norm spends a total of 50 minutes per day commuting, what is the rate of each train?

84. Kristen drove 50 miles to her sister's house, traveling 10 miles in heavy traffic to get out of the city and then 40 miles in less congested traffic. Her average speed in the city was 20 miles per hour less than her speed in light traffic. What was each rate if her trip took 1 hour and 30 minutes?

85. Hattie's annual income from an investment is $32. If she had invested $200 more and the rate had been $\frac{1}{2}\%$ less, then her annual income would have been $35. What are the amount and rate of Hattie's investment?

86. At a constant temperature the pressure, P, and the volume, V, of a gas are related by the equation $PV = K$.   The product of the pressure (in pounds per square inch) and the volume (in cubic inches) of a certain gas is 30 inch-pounds. If the temperature remains constant as the pressure is increased by 4 pounds per square inch, the volume is decreased by 2 cubic inches. Find the original pressure and volume of the gas.

Show that the shaded areas in the figures in Exercises 87 and 88 are equal.

87.

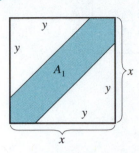

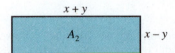

88.

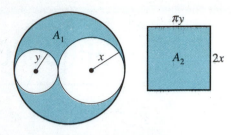

Chapter 9

Sequences and Series

*I*n this chapter we will study functions whose domain is the set of natural numbers, rather than an interval of the real line. We will also learn techniques for summing a sequence and for finding powers of a binomial. These skills are used in discrete mathematics, which has applications in finance, computer science, and many other fields.

9.1 SEQUENCES

Definitions and Notation

Consider the following function. Gwynn would like to compete in a triathlon, but she needs to improve her swimming. She begins a training schedule in which she swims 20 laps a day for the first week and increases that number by 6 laps each week. Table 9.1 gives the first few values of the function.

Week	1	2	3	4	5	6	7	8	n
Number of Laps	20	26	32	38	44	50	56	62	$f(n)$

Table 9.1

This function, f, only makes sense for domain values that are positive integers. We would not ask how many laps Gwynn swims in week 4.63, or in week -6. A function whose domain is a set of successive positive integers is called a **sequence**. Most people think of a sequence as a list of objects in which the order is important, and we often present a mathematical sequence in just that way. The information in Table 9.1 can be displayed in a simpler way by suppressing the domain values and simply listing the range values in order:

$$20, \quad 26, \quad 32, \quad 38, \quad 44, \quad 50, \quad 56, \quad 62$$

When we list them in this way, the range values are called the **terms** of the sequence. The domain values are indicated implicitly by the term's position. For example, the third term of the sequence, 32, is the value of $f(3)$.

We often use the notation a_n instead of $f(n)$ to refer to the terms of a sequence. Thus $a_1 = f(1)$, $a_2 = f(2)$, and so on. In the example above we can find a formula in terms of n for the nth term of the sequence, $f(n) = 14 + 6n$. The expression $14 + 6n$ is called the **general term** for the sequence, and we write

$$a_n = 14 + 6n$$

Example 1

Find the first four terms in each sequence with the given general term.

a. $a_n = \dfrac{n(n + 1)}{2}$

b. $a_n = (-1)^n 2^n$

Solutions Evaluate the general term for successive values of n.

a. $a_1 = \dfrac{1(1 + 1)}{2} = 1 \qquad a_2 = \dfrac{2(2 + 1)}{2} = 3$

$a_3 = \dfrac{3(3 + 1)}{2} = 6 \qquad a_4 = \dfrac{4(4 + 1)}{2} = 10$

The first four terms are 1, 3, 6, and 10.

b. $a_1 = (-1)^1 2^1 = -2 \qquad a_2 = (-1)^2 \, 2^2 = 4$

$a_3 = (-1)^3 2^3 = -8 \qquad a_4 = (-1)^4 \, 2^4 = 16$

The first four terms are -2, 4, -8, and 16.

Example 2

Suppose you deposit $8000 in a savings account that pays 5% interest compounded annually. How much money will be in the account at the end of each of the next 4 years?

Solution During the first year the account will earn 5% of $8000, or $0.05(8000) = 400$ dollars. Thus at the end of the first year the account will contain the original $8000 plus $400 interest for a total of $8400.

At the end of the second year the account will have the $8400 from the previous year plus 5% of $8400, or $0.05\,(8400) = 420$ dollars, in interest. Notice that we can write this sum as

$$8400 + 8400(0.05) = 8400(1 + 0.05) = 8400(1.05) \text{ dollars}$$

Thus at the end of the second year the account will contain $8400(1.05) = 8820$ dollars.

At the end of the third year the account will have the $8820 from the previous year plus the interest, or

$$8820 + 8820(0.05) = 8820(1.05) = 9261 \text{ dollars}$$

At the end of the fourth year there will be the $9261 from the previous year plus the interest, or

$$9261 + 9261(0.05) = 9261(1.05) = 9724.05 \text{ dollars}$$

The dollar amounts at the end of each year form the sequence 8400, 8820, 9261, 9724.05. Notice that each term of the sequence can be found by multiplying the preceding term by 1.05.   ▭

Example 3

Suppose you deposit $8000 in a savings account that pays 5% *simple* annual interest, that is, interest earned only on the initial $8000 and not compounded. How much money will be in the account at the end of each of the next 4 years?

Solution   As in Example 2 the account earns 5% of $8000, or $0.05(8000) = 400$ dollars, during the first year. At the end of the first year the account will contain the original $8000 plus $400 interest for a total of $8400.

During the second year, interest will be earned only on the initial $8000 deposited, for another $400. At the end of the second year the account will have the $8400 from the previous year plus $400 interest for a total of

$$8400 + 400 = 8800 \text{ dollars}$$

At the end of the third year the account will have the $8800 from the previous year plus another $400 simple interest for a total of

$$8800 + 400 = 9200 \text{ dollars}$$

At the end of the fourth year there will be the $9200 from the previous year plus another $400 simple interest for a total of

$$9200 + 400 = 9600 \text{ dollars}$$

The dollar amounts at the end of each year form the sequence 8400, 8800, 9200, 9600. Notice that each term of the sequence can be found by adding 400 to the preceding term.   ▭

Example 4

A regular polygon is a geometric figure in which all the sides have equal length. For example, an equilateral triangle and a square are regular polygons. All the interior angles in a regular polygon are equal also. Find the general term a_n for the sequence that gives the size of an interior angle in a regular polygon of n sides. (See Figure 9.1.)

Solution   Note that this sequence starts with a_3 because we cannot have a polygon with fewer than three sides. We already know the values of a_3 and a_4: the angles in an equilateral triangle are each 60°, and in a square the

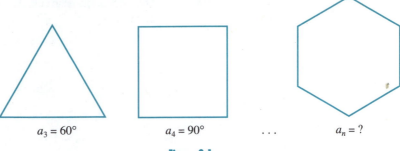

$a_3 = 60°$ $a_4 = 90°$ . . . $a_n = ?$

Figure 9.1

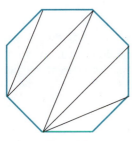

Figure 9.2

angles are each 90°. We would like to find a formula for the size of the angles in any regular polygon.

If we can find the sum of *all* the angles in a regular polygon, then we can simply divide by n to find the size of each. (You can check that this idea works for the equilateral triangle and the square.) To find this sum, notice that any polygon can be partitioned into triangles, as shown in Figure 9.2.

Convince yourself that a polygon of n sides (for $n \geq 3$) can be partitioned into $n - 2$ triangles. Since the angles in every triangle add up to 180°, the sum of the angles in an n-sided polygon is then $(n - 2)$ times 180°. To find the size of just one of the angles, we divide the sum by n, which gives us the general term of the sequence:

$$a_n = \frac{(n - 2)180}{n}$$

Recursively Defined Sequences

A sequence is said to be defined **recursively** if each term of the sequence is defined in terms of its predecessors. For example, the sequence defined by

$$a_1 = 2, \qquad a_{n+1} = 3a_n - 2$$

is a recursive sequence. Its first four terms are

$$a_1 = 2$$
$$a_2 = 3a_1 - 2 = 3(2) - 2 = 4$$
$$a_3 = 3a_2 - 2 = 3(4) - 2 = 10$$
$$a_4 = 3a_3 - 2 = 3(10) - 2 = 28$$

Example 5

Find the first five terms of the recursive sequence

$$a_1 = -1, \qquad a_{n+1} = (a_n)^2 - 4$$

Solution   The first term is given. We find each subsequent term by using the recursive formula:

$$a_2 = (a_1)^2 - 4 = (-1)^2 - 4 = -3$$

$$a_3 = (a_2)^2 - 4 = (-3)^2 - 4 = 5$$

$$a_4 = (a_3)^2 - 4 = 5^2 - 4 = 21$$

$$a_5 = (a_4)^2 - 4 = 21^2 - 4 = 437$$

The first five terms are $-1, -3, 5, 21,$ and 437.   ▭

Example 6

Find a recursive definition for the sequence in Example 2 if the money is kept in the account for n years.

Solution   If we use the letter a for that sequence, we have $a_1 = 8400$.   Each successive term was obtained by multiplying the preceding term by 1.05, which means that $a_n = 1.05 a_{n-1}$.   The sequence is determined by the recursive definition

$$a_1 = 8400, \qquad a_n = 1.05 a_{n-1}$$ ▭

Example 7

Find a recursive definition for the sequence in Example 3 if the money is kept in the account for n years.

Solution   If we use the letter b for that sequence, we have $b_1 = 8400$.   Each successive term was obtained by adding 400 to the preceding term, which means that $b_n = b_{n-1} + 400$.   The sequence is determined by the recursive definition

$$b_1 = 8400, \qquad b_n = b_{n-1} + 400$$ ▭

Example 8

Karen joins a savings plan in which she deposits $200 per month and receives 12% annual interest compounded monthly.
a.   Find a recursively defined sequence that gives the amount of money in Karen's account after n deposits.
b.   Find the first four terms of the sequence.

Solutions   **a.**   In the first month Karen deposits $200, so $a_1 = 200$.   Each month thereafter Karen receives 1% interest (one-twelfth of 12% annual interest) on the previous month's balance and then adds $200 to the total. For example, before she makes her deposit in the second month, the account has

$$200 + 0.01(200) = 1.01(200) \text{ dollars}$$

She then adds $200 to this amount for a total of

$$a_2 = 1.01(200) + 200 \text{ dollars}$$

In general, after the nth deposit, Karen's account contains a_n dollars. In the next month she earns 1% interest on that balance, giving her

$$a_n + 0.01a_n = 1.01a_n \text{ dollars}$$

Then she deposits another $200 for a total of

$$a_{n+1} = 1.01a_n + 200 \text{ dollars}$$

Thus the recursive sequence is defined by

$$a_1 = 200, \qquad a_{n+1} = 1.01a_n + 200$$

b. Evaluate the recursive formula found in part (a) for $n = 1, 2, 3$:

$$a_1 = 200$$

$$a_2 = 1.01a_1 + 200$$

$$= 1.01(200) + 200 = 402$$

$$a_3 = 1.01a_2 + 200$$

$$= 1.01(402) + 200 = 606.02$$

$$a_4 = 1.01a_3 + 200$$

$$= 1.01(606.02) + 200 \approx 812.08$$

Using the TI-82 for Recursively Defined Sequences

The TI-82 has a built-in feature to calculate terms of recursively defined sequences. To use it we must first put the calculator in the sequence mode. Press the $\boxed{\text{MODE}}$ key and use the arrow keys to select **Seq**, as shown in Figure 9.3(a), and then press $\boxed{\text{ENTER}}$.

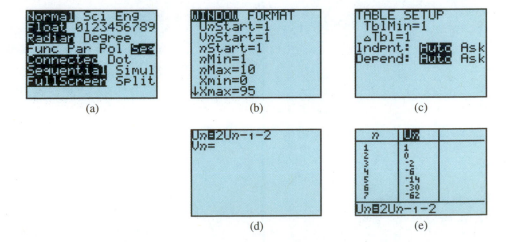

(a)   (b)   (c)

(d)   (e)

Figure 9.3

As an example, we will define the recursive sequence

$$a_1 = 1, \qquad a_n = 2a_{n-1} - 2$$

(The calculator uses U_n to represent the general term.) To define $U_1 = 1$, press the WINDOW key and set **UnStart = 1** and **nStart = 1** as shown in Figure 9.3(b). (The other values do not matter for this application.) Then press 2nd WINDOW and make sure that **TblMin = 1** (the value of **nStart**), that **ΔTbl = 1**, and that **Auto** is selected in both **Indpnt** and **Depend**, as shown in Figure 9.3(c).

To enter the recursive part of the definition of the sequence, press Y= and clear out any old definitions. On the first line set $U_n = 2U_{n-1} - 2$ as shown in Figure 9.3(d). (To enter U_{n-1}, use 2nd and the key for the number 7.)

Finally, enter 2nd GRAPH to see a table of seven values for the sequence, as shown in Figure 9.3(e). If you wish to see more values, you can press the down arrow to scroll farther down the table.

Exercise 9.1

In Exercises 1–14 find the first four terms in the sequence whose general term is given.

1. $a_n = n - 5$

2. $b_n = 2n - 3$

3. $c_n = \dfrac{n^2 - 2}{2}$

4. $d_n = \dfrac{3}{n^2 + 1}$

5. $s_n = 1 + \dfrac{1}{n}$

6. $t_n = \dfrac{n}{2n - 1}$

7. $u_n = \dfrac{n(n - 1)}{2}$

8. $v_n = \dfrac{5}{n(n + 1)}$

9. $w_n = (-1)^n$

10. $A_n = (-1)^{n+1}$

11. $B_n = \dfrac{(-1)^n(n - 2)}{n}$

12. $C_n = (-1)^{n-1} 3^{n+1}$

13. $D_n = 1$

14. $E_n = -1$

Find the first five terms of the recursively defined sequence in Exercises 15–22.

15. $s_1 = 3, \quad s_n = s_{n-1} + 2$

16. $c_1 = 6, \quad c_n = c_{n-1} - 4$

17. $d_1 = 24, \quad d_{n+1} = \dfrac{-1}{2}d_n$

18. $r_1 = 27, \quad r_{n+1} = \dfrac{2}{3}r_n$

19. $t_1 = 1, \quad t_{n+1} = (n + 1)t_n$

20. $x_1 = 1, \quad x_n = \left(\dfrac{n}{n - 1}\right)x_{n-1}$

21. $w_1 = 100, \quad w_n = 1.10w_{n-1} + 100$

22. $q_1 = 100, \quad q_{n+1} = 0.9q_n + 100$

In Exercises 23–30
a. find the first four terms of each sequence.
b. determine equations to define the sequence *recursively*.

23. A new car costs \$14,000 and depreciates in value by 15% each year. How much is the car worth after n years?

24. Krishna takes a job as an executive secretary for \$21,000 per year with a guaranteed 5% raise each year. What will his salary be after n years?

25. A long-distance phone call costs \$1.10 to make the connection and an additional \$0.45 for each minute. What is the cost of a call that lasts n minutes?

26. Bettina earns $1000 per month plus $57 for each satellite dish that she sells. What is her monthly income when she sells n satellite dishes?

27. Geraldo inherits an annuity of $50,000 that earns 12% annual interest compounded monthly. If he withdraws $500 at the end of each month, what is the value of the annuity after n months?

28. Eve borrowed $18,000 for a new car at 6% annual interest compounded monthly. If she pays $400 per month toward the loan, how much does she owe after n months?

29. Majel must take 10 milliliters of a medication directly into her bloodstream at constant intervals. During each time interval, her kidneys filter out 20% of the drug present just after the most recent dose. How much of the drug will be in her bloodstream after n doses?

30. A forest contains 64,000 trees. According to a new logging plan, each year 5% of the trees will be cut down and 16,000 new trees will be planted. How many trees will be in the forest after n years?

31. **a.** Draw three noncollinear points in the plane. (The points should not lie on the same line.) How many distinct lines are determined by the points? (In other words, how many different lines can you draw by choosing two of the points and joining them?)

    b. Add a fourth point to your diagram. Now how many lines are determined?

    c. Let L_n stand for the number of distinct lines determined by n noncollinear points. Write the first five terms of the sequence, and find a recursive formula for the sequence L_n.

32. **a.** Draw two distinct nonparallel lines in the plane. In how many points do the lines intersect?

    b. Add a third line to your diagram that is not parallel to either of the first two lines and does not pass through their intersection point. How many intersection points are there?

    c. Let P_n stand for the maximum number of intersection points determined by n lines in the plane,

no two of which are parallel. Write the first five terms of the sequence, and find a recursive formula for the sequence P_n.

Find the indicated term for each sequence in Exercises 33–40.

33. $D_n = 2^n - n$; D_6

34. $E_n = \sqrt{n + 1}$; E_{11}

35. $x_n = \log n$; x_{26}

36. $y_n = \log (n + 1)$; y_9

37. $z_n = 2\sqrt{n}$; z_{20} 38. $U_n = \dfrac{n + 1}{n - 1}$; U_{17}

39. **a.** The Fibonacci sequence is found throughout nature. For example, the number of spirals visible in a sunflower or on a pineapple are elements of the sequence. The Italian mathematician Leonardo Fibonnaci used this sequence to model the growth of a population of rabbits. The Fibonacci sequence is defined recursively by

$$f_1 = 1, \qquad f_2 = 1, \qquad f_{n+2} = f_n + f_{n+1}$$

Find the first 16 terms of the Fibonacci sequence.

b. Calculate the quotients $\dfrac{f_{n+1}}{f_n}$ for $n = 1$ to $n = 15$. What do you observe? Now find a decimal approximation for the "golden ratio," $\dfrac{1 + \sqrt{5}}{2}$.

40. The Lucas sequence is defined recursively by

$$L_1 = 2, \qquad L_2 = 1, \qquad L_{n+2} = L_n + L_{n+1}$$

   a. Find the first 10 terms of the Lucas sequence.

   b. Calculate $(L_{n+1})^2 - L_n (L_{n+2})$ for $n = 1$ to $n = 8$. What do you notice?

In Exercises 41–46 use a calculator to evaluate a large number of terms for each recursive sequence. What happens to the terms as n gets larger?

41. $a_1 = 1$, $a_n = \dfrac{1}{1 + a_{n-1}} + 1$

42. $b_1 = 1$, $b_n = \dfrac{2}{1 + b_{n-1}} + 1$

43. $c_1 = 3, \quad c_n = \dfrac{\sqrt{1 + c_{n-1}}}{2}$

45. $s_1 = 1, \quad s_n = \dfrac{1}{2}\left(s_{n-1} + \dfrac{4}{s_{n-1}}\right)$

44. $d_1 = 8, \quad d_n = \dfrac{\sqrt{1 + d_{n-1}}}{2}$

46. $t_1 = 1, \quad t_n = \dfrac{1}{2}\left(t_{n-1} + \dfrac{9}{t_{n-1}}\right)$

9.2 ARITHMETIC AND GEOMETRIC SEQUENCES

Arithmetic Sequences

Suppose a charter tour bus service charges $50 plus $15 for each passenger. The cost of a tour is then a function of the number of passengers. Since the number of passengers, n, can only be a natural number, the domain of the function is the set of natural numbers 1, 2, 3, . . . up to the capacity of the tour bus.

If C_n represents the cost of a tour for n passengers, then

$$C_1 = 50 + 15(1) = 65$$

$$C_2 = 50 + 15(2) = 80$$

$$C_3 = 50 + 15(3) = 95$$

and in general

$$C_n = 50 + 15n$$

Note that each term of this sequence can be obtained from the preceding one by adding 15. A sequence in which each term can be obtained from the preceding term by adding a fixed amount is called an **arithmetic sequence**.

The fixed amount we add to each term is the difference between two successive terms and is called the **common difference**. In the example above the common difference is 15. If we denote the first term of an arithmetic sequence by a and the common difference by d, then the sequence can be defined recursively by

$$a_1 = a$$

$$a_{n+1} = a_n + d$$

Example 1

Find the first four terms of an arithmetic sequence with first term 6 and common difference 3.

Solution Since the first term is 6, we have $a_1 = 6$. To find each subsequent term, add 3 to the preceding term:

$$a_2 = a_1 + 3 = 6 + 3 = 9$$

$$a_3 = a_2 + 3 = 9 + 3 = 12$$

$$a_4 = a_3 + 3 = 12 + 3 = 15$$

The first four terms are 6, 9, 12, and 15.

Observe that an arithmetic sequence defines a linear function of n. In Figure 9.4 compare the graph of the linear function $f(x) = 2x + 3$, whose domain is the set of real numbers, and the graph of the arithmetic sequence $a_n = 2n + 3$, whose domain is the set of natural numbers. The common difference, 2, of the sequence corresponds to the slope of the linear function.

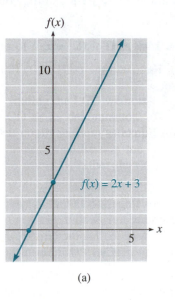

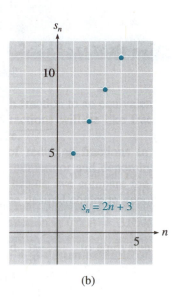

Figure 9.4   (a)   (b)

General Term of an Arithmetic Sequence

We have found a recursive definition for an arithmetic sequence, but we can also find a nonrecursive definition; that is, we can find a formula for the general term. Consider an arithmetic sequence with first term a and common difference d. The

first term is	a
second term is	$a + d$
third term is	$a + d + d = a + 2d$
fourth term is	$a + d + d + d = a + 3d$

$$\vdots$$

nth term is $a + d + d + \cdots + d = a + (n - 1)d$

Thus we have the following property.

> The *n*th term of an arithmetic sequence is
>
> $$a_n = a + (n - 1)d \qquad\qquad (1)$$

We can use Equation (1) to find a particular term of an arithmetic sequence if we know the first term and the common difference.

Example 2

Find the 14th term of the arithmetic sequence $-6, -1, 4, \ldots$.

Solution First find the common difference by subtracting any term from its successor:

$$d = -1 - (-6) = 5$$

Then use Equation (1) with $n = 14$:

$$a_{14} = -6 + (14 - 1)5 = 59$$

Example 3

Suppose you deposit \$8000 in an account that pays 5% *simple* annual interest. (See Example 3 of Section 9.1.)
a.   Find a formula for the amount of money in the account after *n* years.
b.   How much money will be in the account after 20 years?

Solutions   **a.**   Each year $0.05(8000)$ dollars, or \$400, interest is added to the account balance. Thus the annual balances, b_n, form an arithmetic sequence whose first term is 8400 and whose common difference is 400. Hence

$$b_n = 8400 + (n - 1)400$$

b.   Substitute $n = 20$ into the formula for the general term found in part (a).

$$b_{20} = 8400 + (20 - 1)400 = 16{,}000$$

After 20 years the account will contain \$16,000.

Geometric Sequences

Suppose a national junior chess tournament starts with 1024 invited contestants. At the end of each round the winners move on to the next level. Thus after each round there are half as many contestants as before, and the number of remaining contestants is a function of the number of rounds completed. Since the number of rounds, *n*, is a natural number, the domain of the function is a set of natural numbers.

If C_n represents the number of contestants after n rounds of competition, then

$$C_1 = 1024 \left(\frac{1}{2}\right) = 512$$

$$C_2 = 512 \left(\frac{1}{2}\right) = 1024 \left(\frac{1}{2}\right)^2 = 256$$

$$C_3 = 256 \left(\frac{1}{2}\right) = 1024 \left(\frac{1}{2}\right)^3 = 128$$

In general,

$$C_n = 1024 \left(\frac{1}{2}\right)^n$$

Note that each term of this sequence can be obtained from the preceding one by multiplying by $\frac{1}{2}$. A sequence in which each term can be obtained from the preceding term by multiplying by a fixed amount is called a **geometric sequence**.

The fixed amount by which we multiply each term is the ratio of two successive terms and is called the **common ratio**. In the example above the common ratio is $\frac{1}{2}$. If we denote the first term of a geometric sequence by a and the common ratio by r, then the sequence can be defined recursively by

$$a_1 = a$$

$$a_{n+1} = ra_n$$

Example 4

Find the first four terms of a geometric sequence whose first term is 64 and whose common ratio is $\frac{5}{4}$.

Solution Since the first term is 64, we have $a_1 = 64$. To find each subsequent term, multiply the previous term by $\frac{5}{4}$:

$$a_2 = \frac{5}{4} a_1 = \frac{5}{4}(64) = 80$$

$$a_3 = \frac{5}{4} a_2 = \frac{5}{4}(80) = 100$$

$$a_4 = \frac{5}{4} a_3 = \frac{5}{4}(100) = 125$$

The first four terms are 64, 80, 100, and 125.

Observe that a geometric sequence defines an exponential function of n. In Figure 9.5(a) and (b), compare the graphs of the exponential function $f(x) = 100\,(2)^{x-1}$, whose domain is the set of real numbers, and the geometric sequence $a_n = 100\,(2)^{n-1}$, whose domain is the set of natural numbers. The common ratio of the geometric sequence corresponds to the base of the exponential function.

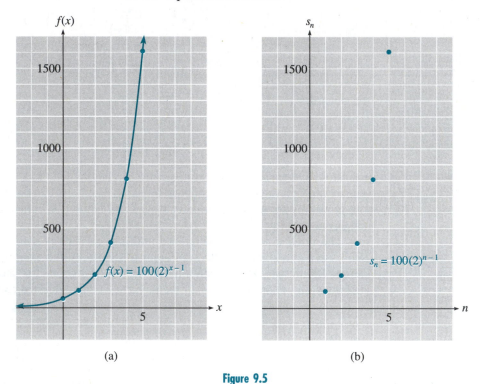

(a)   (b)

Figure 9.5

Recall that an arithmetic sequence defines a linear function whose slope corresponds to the common difference. Just as an exponential function grows much faster in the long run than a linear function, so does a geometric sequence grow much faster than an arithmetic sequence.

Example 5

Identify the following sequences as arithmetic, geometric, or neither.
a. 3, 5, 7, . . .
b. 3, −6, 12, . . .
c. 3, 6, 10, . . .
d. 3, 1, $\dfrac{1}{3}$, . . .

Solutions   **a.** This sequence is arithmetic. Each term is obtained from the preceding term by adding 2.

b. This sequence is geometric. Each term is obtained from the preceding term by multiplying by -2.
c. This sequence is neither arithmetic nor geometric.
d. This sequence is geometric. Each term is obtained from the preceding term by multiplying by $\frac{1}{3}$.

General Term of a Geometric Sequence

We can now find a nonrecursive formula for the general term of a geometric sequence. If we denote the first term of the geometric sequence by a, then the

second term is ar

third term is $ar \cdot r = ar^2$

fourth term is $ar^2 \cdot r = ar^3$

$$\vdots$$

nth term is ar^{n-1}

In general, we have the following property.

> The nth term of a geometric sequence is
> $$a_n = ar^{n-1} \tag{2}$$

We can use Equation (2) to find a particular term of a geometric sequence if we know the first term and the common ratio.

Example 6

Find the ninth term of the geometric sequence $-24, 12, -6, \ldots$.

Solution First find the common ratio by dividing any term by its predecessor:

$$r = \frac{12}{-24} = -\frac{1}{2}$$

Then use Equation (2) with $a = -24$, $r = -\frac{1}{2}$, and $n = 9$:

$$a_9 = -24\left(-\frac{1}{2}\right)^8 = -\frac{3}{32}$$

Example 7

Suppose you deposit \$8000 in an account that pays 5% annual interest compounded annually. (See Example 2 of Section 9.1.)
a. Find a formula for the amount of money in the account after n years.
b. How much money will be in the account after 20 years?

Solutions   **a.**   Each year the account balance is multiplied by 1.05. Thus the annual balances, c_n, form a geometric sequence whose first term is 8400 and whose common ratio is 1.05. Hence

$$c_n = 8400(1.05)^{n-1}$$

b.   Substitute $n = 20$ into the formula for the general term found in part (a).

$$c_{20} = 8400(1.05)^{19} \approx 21,226.38164$$

After 20 years the account will contain $21,226.38.   ◻

Example 8

Find a nonrecursive definition for each sequence.
a. $a_1 = 2,$ $a_n = a_{n-1} + 3$
b. $b_1 = 2,$ $b_{n+1} = 3b_n$

Solutions   **a.**   From the definition we see that the first few terms of the sequence are 2, 5, 8, 11. Since each new term is found by adding 3 to the preceding term, we see that this sequence is arithmetic with a common difference of 3. So the sequence has the form

$$a_n = 2 + (n - 1)3$$

or

$$a_n = 3n - 1$$

b.   The first four terms of this sequence are 2, 6, 18, 54. Since each new term is found by multiplying the previous term by 3, we see that this sequence is geometric with a common ratio of 3. So the sequence has the form

$$b_n = 2 \cdot 3^{n-1}$$ ◻

Exercise 9.2

Identify each sequence in Exercises 1–12 as arithmetic, geometric, or neither.

1. $-2, -6, -18, -54, \ldots$

2. $-2, -6, -10, -14, \ldots$

3. $16, 8, 0, -8, \ldots$ **4.** $16, 8, 4, 2, \ldots$

5. $-1, 1, -1, 1, \ldots$ **6.** $1, 3, 6, 10, \ldots$

7. $1, 4, 9, 16, \ldots$ **8.** $5, -5, 5, -5, \ldots$

9. $27, 9, 3, 1, \ldots$ **10.** $-\dfrac{1}{3}, \dfrac{1}{3}, 1, \dfrac{5}{3}, \ldots$

11. $\dfrac{2}{3}, -1, \dfrac{3}{2}, -\dfrac{9}{4}, \ldots$

12. $2, -8, 32, -124, \ldots$

Find the first four terms of each sequence in Exercises 13–24.

13. $a = 2, \quad d = 4$ **14.** $a = 7, \quad d = 3$

15. $a = \dfrac{1}{2}, \quad d = \dfrac{1}{4}$ **16.** $a = \dfrac{2}{3}, \quad d = \dfrac{1}{3}$

17. $a = 2.7, \quad d = -0.8$

18. $a = 5.9, \quad d = -1.3$

19. $a = 5$, $r = -2$ **20.** $a = -4$, $r = 3$

21. $a = 9$, $r = \dfrac{2}{3}$ **22.** $a = 25$, $r = \dfrac{4}{5}$

23. $a = 60$, $r = 0.4$ **24.** $a = 10$, $r = 0.3$

Find the next three terms and an expression for the general term in Exercises 25–32.

25. $3, 7, 11, \ldots$

26. $-10, -20, -30, \ldots$

27. $-1, -5, -9, \ldots$ **28.** $-6, -1, 4, \ldots$

29. $\dfrac{2}{3}, \dfrac{4}{3}, \dfrac{8}{3}, \ldots$ **30.** $6, 3, \dfrac{3}{2}, \ldots$

31. $4, -2, 1, \ldots$ **32.** $\dfrac{1}{2}, -\dfrac{3}{2}, \dfrac{9}{2}, \ldots$

33. Find the twelfth term in the arithmetic sequence $2, \dfrac{5}{2}, 3, \ldots$.

34. Find the tenth term in the arithmetic sequence $\dfrac{3}{4}, 2, \dfrac{13}{4}, \ldots$.

35. Find the eighth term in the geometric sequence $-3, \dfrac{3}{2}, -\dfrac{3}{4}, \ldots$.

36. Find the sixth term in the geometric sequence $-5, -1, -\dfrac{1}{5}, \ldots$.

37. Find the first term of a geometric sequence with fifth term 48 and common ratio 2.

38. Find the first term of a geometric sequence with fifth term 1 and common ratio $\dfrac{1}{2}$.

39. How many terms are in the sequence $\dfrac{1}{8}, \dfrac{1}{4}, \dfrac{1}{2}, \ldots, 512$?

40. How many terms are in the sequence $\dfrac{27}{64}, \dfrac{9}{16}, \dfrac{3}{4}, \ldots, \dfrac{64}{27}$?

Find a nonrecursive definition for each sequence in Exercises 41–48.

41. $s_1 = 3$, $s_n = s_{n-1} + 2$

42. $c_1 = 6$, $c_n = c_{n-1} - 4$

43. $x_1 = 0$, $x_{n+1} = x_n - 3$

44. $y_1 = -1$, $y_{n+1} = y_n + 5$

45. $d_1 = 24$, $d_{n+1} = \dfrac{-1}{2}d_n$

46. $r_1 = 27$, $r_{n+1} = \dfrac{2}{3}r_n$

47. $w_1 = 1$, $w_n = 2w_{n-1}$

48. $q_1 = 7$, $q_n = 3q_{n-1}$

49. An outdoor theater has 30 seats in the first row, 32 seats in the second row, 34 seats in the third row, and so on, with two more seats in any one row than in the preceding one. Let s_n be the number of seats in the nth row of the theater.
   a. Graph the first five terms of the sequence.
   b. How many seats are in the 50th row?

50. Gwynn is training for a triathlon. In her first week of training she bicycled a total of 20 miles, then increased to 24 miles in her second week, 28 miles in her third week, and so on, increasing her mileage by 4 miles per week. Let m_n be Gwynn's mileage in her nth week of training.
   a. Graph the first five terms of the sequence.
   b. How many miles will she bicycle in her 18th week of training?

51. The cost of drilling a well increases the deeper you drill. Mel's Wells charges $50 for the first 5 feet, $55 for the second 5 feet, $60 for the third 5 feet, and so on. Let d_n be the cost of drilling the nth 5 feet.
   a. Graph the first five terms of the sequence.
   b. How much does Mel charge to drill from a depth of 65 feet to a depth of 70 feet?

52. The cost of having the windows washed in a high-rise office building depends on the height of the window. We Do Windows charges $0.50 per square yard of glass below 10 feet high, $0.60 per square yard for windows between 10 and 20 feet above the ground, $0.70 per square yard for windows between 20 and 30 feet above the ground, and so on. Let w_n be the cost of washing windows on the nth floor if each floor is 10 feet high.

a. Graph the first five terms of the sequence.

b. How much does We Do Windows charge to wash 100 square yards of windows on the 23rd floor?

53. Valerie's grandmother deposited $500 into a college fund for Valerie on the day Valerie was born. The fund earns 5% interest compounded annually. Let V_n be the value of the deposit, including interest, n years later.

a. Graph the first five terms of the sequence.

b. How much will the deposit be worth on Valerie's 18th birthday?

54. When he was 25 years old, Bruce won $2000 on the lottery and deposited the money into a retirement fund. The fund earns 6% interest compounded annually. Let R_n be the value of the deposit, including interest, n years later.

a. Graph the first five terms of the sequence.

b. How much will the deposit be worth when Bruce turns 65 years old?

55. One hundred kilograms of a toxic chemical was dumped illegally into a clean reservoir. A filter can remove 20% of the chemical still present each week (so that 80% of the previous amount remains). Let c_n be the amount of the chemical remaining after n weeks.

a. Graph the first five terms of the sequence.

b. How much of the chemical will remain in the water after 20 weeks?

56. A heart patient is given 40 milliliters of a medication by injection. Each hour 15% of the medicine still present is eliminated from the body (so that 85% of the previous amount remains). Let m_n be the amount of medicine remaining in the patient's body after n hours.

a. Graph the first five terms of the sequence.

b. How much of the medication is left in the patient's body after 10 hours?

 SERIES

Often we are more interested in the sum of a sequence of numbers than in the sequence itself. For example, suppose that for 5 years you have put $100 a month into a savings account that pays 12% annual interest compounded monthly. Then the sequence

$$p_n = 100(1.01)^n$$

gives the current value of the payment you made n months ago (plus interest). However, you are probably more interested in the *total* amount in your savings account, which is the *sum* of the terms p_n from $n = 1$ to $n = 60$.

The sum of the terms of a sequence is called a **series**. Note that although the words *sequence* and *series* are often used interchangeably in everyday English, they have different meanings in mathematics. A sequence is a *list* of numbers, whereas a series is a *single* number obtained by computing a sum. For example, the list of numbers

$$2, \quad 4, \quad 8, \quad 16, \quad 32$$

is a sequence. The sum of those five terms is the series

$$2 + 4 + 8 + 16 + 32 = 62$$

We use the symbol S_n to denote the sum of the first n terms of a sequence. Thus for the sequence above, $S_5 = 62$.

Example 1

Find the series S_6 for the sequence with general term $a_n = 3n + 1$.

Solution   The first six terms of the sequence are 4, 7, 10, 13, 16 and 19. Thus

$$S_6 = 4 + 7 + 10 + 13 + 16 + 19 = 69$$

Arithmetic Series

It is usually very difficult to obtain a formula for the sum of the first n terms of a sequence, but in the special case of an arithmetic sequence, we can. As an example consider the sum of the first 12 terms of the sequence with general term $a_n = 4n + 1$:

$$S_{12} = 5 + 9 + 13 + \cdots + 41 + 45 + 49 \tag{3}$$

Making discoveries in any field, including mathematics, often depends on looking at familiar objects in a new way. To find our formula for arithmetic series, we will write the terms of the series in the opposite order:

$$S_{12} = 49 + 45 + 41 + \cdots + 13 + 9 + 5 \tag{4}$$

By adding Equations (3) and (4) term by term, we find

$$
\begin{aligned}
S_{12} &= 5 + 9 + 13 + \cdots + 41 + 45 + 49 \\
S_{12} &= 49 + 45 + 41 + \cdots + 13 + 9 + 5 \\
\hline
2S_{12} &= 54 + 54 + 54 + \cdots + 54 + 54 + 54
\end{aligned}
$$

Each term of the sum is the same, namely, 54. The term 54 occurs 12 times (since we are adding 12 terms of the original sequence), so

$$2S_{12} = 12(54)$$

$$S_{12} = \frac{12(54)}{2} = 324$$

Notice that the number 54 is the sum of the first term of the arithmetic series, 5, and the last term, 49. In other words, $54 = a_1 + a_n$.   This observation is the key to producing our formula. In general, when we add two copies of S_n (one forwards and one backwards), the sum $a_1 + a_n$ occurs n times and results in twice the series we want. Thus

$$2S_n = n(a_1 + a_n)$$

Dividing both sides by 2 gives us the following formula for computing an arithmetic series.

$$S_n = \frac{n}{2}(a_1 + a_n)$$

Example 2

Find the sum of the first 15 odd integers.

Solution The odd integers form an arithmetic sequence with first term $a_1 = 1$ and common difference 2. Thus the general term of the sequence is $a_n = 1 + (n - 1)2$, and we would like to find the series S_{15}.

We begin by finding the last term of the series, a_{15}:

$$a_{15} = 1 + (15 - 1)2 = 29$$

Next we use the formula for S_n with $n = 15$, $a_1 = 1$, and $a_{15} = 29$:

$$S_{15} = \frac{15}{2}(1 + 29) = 225$$

Example 3

Arlene starts a new job in a print shop at a salary of $800 per month. If she keeps up with the training program, her salary will increase by $35 per month. How much will Arlene have earned at the end of the 18-month training program?

Solution Arlene's monthly salary forms an arithmetic sequence with $a_1 = 800$ and $d = 35$. Thus the general term of the sequence is $a_n = 800 + (n - 1)35$, and we would like to find the series S_{18}.

We begin by finding the last term of the series, a_{18}:

$$a_{18} = 800 + (18 - 1)35 = 1395$$

Next we use the formula for S_n with $n = 18$, $a_1 = 800$, and $a_{18} = 1395$:

$$S_{18} = \frac{18}{2}(800 + 1395) = 19,755$$

Arlene's total earnings for the 18 months will be $19,755.

Geometric Series

We can also find a formula for the sum of the first n terms of a geometric sequence. As an example we will compute the sum of the first nine terms of the sequence $a_n = 3^n$. (This sequence is geometric with $a = 3$ and $r = 3$. Its general term is $a_n = 3(3)^{n-1}$, or 3^n.) Then

$$S_9 = 3 + 3^2 + 3^3 + \cdots + 3^8 + 3^9 \tag{5}$$

Instead of writing the terms in the opposite order as we did for the arithmetic series, here we will multiply each term of S_9 by the common ratio, 3.

$$3S_9 = 3(3) + 3(3)^2 + 3(3)^3 + \cdots + 3(3)^8 + 3(3)^9$$

or

$$3S_9 = 3^2 + 3^3 + 3^4 + \cdots + 3^8 + 3^9 + 3^{10} \qquad (6)$$

Next we subtract Equation (5) from Equation (6). Note that most of the terms will cancel out in the subtraction.

$$
\begin{array}{rl}
3S_9 = & 3^2 + 3^3 + \cdots + 3^8 + 3^9 + 3^{10} \\
-[\ S_9 = & 3 + 3^2 + 3^3 + \cdots + 3^8 + 3^9 \qquad] \\
\hline
2S_9 = & -3 \qquad\qquad\qquad\qquad\qquad + 3^{10}
\end{array}
$$

Thus

$$2S_9 = 3^{10} - 3$$

or

$$S_9 = \frac{3^{10} - 3}{2}.$$

which simplifies to 29,523. Notice that the second term of the numerator is a_1 (3 in this example) and the first term is a_{10} ($3 \cdot 3^9$, or 3^{10}). The denominator of the expression, 2, is equal to $r - 1$. In general, we can derive the following formula for computing a geometric series.

$$S_n = \frac{a_{n+1} - a_1}{r - 1}$$

Think of a_{n+1} as the next term following the last term in the sum. The formula is valid as long as the common ratio, r, is not equal to 1. In Exercise 9.3 we will derive this formula and the formula for arithmetic series.

Example 4

Find the sum of the first five terms of the sequence $5, \dfrac{10}{3}, \dfrac{20}{9}, \dfrac{40}{27}, \ldots$.

Solution   The sequence is geometric with $a_1 = 5$.   We find r by dividing $\dfrac{10}{3}$ by 5:

$$\frac{10}{3} \div 5 = \frac{2}{3}$$

Thus the general term of the series is $a_n = 5 \left(\frac{2}{3}\right)^{n-1}$,   and we would like to find S_5. We can use the formula for a geometric series with $n = 5$,

$$S_5 = \frac{a_6 - a_1}{r - 1}$$

where $a_1 = 5$, $a_6 = 5\left(\frac{2}{3}\right)^5$, and $r = \frac{2}{3}$. Thus

$$S_5 = \frac{5\left(\frac{2}{3}\right)^5 - 5}{\frac{2}{3} - 1} = \frac{5\left(\frac{32}{243}\right) - 5}{-\frac{1}{3}}$$

$$= 5\left(\frac{32}{243} - 1\right)\left(\frac{-3}{1}\right) \approx 13.02$$

Example 5

Payam's starting salary as an engineer is $20,000 with a 5% annual raise for each of the next 5 years, depending on suitable progress. If Payam receives each salary increase, how much will he make during his first 6 years?

Solution Payam's salary is multiplied each year by a factor of 1.05, so its values form a geometric sequence with a common ratio of $r = 1.05$. The general term of the sequence is $a_n = 20{,}000(1.05)^{n-1}$. His total income over the 6 years will be the sum of the first six terms of the sequence. Thus

$$S_6 = \frac{a_7 - a_1}{r - 1} = \frac{20{,}000(1.05)^6 - 20{,}000}{1.05 - 1}$$

$$\approx \frac{20{,}000(0.3401)}{0.05} = 136{,}038.26$$

Payam will earn $136,038.26 over the next 6 years.

Exercise 9.3

In Exercises 1–6 evaluate each arithmetic series.

1. The sum of the first 9 terms of the sequence $a_n = -4 + 3n$

2. The sum of the first 10 terms of the sequence $a_n = 5 - 2n$

3. The sum of the first 16 terms of the sequence $a_n = 18 - \frac{4}{3}n$

4. The sum of the first 13 terms of the sequence $a_n = -6 - \frac{1}{2}n$

5. The sum of the first 30 terms of the sequence $a_n = 1.6 + 0.2n$

6. The sum of the first 25 terms of the sequence $a_n = 2.5 + 0.3n$

In Exercises 7–12 evaluate each geometric series.

7. The sum of the first five terms of $a_n = 2(-4)^{n-1}$

8. The sum of the first eight terms of $a_n = 12(3)^{n-1}$

9. The sum of the first nine terms of
$$a_n = -48\left(\frac{1}{2}\right)^{n-1}$$

10. The sum of the first six terms of
$$a_n = 81\left(\frac{2}{3}\right)^{n-1}$$

11. The sum of the first four terms of
$$a_n = 18(1.15)^{n-1}$$

12. The sum of the first four terms of
$$a_n = 512(0.72)^{n-1}$$

Identify each series in Exercises 13–22 as arithmetic, geometric, or neither, and then evaluate it.

13. $2 + 4 + 6 + \cdots + 96 + 98 + 100$

14. $1 + 3 + 5 + \cdots + 95 + 97 + 99$

15. $2 + 4 + 8 + 16 + \cdots + 256 + 512 + 1024$

16. $1 + 3 + 9 + 27 + \cdots + 6561 + 19{,}683$

17. $1 + 8 + 27 + 64 + 125 + 216 + 343$

18. $1 + 11 + 111 + 1111 + 11{,}111 + 111{,}111$

19. $87 + 84 + 81 + 78 + \cdots + 45 + 42 + 39$

20. $1 + (-2) + (-5) + \cdots + (-41) + (-44)$

21. $6 + 2 + \dfrac{2}{3} + \cdots + \dfrac{2}{81} + \dfrac{2}{243}$

22. $12 + 3 + \dfrac{3}{4} + \cdots + \dfrac{3}{64} + \dfrac{3}{256}$

Write a series to describe each problem in Exercises 23–40, and then evaluate it.

23. Find the sum of all even integers from 14 to 88.

24. Find the sum of all multiples of 7 from 14 to 105.

25. A clock strikes once at one o'clock, twice at two o'clock, and so on. How many times will the clock strike in a 12-hour period?

26. Jessica puts one candle on the cake at her daughter's first birthday, two candles at her second birthday, and so on. How many candles will Jessica have used after her daughter's 16th birthday?

26. A rubber ball is dropped from a height of 24 feet and returns to three-fourths of its previous height on each bounce.

a. How high does the ball bounce after hitting the floor for the third time?

b. How far has the ball traveled vertically when it hits the floor for the fourth time?

28. A Yorkshire terrier can jump 3 feet into the air on his first bounce and five-sixths the height of his previous jump on each successive bounce.

a. How high can the terrier go on his fourth bounce?

b. How far has the terrier traveled vertically when he returns to the ground after his fourth bounce?

29. Sales of Brussels Sprouts dolls peaked at $920,000 in 1991 and began to decline at a steady rate of $40,000 per year. What total revenue should the manufacturer expect to gain from sale of the dolls from 1991 to 2000?

30. It takes Alida 20 minutes to type the first page of her term paper, but each subsequent page takes her 40 seconds less than the previous one. How long will it take her to type a 30-page paper?

31. A computer takes 0.1 second to perform the first iteration of a certain loop, and each subsequent iteration takes 0.05 second longer than the previous one. How long will it take the computer to perform 50 iterations?

32. Richard's water bill was $63.50 last month. If his bill increases by $2.30 per month, how much should he expect to pay for water during the next 10 months?

33. Sales of Energy Ranger dolls peaked at $920,000 in 1991 and began to decline by 8% per year. What total revenue should the manufacturer expect to gain from sale of the dolls from 1991 to 2000?

34. It takes Emily 20 minutes to type the first page of her term paper, but each subsequent page takes only 95% as long as the previous one. How long will it take her to type a 30-page paper?

35. A computer takes 0.1 second to perform the first iteration of a certain loop, and each subsequent iteration requires 20% longer than the previous one. How long will it take the computer to perform 50 iterations?

36. Megan's water bill was $63.50 last month. If her bill increases by 2% per month, how much should she expect to pay for water during the next 10 months?

37. Jim and Nora maintain a college fund for their son David by depositing $500 into an account each year, beginning on the day David was born. If the account earns an interest rate of 5% compounded annually, how much will be in the account on David's 18th birthday?

38. Ben begins an individual retirement account (IRA) when he turns 25 years old, depositing $2000 into the account each year. If the account earns 6% interest compounded annually, how much will he have in his IRA when he turns 65 years old?

39. Suppose you are given 1¢ on the first day of the month, 2¢ on the second day, 4¢ on the third day, and so on, each day's payment being twice the previous day's. What would be your total income on the 30th day?

40. According to legend, a man who had pleased a Persian king asked for the following reward. The man was to receive a single grain of wheat for the first square of a chessboard, two grains for the second square, four grains for the third square, and so on, doubling the amount for each square up to the sixty-fourth square. How many grains would he receive in all?

In Exercises 41–46 find formulas for arithmetic and geometric series by following the indicated steps.

41. In this exercise we will find a formula for the sum of an arithmetic series

$$1 + 2 + 3 + \cdots + (N - 2)$$
$$+ (N - 1) + N$$

the sum of the first N positive integers.

a. How many terms are in the series?

b. If we call this sum S, then

$$S = 1 + 2 + 3 + \cdots + (N - 2)$$
$$+ (N - 1) + N$$

Rewrite the sum by reversing the order of the terms on the right side of the equation and add the result "columnwise" to the equation above. What do the two numbers in any one column on the right side add up to?

c. You know how many columns there are from part (a). Based on that and your answer to part (b), write an expression for $2S$ and solve for S.

d. Use your answer from part (c) to write a formula for the sum of the arithmetic series

$$1 + 2 + 3 + \cdots + (N - 2)$$
$$+ (N - 1) + N$$

42. In this exercise we will find a formula for the sum of an arithmetic series when we are given the first term, F, the last term, L, the common difference, d, and the number of terms, N.

a. If we call this sum S, then

$$S = F + (F + d) + (F + 2d) + \cdots$$
$$+ (L - 2d) + (L - d) + L$$

Rewrite the sum by reversing the order of the terms on the right side of the equation and add the result "columnwise" to the equation above. What do the two numbers in any one column on the right side add up to?

b. You know that the original series has N terms. Based on that and your answer to part (b), write an expression for $2S$ and solve for S.

c. Use your answer from part (c) to write a formula for the sum of an arithmetic series when we are given the first term, F, the last term, L, the common difference, d, and the number of terms, N.

43. In this exercise we will find a formula for the sum of a geometric series

$$1 + r + r^2 + \cdots + r^{N-1} + r^N$$

where r is an arbitrary constant.

a. What is the common ratio of consecutive terms?

b. If we call this sum A, then

$$A = 1 + r + r^2 + \cdots + r^{N-1} + r^N$$

Multiply both sides of the equation by the common ratio. What is the result?

c. Subtract your last equation from the equation for A given above and simplify. What is the result?

d. Starting with your answer to part (c), solve for A. (*Hint:* Start by factoring out an A on the left side of the equation.)

e. Use your answer from part (d) to write a formula for $1 + r + r^2 + \cdots + r^{N-1} + r^N$.

44. In this exercise we will find a formula for the sum of a geometric series

$$a + ar + ar^2 + \cdots + ar^{N-1} + ar^N$$

where a and r are arbitrary constants.

a. What is the common ratio of consecutive terms?

b. If we call this sum A, then

$$A = a + ar + ar^2 + \cdots + ar^{N-1} + ar^N$$

Multiply both sides of the equation by the common ratio. What is the result?

c. Subtract your last equation from the equation for A given above, simplify, and factor out an A on the left side of the equation and an a on the right. What is the result?

d. Starting with your answer to part (c), solve for A.

e. Use your answer from part (d) to write a formula for $a + ar + ar^2 + \cdots + ar^{N-1} + ar^N$.

45. In this exercise we will find a formula for the sum of an arithmetic series when we are given the first term, F, the common difference, d, and the number of terms, N, but we are *not* given the last term.

a. Because the common difference is d, when we add d to a term we get the next term. The first term is F, so the second term is $F + d$, and the third term is $(F + d) + d = F + 2d$. How can you express the fourth term? How can you express the ninth term?

b. How can you express the Nth (or last) term, L, of the series in terms of F, d, and N?

c. Now you know the first term, F, the last term, L, the common difference, d, and the number of terms, N. Use the information in Exercise 42 to find the value of the series.

46. In this exercise we will find a formula for the sum of a geometric series

$$ar^M + ar^{M+1} + ar^{M+2} + \cdots + ar^{N-1} + ar^N$$

where a and r are arbitrary constants. M and N are positive integers with $M < N$.

a. What is the common ratio of consecutive terms?

b. If we call this sum A, then

$$A = ar^M + ar^{M+1} + ar^{M+2} + \cdots + ar^{N-1} + ar^N$$

Multiply both sides of the equation by the common ratio. What is the result?

c. Subtract your last equation from the equation for A given above, simplify, and factor out an A on the left side of the equation and an a on the right. What is the result?

d. Starting with your answer to part (c), solve for A.

e. Use your answer from part (d) to write a formula for the value of the geometric series

$$ar^M + ar^{M+1} + ar^{M+2} + \cdots + ar^{N-1} + ar^N$$

9.4 SIGMA NOTATION AND INFINITE GEOMETRIC SERIES

Summation Notation We can use a convenient notation for representing a series when we know an expression for the general term. For example, suppose we would like to sum the first 15 terms of the sequence

$$4, \quad 7, \quad 10, \ldots, \quad 3n + 1, \ldots$$

Since the terms of the series are obtained by replacing n in the general term $3n + 1$ by the numbers 1 through 15, we might express the sum as

The sum, as n runs from 1 to 15, of $3n + 1$

Thus instead of writing out all the terms of the series, we merely indicate which terms are to be included.

To abbreviate the expression further, we use the Greek letter Σ (called "sigma") to stand for "the sum." We indicate the first and last values of n below and above the summation symbol Σ, as shown here:

$$S_{15} = \sum_{n=1}^{15} (3n + 1)$$

The letter n is called the **index of summation**; it is like a variable because it represents numerical values. Any letter can be used for the index of summation; i, j, and k are other common choices. Of course, the letter used for the index of summation does not affect the sum. This way of expressing a series is sometimes called "using sigma notation."

Example 1 Use sigma notation to represent the sum of the first 20 terms of the sequence

$$-1, \quad 2, \quad 7, \ldots, \quad k^2 - 2, \ldots$$

Solution The general term of the sequence is $k^2 - 2$, and the first term is -1, which is obtained by letting $k = 1$ in the formula for the general term. Thus

$$S_{20} = \sum_{k=1}^{20} (k^2 - 2)$$

The **expanded form** of a series written in sigma notation is obtained by writing out all the terms of the series. For example,

$$\sum_{m=4}^{8} \frac{3}{m} = \frac{3}{4} + \frac{3}{5} + \frac{3}{6} + \frac{3}{7} + \frac{3}{8}$$

Notice that the series above has five terms, which is one more than the difference of the upper and lower limits of summation.

Recall that the general term of an arithmetic series is a linear function of the index, and the general term of a geometric series is an exponential function. If we recognize a given series as one of these two types, we can use the formulas developed in Section 9.3 to evaluate the sum.

Example 2

Compute the value of each series.

a. $\displaystyle\sum_{i=1}^{13} (90 - 5i)$ b. $\displaystyle\sum_{k=0}^{9} 2^k$

Solutions **a.** Since the general term $90 - 5i$ is linear, we see that this series is arithmetic. By writing out the first few terms of the series,

$$85 + 80 + 75 + 70 + \cdots$$

we can verify that the first term of the series is $a_1 = 85$ and the common difference is $d = -5$. In order to use the formula for arithmetic series, we also need to know the last term of the series. We substitute $n = 13$ in the general term to find $a_{13} = 90 - 5(13) = 25$. Thus

$$\sum_{i=1}^{13} (90 - 5i) = \frac{13}{2}(85 + 25) = 715$$

b. The general term of this series is exponential, so the series is geometric. By writing out a few terms of the series,

$$1 + 2 + 4 + 8 + \cdots$$

we confirm that the first term of the series is $a_1 = 1$ and the common ratio is $r = 2$. We also observe that the series has 10 terms, from $k = 0$ to $k = 9$, so $n = 10$. Finally, we substitute these values into the formula for geometric series to obtain

$$\sum_{k=0}^{9} 2^k = \frac{2^{10} - 1}{2 - 1} = 1023$$

If the series is not arithmetic or geometric, we must use direct methods to compute the sum.

Example 3

Compute the value of each series.

a. $\displaystyle\sum_{m=1}^{5} m^2$ b. $\displaystyle\sum_{p=1}^{800} 5$

Solutions **a.** Because the general term, m^2, is neither linear nor exponential, we know that the series is not arithmetic or geometric. However, we can expand the series and evaluate it directly.

$$\sum_{m=1}^{5} m^2 = 1^2 + 2^2 + 3^2 + 4^2 + 5^2$$

$$= 1 + 4 + 9 + 16 + 25 = 55$$

b. The general term is the number 5. Since the index p runs from 1 to 800, we are adding 800 terms, each of which is 5. Thus

$$\sum_{p=1}^{800} 5 = 5 + 5 + 5 + \cdots + 5 + 5 + 5$$

$$= 800(5) = 4000 \qquad \square$$

Infinite Series

A series with infinitely many terms is called an **infinite series**. Is it possible to add infinitely many terms and arrive at a finite sum? If the terms added become small enough, the answer is yes. Consider the infinite geometric series

$$\frac{1}{2} + \frac{1}{4} + \frac{1}{8} + \frac{1}{16} + \cdots$$

The nth **partial sum** of the series is the sum of its first n terms and is denoted by S_n. Thus

$$S_1 = \frac{1}{2}$$

$$S_2 = \frac{1}{2} + \frac{1}{4} = \frac{3}{4}$$

$$S_3 = \frac{1}{2} + \frac{1}{4} + \frac{1}{8} = \frac{7}{8}$$

$$S_4 = \frac{1}{2} + \frac{1}{4} + \frac{1}{8} + \frac{1}{16} = \frac{15}{16}$$

$$\vdots$$

Note that as n increases—as we add more and more terms of the series—the partial sums appear to be approaching 1; that is, as n becomes very large, S_n gets very close to 1. It seems reasonable that the sum of *all* the terms of the series is 1.

We can make this sum more plausible by examining the formula for the nth partial sum of a geometric series,

$$S_n = \frac{a_{n+1} - a_1}{r - 1}$$

$$= \frac{ar^n - a}{r - 1} = \frac{a - ar^n}{1 - r} \qquad (r \neq 1)$$

If r is a fraction between -1 and 1, then r^n gets closer to 0 for increasingly large n. For example, if $r = \frac{1}{2}$, then

$$r^2 = \left(\frac{1}{2}\right)^2 = \frac{1}{4}, \qquad r^3 = \left(\frac{1}{2}\right)^3 = \frac{1}{8}, \qquad r^4 = \left(\frac{1}{2}\right)^4 = \frac{1}{16}$$

and so on, with $\left(\dfrac{1}{2}\right)^n$ becoming smaller and smaller for larger values of n. If we write the formula for S_n in the form

$$S_n = \frac{a}{1 - r}(1 - r^n)$$

we see that the factor $(1 - r^n)$ will get closer and closer to 1 as n grows larger. Consequently, as we compute S_n for larger and larger values of n, the sum approaches the value

$$\frac{a}{1 - r}$$

We now define the sum of an infinite geometric series as follows.

$$S_\infty = \frac{a}{1 - r} \qquad (-1 < r < 1)$$

If $|r| \geq 1$, as in the infinite series

$$3 + 6 + 12 + 24 + \cdots$$

where $r = 2$, the terms become larger as n increases and the sum of the series is not a finite number. In this case the series does not have a sum.

Example 4

Find the sum of each infinite series, if it exists.

a. $\displaystyle\sum_{j=0}^{\infty} 30(0.8)^j$

b. $\displaystyle\sum_{m=0}^{\infty} 3\left(\frac{4}{3}\right)^m$

Solutions a. This series is geometric, with $a = 30$ and $r = 0.8$. The series has a sum because $|r| < 1$. Thus

$$S_\infty = \frac{a}{1 - r} = \frac{30}{1 - 0.8} = 150$$

b. This series is geometric, with $a = 3$ and $r = \dfrac{4}{3}$. The series does not have a sum because $|r| > 1$.

Repeating Decimals

An interesting application of geometric series involves repeating decimals. Recall that the decimal representation of a rational number either terminates, as does 0.75, or repeats a pattern of digits. For example, you probably recognize $0.333\overline{3}$ as the decimal representation of $\frac{1}{3}$. We can easily find the decimal form of a fraction: just divide the denominator into the numerator. Is there a way to find the fractional form of a repeating decimal?

Consider the repeating decimal

$$0.2121\overline{21}$$

We will first write this number as an infinite geometric series:

$$0.21 + 0.0021 + 0.000021 + \cdots$$

The first term of this series is $a = 0.21$, or $\dfrac{21}{100}$, and its common ratio is $r = 0.01$, or $\dfrac{1}{100}$. Since $|r| < 1$, the series has a sum given by

$$S_\infty = \frac{a}{1 - r} = \frac{\dfrac{21}{100}}{1 - \dfrac{1}{100}}$$

$$= \frac{\dfrac{21}{100}}{\dfrac{99}{100}} = \frac{21}{99} = \frac{7}{33}$$

Thus the decimal number $0.2121\overline{21}$ is equal to the fraction $\dfrac{7}{33}$.

Example 5

Find a fraction equivalent to $0.3\overline{7}$.

Solution First observe that the decimal can be written as $0.3 + 0.0\overline{7}$. We will find a fraction equivalent to the repeating decimal $0.0\overline{7}$ and add that to 0.3, or $\dfrac{3}{10}$. Write $0.0\overline{7}$ as a series:

$$0.0\overline{7} = \frac{7}{100} + \frac{7}{1000} + \frac{7}{10,000} + \cdots$$

This is an infinite geometric series with first term $a = \dfrac{7}{100}$ and common ratio $r = \dfrac{1}{10}$. The sum of the series is given by

$$S_\infty = \frac{a}{1 - r} = \frac{\frac{7}{100}}{1 - \frac{1}{10}}$$

$$= \frac{\frac{7}{100}}{\frac{9}{10}} = \frac{7}{100} \cdot \frac{10}{9} = \frac{7}{90}$$

Thus

$$0.3\overline{7} = \frac{3}{10} + \frac{7}{90} = \frac{34}{90}$$

Exercise 9.4

Write each sum in Exercises 1–8 in expanded form.

1. $\displaystyle\sum_{i=1}^{4} i^2$

2. $\displaystyle\sum_{i=1}^{3} (3i - 2)$

3. $\displaystyle\sum_{j=5}^{7} (j - 2)$

4. $\displaystyle\sum_{j=2}^{6} (j^2 + 1)$

5. $\displaystyle\sum_{k=1}^{4} k(k + 1)$

6. $\displaystyle\sum_{k=2}^{6} \frac{k}{2}(k + 1)$

7. $\displaystyle\sum_{m=1}^{4} \frac{(-1)^m}{2^m}$

8. $\displaystyle\sum_{m=3}^{5} \frac{(-1)^{m+1}}{m - 2}$

Write each series in Exercises 9–20 using sigma notation.

9. $1 + 3 + 5 + 7$

10. $2 + 4 + 6 + 8$

11. $5 + 5^3 + 5^5 + 5^7$

12. $4^3 + 4^5 + 4^7 + 4^9 + 4^{11}$

13. $1 + 4 + 9 + 16 + 25$

14. $1 + 8 + 27 + 64 + 125$

15. $\frac{1}{2} + \frac{2}{3} + \frac{3}{4} + \frac{4}{5} + \frac{5}{6}$

16. $\frac{2}{1} + \frac{3}{2} + \frac{4}{3} + \frac{5}{4} + \frac{6}{5}$

17. $\frac{1}{1} + \frac{2}{3} + \frac{3}{5} + \frac{4}{7} + \frac{5}{9} + \frac{6}{11}$

18. $\frac{3}{1} + \frac{5}{3} + \frac{7}{5} + \frac{9}{7} + \frac{11}{9}$

19. $\frac{1}{1} + \frac{2}{2} + \frac{4}{3} + \frac{8}{4} + \cdots$

20. $\frac{1}{2} + \frac{3}{4} + \frac{9}{6} + \frac{27}{8} + \cdots$

Identify each series in Exercises 21–40 as arithmetic, geometric, or neither, and then evaluate it.

21. $\displaystyle\sum_{i=1}^{6} (i^2 + 1)$

22. $\displaystyle\sum_{i=1}^{5} 3i^2$

23. $\displaystyle\sum_{j=1}^{4} \frac{1}{j}$

24. $\displaystyle\sum_{j=0}^{4} \frac{2}{j + 1}$

25. $\displaystyle\sum_{k=1}^{100} 1$

26. $\displaystyle\sum_{k=1}^{300} 3$

27. $\displaystyle\sum_{q=1}^{20} 3^q$

28. $\displaystyle\sum_{p=1}^{30} 2^p$

29. $\displaystyle\sum_{k=1}^{200} k$

30. $\displaystyle\sum_{k=1}^{150} k$

31. $\displaystyle\sum_{n=1}^{6} n^3$

32. $\displaystyle\sum_{n=1}^{7} n^2$

33. $\displaystyle\sum_{n=0}^{30} (3n - 1)$

34. $\displaystyle\sum_{k=0}^{20} (5k + 2)$

35. $\displaystyle\sum_{k=0}^{25} (5-2k)$

36. $\displaystyle\sum_{p=0}^{15} (2-3p)$

37. $\displaystyle\sum_{j=0}^{10} 2 \cdot 5^j$

38. $\displaystyle\sum_{j=0}^{10} 3 \cdot 2^j$

39. $\displaystyle\sum_{m=0}^{12} 50(1.08)^m$

40. $\displaystyle\sum_{m=0}^{18} 300(1.12)^m$

In Exercises 41–48 evaluate each infinite geometric series.

41. $\displaystyle\sum_{n=1}^{\infty} \left(\frac{1}{2}\right)^n$

42. $\displaystyle\sum_{n=1}^{\infty} \left(\frac{1}{3}\right)^n$

43. $\displaystyle\sum_{k=1}^{\infty} 12(0.15)^{k-1}$

44. $\displaystyle\sum_{k=1}^{\infty} 25(0.08)^{k-1}$

45. $\displaystyle\sum_{j=0}^{\infty} 4 \cdot \left(\frac{-3}{5}\right)^j$

46. $\displaystyle\sum_{j=0}^{\infty} 6 \cdot \left(\frac{-2}{5}\right)^j$

47. $\displaystyle\sum_{n=4}^{\infty} 3\left(\frac{1}{2}\right)^n$

48. $\displaystyle\sum_{n=3}^{\infty} 2\left(\frac{1}{3}\right)^n$

Find a fraction equivalent to each of the repeating decimals in Exercises 49–56.

49. $0.\overline{4}$

50. $0.\overline{6}$

51. $0.\overline{31}$

52. $0.\overline{45}$

53. $2.\overline{410}$

54. $3.\overline{027}$

55. $0.12\overline{8}$

56. $0.8\overline{3}$

Solve the word problems in Exercises 57–60.

57. On each swing of a pendulum, the bob traces an arc nine-tenths as long as the previous swing. Approximately how far will the bob move before coming to rest if the first arc length is 12 inches?

58. A force is applied to an object so that each second it moves only one-half of the distance it moved the preceding second. If the object moves 10 centimeters the first second, approximately how far will it move before coming to rest?

59. A ball returns two-thirds of its preceding height on each bounce. If the ball is dropped from a height of 6 feet, approximately what is the total distance the ball travels before coming to rest? (*Hint:* Compute separately the total distance the ball falls from the total distance it moves upward.)

60. If a ball is dropped from a height of 10 feet and returns three-fifths of its preceding height on each bounce, approximately what is the total distance the ball travels before coming to rest? (See the hint for Exercise 59.)

9.5   BINOMIAL EXPANSION

In Chapter 8 we studied products of polynomials, and in particular we found expanded forms for powers of binomials such as

$$(a+b)^2 \quad \text{and} \quad (a+b)^3$$

The amount of work involved in expanding such powers increases as the exponent gets larger. In this section we will learn how to raise any binomial to any positive integer power without having to multiply polynomials.

Investigation #1

Patterns in $(a+b)^n$ for Different Values of n

In this investigation we will look for patterns in the expansion of $(a+b)^n$. We begin by computing a number of such powers.

Expand each power and fill in the blanks. Arrange the terms in each expansion in descending powers of a.

1. $(a + b)^0 =$ _____

2. $(a + b)^1 =$ _____

3. $(a + b)^2 =$ _____

4. $(a + b)^3 =$ _____
   (*Hint:* Start by writing $(a + b)^3 = (a + b)(a + b)^2$ and use your answer to part 3.)

5. $(a + b)^4 =$ _____
   (*Hint:* Start by writing $(a + b)^4 = (a + b)(a + b)^3$ and use your answer to part 4.)

6. $(a + b)^5 =$ _____

7. Do you see a relationship between the exponent n and the number of terms in the expansion of $(a + b)^n$? (Notice that for $n = 0$ we have $(a + b)^0 = 1$, which has one term.) Fill in Table 9.2.

n	Number of Terms in $(a + b)^n$
0	
1	
2	
3	
4	
5	

Table 9.2

8. **First observation:** In general, the expansion of $(a + b)^n$ has _____ terms.

9. Next we will consider the exponents on a and b in each term of the expansions. Refer to your expanded powers in parts 1–5, and fill in Table 9.3.

Table 9.3

n	First Term of $(a + b)^n$	Last Term of $(a + b)^n$	Sum of Exponents on a and b in Each Term
0			
1			
2			
3			
4			
5			

10. Second observation: In any term of the expansion of $(a + b)^n$, the sum of the exponents on a and on b is _____.

11. In fact, we can be more specific in describing the exponents in the expansions. We will use k to label the *terms* in the expansion of $(a + b)^n$, starting with $k = 0$. For example, for $n = 2$ we label the terms as follows:

$$(a + b)^2 = \underset{k\,=\,0}{a^2} + \underset{k\,=\,1}{2ab} + \underset{k\,=\,2}{b^2}$$

We can make a table showing the exponents on a and b in each term of $(a + b)^2$:

Case $n = 2$:

k	0	1	2
Exponent on a	2	1	0
Exponent on b	0	1	2

Table 9.4

Complete the tables shown for the cases $n = 3, 4,$ and 5.

Case $n = 3$:

k	0	1	2	3
Exponent on a	3			
Exponent on b	0			

Table 9.5

Case $n = 4$:

k	0	1	2	3	4
Exponent on a	4				
Exponent on b	0				

Table 9.6

Case $n = 5$:

k	0	1	2	3	4	5
Exponent on a	5					
Exponent on b	0					

Table 9.7

12. Third observation: The variable factors of the kth term in the expansion of $(a + b)^n$ may be expressed as _____. (Fill in the correct powers in terms of n and k for a and b.)

In the next investigation we will look for patterns in the *coefficients* of the terms of the expansions.

Powers of Other Binomials

We can use what we learned from Investigation 1 to raise other binomials to powers.

Example 1

Expand each of the following powers.

a. $(x + 1)^3$ **b.** $(2m - n)^4$

Solutions   **a.** We know from Investigation 1 that

$$(a + b)^3 = a^3 + 3a^2b + 3ab^2 + b^3$$

We can replace a with x and b with 1 to get

$$(x + 1)^3 = (x)^3 + 3(x)^2(1) + 3(x)(1)^2 + (1)^3$$
$$= x^3 + 3x^2 + 3x + 1$$

b. Here we take the expansion

$$(a + b)^4 = a^4 + 4a^3b + 6a^2b^2 + 4ab^3 + b^4$$

from Investigation 1 and let $a = 2m$ and $b = -n$.

$$(2m - n)^4 = (2m)^4 + 4(2m)^3(-n) + 6(2m)^2(-n)^2 + 4(2m)(-n)^3 + (-n)^4$$
$$= 16m^4 + 4(8m^3)(-n) + 6(4m^2)(n^2) + 4(2m)(-n^3) + n^4$$
$$= 16m^4 - 32m^3n + 24m^2n^2 - 8mn^3 + n^4$$

Binomial Coefficient

Even without carrying out the multiplication, we know from Investigation 1 that the expansion of $(a + b)^6$ should have seven terms, with the exponents on a ranging from 6 down to 0 and the exponents on b ranging from 0 up to 6. The expansion should have the form

$$(a + b)^6 = \underline{\hspace{1cm}}a^6 + \underline{\hspace{1cm}}a^5b + \underline{\hspace{1cm}}a^4b^2$$
$$+ \underline{\hspace{1cm}}a^3b^3 + \underline{\hspace{1cm}}a^2b^4 + \underline{\hspace{1cm}}ab^5 + \underline{\hspace{1cm}}b^6$$

We will use the notation ${}_nC_k$ for the coefficient of the kth term in the expansion of $(a + b)^n$. Thus when we expand $(a + b)^6$, the coefficient of a^6 will be denoted by ${}_6C_0$, the coefficient of a^5b^1 will be denoted by ${}_6C_1$, the coefficient of a^4b^2 will be denoted by ${}_6C_2$, and so forth. Note that the 6 before the C indicates that $n = 6$ and the number following the C corresponds to the exponent on b. The symbol ${}_nC_k$ is called the **binomial coefficient**.

> The **binomial coefficient**, ${}_nC_k$, is the coefficient of the term containing b^k in the expansion of $(a + b)^n$.

Example 2

Use your expansions from Investigation 1 to evaluate the following binomial coefficients.

a. ${}_4C_3$ **b.** ${}_5C_2$

Solutions   **a.** $_4C_3$ is the coefficient of the term containing b^3 in the expansion of $(a + b)^4$. Referring to **Step 5** in Investigation 1, we see that the coefficient of the term $4ab^3$ is 4. Thus $_4C_3 = 4$.

b. $_5C_2$ is the coefficient of the term containing b^2 in the expansion of $(a + b)^5$. Referring to **Step 6** in Investigation 1, we see that the coefficient of the term $10a^3b^2$ is 10. Thus $_5C_2 = 10$.   ▭

Pascal's Triangle

To get a clearer picture of the binomial coefficients, consider again the expansions of $(a + b)^n$ you calculated in Investigation 1, but this time look only at the numerical coefficients of each term:

```
n = 0                      1
n = 1                   1     1
n = 2                1     2     1
n = 3             1     3     3     1
n = 4          1     4     6     4     1
n = 5       1     5     10    10    5     1
```

    This triangular array of numbers is known as **Pascal's triangle**. It has many interesting and surprising properties that have been studied extensively. We might first make the following observations.

1. Each row of Pascal's triangle begins with the number _____ and ends with the number _____.

2. Starting with row $n = 1$, the second number and the next-to-last number in the nth row are _____.

    The rest of the numbers in each row follow an interesting pattern. Pick any number in the row $n = 4$ and look at the two closest numbers in the preceding row. (For example, if you picked 6, the two closest numbers in the preceding row are 3 and 3.) Do you see a relationship between the numbers? Try the same thing with several numbers in row $n = 5$ to test your theory.

3. Starting with the row $n = 2$, any number in the triangle (except the first and last 1's in each row) can be found by _____.

4. Using your answer to part 3, continue Pascal's triangle to include the row for $n = 6$.

    The numbers in Pascal's triangle are the binomial coefficients we have been looking for. Specifically, the number in the kth position (starting with $k = 0$) of the nth row of the triangle is $_nC_k$. Thus we can use the numbers in Pascal's triangle to expand $(a + b)^n$.

5. Use Pascal's triangle to find the binomial coefficient $_6C_4$.

6. Expand: $(a + b)^6 =$ _____

7. Expand: $(x - 2)^6 =$ _____ (*Hint:* Use your answer to part 5, replacing a by x and b by -2.)

8. Continue Pascal's triangle to include the row for $n = 7$.

Using Pascal's Triangle

Combining what we learned in Investigation 1 about the form of an expansion with our method for finding the coefficients from Pascal's triangle, we can now write the expanded form of any binomial power without having to perform the multiplication.

Example 3

Find the expanded form of $(3r - q)^6$ without multiplying.

Solution   Start by writing down the expansion of $(a + b)^6$, using your knowledge from Investigation 1.

$$(a + b)^6 = a^6 + \underline{\hspace{1cm}} a^5b + \underline{\hspace{1cm}} a^4b^2 + \underline{\hspace{1cm}} a^3b^3$$
$$+ \underline{\hspace{1cm}} a^2b^4 + \underline{\hspace{1cm}} ab^5 + b^6$$

Next fill in the blanks with the binomial coefficients from Pascal's triangle. Consult Step 4 of Investigation 2 to obtain

$$(a + b)^6 = a^6 + 6a^5b + 15a^4b^2 + 20a^3b^3 + 15a^2b^4 + 6ab^5 + b^6$$

Finally, replace a by $3r$ and replace b by $-q$ and simplify to find

$$(3r - q)^6 = (3r)^6 + 6(3r)^5(-q) + 15(3r)^4(-q)^2 + 20(3r)^3(-q)^3$$
$$+ 15(3r)^2(-q)^4 + 6(3r)(-q)^5 + (-q)^6$$
$$= 729r^6 - 1458r^5q + 1215r^4q^2 - 540r^3q^3 + 135r^2q^4$$
$$- 18rq^5 + q^6$$

Factorial Notation

Although we can find the binomial coefficients in the expansion of $(a + b)^n$ for any exponent n by extending Pascal's triangle far enough, this process is tedious for large values of n. In such cases it would be more convenient to have a formula for calculating the binomial coefficients directly. For this formula we need a special symbol, *n*! (read "*n* factorial"), which is defined for non-negative integers n as follows.

$$n! = n \cdot (n - 1) \cdot (n - 2) \cdot \ \cdots \ \cdot 3 \cdot 2 \cdot 1$$

For example,

$$6! = 6 \cdot 5 \cdot 4 \cdot 3 \cdot 2 \cdot 1 = 720 \quad \text{and} \quad 4! = 4 \cdot 3 \cdot 2 \cdot 1 = 24$$

The factorial symbol applies only to the variable or number it follows; for example, $3 \cdot 4!$ is not equal to $(3 \cdot 4)!$.

Example 4

Write each expression in expanded form and simplify.
a. $2n!$ for $n = 4$ **b.** $(2n - 1)!$ for $n = 4$

Solutions **a.** $2n! = 2 \cdot 4! = 2(4 \cdot 3 \cdot 2 \cdot 1) = 48$
b. $(2n - 1)! = 7! = 7 \cdot 6 \cdot 5 \cdot 4 \cdot 3 \cdot 2 \cdot 1 = 5040$

Example 5

Write $(3n + 1)!$ in factored form, showing the first three factors and the last three factors.

Solution $(3n + 1)! = (3n + 1)(3n)(3n - 1) \cdot \ \cdots \ \cdot 3 \cdot 2 \cdot 1$

Note that

$$7! = 7 \cdot 6 \cdot 5 \cdot 4 \cdot 3 \cdot 2 \cdot 1$$

can be written as

$$7 \cdot 6! = 7 \cdot (6 \cdot 5 \cdot 4 \cdot 3 \cdot 2 \cdot 1)$$

and that

$$5! = 5 \cdot 4 \cdot 3 \cdot 2 \cdot 1$$

can be written as

$$5 \cdot 4! = 5 \cdot (4 \cdot 3 \cdot 2 \cdot 1)$$

In general, we have the following relationship.

$$n! = n(n-1)! \tag{7}$$

This relationship can be helpful in simplifying expressions involving factorials.

Example 6

Write each expression in expanded form and simplify.

a. $\dfrac{6!}{3!}$ **b.** $\dfrac{4!6!}{8!}$

Solutions **a.** $\dfrac{6!}{3!} = \dfrac{6 \cdot 5 \cdot 4 \cdot 3!}{3!} = 6 \cdot 5 \cdot 4 = 120$

b. $\dfrac{4!6!}{8!} = \dfrac{4 \cdot 3 \cdot 2 \cdot 1 \cdot 6!}{8 \cdot 7 \cdot 6!} = \dfrac{3}{7}$

 COMMON ERROR

In Example 6(a) notice that $\dfrac{6!}{3!}$ is *not* equal to $2!$

Binomial Coefficient in Factorial Notation

Our definition for $n!$ only makes sense when n is a positive integer. We will also define $n!$ for $n = 0$. To be consistent with Equation (7), we must have

$$1! = 1 \cdot (1 - 1)!$$

or

$$1! = 1 \cdot 0!$$

This observation leads us to make the following definition.

$$0! = 1$$

Note that both 1! and 0! are equal to 1.

Earlier we introduced the notation $_nC_k$ for the numerical coefficients in the expansion of $(a + b)^n$. These binomial coefficients, which are given by Pascal's triangle, can also be expressed using factorial notation as follows.

$$_nC_k = \frac{n!}{(n - k)! \; k!}$$

Example 7

Evaluate each binomial coefficient.

a. $_6C_2$ **b.** $_9C_8$

Solutions **a.** $_6C_2 = \dfrac{6!}{(6 - 2)! \; 2!}$

$$= \frac{6!}{4! \; 2!} = \frac{6 \cdot 5 \cdot 4!}{(4!)(2 \cdot 1)} = 15$$

b. $_9C_8 = \dfrac{9!}{(9 - 8)! 8!}$

$$= \frac{9!}{1! \; 8!} = \frac{9 \cdot 8!}{1 \cdot 8!} = 9$$

You might want to check that the formula works for $k = 0$ when we use $0! = 1$. We should find that $_nC_0 = 1$, and in fact,

$$_nC_0 = \frac{n!}{(n - 0)! \; (0)!} = \frac{n!}{n!(1)} = 1$$

In other words, the formula correctly gives us the coefficient of a^n in the expansion of $(a + b)^n$.

We now have two methods for computing the binomial coefficient, $_nC_k$: Pascal's triangle and our formula using factorials. When n is small, especially if all the coefficients of the binomial expansion are needed, Pascal's triangle is often the easier method. But when n is large, or if only one coefficient is required, the factorial formula for $_nC_k$ is probably quicker than Pascal's triangle.

Example 8

Find the coefficient of $m^{11}n^3$ in the expansion of $(m + n)^{14}$.

Solution Since we are simply replacing a with m and b with n in the expansion of $(a + b)^{14}$, the term containing $m^{11}n^3$ has $n = 14$ and $k = 3$. Thus the coefficient of $m^{11}n^3$ is $_{14}C_3$.

$$_{14}C_3 = \frac{14!}{(14 - 3)! \, 3!} = \frac{14 \cdot 13 \cdot 12 \cdot 11!}{(11)! \, 3!} = \frac{14 \cdot 13 \cdot 12}{3 \cdot 2 \cdot 1} = 364$$

The coefficient of $m^{11}n^3$ is 364.

Binomial Theorem

Consider the expansion of $(a + b)^6$ written with the $_nC_k$ notation* for the coefficients.

$$(a + b)^6 = \underset{k=0}{_6C_0 a^6 b^0} + \underset{k=1}{_6C_1 a^5 b^1} + \underset{k=2}{_6C_2 a^4 b^2} + \underset{k=3}{_6C_3 a^3 b^3} + \underset{k=4}{_6C_4 a^2 b^4}$$

$$+ \underset{k=5}{_6C_5 ab^5} + \underset{k=6}{_6C_6 b^6}$$

Check that each term can be written in the form

$$_6C_k a^{6-k} b^k$$

for $k = 0$ to $k = 6$. Consequently, we can use sigma notation to write the sum of the terms in the expansion.

$$(a + b)^6 = \sum_{k=0}^{6} {_6C_k} \, a^{6-k} b^k \qquad (8)$$

Notice that the sigma notation indicates terms for $k = 0$ to $k = 6$, so the expansion has seven terms, as it should. Also note that the sum of the exponents on a and b is $(6 - k) + k = 6$.

Equation (8) is a special case of the **binomial theorem**, which uses everything we have learned to write the expanded form of a power of a

---

*Other common notations are used for binomial coefficients. Instead of $_nC_k$ you may see $\binom{n}{k}$ or $C_{n,k}$ or $C(n, k)$ or C_k^n.

binomial in the most compact form possible. We can write the general form of the theorem, for positive integers n, as follows.

> **BINOMIAL THEOREM**
>
> $$(a + b)^n = \sum_{k=0}^{n} {}_nC_k\, a^{n-k}b^k$$

Example 9

a. Use sigma notation to write the expanded form for $(x - 2y)^{10}$.
b. Find the term containing y^7, and simplify.

Solutions **a.** Apply the binomial theorem with $n = 10$, replacing a by x and b by $-2y$ to get

$$(x - 2y)^{10} = \sum_{k=0}^{10} {}_{10}C_k\, x^{10-k}(-2y)^k$$

b. The term containing y^7 corresponds to $k = 7$. The $k = 7$ term looks like

$${}_{10}C_7 x^{10-7}(-2y)^7$$

or

$${}_{10}C_7 x^3(-2y)^7$$

The value of ${}_{10}C_7$ is

$${}_{10}C_7 = \frac{10!}{(10-7)!\,7!} = \frac{10 \cdot 9 \cdot 8 \cdot 7!}{(3)!\,7!} = \frac{10 \cdot 9 \cdot 8}{3 \cdot 2 \cdot 1} = 120$$

and $(-2y)^7 = -128y^7$. Thus the term we want is

$${}_{10}C_7 (x)^3(-2y)^7 = 120x^3(-128y^7) = -15{,}360x^3y^7$$

Using the TI-82 to Compute Factorials and Binomial Coefficients

Both factorials and binomial coefficients can be accessed from the probability submenu of the TI-82's [MATH] menu. (Press [MATH] and then use the ◁ key to highlight **PRB**.) Note that your calculator uses ${}_nC_r$ for the binomial coefficient instead of ${}_nC_k$. (See Figure 9.6.)

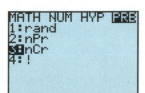

Figure 9.6

| Example 10 | Use your calculator to evaluate each expression. |

a.   10!   **b.** $_{15}C_7$ **c.** $\dfrac{13!}{5!\,8!}$

Solutions   **a.**   To access the factorial command !, we press $\boxed{\text{MATH}}$ $\boxed{\triangleleft}$ 4. Thus to evaluate 10! we key in

$$10 \;\boxed{\text{MATH}}\; \boxed{\triangleleft}\; 4 \;\boxed{\text{ENTER}}$$

and the calculator returns the result: 3,628,800.

b.   To evaluate a binomial coefficient, we first enter n and then access the $_nC_r$ command by pressing $\boxed{\text{MATH}}$ $\boxed{\triangleleft}$ 3. Thus to evaluate $_{15}C_7$ we first key in

$$15 \;\boxed{\text{MATH}}\; \boxed{\triangleleft}\; 3$$

The display will show "15 **nCr**," and the calculator is ready for the value of r. Now key in

$$7 \;\boxed{\text{ENTER}}$$

and the calculator shows 6435.

c.   To evaluate $\dfrac{13!}{5!\,8!}$ we key in the expression

$$13! \;\boxed{\div}\; (5!\,8!)$$

accessing the factorial command each time as shown in part (a). Or we recognize that this fraction is the factorial equivalent of $_{13}C_8$ and access the binomial coefficient command as in part (b), keying in

$$13 \;\textbf{nCr}\; 8 \;\boxed{\text{ENTER}}$$

to obtain 1287.

Exercise 9.5

Answer each question in Exercises 1–4 without expanding the power.

1. How many terms are in the expansion of $(a + b)^{50}$? In the expansion of $(2x + 3y)^{100}$?

2. How many terms are in the expansion of $(a + b)^{75}$? In the expansion of $(5x - 7y)^{200}$?

3. What is the sum of the exponents on x and y in each term of the expansion of $(x + y)^{100}$? In the expansion of $(8x - 7y)^{50}$?

4. What is the sum of the exponents on x and y in each term of the expansion of $(x + y)^{200}$? In the expansion of $(9x - 4y)^{75}$?

In Exercises 5 and 6 find the appropriate entries in Pascal's triangle.

5. Give the portion of Pascal's triangle corresponding to rows from $n = 0$ to $n = 10$. How many rows are involved?

6. Give the portion of Pascal's triangle corresponding to rows from $n = 0$ to $n = 12$. How many rows are involved?

In Exercises 7–22 write each power in expanded form.

7. $(x + 3)^5$ **8.** $(2x + y)^4$

9. $(x - 3)^4$ **10.** $(2x - 1)^5$

11. $\left(2x - \dfrac{y}{2}\right)^3$ **12.** $\left(\dfrac{x}{3} + 3\right)^6$

13. $(x^2 - 3)^7$ **14.** $(1 - y^2)^5$

15. $\left(\dfrac{2}{3} - a^2\right)^4$ **16.** $(x + y)^5$

17. $(x + y)^6$ **18.** $(x - 2y)^4$

19. $(2x - y)^7$ **20.** $\left(b - \dfrac{a}{3}\right)^7$

21. $(a - b)^8$ **22.** $(m - 2n)^8$

Write the expressions in Exercises 23–30 in expanded form and simplify them.

23. $5!$ **24.** $7!$ **25.** $\dfrac{9!}{7!}$

26. $\dfrac{12!}{11!}$ **27.** $\dfrac{5!\,7!}{12!}$ **28.** $\dfrac{12!\,4!}{16!}$

29. $\dfrac{8!}{2!(8 - 2)!}$ **30.** $\dfrac{10!}{4!(10 - 4)!}$

In Exercises 31–38 evaluate each binomial coefficient.

31. ${}_9C_6$ **32.** ${}_8C_5$ **33.** ${}_{12}C_3$ **34.** ${}_{13}C_4$

35. ${}_{20}C_{18}$ **36.** ${}_{18}C_{16}$ **37.** ${}_{14}C_9$ **38.** ${}_{16}C_7$

Find the coefficient of the indicated term in Exercises 39–50.

39. $(x + y)^{20}$; $x^{13}y^7$ **40.** $(x - y)^{15}$; $x^{12}y^3$

41. $(a - 2b)^{12}$; a^5b^7 **42.** $(2a - b)^{12}$; a^8b^4

43. $\left(x - \sqrt{2}\right)^{10}$; x^4 **44.** $\left(\dfrac{x}{2} + 2\right)^8$; x^6

45. $(a - b)^{15}$; $a^{11}b^4$ **46.** $(x + 2)^{12}$; x^8

47. $(a^3 - b^3)^9$; $a^{18}b^9$ **48.** $(x^2 - y^2)^7$; $x^{10}y^4$

49. $\left(x - \dfrac{1}{2}\right)^8$; x^5 **50.** $\left(\dfrac{x}{2} - 2y\right)^{10}$; x^6y^4

Refer to Pascal's triangle and Investigation 2 to answer Exercises 51–52.

51. Write out the terms of the sequence 11^n for $n = 0, 1, 2, 3, 4$. How are these terms related to the rows of Pascal's triangle? Can you explain why? (*Hint:* Rewrite 11 as 10 + 1.)

52. Write out the terms of the sequence 1.1^n for $n = 0, 1, 2, 3, 4$. How are these terms related to the rows of Pascal's triangle? Can you explain why? (*Hint:* Rewrite 1.1 as 1 + 0.1.)

Row n	Sum of Entries
0	
1	
2	
3	
4	
5	
6	

Table 9.8

Investigation #3

Other Properties of Pascal's Triangle

Pascal's triangle has other interesting properties besides providing the binomial coefficients.

1. Add up the numbers in each row of Pascal's triangle and fill in Table 9.8.

2. The sum of the entries in row n of Pascal's triangle is _____.

Now add entries of Pascal's triangle along the diagonal lines indicated on page 560. Fill in two more row of Pascal's triangle and find two more diagonal sums.

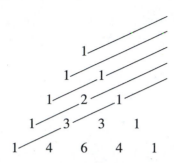

3. Write the sequence of numbers obtained by taking sums along the indicated diagonal lines.

————, ————, ————, ————, ————, ————, ————, . . .

4. What is the name of this sequence? (*Hint:* See Exercise 39 of Section 9.1.)

The binomial coefficient, $_nC_k$, also gives the number of different ways you can choose k distinct items from a set of n items. Finding these numbers is important is determining the probability that certain events will occur.

5. Find the number of different choices you can make for a four-flavor ice cream sundae if there are six flavors to choose from. (*Caution:* When you are deciding which entry in the row corresponds to $k = 4$, remember that the left-most entry corresponds to $k = 0$.)

6. How many ways can you choose a debate team of three members from a debate club of five members?

7. Evaluate the series

$$\sum_{r=0}^{10} {_{10}C_r}$$

(*Hint:* This expression is the sum of all the binomial coefficients in the expansion of $(a + b)^n$. What did you learn in Step 2 about the sum of the terms in a row of Pascal's triangle?)

8. Evaluate the series

$$\sum_{r=0}^{12} {_{12}C_r}$$

(*Hint:* See part 7.)

9. How many ways can you choose 6 objects from a set of 49? (The answer equals the total number of possible choices in a lottery that requires a participant to pick 6 numbers from 1 to 49.)

10. How many ways can you choose 5 cards out of a deck of 52 distinct cards? (The answer equals the total number of different poker hands.)

Chapter 9 Review

In Exercises 1 and 2 find the first four terms in the sequence whose general term is given.

1. $a_n = \dfrac{n}{n^2 + 1}$ **2.** $b_n = \dfrac{(-1)^{n-1}}{n}$

In Exercises 3 and 4 find the first five terms in the recursively defined sequence.

3. $c_1 = 5$; $c_{n+1} = c_n - 3$

4. $d_1 = 1$; $d_{n+1} = -\dfrac{3}{4} d_n$

In Exercises 5–8
a. find the first four terms of each sequence.
b. determine equations to define the sequence *recursively*.

5. Rick purchased a sailboat for $1800. How much is the sailboat worth after n years if it depreciates in value by 12% each year?

6. Sally earns $24,000 per year. If she receives a 6% raise each year, what will her salary be after n years?

7. To fight off an infection, Garrison receives a 30-milliliter dose of an antibiotic followed by doses of 15 milliliters at regular intervals. Between doses, Garrison's body removes 25% of the antibiotic that was present after the previous dose. How much of the antibiotic is present after n doses?

8. Opal joined the Weight Losers Club. At the end of the first meeting she weighed 187 pounds, and she loses 2 pounds from one meeting to the next. How much does Opal weigh after n meetings?

In Exercises 9–18 find the indicated term for each sequence.

9. $x_n = (-1)^n (n - 2)^2$; x_7

10. $y_n = \sqrt{n^3 - 2}$; y_3

11. The tenth term in the arithmetic sequence that begins $-4, 0, \ldots$

12. The sixth term in the arithmetic sequence that begins $x - a, x + a, \ldots$

13. The eighth term in the geometric sequence that begins $\dfrac{16}{27}, \dfrac{-8}{9}, \dfrac{4}{3}, \ldots$

14. The fifth term in the geometric sequence with third term $-\dfrac{2}{3}$ and sixth term $\dfrac{16}{81}$

15. The twenty-third term of the arithmetic sequence $-84, -74, -64, \ldots$

16. The ninth term of the arithmetic sequence $-\dfrac{1}{2}, 1, \dfrac{5}{2}, \ldots$

17. The first term of an arithmetic sequence is 8 and the twenty-eighth term is 89. Find the twenty-first term.

18. What term in the arithmetic sequence $5, 2, -1, \ldots$ is -37?

Identify each sequence in Exercises 19–28 as arithmetic, geometric, or neither. For each arithmetic and geometric sequence, find the next four terms of the sequence and a nonrecursive expression for the general term.

19. $-1, \dfrac{1}{2}, -\dfrac{1}{4}, \dfrac{1}{8}, \ldots$ **20.** $12, 9, 3, 1, \ldots$

21. $6, 1, -4, -9, \ldots$

22. $1, -4, 16, -64, \ldots$

23. $-1, 2, -4, 8, \ldots$ **24.** $\dfrac{2}{3}, \dfrac{1}{2}, \dfrac{3}{8}, \dfrac{9}{32}, \ldots$

25. First term 3, common difference -4

26. First term $\dfrac{1}{4}$, common difference $\dfrac{1}{2}$

27. First term 12, common ratio -4

28. First term 6, common ratio $\dfrac{1}{3}$

Write Exercises 29 and 30 in expanded form.

29. $\displaystyle\sum_{k=2}^{5} k(k-1)$ **30.** $\displaystyle\sum_{j=2}^{\infty} \frac{j}{2j-1}$

Write Exercises 31 and 32 using sigma notation.

31. The sum of the first 12 terms of $1, 3, 7, \ldots ,$ $2^k - 1, \ldots$

32. The sum of the fourth through fifteenth terms of $x, 4x^2, 9x^3, \ldots , k^2x^k, \ldots$

Identify each series in Exercises 33–42 as arithmetic, geometric, or neither, and then evaluate it.

33. The sum of the first 12 terms of the sequence $a_n = 3n - 2$.

34. The sum of the first 20 terms of the sequence $b_n = 1.4 + 0.1n$.

35. $\displaystyle\sum_{i=1}^{6} (3i - 1)$ **36.** $\displaystyle\sum_{k=1}^{12} \left(\frac{2}{3}k - 1\right)$

37. $\displaystyle\sum_{j=1}^{5} \left(\frac{1}{3}\right)^j$ **38.** $\displaystyle\sum_{k=1}^{6} 2^{k-1}$

39. $\displaystyle\sum_{n=1}^{5} (-1)^n (n + 1)$ **40.** $\displaystyle\sum_{n=1}^{4} \frac{n}{n+1}$

41. $-3 + 2 + \left(-\dfrac{4}{3}\right) + \left(\dfrac{8}{9}\right) + \cdots$

42. $\displaystyle\sum_{i=1}^{\infty} 3\left(\frac{1}{3}\right)^{i-1}$

43. A rubber ball is dropped from a height of 12 feet and returns two-thirds of its previous height on each bounce. How high does the ball bounce after hitting the floor for the fourth time?

44. The property taxes on the Hardesty's family home were $840 in 1994. If the taxes increase by 2% each year, what will the taxes be in 2000?

45. a. Find the sum of all integral multiples of 6 between 10 and 100.

b. Write the sum in (a) using sigma notation.

46. Kathy planted a 7-foot silver maple tree in 1994. If the tree grows 1.3 feet each year, in what year will it be 20 feet tall?

47. A rubber ball is dropped from a height of 12 feet and returns three-fourths of its previous height on each bounce. Approximately what is the total distance the ball travels before coming to rest?

48. Suppose you will be paid 1¢ on the first day of June, 2¢ on the second, 4¢ on the third, and so on, so that each new day you are paid twice what your received the preceding day. What will be the total amount you will receive in June?

In Exercises 49 and 50 find a fraction equivalent to each repeating decimal.

49. $3.222\overline{2}$ **50.** $0.41818\overline{18}$

Write each power in Exercises 51 and 52 in expanded form.

51. $(x - 2)^5$ **52.** $\left(\dfrac{x}{2} - y\right)^4$

Evaluate the expressions in Exercises 53–58.

53. $\dfrac{6!}{3!(6-3)!}$ **54.** $\dfrac{9!}{5!(9-5)!}$

55. $_7C_2$ **56.** $_{16}C_{14}$ **57.** $\displaystyle\sum_{k=0}^{5} {_5}C_k$

58. $\displaystyle\sum_{k=0}^{6} {_6}C_k (1.4)^{6-k} (0.6)^k$

In Exercises 59 and 60 find the coefficient of the indicated term.

59. $(x - 2y)^9$; x^6y^3 **60.** $\left(\dfrac{x}{2} - 3\right)^8$; x

Additional Topics

*I*n this chapter we consider some additional topics about functions and graphing, including transformations of basic graphs, inverse functions, and conic sections. We also consider a variation on the method of Gaussian reduction using matrices.

10.1 FURTHER GRAPHING TECHNIQUES

Translations   Many useful functions are variations of the standard functions introduced in Section 5.3. Their graphs can be obtained, without constructing tables of values, by modifying the basic graphs in Figures 5.24 through 5.26 and 5.30. We have already seen some examples of this idea in our study of quadratic functions. Recall the graphs of $f(x) = x^2 + 4$ and $g(x) = x^2 - 4$ shown in Figure 4.23. The graphs are reproduced in Figure 10.1.

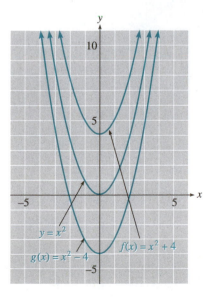

x	f(x)	g(x)
-2	8	0
-1	5	-3
0	4	-4
1	5	-3
2	8	0

Figure 10.1

The graphs of $y = f(x)$ and $y = g(x)$ are said to be **translations** of the graph of $y = x^2$; that is, they are shifted to a different location in the plane but retain the same size and shape as the original graph. In general, we have the following graphing principle.

VERTICAL TRANSLATIONS

Compared with the graph of $y = f(x)$,

1. the graph of $y = f(x) + k$ $(k > 0)$ is shifted *upward k* units.

2. the graph of $y = f(x) - k$ $(k > 0)$ is shifted *downward k* units.

Example 1

Graph the following functions.

a. $g(x) = |x| + 3$ **b.** $h(x) = \dfrac{1}{x} - 2$

Solutions   **a.** The graph of $g(x) = |x| + 3$ is a translation of the basic graph of $y = |x|$, shifted upward three units. (See Figure 10.2.)

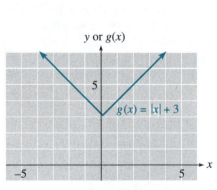

Figure 10.2

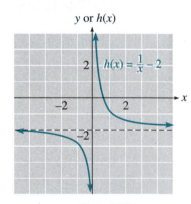

Figure 10.3

b. The graph of $h(x) = \dfrac{1}{x} - 2$ is a translation of the basic graph of $y = \dfrac{1}{x}$, shifted downward two units. We indicate the horizontal asymptote, $y = -2$, by a dotted line. (See Figure 10.3.)

Now consider the graphs of $f(x) = (x + 2)^2$ and $g(x) = (x - 2)^2$ shown in Figure 10.4.

x	f(x)	g(x)
−3	1	25
−2	0	16
−1	1	9
0	4	4
1	9	1
2	16	0
3	25	1

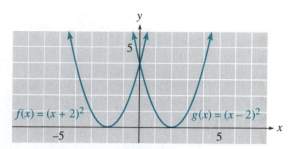

Figure 10.4

Compared with the graph of the basic function $y = x^2$, the graph of $f(x) = (x + 2)^2$ is shifted two units to the left, and the graph of $g(x) = (x - 2)^2$ is shifted two units to the right. In general, we have the following principle.

HORIZONTAL TRANSLATIONS

Compared with the graph of $y = f(x)$,
1. the graph of $y = f(x + h)$ $(h > 0)$ is shifted h units to the *left*.
2. the graph of $y = f(x - h)$ $(h > 0)$ is shifted h units to the *right*.

Example 2

Graph the following functions.

a. $g(x) = \sqrt{x + 1}$ **b.** $h(x) = \dfrac{1}{(x - 3)^2}$

Solutions   a. The graph of $g(x) = \sqrt{x + 1}$ is a translation of the basic graph of $y = \sqrt{x}$, shifted one unit to the left. (See Figure 10.5.)

b. The graph of $h(x) = \dfrac{1}{(x - 3)^2}$ is a translation of the basic graph of

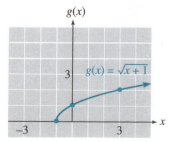

Figure 10.5

$y = \dfrac{1}{x^2}$, shifted three units to the right. We indicate the vertical asymptote, $x = 3$, by a dotted line. (See Figure 10.6.)

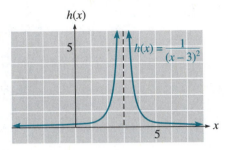

Figure 10.6

The graphs of some functions involve both horizontal and vertical translations.

Example 3

Graph $f(x) = (x + 4)^3 + 2$.

Solution Analyze the function by performing the translations separately. First consider the graph of $y = (x + 4)^3$, which is a translation of the basic graph of $y = x^3$, shifted four units to the left. The graph of $f(x)$ is then obtained by shifting this graph upward two units, as shown in Figure 10.7.

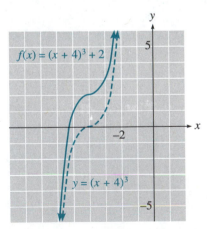

Figure 10.7

Scale Factors We have seen that *adding* a constant to the expression defining a function results in a translation of its graph. We now investigate the effect of *multiplying* by a constant. Consider the graphs of the functions

$$f(x) = 2x^2, \qquad g(x) = \frac{1}{2}x^2, \qquad \text{and} \qquad h(x) = -x^2$$

shown in Figure 10.8.

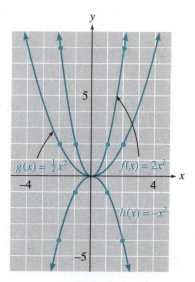

x	f(x)	g(x)	h(x)
−2	8	2	−4
−1	2	$\frac{1}{2}$	−1
0	0	0	0
1	2	$\frac{1}{2}$	−1
2	8	2	−4

Figure 10.8

If we compare each of these graphs with the graph of $y = x^2$, we see that the graph of $f(x) = 2x^2$ is stretched, or expanded, vertically by a factor of 2. The y-coordinate of each point on the graph of $y = x^2$ has been doubled, so each point on the graph of f is twice as far from the x-axis as its counterpart with the same x-coordinate on the basic graph $y = x^2$. The graph of $g(x) = \frac{1}{2}x^2$ is compressed vertically by a factor of $\frac{1}{2}$; each point is half as far from the x-axis as its counterpart on the graph of $y = x^2$. The graph of $h(x) = -x^2$ is reflected about the x-axis.

In general, we have the following graphing principles.

SCALE FACTORS

Compared with the graph of $y = f(x)$, the graph of $y = af(x)$, where $a \neq 0$, is

1. expanded vertically by a factor of a if $|a| > 1$.

2. compressed vertically by a factor of a if $0 < |a| < 1$.

3. reflected about the x-axis if $a < 0$.

The constant a is called the **scale factor** for the graph.

Example 4

Graph the following functions.

a. $g(x) = 3\sqrt[3]{x}$ **b.** $h(x) = -\frac{1}{2}|x|$

Solutions **a.** The graph of $g(x) = 3\sqrt[3]{x}$ is an expansion of the basic graph of $y = \sqrt[3]{x}$ by a factor of 3. Each point on the basic graph has its y-coordinate tripled. (See Figure 10.9.)

b. The graph of $h(x) = -\frac{1}{2}|x|$ is a compression of the basic graph of $y = |x|$ by a factor of $\frac{1}{2}$, combined with a reflection about the x-axis. You may find it helpful to graph the function in two steps, as shown in Figure 10.10.

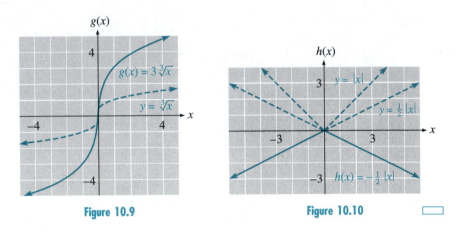

Figure 10.9   **Figure 10.10**

Exploring Translations and Scale Factors with a Graphing Calculator

It is easy to use a graphing calculator to compare the graphs of $y = f(x)$ and $y = f(x) + k$.

Example 5

Compare the graphs of the three equations in each part.
a. $y = x^2$, $y = x^2 + 4$, $y = x^2 - 4$

b. $y = \sqrt{x}$, $y = \sqrt{x} + 4$, $y = \sqrt{x} - 4$

c. $y = |x|$, $y = |x| + 4$, $y = |x| - 4$

Solutions Notice that the three equations in each part are of the forms $y = f(x)$, $y = f(x) + 4$, and $y = f(x) - 4$.

a. Begin by pressing the $\boxed{Y=}$ key and clearing out all old definitions. Next define $Y_1 = X^2$. Define $Y_2 = Y_1 + 4$ and $Y_3 = Y_1 - 4$. (To enter Y_1, we access the $\boxed{Y\text{-VARS}}$ menu. Press $\boxed{2nd}$ $\boxed{VARS}$ to open the menu, press $\boxed{ENTER}$ to choose the function variables, and press $\boxed{ENTER}$ again to select Y_1.) (See Figure 10.11[a].) Finally, press $\boxed{ZOOM}$ $\boxed{6}$ to see the graphs in the standard graphing window, as shown in Figure 10.11(b). We see that the second graph, $y = x^2 + 4$, is just the standard parabola translated four units upward. The third graph, $y = x^2 - 4$, is the standard parabola translated four units downward.

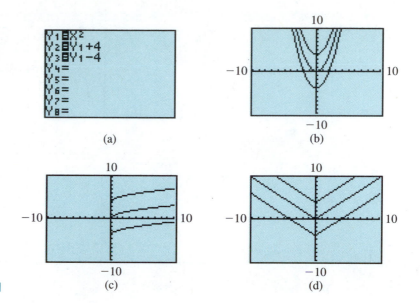

Figure 10.11

b. Keep the definitions of Y_2 and Y_3 as they are and define $Y_1 = \sqrt{X}$. Press $\boxed{\text{GRAPH}}$. The second and third graphs are simply the graph of the square root function shifted four units up and four units down, respectively. The graphs are shown in Figure 10.11(c).

c. Define $Y_1 = \text{abs } X$. (The absolute value function is accessed using $\boxed{\text{2nd}}$ $\boxed{x^{-1}}$.) Press $\boxed{\text{GRAPH}}$. The second and third graphs are simply the graph of the absolute value function shifted four units up and four units down, respectively. The graphs are shown in Figure 10.11(d).

We can also use a graphing calculator to compare the graphs of $y = f(x)$ and $y = f(x + k)$.

Example 6

Compare the graphs of the three equations in each part.

a. $y = x^2$, $y = (x + 4)^2$, $y = (x - 4)^2$

b. $y = \sqrt{x}$, $y = \sqrt{x + 4}$, $y = \sqrt{x - 4}$

c. $y = |x|$, $y = |x + 4|$, $y = |x - 4|$

Solutions Notice that the three equations in each part are of the forms $y = f(x)$, $y = f(x + 4)$, and $y = f(x - 4)$.

a. Define $Y_1 = X^2$, $Y_2 = Y_1(X + 4)$, and $Y_3 = Y_1(X - 4)$. (See Figure 10.12[a].) Press $\boxed{\text{ZOOM}}$ $\boxed{6}$ to see the graphs in the standard graphing window, as shown in Figure 10.12(b). We see that the second graph, $y = (x + 4)^2$, is just the standard parabola shifted four units to the left. The third graph, $y = (x - 4)^2$, is the standard parabola shifted four units to the right.

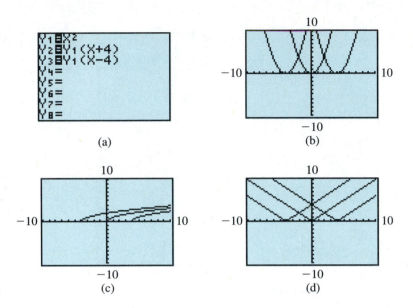

(a)

(b)

(c)

(d)

Figure 10.12

b. Keep the definitions of Y_2 and Y_3 as they are and define $Y_1 = \sqrt{X}$. Press GRAPH. The second and third graphs are simply the graph of the square root function shifted four units left and four units right, respectively. The graphs are shown in Figure 10.12(c).

c. Set $Y_1 = abs\ X$. (The absolute value function is accessed using 2nd x^{-1}.) Press GRAPH. The second and third graphs are simply the graph of the absolute value function shifted four units left and four units right, respectively. The graphs are shown in Figure 10.12(d). ⬜

Graphing calculators can also help us study the effects of scale factors.

Example 7

Compare the graphs of the four equations in each part.
a. $y = x^2$, $y = 2x^2$, $y = -x^2$, $y = -2x^2$
b. $y = \sqrt{x}$, $y = 2\sqrt{x}$, $y = -\sqrt{x}$, $y = -2\sqrt{x}$
c. $y = |x|$, $y = 2|x|$, $y = -|x|$, $y = -2|x|$

Solutions   Here we see that the four equations in each part are of the forms $y = f(x)$, $y = 2f(x)$, $y = -f(x)$, and $y = -2f(x)$.
a. Define $Y_1 = X^2$, $Y_2 = 2Y_1$, $Y_3 = -Y_1$, and $Y_4 = -2Y_1$. (See Figure 10.13[a].) Press ZOOM 6 to see the graphs in the standard graphing window, as shown in Figure 10.13(b). We see that the second graph, $y = 2x^2$, is a parabola that is twice as steep as the standard one. (We say that it is stretched vertically by a factor of 2.) The third graph, $y = -x^2$, is the standard parabola reflected about the x-axis. The fourth graph, $y = -2x^2$, is reflected and stretched vertically.

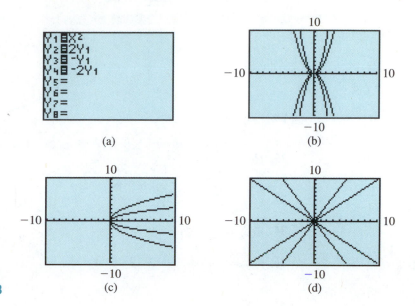

Figure 10.13

b.   Keep Y_2, Y_3, and Y_4 as they are and define $Y_1 = \sqrt{X}$. Press $\boxed{\text{GRAPH}}$.
The second graph is the first one stretched to twice as steep, the third is the
standard parabola reflected about the x-axis, and the fourth is reflected and
stretched. The graphs are shown in Figure 10.13(c).
c.   Set $Y_1 = \text{abs } X$.   The second graph is the first stretched to twice as
steep, the third is the first reflected about the x-axis, and the fourth is reflected
and stretched. The graphs are shown in Figure 10.13(d). $\square$

Exercise 10.1

Graph the functions in Exercises 1–38.

14. $f(x) = (x + 4)^2 - 1$

1. $f(x) = |x| - 4$ **2.** $g(x) = (x + 1)^3$

3. $g(s) = \sqrt[3]{s} - 4$ **4.** $f(s) = s^2 + 3$

15. $g(z) = \dfrac{1}{z + 2} - 3$ **16.** $g(z) = \dfrac{1}{z - 1} + 1$

5. $F(t) = \dfrac{1}{t^2} + 1$ **6.** $G(t) = \sqrt{t} - 2$

17. $F(u) = \sqrt{u + 4} + 4$

18. $F(u) = \sqrt{u - 3} - 5$

7. $G(r) = (r + 2)^3$ **8.** $F(r) = \dfrac{1}{r - 4}$

19. $G(t) = |t - 5| - 1$ **20.** $G(t) = |t + 4| + 2$

9. $H(d) = \sqrt{d} - 3$ **10.** $h(d) = \sqrt[3]{d} + 5$

21. $h(p) = (p + 2)^3 - 5$

11. $h(v) = \dfrac{1}{v + 6}$ **12.** $H(v) = \dfrac{1}{v^2} - 2$

22. $h(p) = (p - 2)^3 + 3$

13. $f(x) = 2 + (x - 3)^2$

23. $H(w) = \dfrac{1}{(w - 1)^2} + 6$

24. $H(w) = \dfrac{1}{(w + 3)^2} - 3$

25. $f(t) = \sqrt[3]{t - 8} - 1$

26. $f(t) = \sqrt[3]{t + 1} + 8$

27. $F(t) = 4t^2$ **28.** $G(t) = \dfrac{1}{3}t^2$

29. $f(x) = \dfrac{1}{3}|x|$ **30.** $H(x) = 3|x|$

31. $h(z) = \dfrac{2}{z^2}$ **32.** $g(z) = \dfrac{2}{z}$

33. $G(v) = -2\sqrt{v}$ **34.** $F(v) = -4\sqrt[3]{v}$

35. $g(s) = \dfrac{-1}{2}s^3$ **36.** $f(s) = \dfrac{-1}{8}s^3$

37. $H(x) = \dfrac{1}{3x}$ **38.** $h(x) = \dfrac{1}{4x^2}$

In Exercises 39–44 give an equation for the function graphed in each figure.

39.

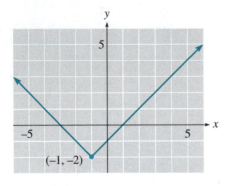

40.

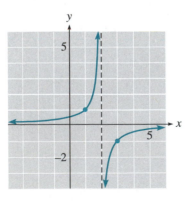

41.

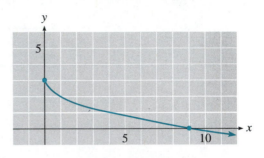

42.

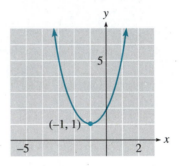

43.

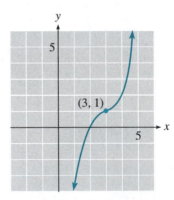

44.

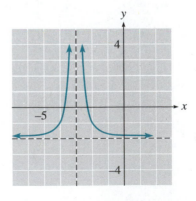

In Exercises 45–48
a. write each equation in the form $y = (x - p)^2 + q$ by completing the square.
b. graph each equation using horizontal and vertical translations.

45. $y = x^2 - 4x + 7$ **46.** $y = x^2 - 2x - 1$

47. $y = x^2 + 2x - 3$ **48.** $y = x^2 + 4x + 5$

10.2 CONIC SECTIONS

In Chapter 2 we found that the graph of any first-degree equation in two variables,

$$Ax + By = C$$

is a line. We now turn our attention to second-degree equations in two variables. The most general form of such an equation is

$$Ax^2 + Bxy + Cy^2 + Dx + Ey + F = 0$$

where A, B, and C cannot all be zero (since in that case the equation would not be of second degree). The graphs of such equations are curves called **conic sections** because they are formed by the intersection of a plane and a cone, as shown in Figure 10.14. Except for a few special cases called *degenerate* conics, which we will describe later, the conic sections fall into four categories called **circles**, **ellipses**, **hyperbolas**, and **parabolas**.

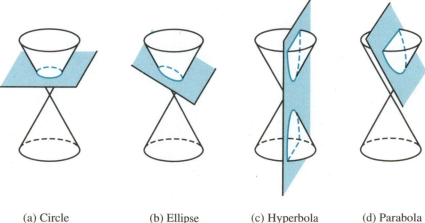

Figure 10.14 (a) Circle   (b) Ellipse   (c) Hyperbola   (d) Parabola

Circles and Ellipses In this section we will consider conic sections whose centers (or vertices, in the case of parabolas) are located at the origin. Such curves are called **central conics**.

The **circle** is the most familiar of the conic sections. In Chapter 6 we used the distance formula to derive the standard equation for a circle of radius r centered at the point (h, k):

$$(x - h)^2 + (y - k)^2 = r^2$$

From this equation we can see that a circle whose center is the origin has equation

$$x^2 + y^2 = r^2$$

We can also rewrite this equation in the form

$$\frac{x^2}{r^2} + \frac{y^2}{r^2} = 1$$

Then the denominator of the x-squared term is the square of the x-intercepts, and the denominator of the y-squared term is the square of the y-intercepts. (Do you see why?) Of course, in this case the x- and y-intercepts are equal because the graph is a circle.

If the denominators of the x-squared and y-squared terms are different, the graph is called an **ellipse**. For example, the graph of

$$\frac{x^2}{9} + \frac{y^2}{4} = 1$$

is shown in Figure 10.15. When $y = 0$, we have $\frac{x^2}{9} = 1$, so the x-intercepts are $(3, 0)$ and $(-3, 0)$. When $x = 0$, we have $\frac{y^2}{4} = 1$, so the y-intercepts are $(0, 2)$ and $(0, -2)$.

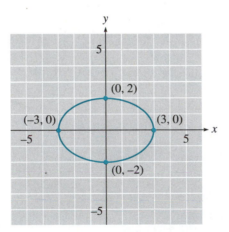

Figure 10.15

In general, an ellipse is defined as follows.

An **ellipse** is the set of points in a plane the sum of whose distances from two fixed points (the **foci**) is a constant.

We can visualize the definition in the following way. Drive two nails into a board to represent the two foci. Attach the two ends of a piece of string to the two nails, and stretch the string taut with a pencil. Trace around the two nails, keeping the string taut, as illustrated in Figure 10.16. The figure described will be an ellipse because the sum of the distances from each point to the two foci is the length of the string, which is constant.

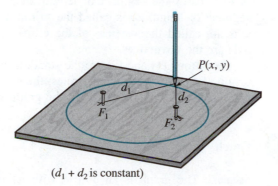

Figure 10.16 $(d_1 + d_2$ is constant)

Ellipses appear in a variety of applications. The orbits of the planets about the sun and of satellites about the earth are ellipses. The arches in some bridges are elliptical in shape, as are certain bicycle gears and the styli in some stereo systems.

Using the distance formula and the definition on page 574, we can show that the equation of an ellipse centered at the origin has the form

$$\frac{x^2}{a^2} + \frac{y^2}{b^2} = 1 \qquad (a > b) \tag{1}$$

when the foci lie on the x-axis. (See Figure 10.17[a].) By setting y equal to zero in Equation (1), we find that the x-intercepts of this ellipse are a and $-a$; by setting x equal to zero, we find that the y-intercepts are b and $-b$.

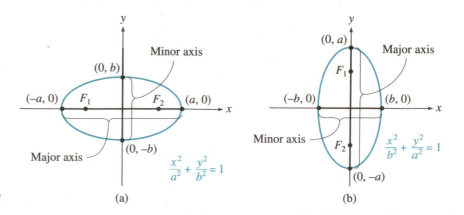

Figure 10.17   (a)   (b)

Figure 10.17(b) shows an ellipse whose foci lie on the y-axis. The equation of this ellipse is

$$\frac{x^2}{b^2} + \frac{y^2}{a^2} = 1 \qquad (a > b) \tag{2}$$

In this case the x-intercepts are b and $-b$ and the y-intercepts are a and $-a$. In both cases the segment of length $2a$ is called the **major axis**, and the segment of length $2b$ is called the **minor axis**. The endpoints of the major axis are called the **vertices** of the ellipse, and the endpoints of the minor axis are the **covertices**.

Equations (1) and (2) are the standard forms for ellipses centered at the origin. As was the case with circles, the standard form of the equation gives us enough information to sketch the graph of the conic section.

Example 1

Graph $\dfrac{x^2}{8} + \dfrac{y^2}{25} = 1$.

Solution The graph is an ellipse with its major axis on the y-axis. Since $a^2 = 25$ and $b^2 = 8$, the vertices are located at $(0, 5)$ and $(0, -5)$, and the covertices lie $\sqrt{8}$ units to the right and left of the center, or approximately at $(2.8, 0)$ and $(-2.8, 0)$. (See Figure 10.18.) To sketch the ellipse, first locate the vertices and covertices. Then draw a smooth curve through the points.

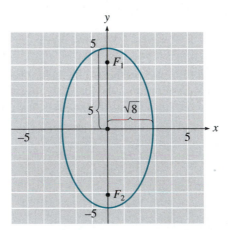

Figure 10.18

The equation of any central ellipse may be written as

$$Ax^2 + By^2 = C$$

where A, B, and C have like signs. The features of the graph are easier to identify if we first convert the equation to standard form.

Example 2

Graph $4x^2 + y^2 = 12$.

Solution First convert the equation to standard form: divide through by the constant term, 12, to obtain

$$\frac{x^2}{3} + \frac{y^2}{12} = 1$$

Since $a^2 = 12$ and $b^2 = 3$, we see that the vertices are $(0, \pm\sqrt{12})$ and the covertices are $(\pm\sqrt{3}, 0)$. Plot points at approximately $(0, \pm3.4)$ and $(\pm1.7, 0)$, then draw a central ellipse through the points. (See Figure 10.19.)

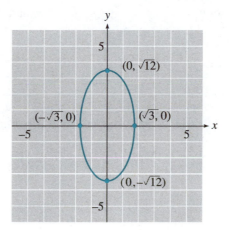

Figure 10.19

Hyperbolas If a cone is cut by a plane parallel to its axis, the intersection is a **hyperbola**, the only conic section made of two separate pieces, or **branches**. (Look back at Figure 10.14.) As do the other conic sections, hyperbolas occur in a number of applied settings. The navigational system called *loran* (long-range navigation) uses radio signals to locate a ship or plane at the intersection of two hyperbolas. Satellites moving with sufficient speed will follow an orbit that is a branch of a hyperbola; for example, a rocket sent to the moon must be fitted with retrorockets to reduce its speed in order to achieve an elliptical, rather than a hyperbolic, orbit about the moon.

The hyperbola is defined as follows.

A **hyperbola** is the set of points in the plane, the difference of whose distances from two fixed points (the **foci**) is a constant.

If the origin is the center of a hyperbola and the foci lie on the *x*-axis, we can show that its equation may be written as

$$\frac{x^2}{a^2} - \frac{y^2}{b^2} = 1 \tag{3}$$

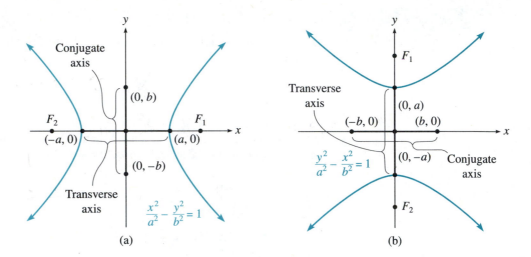

Figure 10.20

(a) (b)

The graph has x-intercepts at a and $-a$ but no y-intercepts. (See Figure 10.20[a].) A hyperbola centered at the origin with foci on the y-axis has the equation

$$\frac{y^2}{a^2} - \frac{x^2}{b^2} = 1 \tag{4}$$

In this case the graph has y-intercepts at a and $-a$ but no x-intercepts. (See Figure 10.20[b].) In both orientations the line segment of length $2a$ is called the **transverse axis**, and the segment of length $2b$ is called the **conjugate axis**. The endpoints of the transverse axis, which lie on the hyperbola, are called the **vertices**. Unlike the equations for an ellipse, a can be less than or equal to b for a hyperbola.

Asymptotes of Hyperbolas

The branches of the hyperbola approach two straight lines that intersect at its center. These lines are asymptotes of the graph, and they are useful as guidelines for sketching the hyperbola. We can obtain the asymptotes by first forming a rectangle (called the "central rectangle") whose sides are parallel to and the same length as the transverse and conjugate axes. The asymptotes are determined by the diagonals of this rectangle.

Example 3

Graph $\dfrac{x^2}{9} - \dfrac{y^2}{16} = 1$.

Solution The graph is a hyperbola with center at the origin. Since the equation is of form (3), the branches of the hyperbola open to the left and right. Also, $a^2 = 9$ and $b^2 = 16$, so $a = 3$ and $b = 4$, and the vertices of the hyperbola are $(3, 0)$ and $(-3, 0)$.

Construct the central rectangle with dimensions $2a = 6$ and $2b = 8$, as shown in Figure 10.21. Draw the asymptotes through the opposite corners of the rectangle, and sketch the branches of the hyperbola through the vertices and approaching the asymptotes to obtain the graph shown.

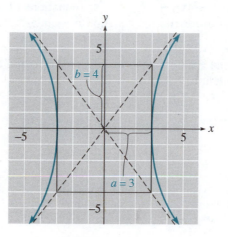

Figure 10.21

Example 4

Graph $\dfrac{y^2}{9} - \dfrac{x^2}{4} = 1$.

Solution The equation is in form (4), so the branches of the hyperbola open upward and downward. Since $a^2 = 9$ and $b^2 = 4$, $a = 3$ and $b = 2$, and the vertices are $(0, 3)$ and $(0, -3)$.

Construct the central rectangle with dimensions $2a = 6$ and $2b = 4$ and draw the asymptotes through its opposite corners. Sketch the branches of the hyperbola through the vertices and approaching the asymptotes as shown in Figure 10.22.

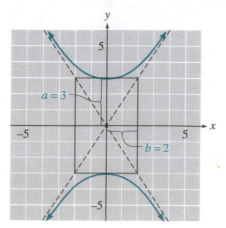

Figure 10.22

The equation of a central hyperbola may be written as

$$Ax^2 + By^2 = C$$

where A and B have *opposite* signs. (For example, $2x^2 - 3y^2 = 6$ and $-4x^2 + y^2 = 3$ are equations of hyperbolas.) As with ellipses, it is best to rewrite the equation in standard form in order to graph it.

Example 5

Write each equation in standard form and describe the important features of its graph.

a. $4y^2 - x^2 = 16$ **b.** $4x^2 = y^2 + 25$

Solutions **a.** First divide each side by 16 to obtain

$$\frac{y^2}{4} - \frac{x^2}{16} = 1 \qquad \leftrightarrow \qquad \frac{y^2}{2^2} - \frac{x^2}{4^2} = 1$$

The graph is a central hyperbola with y-intercepts 2 and -2. (See Figure 10.23(a).)

b. We first write the equation as

$$4x^2 - y^2 = 25$$

and then divide each side by 25 to obtain

$$\frac{4x^2}{25} - \frac{y^2}{25} = 1 \qquad \leftrightarrow \qquad \frac{x^2}{\left(\frac{5}{2}\right)^2} - \frac{y^2}{5^2} = 1$$

The graph is a central hyperbola with x-intercepts $\frac{5}{2}$ and $-\frac{5}{2}$. (See Figure 10.23(b).)

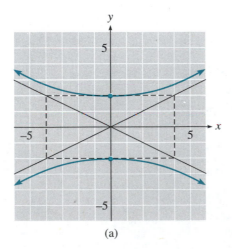

(a)

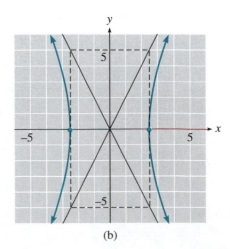

(b)

Figure 10.23

Parabolas   If a cone is cut by a plane parallel to the side of the cone, the intersection is a **parabola**. (Refer to Figure 10.14.) In Section 4.4 we graphed parabolas whose equations were of the form

$$y = ax^2 + bx + c$$

These parabolas opened either up or down and had vertical axes of symmetry. However, parabolas can also open to the left or right. In general, we define a parabola as follows.

> A **parabola** is the set of points in a plane whose distances from a fixed line l and a fixed point F are equal.

The fixed line in the definition is called the **directrix**, and the fixed point is the **focus**. The **axis** of a parabola is the line running through the focus of the parabola and perpendicular to the directrix. (See Figure 10.24.)

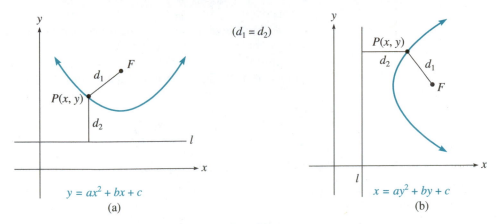

Figure 10.24

Parabolas have many applications in optics and communications. Parabolic mirrors are used in telescopes because the light waves received reflect off the surface and form an image at the focus of the parabola. For similar reasons radio antennae and television dish receivers are parabolic in shape. The parabolic mirrors in searchlights and automobile headlights reflect light from the focus into a beam of parallel rays.

We first consider parabolas with vertex at the origin. Using the distance formula and the definition above, we can show that parabolas that open upward and those that open downward have equations of the forms

$$x^2 = 4py \quad \text{and} \quad x^2 = -4py$$

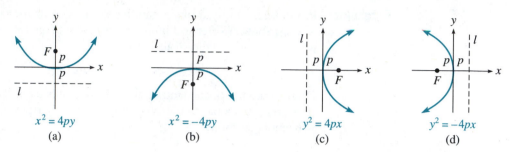

$x^2 = 4py$
(a)

$x^2 = -4py$
(b)

$y^2 = 4px$
(c)

$y^2 = -4px$
(d)

Figure 10.25

respectively, where p is the distance between the vertex of the parabola and its focus or the distance between the vertex and the directrix. (See Figure 10.25.) Parabolas that open to the right and those that open to the left have equations of the forms

$$y^2 = 4px \quad \text{and} \quad y^2 = -4px$$

respectively.

A line drawn through the focus and parallel to the directrix will intersect the parabola at two points. These points are $2p$ units from the directrix, so they must also be $2p$ units from the focus. (Recall that by definition the points of a parabola are equidistant from the focus and the directrix. See

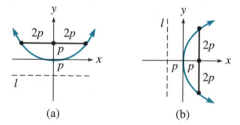

(a)

(b)

Figure 10.26

Figure 10.26.) By locating these two "guide points" and the vertex, we can make a reasonable sketch of the parabola, as illustrated in Example 6.

Example 6

Graph $y^2 = -6x$.

Solution   The parabola has its vertex at the origin and opens to the left, so its axis is the x-axis. Also, $-6 = -4p$, so $p = \frac{3}{2}$, and the focus is the point $(-\frac{3}{2}, 0)$. The directrix is the vertical line $x = \frac{3}{2}$.

To graph the parabola, draw a line segment of length $4p = 6$ perpendicular to the axis and centered at the focus; since $2p = 3$, the endpoints of the segment are $(-\frac{3}{2}, 3)$ and $(-\frac{3}{2}, -3)$. Sketch the parabola through the vertex and the two guide points to obtain the graph shown in Figure 10.27.

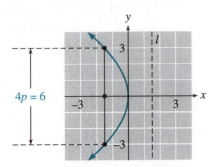

Figure 10.27

The important facts about central conics are summarized in Table 10.1 on page 584.

Example 7

Write each equation in standard form and describe its graph.

a. $x^2 = 9y^2 - 9$ b. $4x^2 + y = 0$

c. $9y^2 = 4 - x^2$ d. $6x^2 = 18 - 6y^2$

e. $x^2 = 6y^2 + 8$ f. $x^2 = \dfrac{4 - y^2}{2}$

Solutions   **a.**   The equation $x^2 = 9y^2 - 9$ is equivalent to

$$9y^2 - x^2 = 9 \quad \text{or} \quad \frac{y^2}{1^2} - \frac{x^2}{3^2} = 1$$

The graph is a hyperbola with the transverse axis on the y-axis.

b.   The equation $4x^2 + y = 0$ is equivalent to

$$y = -4x^2$$

The graph is a parabola that opens downward.

c.   The equation $9y^2 = 4 - x^2$ is equivalent to

$$x^2 + 9y^2 = 4 \quad \text{or} \quad \frac{x^2}{2^2} + \frac{y^2}{\left(\frac{4}{9}\right)^2} = 1$$

The graph is an ellipse with the major axis on the x-axis because $2 > \frac{4}{9}$.

d.   The equation $6x^2 = 18 - 6y^2$ is equivalent to

$$6x^2 + 6y^2 = 18 \quad \text{or} \quad x^2 + y^2 = (\sqrt{3})^2$$

The graph is a circle with radius $\sqrt{3}$.

Table 10.1

Name of Conic	Standard Form of Equation	Graph
Circle	$x^2 + y^2 = r^2$	
Ellipse (a) Major axis on the x-axis	$\dfrac{x^2}{a^2} + \dfrac{y^2}{b^2} = 1 \quad (a > b)$	
(b) Major axis on the y-axis	$\dfrac{x^2}{b^2} + \dfrac{y^2}{a^2} = 1 \quad (a > b)$	
Hyperbola (a) Transverse axis on the x-axis	$\dfrac{x^2}{a^2} - \dfrac{y^2}{b^2} = 1$	
(b) Transverse axis on the y-axis	$\dfrac{y^2}{a^2} - \dfrac{x^2}{b^2} = 1$	
Parabola (a) Opens upward	$x^2 = 4py$	
(b) Opens downward	$x^2 = -4py$	
(c) Opens to the right	$y^2 = 4px$	
(d) Opens to the left	$y^2 = -4px$	

e. The equation $x^2 = 6y^2 + 8$ is equivalent to

$$x^2 - 6y^2 = 8 \quad \text{or} \quad \frac{x^2}{(\sqrt{8})^2} + \frac{y^2}{\left(\dfrac{2}{\sqrt{3}}\right)^2} = 1$$

The graph is a hyperbola with the transverse axis on the x-axis.

f. The equation $x^2 = \dfrac{4 - y^2}{2}$ is equivalent to

$$2x^2 + y^2 = 4 \quad \text{or} \quad \frac{x^2}{(\sqrt{2})^2} + \frac{y^2}{2^2} = 1$$

The graph is an ellipse with the major axis on the y-axis because $2 > \sqrt{2}$.

Conic Sections Using Graphing Calculators

Graphing calculators are designed to graph functions. Circles, ellipses, hyperbolas, and some parabolas are not the graphs of functions because most values of x correspond to two different values of y. We must do a little more work to obtain these graphs on a calculator.

Example 8

Use a calculator to graph each conic section.
a. $x^2 + y^2 = 9$
b. $x^2 - y^2 = 4$
c. $x - y^2 = 1$

Solutions a. We will first solve for y in terms of x.

$$x^2 + y^2 = 9$$
$$y^2 = 9 - x^2$$
$$y = \pm\sqrt{9 - x^2}$$

We see that there are two values of y associated with a single value of x; that is, the graph of the circle is actually the graph of two functions,

$$Y_1 = \sqrt{9 - x^2} \quad \text{and} \quad Y_2 = -Y_1$$

Enter these expressions and press $\boxed{\text{ZOOM}}$ $\boxed{6}$ for the standard graphing window.

The graph appears to be part of an ellipse, not a circle, as shown in Figure 10.28(a). This distortion in shape occurs because the scales on the two axes are different. One way to remedy this distortion is to use the "square" window setting: press $\boxed{\text{ZOOM}}$ $\boxed{5}$. We now see that the conic section is a circle, as in Figure 10.28(b). (What is the circle's radius?)

If we look closely, we see gaps at the left and right of the circle. These

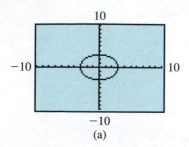

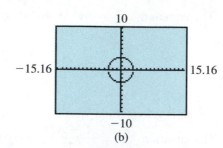

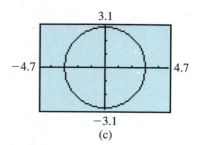

(a)

(b)

(c)

Figure 10.28

gaps occur because the calculator plots only a selection of points (depending on the window chosen) and then tries to "connect the dots." We get a better picture with the **ZDecimal** window ([ZOOM] [4]). (See Figure 10.28[c].)

b.   Begin by solving for y in terms of x.

$$x^2 - y^2 = 4$$
$$x^2 - 4 = y^2$$
$$\pm\sqrt{x^2 - 4} = y$$

Enter

$$Y_1 = \sqrt{x^2 - 4} \quad \text{and} \quad Y_2 = -Y_1$$

Press [GRAPH] to see the hyperbola in Figure 10.29(a).

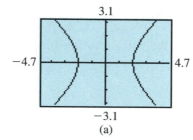

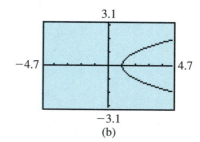

Figure 10.29   (a)   (b)

c.   Solve for y in terms of x.

$$x - y^2 = 1$$
$$x - 1 = y^2$$
$$\pm\sqrt{x - 1} = y$$

Enter

$$Y_1 = \sqrt{x - 1} \quad \text{and} \quad Y_2 = -Y_1$$

The graph is a parabola that opens to the right, as shown in Figure 10.29(b).

Exercise 10.2

Graph each circle or ellipse in Exercises 1–16.

1. $x^2 + y^2 = 25$

2. $x^2 + y^2 = 16$

3. $4x^2 = 16 - 4y^2$

4. $2x^2 = 18 - 2y^2$

5. $\dfrac{x^2}{16} + \dfrac{y^2}{4} = 1$

6. $\dfrac{x^2}{9} + \dfrac{y^2}{16} = 1$

7. $\dfrac{x^2}{10} + \dfrac{y^2}{25} = 1$

8. $\dfrac{x^2}{16} + \dfrac{y^2}{12} = 1$

9. $x^2 + \dfrac{y^2}{14} = 1$

10. $\dfrac{x^2}{8} + y^2 = 1$

11. $3x^2 + 4y^2 = 36$

12. $5x^2 + 2y^2 = 20$

13. $x^2 = 36 - 9y^2$

14. $4x^2 = 36 - y^2$

15. $3y^2 = 30 - 2x^2$

16. $5y^2 = 30 - 3x^2$

Graph each hyperbola in Exercises 17–28.

17. $\dfrac{x^2}{25} - \dfrac{y^2}{9} = 1$

18. $\dfrac{y^2}{4} - \dfrac{x^2}{16} = 1$

19. $\dfrac{y^2}{12} - \dfrac{x^2}{8} = 1$

20. $\dfrac{x^2}{15} + \dfrac{y^2}{10} = 1$

21. $9x^2 - 4y^2 = 36$

22. $4x^2 - 9y^2 = 36$

23. $y^2 - 9x^2 = 36$

24. $4y^2 - x^2 = 36$

25. $3x^2 = 4y^2 + 24$

26. $4x^2 = 3y^2 + 24$

27. $\dfrac{1}{2}x^2 = y^2 - 12$

28. $y^2 = \dfrac{1}{2}x^2 - 16$

Graph each parabola in Exercises 29–38.

29. $x^2 = 2y$

30. $y^2 = 4x$

31. $x = -\dfrac{1}{16}y^2$

32. $y = -\dfrac{1}{18}x^2$

33. $y^2 = 12x$

34. $4x^2 = 3y$

35. $x^2 + 8y = 0$

36. $4y^2 - 2x = 0$

37. $2y^2 - 3x = 0$

38. $3x^2 + 5y = 0$

The graphs of the equations in Exercises 39–50 are circles, ellipses, hyperbolas, or parabolas. Name the graph and describe its main features.

39. $y^2 = 4 - x^2$

40. $y^2 = 6 - 4x^2$

41. $4y^2 = x^2 - 8$

42. $x^2 + 2y - 4 = 0$

43. $4x^2 = 12 - 2y^2$

44. $6x^2 = 8 - 6y^2$

45. $4x^2 = 6 + 4y$

46. $2x^2 = 5 + 4y^2$

47. $6 + \dfrac{x^2}{4} = y^2$

48. $y^2 = 6 - \dfrac{2x^2}{3}$

49. $\dfrac{1}{2}x^2 - y = 4$

50. $\dfrac{x^2}{4} = 4 + 6y^2$

51. An ellipse is centered at the origin with a vertical major axis of length 16 and a horizontal minor axis of length 10.

 a. Find the equation of the ellipse.

 b. What are the values of y when $x = 4$?

52. An ellipse is centered at the origin with a horizontal major axis of length 26 and a vertical minor axis of length 18.

 a. Find the equation of the ellipse.

 b. What are the values of y when $x = 12$?

53. The arch of a bridge is in the shape of the top half of an ellipse with a horizontal major axis. (See Figure 10.30.) The arch is 7 feet high and 20 feet wide. How high is the arch at a distance of 8 feet from the peak?

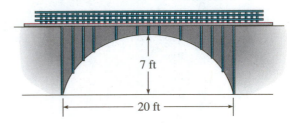

7 ft

|◄———— 20 ft ————►|

Figure 10.30

54. A doorway is topped by a semielliptical arch. (See Figure 10.31.) The doorway is 230 centimeters high at its highest point and 200 centimeters high at its lowest point. It is 80 centimeters wide. How high is the doorway 8 centimeters from the left side?

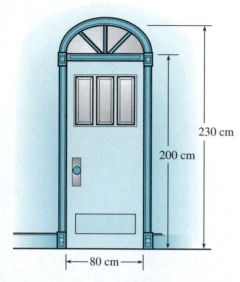

Figure 10.31

55. The trailing edge of a racing sailboat's keel is semielliptical. The major axis of the ellipse is 360 centimeters. The minor axis is 100 centimeters, but part of the ellipse has been cut off by a straight-line leading edge. The leading edge is parallel to the major axis and 330 centimeters long. (See Figure 10.32.) What is the width of the keel at its widest point? Round your answer to two decimal places.

56. The trailing edge of the wing of the World War II British Spitfire has a semielliptical shape. The major axis of the ellipse is 48 feet. The minor axis is 16 feet, but part of the top half of the ellipse is cut off by a straight-line leading edge. (See Figure 10.33.) This edge is parallel to the major axis and 46 feet long. How wide is the wing at the center from leading edge to trailing edge? Round your answer to two decimal places.

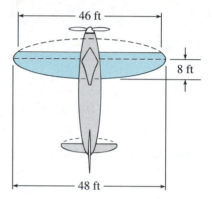

Figure 10.33

57. A reflecting telescope has a parabolic primary mirror to collect light. Suppose you want to build a primary mirror with a diameter of 72 inches but only 3 inches deep. (See Figure 10.34.) How far will the focus be from the vertex of the parabola?

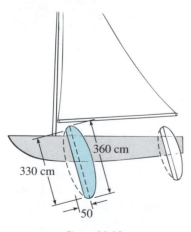

Figure 10.32

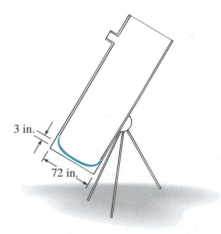

Figure 10.34

58. A parabolic sonar dish has its receiver at the focus of a parabola at a distance of 12 inches from the vertex. The dish is 12 inches tall. (See Figure 10.35.) How deep is the dish?

is the diameter of the base of the tower if the base is 360 feet lower than the center of the hyperbola?

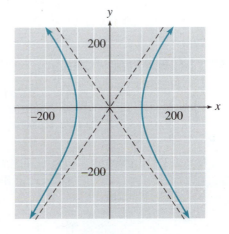

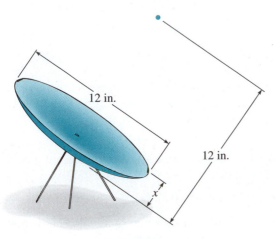

Figure 10.35

59. Tien-Ying wants to build a parabolic satellite dish. The diameter will be 60 centimeters and the dish will be 18 centimeters deep. (See Figure 10.36.) She must place the receiver at the focus of the parabola. How far is the receiver from the vertex of the parabola?

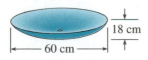

Figure 10.36

Figure 10.37

60. A flashlight has a parabolic mirror that has an 8-centimeter diameter and is 4 centimeters deep. How far should the bulb be from the vertex if the bulb is to be positioned at the focus?

61. The cooling tower for an electricity-generating plant has the shape obtained by revolving a portion of the hyperbola $\dfrac{x^2}{100^2} - \dfrac{y^2}{150^2} = 1$ around the y-axis, as shown in Figure 10.37. What

62. What is the diameter at the top of the cooling tower in Exercise 61 if the top is 200 feet higher than the center of the hyperbola?

63. How far below the top of the cooling tower in Exercise 61 is the highest point where the diameter of the tower is 250 feet?

64. How far above the bottom of the cooling tower in Exercise 61 is the lowest point where the diameter of the tower is 200 feet?

65. Graph $x^2 - y^2 = 0$. (*Hint:* First write as $(x - y)(x + y) = 0$ and then graph $y = x$ and $y = -x$.)

66. Graph $4x^2 - y^2 = 0$. (See hint for Exercise 65.)

67. Graph $x^2 - y^2 = 4$, $x^2 - y^2 = 1$, and $x^2 - y^2 = 0$ on the same set of axes.

68. Graph $4x^2 - y^2 = 16$, $4x^2 - y^2 = 4$, and $4x^2 - y^2 = 0$ on the same set of axes.

10.3   TRANSLATED CONICS

In this section we consider ellipses and hyperbolas that are not centered at the origin and parabolas whose vertices are not at the origin. These figures are **translations** of the central conics we studied in Section 10.2.

Ellipses   In Section 6.6 we used the distance formula to derive the equation of a circle centered at the point (h, k). The standard form for this equation is

$$(x - h)^2 + (y - k)^2 = r^2$$

The standard forms for the equation of an ellipse centered at any point (h, k) can also be derived from the distance formula.

$$\frac{(x - h)^2}{a^2} + \frac{(y - k)^2}{b^2} = 1 \qquad (a > b) \qquad (5)$$

$$\frac{(x - h)^2}{b^2} + \frac{(y - k)^2}{a^2} = 1 \qquad (a > b) \qquad (6)$$

Equation (5) describes an ellipse whose major axis is parallel to the x-axis, as shown in Figure 10.38(a). Its vertices lie a units to the left and

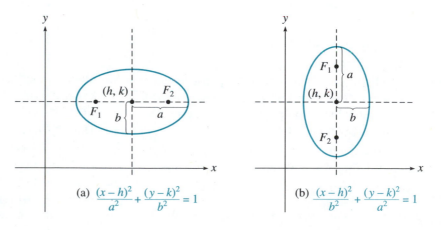

(a) $\dfrac{(x - h)^2}{a^2} + \dfrac{(y - k)^2}{b^2} = 1$

(b) $\dfrac{(x - h)^2}{b^2} + \dfrac{(y - k)^2}{a^2} = 1$

Figure 10.38

right of the center, and its covertices lie b units above and below the center. The ellipse described by Equation (6) and shown in Figure 10.38(b) has its major axis parallel to the y-axis; its vertices lie a units above and below the center, and its covertices lie b units to the left and right of the center.

Example 1   Graph $\dfrac{(x + 2)^2}{16} + \dfrac{(y - 1)^2}{5} = 1$.

Solution   The graph is an ellipse with center at $(-2, 1)$. Since the equation is in standard form (5), we have $a^2 = 16$ and $b^2 = 5$, and the major axis is parallel to the x-axis. The vertices lie four units to the left and right of the center, at $(-6, 1)$ and $(2, 1)$. The covertices lie $\sqrt{5}$ units above and below the center, at approximately $(-2, 3.2)$ and $(-2, -1.2)$. The graph is shown in Figure 10.39.

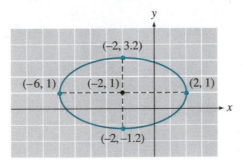

Figure 10.39

Second-degree equations in which the coefficients of x^2 and y^2 have the *same* sign can be written in one of the standard forms for an ellipse by completing the square. As with circles, the equation can be graphed easily from the standard form.

Example 2   **a.**   Write the equation

$$4x^2 + 9y^2 - 16x - 18y - 11 = 0$$

in standard form.

b.   Graph the equation.

Solutions   **a.**   First prepare to complete the square in both x and y by writing

$$4(x^2 - 4x \quad) + 9(y^2 - 2y \quad) = 11$$

Note that the coefficients of x^2 and y^2 have been factored from their respective terms. Complete the square in x by adding $4 \cdot 4$, or 16, to each side of the equation, and complete the square in y by adding $9 \cdot 1$, or 9, to each side, to obtain

$$4(x^2 - 4x + 4) + 9(y^2 - 2y + 1) = 11 + 16 + 9$$

Write each term on the left side as a perfect square to get

$$4(x - 2)^2 + 9(y - 1)^2 = 36$$

To write this equation in standard form, divide each side by 36 to get

$$\frac{(x - 2)^2}{9} + \frac{(y - 1)^2}{4} = 1$$

b. The graph is an ellipse with center at $(2, 1)$, $a^2 = 9$, and $b^2 = 4$. The vertices lie three units to the right and left of the center at $(5, 1)$ and $(-1, 1)$, and the covertices lie two units above and below the center at $(2, 3)$ and $(2, -1)$. The graph is shown in Figure 10.40.

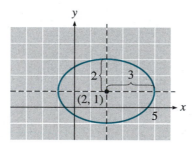

Figure 10.40

To write the equation of an ellipse from a description of its properties, we must find the center of the ellipse and the lengths of its axes. We can then substitute this information into the appropriate equation in standard form.

Example 3

Find the equation of the ellipse with vertices at $(3, 3)$ and $(3, -5)$ and covertices at $(1, -1)$ and $(5, -1)$.

Solution   To find the center of the ellipse, find the midpoint of the major or minor axis:

$$h = \bar{x} = \frac{3 + 3}{2} = 3$$

$$k = \bar{y} = \frac{3 - 5}{2} = -1$$

The center of the ellipse is the point $(3, -1)$. The value of a is half the length of the major axis, or the distance from the center, $(3, -1)$, to one of the vertices, say $(3, 3)$:

$$a = |3 - (-1)| = 4$$

The value of b is half the length of the minor axis, or the distance from the center, $(3, -1)$, to one of the covertices, say $(5, -1)$:

$$b = |5 - 3| = 2$$

Since the major axis is a vertical line segment, the ellipse has an equation of standard form (6), or

$$\frac{(x - h)^2}{b^2} + \frac{(y - k)^2}{a^2} = 1$$

with $h = 3$, $k = -1$, $a = 4$, and $b = 2$. Thus the equation is

$$\frac{(x - 3)^2}{2^2} + \frac{(y + 1)^2}{4^2} = 1$$

or

$$\frac{(x - 3)^2}{4} + \frac{(y + 1)^2}{16} = 1$$

$$4(x - 3)^2 + (y + 1)^2 = 16$$

$$4(x^2 - 6x + 9) + (y^2 + 2y + 1) = 16$$

$$4x^2 - 24x + 36 + y^2 + 2y + 1 = 16$$

$$4x^2 + y^2 - 24x + 2y + 21 = 0$$

Hyperbolas The following standard forms for equations of hyperbolas centered at the point (h, k) can be derived using the distance formula and the definition of hyperbola.

$$\frac{(x - h)^2}{a^2} - \frac{(y - k)^2}{b^2} = 1 \tag{7}$$

$$\frac{(y - k)^2}{a^2} - \frac{(x - h)^2}{b^2} = 1 \tag{8}$$

Equation (7) describes a hyperbola whose transverse axis is parallel to the x-axis, and Equation (8) describes a hyperbola whose transverse axis is parallel to the y-axis. (See Figure 10.41.)

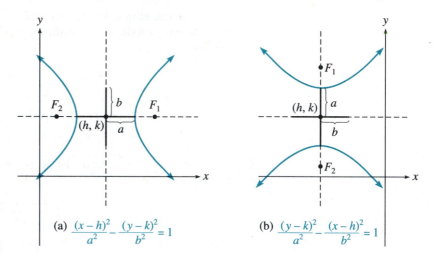

Figure 10.41

(a) $\dfrac{(x-h)^2}{a^2} - \dfrac{(y-k)^2}{b^2} = 1$

(b) $\dfrac{(y-k)^2}{a^2} - \dfrac{(x-h)^2}{b^2} = 1$

Example 4

Graph $\dfrac{(x-3)^2}{8} - \dfrac{(y+2)^2}{10} = 1$.

Solution The graph is a hyperbola with center at $(3, -2)$. Since the equation is in standard form (7), the transverse axis is parallel to the x-axis, and

$$a = \sqrt{8}, \qquad b = \sqrt{10}$$

The coordinates of the vertices are thus $(3 + \sqrt{8}, -2)$ and $(3 - \sqrt{8}, -2)$, or approximately $(5.8, -2)$ and $(0.2, -2)$. The ends of the conjugate axes are $(3, -2 + \sqrt{10})$ and $(3, -2 - \sqrt{10})$, or approximately $(3, 1.2)$ and $(3, -5.2)$.

The central rectangle is centered at the point $(3, -2)$ and extends to the vertices in the horizontal direction and to the ends of the conjugate axis in the vertical direction, as shown in Figure 10.42. Draw the asymptotes through the opposite corners of the central rectangle, and sketch the hyperbola through the vertices and approaching the asymptotes to obtain the graph in Figure 10.42.

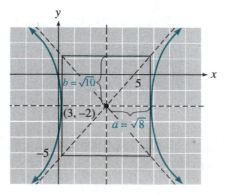

Figure 10.42

Quadratic equations in which the x^2 term and the y^2 term have *opposite* signs describe hyperbolas. We can write such equations in one of the standard forms (7) or (8) by completing the square in x and in y.

Example 5

a. Write the equation

$$y^2 - 4x^2 + 4y - 8x - 9 = 0$$

in standard form.

b. Graph the equation.

Solutions a. First prepare to complete the square by writing

$$(y^2 + 4y \quad) - 4(x^2 + 2x \quad) = 9$$

Complete the square in y by adding 4 to each side of the equation, and complete the square in x by adding $-4 \cdot 1$, or -4, to each side, to get

$$(y^2 + 4y + 4) - 4(x^2 + 2x + 1) = 9 + 4 - 4$$

and then

$$(y + 2)^2 - 4(x + 1)^2 = 9$$

Divide each side by 9 to obtain the standard form

$$\frac{(y + 2)^2}{9} - \frac{(x + 1)^2}{\dfrac{9}{4}} = 1$$

b. The graph is a hyperbola with center at $(-1, -2)$. Since the equation is in standard form (8), the transverse axis is parallel to the y-axis, and

$$a^2 = 9, \qquad b^2 = \frac{9}{4}$$

Thus $a = 3$ and $b = \frac{3}{2}$, and the vertices are $(-1, 1)$ and $(-1, -5)$. The ends of the conjugate axis are $(-\frac{5}{2}, -2)$ and $(\frac{1}{2}, -2)$.

The central rectangle is centered at $(-1, -2)$ and extends to the vertices in the vertical direction and to the ends of the conjugate axis in the horizontal direction, as shown in Figure 10.43. Draw the asymptotes through the corners of the rectangle, and sketch the hyperbola through the vertices and approaching the asymptotes to obtain the graph in Figure 10.43.

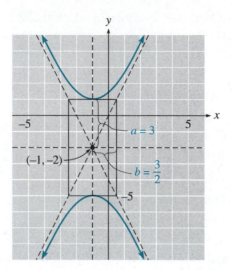

Figure 10.43

Parabolas The equation of a parabola whose vertex is located at the point (h, k) has one of the following four standard forms.

$$(x - h)^2 = \quad 4p(y - k) \quad \text{opens upward} \qquad (9)$$

$$(x - h)^2 = -4p(y - k) \quad \text{opens downward} \qquad (10)$$

$$(y - k)^2 = \quad 4p(x - h) \quad \text{opens to the right} \qquad (11)$$

$$(y - k)^2 = -4p(x - h) \quad \text{opens to the left} \qquad (12)$$

Example 6

Graph $(x + 2)^2 = -8(y - 3)$.

Solution The equation is in standard form (10), so the parabola opens downward. The vertex is the point $(-2, 3)$, and the axis of the parabola is the vertical line through $(-2, 3)$. Since $-4p = -8$, $p = 2$; and the focus, which lies two units below the vertex, is the point $(-2, 1)$. The directrix is the horizontal line $y = 5$.

To graph the parabola, draw a line segment of length $4p = 8$ centered at the focus and perpendicular to the axis; since $2p = 4$, the endpoints of the segment are $(-6, 1)$ and $(2, 1)$. Sketch a parabola through the vertex, $(-2, 3)$, and the two guide points to obtain the graph in Figure 10.44.

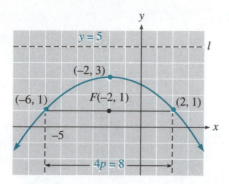

Figure 10.44

A quadratic equation that includes either an x^2 term or a y^2 term but not both can be put into one of the standard forms (9) through (12) by completing the square.

Example 7

Write the equation

$$y^2 - 4y - 4x + 8 = 0$$

in standard form.

Solution First prepare to complete the square in y by writing

$$(y^2 - 4y \qquad) = 4x - 8$$

Complete the square by adding 4 to each side of the equation to get

$$(y^2 - 4y + 4) = 4x - 8 + 4$$

or

$$(y - 2)^2 = 4(x - 1)$$

To write the equation of a given parabola, we must find the coordinates of the vertex and the value of the constant, p.

Example 8

Find an equation for the parabola with vertex $(4, -1)$ and focus $(2, -1)$.

Solution The axis of the parabola is the horizontal line $y = -1$, which passes through the vertex and the focus. The parabola opens to the left, and its equation has the standard form (12), so

$$(y - k)^2 = -4p(x - h)$$

The value of p is the distance from the vertex to the focus, so $p = 4 - 2 = 2$. Since the vertex is the point $(4, -1)$, the equation of the parabola is

$$(y + 1)^2 = -4(2)(x - 4)$$

or

$$(y + 1)^2 = -8(x - 4)$$

Or, alternatively, removing parentheses yields

$$y^2 + 2y + 8x - 31 = 0$$

The graph is shown in Figure 10.45.

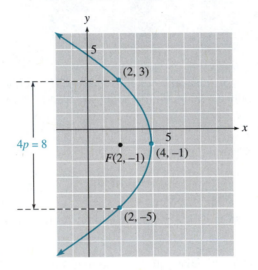

Figure 10.45

General Quadratic Equation in Two Variables

We have considered graphs of second-degree equations in two variables,

$$Ax^2 + Bxy + Cy^2 + Dx + Ey + F = 0$$

for which *B,* the coefficient of the *xy* term, is zero. Such graphs are conic sections with axes parallel to one or both of the coordinate axes. If *B* does not equal zero, the axes of the conic section are rotated with respect to the coordinate axes. The graphing of such equations is taken up in more advanced courses in analytic geometry.

The graph of a second-degree equation can also be a point, a line, a pair of lines, or no graph at all, depending on the values of the coefficients *A* through *F.* For example, the graph of the equation

$$x^2 + y^2 - 2x + 4y + 5 = 0$$

or

$$(x - 1)^2 + (y + 2)^2 = 0$$

Is not a circle because $r^2 = 0$. The graph consists of the single point $(1, -2)$. Such graphs are called **degenerate conics**, and we will not include them in our discussion.

Given an equation of the form

$$Ax^2 + Cy^2 + Dx + Ey + F = 0$$

we can determine the nature of the graph from the coefficients of the quadratic terms. If the graph is not a degenerate conic, the following criteria apply.

The graph of $Ax^2 + Cy^2 + Dx + Ey + F = 0$ is
1. a circle if $A = C$.
2. a parabola if $A = 0$ or $C = 0$ (but not both).
3. an ellipse if $A \neq C$ and they have the same sign.
4. a hyperbola if A and C have opposite signs.

Example 9

Name the graph of each equation, assuming that the graph is not degenerate.
a. $3x^2 + 3y^2 - 2x + 4y - 6 = 0$
b. $4y^2 + 8x^2 - 3y = 0$
c. $4x^2 - 6y^2 + x - 2y = 0$
d. $y + x^2 - 4x + 1 = 0$

Solutions   a.   The graph is a circle because the coefficients of x^2 and y^2 are equal.
b.   The graph is an ellipse because the coefficients of x^2 and y^2 are both positive.
c.   The graph is a hyperbola because the coefficients of x^2 and y^2 have opposite signs.
d.   The graph is a parabola because y is of first degree and x is of second degree.

Note that the coefficients D, E, and F do not figure in determining the *type* of conic section the equation represents. They do, however, determine the *position* of the graph relative to the origin. Once we recognize the form of the graph, it is helpful to write the equation in the appropriate standard form in order to discover more specific information about the graph. The standard forms for the conic sections are summarized in Table 10.2 on page 600. For the parabola, (h, k) is the vertex of the graph, whereas for the other conics, (h, k) is the center.

Example 10

Describe the graph of each equation without graphing.
a. $x^2 = 9y^2 - 9$
b. $x^2 - y = 2x + 4$
c. $x^2 + 9y^2 + 4x - 18y + 9 = 0$

Table 10.2

Name of Curve	Standard Form of Equation	Graph
Circle	$(x - h)^2 + (y - k)^2 = r^2$	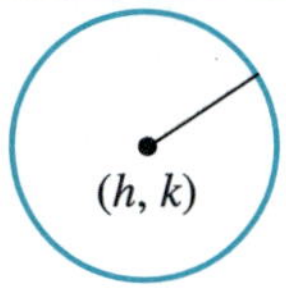
Ellipse (a) Major axis parallel to x-axis	$\dfrac{(x - h)^2}{a^2} + \dfrac{(y - k)^2}{b^2} = 1$	
(b) Major axis parallel to y-axis	$\dfrac{(x - h)^2}{b^2} + \dfrac{(y - k)^2}{a^2} = 1$	
Hyperbola (a) Transverse axis parallel to x-axis	$\dfrac{(x - h)^2}{a^2} - \dfrac{(y - k)^2}{b^2} = 1$	
(b) Transverse axis parallel to y-axis	$\dfrac{(y - k)^2}{a^2} - \dfrac{(x - h)^2}{b^2} = 1$	
Parabola (a) Opens upward	$(x - h)^2 = 4p(y - k)$	
(b) Opens downward	$(x - h)^2 = -4p(y - k)$	
(c) Opens to the right	$(y - k)^2 = 4p(x - h)$	
(d) Opens to the left	$(y - k)^2 = -4p(x - h)$	

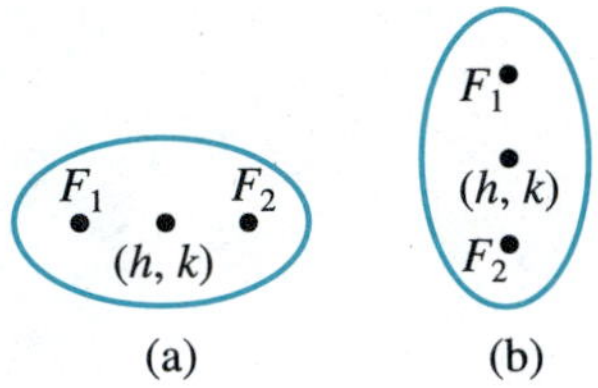

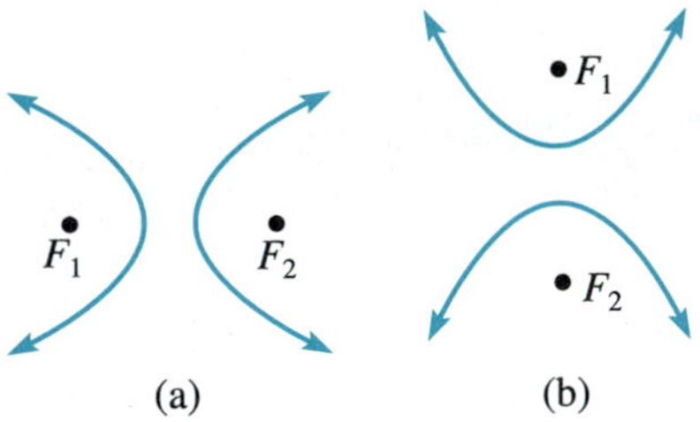

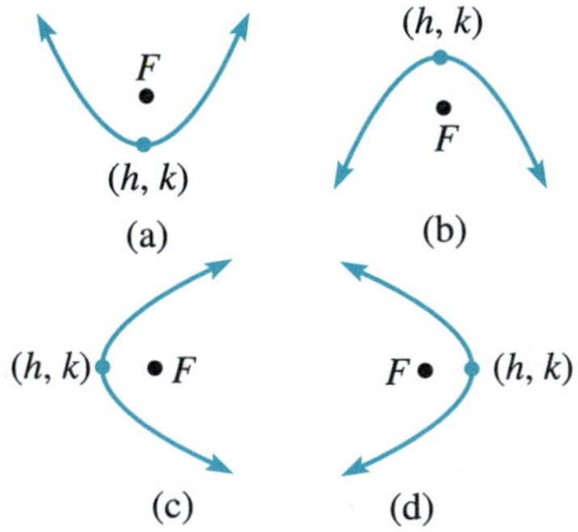

Solutions   **a.** The equation $x^2 = 9y^2 - 9$ is equivalent to

$$9y^2 - x^2 = 9$$

or

$$\frac{y^2}{1} - \frac{x^2}{9} = 1$$

The graph is a hyperbola centered at the origin with vertices $(0, 1)$ and $(0, -1)$.

b. The equation $x^2 - y = 2x + 4$ is equivalent to

$$(x^2 - 2x + 1) = y + 4 + 1$$

or

$$(x - 1)^2 = (y + 5)$$

The graph is a parabola that opens upward from the vertex, $(1, -5)$.

c. The equation $x^2 + 9y^2 + 4x - 18y + 9 = 0$ is equivalent to

$$(x^2 + 4x + 4) + 9(y^2 - 2y + 1) = -9 + 4 + 9$$

or

$$(x + 2)^2 + 9(y - 1)^2 = 4$$

or

$$\frac{(x + 2)^2}{4} + \frac{(y - 1)^2}{\dfrac{4}{9}} = 1$$

The graph is an ellipse centered at $(-2, 1)$ with a major axis of length $2a = 4$ parallel to the x-axis. It has vertices at $(0, 1)$ and $(-4, 1)$ and covertices at $\left(-2, \frac{5}{3}\right)$ and $\left(-2, \frac{1}{3}\right)$.

Using Graphing Calculators

As we saw in Section 10.2, the graph of a conic section is actually the graph of two functions. We can find these two functions by solving the quadratic equation that defines the conic section for y in terms of x.

Example 11

Use a calculator to graph each conic section.
a. $3y - 4x^2 + 6x - 2 = 0$
b. $y^2 - 3y + x^2 = 0$

Solutions **a.** Solve for y in terms of x.

$$3y = 4x^2 - 6x + 2$$

$$y = \frac{4x^2 - 6x + 2}{3} = \frac{4x^2}{3} - 2x + \frac{2}{3}$$

This is the equation of a parabola that opens upward. Figure 10.46(a) shows the graph in the **ZDecimal** ([ZOOM] [4]) window.

b. The equation $y^2 - 3y + x^2 = 0$ is quadratic in y. To solve for y, we will use the quadratic formula. Because we are regarding y (not x) as the variable, we have

$$a = 1, \qquad b = -3, \qquad \text{and} \qquad c = x^2$$

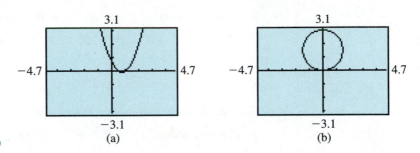

Figure 10.46

Thus

$$y = \frac{-(-3) \pm \sqrt{(-3)^2 - 4(1)(x^2)}}{2(1)}$$

$$= \frac{3 \pm \sqrt{9 - 4x^2}}{2}$$

The two functions defined by this equation are

$$Y_1 = \frac{3 + \sqrt{9 - 4x^2}}{2}$$

and

$$Y_2 = \frac{3 - \sqrt{9 - 4x^2}}{2}$$

Use the **ZDecimal** window (ZOOM 4) to obtain the graph shown in Figure 10.46(b). By tracing values on the two curves, we can see that this circle has a diameter of 3.

Exercise 10.3

Graph each ellipse in Exercises 1–16.

1. $\dfrac{(x-3)^2}{16} + \dfrac{(y-4)^2}{9} = 1$

2. $\dfrac{(x-2)^2}{4} + \dfrac{(y-5)^2}{25} = 1$

3. $\dfrac{(x+2)^2}{6} + \dfrac{(y-5)^2}{12} = 1$

4. $\dfrac{(x-5)^2}{15} + \dfrac{(y+3)^2}{8} = 1$

5. $\dfrac{x^2}{16} + \dfrac{(y+4)^2}{6} = 1$

6. $\dfrac{(x-5)^2}{15} + \dfrac{y^2}{25} = 1$

7. $9x^2 + 4y^2 - 16y = 20$

8. $x^2 + 16y^2 + 6x = 7$

9. $9x^2 + 16y^2 - 18x + 96y + 9 = 0$

10. $16x^2 + 9y^2 + 64x - 18y - 71 = 0$

11. $x^2 + 4y^2 + 4x - 16y + 4 = 0$

12. $2x^2 + y^2 - 16x + 6y + 11 = 0$

13. $6x^2 + 5y^2 - 12x + 20y - 4 = 0$

14. $5x^2 + 8y^2 - 20x + 16y - 12 = 0$

15. $8x^2 + y^2 - 48x + 4y + 68 = 0$

16. $x^2 + 10y^2 + 4x + 20y + 4 = 0$

In Exercises 17–22 write an equation for the ellipse with the properties given.

17. Center at $(1, 6)$, $a = 3$, $b = 2$, major axis horizontal

18. Center at $(2, 3)$, $a = 4$, $b = 3$, major axis vertical

19. Vertices at $(3, 2)$ and $(-7, 2)$, minor axis of length 6

20. Covertices at $(3, 7)$ and $(3, -1)$, major axis of length 10

21. Vertices at $(-4, 9)$ and $(-4, -3)$, covertices at $(-7, 3)$ and $(-1, 3)$

22. Vertices at $(-3, -5)$ and $(9, -5)$, covertices at $(3, 0)$ and $(3, -10)$

Graph each hyperbola in Exercises 25–38.

23. $\dfrac{(x - 4)^2}{9} - \dfrac{(y + 2)^2}{16} = 1$

24. $\dfrac{(y + 4)^2}{25} - \dfrac{(x - 3)^2}{4} = 1$

25. $\dfrac{x^2}{4} - \dfrac{(y - 3)^2}{8} = 1$ 26. $\dfrac{y^2}{9} - \dfrac{(x + 4)^2}{12} = 1$

27. $\dfrac{(y + 2)^2}{6} - \dfrac{(x + 2)^2}{10} = 1$

28. $\dfrac{(x - 4)^2}{5} - \dfrac{(y - 4)^2}{8} = 1$

29. $\dfrac{y^2}{6} - \dfrac{(x - 3)^2}{15} = 1$ 30. $\dfrac{x^2}{12} - \dfrac{(y + 2)^2}{7} = 1$

31. $9x^2 - 4y^2 - 36x - 24y - 36 = 0$

32. $9y^2 - 4x^2 - 72y - 24x + 72 = 0$

33. $16y^2 - 4x^2 + 32x - 128 = 0$

34. $16x^2 - 9y^2 + 54y - 225 = 0$

35. $4x^2 - 6y^2 - 32x - 24y + 16 = 0$

36. $9y^2 - 8x^2 + 72y + 16x + 64 = 0$

37. $12x^2 - 3y^2 + 24y - 84 = 0$

38. $10y^2 - 5x^2 + 30x - 95 = 0$

In Exercises 39–42 find an equation for the hyperbola with the properties given.

39. Center at $(-1, 5)$, $a = 8$, $b = 6$, transverse axis vertical

40. Center at $(6, -2)$, $a = 1$, $b = 4$, transverse axis horizontal

41. One vertex at $(-1, 3)$, one end of the horizontal conjugate axis at $(-5, 1)$

42. One vertex at $(1, -2)$, one end of the vertical conjugate axis at $(-5, -4)$

Graph each parabola in Exercises 43–54.

43. $2x = (y + 3)^2$ 44. $-3y = (x - 2)^2$

45. $-6(y + 4) = (x - 3)^2$

46. $4(x - 5) = (y + 1)^2$

47. $(y - 4)^2 + 3 = x$

48. $(x + 3)^2 - 2 = y$

49. $y^2 - 6y + 10x + 4 = 0$

50. $y^2 - 4y + 8x + 6 = 0$

51. $4x^2 - 4x = 23 - 12y$

52. $4x^2 + 4x = 8y - 5$

53. $9y^2 = 6y + 12x - 1$

54. $9y^2 + 12y - 12x = 0$

In Exercises 55–62 find an equation for the parabola with the properties given.

55. Vertex at $(1, 2)$, focus at $(1, 1)$

56. Vertex at $(-1, 3)$, focus at $(-1, 5)$

57. Vertex at $(-4, -2)$, focus at $(-2, -2)$

58. Vertex at $(6, -3)$, focus at $(3, -3)$

59. Vertex at $(2, 5)$, directrix the line $x = 3$

60. Vertex at $(-4, 1)$, directrix the line $y = 3$

61. Focus at $(3, -2)$, directrix the line $y = 2$

62. Focus at $(-2, -2)$, directrix the line $x = -6$

In Exercises 63–82 name the graph of each equation and describe its main features.

63. $y^2 = 4 - x^2$ 64. $y^2 = 6 - 4x^2$

65. $4y^2 = x^2 - 8$ 66. $x^2 + 2y - 4 = 0$

67. $4x^2 = 12 - 2y^2$

68. $6x^2 = 8 - 6y^2$

69. $4x^2 = 6 + 4y$

70. $2x^2 = 5 + 4y^2$

71. $6 + \dfrac{x^2}{4} = y^2$

72. $y^2 = 6 - \dfrac{2x^2}{3}$

73. $\dfrac{1}{2}y^2 - x = 4$

74. $\dfrac{x^2}{4} = 4 + 6y^2$

75. $y = (x - 3)^2 + 2$

76. $\dfrac{(y - 2)^2}{4} - \dfrac{(x + 3)^2}{8} = 1$

77. $(x + 1)^2 + (y - 4)^2 = 16$

78. $\dfrac{(x + 3)^2}{4} + \dfrac{y^2}{12} = 1$

79. $2x^2 + y^2 + 4x = 2$

80. $y^2 - 4x^2 + 2y - x = 0$

81. $x^2 + 6x = 4 - y^2$

82. $y - 2 = \dfrac{(x + 4)^2}{4}$

10.4   INVERSE FUNCTIONS

It is often convenient to think of a function as a machine or process that acts on the elements of the domain (the "input" values) and produces the elements of the range (the "output" values). (See Figure 10.47.) For example, from the input values 1, 2, 3, and 4, the function $f(x) = 2x$ produces the output values 2, 4, 6, and 8, respectively. This is just another way of saying that $f(1) = 2$, $f(2) = 4$, $f(3) = 6$, and $f(4) = 8$.

The **inverse** of a function f is obtained by running the function machine in reverse: the elements of the *range* of f are used as the input values, and the output values produced by the machine are the corresponding elements of the domain of f. If g is the inverse of the function f above, then g turns 2 into 1, 4 into 2, 6 into 3, and 8 into 4. (See Figure 10.48.) In other words, $g(2) = 1$, $g(4) = 2$, $g(6) = 3$, and $g(8) = 4$. Thus the domain of g is the range of f, and the range of g is the domain of f.

Notice that if (a, b) is an ordered pair obtained from the function f, then (b, a) will be an ordered pair from the inverse function g. The function $f(x) = 2x$ produces the ordered pairs

$$(1, 2), \quad (2, 4), \quad (3, 6), \quad \text{and} \quad (4, 8)$$

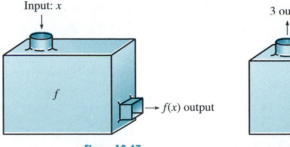

Figure 10.47

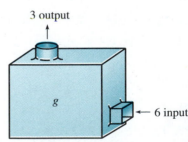

Figure 10.48

and its inverse function g produces the ordered pairs

$$(2, 1), \quad (4, 2), \quad (6, 3), \quad \text{and} \quad (8, 4)$$

To obtain the ordered pairs of the inverse function g, we need only interchange the components of the ordered pairs of f.

   This observation suggests that the inverse of a function defined by an equation can be obtained by interchanging the variables in the equation. Thus if the function f is defined by the equation $\quad y = 2x$, then an equation for the inverse function g is found by interchanging the variables x and y to obtain $\quad x = 2y$. Solving this new equation for y, we find

$$y = \frac{x}{2} \quad \text{or} \quad g(x) = \frac{x}{2}.$$

Example 1

Find the inverse of the function $\quad f(x) = 4x - 3$.

Solution   Write the equation for f in the form

$$y = 4x - 3$$

Interchange the variables to obtain

$$x = 4y - 3$$

and solve for y:

$$y = \frac{x}{4} + \frac{3}{4}$$

The inverse function is $\quad g(x) = \frac{x}{4} + \frac{3}{4}$

   The inverse function g "undoes" the effect of the function f. For example, the function $\quad f(x) = 2x \quad$ doubles the elements of its domain, and its inverse function, $\quad g(x) = \frac{x}{2}, \quad$ halves them. If we apply the function f to a given input value and then apply the function g to the output from f, the end result will be the original input value. This is like feeding a number into the function machine and then feeding the result back in with the machine running in reverse. For the function $\quad f(x) = 2x \quad$ and its inverse, $\quad g(x) = \frac{x}{2}, \quad$ if we choose $\quad x = 5 \quad$ as an input value, we find that $\quad f(5) = 2(5) = 10, \quad$ and $g(10) = \dfrac{10}{2} = 5.$ (See Figure 10.49 on page 606.)

Example 2

a.   Find the inverse of the function $\quad f(x) = x^3 + 2$.
b.   Show that the inverse function "undoes" the effect of f on $\quad x = 2$.
c.   Show that f "undoes" the effect of the inverse function on $\quad x = -25$.

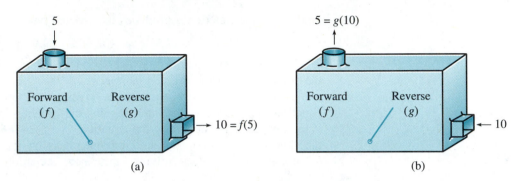

(a)   (b)

Figure 10.49

Solutions   **a.**   Write the equation for f in the form

$$y = x^3 + 2$$

Interchange the variables to obtain

$$x = y^3 + 2$$

and solve for y:

$$y^3 = x - 2$$

$$y = \sqrt[3]{x - 2}$$

The inverse function is $g(x) = \sqrt[3]{x - 2}.$

b. $f(2) = 2^3 + 2 = 10$ and $g(10) = \sqrt[3]{10 - 2} = 2.$

c. $g(-25) = \sqrt[3]{-25 - 2} = -3$ and $f(-3) = (-3)^3 + 2 = -25.$

Graph of the Inverse   The graphs of a function and its inverse are related in an interesting way. To see this relationship, we first observe in Figure 10.50 that the graphs of the ordered pairs (a, b) and (b, a) are always located symmetrically with respect to the graph of $y = x.$ Now, for every ordered pair (a, b) in f, the ordered pair (b, a) is in the inverse of f. Thus the graphs of $y = f(x)$ and its inverse, $y = g(x),$ are reflections of each other about the line

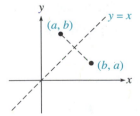

Figure 10.50

$y = x$.   Figure 10.51 shows the graphs of $y = 4x - 3$ and its inverse,
$y = \dfrac{x}{4} + \dfrac{3}{4}$,   from Example 1, along with the graph of $y = x$.

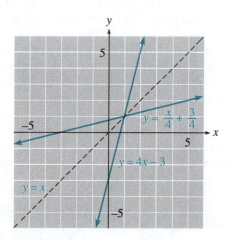

Figure 10.51

Example 3

Graph the function

$$f(x) = x^3 + 2$$

and its inverse,

$$g(x) = \sqrt[3]{x - 2}$$

which we found in Example 2, on the same set of axes.

Solution   The graph of f is the graph of $y = x^3$ translated two units upward.
The graph of g is the graph of $y = \sqrt[3]{x}$ translated two units to the right.
The two graphs are symmetric bout the line $y = x$. (See Figure 10.52.)

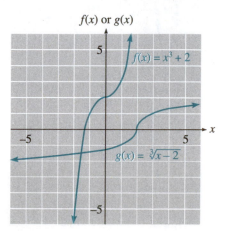

Figure 10.52

Horizontal Line Test   In some cases the inverse of a function is not itself a function. For example, consider the function $f(x) = x^2$, or $y = x^2$. To find its inverse, we first interchange x and y to obtain $x = y^2$ and then solve for y to get $y = \pm\sqrt{x}$. The graphs of f and its inverse are shown in Figure 10.53. Since the graph of the inverse does not pass the vertical line test, it is *not* a function.

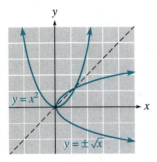

Figure 10.53

For many applications it is important to know whether or not the inverse of f is a function. This can be determined from the graph of f. Note that when we interchange x and y to find a formula for the inverse, horizontal lines of the form $y = k$ become vertical lines of the form $x = k$. Thus if the graph of the *inverse* is to pass the vertical line test, then the graph of the *original function* must pass the horizontal line test, namely, that no horizontal line should intersect the graph in more than one point. Notice that the graph of $f(x) = x^2$ does *not* pass the horizontal line test, so we would not expect its inverse to be a function.

> **HORIZONTAL LINE TEST**
>
> If no horizontal line intersects the graph of a function more than once, then the inverse is also a function.

Example 4   Which of the functions in Figure 10.54 have inverses that are also functions?

Solutions   In each case, to determine whether the inverse is a function, apply the horizontal line test. Since no horizontal line intersects their graphs more than once, the functions pictured in Figure 10.54(a) and (c) have inverses that are also functions.

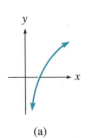

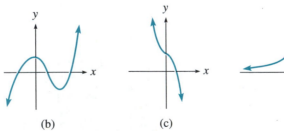

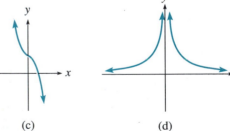

Figure 10.54   (a)   (b)   (c)   (d)

Inverse Notation   If the inverse of a function f is also a function, then the inverse is often denoted by the symbol f^{-1}, read "f inverse." For example, since the function $f(x) = x^3 + 2$ passes the horizontal line test (see Example 3), its inverse is a function and can be denoted by $f^{-1}(x) = \sqrt[3]{x - 2}$.

Example 5

If $h(x) = 2x - 6$, find $h^{-1}(10)$.

Solution   First find the inverse function for $y = 2x - 6$. Interchange x and y to get

$$x = 2y - 6$$

and solve for y:

$$2y = x + 6$$

$$y = \frac{x}{2} + 3$$

The inverse function is

$$h^{-1}(x) = \frac{x}{2} + 3$$

Now evaluate the inverse function at $x = 10$:

$$h^{-1}(10) = \frac{10}{2} + 3 = 8$$

COMMON ERROR

Although the same symbol, $^{-1}$, is used for both reciprocals and inverse functions, the two notions are not equivalent. In Example 5 note that $h^{-1}(x)$ does *not* indicate $\dfrac{1}{2x - 6}$. To avoid confusion, we use the notation $\dfrac{1}{h}$ to refer to the reciprocal of the function h.

Exercise 10.4

Find the inverse of each function in Exercises 1–16.

1. $f(x) = x + 2$

2. $f(x) = x - 3$

3. $f(x) = 2x$

4. $f(x) = \dfrac{x}{5}$

5. $f(x) = 2x - 6$

6. $f(x) = 3x + 1$

7. $f(x) = \dfrac{3 - x}{2}$

8. $f(x) = \dfrac{5 - x}{3}$

9. $f(x) = x^3 + 1$

10. $f(x) = x^3 - 8$

11. $f(x) = \sqrt[3]{x}$

12. $f(x) = \dfrac{1}{x}$

13. $f(x) = \dfrac{1}{x - 1}$

14. $f(x) = \sqrt[3]{x + 1}$

15. $f(x) = \sqrt[3]{x} + 4$

16. $f(x) = \dfrac{1}{x} - 3$

17. a. Find the inverse g of the function $f(x) = (x - 2)^3$.
b. Show that g "undoes" the effect of f on $x = 4$.
c. Show that f "undoes" the effect of g on $x = -8$.

18. a. Find the inverse g of the function $f(x) = \dfrac{2}{x + 1}$.
b. Show that g "undoes" the effect of f on $x = 3$.
c. Show that f "undoes" the effect of g on $x = -1$.

For Exercises 19–34 graph the function and its inverse on the same set of axes, along with the graph of y = x, for each of Exercises 1–16.

Which of the functions in Exercises 35–40 have inverses that are also functions?

35. a.

b.

c.
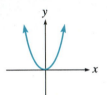

d.

36. a.
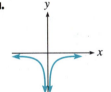

b.

c.
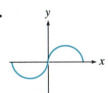

d.
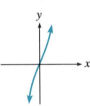

37. a. $f(x) = x$ **b.** $f(x) = x^2$
38. a. $f(x) = x^3$ **b.** $f(x) = x^4$
39. a. $f(x) = \dfrac{1}{x}$ **b.** $f(x) = \dfrac{1}{x^2}$
40. a. $f(x) = \sqrt{x}$ **b.** $f(x) = \sqrt[3]{x}$

41. If $F(t) = \dfrac{2}{3}t + 1$, find $F^{-1}(5)$.

42. If $G(s) = \dfrac{s - 3}{4}$, find $G^{-1}(-2)$.

43. If $m(v) = 6 - \dfrac{2}{v}$, find $m^{-1}(-3)$.

44. If $p(z) = 1 - 2z^3$, find $p^{-1}(7)$.

45. $f(x) = \dfrac{x + 2}{x - 1}$.   Find $f^{-1}(x)$.

46. $f(x) = \dfrac{3x + 1}{x - 3}$.   Find $f^{-1}(x)$.

47. $f(x) = x^3 + x + 1$.   Find
   a. $f^{-1}(1)$ **b.** $f^{-1}(3)$

48. $f(x) = x^5 + x^3 + 7$.   Find
   a. $f^{-1}(7)$ **b.** $f^{-1}(5)$

49. If $f(-1) = 0$, $f(0) = 1$, $f(1) = -2$, and $f(2) = -1$,   find
   a. $f^{-1}(1)$ **b.** $f^{-1}(-1)$

50. If $f^{-1}(-2) = 1$, $f^{-1}(-1) = -2$, $f^{-1}(0) = 0$, and $f^{-1}(1) = -1$,   find
   a. $f(-1)$ **b.** $f(1)$

51. Let $f(x) = 2^x$.
   a. Fill in the table for f.
   b. Graph f.
   c. By interchanging the components of the ordered pairs in (a), make a table for f^{-1}.
   d. Graph f^{-1} on the same set of axes with f.

x	$f(x)$
-2	
-1	
0	
1	
2	

52. Repeat Exercise 51 with $f(x) = \left(\dfrac{1}{2}\right)^x$.

10.5 SOLUTION OF LINEAR SYSTEMS USING MATRICES

In Chapter 3 we solved linear systems by using linear combinations and Gaussian reduction. In this section we consider another mathematical tool called a matrix (plural: matrices) that has wide application in mathematics. In particular, we will see how matrices can be used to solve linear systems.

A **matrix** is a rectangular array of elements or **entries**. (In this book the entries will be real numbers.) These entries are ordinarily displayed in rows and columns, and the entire matrix is enclosed in brackets or parentheses. Thus

$$\begin{bmatrix} 1 & 2 & 3 \\ 4 & 5 & 6 \\ 7 & 8 & 9 \end{bmatrix}, \qquad \begin{bmatrix} 2 & -1 & 3 \\ 4 & 0 & 2 \end{bmatrix}, \qquad \text{and} \qquad \begin{bmatrix} 4 \\ 5 \\ 6 \end{bmatrix}$$

are matrices. A matrix of **order**, or **dimension**, $n \times m$ (read "n by m") has n (horizontal) rows and m (vertical) columns. The matrices above are 3×3, 2×3, and 3×1, respectively. The first matrix—in which the number of rows is equal to the number of columns—is an example of a **square matrix**.

Coefficient Matrix and Augmented Matrix of a System

For a system of linear equations of the form

$$a_1 x + b_1 y + c_1 z = d_1$$
$$a_2 x + b_2 y + c_2 z = d_2$$
$$a_3 x + b_3 y + c_3 z = d_3$$

the matrices

$$\begin{bmatrix} a_1 & b_1 & c_1 \\ a_2 & b_2 & c_2 \\ a_3 & b_3 & c_3 \end{bmatrix} \qquad \text{and} \qquad \left[\begin{array}{ccc|c} a_1 & b_1 & c_1 & d_1 \\ a_2 & b_2 & c_2 & d_2 \\ a_3 & b_3 & c_3 & d_3 \end{array} \right]$$

are called the **coefficient matrix** and the **augmented matrix**, respectively. Notice that each *row* of the augmented matrix represents one of the equations of the system. For example, the augmented matrix of the system

$$\begin{array}{rrrr} 3x & - & 4y & + & z & = & 2 \\ -x & + & 2y & & & = & -1 \\ 2x & - & y & - & 3z & = & 4 \end{array} \qquad \text{is} \qquad \left[\begin{array}{ccc|c} 3 & -4 & 1 & 2 \\ -1 & 2 & 0 & -1 \\ 2 & -1 & -3 & 4 \end{array} \right]$$

and the augmented matrix of the system

$$\begin{array}{rrrr} x & - & 3y & + & 2z & = & 5 \\ & & 2y & - & z & = & 4 \\ & & & & 4z & = & 8 \end{array} \qquad \text{is} \qquad \left[\begin{array}{ccc|c} 1 & -3 & 2 & 5 \\ 0 & 2 & -1 & 4 \\ 0 & 0 & 4 & 8 \end{array} \right]$$

The augmented matrix of this last system, which has all zero entries in the lower left corner, is said to be in **upper triangular form**. As we saw in Section 3.3, it is easy to find the solution of such a system by back-substitution. If we can change a system of linear equations into an equivalent system in upper triangular form, we can then use back-substitution to find the solution of the system.

Elementary Transformations

In Section 3.2 we used the two properties on pages 149 and 150 to change a given linear system into an equivalent one, that is, one that has the same solution as the original system. The properties allowed us to perform the following operations on the equations of a system.

1. Multiply both sides of an equation by a nonzero real number.

2. Add a constant multiple of one equation to another equation.

Clearly, we can also perform the following operation without changing the solution of the system.

3. Interchange two equations.

Since each equation of the system corresponds to a row in the augmented matrix for the system, the three operations above correspond to the following **elementary row operations** on the augmented matrix.

> **ELEMENTARY ROW OPERATIONS**
> **1.** Multiply the entries of any row by a nonzero number.
> **2.** Add a constant multiple of one row to another.
> **3.** Interchange two rows.

Example 1

a. $A = \begin{bmatrix} 1 & 3 & -1 & -1 \\ 2 & 1 & 4 & 5 \\ 6 & 2 & -1 & -12 \end{bmatrix}$ and $B = \begin{bmatrix} 1 & 3 & -1 & -1 \\ 6 & 3 & 12 & 15 \\ 6 & 2 & -1 & -12 \end{bmatrix}$

represent equivalent systems because we can multiply each entry in the second row of A by 3 to obtain B.

b. $A = \begin{bmatrix} 3 & -1 & 2 & 7 \\ 2 & 1 & 4 & -5 \\ 3 & 1 & 9 & -16 \end{bmatrix}$ and $B = \begin{bmatrix} 3 & 1 & 9 & -16 \\ 2 & 1 & 4 & -5 \\ 3 & -1 & 2 & 7 \end{bmatrix}$

represent equivalent systems because we can interchange the first and third rows of A to obtain B.

c. $A = \begin{bmatrix} 1 & 2 & 1 & -3 \\ 2 & 0 & -1 & 7 \\ 3 & 1 & 2 & 10 \end{bmatrix}$ and $B = \begin{bmatrix} 1 & 2 & 1 & -3 \\ 0 & -4 & -3 & 13 \\ 3 & 1 & 2 & 10 \end{bmatrix}$

represent equivalent systems because we can add -2 times each entry of the first row of A to the corresponding entry of the second row of A to obtain B.   ☐

Example 2

Use row operations to form an equivalent matrix with the given elements:

$$\begin{bmatrix} 1 & -4 & -5 \\ 3 & 6 & 3 \end{bmatrix} \quad \rightarrow \quad \begin{bmatrix} 1 & -4 & -5 \\ 0 & ? & ? \end{bmatrix}$$

Solution   To obtain 0 as the first entry in the second row, add -3(row 1) to row 2:

$$-3(\text{row 1}) + \text{row 2} \begin{bmatrix} 1 & -4 & -5 \\ 3 & 6 & 3 \end{bmatrix} \quad \rightarrow \quad \begin{bmatrix} 1 & -4 & -5 \\ 0 & 18 & 18 \end{bmatrix} \quad ☐$$

It is often convenient to perform more than one row operation at a time on a given matrix.

Example 3

Use row operations on the first matrix to form an equivalent matrix with the given elements:

$$\begin{bmatrix} 1 & -3 & 1 & -4 \\ 3 & -1 & -1 & 8 \\ 2 & -2 & 3 & -1 \end{bmatrix} \quad \rightarrow \quad \begin{bmatrix} 1 & -3 & 1 & -4 \\ 0 & ? & ? & ? \\ 0 & 0 & ? & ? \end{bmatrix}$$

Solution   Perform the transformation in two steps: first obtain zeros in the second and third entries of the first column by adding suitable multiples of the first row to the second and third rows:

$$\begin{matrix} \\ -3(\text{row 1}) + \text{row 2} \\ -2(\text{row 1}) + \text{row 3} \end{matrix} \begin{bmatrix} 1 & -3 & 1 & -4 \\ 3 & -1 & -1 & 8 \\ 2 & -2 & 3 & -1 \end{bmatrix} \quad \rightarrow \quad \begin{bmatrix} 1 & -3 & 1 & -4 \\ 0 & 8 & -4 & 20 \\ 0 & 4 & 1 & 7 \end{bmatrix}$$

Now obtain a zero as the second entry of the third row by adding $-\frac{1}{2}$ (row 2) to row 3:

$$-\frac{1}{2}(\text{row 2}) + \text{row 3} \begin{bmatrix} 1 & -3 & 1 & -4 \\ 0 & 8 & -4 & 20 \\ 0 & 4 & 1 & 7 \end{bmatrix} \quad \rightarrow \quad \begin{bmatrix} 1 & -3 & 1 & -4 \\ 0 & 8 & -4 & 20 \\ 0 & 0 & 3 & -3 \end{bmatrix}$$

Note that the last matrix is in upper triangular form.   ☐

Gaussian Reduction   It is now a simple matter to adapt the method of Gaussian reduction, which we learned in Section 3.3, to the augmented matrix of a linear system. This method can be used to solve linear systems of any size and is well suited for implementation as a computer program.

The method has three steps.

> **SOLVING A LINEAR SYSTEM WITH GAUSSIAN REDUCTION**
> **1.** Write the augmented matrix for the system.
> **2.** Using row operations, transform the matrix into an equivalent one in upper triangular form.
> **3.** Use back-substitution to find the solution to the system.

Example 4   Use Gaussian reduction to solve the system

$$x - 2y = -5$$
$$2x + 3y = 11$$

Solution   The augmented matrix is

$$\begin{bmatrix} 1 & -2 & -5 \\ 2 & 3 & 11 \end{bmatrix}$$

Use row operations to obtain 0 in the first entry of the second row:

$$-2(\text{row } 1) + \text{row } 2 \begin{bmatrix} 1 & -2 & -5 \\ 2 & 3 & 11 \end{bmatrix} \rightarrow \begin{bmatrix} 1 & -2 & -5 \\ 0 & 7 & 21 \end{bmatrix}$$

The last matrix corresponds to the system

$$x - 2y = -5 \qquad (1)$$
$$7y = 21 \qquad (2)$$

Use back-substitution to solve the system. From Equation (2), $y = 3$.   Substitute 3 for y in Equation (1) to find

$$x - 2(3) = -5$$
$$x = 1$$

The solution is the ordered pair $(1, 3)$.

To transform the augmented matrix of a 3×3 system into upper triangular form, we can use the following sequence of row operations.

1. Obtain zeros in the *first* entries of the second and third rows by adding suitable multiples of the *first* row to the second and third rows.
2. Obtain a zero in the *second* entry of the third row by adding a suitable multiple of the *second* row to the third row.

Example 5   Use Gaussian reduction to solve the system

$$2x - 4y \quad\;\;\; = \quad\; 6$$
$$3x - 4y + z = \quad\; 8$$
$$2x \quad\quad\; - 3z = -11$$

Solution   The augmented matrix for the system is

$$\begin{bmatrix} 2 & -4 & 0 & 6 \\ 3 & -4 & 1 & 8 \\ 2 & 0 & -3 & -11 \end{bmatrix}$$

For each transformation of the augmented matrix, we show the corresponding system of equations. At each step the new system is equivalent to the original one. Begin by obtaining a 1 in the first entry of the first row by multiplying each entry of the first row by $\frac{1}{2}$ (this will make it easier to obtain zeros in the first entries of the second and third rows):

$$\tfrac{1}{2}(\text{row } 1) \begin{bmatrix} 2 & -4 & 0 & 6 \\ 3 & -4 & 1 & 8 \\ 2 & 0 & -3 & -11 \end{bmatrix} \quad \begin{aligned} 2x - 4y + 0z &= \quad 6 \\ 3x - 4y + z &= \quad 8 \\ 2x + 0y - 3z &= -11 \end{aligned}$$

$$\downarrow$$

$$\begin{bmatrix} 1 & -2 & 0 & 3 \\ 3 & -4 & 1 & 8 \\ 2 & 0 & -3 & -11 \end{bmatrix} \quad \begin{aligned} x - 2y + 0z &= \quad 3 \\ 3x - 4y + z &= \quad 8 \\ 2x + 0y - 3z &= -11 \end{aligned}$$

Next obtain zeros in the first entries of the second and third rows by adding suitable multiples of the first row to the second and third rows:

$$\begin{matrix} \\ -3(\text{row } 1) + \text{row } 2 \\ -2(\text{row } 1) + \text{row } 3 \end{matrix} \begin{bmatrix} 1 & -2 & 0 & 3 \\ 3 & -4 & 1 & 8 \\ 2 & 0 & -3 & -11 \end{bmatrix} \quad \begin{aligned} x - 2y + 0z &= \quad 3 \\ 3x - 4y + z &= \quad 8 \\ 2x + 0y - 3z &= -11 \end{aligned}$$

$$\downarrow$$

$$\begin{bmatrix} 1 & -2 & 0 & 3 \\ 0 & 2 & 1 & -1 \\ 0 & 4 & -3 & -17 \end{bmatrix} \quad \begin{aligned} x - 2y + 0z &= \quad 3 \\ 0x + 2y + z &= -1 \\ 0x + 4y - 3z &= -17 \end{aligned}$$

Finally, obtain a zero in the second entry of the third row by adding $-2(\text{row } 2)$ to row 3:

$$\begin{matrix} \\ \\ -2(\text{row } 2) + \text{row } 3 \end{matrix} \begin{bmatrix} 1 & -2 & 0 & 3 \\ 0 & 2 & 1 & -1 \\ 0 & 4 & -3 & -17 \end{bmatrix} \quad \begin{aligned} x - 2y + 0z &= \quad 3 \\ 0x + 2y + z &= -1 \\ 0x + 4y - 3z &= -17 \end{aligned}$$

$$\downarrow$$

$$\begin{bmatrix} 1 & -2 & 0 & 3 \\ 0 & 2 & 1 & -1 \\ 0 & 0 & -5 & -15 \end{bmatrix} \quad \begin{aligned} x - 2y + 0z &= \quad 3 \\ 0x + 2y + z &= -1 \\ 0x + 0y - 5z &= -15 \end{aligned}$$

The system is now in upper triangular form; use back-substitution to find the solution. Solve the last equation to get $z = 3$ and substitute 3 for z in the second equation to find $y = -2$. Finally, substitute 3 for z and -2 for y in the first equation to find $x = -1$. The solution is the ordered triple $(-1, -2, 3)$.

If any step in this procedure results in the equation $0x + 0y + 0z = 0$, or in a contradiction such as $0x + 0y + 0z = d$, $d \neq 0$, then the system is either dependent or inconsistent, respectively. In either case the system does not have a unique solution.   ▭

Exercise 10.5

Perform the given elementary row operation on the matrices in Exercises 1–8.

1. Multiply row 2 by -3:

$$\begin{bmatrix} -2 & 1 & | & 0 \\ 3 & -1 & | & 2 \end{bmatrix}$$

2. Multiply row 1 by $\dfrac{1}{4}$:

$$\begin{bmatrix} 2 & 0 & | & 3 \\ -1 & 5 & | & 4 \end{bmatrix}$$

3. Add 2(row 1) to row 2:

$$\begin{bmatrix} 1 & -3 & | & 6 \\ -2 & 4 & | & -1 \end{bmatrix}$$

4. Add -3(row 1) to row 2:

$$\begin{bmatrix} 1 & -4 & | & 8 \\ 3 & -2 & | & 10 \end{bmatrix}$$

5. Interchange row 1 and row 3:

$$\begin{bmatrix} 0 & -3 & 2 & | & -3 \\ 2 & 6 & -1 & | & 4 \\ 1 & 0 & -2 & | & 5 \end{bmatrix}$$

6. Interchange row 2 and row 3:

$$\begin{bmatrix} 1 & 6 & 0 & | & -2 \\ 0 & 0 & 5 & | & -10 \\ 0 & 3 & -2 & | & 8 \end{bmatrix}$$

7. Add -4(row 1) to row 3:

$$\begin{bmatrix} 1 & 2 & 1 & | & -5 \\ 0 & 4 & -2 & | & 3 \\ 4 & -1 & 6 & | & -8 \end{bmatrix}$$

8. Add 2(row 2) to row 3:

$$\begin{bmatrix} 1 & -7 & 5 & | & 2 \\ 0 & 1 & -3 & | & -1 \\ 0 & -2 & -3 & | & 4 \end{bmatrix}$$

In Exercises 9–18 use row operations on the first matrix to form an equivalent matrix with the given elements.

9. $\begin{bmatrix} 1 & -3 & | & 2 \\ 2 & 1 & | & 4 \end{bmatrix} \rightarrow \begin{bmatrix} 1 & -3 & | & 2 \\ 0 & ? & | & ? \end{bmatrix}$

10. $\begin{bmatrix} -2 & 3 & | & 0 \\ 4 & 1 & | & 6 \end{bmatrix} \rightarrow \begin{bmatrix} -2 & 3 & | & 0 \\ 0 & ? & | & ? \end{bmatrix}$

11. $\begin{bmatrix} 2 & 6 & | & -4 \\ 5 & 3 & | & 1 \end{bmatrix} \rightarrow \begin{bmatrix} 2 & 6 & | & -4 \\ ? & 0 & | & ? \end{bmatrix}$

12. $\begin{bmatrix} 6 & 4 & | & -2 \\ -1 & -2 & | & -3 \end{bmatrix} \rightarrow \begin{bmatrix} 6 & 4 & | & -2 \\ ? & 0 & | & ? \end{bmatrix}$

13. $\begin{bmatrix} 1 & -2 & 2 & | & 1 \\ 2 & 3 & -1 & | & 6 \\ 4 & 1 & -3 & | & 3 \end{bmatrix} \rightarrow \begin{bmatrix} 1 & -2 & 2 & | & 1 \\ 0 & ? & ? & | & ? \\ 0 & ? & ? & | & ? \end{bmatrix}$

14. $\begin{bmatrix} 2 & -1 & 3 & | & -1 \\ -4 & 0 & 4 & | & 5 \\ 6 & 2 & -1 & | & -2 \end{bmatrix} \rightarrow \begin{bmatrix} 2 & -1 & 3 & | & -1 \\ 0 & ? & ? & | & ? \\ 0 & ? & ? & | & ? \end{bmatrix}$

15. $\begin{bmatrix} -1 & 4 & 3 & | & 2 \\ 2 & -2 & -4 & | & 6 \\ 1 & 2 & 3 & | & -3 \end{bmatrix} \rightarrow \begin{bmatrix} -1 & 4 & 3 & | & 2 \\ ? & 0 & ? & | & ? \\ ? & 0 & ? & | & ? \end{bmatrix}$

16. $\begin{bmatrix} 3 & -2 & 4 & | & -4 \\ 2 & 2 & 1 & | & 2 \\ -1 & 1 & 5 & | & -1 \end{bmatrix} \rightarrow \begin{bmatrix} 3 & -2 & 4 & | & -4 \\ ? & ? & 0 & | & ? \\ ? & ? & 0 & | & ? \end{bmatrix}$

17. $\begin{bmatrix} -2 & 1 & -3 & | & -2 \\ 4 & 2 & 0 & | & 2 \\ 6 & -1 & 2 & | & 0 \end{bmatrix} \rightarrow \begin{bmatrix} -2 & 1 & -3 & | & -2 \\ 0 & ? & ? & | & ? \\ 0 & 0 & ? & | & ? \end{bmatrix}$

18. $\begin{bmatrix} -1 & 2 & 3 & | & 3 \\ 4 & 0 & 1 & | & -6 \\ 2 & 2 & -3 & | & -2 \end{bmatrix} \rightarrow \begin{bmatrix} -1 & 2 & 3 & | & 3 \\ 0 & ? & ? & | & ? \\ 0 & 0 & ? & | & ? \end{bmatrix}$

Use Gaussian reduction on the augmented matrix to solve each system in Exercises 19–26.

19. $x + 3y = 11$
$2x - y = 1$

20. $x - 5y = 11$
$2x + 3y = -4$

21. $x - 4y = -6$
$3x + y = -5$

22. $x + 6y = -14$
$5x + 3y = -4$

23. $2x + y = 5$
$3x - 5y = 14$

24. $3x - 2y = 16$
$4x + 2y = 12$

25. $x - y = -8$
$x + 2y = 9$

26. $4x - 3y = 16$
$2x + y = 8$

Use Gaussian reduction to solve each system in Exercises 27–34.

27. $x + 3y - z = 5$
$3x - y + 2z = 5$
$x + y + 2z = 7$

28. $x - 2y + 3z = -11$
$2x + 3y - z = 6$
$3x - y - z = 3$

29. $2x - y + z = 5$
$x - 2y - 2z = 2$
$3x + 3y - z = 4$

30. $x - 2y - 2z = 4$
$2x + y - 3z = 7$
$x - y - z = 3$

31. $2x - y - z = -4$
$x + y + z = -5$
$x + 3y - 4z = 12$

32. $x - 2y - 5z = 2$
$2x + 3y + z = 11$
$3x - y - z = 11$

33. $2x - y = 0$
$3y + z = 7$
$2x + 3z = 1$

34. $3x - z = 7$
$2x + y = 6$
$3y - z = 7$

Chapter 10 Review

Graph each function in Exercises 1–10.

1. $g(x) = |x| + 2$

2. $F(t) = \dfrac{1}{t} - 2$

3. $f(s) = \sqrt{s} + 3$

4. $h(r) = (r - 2)^3$

5. $f(x) = (x - 2)^2 - 4$

6. $g(u) = \sqrt{u + 2} - 3$

7. $G(t) = |t + 2| - 3$

8. $H(t) = \dfrac{1}{(t - 2)^2} + 3$

9. $h(s) = -2\sqrt{s}$

10. $g(s) = \dfrac{1}{2|s|}$

Give an equation for the function graphed in Exercises 11 and 12.

11.

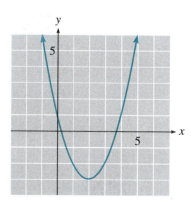

12.

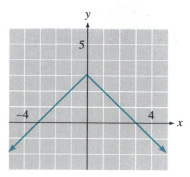

In Exercises 13–34 graph each conic section and describe its main features.

13. $x^2 + y^2 = 9$

14. $(x - 2)^2 + (y + 3)^2 = 16$

15. $x^2 + y^2 - 4x + 2y - 4 = 0$

16. $x^2 + y^2 - 6y - 4 = 0$

17. $\dfrac{x^2}{9} + y^2 = 1$

18. $\dfrac{x^2}{4} + \dfrac{y^2}{16} = 1$

19. $\dfrac{(x - 2)^2}{4} + \dfrac{(y + 3)^2}{9} = 1$

20. $\dfrac{(x + 4)^2}{12} + \dfrac{(y - 2)^2}{6} = 1$

21. $4x^2 + y^2 - 16x + 4y + 4 = 0$

22. $8x^2 + 5y^2 + 16x - 20y - 12 = 0$

23. $4(y - 2) = (x + 3)^2$

24. $(x - 2)^2 + 4y = 4$

25. $x^2 - 8x - y + 6 = 0$

26. $y^2 + 6y + 4x + 1 = 0$

27. $x^2 + y = 4x - 6$

28. $y^2 = 2y + 2x + 2$

29. $\dfrac{y^2}{6} - \dfrac{x^2}{8} = 1$

30. $\dfrac{(x - 2)^2}{4} - \dfrac{(y + 3)^2}{9} = 1$

31. $2y^2 - 3x^2 - 16y - 12x + 8 = 0$

32. $9x^2 - 4y^2 - 72x - 24y + 72 = 0$

33. $2x^2 - y^2 + 6y - 19 = 0$

34. $4y^2 - x^2 + 8x - 28 = 0$

In Exercises 35–42 write an equation for the conic section with the given properties.

35. Circle: center at $(-4, 3)$, radius $2\sqrt{5}$

36. Circle: endpoints of a diameter at $(-5, 2)$ and $(1, 6)$

37. Ellipse: center at $(-1, 4)$, $a = 4$, $b = 2$, major axis vertical

38. Ellipse: vertices at (3, 6) and (3, −4), covertices at (1, 1) and (5, 1)

39. Parabola: vertex at (0, 0), focus at (0, 4)

40. Parabola: vertex at (2, 4), focus at (−2, 4)

41. Parabola: vertex at (2, −3), directrix line $y = -5$

42. Parabola: focus at (−4, 1), directrix line $x = 2$

Find the inverse f^{-1} of each function in Exercises 43–50.

43. $f(x) = x + 4$

44. $f(x) = \dfrac{x - 2}{4}$

45. $f(x) = x^3 - 1$

46. $f(x) = \dfrac{1}{x + 2}$

47. $f(x) = \dfrac{1}{x} + 2$

48. $f(x) = \sqrt[3]{x} - 2$

49. If $F(t) = \dfrac{3}{4}t + 2$, find $F^{-1}(2)$.

50. If $G(x) = \dfrac{1}{x} - 4$, find $G^{-1}(3)$.

Exercises 51–56: For Exercises 43–48, graph the function and its inverse on the same set of axes, along with the graph of $y = x$.

Use Gaussian reduction to solve each system in Exercises 57–62.

57. $x - 2y = 5$
$2x + y = 5$

58. $4x - 3y = 16$
$2x + y = 8$

59. $2x - y = 7$
$3x + 2y = 14$

60. $2x - y + 3z = -6$
$x + 2y - z = 7$
$3x + y + z = 2$

61. $x + 2y - z = -3$
$2x - 3y + 2z = 2$
$x - y + 4z = 7$

62. $x + y + z = 1$
$2x - y - z = 2$
$2x - y + 3z = 2$

Solve Exercises 63–66 using a system with two or three variables.

63. A collection of coins consisting of dimes and quarters has a value of $4.95. How many dimes are in the collection if there are 25 more dimes than quarters?

64. The first-class fare on an airplane flight is $280 and the tourist fare is $160. If 64 passengers paid a total of $12,160 for the flight, how many of each type of ticket were sold?

65. The three points (0, 3), (1, 8), and (3, 30) all lie on a parabola with a vertical axis. What is the equation of the parabola?

66. Juan is preparing a quart of a fruit punch consisting of cranberry juice, apricot nectar, and club soda. Each quart of cranberry juice has 1200 calories and costs 25 cents. The apricot nectar has 1600 calories and costs 20 cents per quart. Club soda has no calories and costs 5 cents per quart. How much of each ingredient should Juan use to make a quart of fruit punch that will have 800 calories and cost 16 cents?

Appendix

 Subsets and Properties of the Real Numbers

Subsets of the Real Numbers

In algebra we work primarily with collections, or **sets**, of numbers. Our study of algebra will deal almost exclusively with the **real numbers**, denoted by *R*. The real numbers are classified according to their properties:

- The set *N* of **natural**, or **counting, numbers**, as its name suggests, consists of the numbers 1, 2, 3, 4, . . . , where ". . ." indicates that the list continues without end.

- The set *W* of **whole numbers** consists of the natural numbers and zero: 0, 1, 2, 3, ...

- The set *J* of **integers** consists of the natural numbers, their negatives, and zero: . . . , −3, −2, −1, 0, 1, 2, 3, ...

- The set *Q* of **rational numbers** includes all numbers that can be represented in the form *a*/*b*, where *a* and *b* are integers and *b* does not equal zero. Examples are −3/4, 18/27, 3, and −6. (The integers 3 and −6 are rational because they can be written in the form 3/1 and −6/1.) When represented in decimal form, all rational numbers either terminate or exhibit a repeating pattern. For example, −3/4 is equivalent to the terminating decimal −0.75, and 2/3 is equivalent to the repeating decimal 0.666. . . . (A repeating decimal is denoted by a bar over the repeating digit or block of digits; thus $0.666 \ldots = 0.\bar{6}$.)

- The set *H* of **irrational numbers** includes all numbers whose decimal representations are nonterminating and nonrepeating. An irrational number *cannot* be written in the form *a*/*b*, where *a* and *b* are integers. Examples of irrational numbers are $\sqrt{15}$, π, and $-\sqrt[3]{7}$. (Note that $\sqrt{9}$ is *not* irrational, since $\sqrt{9} = 3$, a rational number.)

Each item in a set is called an **element** or **member** of the set. For example, −4 is an element of the set of integers. We use the symbol ∈ to

mean "is an element of"; thus we might write $-4 \in J$ for the statement
" -4 is an element of J."

If all the elements of a set are also members of a larger set, that set is
a **subset** of the larger set. The different subsets of the real numbers are related
as shown in Figure A.1. Notice that the natural numbers are a subset of the
whole numbers, the whole numbers are a subset of the integers, and the
integers are a subset of the rational numbers. Also observe that every real
number is either rational or irrational.

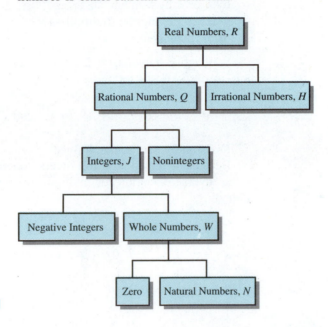

Figure A.1

Example 1

a.   2 is an element of N, W, J, Q, and R.
b. $\sqrt{15}$ is an element of H and R.
c. -5.73 is an element of Q and R.
d.   The number π, whose decimal representation begins $3.14159 \ldots$, is
an element of H and R.

**Properties of the
Real Numbers**

The four arithmetic operations (addition, subtraction, multiplication, and division) and the notions of equality and order in the set of real numbers are
governed by a number of properties. Several important properties are given
here:

$$a + b = b + a$$
$$ab = ba$$

Commutative properties

$$(a + b) + c = a + (b + c)$$
$$(ab)c = a(bc)$$

Associative properties

$$a(b + c) = ab + ac \qquad \text{Distributive property}$$

$$a + 0 = a$$
$$a \cdot 1 = a \qquad \text{Identity properties}$$

If $a = b$ and $b = c$, then $a = c$.
If $a < b$ and $b < c$, then $a < c$.

Transitive property

If $a = b$, then b may be replaced
by a or a by b in any statement
without altering the truthfulness
of the statement.

Substitution property

Subtraction and division are defined in terms of addition and multiplication, respectively, as follows:

$$a - b = a + (-b) \qquad \text{and} \qquad \frac{a}{b} = a\left(\frac{1}{b}\right) \qquad (b \neq 0)$$

Division by zero is not defined.

Because subtraction is the inverse operation for addition and division is the inverse operation for multiplication, we have the following inverse properties:

$$a + (-a) = 0 \qquad \text{Negative (or additive-inverse) property}$$

$$a \cdot \frac{1}{a} = 1 \qquad (a \neq 0) \qquad \text{Reciprocal (or multiplicative-inverse) property}$$

Exercise A.1

Name the subsets of the real numbers to which each of the numbers in Exercises 1–12 belongs.

1. $-\dfrac{5}{8}$

2. 137

3. $\sqrt{8}$

4. $2.71828\ldots$

5. -36

6. $\sqrt{49}$

7. 0

8. $0.\overline{357}$

9. $13.\overline{289}$

10. $\sqrt{\dfrac{4}{9}}$

11. $6.468725\ldots$

12. $\dfrac{13}{7}$

A.2 REVIEW OF PRODUCTS AND FACTORING

In Section 6.1 we used the first law of exponents to multiply two or more monomials. In this section we review techniques for multiplying and factoring polynomials of several terms.

Products of Polynomials To multiply polynomials containing more than one term, we use a generalized form of the distributive property:

$$a(b + c + d + \cdots) = ab + ac + ad + \cdots$$

Thus to find the product of a monomial and a polynomial, we multiply each term of the polynomial by the monomial.

Example 1

a. $3x(x + y + z) = 3x(x) + 3x(y) + 3x(z)$

$$= 3x^2 + 3xy + 3xz$$

b. $-2ab^2(3a^2b - ab + 2ab^2)$

$$= -2ab^2(3a^2b) - 2ab^2(-ab) - 2ab^2(2ab^2)$$

$$= -6a^3b^3 + 2a^2b^3 - 4a^2b^4$$

Products of Binomials Products of binomials occur so frequently that it is worthwhile to learn a shortcut for this type of multiplication. We can use the following scheme to perform the multiplication mentally:

$$(3x - 2y)(x + y) = 3x^2 + 3xy - 2xy - 2y^2$$

$$= 3x^2 + xy - 2y^2$$

This process is sometimes called the **FOIL** method, where FOIL represents

the product of the **F**irst terms
the product of the **O**uter terms
the product of the **I**nner terms
the product of the **L**ast terms

Example 2

a. $(2x - 1)(x + 3)$

$$= 2x^2 + 6x - x - 3$$

$$= 2x^2 + 5x - 3$$

b. $(3x + 1)(2x - 1)$

$$= 6x^2 - 3x + 2x - 1$$

$$= 6x^2 - x - 1$$

Factoring We sometimes find it useful to write a polynomial as a single *term* composed of two or more *factors*. This process is the reverse of multiplication and is called **factoring**. For example, observe that

$$3x^2 + 6x = 3x(x + 2)$$

We will consider only factorizations in which the factors are polynomials with integer coefficients.

Common Factors We can factor a common factor from a polynomial by using the distributive property in the form

$$ab + ac = a(b + c)$$

We first identify the common factor. For example, observe that the polynomial

$$6x^3 + 9x^2 - 3x$$

contains the monomial $3x$ as a factor of each term. We therefore write

$$6x^3 + 9x^2 - 3x = 3x(\qquad)$$

and then insert the appropriate polynomial factor within the parentheses. This factor can be determined by inspection. We ask ourselves for monomials that multiply $3x$ to yield $6x^3$, $9x^2$, and $-3x$, respectively, to obtain

$$6x^3 + 9x^2 - 3x = 3x(2x^2 + 3x - 1)$$

We can always check the result of factoring an expression by multiplying the factors. In the example above,

$$3x(2x^2 + 3x - 1) = 6x^3 + 9x^2 - 3x$$

Example 3

a. $18x^2y - 24xy^2 = 6xy(? - ?)$
$$= 6xy(3x - 4y)$$

because

$$6xy(3x - 4y) = 18x^2y - 24xy^2$$

b. $y(x - 2) + z(x - 2) = (x - 2)(? + ?)$
$$= (x - 2)(y + z)$$

because

$$(x - 2)(y + z) = y(x - 2) + z(x - 2)$$

It is often useful to factor -1 from the terms of a binomial:

$$a - b = (-1)(-a + b)$$
$$= (-1)(b - a)$$
$$= -(b - a)$$

Hence we have the following important relationship:

$$a - b = -(b - a)$$

that is, $a - b$ and $b - a$ are negatives of each other.

Example 4

a. $3x - y = -(y - 3x)$
b. $a - 2b = -(2b - a)$

Factoring Quadratic Polynomials

A useful type of factoring involves quadratic (second-degree) binomials or trinomials. For example, consider the trinomial

$$x^2 + 6x - 16 \qquad (1)$$

We desire, if possible, to find two binomial factors,

$$(x + a)(x + b)$$

whose product is the given trinomial. Now,

$$(x + a)(x + b) = x^2 + (a + b)x + ab \qquad (2)$$

Comparing Equations (1) and (2), we seek two integers a and b such that $a + b = 6$ and $ab = -16$; that is, their sum must be the coefficient of the linear term, $6x$, and their product must be the constant term -16. By inspection, or by trial and error, we determine that the two numbers are 8 and -2, so that

$$x^2 + 6x - 16 = (x + 8)(x - 2)$$

Checking, we note that $(x + 8)(x - 2) = x^2 + 6x - 16$.

Example 5

Factor.
a. $x^2 - 7x + 12$ b. $x^2 - x - 12$

Solutions   **a.** Find two numbers whose product is 12 and whose sum is -7. Since the product is positive and the sum is negative, the two numbers must both be negative. By inspection, or by trial and error, we find that the two numbers are -4 and -3. Hence

$$x^2 - 7x + 12 = (x - 4)(x - 3)$$

b. Find two numbers whose product is -12 and whose sum is -1. Since the product is negative, the two numbers must be of opposite sign and their sum must be -1. By inspection, or by trial and error, we find that the two numbers are -4 and 3. Hence

$$x^2 - x - 12 = (x - 4)(x + 3)$$

Although we have not specifically noted the check in the examples above, you should do the check mentally for each factorization.

Often we are confronted with a quadratic trinomial in which the coefficient of the second-degree term is other than 1.

Example 6

Factor $8x^2 - 9 - 21x$.

Solution

1. Write the trinomial in decreasing powers of x.

$$8x^2 - 21x - 9$$

2. Consider possible combinations of first-degree factors of the first term:

$$(8x \quad)(x \quad)$$
$$(4x \quad)(2x \quad)$$

3. Consider possible factorizations of the last term: 9 may be factored as $9 \cdot 1$ or as $3 \cdot 3$. Form all possible pairs of binomial factors using these factorizations:

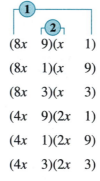

$$(8x \quad 9)(x \quad 1)$$
$$(8x \quad 1)(x \quad 9)$$
$$(8x \quad 3)(x \quad 3)$$
$$(4x \quad 9)(2x \quad 1)$$
$$(4x \quad 1)(2x \quad 9)$$
$$(4x \quad 3)(2x \quad 3)$$

4. Select the combinations of products **1** and **2** whose sum or difference could be the second term, $-21x$:

$$(8x \quad 3)(x \quad 3)$$

5. Insert the proper signs:

$$(8x + 3)\,(x - 3) \qquad \qquad \square$$

With practice, you can usually mentally factor trinomials of the form $Ax^2 + Bx + C$.   The following observations may help:

1. If both B and C are positive, both signs in the factored form are positive. For example, as a first step in factoring $6x^2 + 11x + 4$ we could write

$$(\quad + \quad)(\quad + \quad)$$

2. If B is negative and C is positive, both signs in the factored form are negative. Thus as the first step in factoring $6x^2 - 11x + 4$ we could write

$$(\ \ -\ \ )(\ \ -\ \ )$$

3. If C is negative, the signs in the factored form are opposite. Thus as a first step in factoring $6x^2 - 5x - 4$ we could write

$$(\ \ +\ \ )(\ \ -\ \ ) \quad \text{or} \quad (\ \ -\ \ )(\ \ +\ \ )$$

Example 7

a. $6x^2 + 5x + 1$

$\qquad = (\ \ +\ \ )(\ \ +\ \ )$

$\qquad = (3x + 1)(2x + 1)$

c. $6x^2 - x - 1$

$\qquad = (\ \ +\ \ )(\ \ -\ \ )$

$\qquad = (3x + 1)(2x - 1)$

b. $6x^2 - 5x + 1$

$\qquad = (\ \ -\ \ )(\ \ -\ \ )$

$\qquad = (3x - 1)(2x - 1)$

d. $6x^2 - xy - y^2$

$\qquad = (\ \ +\ \ )(\ \ -\ \ )$

$\qquad = (3x + y)(2x - y)$

Special Products and Factors

The products below are special cases of the multiplication of binomials. They occur so often that you should learn to recognize them on sight.

> **I.** $(x + a)^2 = (x + a)(x + a) = x^2 + 2ax + a^2$
>
> **II.** $(x - a)^2 = (x - a)(x - a) = x^2 - 2ax + a^2$
>
> **III.** $(x + a)(x - a) = x^2 - a^2$

⚠ COMMON ERRORS

Note that in (I) $(x + a)^2 \neq x^2 + a^2$ and in (II) $(x - a)^2 \neq x^2 - a^2$.

Example 8

a. $3(x + 4)^2$

$\qquad = 3(x^2 + 2 \cdot 4x + 4^2)$

$\qquad = 3x^2 + 24x + 48$

b. $(y + 5)(y - 5)$

$\qquad = y^2 - 5^2$

$\qquad = y^2 - 25$

c. $(3x - 2y)^2 = (3x)^2 - 2(3x)(2y) + (2y)^2$

$\qquad\qquad\qquad = 9x^2 - 12xy + 4y^2$

Each of the formulas on page 627, when viewed from right to left, also represents a special case of factoring quadratic polynomials.

$$\textbf{I. } x^2 + 2ax + a^2 = (x + a)^2$$
$$\textbf{II. } x^2 - 2ax + a^2 = (x - a)^2$$
$$\textbf{III. } x^2 - a^2 = (x + a)(x - a)$$

The trinomials in (I) and (II) are sometimes called **perfect-square trinomials** because they are squares of binomials. Note that the expression $x^2 + a^2$ *cannot* be factored.

Example 9

Factor.

a. $x^2 + 8x + 16$ **b.** $x^2 - 10x + 25$
c. $4a^2 - 12ab + 9b^2$ **d.** $25a^2b^2 + 20ab + 4$

Solutions   **a.**   Observe that 16 is equal to 4^2 and 8 is equal to $2 \cdot 4$. Thus

$$x^2 + 8x + 16 = x^2 + 2 \cdot 4x + 4^2$$
$$= (x + 4)^2$$

b.   Observe that 25 is equal to 5^2 and 10 is equal to $2 \cdot 5$. Thus

$$x^2 - 10x + 25 = x^2 - 2 \cdot 5x + 5^2$$
$$= (x - 5)^2$$

c.   Observe that $4a^2 = (2a)^2$, $9b^2 = (3b)^2$,   and $12ab = 2(2a)(3b)$. Thus

$$4a^2 - 12ab + 9b^2 = (2a)^2 - 2(2a)(3b) + (3b)^2$$
$$= (2a - 3b)^2$$

d.   Observe that $25a^2b^2 = (5ab)^2$, $4 = 2^2$,   and $20ab = 2(5ab)(2)$. Thus

$$25a^2b^2 + 20ab + 4 = (5ab)^2 + 2(5ab)(2) + 2^2$$
$$= (5ab + 2)^2$$

Binomials of the form $x^2 - a^2$ are often called the **difference of two squares**.

Example 10

Factor if possible.

a. $x^2 - 81$ **b.** $4x^2 - 9y^2$ **c.** $x^2 + 81$

Solutions   **a.**   The expression $x^2 - 81$ can be written as the difference of two squares, $x^2 - 9^2$,   and thus can be factored according to (III) above:

$$x^2 - 81 = x^2 - 9^2$$
$$= (x + 9)(x - 9)$$

b. The expression $4x^2 - 9y^2$ can be written as the difference of two squares, $(2x)^2 - (3y)^2$, and thus can be factored as

$$4x^2 - 9y^2 = (2x)^2 - (3y)^2$$
$$= (2x + 3y)(2x - 3y)$$

c. The expression $x^2 + 81$, equivalent to $x^2 + 0x + 81$, is *not* of form (III). It is *not* factorable, because no two real numbers have a product of 81 and a sum of 0.

COMMON ERROR

$x^2 + 81 \neq (x + 9)(x + 9)$, which you can verify by performing the indicated multiplication.

The factors $x + 9$ and $x - 9$ in Example 10(a) are called **conjugates** of each other. In general, any binomials of the form $a - b$ and $a + b$ are called a **conjugate pair**.

Exercise A.2

Write each product in Exercises 1–22 as a polynomial and simplify.

1. $4y(x - 2y)$ **2.** $3x(2x + y)$

3. $-6x(2x^2 - x + 1)$ **4.** $-2y(y^2 - 3y + 2)$

5. $a^2b(3a^2 - 2ab - b)$

6. $ab^3(-a^2b^2 + 4ab - 3)$

7. $2x^2y^3(4xy^4 - 2x^2y - 3x^3y^2)$

8. $5x^2y^2(3x^4y^2 + 3x^2y - xy^6)$

9. $(n + 2)(n + 8)$ **10.** $(r - 1)(r - 6)$

11. $(r + 5)(r - 2)$ **12.** $(z - 3)(z + 5)$

13. $(2z + 1)(z - 3)$ **14.** $(3t - 1)(2t + 1)$

15. $(4r + 3s)(2r - s)$ **16.** $(2z - w)(3z + 5w)$

17. $(2x - 3y)(3x - 2y)$

18. $(3a + 5b)(3a + 4b)$

19. $(3t - 4s)(3t + 4s)$ **20.** $(2x - 3z)(2x + 3z)$

21. $(2a^2 + b^2)(a^2 - 3b^2)$

22. $(s^2 - 5t^2)(3s^2 + 2t^2)$

Factor Exercises 23–40 completely. Check your answers by multiplying factors.

23. $4x^2z + 8xz$ **24.** $3x^2y + 6xy$

25. $3n^4 - 6n^3 + 12n^2$ **26.** $2x^4 - 4x^2 + 8x$

27. $15r^2s + 18rs^2 - 3r$ **28.** $2x^2y^2 - 3xy + 5x^2$

29. $3m^2n^4 - 6m^3n^3 + 14m^3n^2$

30. $6x^3y - 6xy^3 + 12x^2y^2$

31. $15a^4b^3c^4 - 12a^2b^2c^5 + 6a^2b^3c^4$

32. $14xy^4z^3 + 21x^2y^3z^2 - 28x^3y^2z^5$

33. $a(a + 3) + b(a + 3)$

34. $b(a - 2) + a(a - 2)$

35. $y(y - 2) - 3x(y - 2)$

36. $2x(x + 3) - y(x + 3)$

37. $4(x - 2)^2 - 8x(x - 2)^3$

38. $6(x + 1) - 3x(x + 1)^2$

39. $x(x - 5)^2 - x^2(x - 5)^3$

40. $x^2(x + 3)^3 - x(x + 3)^2$

Supply the missing factors or terms in Exercises 41–48.

41. $3m - 2n = -(?)$ **42.** $2a - b = -(?)$

43. $-2x + 2 = -2(?)$ **44.** $-6x - 9 = -3(?)$

45. $-ab - ac = ?(b + c)$

46. $-a^2 + ab = ?(a - b)$

47. $2x - y + 3z = -(?)$

48. $3x + 3y - 2z = -(?)$

Factor Exercises 49–84 completely.

49. $x^2 + 5x + 6$ **50.** $x^2 + 5x + 4$

51. $y^2 - 7y + 12$ **52.** $y^2 - 7y + 10$

53. $x^2 - 6 - x$ **54.** $x^2 - 15 - 2x$

55. $2x^2 + 3x - 2$ **56.** $3x^2 - 7x + 2$

57. $7x + 4x^2 - 2$ **58.** $1 - 5x + 6x^2$

59. $9y^2 - 21y - 8$ **60.** $10y^2 - 3y - 18$

61. $10u^2 - 3 - u$ **62.** $8u^2 - 3 + 5u$

63. $21x^2 - 43x - 14$ **64.** $24x^2 - 29x + 5$

65. $5a + 72a^2 - 12$

66. $-30a + 72a^2 - 25$

67. $12 - 53x + 30x^2$ **68.** $39x + 80x^2 - 20$

69. $-30t - 44 + 54t^2$ **70.** $48t^2 - 122t + 39$

71. $3x^2 - 7ax + 2a^2$ **72.** $9x^2 + 9ax - 10a^2$

73. $15x^2 - 4xy - 4y^2$ **74.** $12x^2 + 7xy - 12y^2$

75. $18u^2 + 20v^2 - 39uv$

76. $24u^2 - 20v^2 + 17uv$

77. $12a^2 - 14b^2 - 13ab$

78. $24a^2 - 15b^2 - 2ab$

79. $10a^2b^2 - 19ab + 6$ **80.** $12a^2b^2 - ab - 20$

81. $56x^2y^2 - 2xy - 4$ **82.** $54x^2y^2 + 3xy - 2$

83. $22a^2z^2 - 21 - 19az$

84. $26a^2z^2 - 24 + 23az$

Write each expression in Exercises 85–96 as a polynomial and simplify.

85. $(x + 3)^2$ **86.** $(y - 4)^2$

87. $(2y - 5)^2$ **88.** $(3x + 2)^2$

89. $(x + 3)(x - 3)$ **90.** $(x - 7)(x + 7)$

91. $(3t - 4s)(3t + 4s)$ **92.** $(2x + a)(2x - a)$

93. $(5a - 2b)^2$ **94.** $(4u + 5v)^2$

95. $(8xz + 3)^2$ **96.** $(7yz - 2)^2$

Factor Exercises 97–116 completely.

97. $x^2 - 25$ **98.** $x^2 - 36$

99. $x^2 - 24x + 144$ **100.** $x^2 + 26x + 169$

101. $x^2 - 4y^2$ **102.** $9x^2 - y^2$

103. $4x^2 + 12x + 9$ **104.** $4y^2 + 4y + 1$

105. $9u^2 - 30uv + 25v^2$

106. $16s^2 - 56st + 49t^2$

107. $4a^2 - 25b^2$ **108.** $16a^2 - 9b^2$

109. $x^2y^2 - 81$ **110.** $x^2y^2 - 64$

111. $9x^2y^2 + 6xy + 1$ **112.** $4x^2y^2 + 12xy + 9$

113. $16x^2y^2 - 1$ **114.** $64x^2y^2 - 1$

115. $(x + 2)^2 - y^2$ **116.** $x^2 - (y - 3)^2$

A.3   COMPLEX NUMBERS

In Section 6.3 we noted that the square root of a negative number is not a real number. However, for many applications it is necessary to consider such square roots. In this section we introduce a set of numbers, **C**, called **complex numbers**, which includes all the real numbers and also square roots of negative real numbers.

Imaginary Numbers   We begin by defining a new number, i, whose square is -1.

$$i^2 = -1 \quad \text{or} \quad i = \sqrt{-1}$$

Furthermore, we define the principal square root of *any* negative real number in the following way.

For $a > 0$,
$$\sqrt{-a} = \sqrt{-1}\,\sqrt{a} = i\sqrt{a}$$

Example 1

a. $\sqrt{-4} = \sqrt{-1}\sqrt{4}$
$\qquad = i\sqrt{4} = 2i$

b. $\sqrt{-3} = \sqrt{-1}\sqrt{3}$
$\qquad = i\sqrt{3}$

Thus a square root of a negative real number can be represented as the product of a real number and the number $\sqrt{-1}$, or i. For historical reasons such numbers are called **imaginary numbers**.

Each negative real number $-a$, for $a > 0$, has *two* imaginary square roots, $i\sqrt{a}$ and $-i\sqrt{a}$, since

$$\left(i\sqrt{a}\right)^2 = i^2\left(\sqrt{a}\right)^2 = i^2 a = -a$$

and

$$\left(-i\sqrt{a}\right)^2 = i^2\left(-\sqrt{a}\right)^2 = i^2 a = -a$$

For example, the two square roots of -9 are

$$\sqrt{-9} = \sqrt{-1}\sqrt{9} = 3i \quad \text{and} \quad -\sqrt{-9} = -\sqrt{-1}\sqrt{9} = -3i$$

Complex Numbers   Now consider all possible expressions of the form $a + bi$, where a and b are real numbers and $i = \sqrt{-1}$. Such an expression represents a complex number, that is, a number in the set C. Here a is called the **real part** of the number and b is the **imaginary part**. If $b = 0$, then $a + bi = a$, and it is evident that the set R of real numbers is contained in the set C of complex numbers. The relationships among the subsets of C are shown in Figure A.2.

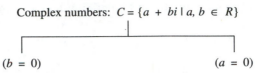

Complex numbers: $C = \{a + bi \mid a, b \in R\}$

$(b = 0)$
Real numbers: $a + bi = a$

$(a = 0)$
Imaginary numbers: $a + bi = bi$

Example 2

Write each expression in the form $a + bi$.

a. $3\sqrt{-18}$ **b.** $2 - 3\sqrt{-16}$

Solutions

a.
$$
\begin{aligned}
3\sqrt{-18} &= 3\sqrt{-1 \cdot 9 \cdot 2} \\
&= 3\sqrt{-1}\sqrt{9}\sqrt{2} \\
&= 3i(3)\sqrt{2} \\
&= 9i\sqrt{2}
\end{aligned}
$$

b.
$$
\begin{aligned}
2 - 3\sqrt{-16} &= 2 - 3\sqrt{-1 \cdot 16} \\
&= 2 - 3\sqrt{-1}\sqrt{16} \\
&= 2 - 3i(4) \\
&= 2 - 12i
\end{aligned}
$$

Operations on Complex Numbers

To add or subtract complex numbers, we simply add or subtract their real parts and their imaginary parts.

Example 3

a.
$$
\begin{aligned}
(2 + 3i) + (5 - 4i) &= (2 + 5) + (3 - 4)i \\
&= 7 - i
\end{aligned}
$$

b.
$$
\begin{aligned}
(2 + 3i) - (5 - 4i) &= (2 - 5) + [3 - (-4)]i \\
&= -3 + 7i
\end{aligned}
$$

To multiply complex numbers, we treat them as though they were binomials and replace i^2 with -1.

Example 4

a.
$$
\begin{aligned}
(2 - i)(1 + 3i) &= 2 + 6i - i - 3i^2 \\
&= 2 + 6i - i - 3(-1) \\
&= 2 + 6i - i + 3 \\
&= 5 + 5i
\end{aligned}
$$

b.
$$
\begin{aligned}
(3 - i)^2 &= (3 - i)(3 - i) \\
&= 9 - 3i - 3i + i^2 \\
&= 9 - 6i + (-1) \\
&= 8 - 6i
\end{aligned}
$$

The quotient of two complex numbers can be found by using the following property, which is analogous to the fundamental principle of fractions in the set of real numbers. Recall from Section 6.5 that for $b > 0$ the conjugate of $a + \sqrt{b}$ is $a - \sqrt{b}$. Similarly, the conjugate of $a + bi$ is $a - bi$.

Example 5

a. The conjugate of $2 + 3i$ is $2 - 3i$.
b. The conjugate of $-3 - i$ is $-3 + i$.
c. The conjugate of $2i$ is $-2i$.
d. The conjugate of $-4 + i$ is $-4 - i$.

The quotient

$$\frac{a + bi}{c + di}$$

of two complex numbers can be simplified by multiplying the numerator and denominator by $c - di$, the conjugate of the denominator; that is,

$$\frac{a + bi}{c + di} = \frac{(a + bi)(c - di)}{(c + di)(c - di)}$$

If the divisor is of the form bi, we need only multiply the numerator and denominator by i.

Example 6

a.

$$\frac{4 - i}{-2i} = \frac{(4 - i)i}{-2i \cdot i}$$

$$= \frac{4i - i^2}{-2i^2}$$

$$= \frac{4i - (-1)}{-2(-1)}$$

$$= \frac{4i + 1}{2}$$

$$= \frac{1}{2} + 2i$$

b.

$$\frac{4 + i}{2 + 3i} = \frac{(4 + i)(2 - 3i)}{(2 + 3i)(2 - 3i)}$$

$$= \frac{8 - 10i - 3i^2}{4 - 9i^2}$$

$$= \frac{8 - 10i - 3(-1)}{4 - 9(-1)}$$

$$= \frac{8 - 10i + 3}{4 + 9}$$

$$= \frac{11}{13} - \frac{10}{13}i$$

Using Radical Notation   The symbol $\sqrt{-b}$ $(b > 0)$ should be used with care, since certain properties involving square root symbols are valid for real numbers but are *not* valid when the symbols do not represent real numbers.

 COMMON ERROR

$$\sqrt{-2}\,\sqrt{-3} \neq \sqrt{(-2)(-3)} = \sqrt{6}$$

$$\sqrt{-2}\,\sqrt{-3} = (i\sqrt{3})(i\sqrt{3}) = i^2\sqrt{6} = -\sqrt{6}$$

To avoid difficulty with this point, rewrite all expressions of the form $\sqrt{-b}$ $(b > 0)$ in the form $i\sqrt{b}$ before performing any computations.

Example 7

a. $\sqrt{-2}(3 - \sqrt{-5})$

$= i\sqrt{2}(3 - i\sqrt{5})$

$= 3i\sqrt{2} - i^2\sqrt{10}$

$= 3i\sqrt{2} - (-1)\sqrt{10}$

$= \sqrt{10} + 3i\sqrt{2}$

b. $(2 + \sqrt{-3})(2 - \sqrt{-3})$

$= (2 + i\sqrt{3})(2 - i\sqrt{3})$

$= 4 - 3i^2$

$= 4 - 3(-1)$

$= 7$

c. $\dfrac{2}{\sqrt{-3}} = \dfrac{2}{i\sqrt{3}} \cdot \dfrac{i\sqrt{3}}{i\sqrt{3}}$

$= \dfrac{2i\sqrt{3}}{3i^2}$

$= \dfrac{2i\sqrt{3}}{3(-1)} = \dfrac{-2i\sqrt{3}}{3}$

d. $\dfrac{1}{3 - \sqrt{-1}} = \dfrac{1}{3 - i}$

$= \dfrac{1 \cdot (3 + i)}{(3 - i)(3 + i)}$

$= \dfrac{3 + i}{9 - i^2}$

$= \dfrac{3 + i}{9 - (-1)} = \dfrac{3}{10} + \dfrac{1}{10}i$

Exercise A.3

Write each expression in Exercises 1–18 in the form $a + bi$.

1. $\sqrt{-4}$

2. $\sqrt{-9}$

3. $\sqrt{-32}$

4. $\sqrt{-50}$

5. $3\sqrt{-8}$

6. $4\sqrt{-18}$

7. $3\sqrt{-24}$

8. $2\sqrt{-40}$

9. $5\sqrt{-64}$

10. $7\sqrt{-81}$

11. $-2\sqrt{-12}$

12. $-3\sqrt{-75}$

13. $4 + 2\sqrt{-1}$

14. $5 - 3\sqrt{-1}$

15. $3\sqrt{-50} + 2$

16. $5\sqrt{-12} - 1$

17. $\sqrt{4} + \sqrt{-4}$

18. $\sqrt{20} - \sqrt{-20}$

Add or subtract in Exercises 19–24.

19. $(2 + 4i) + (3 + i)$

20. $(2 - i) + (3 - 2i)$

21. $(4 - i) - (6 - 2i)$

22. $(2 + i) - (4 - 2i)$

23. $3 - (4 + 2i)$

24. $(2 - 6i) - 3$

Multiply in Exercises 25–34.

25. $(2 - i)(3 + 2i)$

26. $(1 - 3i)(4 - 5i)$

27. $(3 + 2i)(5 + i)$

28. $(-3 - i)(2 - 3i)$

29. $(6 - 3i)(4 - i)$

30. $(7 + 3i)(-2 - 3i)$

31. $(2 - i)^2$

32. $(2 + 3i)^2$

33. $(2 - i)(2 + i)$

34. $(1 - 2i)(1 + 2i)$

Divide in Exercises 35–46.

35. $\dfrac{1}{3i}$

36. $\dfrac{-2}{5i}$

37. $\dfrac{3 - i}{5i}$

38. $\dfrac{4 + 2i}{3i}$

39. $\dfrac{2}{1 - i}$

40. $\dfrac{-3}{2 + i}$

41. $\dfrac{2 + i}{1 + 3i}$

42. $\dfrac{3 - i}{1 + i}$

43. $\dfrac{2 - 3i}{3 - 2i}$

44. $\dfrac{6 + i}{2 - 5i}$

45. $\dfrac{3 + 2i}{5 - 3i}$

46. $\dfrac{-4 - 3i}{2 + 7i}$

Simplify Exercises 47–54. Write each expression in the form $a + bi$.

47. $\sqrt{-4}(1 - \sqrt{-4})$

48. $\sqrt{-9}(3 + \sqrt{-16})$

49. $(2 + \sqrt{-9})(3 - \sqrt{-9})$

50. $(4 - \sqrt{-2})(3 + \sqrt{-2})$

51. $\dfrac{3}{\sqrt{-4}}$

52. $\dfrac{-1}{\sqrt{-25}}$

53. $\dfrac{2 - \sqrt{-1}}{2 + \sqrt{-1}}$

54. $\dfrac{1 + \sqrt{-2}}{3 - \sqrt{-2}}$

55. For what values of x will $\sqrt{x - 5}$ be real? Imaginary?

56. For what values of x will $\sqrt{x + 3}$ be real? Imaginary?

57. Simplify. (*Hint:* $i^2 = -1$ and $i^4 = 1$.)

a. i^6 **b.** i^{12} **c.** i^{15} **d.** i^{102}

58. Express with a positive exponent and simplify.

a. i^{-1} **b.** i^{-2} **c.** i^{-3} **d.** i^{-6}

59. Evaluate $x^2 + 2x + 3$ for $x = 1 + i$.

60. Evaluate $2y^2 - y + 2$ for $y = 2 - i$.

Answers

Answers to Odd-Numbered Exercises in Each Section and to All Chapter Review Exercises

Exercise 1.2 [Page 14]

1a. $j = a + 27$ **1b.** 49   **3a.** $h = \dfrac{1260}{r}$

3b. 28 hr   **5a.** $y = \dfrac{f}{3}$ **5b.** $c = 5.79\left(\dfrac{f}{3}\right)$

5c. \$19.30   **7a.** $A = \pi r^2$ **7b.** 78.54 sq cm

9a. $n = 20 + m$ **9b.** $a = \dfrac{198}{20 + m}$ **9c.** 7.92

11a. $t = 0.079p$ **11b.** $b = 1.079p$ **11c.** \$528.71

13a. $C = 1.97 + 0.39m$ **13b.** \$12.50

15a. $c = 0.4w$ **15b.** $r = 50 - 0.4w$ **15c.** 47.6

17a. $u = \dfrac{m}{20}$ **17b.** $g = 14.6 - \dfrac{m}{20}$ **17c.** 9.1

19. 6   **21.** -42 **23.** 5   **25.** -2 **27.** -2
29. -60 **31.** -25 **33.** 81   **35.** -64
37. -32 **39.** 50   **41.** -2 **43.** -5 **45.** -19
47. 98.4   **49.** 36   **51.** -3 **53.** 45   **55.** -5

57a. $2 + \dfrac{3}{4}$ **57b.** $\dfrac{2 + 3}{4}$ **59a.** -23^2

59b. $(-23)^2$ **61a.** $\sqrt{9 + 16}$ **61b.** $\sqrt{9} + 16$
63. 100   **65.** 1080   **67.** 0   **69.** 72.5904

Exercise 1.3 [Page 24]

1. 7   **3.** $\dfrac{31}{3}$ **5.** $\dfrac{2}{3}$ **7.** $-4.8\overline{3}$ **9.** 34.29

11. \$18   **13a.** $t = $ time wife drives   **13b.** $45t$
13c. $16(t + 6)$ **13d.** 3.3 hr
15a. $n = $ number of copies
15b. $20{,}000 + 0.02n$; $17{,}500 + 0.025n$
15c. 500,000 copies   **17a.** 145,800   **17b.** 125,000

19. 7.53%   **21.** 87
23d. $7.\overline{7}$ lb of 6%; $2.\overline{2}$ lb of 15%   **25d.** \$16,600

27. $\dfrac{v - k}{g}$ **29.** $\dfrac{S - 2w^2}{4w}$ **31.** $\dfrac{P - a + d}{d}$

33. $\dfrac{A - \pi r^2}{\pi r}$

Exercise 1.4 [Page 34]

1. $\dfrac{\pm 5}{3}$ **3.** $\pm\sqrt{7}$ **5.** $\pm\sqrt{6}$ **7.** $\pm\sqrt{6}$

9. 11.8%   **11.** 64   **13.** -2 **15.** $\dfrac{9}{4}$

17. -5 **19.** 90 ft   **21.** $\dfrac{-1}{2}$ **23.** $\dfrac{13}{8}$

25. $\pm\sqrt{\dfrac{15}{8}}$ **27.** 0.97   **29.** 37 ft   **31.** 4   **33.** 40

35. \$6187.50   **37.** \$37,500   **39.** 45 mi   **41.** 689

43. $\dfrac{S - a}{S}$ **45.** $\dfrac{Hy}{2y - H}$ **47.** $\pm\sqrt{\dfrac{Fr}{m}}$

49. $\pm\sqrt{\dfrac{Gm_1m_2}{F}}$ **51.** $\dfrac{gT^2}{4\pi^2}$ **53.** $\pm\sqrt{t^2 - r^2}$

Exercise 1.5 [Page 43]

1. 47°, 57°, 76°   **3.** 30°, 60°   **5.** 50°, 50°, 80°
7. 15 cm   **9.** 21 in.   **11.** 42.4 m   **13.** 11.3 in.

15. $x > 3$ **17.** $x > 0$ **19.** $x \le \dfrac{6}{13}$

21. $4 < x < 16$ **23.** 14 ft   **25.** 17.1 sq ft
27. 89.23 ft   **29a.** 7.24 cu m   **29b.** 6.16 sq cm
31a. 2623.86 cu m   **31b.** 1903.43 sq in.

33a. $72h + 640$ **33b.** 8 in. **35a.** $8.8r^2$
35b. 5.86 cm

Exercise 1.6 [Page 57]

1a. High: 7°F; low: −19°F

1b. Above 5°F from noon to 3:00 P.M.; below −5°F from midnight to 9:00 A.M. and from 7:00 P.M. to midnight

1c. 7:00 A.M.: −10°F; 2:00 P.M.: 6°F; 10:00 A.M. and 5:00 P.M.: 0°F; 6:00 A.M. and 10:00 P.M.: −12°F

1d. Between 3:00 A.M. and 6:00 A.M.: 6°F; between 9:00 A.M. and noon: 10°F; between 6:00 P.M. and 9:00 P.M.: 9°F

1e. Increased most rapidly: 9:00 A.M. to noon; decreased most rapidly: 6:00 P.M. to 9:00 P.M.

3a. At 43 mph: 28 mpg **3b.** 34 mpg at 52 mph

3c. Best gas mileage: at 70 mph. The graph seems to be "leveling off" for higher speeds; any improvement in mileage probably would not be appreciable, and the mileage might in fact deteriorate.

3d. Road condition, weather conditions, traffic, weight in the car

5a. 12 min **5b.** First 38 min

5c. Approximately from 38 min to 55 min

7. $(-3, -2), (1, 6), (-2, 0), (0, 4)$

9. $(-2, 6), (2, 6), (1, 3)$ or $(-1, 3), (0, 2)$

11. $\left(-1, -\dfrac{1}{2}\right), \left(\dfrac{1}{2}, -2\right), \left(4, \dfrac{1}{3}\right), (0, -1)$

13. $(-2, -8), \left(\dfrac{1}{2}, \dfrac{1}{8}\right), (0, 0), (-1, -1)$

15. 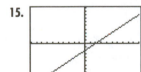 **17.**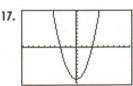

19. 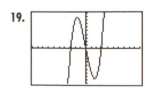 **21.**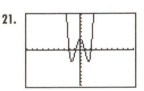

23. 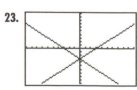 **25.**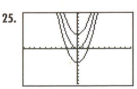

27. 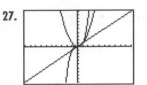 **29.**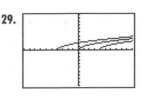

31. $(-6, -3), (-2, -3)$ **33.** $(5, -1)$
35. $(5.4, 4.8)$ **37.** $(1.2, 5.1), (-1.2, -1.1)$

39a. 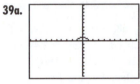 **39b.**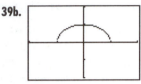

41a. 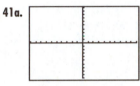 **41b.**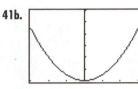

43a. 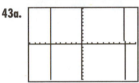 **43b.**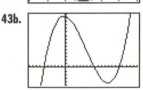

CHAPTER 1 REVIEW [Page 63]

1a. $48t$ **1b.** $2400 - 48t$ **2a.** $0.07s$ **2b.** $1.07s$

3a. $\dfrac{kT}{V}$ **3b.** 40 lb per sq in. **4a.** $kL(T - 65)$

4b. 0.48 ft **5.** -36 **6.** 36 **7.** 10 **8.** $\dfrac{-5}{2}$

9. -9 **10.** 13 **11.** 19.900 **12.** 0.014 **13.** 52

14. 3.675 **15.** 1 **16.** 600 **17.** $\dfrac{2}{3}$ **18.** -5

19. 5 **20.** 200 **21.** $\pm\sqrt{6}$ **22.** $\pm\sqrt{2}$

23. $\pm\sqrt{40}$ **24.** ± 4 **25.** -2 **26.** 2 **27.** ± 2

28. ± 2 **29.** 169 **30.** 49 **31.** 16 **32.** -16

33. $\dfrac{3N + 3c}{5}$ **34.** $\dfrac{2p + 10 - C}{2}$ **35.** $\dfrac{S + 2\pi r}{2\pi}$

36. $\dfrac{9C + 160}{5}$ **37.** $n\left(1 - \dfrac{V}{C}\right)$ **38.** $\dfrac{rp}{r - p}$

39. $\dfrac{2v}{t^2}$ **40.** πr^2 **41.** $x < \dfrac{2}{3}$ **42.** $x \geq 11$

43. $x < -4.7$ **44.** $x \geq \dfrac{-8}{5}$ **45a.** 2000 **45b.** 3112

46a. 5334   **46b.** 2000   **47.** 5 ft by 9 ft

48. 27 ft by 78 ft   **49.** 28°, 56°, 96°

50. 35°, 60°, 85°   **51.** 11.5 cm, 27.1 cm, 27.1 cm

52. 40.1 cm, 112.5 cm, 120.3 cm   **53.** 800   **54.** $150

55. 1800   **56.** 22,000   **57.** 22.08%   **58.** 24.2%

59. 20 L   **60.** 4 qt   **61.** $\frac{5}{9}$ cup   **62.** $12\frac{4}{9}$ lb

63. $2\frac{2}{9}$ km   **64.** 18.75 lb   **65.** 2.3 ft

66. 17.04 ft   **67.** 18.04 ft   **68.** 3.02 mi   **69.** 4 ft

70. 40 in.   **71.** 3,333,333.$\overline{3}$ cu yd   **72a.** 262.49 yd

72b. 83,996.8 sq yd   **73a.** 9 ft   **73b.** 10 days

73c. 3 ft   **73d.** Jan. 8–Jan. 9; Jan. 13–Jan. 14

74a. 32°F, 11°F   **74b.** 4 days   **74c.** Jan. 9

74d. 15°F   **75.** $(2, -6), (-1, 3), (0, 2), (1, 1)$

76. $(1, 0), (-3, 2), (0, 1), (-8, 3)$

77a. $(-3, 0), (-1, 0), (1, 0), (3, 0)$

77b. $(-3.3, 8), (3.3, 8)$ **77c.** Four

77d. $(0, 4.5), (-2.1, 7.9), (2.1, 7.9)$ **78a.** $(1, 2)$

78b. $(-1, 8)$ **78c.** $(-2.1, 6.8), (1.4, 2.1)$

Exercise 2.1 [Page 75]

1a. $h = 6 + 2t$

1b.

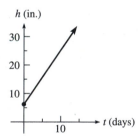

1c. 48 in., or 4 ft   **1d.** 33 days

3a. $A = 250 - 15w$

3b.

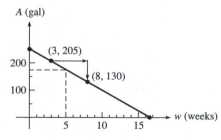

3c. 75 gal   **3d.** Up to the fifth week

5a. $P = -800 + 40t$

5b.

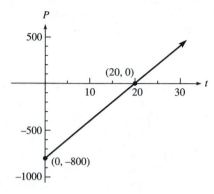

5c. The P-intercept, -800, is the initial ($t = 0$) value of the profit. Phil and Ernie start out $800 in debt. The t-intercept, 20, is the number of hours required for Phil and Ernie to break even.

7a. $0.60x + 0.80y = 4800$

7b.

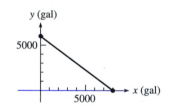

7c. The y-intercept, 6000, is the amount of premium that the gas station owner can buy if he buys no regular. The x-intercept, 8000, is the amount of regular he can buy if he buys no premium.

9a. $I = 10,000 + 0.03s$

9b.

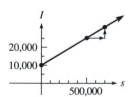

9c. Sales between $200,000 and $400,000

9d. Her salary will increase by $6000.

11.

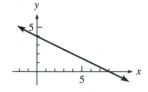

13.

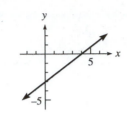

15.

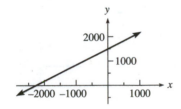

17.

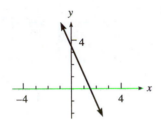

19.

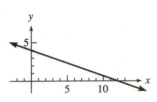

21.

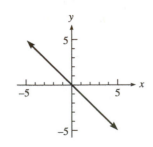

23.

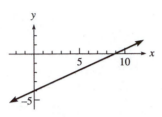

25.

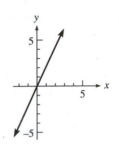

27.

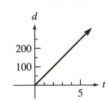

29a. $d = 50t$

29b.

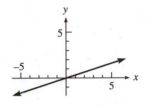

31a. $T = \dfrac{h}{24}$

31b.

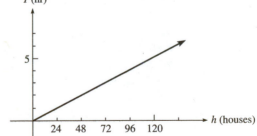

Exercise 2.2 [Page 84]

1. Anthony   **3.** Øksendahl's   **5.** Los Angeles

7. Highway 33   **9.** Acme Movers' ramp   **11.** $\dfrac{1}{4}$

13. $-\dfrac{2}{3}$ **15.** -1 **17a.** 14   **17b.** $23\dfrac{1}{3}$ **17c.** -56

19. 14.29 ft

21. $m = \dfrac{3}{4}$

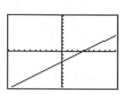

23. $m = -3$

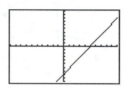

25. $m = \dfrac{8}{5}$

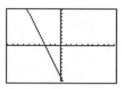

27. a and c

29a.

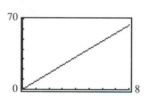

29b. $m = 8$ dollars per hour
29c. The slope gives the typist's rate of pay, in dollars per hour.
31a. $m = 1.5$ m per minute   **31b.** Speed of the train
33a. $m = 1250$ barrels per day   **33b.** Rate of pumping
35a. $m = -6$ L per day
35b. Rate of water consumption
37a. $m = 12$ in. per kilogram
37b. Conversion rate from feet to inches
39a. $m = 4$ dollars per kilogram
39b. Unit cost of coffee beans, per kilogram

Exercise 2.3 [Page 96]

1. $y = -2x + 6$ **3.** $y = 3x + 8$

5. $y = -\dfrac{x}{2} + \dfrac{5}{2}$ **7.** $y = \dfrac{3}{4}x - \dfrac{3}{2}$

9. $y = -0.4x + 2$ **11.** $y = -\dfrac{7}{2}x - 16$

13a. $(100, 0), (0, 100)$

13b. Xmin $= -20$, Xmax $= 120$
     Ymin $= -20$, Ymax $= 120$
13c. $y = -x + 100$

13d.

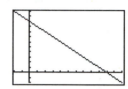

15a. $(0.04, 0), (0, -0.028)$
15b. Xmin $= -0.1$, Xmax $= 0.1$
     Ymin $= -0.1$, Ymax $= 0.1$
15c. $y = (25x - 1)/36$

15d.

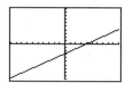

17a. $(-47, 0), (0, 12)$
17b. Xmin $= -50$, Xmax $= 10$
     Ymin $= -5$,   Ymax $= 15$
17c. $y = 12(x/47 + 1)$

17d.

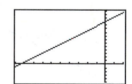

19a. $(-42, 0), (0, -28)$
19b. Xmin $= -50$, Xmax $= 10$
     Ymin $= -30$, Ymax $= 5$

19c. $y = -\dfrac{2}{3}x - 28$

19d.

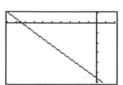

21a. $(-114, 0), (0, 38)$
21b. Xmin $= -120$, Xmax $= 10$
     Ymin $= -10$,   Ymax $= 40$

21c. $y = \dfrac{1}{3}x + 38$

21d.

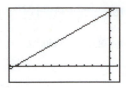

23a. $\left(\dfrac{1}{13}, 0\right), \left(0, \dfrac{1}{22}\right)$

23b. Xmin $= -0.1$,   Xmax $= 0.1$
Ymin $= -0.1$,   Ymax $= 0.1$

23.c $y = -\dfrac{13}{22}x + \dfrac{1}{22}$

23d.

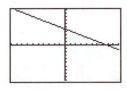

25. $\dfrac{12}{5}$ **27.** $-\dfrac{1}{2}$ **29.** Undefined   **31.** -1.028

33. $-\dfrac{9}{2}$ **35.** $-\dfrac{9}{19}$ **37.** 0.387 **39.** Undefined

41. $m = 2.5$, y-int. $= 6.25$ **43.** $m = -8.4$, y-int. $= 63$
45a. Yes   **45b.** 2π

47a.

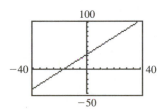

47b. $50°F$ **47c.** $-20°C$ **47d.** $m = \dfrac{9}{5}$

47e. The Fahrenheit temperature increases $\dfrac{9}{5}$ of a degree
for every degree increase in the Celsius temperature.
47f. The C-intercept gives the Celsius temperature at $0°F$,
and the F-intercept gives the Fahrenheit temperature at
$0°C$.

49a.

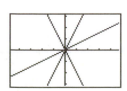

49b. $m = \dfrac{1}{2}, 2, -2$; y-int. $= 0, 0, 0$

51a.

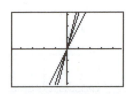

51b. $m = 2, 3, 4$; y-int. $= 0, 0, 0$

53a.

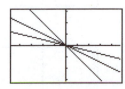

53b. $m = -1, -\dfrac{1}{2}, -\dfrac{1}{4}$; $b = 0, 0, 0$

55a.

55b. $m = -2, -2, -2$; y-int. $= 0, 10, -25$
55c. The coefficient of x in each equation gives the slope
of the line. The constant term is the y-intercept.

57a.

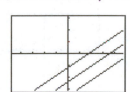

57b. $m = \dfrac{2}{3}, \dfrac{2}{3}, \dfrac{2}{3}$; y-int. $= -12, -24, -36$

Exercise 2.4 [Page 107]

1a. $y = -\dfrac{3}{2}x + \dfrac{1}{2}$ **1b.** $m = -\dfrac{3}{2}$, $b = \dfrac{1}{2}$

3a. $y = \dfrac{1}{3}x - \dfrac{2}{3}$ **3b.** $m = \dfrac{1}{3}$, $b = -\dfrac{2}{3}$

5a. $y = -\dfrac{1}{6}x + \dfrac{1}{9}$ **5b.** $m = -\dfrac{1}{6}$, $b = \dfrac{1}{9}$

7a. $y = 14x - 22$ **7b.** $m = 14$, $b = -22$

9a. $y = -29$ **9b.** $m = 0, \quad b = -29$

11a. $y = -\dfrac{5}{3}x + 16\dfrac{1}{3}$

11b. $m = -\dfrac{5}{3}, \quad b = 16\dfrac{1}{3}$

13a.

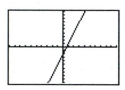

13b. $y = 3x - 2$

15a.

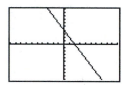

15b. $y = -2x + 4$

17a.

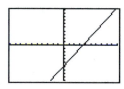

17b. $y = \dfrac{5}{3}x - 6$

19a.

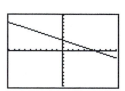

19b. $y = -\dfrac{1}{2}x + 3$ **21.** $d = 1.5t + 5$

23. $B = 1250t + 2000$ **25.** $W = -6t + 48$

27. $i = 12f$ **29.** $C = 4b$ **31a.** $m = 4, \quad b = 40$

31b. $y = 4x + 40$ **33a.** $m = -80, \quad b = -2000$

33b. $P = -80 - 2000$ **35a.** $m = \dfrac{1}{4}, \quad b = 0$

35b. $V = \dfrac{1}{4}d$ **37.** $y = -3x + 1$

39. $y = \dfrac{5}{3}x - \dfrac{13}{3}$ **41.** $y = -0.27x - 5.228$

43. $y = \dfrac{x}{7} + \dfrac{18}{7}$ **45.** $y = \dfrac{11}{4}x - \dfrac{1}{2}$

47. $y = 1.15x - 8.03$ **49a.** $C = 80x + 5000$

49b.

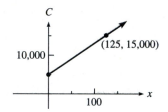

49c. $m = 80$ dollars/bike is the cost of making each bike.

51a. $p = 2.2k$

51b.

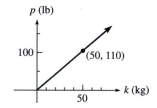

51c. $m = 2.2$ pounds/kilogram is the conversion factor from kilograms to pounds.

53a. $d = 65t + 265$

53b.

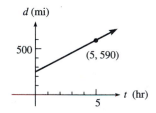

53c. $m = 65$ miles/hour is their average speed.

Exercise 2.5 [Page 116]

1a.

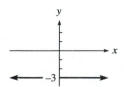

1b. $m = 0$

3a.

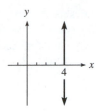

3b. Undefined

5a.

5b. Undefined
7a. l_1 negative, l_2 negative, l_3 positive, l_4 zero
7b. l_1, l_2, l_4, l_3 **9a.** Parallel   **9b.** Neither
9c. Neither   **9d.** Parallel

11b. Slope $\overline{AB}$: -1; slope $\overline{BC}$: 1; slope $\overline{AC}$: $\frac{1}{4}$. Hence $\overline{AB} \perp \overline{BC}$, so the triangle is a right triangle.

13b. Slope $\overline{PQ}$: 4; slope $\overline{QR}$: $-\frac{7}{2}$; slope $\overline{RS}$: 4; slope $\overline{SP}$: $-\frac{7}{2}$. Hence $\overline{PQ} \parallel \overline{RS}$ and $\overline{QR} \parallel \overline{SP}$, so the points are the vertices of a parallelogram.
15a. $x - 2y = 4$

15b.

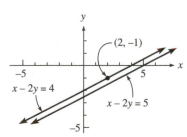

17a. $2x + 3y = 14$

17b.

y
$2y - 3x = 5$
5
(1, 4)
$2x + 3y = 14$
5
x

19a. $y = -2x - 8$ **19b.** $y = \frac{1}{2}x - 3$

21. $y = \frac{3}{2}x + \frac{15}{2}$ **23a.** Right angles are equal.

23b. Alternate interior angles are equal.
23c. Two angles of one triangle equal two angles of the other.
23d. Definition of slope
23e. Corresponding sides of similar triangles are proportional.

Exercise 2.6 [Page 126]
  1a. $x = 0.6$ **1b.** $x = -0.4$ **1c.** $x > 0.6$
  1d. $x < -0.4$ **3a.** $x = 6$ **3b.** $x = 1$ **3c.** $x \geq 6$
  3d. $x \leq 1$ **5a.** $x \approx 12$ **5b.** $x \approx 18$ **5c.** $x < 9$
  5d. $x > 3$ **7a.** $t \approx -3$ **7b.** $t \approx 1.5$ **7c.** $t < 0.8$
  7d. $-2.5 \leq t \leq 0.5$ **7e.** all t
  9a. $q = -2$, $q = 2$
  9b. $q \approx -2.8$, $q = 0$, $q \approx 2.8$
  9c. $-2.5 < q < -1.25$, $1.25 < q < 2.5$
  9d. $-2 < q < 0$, $2 < q$ **9e.** $0 \leq M \leq 26$
  11a. $u = -1$ **11b.** $u = 0$ **11c.** $u < 1$
  11d. $-0.5 < u < 0$ **13a.** 1991   **13b.** 1 yr
  13c. 1 yr   **13d.** 7000   **15a.** Approximately \$365
  15b. \$2 or \$8   **15c.** $3.25 < d < 6.75$ **17a.** $x = 4$
  17b. $x = -5$ **17c.** $x > 1$ **17d.** $x < 14$
  19a. $x = 11$ **19b.** $x = -10$ **19c.** $x \geq -5$
  19d. $x \leq 8$ **21a.** $x = 20$ **21b.** $x \leq 7$
  23a. $x = 4$ **23b.** $x < 22$ **25a.** $x = 41$
  25b. $29 < x \leq 61$ **27a.** $x = -5$ or $x = 17$
  27b. $-1 < x < 13$ **29a.** $x = -15$, $x = 5$, $x = 20$
  29b. Three solutions: approximately -13, 2, 22

CHAPTER 2 REVIEW [Page 133]
  1a. $C = 20n + 2000$

1b.

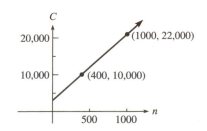

1c. $22,000   **1d.** 400   **2a.** $W = 18m + 80$

2b.

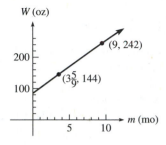

2c. 242 oz, or 15 lb 2 oz

2d. $3\frac{5}{9}$ months, or 3 months, 17 days

3a. $R = 1660 - 20t$

3b.

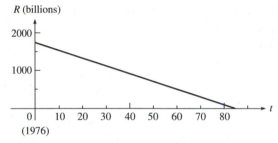

3c. When $t = 0$ (1976), R was 1660. R will be 0 when $t = 83$ (in 2059).   **4a.** $R = 500 - 8t$

4b.

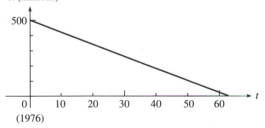

4c. When $t = 0$ (1976), R was 500. R will be 0 when $t = 62.5$ (in 2038).   **5a.** $5A + 2C = 1000$

5b.

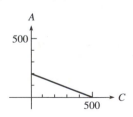

5c. 200 children's tickets

5d. x-intercept (500): only children's tickets are sold. y-intercept (200): only adults' tickets are sold.

6a. $60A + 100S = 1200$

6b.

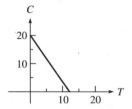

6c. 6 days

6d. x-intercept: all days in Saint-Tropez y-intercept: all days in Atlantic City

7.

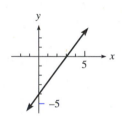

8.

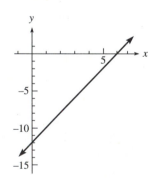

9.

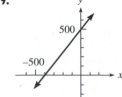

10.

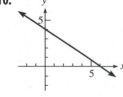

11.

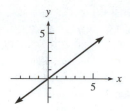

12.

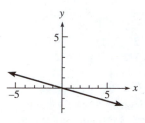

13.

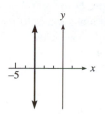

14.

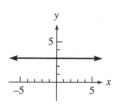

15a. $B = 800 - 5t$

15b.

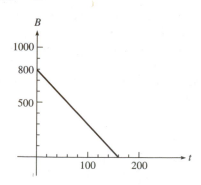

15c. $m = -5$ barrels per minute gives the rate at which the oil is leaking.

16a. $h = 12 - 0.05t$

16b.

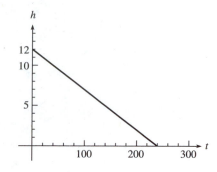

16c. $m = -0.05$ in. per minute gives the rate at which the candle is growing shorter.

17a. $F = 500 + 0.10C$

17b.

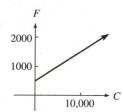

17c. $m = 0.10$ The decorator charges 10% of the cost of the job (plus a flat $500 fee).

18a. $R = 35 + 0.02V$

18b.

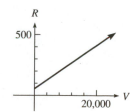

18c. $m = 0.02$ The registration fee is 2% of the value of the car (plus a flat $35 fee).

19. $\dfrac{-3}{2}$ **20.** 2 **21.** -0.4 **22.** -1.7

23. $d = 1$; $V = 4.2$ **24.** $S = 168$; $q = 5$

25. 80 ft **26.** 24 ft **27.** $m = \dfrac{1}{2}$, $b = \dfrac{-5}{4}$

28. $m = \dfrac{-3}{4}$, $b = \dfrac{5}{4}$ **29.** $m = -4$; $b = 3$

30. $m = 0$; $b = 3$

31a.

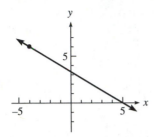

31b. $y = \dfrac{-2}{3}x + \dfrac{10}{3}$

32a.

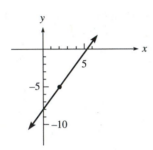

32b. $y = \dfrac{3}{2}x - 8$ **33.** $y = \dfrac{-9}{5}x + \dfrac{2}{5}$

34. $y = \dfrac{5}{2}x + 8$ **35a.** $P = 4800 + 132t$

35b. $m = 132$ people per year gives the rate of population growth.

36a. $M = 112 + 28g$

36b. $m = 28$ mpg gives the gas mileage (rate of gas consumption).

37a. $m = -2$, $b = 3$ **37b.** $y = -2x + 3$

38a. $m = \dfrac{3}{2}$, $b = -5$ **38b.** $y = \dfrac{3}{2}x - 5$

39. Parallel   **40.** Perpendicular

41. $y = \dfrac{-2}{3}x + \dfrac{14}{3}$ **42.** $y = \dfrac{3}{2}x + \dfrac{5}{2}$

43. $y = \dfrac{2}{3}x - \dfrac{26}{3}$ **44.** $y = \dfrac{3}{2}x - 2$

45a. $-5, 7$ **45b.** $-10 \le x \le 12$

46a. $10, 30$ **46b.** $x < 15$ or $x > 25$

47a. $-1, 1$ **47b.** $-3 < x < 2$ or $2 < x < 3$

48a. 1 **48b.** $x < \dfrac{1}{4}$ or $x > 4$

Exercise 3.1 [Page 146]

1. $(3, 0)$ **3.** $(50, 70)$ **5.** $(7, 2)$ **7.** $(-2, 3)$
9. $(2, 3)$ **11.** $(-80, 70)$ **13.** $(-4.6, 52)$
15. $(7.15, 4.3)$ **17.** Inconsistent   **19.** Dependent
21. Consistent   **23.** Inconsistent
25. 12 hardbacks, 36 paperbacks
27. 30 cents per bushel, 1500 bushels   **29.** 25 pendants
31. 125 min   **33.** 36 adults, 46 students

Exercise 3.2 [Page 155]

1. $(3, 0)$ **3.** $(2, 1)$ **5.** $(1, 2)$ **7.** $(1, -2)$
9. $(6, 0)$ **11.** $(1, 2)$ **13.** $(1, 2)$ **15.** $(2.3, 1.6)$
17. $(182, 134)$ **19.** Dependent   **21.** Inconsistent
23. Consistent   **25.** Dependent
27. 3595 for winner, 3584 for loser
29. \$800 at 8%, \$1200 at 10%   **31.** 32 lb
33. Leon: 180 mph; Marlene: 60 mph
35. Airplane: 300 mph; wind: 20 mph
37. 0.6 cup oats, 0.4 cup wheat
39. \$34 per pair, 1190 pairs

Exercise 3.3 [Page 166]

1. $(1, 2, -1)$ **3.** $(2, -1, -1)$ **5.** $(4, 4, -3)$
7. $(2, -2, 0)$ **9.** $(2, 2, 1)$ **11.** $(0, -2, 3)$
13. $(-1, 1, -2)$ **15.** $\left(\dfrac{1}{2}, \dfrac{2}{3}, -3\right)$ **17.** $(4, -2, 2)$
19. $(1, 1, 0)$ **21.** $\left(\dfrac{1}{2}, -\dfrac{1}{2}, \dfrac{1}{3}\right)$ **23.** Inconsistent
25. $\left(\dfrac{1}{2}, 0, 3\right)$ **27.** Dependent   **29.** $(-1, 3, 0)$
31. Inconsistent   **33.** $\left(\dfrac{1}{2}, \dfrac{1}{2}, 3\right)$
35. 60 nickels, 20 dimes, 5 quarters
37. $x = 40$ in., $y = 60$ in., $z = 55$ in.
39. 0.3 cup carrots, 0.4 cup green beans, 0.3 cup cauliflower
41. 40 score only, 20 evaluation, 80 narrative report
43. 20 tennis, 15 Ping-Pong, 10 squash

Exercise 3.4 [Page 175]

1. 3:00 A.M.: 77°; 4:00 A.M.: 75°
3. Third day: $3\dfrac{4}{5}$ cm; fifth day: $6\dfrac{3}{5}$ cm
5. 2 s: 20 mph; 4 s: 40 mph
7. Midnight: 20.5°C; 2:00 A.M.: 18°C

9. 2 min: 21°C; 2 hr: 729°C   **11.** 128 lb
13b. 60, 10   **13c.** $y = -10x + 90$ **13d.** 45
15b. 129 lb, 145 lb   **15c.** $y = 2.\overline{6}x - 44.\overline{3}$
15d. 137 lb   **17b.** 275, 450   **17c.** $y = 1.75x - 92.5$
17d. 293   **19.** $y = -9.97x + 90.23$, 45.37
21. $y = 2.84x - 55.74$, 137.33 lb
23. $y = 1.75x - 95.92$, 288

Exercise 3.5 [Page 186]

1.

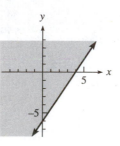

3.

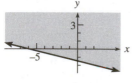

5.

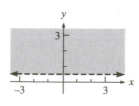

7.

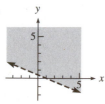

9.

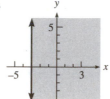

11.

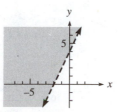

13.

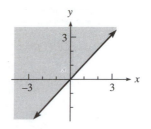

15.

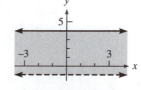

17.

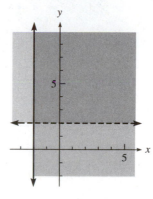

19.

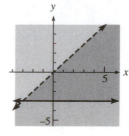

21.

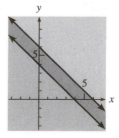

23.

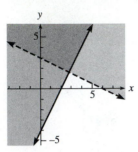

25.

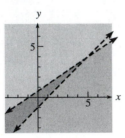

27.

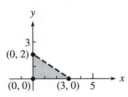

29.

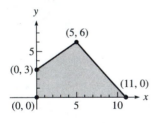

31.

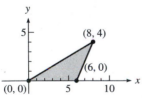

33.

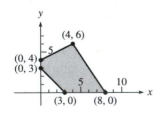

35.

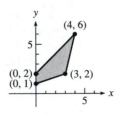

37.

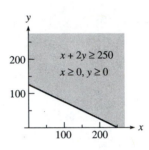

$x + 2y \geq 250$
$x \geq 0, y \geq 0$

39.

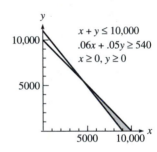

$x + y \leq 10,000$
$.06x + .05y \geq 540$
$x \geq 0, y \geq 0$

41.

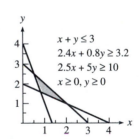

$x + y \leq 3$
$2.4x + 0.8y \geq 3.2$
$2.5x + 5y \geq 10$
$x \geq 0, y \geq 0$

Exercise 3.6 [Page 195]

1. The graph of $12 = 3x + 4y$ does not intersect the set of feasible solutions. **3.** (8, 2); $32

5. (2, 0)   **7a.** $22   **7b.** (8, 0)   **7c.** $32
9a. (1, 4)   **9b.** 7   **9c.** (4, 5)   **9d.** 17
11a. (0, 5)   **11b.** −10   **11c.** (5, 0)   **11d.** 25
13b. (0, 0); 0   **13c.** (3, 2); 13   **15b.** (0, 14); −14
15c. (10, 0); 30   **17b.** (0, 8); −160   **17c.** (8, 0); 1600
19b. (1, 0); 18   **19c.** (5, 2); 186   **21.** 250

23. $5040
25. 986.4 calories
27. Maximum: 17.4; minimum: -8.4
29. Maximum: 4112; minimum: 0
31. Maximum: 1908; minimum: 0

CHAPTER 3 REVIEW [Page 199]

1. $x = -1, \quad y = 2$

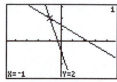

2. $x = 1.9, \quad y = -0.8$

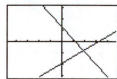

3. $x = \frac{1}{2}, \quad y = \frac{7}{2}$ **4.** $x = 1, \quad y = 2$

5. $x = 12, \quad y = 0$ **6.** $x = \frac{1}{2}, \quad y = \frac{3}{2}$

7. Consistent **8.** Inconsistent **9.** Dependent
10. Consistent **11.** $x = 2, \quad y = 0, \quad z = -1$
12. $x = 2, \quad y = 1, \quad z = -1$
13. $x = 2, \quad y = -5, \quad z = 3$
14. $x = -2, \quad y = \frac{3}{2}, \quad z = -1$
15. $x = -2, \quad y = 1, \quad z = 3$
16. $x = 2, \quad y = -1, \quad z = 0$
17. 26 **18.** 17
19. $3181.82 at 8%, $1818.18 at 13.5%
20. $4000 **21.** 5 cm, 12 cm, 13 cm
22. 25 crates to Chicago, 20 crates to Boston, 10 crates to Los Angeles
23.

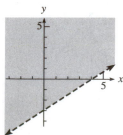

24.

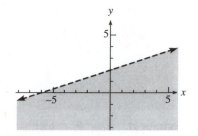

25.

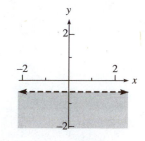

26.

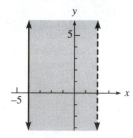

27.

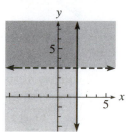

28.

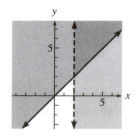

29.

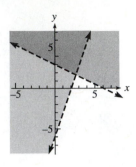

30.

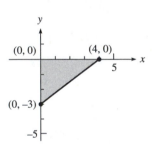

31.

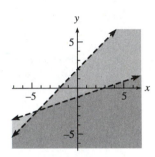

Vertices are (0, 0), (0, −3), (4, 0)

32.

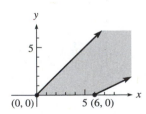

Vertices are (0, 0), (6, 0)

33.

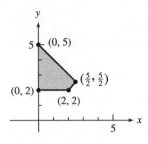

Vertices are $(0, 2)$, $(0, 5)$, $(2, 2)$, $\left(\dfrac{5}{2}, \dfrac{5}{2}\right)$

34.

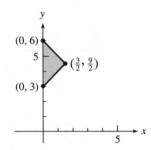

Vertices are $(0, 3)$, $(0, 6)$, $\left(\dfrac{3}{2}, \dfrac{9}{2}\right)$

35.

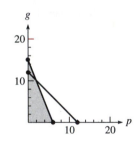

$$20p + 8g \le 120$$
$$10p + 10g \le 120$$

36.

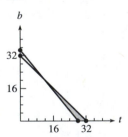

$$t + b \le 32$$
$$2t + 1.6b \ge 56$$

37. 2 batches of peanut butter cookies, 10 batches of granola cookies

38. 28 g of tofu, 0 g of brown rice **39.** 6

40. 26 min **41a.** 850 **41b.** 1450

41c. $y = 30x - 50$ **41d.** 1300

41e. $y = 31.10x - 93.91$; 1305.59 **42a.** 45 cm

42b. 87 cm **42c.** $y = 1.2x - 3$ **42d.** 69 cm

42e. $y = 1.197x - 3.660$; 68.16

Exercise 4.0 [Page 202]

1. $3x^2 - 15x$ **3.** $2b^2 + 9b - 18$

5. $16w^2 - 24w + 9$ **7.** $6p^3 - 33p^2 + 45p$

9. $-50 - 100r - 50r^2$ **11.** $12q^4 - 36q^3 + 27q^2$

13. $(x - 5)(x - 2)$ **15.** $(x - 15)(x + 15)$

17. $(w - 8)(w + 4)$ **19.** $(2z - 5)(z + 8)$

21. $(3n + 4)^2$ **23.** $3a^2(a + 1)^2$

25. $4h^2(h - 3)(h + 3)$ **27.** $-10(u + 13)(u - 3)$

29. $6t^2(4t^2 + 1)$

Exercise 4.1 [Page 208]

1c. Base: $36 - 2x$; perimeter: $72 - 2x$; area: $36x - 2x^2$

1f. Area: 162 sq in.; 9 in. by 18 in.

1g. Two possible rectangles, one with height 11.5 in. and the other with height 6.5 in.

3b. 306.25 ft at 0.625 s **3c.** 1.25 s **3d.** 5 s

5b. Price of a room: $20 + 2x$; rooms rented: $60 - 3x$; revenue: $1200 + 60x - 6x^2$

5d. 20 **5f.** $24; $36 **5g.** $1350; $30; 45 rooms

Exercise 4.2 [Page 215]

1. $x = 2, -2.5$ **3.** $x = 0, -\dfrac{10}{3}$ **5.** $x = 3, -\dfrac{3}{2}$

7. $x = -\dfrac{3}{4}, -8$ **9.** $x = 4, 4$ **11.** $a = \dfrac{1}{2}, -3$

13. $x = 0, 3$ **15.** $y = 1, 1$ **17.** $x = \dfrac{1}{2}, 1$

19. $t = 2, 3$ **21.** $z = -1, 2$ **23.** $v = -3, 6$

25 and 27. The graphs have the same x-intercepts. In general, the graph of $y = ax^2 + bx + c$ has the same x-intercepts as the graph of $y = k(ax^2 + bx + c)$.

29. $x^2 + x - 2 = 0$ **31.** $x^2 + 5x = 0$

33. $2x^2 + 5x - 3 = 0$ **35.** $8x^2 - 10x - 3 = 0$

37. $0.1(x - 18)(x + 15)$

39. $-0.08(x - 18)(x + 32)$ **41.** 5, -1

43. $\dfrac{5}{2}, -\dfrac{3}{2}$ **45.** $-2 \pm \sqrt{3}$ **47.** $\dfrac{1}{2} \pm \dfrac{\sqrt{3}}{2}$

49. $\dfrac{-2}{9}, \dfrac{-4}{9}$ **51.** $\dfrac{7}{8} \pm \dfrac{\sqrt{8}}{8}$ **53.** 5.26 cm

55. 24 ft **57a.** $h = -16t^2 + 16t + 8$

57b. 12 ft; 8 ft **57c.** At $\dfrac{1}{4}$ s and $\dfrac{3}{4}$ s

59a. $l = 180 - x$; $A = 180x - x^2$

59c. 100 yd by 80 yd **61a.** $2(x - 4)^2$

61b. V increases as x increases for $x > 4$.

61c. 9 in. by 9 in. **63a.** $20 + x$; $600 - 10x$

63b. $12,000 + 400x - 10x^2$ **63c.** $13,750; $15,000

63d. 35 or 45

Exercise 4.3 [Page 226]

1. $(x + 4)^2$ **3.** $\left(x - \dfrac{7}{2}\right)^2$ **5.** $\left(x + \dfrac{3}{4}\right)^2$

7. $\left(x - \dfrac{2}{5}\right)^2$ **9.** 1, 1 **11.** $-4, -5$

13. $\dfrac{-3 \pm \sqrt{21}}{2}$ **15.** $1 \pm \sqrt{\dfrac{5}{2}}$ **17.** $\dfrac{1 \pm \sqrt{13}}{4}$

19. $\dfrac{4}{3}, -1$ **21.** 1.618, -0.618 **23.** 1.449, -3.449

25. 1.695, -0.295 **27.** 1.434, 0.232

29. $-5.894, 39.740$ **31.** 29.16 mph **33a.** 24.5 s

33b. 1.2 s **35.** 11.8%

37. 18.09 ft by 4.61 ft, or 6.91 ft by 12.06 ft

39a. 45 mi **39b.** 1.26 mi **41.** $\dfrac{-b \pm \sqrt{b^2 - 4c}}{2}$

43. $\dfrac{-b}{a}$

Exercise 4.4 [Page 238]

For Exercises 1–15 each graph is shown compared to the standard parabola.

1.
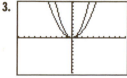

3.

5.
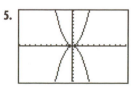

7.
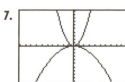

9.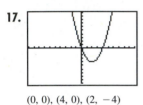

11.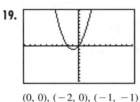

13.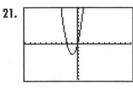

15.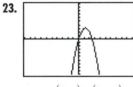

17.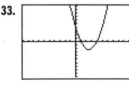

$(0, 0), (4, 0), (2, -4)$

19.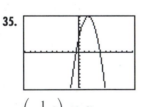

$(0, 0), (-2, 0), (-1, -1)$

21.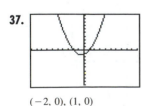

$(0, 0), (-2, 0), (-1, -3)$

23.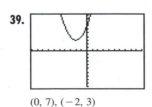

$(0, 0), \left(\frac{5}{2}, 0\right), \left(\frac{5}{4}, 3\frac{1}{8}\right)$

25. $(1, 1)$ **27.** $(1.5, 4.25)$ **29.** $\left(\frac{2}{3}, \frac{1}{9}\right)$

31. $(-4.5, 18.5)$

33.

$(1, 0), (4, 0)$

$(0, 4), \left(\frac{5}{2}, -2\frac{1}{4}\right)$

35.

$\left(-\frac{1}{2}, 0\right), (4, 0)$

$(0, 4), \left(\frac{7}{4}, 10\frac{1}{8}\right)$

37.

$(-2, 0), (1, 0)$
$(0, -1.2), (-0.5, -1.35)$

39.

$(0, 7), (-2, 3)$

41.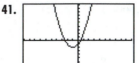

$(0.41, 0), (-2.41, 0)$
$(0, -1), (-1, -2)$

43.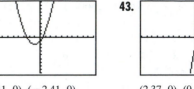

$(2.37, 0), (0.63, 0)$
$(0, -3), (1.5, 1.5)$

45a. $y = x^2 + x - 6$ **45b.** $y = 2x^2 + 2x - 12$
47. $y = x^2 + x + 2; \quad y = 3x^2 - 6x + 2$
49a. $(3, 4)$ **49b.** $y = 2x^2 - 12x + 22$

51a. $(-4, -3)$ **51b.** $y = -\frac{1}{2}x^2 - 4x - 11$

53. We can substitute the coordinates of the vertex and need only to determine the value of a.

55a. $y = a(x + 2)^2 + 6$ **55b.** 3   **57.** $y = ax^2 - 3$

Exercise 4.5 [Page 247]

1. $x < -12, \quad$ or $\quad x > 15$ **3.** $0.3 < x < 0.5$
5. $x < -2 \quad$ or $\quad x > 3$ **7.** $0 \le k \le 4$
9. $p < -1 \quad$ or $\quad p > 6$ **11.** $-4.8 < x < 6.2$
13. $x \le -7.2 \quad$ or $\quad x \ge -0.6$
15. $x < 5.2 \quad$ or $\quad x > 8.8$ **17.** $x < -3.5 \quad$ or $\quad x > 3.5$
19. All x **21.** $-10.6 < x < 145.6$ **23.** $(-5, 3]$
25. $[-4, 0]$ **27.** $(-6, \infty)$
29. $(-\infty, -3) \cup [-1, \infty)$ **31.** $[-6, -4) \cup (-2, 0]$
33. $\left(-\infty, -\frac{1}{2}\right) \cup (4, \infty)$ **35.** $(-2.24, 2.24)$
37. $(-\infty, 0.4] \cup [6, \infty)$ **39.** $4 < t < 16$
41. $0 \le x < 100 \quad$ or $\quad 600 < x \le 700$
43. $10 < p < 30$

Exercise 4.6 [Page 256]

1. 3 s, 144 ft   **3a.** $50w - w^2$ **3b.** 625 sq in.
5a. $300w - 2w^2$ **5b.** 11,250 sq yd
7a. $16 + x; \quad 2400 - 100x; \quad 38,400 + 800x - 100x^2$
7b. 20   **9.** 100; $8   **11.** $a = 0.9; \quad I = \$865.8$
13. $a = 3, \quad b = 1, \quad c = -2$
15a. $P = -0.16x^2 + 7.4x - 71$ **15b.** 14%

15c.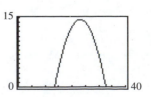

17a. $C = 0.75t^2 - 1.85t + 56.2$ **17b.** 65.7 lb

17c.

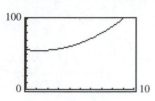

19. $D = \dfrac{1}{2}n^2 - \dfrac{3}{2}n$

21. $y = 0.00012x^2 - 0.48x + 500$

23. $(-1, 12), (4, 7)$ **25.** $(-2, 7)$ **27.** None

29. $(-2, -5), (5, 16)$ **31.** $(1, 4)$ **33.** $(3, 1)$

35. $(200, 5600), (1000, 12{,}000)$, maximum revenue: $12,800

37. $(50, 30{,}625), (250, 65{,}625)$, maximum revenue: $70,000

CHAPTER 4 REVIEW [Page 261]

1. $x = -1, \quad x = 2$ **2.** $y = 2$

3. $x = -2, \quad x = 3$ **4.** $x = \pm 1$

5. $4x^2 - 29x - 24 = 0$ **6.** $9x^2 - 30x + 25 = 0$

7. $y = (x - 3)(x + 2.4)$ **8.** $y = -(x + 1.3)(x - 2)$

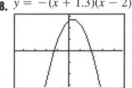

9. $x = 1, \quad x = 4$ **10.** $x = \dfrac{1 \pm \sqrt{15}}{7}$

11. $x = 2 \pm \sqrt{10}$ **12.** $x = \dfrac{-3 \pm \sqrt{21}}{2}$

13. $x = \dfrac{3 \pm \sqrt{3}}{2}$ **14.** $x = \dfrac{1 \pm \sqrt{10}}{3}$

15. $x = 1, \quad x = 2$ **16.** $x = 2.62, x = 0.38$

17. $x = 3.41, x = 0.59$ **18.** $x = 0.82, x = -1.82$

19. 9   **20.** 13   **21.** 11%   **22.** 8.5%   **23.** 120

24. 10 ft by 18 ft or 12 ft by 15 ft

25. Vertex: $(0, 0)$; x-intercept: $(0, 0)$; y-intercept: $(0, 0)$

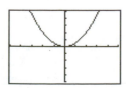

26. Vertex: $(0, -4)$; x-intercepts: $(2, 0)$ and $(-2, 0)$; y-intercept: $(0, -4)$

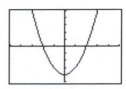

27. Vertex: $\left(\dfrac{9}{2}, -\dfrac{81}{4} \right)$; x-intercepts: $(0, 0)$ and $(9, 0)$; y-intercept: $(0, 0)$

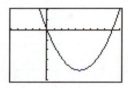

28. Vertex: $(-1, 2)$; x-intercepts: $(0, 0)$ and $(-2, 0)$; y-intercept: $(0, 0)$

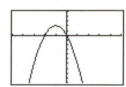

29. Vertex: $\left(\dfrac{1}{2}, -\dfrac{49}{4} \right)$; x-intercepts: $(-3, 0)$ and $(4, 0)$; y-intercept: $(0, -12)$

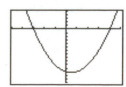

30. Vertex: $\left(\dfrac{1}{4}, -\dfrac{31}{8} \right)$; no x-intercepts; y-intercept: $(0, -4)$

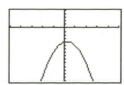

31. Vertex: $(1, 5)$; x-intercepts: $(-1.24, 0)$ and $(3.24, 0)$; y-intercept: $(0, 4)$

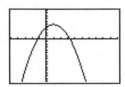

32. Vertex: $\left(\dfrac{3}{2}, \dfrac{7}{4}\right)$; no x-intercepts; y-intercept: $(0, 4)$

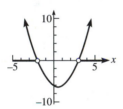

33. $(-\infty, -2) \cup (3, \infty)$

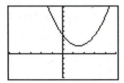

34. $[-3, 4]$

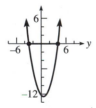

35. $\left[-1, \dfrac{3}{2}\right]$

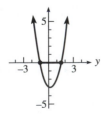

36. $\left(-\infty, -\dfrac{1}{3}\right) \cup (2, \infty)$

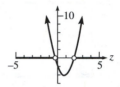

37. $[-2, 2]$

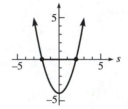

38. $(-\infty, -\sqrt{3}) \cup (\sqrt{3}, \infty)$

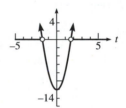

39. $200 < p < 280$ **40.** $20 < p < 40$
41. 18   **42.** \$35
43. $(1, 3), (-1, 3)$

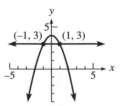

44. $(1, 2), (4, -13)$

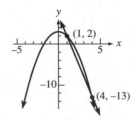

45. $(-1, -4)$, $(5, 20)$

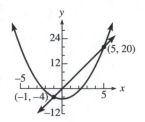

46. $(-1, 4)$, $(2, 1)$

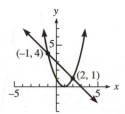

47. $(-9, 155)$, $(5, 15)$

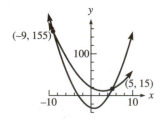

48. $\left(-\frac{7}{2}, \frac{7}{4}\right)$, $(8, -4)$

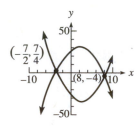

49. $a = 1$, $b = -1$, $c = -6$

50. $y = -\frac{1}{2}x^2 - 4x + 10$

Exercise 5.1 [Page 271]

1. Function; tax determined by price of item

3. Not a function; incomes may differ for same number of years of education

5. Function; weight determined by volume

7. Independent: items purchased; dependent: price of item. Yes, a function.

9. Independent: topics; dependent: page or pages on which topic occurs. No, not a function.

11. Independent: students' names; dependent: students' scores on tests, quizzes, etc. No, not a function.

13. Independent: person stepping on scales; dependent: person's weight. Yes, a function.

15. No   **17.** Yes   **19.** Yes   **21.** Yes   **23.** No

25. Yes   **27a.** 0   **27b.** 10   **27c.** -19.4

27d. $\frac{14}{3}$ **29a.** 1   **29b.** 6   **29c.** $\frac{3}{8}$ **29d.** 96.48

31a. $\frac{5}{6}$ **31b.** 9   **31c.** $-\frac{1}{10}$ **31d.** ≈ 0.923

33a. $\sqrt{12}$ **33b.** 0   **33c.** $\sqrt{3}$ **33d.** ≈ 0.447

35a. 60   **35b.** 37.5   **35c.** 30   **37a.** 15%

37b. 14%   **37c.** $7010–$9169

39a. Independent: time in years; dependent: value of computer

39c. $V(10) = 11,200$ is the value of the computer after 10 years.

41a. Independent: number of clients; dependent: revenue

41c. $R(40) = 72,800$ thousand dollars is the revenue generated from 40 clients.

43a. Independent: price of a car; dependent: number of cars sold

43c. $N(6000) = 2000$ is the number of cars the dealership will sell at $6000 apiece.

45a. Independent: length of skid marks; dependent: velocity before braking

45c. $v(250) \approx 54.77$ mph is the velocity of a car that left skid marks 250 ft long.

47a. $27a^2 - 18a$ **47b.** $3a^2 + 6a$

47c. $3a^2 - 6a + 2$ **47d.** $3a^2 + 6a$ **49a.** 8

49b. 8   **49c.** 8   **49d.** 8   **51a.** $8x^3 - 1$

51b. $2x^3 - 2$ **51c.** $x^6 - 1$ **51d.** $x^6 - 2x^3 + 1$

53a. 11   **53b.** 13   **53c.** $3a + 3b - 4$

53d. $3a + 3b - 2$; $f(a) + f(b) \neq f(a + b)$

55a. 19   **55b.** 28   **55c.** $a^2 + b^2 + 6$

55d. $a^2 + 2ab + b^2 + 3$; $f(a) + f(b) \neq f(a + b)$

57a. $\sqrt{3} + 2$ **57b.** $\sqrt{6}$

57c. $\sqrt{a + 1} + \sqrt{b + 1}$

57d. $\sqrt{a + b + 1}$; $f(a) + f(b) \neq f(a + b)$

59a. $-\dfrac{5}{3}$ **59b.** $-\dfrac{2}{5}$ **59c.** $-\dfrac{2}{a} - \dfrac{2}{b}$

59d. $-\dfrac{2}{a + b}$; $f(a) + f(b) \neq f(a + b)$

61a. $0, 5$ **61b.** $0, -\dfrac{3}{2}$ **61c.** $\dfrac{5}{6}$ **61d.** $-5, \dfrac{1}{2}$

63a. $\sqrt{2}, -4$ **63b.** -2 **63c.** $\dfrac{4}{3}$ **63d.** $2, \dfrac{7}{9}$

65. $f(x) = 3x - 2$ **67.** $G(t) = t^2 + 1$

Exercise 5.2 [Page 284]

1a. $-2, 0, 5$ **1b.** 2
1c. $h(-2) = 0, \quad h(0) = -2, \quad h(1) = 0$ **1d.** 5
1e. 3 **1f.** Domain: $[-4, 3]$; range: $[-5, 5]$
3a. $-1, 2$ **3b.** $3, \approx -1.25$
3c. $R(-2) = 0, \quad R(0) = 4, \quad R(2) = 0, \quad R(4) = 0$
3d. Maximum: 4; minimum: -5
3e. Maximum at $p = 0$;   minimum at $p = 5$
3f. Domain: $[-3, 5]$; range: $[-5, 4]$ **5a.** $0, \dfrac{1}{2}, 0$

5b. $\approx \dfrac{5}{6}$ **5c.** $-\dfrac{5}{6}, -\dfrac{1}{6}, \dfrac{7}{6}, \dfrac{11}{6}$

5d. Maximum: 1; minimum: -1
5e. Maximum at $x = -1.5, 0.5$;   minimum at $x = -0.5, 1.5$
5f. Domain: $[-2, 2]$; range: $[-1, 1]$ **7a.** $2, 2, 1$
7b. $-6 \leq s < -4$ or $0 \leq s < 2$
7c. Maximum: 2; minimum: -1
7d. Maximum for $-3 \leq s < -1$ or $3 \leq s < 5$; minimum for $-6 \leq s < -4$ or $0 \leq s < 2$
7e. Domain: $[-6, 5)$; range: $\{-1, 1, 2\}$
9. Domain: $[-2, 6]$; range: $[-4, 12]$
11. Domain: $[-5, 3]$; range: $[-15, 1]$
13. Domain: $[-2, 2]$; range: $[-9, 7]$
15. Domain: $[-1, 8]$; range: $[0, 3]$

17. Domain: $[-1.25, 2.75]$; range: $\left[\dfrac{4}{17}, 4\right]$

19. Domain: $(3, 6]$; range: $\left(-\infty, -\dfrac{1}{3}\right]$ **21.** Yes

23. No   **25.** No   **27.** Yes   **29.** No

Exercise 5.3 [Page 295]

1.

3.
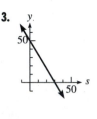

5.
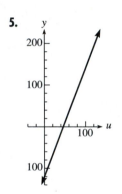

7.
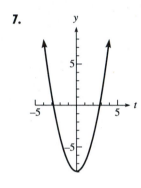

9.
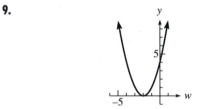

11.
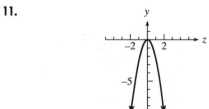

13. 8: $8^3 = 512$ **15.** -0.4: $(-0.4)^3 = -0.064$
17. $2.080^3 \approx 8.9989$
19. -0.126: $(-0.126)^3 \approx -0.0020004$ **21a.** -9
21b. 9 **23a.** -4 **23b.** 20 **25.** -50 **27.** 144
29. 1 **31–35.** See text.   **37a.** $3.2, 6.3$
37b. $\pm 2.6, \pm 3.5$ **37c.** $3.2, -2.5$ **37d.** $\pm 3.9, \pm 1.9$
37e. $10.2, 5.3$ **37f.** $2 \leq x \leq 3.2, \quad -3.2 \leq x \leq -2$

39a. 0.3, −1.3   **39b.** −0.4, 0.9   **39c.** 0.2, −5
39d. 0.2 ≤ h ≤ 3.3

41.

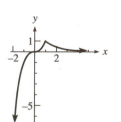

43.

45.

47.

49.

51.

53. $y = \sqrt{x}$ **55.** $y = |x|$ **57.** $y = x^3$
59. $y = \sqrt[3]{x}$ **61.** $y = \dfrac{1}{x}$ **63.** $y = \dfrac{1}{x^2}$
65. $f(x)$ **67.** $g(x)$ **69.** $g(x)$
71. Increasing: x^3, $\sqrt{x}$, $\sqrt[3]{x}$, x; decreasing: $\dfrac{1}{x}$
73. x^3, $|x|$, $\sqrt{x}$, $\sqrt[3]{x}$, x, x^2

Exercise 5.4 [Page 307]

1a. $y = 0.3x$ **1b.** 0.6, 8, 0.3, 15
1d. y doubles also   **3a.** $y = \dfrac{2}{3}x^2$ **3b.** 6, 9, 96, 15
3d. y is quadrupled   **5a.** $y = \dfrac{120}{x}$
5b. 30, 8, 4, 40   **5d.** y is halved   **7.** b   **9.** c
11. b: $k = 0.5$ **13.** c   **15.** b: $k = 72$ **17.** c
19b. No   **19c.** $P = 2l + 16$ **19d.** Yes
19e. $A = 8l$ **21a.** $m = 0.165w$ **21b.** 19.8 lb
21c. 303.03 lb   **23a.** $L = 0.8125T^2$
23b. 234.8125 ft   **23c.** 0.96 s   **25a.** $F = \dfrac{88}{d}$
25b. 8.8 milligauss   **25c.** More than 20.47 in.
27a. $P = 0.228w^3$ **27b.** 752 kilowatts
27c. 35.54 mph   **29a.** $d = 0.005v^2$ **29b.** 50m
31a. $m = \dfrac{8}{p}$ **31b.** 0.8 ton   **33a.** $T = \dfrac{6}{d}$ **33b.** 1°C
35a. $W = 600d^2$ **35b.** 864 newtons
37. One-fourth of original illumination
39. 81% of original resistance
41. If $y_1 = kx_1$, then $y_2 = k(cx_1) = ky_1$.
43. If $y = kx^2$, then $\dfrac{y}{x^2} = k$.   **45.** Yes

Exercise 5.5 [Page 325]

1. b   **3.** a

5.

7.

9.

11a: Table (4), Graph (C)   **11b.** Table (3), Graph (B)
11c. Table (1), Graph (D)   **11d.** Table (2), Graph (A)
13. $|x| = 6$ **15.** $|p + 3| = 5$ **17.** $|t - 6| \leq 3$
19. $|b + 1| \geq 0.5$ **21a.** $|x + 12|, |x + 4|, |x - 24|$
21b. $f(x) = |x + 12| + |x + 4| + |x - 24|$
21c. She should stand at x-coordinate -4.
23. They should find an apartment as close to the health club as possible.
25a. $x = -5, \quad x = -1$ **25b.** $-7 \leq x \leq 1$
25c. $x < -8 \quad$ or $\quad x > 2$ **27.** $x = -\dfrac{3}{2}, \quad x = \dfrac{5}{2}$
29. $\dfrac{-9}{2} < x < \dfrac{-3}{2}$ **31.** $x \leq -2 \quad$ or $\quad x \geq 5$
33a. $k = 50; \quad s = 50\sqrt{d}$ **33b.** 30 cm/s
35a. $k = 1.225; \quad d = 1.225\sqrt{h}$ **35b.** 173.24 mi
37a. $2a$ **37b.** $\sqrt{16 - a^2}$ **37c.** $2a\sqrt{16 - a^2}$
37d. $0 < a < 4$ **37e.** 13.86, 1.32, or 3.77   **37f.** 2.85
39a. $\dfrac{1}{2 - a}$ **39b.** $\dfrac{a}{a - 2}$ **39c.** $\dfrac{a^2}{2(a - 2)}$
39d. $a > 2$ **39e.** 4.5; 2.4 and 12
39f. The smallest area (2 square units) occurs when $a = 4$. The triangle is undefined when $a = 2$, and for values just larger than 2, the area of the triangle becomes smaller as a increases to 4. After that, the area increases again.
41a. $r = \dfrac{C}{2\pi}$ **41b.** $A = \dfrac{C^2}{4\pi}$
41c. Area is a quadratic polynomial (or monomial) of C. The domain is the set of positive reals.
41d. $\dfrac{2500}{\pi}$ sq yd
43.

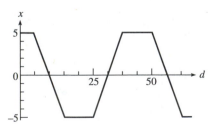

CHAPTER 5 REVIEW [Page 332]

1. Function; there is a unique value of the dependent variable for each value of the independent variable.
2. Not a function; there is not a unique value of the dependent variable for each value of the independent variable.
3. Not a function; there is not a unique value of the dependent variable for each value of the independent variable.
4. Function; there is a unique value of the dependent variable for each value of the independent variable.
5. $N(10) = 7000$; this is the number of barrels of oil 10 days after a new well is opened.
6. $H(16) = 3$; this is the number of hours it takes for the boat to travel upstream between the two cities if the boat's top speed is 16 mph.
7. $F(0) = 1; \quad F(-3) = \sqrt{37} \approx 6.08$
8. $H(2a) = 4a^2 + 4a; \quad H(a + 1) = a^2 + 4a + 3$
9. $f(2) + f(3) = -11; \quad f(2 + 3) = -13$
10. $f(a) + f(b) = 2a^2 + 2b^2 - 8;$
$f(a + b) = 2a^2 + 4ab + 2b^2 - 4$
11a. $P(0) = 5$ **11b.** $x = 1, \quad x = 5$
12a. $R(0) = 2$ **12b.** $x = \pm 2$
13a. $f(-2) = 3, \quad f(2) = 5$ **13b.** $t = 1, 3$
13c. t-intercepts: $-3, 4; f(t)$-intercept: 2
13d. Maximum value of f is 5 for $t = 2$
14a. $P(-3) = -2, \quad P(3) = 3$
14b. $z = -5, -\dfrac{1}{2}, 4$
14c. z-intercepts: $-4, -1, 5; P(z)$-intercept: 3
14d. Minimum value of P is -3 for $z = -2$
15. Domain: $[-2, -4]$; range: $[-10, 2.25]$

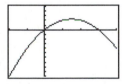

16. Domain: $[2, 6]$; range: $[0, 2]$

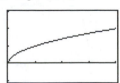

17. Domain: $(-2, 4]$; range: $\left[\dfrac{1}{6}, \infty\right)$

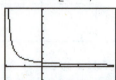

18. Domain: $[-4, 2)$; range: $\left[\dfrac{1}{6}, \infty\right)$

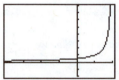

19. Function   **20.** Not a function   **21.** Not a function
22. Function
23.

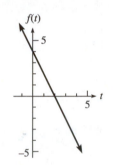

24.

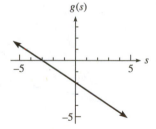

25.

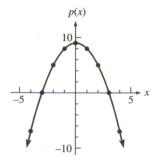

26.

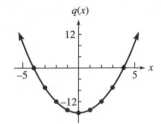

27.

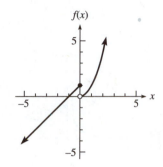

28.

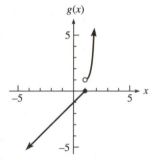

29.

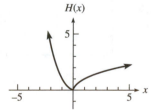

30.

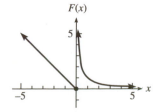

31.

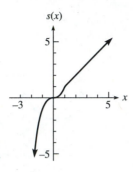

32.

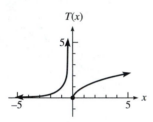

33a. $x = 0.5$ **33b.** $x = 3.4$ **33c.** $x > 4.9$
33d. $x \leq 2.0$

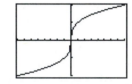

34a. $x = 0.4$ **34b.** $x = 3.2$ **34c.** $x \leq 4.5$
34d. $x > 0.2$ or $x < 0$

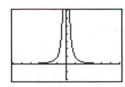

35a. $x = \pm 5.8$ **35b.** $x = \pm 0.4$
35c. $-2.5 < x < 2.5$ **35d.** $x \leq -0.5$ or $x \geq 0.5$

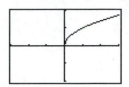

36a. $x = 0.5$ **36b.** $x = 2.9$ **36c.** $x < 2.25$
36d. $x \geq 1.7$ **37.** $y = 1.2x^2$ **38.** $y = 54x$

39. $y = \dfrac{20}{x}$ **40.** $y = \dfrac{720}{x^2}$ **41a.** $s = 1.75t^2$

41b. 63 cm   **42a.** $V = \dfrac{4T}{P}$ **42b.** 32   **43.** 480

44. 14.0625 lumens   **45a.** $w = \dfrac{k}{r^2}$

45b.

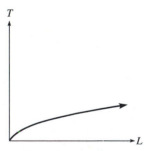

45c. $3960\sqrt{3} \approx 6858.92$ mi   **46a.** $T = k\sqrt{L}$

46b.

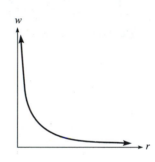

46c. The period is multiplied by a factor $\dfrac{2}{\sqrt{5}}$, or $\dfrac{2\sqrt{5}}{5}$.

47.

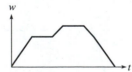

48.

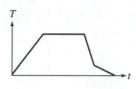

49a.

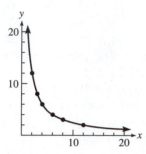

49b. $y = \dfrac{24}{x}$

50a.

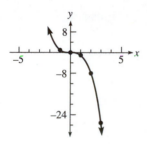

50b. $y = -x^3$ **51.** $|x| = 4$ **52.** $|y + 5| = 3$

53. $|p - 7| \le 4$ **54.** $|q + 4| \ge 0.3$

55. $-\dfrac{2}{3} < x < 2$ **56.** $-0.4 \le x \le 0.1$

57. $y \le -0.9$ or $y \ge 0.1$

58. $z < -\dfrac{5}{18}$ or $z > -\dfrac{1}{18}$ **59a.** $\dfrac{s\sqrt{3}}{2}$

59b. $\dfrac{s^2\sqrt{3}}{4}$

59c.

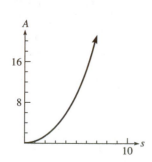

59d. $4\sqrt{3}$ sq cm   **59e.** 2.5 ft

60a. Length $= a$; height $= 10 - \dfrac{1}{2}a$

60b. $A = a\left(10 - \dfrac{1}{2}a\right)$ **60c.** $0 < a < 20$

60d.

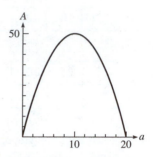

60e. $P = a + 20$

Exercise 6.1 [Page 343]

 1. b^9 **3.** 6^8 **5.** q^9 **7.** $-24y^2$ **9.** $-16wz^4$

11. $-12x^3y^4$ **13.** $21a^2b^5$ **15.** $8x^5y^3$ **17.** $3x^2y^6z^5$

19. w^3 **21.** 2^5 **23.** $\dfrac{1}{z^3}$ **25.** $\dfrac{1}{4ab^4}$ **27.** $\dfrac{-3w^2}{2}$

29. $\dfrac{7z^4}{5x^2y}$ **31.** $\dfrac{5}{c^2}$ **33.** d^{15} **35.** 5^{12} **37.** t^{32}

39. $216x^3$ **41.** $32t^{15}$ **43.** $256a^8b^{16}$ **45.** b^{13}

47. p^7q^6 **49.** $4x^{10}y^{14}$ **51.** $-2y^4z^8w^4$ **53.** $-a^4$

55. $-x^{12}y^6$ **57.** $\dfrac{w^6}{64}$ **59.** $\dfrac{-64}{p^{15}}$ **61.** $\dfrac{h^8}{m^{12}}$

63. $\dfrac{-8x^3}{27y^6}$ **65.** $\dfrac{16}{x}$ **67.** $-x^4y$ **69.** $\dfrac{-8x}{9y^2}$

71a. $2w$ **71b.** w^2 **73a.** $-2z^2$ **73b.** $-24z^4$

75a. Impossible   **75b.** $12p^5$ **77a.** 3^{17}

77b. Impossible   **79a.** a^6 **79b.** a^{12} **79c.** a^3b^3

79d. $(a + b)^3$ **81a.** $96a^5$ **81b.** $6a^5$ **81c.** $3a^{10}$

81d. $9a^{10}$

Exercise 6.2 [Page 350]

 1. $\dfrac{1}{2}$ **3.** $\dfrac{1}{25}$ **5.** 27   **7.** 16   **9.** 320

11. $\dfrac{1}{32q^5}$ **13.** $\dfrac{-4}{x^2}$ **15.** $\dfrac{1}{y^2} + \dfrac{1}{y^3}$ **17.** $\dfrac{1}{(m - n)^2}$

19. $\dfrac{-5x^5}{y^2}$ **21.** 5^{-13} **23.** $7(3^{-24})$ **25.** $3r^{-4}$

27. $\frac{2}{5}w^{-3}$ **29a.** 8 **29b.** -8 **29c.** $\frac{1}{8}$

29d. $-\frac{1}{8}$ **31a.** $\frac{1}{8}$ **31b.** $-\frac{1}{8}$ **31c.** 8 **31d.** -8

33a.

x	1.2	5.4	12.1	63.4	128
$f(x)$	0.69	0.03	0.01	0.00	0.00

33b. They decrease.

35a.

x	2.5	1.3	0.8	0.02	0.004
x^{-2}	0.16	0.59	1.56	2500	62,500

35b. They increase.
37b, 37c, and 37d are all the same graph. **39a.** 2.04
39b. 0.069 **39c.** 6.25 **39d.** 0.174 **41.** a^5

43. $\frac{1}{5^7}$ **45.** $\frac{20}{x^3}$ **47.** $\frac{1}{p^3}$ **49.** $\frac{1}{3u^{12}}$ **51.** $5^8 t$

53. $\frac{1}{7^{10}}$ **55.** $\frac{x^4}{9y^6}$ **57.** $\frac{a^6 b^4}{36}$ **59.** $\frac{5}{6h^6}$

61. $x - 3 + 2x^{-1}$ **63.** $-3 + 6t^{-2} + 12t^{-4}$

65. $-4 - 2u^{-1} + 6u^{-2}$ **67.** $\frac{4(x^4 + 4)}{x^2}$

69. $\frac{6p^4 - 3p - 12}{p}$ **71.** $\frac{3 - 6a^4 + 2a^6}{2a^3}$

73. $\frac{1}{x^2} + \frac{1}{y^2}$ **75.** $\frac{2}{w} - \frac{1}{4w^2}$ **77.** $\frac{b}{a} - \frac{a}{b}$

79. $\dfrac{1}{\frac{1}{x} + \frac{1}{y}}$ **81.** $\dfrac{x + \frac{1}{x^2}}{x}$ **83.** $\dfrac{\frac{1}{a} + \frac{1}{b}}{\frac{1}{ab}}$

85. 2.85×10^2 **87.** 8.372×10^6 **89.** 2.4×10^{-2}
91. 5.23×10^{-4} **93.** 240
95. 6,870,000,000,000,000 **97.** 0.005
99. 0.000202 **101.** 3,000,000 **103.** 112,500
105a. 4.3512×10^{12} **105b.** $16,871.13
107a. 5.866×10^{12}
107b. 7.854×10^8 hr, or 89,658 yr
109a. 4.00338×10^{20} **109b.** 4.17×10^{19} sq ft
109c. Over 7595 times

Exercise 6.3 [Page 359]
1. 11 **3.** -3 **5.** Not real **7.** -2 **9.** 2
11. 9 **13.** 3 **15.** Not real **17.** -2 **19.** -2
21. $\frac{-1}{2}$ **23.** $\frac{1}{8}$ **25.** $\sqrt{3}$ **27.** $4\sqrt[3]{x}$

29. $\sqrt[5]{4x}$ **31.** $\frac{1}{\sqrt[3]{8}}$ **33.** $\frac{3}{\sqrt[3]{xy}}$
35. $\sqrt[4]{x - 2}$
37. $7^{1/2}$ **39.** $(2x)^{1/3}$ **41.** $2z^{1/5}$
43. $-3(6^{-1/4})$ **45.** $(x - 3y)^{1/4}$
47. $-(1 + 3b)^{-1/3}$ **49.** 0.375 **51.** $0.8\overline{3}$
53. $0.\overline{285714}$ **55.** $3.\overline{90}$ **57.** 1.414 **59.** 4.217
61. -2.122 **63.** 1.125 **65.** 0.140 **67.** 2.782
69. 132.6 km **71a.** 287; 343 **71b.** 1985; 2028
73b. 10, 4.64, 3.16, 2.51
73c. 1.58, 1.05, 1.005; $100^{1/n}$ gets closer to 1.
75 and 77. y_1 and y_2 are symmetrical about $y_3 = x$.
79. $\sqrt{\sqrt{x}} = (x^{1/2})^{1/2} = x^{1/4} = \sqrt[4]{x}$ **81.** 125
83. 2 **85.** 63 **87.** $-2x^7$ **89.** -216 **91.** 17
93. 241 **95.** 9.5 **97.** ± 5.477 **99.** -5, 12.5
101a. 1 **101b.** 14 **101c.** -0.098 **101d.** 45.12

103. 6.38 m **105.** 4840°K **107.** $\frac{4\pi r^3}{3}$

109. $\frac{8Lvf}{\pi R^4}$

Exercise 6.4 [Page 366]
1. 27 **3.** -4 **5.** $\frac{1}{64}$ **7.** $\frac{-1}{625}$ **9.** 125

11. $\frac{1}{256}$ **13.** $\sqrt[5]{x^4}$ **15.** $3\sqrt[5]{x^2}$ **17.** $\frac{1}{\sqrt[6]{b^5}}$

19. $\frac{1}{\sqrt[3]{(pq)^2}}$ **21.** $\frac{4}{\sqrt[3]{z^2}}$ **23.** $-2\sqrt[4]{xy^3}$
25. $x^{2/3}$ **27.** $(ab)^{2/3}$ **29.** $2a^{1/5}b^{3/5}$ **31.** $8x^{-3/4}$

33. $-4mp^{-7/6}$ **35.** $\frac{R}{3}T^{-1/2}K^{-5/2}$ **37.** 8 **39.** -81

41. $2y^3$ **43.** $-a^4 b^8$ **45.** $2x^3 y^9$ **47.** $-3a^2 b^3$
49. 7.931 **51.** 10.903 **53.** 0.090 **55.** 35.142
57. $4a^2$ **59.** $16m^{8/3}$ **61.** $4w^{3/2}$ **63.** $-15u$

65. $\frac{1}{2k^{1/4}}$ **67.** $\frac{2}{3}c^{4/3}$ **69.** $2x^{3/2} - 2x$

71. $\frac{1}{2}y^{1/3} + \frac{3}{2}y^{-7/6}$ **73.** $2x^{1/2} - x^{1/4} - 1$

75. $a^{3/2} - 4a^{3/4} + 4$ **77.** $x(x^{1/2} + 1)$ **79.** $\frac{y - 1}{y^{1/4}}$

81. $\frac{a^{2/3} + 3a^{1/3} - 1}{a^{1/3}}$ **83a.** 4096 **83b.** $\frac{1}{8}$

83c. 62.35 **83d.** 400,000 **85.** 64 **87.** $\frac{1}{243}$

89. 2.466   **91.** $\dfrac{13}{3}$ **93.** 0.665   **95a.** 131, 199, 254

95b. ≈20 days   **97.** 1.88 yr   **99.** $7114.32

101a. 2174.34, 4995.32, 8125.92   **101b.** 40,031

Exercise 6.5 [Page 374]

1. $3\sqrt{2}$ **3.** $2\sqrt[3]{3}$ **5.** $-2\sqrt[4]{4}$, or $-2\sqrt{2}$

7. $100\sqrt{6}$ **9.** $10\sqrt[3]{900}$ **11.** $\dfrac{-2}{3}\sqrt[3]{5}$

13. $x^3\sqrt[3]{x}$ **15.** $3z\sqrt{3z}$ **17.** $2a^2b^3\sqrt[4]{3a}$

19. $-2pq\sqrt[5]{3p^2q^4}$ **21.** $-6s^2$ **23.** $-7h$

25. $2\sqrt{4-x^2}$ **27.** $A\sqrt[3]{8+A^3}$ **29.** $\dfrac{5p^6\sqrt{5p}}{a^2}$

31. $\dfrac{2\sqrt[3]{7v^2}}{w^2}$ **33.** a^2b **35.** $7y\sqrt{2x}$

37. $\dfrac{2b\sqrt[3]{b^2}}{a^2}$ **39.** $\sqrt[5]{b}$ **41.** $2|x|$ **43.** $|x-5|$

45. $|x-3|$ **47a.** $y=\sqrt{x^2}=|x|$

47b. $y=\sqrt[3]{x^3}=x$ **49.** $5\sqrt{7}$ **51.** $\sqrt{3}$

53. $9\sqrt{2x}$ **55.** $-\sqrt[3]{2}$ **57.** $5\sqrt[3]{5}+9\sqrt{5}$

59. $-6y\sqrt{x}+4x\sqrt{y}$ **61.** $6-2\sqrt{5}$

63. $2\sqrt{3}+2\sqrt{5}$ **65.** $2\sqrt[3]{5}-4\sqrt[3]{3}$

67. $4x\sqrt{6}+4\sqrt{3x}$ **69.** $x-9$ **71.** $-4+\sqrt{6}$

73. $7-2\sqrt{10}$ **75.** $6x-7\sqrt{2xy}-6y$

77. $a-4\sqrt{ab}+4b$ **79.** $2(1+\sqrt{3})$

81. $6(\sqrt{3}+1)$ **83.** $4(1+\sqrt{y})$

85. $\sqrt{2}(1-\sqrt{3})$ **87.** $\sqrt{x}(2y+3\sqrt{y})$

89. $2\sqrt{x}(2\sqrt{x}-\sqrt{3})$ **91.** $2\sqrt{3}$ **93.** $\dfrac{-\sqrt{21}}{7}$

95. $\dfrac{\sqrt{14x}}{6}$ **97.** $\dfrac{\sqrt{2ab}}{b}$ **99.** $\dfrac{\sqrt{6k}}{k}$

101. $\dfrac{-3x^3\sqrt{30}}{4}$ **103.** $-2(1-\sqrt{3})$

105. $\dfrac{x(x+\sqrt{3})}{x^2-3}$ **107.** $\dfrac{\sqrt{6}}{2}$ **109.** $\dfrac{\sqrt{15}-3}{6}$

111. $\dfrac{\sqrt[3]{4x^2}}{2x}$ **113.** $\dfrac{\sqrt[3]{x}}{x}$ **115.** $\dfrac{\sqrt[3]{18y^2}}{3y}$ **117.** $\dfrac{\sqrt[4]{2xy}}{2y}$

119. $3x^2\sqrt[4]{3x^3}$

Exercise 6.6 [Page 385]

1. $\dfrac{-1}{3}$ **3.** $\dfrac{-1}{2}, 2$ **5.** 12   **7.** 5   **9.** 4   **11.** 5

13. 0   **15.** 4   **17a.** t mi   **17b.** $d=\sqrt{64+t^2}$

17d. 12.7 min   **19a.** $700-10t$ **19b.** $15t$

19d. $D=\sqrt{(15t)^2+(700-10t)^2}$

19e. 618.5 ft   **19f.** 719.8 ft   **19g.** 22 s

21a. 1 hr 30 min   **21b.** 1 hr 42 min

21c. $t=\dfrac{\sqrt{1+x^2}}{3}+\dfrac{5-x}{4}$

21f. $x=1.15$ mi; $t=1$ hr 28.2 min   **23.** 5; $\left(\dfrac{5}{2}, 3\right)$

25. $2\sqrt{5}$; $(0,-2)$ **27.** 5; $\left(\dfrac{1}{2}, 5\right)$

29. $7+\sqrt{89}+2\sqrt{17}$

31. The rectangle is a square because all the sides are equal in length to $\sqrt{61}$.

33.

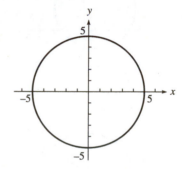

35.

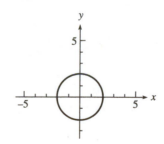

37.

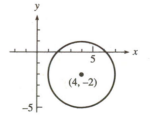

(4, –2)

39.

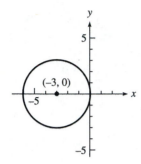

41.

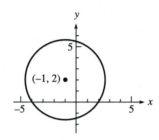

43.

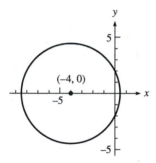

45. $x^2 + y^2 + 4x - 10y + 17 = 0$
47. $x^2 + y^2 - 3x + 8y + 11 = 0$
49. $x^2 + y^2 - 4x - 4y - 2 = 0$
51. $x^2 + y^2 + 6x + 2y + 9 = 0$
53. $x^2 + y^2 + 2x + 2y = 23$
55. $MD \parallel AC$; therefore, $\angle BAC = \angle BMD$ and $\triangle BAC \sim \triangle BMD$. Hence $\dfrac{AB}{AM} = \dfrac{AC}{AE}$. Since M is the midpoint of AB, $AB = 2AM$. Therefore, $AC = 2AE$ as well. Similarly, $\dfrac{AB}{MB} = \dfrac{BC}{BD}$. Hence $BC = 2BD$.

CHAPTER 6 REVIEW [Page 664]

1. $10x^7$ **2.** $21m^9n^6$ **3.** $-4a^3b^4$ **4.** $-10s^7t^3$

5. $\dfrac{u^3}{2v^5}$ **6.** $-\dfrac{4}{z^2}$ **7.** $\dfrac{-3s^6t^2}{r}$ **8.** $\dfrac{4a^2}{3b^3c}$ **9.** $9x^6$

10. $-64y^{12}$ **11.** $\dfrac{-27n^3}{8m^3}$ **12.** $\dfrac{w^{20}}{81v^{32}}$ **13.** $\dfrac{1}{81}$

14. $\dfrac{1}{64}$ **15.** 9 **16.** 75 **17.** $\dfrac{1}{243m^5}$ **18.** $-\dfrac{7}{y^8}$

19. $\dfrac{1}{a} + \dfrac{1}{a^2}$ **20.** $\dfrac{3r^2}{q^9}$ **21.** $\dfrac{2}{c^3}$ **22.** $\dfrac{99}{z^2}$

23. $\dfrac{d^8}{16k^{12}}$ **24.** $\dfrac{2w^{14}}{5}$ **25.** $\dfrac{(5n^2 + 6n - 3)}{3n^3}$

26. $\dfrac{3p^2 - 14p + 2}{7p}$ **27.** $7[1 + (5n)^{1/3}]$

28. $4(3 - 2y^{1/4})$ **29.** $x^{1/2}(3x^{1/2}y^{1/2} - 2y^{1/2})$

30. $\sqrt{d}(3\sqrt{2}c + \sqrt{3}c)$ **31.** $\dfrac{(1 - q^{2/3})}{q^{1/3}}$

32. $\dfrac{5(x + 2)^{3/2} - 1}{(x + 2)^{3/4}}$ **33.** $3\sqrt{t}(3\sqrt{t} + \sqrt{3})$

34. $2\sqrt{w}(3\sqrt{w} + \sqrt{2})$ **35.** $25\sqrt{m}$ **36.** $8\sqrt[3]{n}$

37. $\sqrt[3]{(13d)^2}$ **38.** $6\sqrt[5]{x^2y^3}$ **39.** $\dfrac{1}{\sqrt[4]{(3q)^3}}$

40. $7\sqrt{u^3v^3}$ **41.** $\sqrt{a^2 + b^2}$ **42.** $\sqrt[4]{16 - x^2}$

43. $\dfrac{7\sqrt{5y}}{5y}$ **44.** $3\sqrt{2d}$ **45.** $\dfrac{\sqrt{33rs}}{11s}$ **46.** $\dfrac{\sqrt{13m}}{m}$

47. $\dfrac{-3\sqrt{a} + 6}{a - 4}$ **48.** $\dfrac{-3\sqrt{z} - 12}{z - 16}$

49. $\dfrac{2x^2 + x\sqrt{3} - 3}{x^2 - 3}$ **50.** $\dfrac{5m^2 - 7\sqrt{3}m + 6}{25m^2 - 12}$

51. $\sqrt[3]{2}$ **52.** $\dfrac{9\sqrt[4]{27}}{3}$ **53.** $\dfrac{\sqrt[4]{2qw}}{2w}$ **54.** $\dfrac{\sqrt[3]{35t^2}}{7t}$

55. $1, 4$ **56.** 8 **57.** 9 **58.** 12 **59.** 7

60. ± 8 **61.** 5 **62.** 4 **63.** $g = \dfrac{2v}{t^2}$

64. $r = \dfrac{\pm\sqrt{3q^2 - 6q + 7}}{2}$ **65.** $p = \pm 2\sqrt{R^2 - R}$

66. $r = \pm\sqrt{2q^3 - 1}$

67. 1.018×10^{-9} s   or 0.000000001018 s

68. $250{,}000{,}000{,}000$ hr   or $28{,}538{,}812.79$ yr

69a. 5.72674×10^7 sq mi, 6.1×10^9 **69b.** 107

70a. Distance from sun to earth $= 9.2956 \times 10^7$ mi; speed of light $= 1.86 \times 10^5$ miles per second

70b. 499.76 s or 8.33 min   **71.** 112 kg

72. Radius = height = 2.67 in.   **73a.** 283   **73b.** 2051

74a. 123   **74b.** 2017   **75a.** 2699   **75b.** 50 hr

76a. 293.85 sq in.   **76b.** 89.96 lb   **77.** 3.07 hr

78. 21.82 ft or 60 ft   **79.** 21.59; yes

80. $2\sqrt{10} + 2\sqrt{5} \approx 10.80$

81.

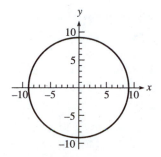

82.

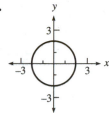

83.

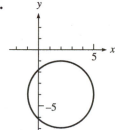

84.

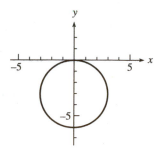

85. $(x - 5)^2 + (y + 2)^2 = 32$

86. $(x - 7)^2 + (y + 1)^2 = 41$

87. $(x - 1)^2 + (y - 4)^2 = 10$

88. $(x - 3)^2 + (y + 2)^2 = 25$

Exercise 7.1 [Page 402]

1b. $P(t) = 300(2)^t$

1c.

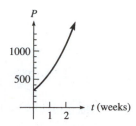

1d. 76,800; 492   **3b.** $P(t) = 20,000(2.5)^{t/6}$

3c.

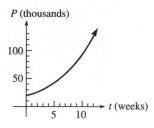

3d. 36,840; 424,127   **5b.** $A(t) = 4000(1.08)^t$

5c.

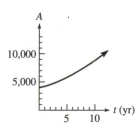

5d. $4665.60; $8635.70   **7b.** $P(t) = 20,000(1.05)^t$

7c.

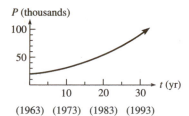

(1963) (1973) (1983) (1993)

7d. $35,917.13; $74,669.13   **9b.** $S(t) = 1500(1.12)^t$

9c.

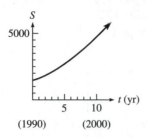

9d. 2960; 5844   **11b.** $P(t) = 250,000(0.75)^{t/2}$

11c.

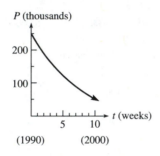

11d. 162,379; 79,101   **13b.** $L(d) = (0.85)^{d/4}$

13c.

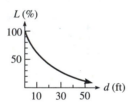

13d. 44%; 16%   **15b.** $P(t) = 50(0.992)^t$

15c.

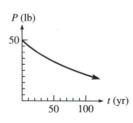

15d. 46.1 lb; 22.4 lb
17. Species A: $r \approx 0.0447$; Species B: $r \approx 0.0466$; Species B
19a. $L(t) = 1.5t + 6$ **19b.** $E(t) = 6(1.5)^{t/2}$

19c.

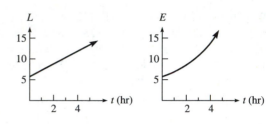

21b. $W(t) = 2^t$ cents   **21c.** $327.68; $10,737,418.24
23. 1.57%   **25a.** 3.53%   **25b.** 3.53%   **25c.** No
25d. 3.53%   **27a.** $466,560   **27b.** 279,936
27c. People lower on the list do not make money—there are not enough victims to go around!

Exercise 7.2 [Page 411]
   1. 4.729   **3.** 19.470   **5.** −0.170   **7.** 0.018
   9. 937.230   **11.** −140.169

13.

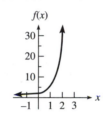

15.

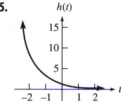

17.

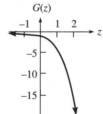

19.

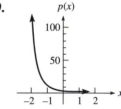

21.

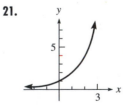

23.

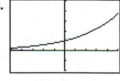

25.

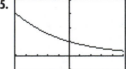

27.

29.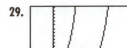

31.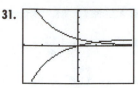

35. 5   **37.** $\dfrac{2}{3}$ **39.** $-\dfrac{1}{4}$ **41.** $\dfrac{1}{7}$ **43.** $-\dfrac{5}{4}$

45. ± 2 **47a.** $N(t) = 100 \cdot 8^{t/4}$ **47c.** 12 days

49a. $N(t) = 26 \cdot 2^{t/6}$ **49c.** 72 days later

51a. $V(t) = 700(0.7)^{t/2}$ **51c.** 4 yr   **53.** 1.58

55. 0.43   **57.** 2.26   **59.** -1.40 **61.** a, d

63a. $P_0 = 300$ **63b.** $a = 2$ **63c.** $f(x) = 300 \cdot 2^x$

65a. $S_0 = 150$ **65b.** $a = 0.55$

65c. $S(d) = 150(0.55)^d$

67.

x	-2	-1	0	1	2	3	4	5	6
$f(x)$	4	1	0	1	4	9	16	25	36
$g(x)$	$\dfrac{1}{4}$	$\dfrac{1}{2}$	1	2	4	8	16	32	64

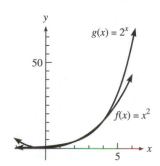

Exercise 7.3 [Page 420]

1. 2   **3.** 3   **5.** $\dfrac{1}{2}$ **7.** -1 **9.** 1   **11.** 0

13. 5   **15.** -4 **17.** 4   **19.** -1

21. $16^w = 256$ **23.** $b^{-2} = 9$ **25.** $10^{-2.3} = A$

27. $4^{2q-1} = 36$ **29.** $u^w = v$ **31.** $\log_8 \dfrac{1}{2} = -\dfrac{1}{3}$

33. $\log_t 16 = \dfrac{3}{2}$ **35.** $\log_{0.8} M = 1.2$

37. $\log_x(W - 3) = 5t$ **39.** $\log_3 2N_0 = -0.2t$

41. 2   **43.** 64   **45.** -1 **47.** 100   **49.** 11

51. $7^{2/3}$ **53.** 4.64   **55.** 1.70   **57.** 0.77   **59.** 3.84

61. 1.7348   **63.** 3.3700   **65.** -1.1367

67. -0.1585 **69.** 2.30   **71.** -0.23 **73.** 2.53

75. 0.77   **77.** -0.68 **79.** 3.63   **81.** 9.60 in.

83. 1.91 mi   **85.** 3.34 mi   **87a.** 19,969,613

87b. 25,372,873; 32,238,116; 40,960,915   **87c.** 1970

87d. 1987   **89.** 1   **91.** 0   **93.** 1   **95.** 0

Exercise 7.4 [Page 429]

1.

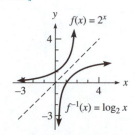

3.

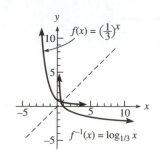

5. 2.688   **7.** 0.334   **9.** Undefined   **11.** 8.683

13. 15.614   **15.** 0.419   **17.** 12.04

19. -1.58×10^{-5} **21.** 1.76   **23.** 25.70   **25.** 3.31

27. 0.05   **29.** 3.2   **31.** 1.3×10^{-2} **33.** 100 dB

35. 6.31×10^6 W per sq m   **37.** 1000 times

39. 25,119 times   **41.** 4.7   **43a.** 81   **43b.** 4

43c. $\log_3 x$ and 3^x are inverse functions.   **43d.** 1.8; a

45. "Take the log base 6 of x."   **47.** 243   **49.** 100,000

51. $x > 9$

53a.

x	x^2	$\log_{10} x$	$\log_{10} x^2$
1	1	0	0
2	4	0.3010	0.6020
3	9	0.4771	0.9542
4	16	0.6021	1.2041
5	25	0.6990	1.3979
6	36	0.7782	1.5563

53b. $2 \log_{10} x = \log_{10} x^2$

55.

x	$y = \log_c x$
1	0
2	0.693
4	1.386
16	2.773
$\frac{1}{2}$	-0.693
$\frac{1}{4}$	-1.386
$\frac{1}{16}$	-2.773

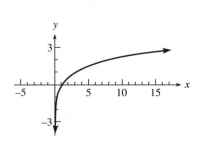

Exercise 7.5 [Page 437]

1. $\log_b 2 + \log_b x$ **3.** $\log_b x - \log_b y$

5. $\log_b x + \log_b y - \log_b z$ **7.** $3 \log_b x$

9. $\frac{1}{2} \log_b x$ **11.** $\frac{2}{3} \log_b x$ **13.** $2 \log_b x + 3 \log_b y$

15. $\frac{1}{2} \log_b x + \log_b y - 2 \log_b z$

17. $\frac{1}{3} (\log_{10} x + 2 \log_{10} y - \log_{10} z)$

19. $\log_{10} 2 + \log_{10} \pi + \frac{1}{2} \log_{10} l - \frac{1}{2} \log_{10} g$

21. $\frac{1}{2} [\log_{10}(s - a) + \log_{10}(s - b)]$ **23.** 1.7917

25. -0.9163 **27.** 2.1972   **29.** 2.0149

31. -11.7357 **33.** 4.3174   **35.** $\log_b 4$

37. $\log_b x^2 y^3$ **39.** $\log_b \frac{1}{x^2}$ **41.** $\log_{10} \sqrt{\frac{xy}{z^3}}$

43. $\log_b \frac{1}{4}$ **45.** 500   **47.** 4   **49.** 11   **51.** 3

53. No solution   **55.** 2.8074   **57.** 0.8928

59. ± 1.3977 **61.** -1.6092 **63.** 0.2736

65. -12.4864 **67a.** $W(t) = 0.01 \cdot 3^{t/16}$ **67b.** 56.97 hr

69a. $P(t) = P_0(1.037)^t$ **69b.** 19.08 yr

71a. $C(t) = 0.7(0.8)^t$ **71b.** 2.5 hr later

73a. 12.49 hr   **73b.** 24.97 hr; 37.46 hr

75. \$15,529.24; \$16,310.19   **77.** 12 yr; $11\frac{3}{4}$ yr

79. 12.9%   **81.** 10 yr, 361 days

83. $t = \frac{1}{k} \log_{10}\left(\frac{A}{A_0} + 1\right)$ **85.** $q = \log_v \left(\frac{w}{p}\right)$

87. $A = k(10^{t/T} - 1)$ **89.** $\log_b(4 \cdot 8) \overset{?}{=} \log_b \frac{64}{2}$

$\log_b 32 = \log_b 32$

91. $\log_b \frac{6^2}{9} \overset{?}{=} \log_b 2^2$ **93.** $\log_b \left(\frac{12}{3}\right)^{1/2} \overset{?}{=} \log_b 8^{1/3}$

$\log_b 4 = \log_b 4$ $\log_b 2 = \log_b 2$

95. $\log_{10}(10 + 100) \overset{?}{=} \log_{10} 10 + \log_{10} 100$

$\log_{10} 110 \neq 1 + 2$

97a. $x = b^m, \quad y = b^n$ **97b.** $\log_b (b^m \cdot b^n)$

97c. $\log_b(b^{m + n})$ **97d.** $m + n$

97e. $\log_b(xy) = \log_b x + \log_b y$ **99a.** $x = b^m$

99b. $\log_b (b^m)^k$ **99c.** $\log_b b^{mk}$ **99d.** mk

99e. $\log_b (x^k) = k \log_b x$

Exercise 7.6 [Page 446]

1. 1.3610   **3.** 2.7726   **5.** -1.2040 **7.** 0

9. 1.4918   **11.** 10.3812   **13.** 0.3012   **15.** 0.6703

17. 0.642   **19.** 3.807   **21.** -1.204 **23.** 4.137

25. 1.878   **27.** 0.0743   **29.** 0.8277   **31.** 7.5204

33. -2.9720 **35.** 1.6451   **37.** -3.0713

39. $t = \frac{1}{k} \ln y$ **41.** $t = \ln \left(\frac{k}{k - y}\right)$

43. $k = e^{T/T_0} - 10$ **45.** $N(t) = 100e^{0.6931t}$

47. $N(t) = 1200e^{-0.5108t}$ **49.** $N(t) = 10e^{0.1398t}$

51b. 8950   **51c.** 12.8 hr   **53a.** 941.8 lumens

53b. 2.2 cm   **55b.** \$3888.98   **55c.** 9.6 yr   **57.** 7.5%

59a. 1921 yr   **59b.** 5589.9 yr

61b. $P(t) = 20,000e^{0.05596t}$ **61c.** 107,182

63. $N(t) = N_0 e^{-0.0866t}$ **65a.** $8^x = 20$

65b. $\log_{10} 8^x = \log_{10} 20$

65c. $x = \frac{\log_{10} 20}{\log_{10} 8} = 1.4406$

67. $\log_8 Q = \frac{\log_{10} Q}{\log_{10} 8}$

69. $\ln Q = \frac{\log_{10} Q}{\log_{10} e} = 2.3 \log_{10} Q$

CHAPTER 7 REVIEW [Page 449]

1a.

Years after 1974	Number of Degrees
0	8
5	12
10	18
15	27
20	40.5

1b. $N(t) = 8(1.5)^{t/5}$

1c.

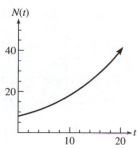

1d. 18; $43.9 \approx 44$

2a.

Years after 1975	Price (dollars)
0	0.25
1	0.28
2	0.30
3	0.33
4	0.37
5	0.40

2b. $P(t) = 0.25(1.1)^t$

2c.

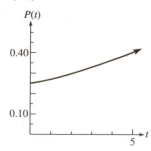

2d. $0.65; $2.71

3a.

Hours after 8:00 A.M.	Amount in Body (mg)
0	100
1	85
2	72.25
3	61.41
4	52.20
5	44.37

3b. $A(t) = 100(0.85)^t$ **3c.**

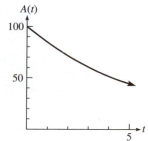

3d. 52.20 mg; 19.69 mg

4a.

Weeks after the Series	Sales (dollars)
0	200,000
1	140,000
2	98,000
3	68,600
4	48,020
5	33,614

4b. $S(t) = 200,000(0.7)^t$

4c.

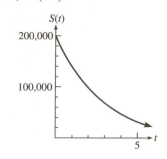

4d. $48,020; $23,529.80   **5.** 15.673   **6.** 94.633
7. 3.248   **8.** -0.258

9.

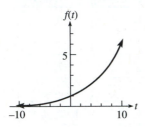

10.

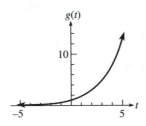

11.

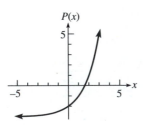

12.

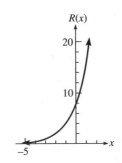

13. $x = -\dfrac{4}{3}$ **14.** $x = \dfrac{1}{7}$ **15.** $x = -11$

16. $x = \pm\sqrt{7}$ **17.** 4   **18.** $\dfrac{1}{2}$ **19.** -1

20. 1   **21.** -3 **22.** -4 **23.** $2^{x-2} = 3$
24. $n^{p-1} = q$ **25.** $\log_{0.3}(x + 1) = -2$
26. $\log_4 3N_0 = 0.3t$ **27.** $y = -1$ **28.** $x = 81$

29. $b = 4$ **30.** $x = 3$ **31.** $x = 0.544$
32. $x = 3.716$ **33.** $x = 0.021$ **34.** $x = 0.502$
35. $k = 0.0543$ **36.** $P = 2.959$ **37.** $h = 0.195$

38. $Q = 2.823$ **39.** $\log_b x + \dfrac{1}{3}\log_b y - 2\log_b z$

40. $\log_b L - \dfrac{1}{2}(\log_b 2 + \log_b R)$

41. $\dfrac{1}{3}(4\log_{10} x - \log_{10} y)$

42. $\dfrac{1}{2}\log_{10}(s - a) + \log_{10}(s - g)$

43. $\log_{10}\sqrt[3]{\dfrac{x}{y^2}}$ **44.** $\log_{10}\dfrac{\sqrt{3x}}{\sqrt[3]{y^2}}$ **45.** $\log_{10}\dfrac{1}{8}$

46. $\log_{10} 300$ **47.** $x = \dfrac{9}{4}$ **48.** $x = 190$

49. $x = 31.16$ or $x = -32.16$ **50.** $x = \dfrac{32}{9}$

51. $x = 3.77$ **52.** $x = 3.\overline{33}$ **53.** $x = -7.278$

54. $x = 3.04$ **55.** $t = \dfrac{1}{k}\log_{10}\left(\dfrac{N}{N_0}\right)$

56. $t = \dfrac{1}{K}10^{(Q - R_0)/R}$ **57a.** 238   **57b.** 2010

58a. 94,757   **58b.** 2001   **59a.** $94.48
59b. About 5 years   **60a.** $1364.23   **60b.** 35 months
61. $x = 1.548$ **62.** $x = -0.693$
63. $x = 411.58$ **64.** $x = 0.247$ **65.** $x = 2.286$

66. $x = -1.470$ **67.** $t = \dfrac{1}{k}\ln\left(\dfrac{12}{y - 6}\right)$

68. $k = e^{(N - N_0)/4} - 10$ **69.** $N(t) \approx 600e^{-0.9163t}$
70. $N(t) \approx 100e^{0.0583t}$

Exercise 8.1 [Page 458]

1. Binomial; 3   **3.** Monomial; 4   **5.** Trinomial; 2
7. Trinomial; 3   **9.** b and c   **11.** a, b, c
13a. -1 **13b.** -21 **13c.** $8b^3 - 12b^2 + 2b + 1$
15a. $\dfrac{11}{4}$ **15b.** $\dfrac{1}{9}$ **15c.** $w^2 - 3w + 1$
17a. 28.0128   **17b.** 126.5728
17c. $3k^4 - 12k^3 + 16k^2 - 8k + 4$ **19a.** 2   **19b.** 96
19c. $\dfrac{m^6}{729} - \dfrac{m^5}{243}$ **21.** $y^3 - y + 6$
23. $12x^3 - 5x^2 - 8x + 4$ **25.** $x^3 - 6x^2 + 11x - 6$
27. $6a^4 - 5a^3 - 5a^2 + 5a - 1$
29. $y^4 + 5y^3 - 20y - 16$ **31.** 4   **33.** 5   **35.** 7

37. 4   **39.** 6

41. $(x + y)^3 = (x + y)(x + y)^2$
$= (x + y)(x^2 + 2xy + y^2)$
$= x(x^2 + 2xy + y^2) + y(x^2 + 2xy + y^2)$
$= x^3 + 2x^2y + xy^2 + x^2y + 2xy^2 + y^3$
$= x^3 + 3x^2y + 3xy^2 + y^3$

43. $(x + y)(x^2 - xy + y^2)$
$= x(x^2 - xy + y^2) + y(x^2 - xy + y^2)$
$= x^3 - x^2y + xy^2 + x^2y - xy^2 + y^3$
$= x^3 + y^3$

45. $x^3 - 1$ **47.** $8x^3 + 1$ **49.** $27a^3 - 8b^3$

51. $(x + 3)(x^2 - 3x + 9)$

53. $(2x - y)(4x^2 + 2xy + y^2)$

55. $(a - 2b)(a^2 + 2ab + 4b^2)$

57. $(xy - 1)(x^2y^2 + xy + 1)$

59. $(3a + 4b)(9a^2 - 12ab + 16b^2)$

61. $(5ab - 1)(25a^2b^2 + 5ab + 1)$ **63a.** $2x^2 + 32x$

63b. 1224 sq in.   **65a.** $\left(6 - \dfrac{5\pi}{4}\right)x^2$

65b. 132.67 sq in.   **67a.** $\dfrac{2}{3}\pi r^3 + \pi r^2 h$ **67b.** $\dfrac{14}{3}\pi r^3$

69a. $500(1 + r)^2$; $500(1 + r)^3$; $500(1 + r)^4$

69b. $500r^2 + 1000r + 500$;
$500r^3 + 1500r^2 + 1500r + 500$;
$500r^4 + 2000r^3 + 3000r^2 + 2000r + 500$

69c. $583.20; $629.86; $680.24; yes

71a. $16 - 2x$, $12 - 2x$, x

71b. $V = x(16 - 2x)(12 - 2x)$

71c. $0 < x < 6$ **71d.** 140, 192, 180

71e.

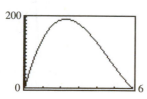

71f. $x = 2.26$ in.; maximum volume $= 194.07$ cu in.

73a. 0, 9

73b.

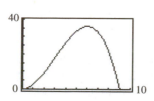

73c. $\dfrac{28}{3}$ points   **73d.** 36 points   **73e.** 3 mL or 8.2 mL

75a.

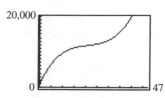

75b. 900; 11,145; 15,078   **75c.** 1341; 171; 627

75d. 1981   **77a.** 20 cm   **77b.** 100 cm

79a.

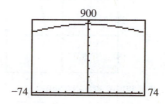

79b. 864 min   **79c.** 859.8 min

79d. For $-34 \le t \le 34$

79e. For $t > 66$ or $t < -66$

Exercise 8.2 [Page 468]

1.

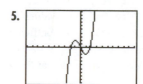

3.

5.

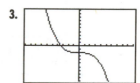

7.

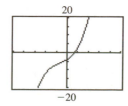

9.

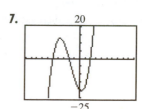

If the coefficient of x^3 is positive, the y-values increase from $-\infty$ as x increases from $-\infty$, and the y-values increase toward $+\infty$ as x increases to $+\infty$. If the coefficient of x^3 is negative, the y-values decrease from $+\infty$ as x increases from $-\infty$, and the y-values decrease toward $-\infty$ as x increases to $+\infty$.

11a. **11b.** **11c.**

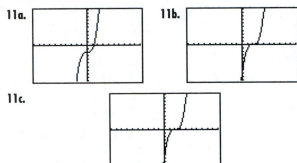

The graphs of b and c are the same.

13. **15.** **17.** **19.**

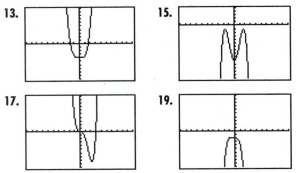

If the coefficient of x^4 is positive, the y-values decrease from $+\infty$ as x increases from $-\infty$, and the y-values increase toward $+\infty$ as x increases to $+\infty$. If the coefficient of x^4 is negative, the y-values increase from $-\infty$ as x increases from $-\infty$, and the y-values decrease toward $-\infty$ as x increases to $+\infty$.

21.

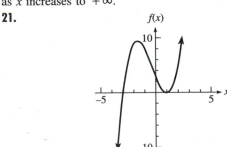

23.

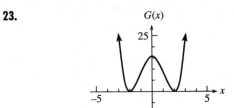

25.

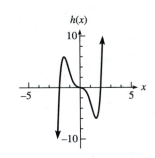

27.

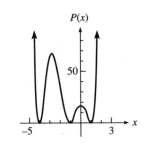

29.

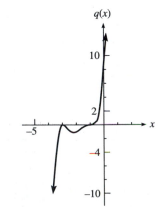

31a. $(-2,0), (-1,0), (3,0)$
31b. $P(x) = (x + 2)(x + 1)(x - 3)$
33a. $(-2,0), (0,0), (1,0), (2,0)$
33b. $R(x) = x(x - 2)(x - 1)(x + 2)$
35a. $(-2,0), (1,0), (4,0)$

35b. $p(x) = (x - 4)(x - 1)(x + 2)$
37a. $(-2,0), (2,0), (3,0)$
37b. $r(x) = (x + 2)^2(x - 2)(x - 3)$
39a. 0 (multiplicity 2)
39b.

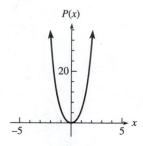

41a. $0, \pm 2\sqrt{2}$
41b.

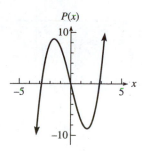

43a. 0 (multiplicity 2), -2 (multiplicity 2)
43b.

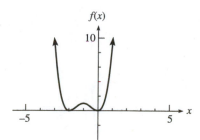

45a. $0, \pm 4$
45b.

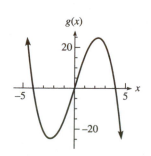

47a. $\pm\sqrt{2}, \pm\sqrt{8}$

47b.

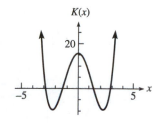

49a. $-3, \pm 1$

49b.

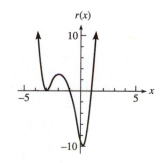

51. $P(x) = (x + 2)(x - 1)(x - 4)$

53. $P(x) = (x + 3)^2(x - 2)$

55. $P(x) = (x + 2)(x - 2)^3$

57a. $y = x^3 - 4x + 3$, graph is shifted upward three units.

57b. $y = x^3 - 4x - 5$, graph is shifted downward five units.

57c. $y = (x - 2)^3 - 4(x - 2)$, graph is shifted two units to the right.

57d. $y = (x + 3)^3 - 4(x + 3)$, graph is shifted three units to the left.

59a. $y = x^4 - 4x^2 + 6$, graph is shifted six units upward.

59b. $y = x^4 - 4x^2 - 2$, graph is shifted two units downward.

59c. $y = (x - 4)^4 - 4(x - 4)^2$, graph is shifted four units to the right.

59d. $y = (x + 2)^4 - 4(x + 2)$, graph is shifted two units to the left.

Exercise 8.3 [Page 481]

1.

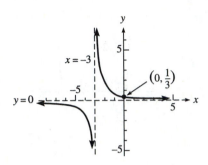

3.

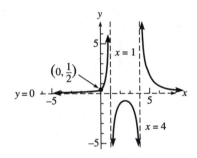

5.

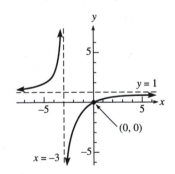

7.

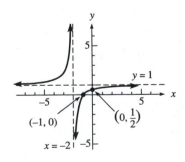

9.

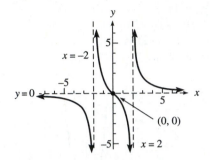

11.

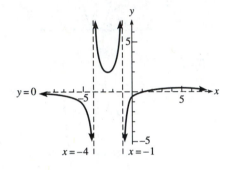

13.

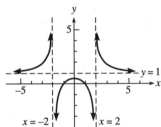

15.

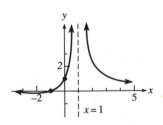

17.

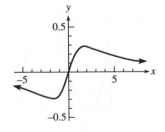

19a. $t = \dfrac{150}{50 - v}$

19b. 3 hr 20 min; travel time will increase

19c.

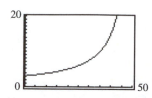

Horizontal asymptote: $t = 0$; vertical asymptote: $v = 50$; if the wind speed is 50 mph, the ducks will never reach their destination.

21a. Average cost $= 8 + \dfrac{20,000}{n}$

21b. $18 = 8 + \dfrac{20,000}{n}, \quad n = 2000$

21c.

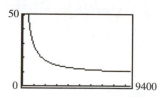

21d. For $n > 5000$ **21e.** $C = 8$; as n gets larger, the average cost per calculator approaches \$8.

23a. Cost of reordering $= 4500 + \dfrac{3000}{x}$

$C = 6x + 4500 + \dfrac{3000}{x}$

23b.

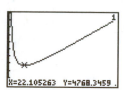

Minimum possible value for $C = \$4768.33$

23c. 22; 14

23d.

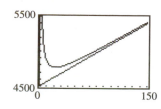

The graph of C approaches the line as an asymptote.

25a. $V = 24x - \dfrac{1}{2}x^3$

25b.

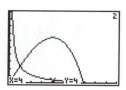

Maximum possible volume is 64 cm³.

25c. $x = 4$ cm

25d.

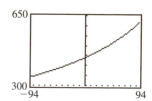

4 cm by 4 cm by 4 cm; $h = 4$ cm

27a. $0 \le p < 100$ **27b.** \$48,000; \$72,000; \$216,000

27c. 60%

27d.

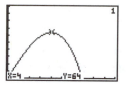

For $p > 96\%$ **27e.** $p = 100$; as the percentage immunized approaches 100, the cost becomes infinitely large.

29a. 468.2 Hz; 365.2 Hz   **29b.** -20 m per s; 68 m per s

29c.

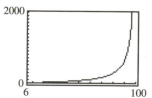

$v > 12$ m per s (approaching at more than 12 m per s)

29d. $v = 332$ m per s; as v approaches 332 m per s, the pitch becomes infinitely large.

31.

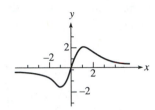

Exercise 8.4 [Page 491]

1. $-\dfrac{2}{d}$ **3.** $\dfrac{2r}{t}$ **5.** $\dfrac{2x+3}{3}$ **7.** $\dfrac{3a^2-2a}{2}$

9. $\dfrac{6(1+t)}{1-t}$ **11.** $y-2$ **13.** $\dfrac{-2}{v^2+3v+9}$

15. $\dfrac{2(x^2+9)}{3(x+3)}$ **17.** $\dfrac{y+3x}{y-3x}$ **19.** $\dfrac{2x-3}{x-1}$

21. $\dfrac{2x+3}{3-2x}$ **23.** $\dfrac{4z^2+6z+9}{2z+3}$ **25.** b   **27.** None

29. $\dfrac{25p}{n}$ **31.** $\dfrac{-np^2}{2}$ **33.** $\dfrac{5}{ab}$ **35.** 5

37. $\dfrac{a(2a-1)}{a+4}$ **39.** $\dfrac{x-2}{x+1}$

41. $\dfrac{6x(x-2)(x-1)^2}{(x^2-8)(x^2-2x+4)}$ **43.** $\dfrac{2}{9}$ **45.** $\dfrac{a+1}{a-2}$

47. $3(x^2-xy+y^2)$ **49.** $\dfrac{x+2}{x^2-1}$ **51.** $\dfrac{x^2(x-4)}{(x+1)}$

53. $\dfrac{x+3}{6y}$ **55.** $6rs-5+\dfrac{2}{rs}$

57. $-5s^8+7s^3-\dfrac{2}{s^2}$ **59.** $2y+5+\dfrac{2}{2y+1}$

61. $x^2+4x+9+\dfrac{19}{x-2}$

63. $4z^3-2z^2+3z+1+\dfrac{2}{2z+1}$

65. $x^3+2x^2+4x+8+\dfrac{15}{x-2}$

67. $P(x)=(x-1)(x^2-x-1)$ so 1 is a zero of $P(x)$.
The other zeros are $\dfrac{1\pm\sqrt{5}}{2}$.

69. $P(x)=(x+3)(x^3-6x^2+8x)$ so -3 is a zero of $P(x)$. The other zeros are 0, 2, and 4.

71a. All real numbers except 2   **71b.** $y=x+2,\ x\neq2$

71c.

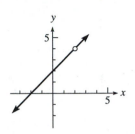

73a. All real numbers except 1 and -1

73b. $y=\dfrac{1}{x-1},\ x\neq-1$

73c.

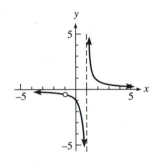

Exercise 8.5 [Page 500]

1. $\dfrac{x-3}{2}$ **3.** $\dfrac{a+b-5c}{6}$ **5.** $\dfrac{2x-1}{2y}$

7. $\dfrac{-2x+7}{x+2y}$ **9.** $\dfrac{4}{a-1}$ **11.** $12xy^2(x+y)^2$

13. $(a+4)(a+1)^2$ **15.** $x(x-1)^3$

17. $12x^3(x-1)^2$ **19.** $\dfrac{6}{18x}$ **21.** $\dfrac{xy^2}{xy}$

23. $\dfrac{3y^2-9y}{y^2-y-6}$ **25.** $\dfrac{3a+3b}{a^2-b^2}$ **27.** $\dfrac{7x}{6}$ **29.** $\dfrac{y}{12}$

31. $\dfrac{7x+1}{6x}$ **33.** $\dfrac{8x-2}{(x+1)(x-1)}$

35. $\dfrac{3y-3y^2}{(y+1)(2y-1)}$ **37.** $\dfrac{y^2-4y+5}{(y+1)(2y-3)}$

39. $\dfrac{-4}{15(x-2)}$ **41.** $\dfrac{4x-2}{(x-2)(x+1)^2}$

43. $\dfrac{3y+1}{y(y-3)(y+2)}$ **45.** $\dfrac{x^2-1}{x}$

47. $\dfrac{x^3-2x^2+2x-2}{(x-1)^2}$ **49.** $\dfrac{y^3-2y^2-y}{(y+1)(y-1)}$

51. $\dfrac{x^2+x+1}{x+2}$ **53.** $\dfrac{1}{10}$ **55.** $\dfrac{7}{10a+2}$

57. $\dfrac{a}{a-2}$ **59.** x **61.** $\dfrac{x}{x-1}$ **63.** $\dfrac{y}{y+2}$

65. xy **67.** $\dfrac{x^2(y-1)}{y^2(x+1)}$ **69.** $\dfrac{-2(x+z)}{xz}$

71. $\dfrac{y^2}{(y+1)^2(y-1)}$ **73a.** $\dfrac{25}{s+8}$ **73b.** $\dfrac{25}{s-8}$

73c. $\dfrac{50s}{s^2-64}$ **75a.** $\dfrac{900}{400+w}$ **75b.** $\dfrac{900}{400-w}$

75c. Orville by $\dfrac{1800w}{160,000-w^2}$

77a. $\dfrac{1}{f}=\dfrac{1}{q+60}+\dfrac{1}{q}$ **77b.** $f=\dfrac{q^2+60q}{2q+60}$

79a. $T_1=\dfrac{d}{r_1},\quad T_2=\dfrac{d}{r_2}$ **79b.** $D=2d,\quad T=\dfrac{d}{r_1}+\dfrac{d}{r_2}$

79c. $r=\dfrac{D}{T}=\dfrac{2d}{\dfrac{d}{r_1}+\dfrac{d}{r_2}}$ **79d.** $r=\dfrac{2r_1r_2}{r_1+r_2}$

79e. $r=58\tfrac{1}{3}$ mph

Exercise 8.6 [page 510]

1. $\dfrac{1}{2}$ **3.** 5   **5.** 1   **7.** $-\dfrac{14}{5}$

9. $-\dfrac{1}{6},\ -\dfrac{4}{3}$ **11.** $2\pm\sqrt{5}$ **13.** ±2 **15.** $-\tfrac{1}{2}$

17a.

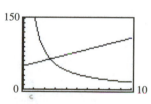

$2.50

17b. $\dfrac{160}{x}=6x+49,\quad x=2.5$ **19a.** $L=\dfrac{3200}{w}$

19b. $P=\dfrac{6400}{w}+2w$

19c.

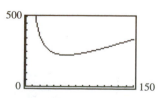

Lowest point: (56.6, 226); the minimum perimeter is 226 ft for a width of 56.6 ft.

19d. $240=\dfrac{6400}{w}+2w$ **19e.** $w=40$ ft, $l=80$ ft

21a. $t_1=\dfrac{500}{180+w}$ **21b.** $t_2=\dfrac{400}{180-w}$

21c.

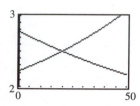

Intersection point: (20, 2.5); if the wind speed is 20 mph, both pilots take the same amount of time: 2.5 hr.

21d. $\dfrac{500}{180+w}=\dfrac{400}{180-w}$ **21e.** 20 mph

23a. $t_1=\dfrac{144}{s-20}$ **23b.** $t_2=\dfrac{144}{s+20}$

23c.

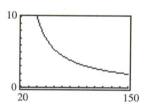

(100, 3): If the airspeed of the plane is 100 mph, the round trip will take 3 hr.

23d. $\dfrac{144}{s-20}+\dfrac{144}{s+20}=3$ **23e.** 100 mph

25.

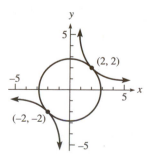

(2, 2), (−2, −2)

27.

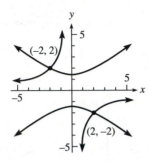

$(2, -2), (-2, 2)$

29.

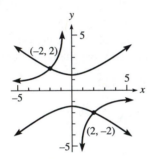

$(3\sqrt{2}, \sqrt{2}), (-3\sqrt{2}, -\sqrt{2})$

CHAPTER 8 REVIEW [Page 513]

1. $\dfrac{a}{2(a-1)}$　　**2.** $-\dfrac{a}{4}$　　**3.** $\dfrac{2y-3}{3}$　　**4.** $\dfrac{y^2-2x}{2}$

5. $\dfrac{x}{x+3}$　　**6.** $\dfrac{2y-x}{2y+x}$　　**7.** $\dfrac{a-3}{2a+6}$

8. $\dfrac{2xy+1}{2xy-1}$　　**9.** $10ab$　　**10.** $-\dfrac{1}{4}a^4b^3$

11. $\dfrac{6x}{2x+3}$　　**12.** $\dfrac{2x^2+5x+3}{6}$

13. $\dfrac{a^2-2a}{a^2+3a+2}$　　**14.** $\dfrac{a^2b+2ab^2+4b^3}{a^2-2ab}$

15. $\dfrac{1}{2x-1}$　　**16.** $\dfrac{y+2}{12x}$　　**17.** $4yz+2-\dfrac{1}{yz}$

18. $9x^2-7+\dfrac{4}{x^2}-\dfrac{1}{x^4}$　　**19.** y^2+2y-4

20. $x^2-2x-2-\dfrac{1}{x-2}$

21. $x^2+x+\dfrac{1}{2}-\dfrac{1}{2(2x-1)}$

22. $-y^2+\dfrac{y}{3}-\dfrac{4}{9}+\dfrac{40}{9(3y+1)}$　　**23.** $\dfrac{2}{x}$

24. $\dfrac{y-2}{y+3}$　　**25.** $-\dfrac{a}{6}$　　**26.** $\dfrac{5b}{4}$

27. $\dfrac{3x+1}{2(x-3)(x+3)}$　　**28.** $\dfrac{4(y+1)}{(y+2)^2(y-2)}$

29. $\dfrac{2a^2-a+1}{(a-3)(a-1)}$　　**30.** $\dfrac{a^4+a^3-a^2+a+4}{(a+1)^2(a-1)}$

31. $\dfrac{1}{5}$　　**32.** $\dfrac{y(x-2)}{x+2}$　　**33.** $\dfrac{x}{x+4}$

34. $\dfrac{x^2}{(x+1)(x-1)^2}$　　**35.** No solution

36. $x=\dfrac{3}{2}$　　**37.** $x=-1,\ x=2$

38. $x=\dfrac{2\pm\sqrt{22}}{3}$　　**39.** $x=0,\ x=\pm2$

40. $x=0,\ x=-2,\ x=\dfrac{1}{2}$

41a. $x=2,\ x=-1$

41b.

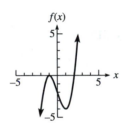

42a. $x=-2,\ x=3$

42b.

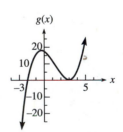

43a. $x=0,\ x=1,\ x=-3$

43b.

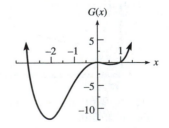

44a. $x = -1$, $x = 2$

44b.

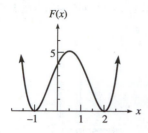

45a. $x = 0$, $x = -1$, $x = 1$

45b.

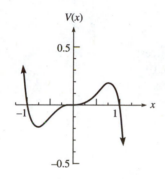

46a. $x = 0$, $x = -3$, $x = 3$

46b.

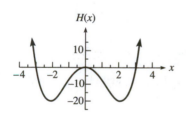

47a. $x = -1$, $x = 1$

47b.

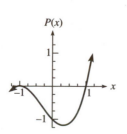

48a. $x = -2$, $x = 1$, $x = 2$

48b.

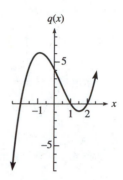

49a. $x = -2$, $x = 0$, $x = 1$

49b.

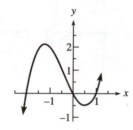

50a. $x = 2$

50b.

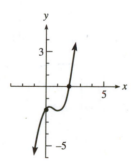

51a. $x = -1$, $x = 1$, $x = -\sqrt{6}$, $x = \sqrt{6}$

51b.

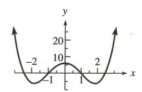

52a. $x = 0$, $x = 1$, $x = \pm\sqrt{3}$

52b.

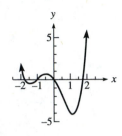

53a. Asymptotes: $x = 4, y = 0$; y-intercept: $\left(0, -\dfrac{1}{4}\right)$

53b.

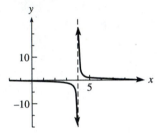

54a. Asymptotes: $x = 5$, $x = -2$, $y = 0$; y-intercept: $\left(0, -\dfrac{1}{5}\right)$

54b.

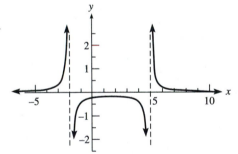

55a. Asymptotes: $x = -3$, $y = 1$; x-intercept: $(2, 0)$; y-intercept: $\left(0, -\dfrac{2}{3}\right)$

55b.

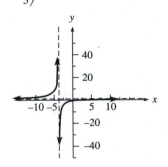

56a. Asymptotes: $x = 3$, $x = -1$, $y = 0$; x-intercept: $(1, 0)$; y-intercept: $\left(0, \dfrac{1}{3}\right)$

56b.

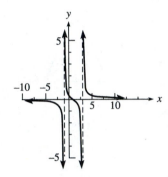

57a. Asymptotes: $x = -2$, $x = 2$, $y = 3$; x and y intercept: $(0, 0)$

57b.

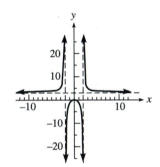

58a. Asymptotes: $x = -3$, $x = 3$, $y = 2$; x-intercepts: $(-1, 0)$ and $(1, 0)$; y-intercept: $\left(0, \dfrac{2}{9}\right)$

58b.

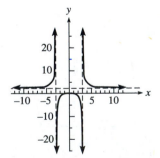

59. $\dfrac{x^3 + y}{x^3 y}$ **60.** $\dfrac{-x^2 y + y - 1}{xy(y - 1)}$ **61.** $\dfrac{1 - xy}{x(y - 1)}$

62. $\dfrac{x + y}{y}$　**63.** $\dfrac{-(x - y)^2}{xy}$　**64.** $\dfrac{1}{y - x}$

65a. $\dfrac{1}{6}n^3 - \dfrac{1}{2}n^2 + \dfrac{1}{3}n$　**65b.** 220　**65c.** 20

66a. $n^3 - 3n^2 + 2n$　**66b.** 7980　**66c.** 23

67a. $A = (x + y)^2$　**67b.** $A = x^2 + 2xy + y^2$

67c. x^2, xy, xy, y^2

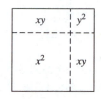

68a. $A = x^2 - y^2$　**68b.** $(x + y)(x - y)$

68c.

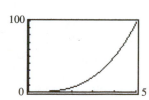

69a. $A = \dfrac{1}{2}(x^2 - y^2)$　**69b.** $A = \dfrac{1}{2}(x - y)(x + y)$

69c. $A = 18$ sq ft　**70a.** $A = \pi x^2 - \pi y^2$

70b. $A = \pi(x - y)(x + y)$　**70c.** $A = 66\pi$ sq in.

71a. $V = \dfrac{\pi h^3}{4}$　**71b.** $2\pi \approx 6.28$ cm³; $16\pi \approx 50.27$ cm³

71c.

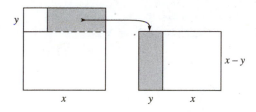

If $V = 100$ cm³, $h \approx 5$ cm.

72a. $V = \dfrac{\pi d^3}{8} \approx 0.39d^3$　**72b.** 10.6 cu in.; 25.1 cu in.

72c.

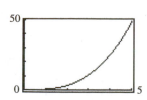

If $V = 5$ cu in., $d \approx 2.3$ in.

73a. $P = -0.02x^2 + 24x - 4000$

73b.

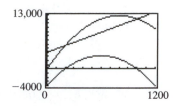

73c. Profit is negative for $x < 200$ and $x > 1000$.
73d. Maximum revenue is \$12,800 for 800 pens; maximum profit is \$3200 for 600 pens.

74a. $P = -0.03x^2 + 33x - 3000$

74b.

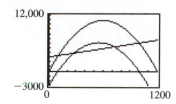

74c. Profit is negative for $x < 100$ or $x > 1000$.
74d. Maximum revenue is \$10,800 for 600 chocolate creams; maximum profit is \$6075 for 550 chocolate creams.

75a. $P = -1.75x^2 + 525x - 21,875$

75b.

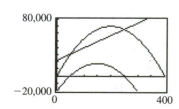

75c. Profit is negative for $x < 50$ and $x > 250$.
75d. Maximum revenue is \$70,000 for 200 machines; maximum profit is \$17,500 for 150 machines.

76a. $P = -2x^2 + 400x - 12,800$

76b.

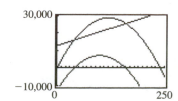

76c. Profit is negative for $x < 40$ and $x > 160$.

76d. Maximum revenue is $28,800 for 120 lawn mowers; maximum profit is $7200 for 100 lawn mowers.

77. Leon flies at 240 mph; Marlene drives at 60 mph.

78. Moia drives at 45 mph; Fran drives at 50 mph.

79. 1 in. by 12 in.   **80.** 7 cm by 10 cm

81. 2 ft by 7 ft   **82.** 12 ft by 18 ft

83. Morning train: 20 mph; evening train: 30 mph

84. Rate in the city: 10 mph; rate in light traffic: 30 mph

85. Amount: $800; rate: 4%

86. Pressure: 6 lb per sq in.; volume: 5 cu in.

87. $A_1 = $ area of square $-$ area of two triangles

$$= x \cdot x - \left[\left(\frac{1}{2}y \cdot y \right) + \left(\frac{1}{2}y \cdot y \right) \right]$$

$$= x^2 - \left[\frac{1}{2}y^2 + \frac{1}{2}y^2 \right]$$

$$= x^2 - y^2$$

$A_2 = $ (length)(width)

$$= (x + y)(x - y)$$

$$= x^2 - y^2$$

Therefore, $A_1 = A_2$.

88. Radius of large circle $= x + y$

$A_1 = $ area of large circle $-$ area of circle radius y
$\quad\quad -$ area of circle radius x

$A_1 = \pi(x + y)^2 - \pi y^2 - \pi x^2$

$\quad = \pi x^2 + 2\pi xy + \pi y^2 - \pi y^2 - \pi x^2$

$\quad = 2\pi xy$

$A_2 = $ (length)(width)

$\quad = (\pi y)(2x) = 2\pi xy$

Therefore, $A_1 = A_2$.

Exercise 9.1 [Page 524]

1. $-4, -3, -2, -1$ **3.** $\frac{-1}{2}, 1, \frac{7}{2}, 7$

5. $2, 1.5, 1.\overline{3}, 1.25$ **7.** $0, 1, 3, 6$ **9.** $-1, 1, -1, 1$

11. $1, 0, \frac{-1}{3}, \frac{1}{2}$ **13.** $1, 1, 1, 1$ **15.** $3, 5, 7, 9, 11$

17. $24, -12, 6, -3, 1.5$ **19.** $1, 2, 6, 24, 120$

21. $100, 210, 331, 464.1, 610.51$

23a. $14,000; 11,900; 10,115; 8597.75$

23b. $v_1 = 14,000; v_{n+1} = 0.85v_n$

25a. $1.55, 2.00, 2.45, 2.90$

25b. $c_1 = 1.55; c_{n+1} = c_n + 0.45$

27a. $50,000; 50,000; 50,000; 50,000$

27b. $v_1 = 50,000; v_{n+1} = v_n$ **29a.** $10, 18, 24.4, 29.52$

29b. $d_1 = 10; d_{n+1} = 0.8d_n + 10$ **31a.** 3   **31b.** 6

31c. $0, 1, 3, 6, 10; L_{n+1} = n + L_n$ **33.** 58

35. 1.415 **37.** 8.944

39a. $1, 1, 2, 3, 5, 8, 13, 21, 34, 55, 89, 144, 233, 377, 610, 987$

39b. $1, 2, 1.5, 1.\overline{6}, 1.6, 1.625, 1.615, 1.619, 1.618, 1.618, 1.618, 1.618, 1.618, 1.618, 1.618$. The quotients approach a limit near 1.618. $\frac{1 + \sqrt{5}}{2} \approx 1.618033989$, about the same as the limit above.

41. a_n approaches 1.4142, or $\sqrt{2}$

43. c_n approaches 0.64039, or $\frac{1 + \sqrt{17}}{8}$

45. s_n approaches 2

Exercise 9.2 [Page 532]

1. Geometric   **3.** Arithmetic   **5.** Geometric

7. Neither   **9.** Geometric   **11.** Geometric

13. $2, 6, 10, 14$ **15.** $\frac{1}{2}, \frac{3}{4}, 1, \frac{5}{4}$ **17.** $2.7, 1.9, 1.1, 0.3$

19. $5, -10, 20, -40$ **21.** $9, 6, 4, \frac{8}{3}$

23. $60, 24, 9.6, 3.84$

25. $15, 19, 23, \ldots, 3 + (n - 1)4$

27. $-13, -17, -21, \ldots, -1 - 4(n - 1)$

29. $\frac{16}{3}, \frac{32}{3}, \frac{64}{3}, \ldots, \frac{2}{3}(2)^{n-1}$

31. $\frac{-1}{2}, \frac{1}{4}, \frac{-1}{8}, \ldots, 4\left(\frac{-1}{2}\right)^{n-1}$ **33.** 7.5

35. $\frac{3}{128}$ **37.** 3   **39.** 13   **41.** $s_n = 3 + 2(n - 1)$

43. $x_n = -3(n - 1)$ **45.** $d_n = 24\left(\frac{-1}{2}\right)^{n-1}$

47. $w_n = 2^{n-1}$ **49b.** 128   **51b.** $115

53b. $1203.31   **55b.** 1.15 kg

Exercise 9.3 [Page 538]

1. 99 **3.** $106\frac{2}{3}$ **5.** 141 **7.** 410 **9.** -95.8125

11. 89.88075 **13.** Arithmetic; 2550

15. Geometric; 2046 **17.** Neither; 784

19. Arithmetic; 1071 **21.** Geometric; 8.996

23. 1938 **25.** 78 **27a.** 10.125 ft **27b.** 107.25 ft

29. $7,400,000 **31.** 66.25 s **33.** $6,504,532.78

35. 4549.7 s **37.** $14,066.19 **39.** $10,737,418.23

41a. N **41b.** $N + 1$

41c. $2S = N \cdot (N + 1)$; $S = \dfrac{N(N + 1)}{2}$

41d. $\dfrac{N(N + 1)}{2}$ **43a.** r

43b. $Ar = r + r^2 + r^3 + \cdots + r^N + r^{N+1}$

43c. $A - Ar = 1 - r^{N+1}$

$\quad A(1 - r) = 1 - r^{N+1}$

43d. $A = \dfrac{1 - r^{N+1}}{1 - r}$

45a. $F + (4 - 1)d$, $F + (9 - 1)d$

45b. $F + (N - 1)d$

45c. $\dfrac{N}{2}[2F + (N - 1)d]$

Exercise 9.4 [Page 547]

1. $1^2 + 2^2 + 3^2 + 4^2$ **3.** $3 + 4 + 5$

5. $1(2) + 2(3) + 3(4) + 4(5)$

7. $\dfrac{-1}{2} + \dfrac{1}{2^2} - \dfrac{1}{2^3} + \dfrac{1}{2^4}$ **9.** $\displaystyle\sum_{k=1}^{4}(2k - 1)$

11. $\displaystyle\sum_{k=1}^{4} 5^{2k-1}$ **13.** $\displaystyle\sum_{k=1}^{5} k^2$ **15.** $\displaystyle\sum_{k=1}^{5} \dfrac{k}{k + 1}$

17. $\displaystyle\sum_{k=1}^{6} \dfrac{k}{2k - 1}$ **19.** $\displaystyle\sum_{k=1}^{\infty} \dfrac{2^{k-1}}{k}$ **21.** Neither; 97

23. Neither; $\dfrac{25}{12}$ **25.** Neither; 100

27. Geometric; 5,230,176,600 **29.** Arithmetic; 20,100

31. Neither; 441 **33.** Arithmetic; 1364

35. Arithmetic; -520 **37.** Geometric; 24,414,062

39. Geometric; 1074.76 **41.** 1 **43.** 14.12

45. 2.5 **47.** 0.375 **49.** $\dfrac{4}{9}$ **51.** $\dfrac{31}{99}$ **53.** $2\dfrac{410}{999}$

55. $\dfrac{29}{225}$ **57.** 120 in. **59.** 30 ft

Exercise 9.5 [Page 558]

1. 51; 101 **3.** 100; 50

5. 11 rows:

```
                    1
                  1   1
                1   2   1
              1   3   3   1
            1   4   6   4   1
          1   5  10  10   5   1
        1   6  15  20  15   6   1
      1   7  21  35  35  21   7   1
    1   8  28  56  70  56  28   8   1
  1   9  36  84 126 126  84  36   9   1
1  10  45 120 210 252 210 120  45  10   1
```

7. $x^5 + 15x^4 + 90x^3 + 270x^2 + 405x + 243$

9. $x^4 - 12x^3 + 54x^2 - 108x + 81$

11. $8x^3 - 6x^2y + \dfrac{3}{2}xy^2 - \dfrac{1}{8}y^3$

13. $x^{14} - 21x^{12} + 189x^{10} - 945x^8 + 2835x^6 - 5103x^4$
$\quad + 5103x^2 - 2187$

15. $\dfrac{16}{81} - \dfrac{32}{27}a^2 + \dfrac{8}{3}a^4 - \dfrac{8}{3}a^6 + a^8$

17. $x^6 + 6x^5y + 15x^4y^2 + 20x^3y^3 + 15x^2y^4$
$\quad + 6xy^5 + y^6$

19. $128x^7 - 448x^6y + 672x^5y^2 - 560x^4y^3 + 280x^3y^4$
$\quad - 84x^2y^5 + 14xy^6 - y^7$

21. $a^8 - 8a^7b + 28a^6b^2 - 56a^5b^3 + 70a^4b^4 - 56a^3b^5$
$\quad + 28a^2b^6 - 8ab^7 + b^8$

23. $5 \cdot 4 \cdot 3 \cdot 2 \cdot 1 = 120$

25. $\dfrac{9 \cdot 8 \cdot 7 \cdot 6 \cdot 5 \cdot 4 \cdot 3 \cdot 2 \cdot 1}{7 \cdot 6 \cdot 5 \cdot 4 \cdot 3 \cdot 2 \cdot 1} = 72$

27. $\dfrac{(5 \cdot 4 \cdot 3 \cdot 2 \cdot 1)(7 \cdot 6 \cdot 5 \cdot 4 \cdot 3 \cdot 2 \cdot 1)}{12 \cdot 11 \cdot 10 \cdot 9 \cdot 8 \cdot 7 \cdot 6 \cdot 5 \cdot 4 \cdot 3 \cdot 2 \cdot 1} = \dfrac{1}{792}$

29. $\dfrac{8 \cdot 7 \cdot 6 \cdot 5 \cdot 4 \cdot 3 \cdot 2 \cdot 1}{(2 \cdot 1)(6 \cdot 5 \cdot 4 \cdot 3 \cdot 2 \cdot 1)} = 28$

31. 84 **33.** 220 **35.** 190 **37.** 2002

39. 77,520 **41.** $-101,376$ **43.** 1680 **45.** 1365

47. -84 **49.** -7

51. 1, 11, 121, 1331, 14,641. The digits in the terms of the sequence correspond to the numbers in the first five rows of Pascal's triangle. If $11^n = (10 + 1)^n$ is expanded as a binomial, each term is the product of a number from Pascal's triangle times a power of 10 times a power of 1.

CHAPTER 9 REVIEW [Page 561]

1. $\dfrac{1}{2}, \dfrac{2}{5}, \dfrac{3}{10}, \dfrac{4}{17}$ **2.** $1, -\dfrac{1}{2}, \dfrac{1}{3}, -\dfrac{1}{4}$

3. $5, 2, -1, -4, -7$ **4.** $1, -\dfrac{3}{4}, \dfrac{9}{16}, -\dfrac{27}{64}, \dfrac{81}{256}$

5a. 1584, 1393.92, 1226.6496, 1079.451648
5b. $a_1 = 1584, \quad a_{n+1} = 0.88a_n$
6a. 25,440; 26,966.4; 28,584.384; 30,299.44704
6b. $a_1 = 25,440, \quad a_{n+1} = 1.06a_n$
7a. 30; 37.5; 43.125; 47.34375
7b. $a_1 = 30, \quad a_{n+1} = 0.75a_n + 15$
8a. 187, 185, 183, 181   **8b.** $a_1 = 187, \quad a_{n+1} = a_n - 2$
9. $x_7 = -25$ **10.** $y_3 = 5$ **11.** 32   **12.** $x + 9a$

13. $-\dfrac{81}{8}$ **14.** $-\dfrac{8}{27}$ **15.** 136   **16.** $\dfrac{23}{2}$ **17.** 68

18. The 15th term

19. Geometric; $-\dfrac{1}{16}, \dfrac{1}{32}, -\dfrac{1}{64}, \dfrac{1}{128}$;

$$a_n = (-1)\left(-\dfrac{1}{2}\right)^{n-1}$$

20. Neither
21. Arithmetic; $-14, -19, -24, -29; \quad a_n = 11 - 5n$
22. Geometric; $256, -1024, 4096, -16,384$;

$$a_n = (-4)^{n-1}$$

23. Geometric; $-16, 32, -64, 128; \quad a_n = (-1)(-2)^{n-1}$

24. Geometric; $\dfrac{27}{128}, \dfrac{81}{512}, \dfrac{243}{2048}, \dfrac{729}{8192}$;

$$a_n = \left(\dfrac{2}{3}\right)\left(\dfrac{3}{4}\right)^{n-1}$$

25. Arithmetic; $-1, -5, -9, -13; \quad a_n = 7 - 4n$

26. Arithmetic; $\dfrac{3}{4}, \dfrac{5}{4}, \dfrac{7}{4}, \dfrac{9}{4}; \quad a_n = \dfrac{1}{2}n - \dfrac{1}{4}$

27. Geometric; $-48, 192, -768, 3072$;

$$a_n = 12(-4)^{n-1}$$

28. Geometric; $2, \dfrac{2}{3}, \dfrac{2}{9}, \dfrac{2}{27}; \quad a_n = \dfrac{2}{3^{n-2}}$

29. $\displaystyle\sum_{k=2}^{5} k(k-1) = 2(1) + 3(2) + 4(3) + 5(4)$

30. $\displaystyle\sum_{j=2}^{\infty} \dfrac{j}{2j-1} = \dfrac{2}{3} + \dfrac{3}{5} + \dfrac{4}{7} + \dfrac{5}{9} + \dfrac{6}{11} + \cdots$

31. $\displaystyle\sum_{k=1}^{12} 2^k - 1$ **32.** $\displaystyle\sum_{k=4}^{15} k^2 x^k$ **33.** Arithmetic; 210

34. Arithmetic; 49   **35.** Arithmetic; 57

36. Arithmetic; 40   **37.** Geometric; $\dfrac{121}{243}$

38. Geometric; 63   **39.** Neither; -4

40. Neither; $\dfrac{163}{60}$ **41.** Geometric; $-\dfrac{9}{5}$

42. Geometric; $\dfrac{9}{2}$ **43.** $\dfrac{64}{27}$ ft   **44.** \$945.98

45a. 810

45b. $\displaystyle\sum_{n=2}^{16} 6n$ **46.** 2004   **47.** 84 ft

48. \$10,737,418.23   **49.** $\dfrac{29}{9}$ **50.** $\dfrac{23}{55}$

51. $x^5 - 10x^4 + 40x^3 - 80x^2 + 80x - 32$

52. $\dfrac{x^4}{16} - \dfrac{1}{2}x^3 y + \dfrac{3}{2}x^2 y^2 - 2xy^3 + y^4$ **53.** 20

54. 126   **55.** 21   **56.** 120   **57.** 32   **58.** 64
59. -672 **60.** -8748

Exercise 10.1 [Page 571]

1.

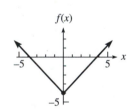

3.

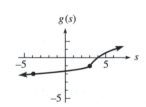

5.

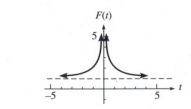

7.
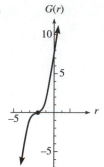
$G(r)$

9.
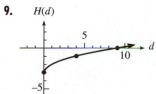
$H(d)$

11.
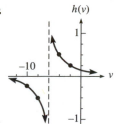
$h(v)$

13.
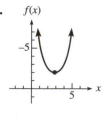
$f(x)$

15.
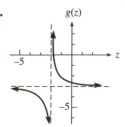
$g(z)$

17.
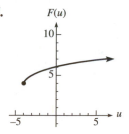
$F(u)$

19.
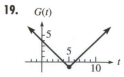
$G(t)$

21.
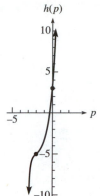
$h(p)$

23.
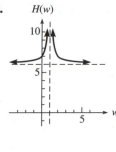
$H(w)$

25.
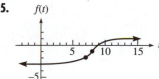
$f(t)$

27.
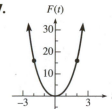
$F(t)$

29.
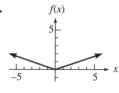
$f(x)$

31.
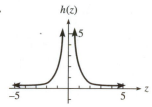
$h(z)$

33.
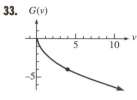
$G(v)$

35.
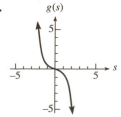
$g(s)$

37.

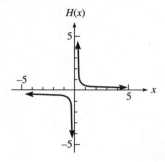

3.

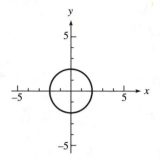

39. $y = |x + 1| - 2$ **41.** $y = -\sqrt{x} + 3$
43. $y = (x - 3)^3 + 1$
45. $y = (x - 2)^2 + 3$

5.

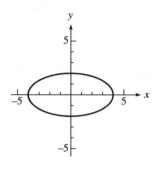

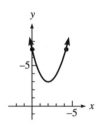

7.

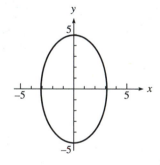

47. $y = (x + 1)^2 - 4$

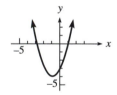

Exercise 10.2 [Page 587]

1.

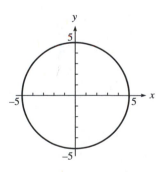

9.

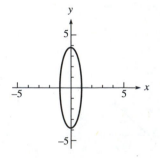

11.

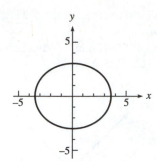

13.

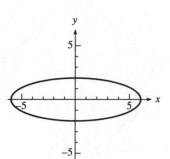

15.

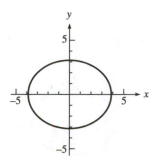

17.

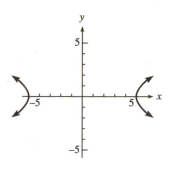

19.

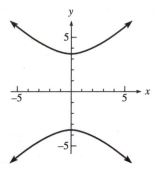

21.

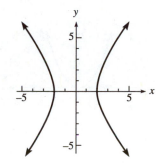

23.

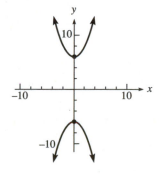

25.

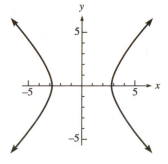

27.

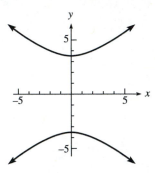

29.

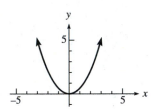

31.

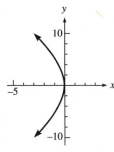

33.

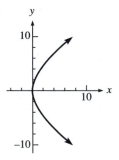

35.

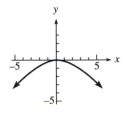

37.

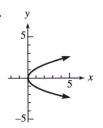

39. Circle; radius: 2
41. Hyperbola; x-intercepts: $\pm\sqrt{8}$
43. Ellipse; x-intercepts: $\pm\sqrt{3}$; y-intercepts: $\pm\sqrt{6}$
45. Parabola; vertex: $\left(0, -\dfrac{3}{2}\right)$, opens up
47. Hyperbola; y-intercepts: $\pm\sqrt{6}$
49. Parabola; vertex: $(0, -4)$, opens up

51a. $\dfrac{x^2}{25} + \dfrac{y^2}{64} = 1$ **51b.** $y = \pm\dfrac{24}{5}$ **53.** $\dfrac{21}{5}$ ft
55. 69.98 cm   **57.** 108 in.   **59.** 12.5 cm
61. 520 ft   **63.** 87.5 ft

65.

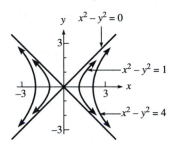

67.

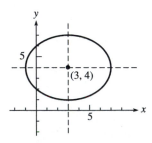

Exercise 10.3 [Page 602]

1.

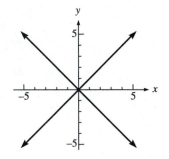

3.

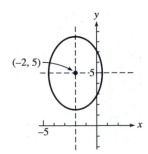

5.

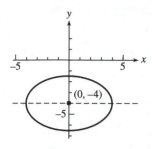

7.

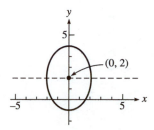

9.

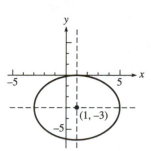

11.

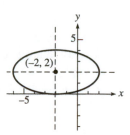

13.

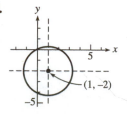

15.

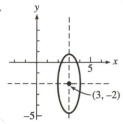

17. $4x^2 + 9y^2 - 8x - 108y + 292 = 0$
19. $9x^2 + 25y^2 + 36x - 100y - 89 = 0$
21. $4x^2 + y^2 + 32x - 6y + 37 = 0$

23.

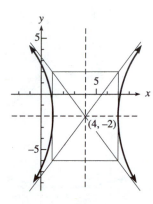

25.

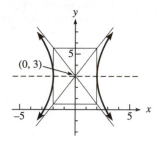

27.

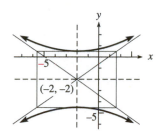

29.

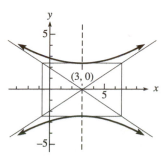

31.

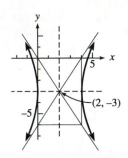

(2, –3)

33.

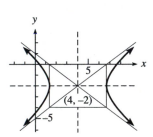

(4, 0)

35.

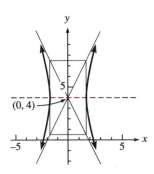

(4, –2)

37.

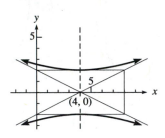

(0, 4)

39. $9y^2 - 16x^2 - 90y - 32x - 367 = 0$
41. $4y^2 - x^2 - 8y - 2x - 13 = 0$

43.

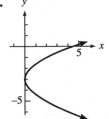

45.

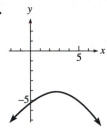

47.

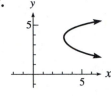

49.

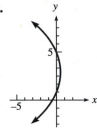

51.

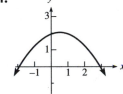

53.

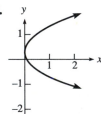

55. $x^2 - 2x + 4y - 7 = 0$
57. $y^2 + 4y - 8x - 28 = 0$
59. $y^2 - 10y + 4x + 17 = 0$
61. $x^2 - 6x + 8y + 9 = 0$
63. Circle; center (0, 0), radius 2
65. Hyperbola; center (0, 0), transverse axis on the x-axis, $a^2 = 8$, $b^2 = 2$
67. Ellipse; center (0, 0), major axis vertical, $a^2 = 6$, $b^2 = 3$
69. Parabola; vertex $\left(0, -\dfrac{3}{2}\right)$, opens upward, $p = \dfrac{1}{4}$
71. Hyperbola; center (0, 0), transverse axis on the y-axis, $a^2 = 6$, $b^2 = 24$
73. Parabola; vertex $(-4, 0)$, opens to the right, $p = \dfrac{1}{2}$
75. Parabola; vertex (3, 2), opens upward, $p = \dfrac{1}{4}$

77. Circle; center $(-1, 4)$, radius 4

79. Ellipse; center $(-1, 0)$, major axis vertical, $a^2 = 4$, $b^2 = 2$

81. Circle; center $(-3, 0)$, radius $\sqrt{13}$

Exercise 10.4 [Page 609]

1. $g(x) = x - 2$ **3.** $g(x) = \dfrac{1}{2}x$

5. $g(x) = \dfrac{1}{2}x + 3$ **7.** $g(x) = 3 - 2x$

9. $g(x) = \sqrt[3]{x - 1}$ **11.** $g(x) = x^3$

13. $g(x) = \dfrac{x + 1}{x}$ **15.** $g(x) = (x - 4)^3$

17a. $f(x) = (x - 2)^3$; $g(x) = \sqrt[3]{x} + 2$

17b. $f(4) = (4 - 2)^3 = 8$; $g(8) = \sqrt[3]{8} + 2 = 4$

17c. $g(-8) = \sqrt[3]{-8} + 2 = 0$;
$f(0) = (0 - 2)^3 = -8$

19.

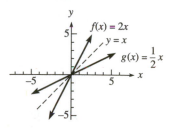

21.

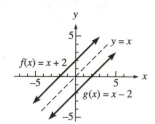

23.

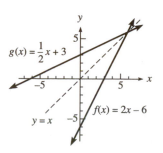

25.

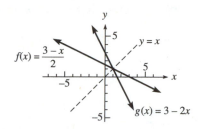

27.

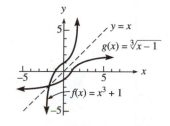

29.

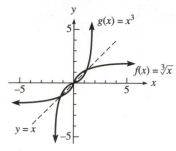

31.

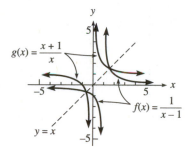

33.

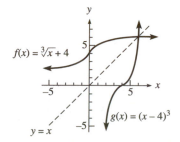

35a. Yes   **35b.** No   **35c.** No   **35d.** Yes   **37a.** Yes

37b. No   **39a.** Yes   **39b.** No   **41.** 6   **43.** $\dfrac{2}{9}$

45. $f^{-1}(x) = \dfrac{x + 2}{x - 1}$ **47a.** 0   **47b.** 1   **49a.** 0

49b. 2

51a.

x	$f(x)$
-2	$\dfrac{1}{4}$
-1	$\dfrac{1}{2}$
0	1
1	2
2	4

51b and d.

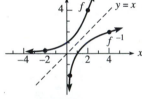

51c.

x	$f^{-1}(x)$
$\dfrac{1}{4}$	-2
$\dfrac{1}{2}$	-1
1	0
2	1
4	2

Exercise 10.5 [Page 616]

1. $\begin{bmatrix} -2 & 1 & | & 0 \\ -9 & 3 & | & -6 \end{bmatrix}$ **3.** $\begin{bmatrix} 1 & -3 & | & 6 \\ 0 & -2 & | & 11 \end{bmatrix}$

5. $\begin{bmatrix} 1 & 0 & -2 & | & 5 \\ 2 & 6 & -1 & | & 4 \\ 0 & -3 & 2 & | & -3 \end{bmatrix}$ **7.** $\begin{bmatrix} 1 & 2 & 1 & | & -5 \\ 0 & 4 & -2 & | & 3 \\ 0 & -9 & 2 & | & 12 \end{bmatrix}$

9. $\begin{bmatrix} 1 & -3 & | & 2 \\ 0 & 7 & | & 0 \end{bmatrix}$ **11.** $\begin{bmatrix} 2 & 6 & | & -4 \\ 4 & 0 & | & 3 \end{bmatrix}$

13. $\begin{bmatrix} 1 & -2 & 2 & | & 1 \\ 0 & 7 & -5 & | & 4 \\ 0 & 9 & -11 & | & -1 \end{bmatrix}$ **15.** $\begin{bmatrix} -1 & 4 & 3 & | & 2 \\ 3 & 0 & -5 & | & 14 \\ -3 & 0 & -3 & | & 8 \end{bmatrix}$

17. $\begin{bmatrix} -2 & 1 & -3 & | & -2 \\ 0 & 4 & -6 & | & -2 \\ 0 & 0 & -4 & | & -5 \end{bmatrix}$ **19.** $(2, 3)$ **21.** $(-2, 1)$

23. $(3, -1)$ **25.** $\left(-\dfrac{7}{3}, \dfrac{17}{3}\right)$ **27.** $(1, 2, 2)$

29. $\left(2, -\dfrac{1}{2}, \dfrac{1}{2}\right)$ **31.** $(-3, 1, -3)$

33. $\left(\dfrac{5}{4}, \dfrac{5}{2}, -\dfrac{1}{2}\right)$

CHAPTER 10 REVIEW [Page 618]

1.

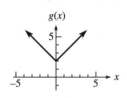

2.

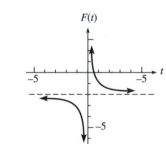

3.

4.

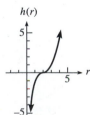

5.

6.

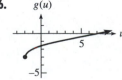

7.

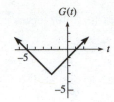

$G(t)$

8.

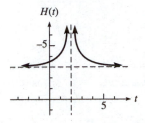

$H(t)$

9. $h(s)$ **10.** $g(s)$

          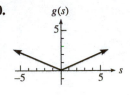

11. $y = (x - 2)^2 - 4$ **12.** $y = -|x| + 3$
13. Circle; center (0, 0), radius 3

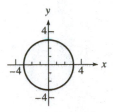

14. Circle; center (2, −3), radius 4

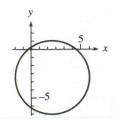

15. Circle; center (2, −1), radius 3

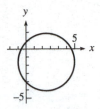

16. Circle; center (0, 3), radius $\sqrt{13}$

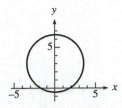

17. Ellipse; center (0, 0), major axis horizontal,
$a^2 = 9$, $b^2 = 1$

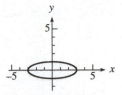

18. Ellipse; center (0, 0), major axis vertical,
$a^2 = 16$, $b^2 = 4$

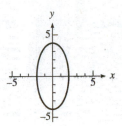

19. Ellipse; center (2, −3), major axis vertical,
$a^2 = 9$, $b^2 = 4$

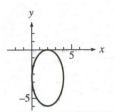

20. Ellipse; center $(-4, 2)$, major axis horizontal,
$a^2 = 12$, $b^2 = 6$

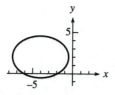

21. Ellipse; center $(2, -2)$, major axis vertical,
$a^2 = 16$, $b^2 = 4$

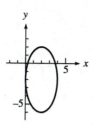

22. Ellipse; center $(-1, 2)$, major axis vertical,
$a^2 = 8$, $b^2 = 5$

23. Parabola, vertex $(-3, 2)$, opens upward, $p = 1$

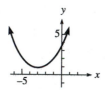

24. Parabola, vertex $(2, 1)$, opens downward, $p = 1$

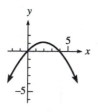

25. Parabola, vertex $(4, -10)$, opens upward, $p = \dfrac{1}{4}$

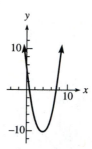

26. Parabola, vertex $(2, -3)$, opens to the left, $p = 1$

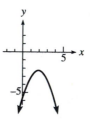

27. Parabola, vertex $(2, -2)$, opens downward, $p = \dfrac{1}{4}$

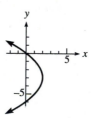

28. Parabola, vertex $\left(-\dfrac{3}{2}, 1\right)$, opens to the right,
$p = \dfrac{1}{2}$

29. Hyperbola; center $(0, 0)$, transverse axis on the y-axis, $a^2 = 6$, $b^2 = 8$

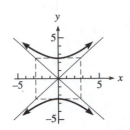

30. Hyperbola; center $(2, -3)$, transverse axis parallel to the x-axis, $a^2 = 4$, $b^2 = 9$

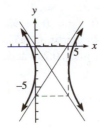

31. Hyperbola; center $(-2, 4)$, transverse axis parallel to the y-axis, $a^2 = 6$, $b^2 = 4$

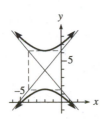

32. Hyperbola; center $(4, -3)$, transverse axis parallel to the x-axis, $a^2 = 4$, $b^2 = 9$

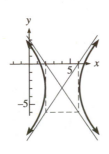

33. Hyperbola; center $(0, 3)$, transverse axis parallel to the x-axis, $a^2 = 5$, $b^2 = 10$

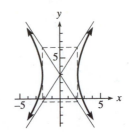

34. Hyperbola; center $(4, 0)$, transverse axis parallel to the y-axis, $a^2 = 3$, $b^2 = 12$

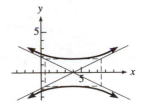

35. $(x + 4)^2 + (y - 3)^2 = 20$
36. $(x + 2)^2 + (y - 4)^2 = 13$
37. $\dfrac{(x + 1)^2}{4} + \dfrac{(y - 4)^2}{16} = 1$
38. $\dfrac{(x - 3)^2}{4} + \dfrac{(y - 1)^2}{25} = 1$ **39.** $x^2 = 16y$
40. $(y - 4)^2 = -16(x - 2)$
41. $(x - 2)^2 = 8(y + 3)$
42. $(y - 1)^2 = -12(x + 1)$ **43.** $f^{-1}(x) = x - 4$
44. $f^{-1}(x) = 4x + 2$ **45.** $f^{-1}(x) = \sqrt[3]{x + 1}$
46. $f^{-1}(x) = \dfrac{1}{x} - 2$ **47.** $f^{-1}(x) = \dfrac{1}{x - 2}$
48. $f^{-1}(x) = (x + 2)^3$ **49.** 0 **50.** $\dfrac{1}{7}$

51.

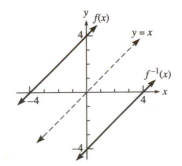

52.

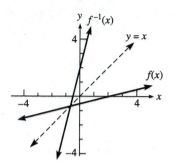

53.

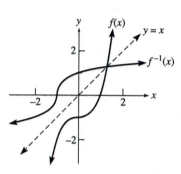

54.

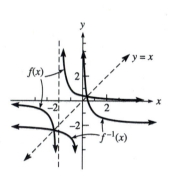

55.

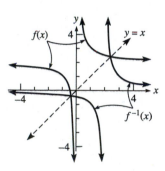

56.

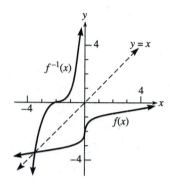

57. $x = 3,\ y = -1$ **58.** $x = 4,\ y = 0$
59. $x = 4,\ y = 1$ **60.** $x = 0,\ y = 3,\ z = -1$
61. $x = -1,\ y = 0,\ z = 2$
62. $x = 1,\ y = 0,\ z = 0$ **63.** 32 dimes
64. 16 first-class, 48 tourist   **65.** $y = 2x^2 + 3x + 3$
66. 0.4 qt cranberry juice, 0.2 qt apricot nectar, 0.4 qt club soda

Exercise A.1 [Page 622]
  1. Q, R **3.** H, R **5.** J, Q, R **7.** W, J, Q, R
  9. Q, R **11.** H, R

Exercise A.2 [Page 629]
  1. $4xy - 8y^2$ **3.** $-12x^3 + 6x^2 - 6x$
  5. $3a^4b - 2a^3b^2 - a^2b^2$ **7.** $8x^3y^7 - 4x^4y^4 - 6x^5y^5$
  9. $n^2 + 10n + 16$ **11.** $r^2 + 3r - 10$
  13. $2z^2 - 5z - 3$ **15.** $8r^2 + 2rs - 3s^2$
  17. $6x^2 - 13xy + 6y^2$ **19.** $9t^2 - 16s^2$
  21. $2a^4 - 5a^2b^2 - 3b^4$ **23.** $4xz(x + 2)$
  25. $3n^2(n^2 - 2n + 4)$ **27.** $3r(5rs + 6s^2 - 1)$
  29. $m^2n^2(3n^2 - 6mn + 14m)$
  31. $3a^2b^2c^4(5a^2b - 4c + 2b)$ **33.** $(a + b)(a + 3)$
  35. $(y - 3x)(y - 2)$
  37. $4(x - 2)^2(-2x^2 + 4x + 1)$
  39. $x(x - 5)^2(-x^2 + 5x + 1)$ **41.** $-(2n - 3m)$
  43. $-2(x - 1)$ **45.** $-a(b + c)$
  47. $-(-2x + y - 3z)$ **49.** $(x + 2)(x + 3)$
  51. $(y - 3)(y - 4)$ **53.** $(x - 3)(x + 2)$
  55. $(2x - 1)(x + 2)$ **57.** $(4x - 1)(x + 2)$
  59. $(3y + 1)(3y - 8)$ **61.** $(2u + 1)(5u - 3)$
  63. $(3x - 7)(7x + 2)$ **65.** $(9a + 4)(8a - 3)$
  67. $(2x - 3)(15x - 4)$ **69.** $2(3t + 2)(9t - 11)$
  71. $(x - 2a)(3x - a)$ **73.** $(3x - 2y)(5x + 2y)$

75. $(3u - 4v)(6u - 5v)$ **77.** $(3a + 2b)(4a - 7b)$
79. $(5ab - 2)(2ab - 3)$ **81.** $2(4xy + 1)(7xy - 2)$
83. $(2az - 3)(11az + 7)$ **85.** $x^2 + 6x + 9$
87. $4y^2 - 20y + 25$ **89.** $x^2 - 9$ **91.** $9t^2 - 16s^2$
93. $25a^2 - 20ab + 4b^2$ **95.** $64x^2z^2 + 48xz + 9$
97. $(x + 5)(x - 5)$ **99.** $(x - 12)^2$
101. $(x + 2y)(x - 2y)$ **103.** $(2x + 3)^2$
105. $(3u - 5v)^2$ **107.** $(2a + 5b)(2a - 5b)$
109. $(xy + 9)(xy - 9)$ **111.** $(3xy + 1)^2$
113. $(4xy - 1)(4xy + 1)$
115. $(x + 2 - y)(x + 2 + y)$

Exercise A.3 [Page 634]
 1. $2i$ **3.** $4i\sqrt{2}$ **5.** $6i\sqrt{2}$ **7.** $6i\sqrt{6}$ **9.** $40i$
11. $-4i\sqrt{3}$ **13.** $4 + 2i$ **15.** $2 + 15i\sqrt{2}$

17. $2 + 2i$ **19.** $5 + 5i$ **21.** $-2 + i$
23. $-1 - 2i$ **25.** $8 + i$ **27.** $13 + 13i$
29. $21 - 18i$ **31.** $3 - 4i$ **33.** 5 **35.** $-\frac{1}{3}i$
37. $-\frac{1}{5} - \frac{3}{5}i$ **39.** $1 + i$ **41.** $\frac{1}{2} - \frac{1}{2}i$
43. $\frac{12}{13} - \frac{5}{13}i$ **45.** $\frac{9}{34} + \frac{19}{34}i$ **47.** $4 + 2i$
49. $15 + 3i$ **51.** $-\frac{3}{2}i$ **53.** $\frac{3}{5} - \frac{4}{5}i$
55. $x \geq 5$; $x < 5$ **57a.** -1
57b. 1 **57c.** $-i$ **57d.** -1 **59.** $5 + 4i$

Index

CALCULATOR KEYSTROKE INDEX

ABS 569

ALPHA 195

CALC 256

CLEAR 52

DRAW 181

GRAPH 52

MATH 291

MODE 410

QUIT 54

STAT 173

STAT PLOT 174

STO▷ 174

TABLE 207

TblSet 207

TRACE 52

WINDOW 56

X,T,θ 52

Y-VARS 568

ZOOM 51